# Structural Dynamics

# Structural Dynamics

## Concepts and Applications

Henry R. Busby
George H. Staab

CRC Press
Taylor & Francis Group
Boca Raton London New York

CRC Press is an imprint of the
Taylor & Francis Group, an **informa** business

CRC Press
Taylor & Francis Group
6000 Broken Sound Parkway NW, Suite 300
Boca Raton, FL 33487-2742



Printed on acid-free paper

International Standard Book Number-13: 978-1-4987-6594-7 (Hardback)



**Visit the Taylor & Francis Web site at**
**http://www.taylorandfrancis.com**

**and the CRC Press Web site at**
**http://www.crcpress.com**

**Visit the eResources: https://www.crcpress.com/9781498765947**

MIX
Paper from responsible sources
FSC www.fsc.org FSC® C013056

Printed and bound in Great Britain by
TJ International Ltd, Padstow, Cornwall

*Henry R. Busby dedicates this book to his wife Nancy, daughters Corinne, Marlene, and grandchildren Anna, Thomas, Abigail, Samuel, and Prairie Rain.*

*George H. Staab dedicates this text to his wife Ellen, children Dan and Ben, daughter-in-law Jen, and granddaughter Claire.*

# Contents


Preface ........ xiii
Authors ........ xv

**1. Single-Degree-of-Freedom Systems** ........ 1
1.1 Introduction ........ 1
1.2 One Degree of Freedom ........ 2
1.3 Equivalent Spring Constant ........ 4
1.4 Free Vibrations ........ 7
1.5 Damped Free Vibration ........ 13
1.6 Forced Vibration under Harmonic Force ........ 21
1.6.1 Forced Undamped Harmonic Motion ........ 22
1.6.2 Forced Damped Harmonic Motion ........ 27
1.6.3 Forced Damped Harmonic Motion Using Complex Format ........ 30
1.6.4 Forced Harmonic Motion of the Support ........ 32
1.6.5 Force Transmitted to Base ........ 37
1.7 Forced Response to Periodic Loading ........ 37
1.7.1 Trigonometric Functions and Fourier Series ........ 38
1.7.2 Alternate Forms of Fourier Series ........ 42
1.7.3 Complex Form of Fourier Series ........ 42
1.7.4 Response to General Periodic Forces ........ 44
1.8 Work Performed by External Forces and Energy Dissipation ........ 47
1.8.1 Material Damping ........ 49
1.8.2 Elastic–Plastic Materials ........ 50
1.8.3 Viscoelastic Materials ........ 51
1.8.4 Using Viscoelastic Relations (for Harmonic Functions) ........ 53
1.9 Response to General Forcing Function ........ 55
1.9.1 Dirac Delta or Impulse Function ........ 56
1.9.2 Response to Dirac Delta Function ........ 57
1.10 Response to Arbitrary Forcing Function ........ 58
1.10.1 Step Response of Undamped System ........ 60
1.10.2 Response to External Force That Varies Linearly with Time ........ 63
1.10.3 Exponentially Decaying Function ........ 65
1.10.4 Asymptotic Step Forcing Function ........ 66
1.10.5 Response to a Ramp-Step Function ........ 67
1.10.6 Rectangular Impulse ........ 70
1.10.7 Triangular Impulse ........ 73
1.10.8 Half-Cycle Sine Impulse ........ 74
1.11 Integral Transformations ........ 77
1.11.1 Application of the Fourier Transformations to the Viscoelastic Relations ........ 80
1.11.2 Application of the Laplace Transformations to the Viscoelastic Relations ........ 80
1.11.3 Dirac and Heaviside Functions ........ 81

1.11.4 Application of Laplace and Fourier Transforms to ODOF Systems with Material Damping ..... 82
1.11.5 An Application of $U^*$ ..... 86
1.11.6 Experimental Determination of $U^*$ ..... 87
Problems ..... 90
References ..... 92

**2. Random Vibrations** ..... 93
2.1 Introduction ..... 93
2.2 Probability, Probability Distribution, and Probability Density ..... 93
2.3 Mean, Variance, Standard Deviation, and Distributions ..... 96
2.4 Combined Probabilities ..... 98
2.5 Random Functions ..... 101
2.5.1 Correlation Function and Spectral Densities ..... 103
2.5.2 Combinations of Random Processes ..... 111
2.5.3 Level Crossings ..... 112
2.6 Dynamic Characteristics of Linear Systems ..... 115
2.7 Input–Output Relations for Stationary Random Processes ..... 117
2.8 Input–Output Relations for Nonstationary Random Processes ..... 121
Problems ..... 125
References ..... 126

**3. Dynamic Response of SDOF Systems Using Numerical Methods** ..... 127
3.1 Introduction ..... 127
3.2 Interpolating the Excitation Function ..... 128
3.3 Finite Differences ..... 133
3.3.1 Euler Method ..... 135
3.3.2 Modified Euler or Heun's Method ..... 138
3.3.3 Runge–Kutta Method ..... 139
3.3.4 Central Difference Method ..... 142
3.4 Newmark Method ..... 145
3.4.1 Constant Acceleration Method ..... 147
3.4.2 Average Acceleration Method ..... 148
3.4.3 Linear Acceleration Method ..... 149
3.5 Wilson-Theta Method ..... 150
3.6 HHT-Alpha Method ..... 153
Problems ..... 157
References ..... 158

**4. Systems with Several Degrees of Freedom** ..... 159
4.1 Introduction ..... 159
4.2 Equations of Motion ..... 159
4.3 Lagrange's Equations ..... 166
4.3.1 Generalized Forces ..... 169
4.4 Potential Energy ..... 171
4.4.1 Application of Lagrange's Equation ..... 174
4.5 Free Vibrations ..... 175
4.6 Frequency Response Function ..... 181
4.6.1 Application of Frequency Response Function ..... 184

4.6.2 Transfer Function (Response to Unit Impulse) ... 184
4.6.3 Arbitrary Forcing Function $P(t)$ ... 186
4.7 Damped Free Vibrations ... 186
4.8 Damped Forced Vibration ... 191
4.9 General Theory of Multi-Degree-of-Freedom Systems ... 194
4.9.1 Matrix Notation for MDF Systems ... 196
4.10 Free Undamped MDOF Systems ... 198
4.10.1 Orthogonality of Natural Modes ... 199
4.10.2 Normalized Modes ... 200
4.10.3 Free Vibrations with Give Initial Conditions ... 202
4.11 Forced Undamped Vibrations ... 208
4.11.1 Steady-State Harmonic Motion ... 208
4.11.2 Arbitrary Forcing Function ... 210
4.12 Forced Damped MDOF Systems ... 211
4.12.1 Proportional Damping ... 212
4.12.2 Arbitrary Viscous Damping ... 213
Problems ... 214
References ... 216

**5. Equations of Motion of Continuous Systems** ... 217
5.1 Introduction ... 217
5.2 Forces and Stresses ... 217
5.3 Equations of Equilibrium ... 219
5.4 Plane Stress ... 222
5.5 Displacement and Strain Relations ... 224
5.6 Stress–Strain Relations ... 228
5.6.1 Special Cases ... 232
5.6.1.1 Plane Stress and Strain ... 232
5.6.1.2 Uniaxial Stress ... 232
5.6.1.3 Anisotropic Materials ... 233
5.7 Displacement Equations for Elastic Bodies ... 233
5.8 Boundary Conditions ... 234
5.9 Work and Energy ... 235
5.9.1 Strain Energy ... 236
5.9.2 Kinetic Energy ... 238
5.10 Principle of Virtual Work ... 238
5.11 Hamilton's Principle ... 241
5.12 General Energy Theorem ... 247
5.13 Rayleigh's Method ... 248
5.14 Ritz Method ... 253
5.14.1 Property of Ritz Method ... 256
5.14.2 Another Approach to Ritz's Method ... 257
5.15 Galerkin's Method ... 258
Problems ... 261
References ... 263

**6. Vibration of Strings and Bars** ... 265
6.1 Introduction ... 265
6.2 Transverse String Vibration ... 265

6.2.1 Initial and Boundary Conditions ..... 266
6.3 General Solution of the Wave Equation ..... 267
6.3.1 Traveling Wave Solution ..... 268
6.3.2 Fourier Transformation Solution ..... 268
6.4 Free Vibrations of Finite Length Strings ..... 271
6.4.1 Discontinuous Strings ..... 274
6.5 Forced Vibrations of Finite Length Strings ..... 277
6.6 Longitudinal Vibrations of Bars ..... 279
6.7 Free Vibrations of Bars ..... 280
6.7.1 Orthogonality of Natural Modes ..... 283
6.7.2 Free Vibrations with Given Initial Conditions ..... 285
6.8 Forced Vibrations of Bars ..... 287
6.9 Material with Damping ..... 289
6.10 Forced Vibrations and Natural Mode Expansion Method ..... 289
6.11 Concentrated Force ..... 293
6.12 Bar with Concentrated End Load and Associated Expressions for $p(x, t)$ ..... 294
Problems ..... 296
References ..... 298

7. **Beam Vibrations** ..... 299
7.1 Introduction ..... 299
7.2 Shear Beam ..... 299
7.3 Euler–Bernoulli Theory ..... 301
7.3.1 Free Vibrations ..... 303
7.3.2 Free Vibration with Given Initial Conditions ..... 309
7.3.3 Forced Vibrations ..... 311
7.3.4 Application of the Laplace Transformations ..... 318
7.3.5 Frequency Response Function ..... 319
7.3.6 Effect of Axial Force ..... 324
7.4 Timoshenko Beam Theory (Effects of Shear Deformation and Rotary Inertia) ..... 326
7.4.1 Equations of Motion ..... 326
7.4.2 Free Vibrations ..... 329
7.4.2.1 Effect of Shear Deformation and Rotary Inertia ..... 332
7.4.3 Forced Vibration ..... 334
Problems ..... 334
References ..... 337

8. **Continuous Beams and Frames** ..... 339
8.1 Introduction ..... 339
8.2 Slope–Deflection Method ..... 339
8.2.1 Forced Vibrations ..... 344
8.3 Vibrations of Frames with Axial Forces ..... 345
Problems ..... 348
References ..... 349

9. **Vibrations of Plates** ..... 351
9.1 Introduction ..... 351

9.2 Equations of Motion ..... 351
9.2.1 Solution to Plate Equations ..... 358
9.2.2 Solutions for Other Boundary Conditions ..... 361
9.2.3 Forced Vibrations ..... 365
9.2.4 Composite Plates ..... 369
9.2.4.1 Free Vibrations of a Simply Supported Plate ..... 370
9.2.4.2 Forced Vibrations ..... 373
9.3 Circular Plates ..... 373
9.3.1 Equations of Motion ..... 374
9.3.2 Forced Vibrations ..... 378
9.4 Approximate Solutions ..... 379
9.4.1 Rayleigh Method ..... 380
9.4.2 Ritz Method ..... 382
9.4.3 Galerkin's Method ..... 382
9.5 Sandwich Plates ..... 382
9.5.1 Equations of Motion ..... 383
9.5.2 Free Vibrations ..... 388
9.5.3 Forced Vibrations ..... 389
9.6 Equations for Plates of Variable Thickness ..... 391
Problems ..... 392
References ..... 393

**10. Vibration of Shells** ..... 395
10.1 Introduction ..... 395
10.2 Cylindrical Shells ..... 395
10.2.1 Equations of Motion ..... 396
10.2.2 Simplified System of Equations ..... 401
10.2.3 Solutions for Cylindrical Shells ..... 403
10.2.3.1 Free Vibrations ..... 403
10.2.3.2 Forced Vibrations ..... 406
10.2.3.3 Other Boundary Conditions ..... 408
10.2.4 Membrane Theory of Cylindrical Shells ..... 409
10.2.4.1 Free Vibrations ..... 411
10.2.4.2 Forced Vibrations ..... 412
10.3 Shells of Revolution ..... 414
10.3.1 Spherical Shell ..... 414
10.3.2 Shallow Spherical Shells ..... 417
10.4 Composite Shells ..... 421
10.4.1 Equations of Motion ..... 422
10.4.2 Free Vibrations ..... 423
10.4.3 Forced Vibrations ..... 425
Problems ..... 425
References ..... 426

**11. Finite Elements and Time Integration Numerical Techniques** ..... 429
11.1 Introduction ..... 429
11.2 Basic Finite Element Approach ..... 430
11.3 Interpolation or Shape Functions ..... 435

11.3.1 One-Dimensional Interpolation Formula ... 436
11.3.1.1 Lagrange's Interpolation Formula ... 438
11.3.1.2 Hermitian Interpolation Function ... 439
11.3.2 Two-Dimensional Interpolation Formula ... 441
11.3.2.1 Triangular Elements ... 441
11.3.2.2 Rectangular Elements ... 446
11.3.2.3 Tetrahedral Elements ... 451
11.3.2.4 Solid Rectangular Hexahedron Elements ... 454
11.3.2.5 Isoparametric Elements ... 455
11.3.2.6 Plate Elements ... 456
11.3.3 Element Properties ... 464
11.4 Element Assembly ... 489
11.4.1 Boundary Conditions ... 495
11.5 Free Response of Finite Element Systems (Eigenvalue Analysis) ... 496
11.5.1 Orthogonality for Eigenvectors of Symmetric Matrices ... 502
11.5.2 Rayleigh Quotient ... 502
11.5.3 Reduction to Standard Form ... 503
11.6 Solution Methods for Calculating Eigenvalues and Eigenvectors ... 504
11.6.1 Vector Iteration Methods ... 505
11.6.1.1 Inverse Iteration ... 507
11.6.1.2 Forward Iteration ... 509
11.6.2 Vector Iteration with Shifts ... 511
11.6.3 Subspace Iteration ... 513
11.7 Transformation Methods ... 515
11.7.1 Generalized Jacobi Method ... 516
11.8 Multiple-Degrees-of-Freedom Numerical Techniques ... 521
11.8.1 Central Difference Method ... 521
11.8.2 The Houbolt Method ... 524
11.8.3 Newmark Method ... 527
11.8.4 Wilson-Theta Method ... 531
11.8.5 HHT-Alpha Method ... 536
Problems ... 539
References ... 545

**12. Shock Spectra** ... 547
12.1 Introduction ... 547
12.2 One-Degree-of-Freedom System ... 547
12.3 Several Degrees of Freedom Systems ... 548
12.4 Random Loading ... 549
12.5 Input–Output Relations ... 552
References ... 556

**Appendix A: Introduction to Composite Materials** ... 557

**Appendix B: Additional References** ... 567

**Index** ... 571

# Preface

The structural dynamics texts currently available present material in a variety of manners and at various technical levels. Some are intended for use by undergraduate students, others for graduate students, and some primarily address specific areas such as continuous systems or thin plates and shells. This text is intended for use by advanced undergraduate and beginning graduate students, and it includes material deemed most pertinent to the anticipated audience by presenting a variety of related topics for single- and multiple-degree-of-freedom systems. No previous knowledge of structural dynamics is required and all necessary background is assumed to come from required undergraduate engineering courses, making the text suitable for self-study. This text is intended to provide the knowledge necessary for establishing the equations of motion and determining the structural responses of systems resulting from dynamic loads. It is applicable to defining and understanding problems relevant to civil, mechanical, and aerospace engineering. Practicing engineers should have no problems using the material in this text as a reference.

In selecting topics for this text, emphasis was placed on the fundamentals of the subject, which include classical analytical methods and modern numerical solution techniques. It is structured so that students can learn the material in a clear forthright manner and develop a firm grasp of mathematical modeling, formulation, and solution of the equations of motion. The text presents an elementary introduction to time-dependent problems using the basic concepts of the single-degree-of-freedom spring–mass systems. Forced and free vibrations as well as damped and undamped systems are considered. Responses to general and arbitrary forcing functions are discussed along with material models (elastic, elastic–plastic, and viscoelastic) and integral transformations (Laplace, Fourier, Dirac, and Heaviside). Random vibrations and their related stochastic processes are presented in an early chapter so that these concepts can be used in subsequent chapters.

Two chapters are dedicated to numerical solution procedures, many of which are addressed using MATLAB® or similar commercially available programs. An early chapter is dedicated to single-degree-of-freedom systems in which finite difference techniques (Euler, Runge–Kutta, etc.), Newmark, Wilson-theta, and HHT-alpha methods are introduced. A later chapter is dedicated to multiple-degree-of-freedom systems, and the finite element method is introduced. Due to the abundance of finite element techniques available for different applications, only selected elements are presented. In addition to finite elements, the numerical techniques presented for one-dimensional problems are expanded upon.

Multiple-degree-of-freedom and continuous systems are presented in Chapters 4 and 5 with discussions of work and energy methods. The subsequent five chapters present topics relevant to specific structural members: strings and bars, beams (Euler–Bernoulli and Timoshenko), frames, plates (including composites), and shells. A brief discussion of continuous fiber-composite laminates is presented in Appendix A, and a relatively comprehensive list of additional references pertaining to structural dynamics is presented in Appendix B.

MATLAB® is a registered trademark of The MathWorks, Inc. For product information, please contact:

The MathWorks, Inc.
3 Apple Hill Drive
Natick, MA 01760-2098, USA
Tel: 508 647 7000
Fax: 508-647-7001
E-mail: info@mathworks.com
Web: www.mathworks.com

# *Authors*

**Henry R. Busby** earned an undergraduate degree in mechanical engineering at California State University, Long Beach, and his MS and PhD from the University of Southern California. Prior to his career in academia, Professor Busby had complied more than 15 years of experience in industry working at companies such as Rocketdyne, Canoga Park, California, TRW, Redondo Beach, California, Aerospace Corporation, El Segundo, California, PDA Engineering, Santa Ana, California, and Holmes & Narver, Inc., Orange, California.

He is currently emeritus faculty of mechanical and aerospace engineering at The Ohio State University, where he has taught machine design, computer-aided design, optimization, composite materials, and advanced strength of materials for the past 30 years. He is coauthor of three books, *Introductory Engineering Modeling Emphasizing Differential Models and Computer Simulations, Practical Inverse Analysis in Engineering*, and *Mechanical Design of Machine Elements and Machines*.

A life member of ASME, Professor Busby has authored or coauthored more than 35 papers in numerical methods and inverse problem as well as a consultant with government and various engineering firms.

**George H. Staab** is an emeritus faculty member in mechanical engineering and aerospace engineering at The Ohio State University. He earned BS and MS in aeronautical engineering from Purdue University in 1972 and 1973, respectively. He was employed as a rotor head and blade analyst by Sikorsky aircraft from 1973 to 1976, when he returned to Purdue in pursuit of his doctorate, which was earned in 1979. He then became an engineering mechanics faculty member at The Ohio State University. His 35-year teaching career ended in mechanical engineering and aerospace engineering. He was the faculty advisor for the FIRST Robotics and Formula SAE student project teams for many years. Additionally, he co-developed educational software accompanying Beer and Johnston's Statics, Dynamics, and Mechanics of Materials texts; is the sole author of *Laminar Composites*; and is the coauthor of *Mechanical Design of Machine Elements and Machines*.

# 1

# Single-Degree-of-Freedom Systems

## 1.1 Introduction

Classical structures such as buildings and bridges have been in the existence for thousands of years. In order for them to perform as intended, they have to be analyzed. The primary purpose of structural analysis is finding out the behavior of a physical structure when subjected to a force or forces. The study of dynamics can be traced back to the time of Aristotle (384–322 BC) and has evolved to a high level of sophistication. The forces applied to a structure can be in the form of a load due to weight, wind, snow, etc. or some other kind of excitation such as an earthquake, or nearby blast loading. It is noted that all these loads are dynamic in nature, including the self-weight of the structure. The distinction is made between dynamic and static analysis on the basis of whether all the external forces are applied so slowly that the loads and the resulting deformations and stresses are independent of time. These are called static loads. However, when the loads are applied rapidly, called dynamic loads, the structure responds and motion occurs. A dynamic load is one that changes with time fairly quickly in comparison to the structure's natural frequency. The study of this motion is called structural dynamics or the vibration of structures. The link between dynamics and structures (in particular structural behavior) became a more focused field of study during the industrial revolution in the late 1700s. Engineers and scientists began to see the effect of rapidly applied or repeated loads on machines. They began to formulate analysis procedures for identifying the relationship between rapidly applied dynamic loads and the effect they have on a structure. As technology advanced, many researchers have added to the knowledge base and today's modern computational tools have made the solution of structural dynamics problems far easier to achieve. A historical overview of many of the contributions is given by Corradi (2006).

Dynamic analysis of simple structures can be performed manually. For more complex structures, some type of numerical analysis such as finite element methods must be performed to determine frequencies, mode shapes, deformations, and stresses. The equations of motion can be formulated in terms of linear or nonlinear ordinary differential equations. We refer to theses as *discrete systems.*

Newton's law of motion forms the basis of structural dynamics. Newton's law states that when a material mass is acted upon by a force the momentum of the mass will change in proportion to the force and in the same direction as the force. That is,

$$\frac{d}{dt}(m\dot{u}) = p(t)$$

Both the momentum $m\dot{u}$ and the external driving force $p(t)$ are functions of time. This is Newton's second law and the basis of structural dynamics. In most structural dynamics problems the mass is constant, consequently Newton's second law becomes

$$m\frac{d\dot{u}}{dt} = m\frac{d^2u}{dt^2} = p(t)$$

The rotational version of Newton's law states that the angular momentum $I\dot{\theta}$ and the external torque $T(t)$ are functions of time. Thus, the time rate of change of the angular momentum equals the torque. Therefore,

$$\frac{d}{dt}(I\dot{\theta}) = T(t)$$

where $I$ is the mass moment of inertia. For most structural dynamic problems, the mass moment of inertia is constant leading to

$$I\ddot{\theta} = T(t)$$

The equations of motion will be investigated for a general elastic solid for various types of motion, such as free, forced, and transient. Free motion can occur with no external applied force but with initial conditions applied such as an initial displacement or velocity. Forced motion as the name implied is due to some type of external load. The response due to forced motion can be either harmonic or transient.

Time-dependent coordinates $u^T(t) = [u_1(t), \ldots, u_n(t)]$ are used to describe the displacement and rotational components of a system of mass particles or rigid bodies from a known static equilibrium state. In most structural dynamics problems, we deal with the number of dynamic degrees of freedom. This is the least number of independent displacements needed to define the displaced position of all the masses relative to their original position. Also, there are as many natural modes, or types of vibration, as there are degrees of freedom. If $n = 1$, we have the so-called single-degree-of-freedom (SDOF) system. For $n > 1$, we have multi-degree-of-freedom (MDOF) systems and for $n = \infty$ we have what is called continuous systems.

## 1.2 One Degree of Freedom

When considering SDOF, a single parameter ($u$) is used to describe the state of the system. The traditional model used to describe an SODF system consists of a mass $m$ suspended from a spring of stiffness $k$ acted on by a time-dependent force $p(t)$. This results in a deflection $u$ in the direction of the applied force. The simplest model for this type of motion is a mass supported by a spring as illustrated in Figure 1.1a and b.

Other SDOF systems often encountered in engineering are illustrated in Figure 1.2.

Application of Newton's second law to the system in Figure 1.1b results in the familiar expression

$$\downarrow + \sum F_{vert} = m\ddot{u} \tag{1.1}$$

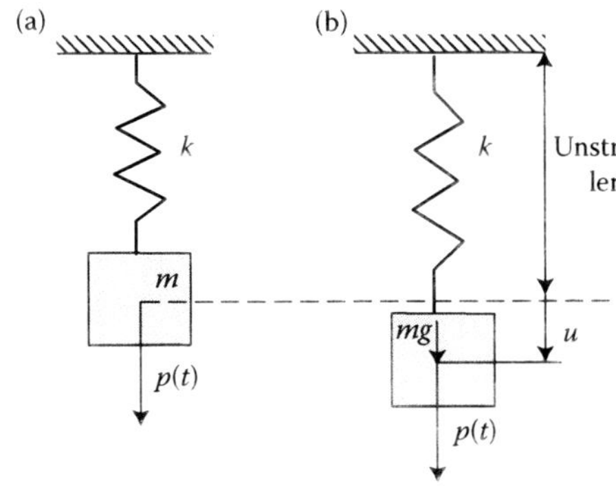


**FIGURE 1.1**
Typical model of an SDOF system (a) undeformed, and (b) deformed.

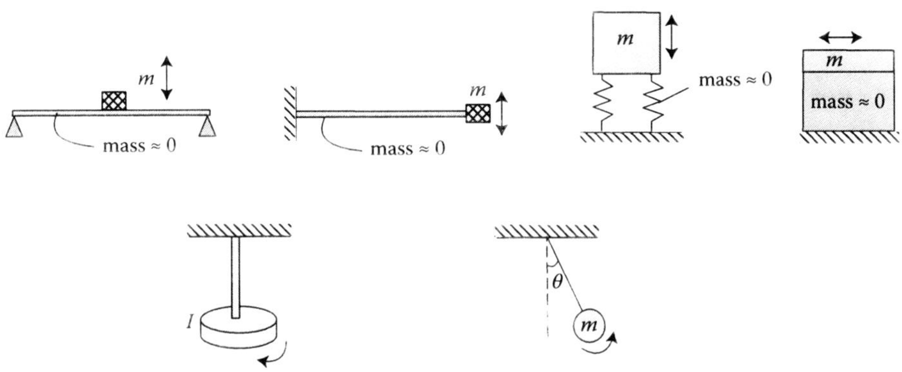


**FIGURE 1.2**
Additional SDOF systems.

$$m\frac{d^2u}{dt^2} = p(t) + mg - ku \Rightarrow m\frac{d^2u}{dt^2} + ku = mg + p(t)$$

or

$$m\ddot{u} + ku = mg + p(t) \tag{1.2}$$

Since the linear spring with the mass is in static equilibrium due to the gravity force, its displacement is given by

$$u_{static} = \frac{mg}{k}$$

Equation 1.2 can then be written as

$$m\frac{d^2}{dt^2}(u(t) + u_{static}) + k(u + u_{static}) = mg + p(t) \tag{1.3}$$

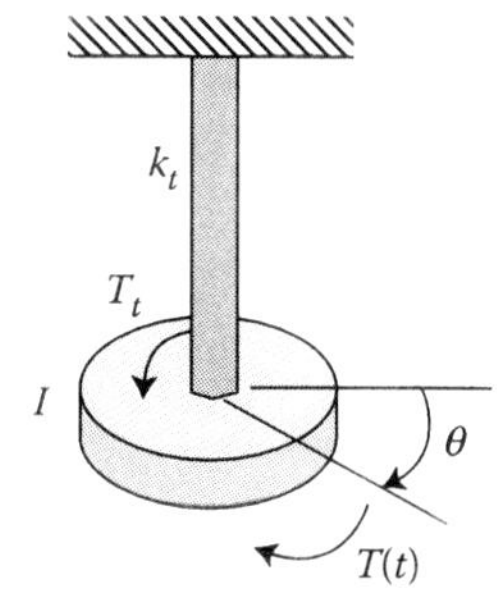


**FIGURE 1.3**
Rotational motion model.

Since $u_{static}$ is a constant, Equation 1.3 becomes

$$m\ddot{u} + ku = p(t) \tag{1.4}$$

Equation 1.4 is an ordinary differential equation whose solution depends upon the loading conditions and the boundary conditions. Depending on the form of the loading condition $p(t)$, the solution can be simple with $p(t) = 0$ or complex with $p(t) \neq 0$.

For rotational motion, consider the system shown in Figure 1.3.

Applying Newton's second law for rotation about a fixed axis gives

$$\sum T = I\ddot{\theta} \tag{1.5}$$

$$I\frac{d^2\theta}{dt^2} = -T_r + T(t) = -k_t\theta + T(t) \Rightarrow I\frac{d^2\theta}{dt^2} + k_t\theta = T(t)$$

or

$$I\ddot{\theta} + k_t\theta = T(t) \tag{1.6}$$

## 1.3 Equivalent Spring Constant

In an elastic system composed of many springs in various arrangements, it is convenient to define an equivalent spring constant to obtain a single force that corresponds to the effect of these springs put together. The springs in a spring mass system will either be in parallel or in series as illustrated in Figure 1.4.

For this system, let $u$ represent the displacement due to the external load $P$. Each spring will see the same displacement $u$, and $P$ will equal the sum of the forces exerted by each of the springs. Therefore,

$$P = k_1u + k_2u + k_3u$$

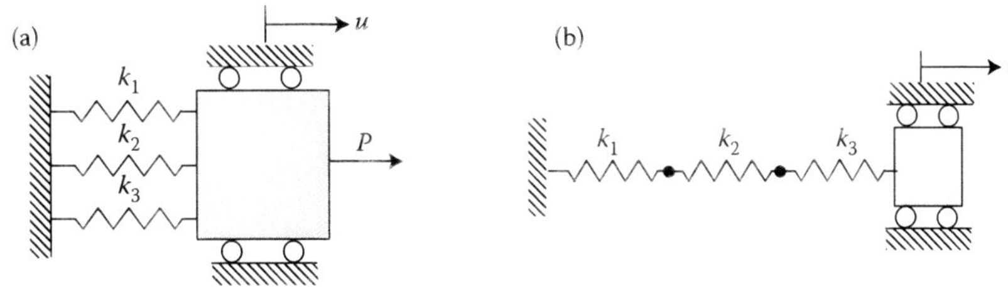


**FIGURE 1.4**
Spring–mass system with multiple springs in (a) parallel (b) series.

For a single equivalent spring that would replace the set of three springs, we define this as

$$P = k_{eq}u$$

Hence, for springs in parallel, we have

$$k_{eq} = k_1 + k_2 + k_3 \tag{1.7}$$

For a system with springs in series as shown in Figure 1.4b, each spring sees the same force but has a different displacement. Hence,

$$P = k_1u = k_2u = k_3u$$

For a single equivalent spring that would replace the set of springs in series, we have

$$P = k_{eq}u$$

Now

$$u = u_1 + u_2 + u_3$$

Hence,

$$\frac{P}{k_{eq}} = \frac{P}{k_1} + \frac{P}{k_2} + \frac{P}{k_3}$$

and

$$\frac{1}{k_{eq}} = \frac{1}{k_1} + \frac{1}{k_2} + \frac{1}{k_3} \tag{1.8}$$

Equations 1.7 and 1.8 can be extended to *n* springs in parallel or series.

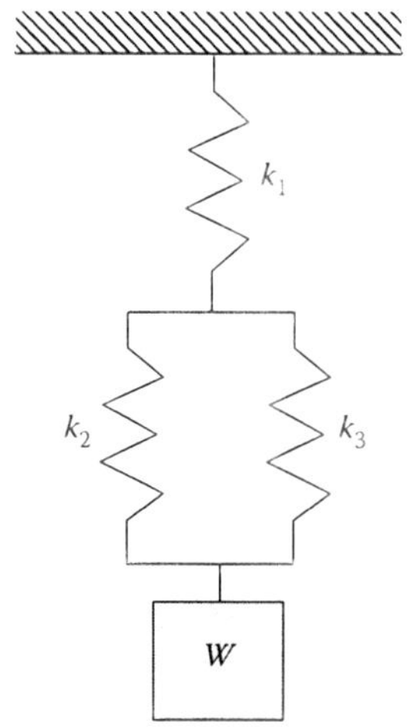


FIGURE 1.5
Spring–mass system.

**EXAMPLE 1.1**

A weight $W$ is suspended from a spring system as shown in Figure 1.5. Determine the equivalent spring constant of this system. Assume linear motion for all parts of the system.

**Solution**

Since springs $k_2$ and $k_3$ are in parallel, we have

$$k' = k_2 + k_3$$

Springs $k_1$ and $k'$ are in series that can be written as

$$\frac{1}{k_{eq}} = \frac{1}{k_1} + \frac{1}{k'} \quad \text{or} \quad k_{eq} = \frac{1}{(1/k_1) + (1/k')} = \frac{k_1 k'}{k_1 + k'} = \frac{k_1(k_2 + k_3)}{k_1 + k_2 + k_3}$$

**EXAMPLE 1.2**

Determine the spring constant $k$ when a load $W$ is applied at the cross-point of two identical beams as shown in Figure 1.6. The deflection of a single beam is $WL^3/48EI$.

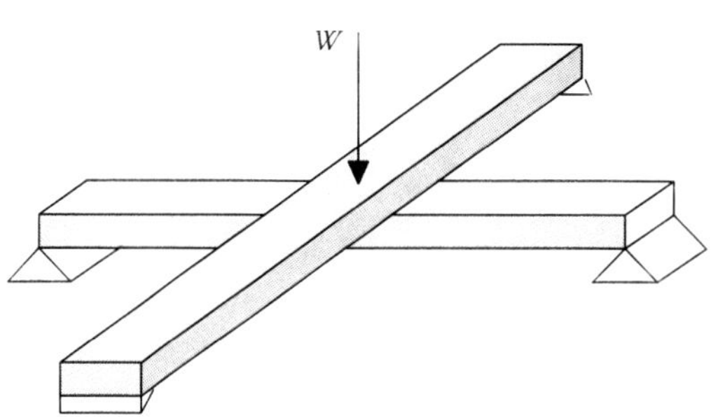


FIGURE 1.6
Cross-beam configuration.

**Solution**

Since the beams see the same deflection due to the load $W$, they are in parallel. Therefore,

$$k_{eq} = k_1 + k_2 = 2k$$

Since $k_1 = k_2$ and due to the fact that $k = W/\delta$, we have

$$k_{eq} = 2k = \frac{2W}{WL^3/48EI} = \frac{96EI}{L^3}$$

## 1.4 Free Vibrations

In the case of free vibrations, the applied force $p(t) = 0$ and in this case Equation 1.4 reduces to

$$m\ddot{u} + ku = 0 \tag{1.9}$$

Rewriting the above Equation 1.9 as

$$\ddot{u} + \frac{k}{m}u = 0$$

and introducing $\omega_0^2 = k/m$, we have

$$\ddot{u} + \omega_0^2 u = 0 \tag{1.10}$$

The initial conditions that must be satisfied at time $t = 0$ are $u = u_0$ and $\dot{u} = \dot{u}_0$, where $u_0$ and $\dot{u}_0$ are given. The elementary functions of a real variable that have this specific property are the sine and cosine functions. The general solution for $u(t)$ is

$$u(t) = C_1 \cos\omega_0 t + C_2 \sin\omega_0 t \tag{1.11}$$

where $C_1$ and $C_2$ are constants of integration. The constants $C_1$ and $C_2$ are determined from the initial conditions

$$u(t=0) = u_0 \rightarrow u_0 = C_1 \cos(\omega_0(0))$$

and

$$\dot{u}(t=0) = \dot{u}_0 \rightarrow \dot{u}_0 = -C_1\omega_0 \sin(\omega_0(0)) + C_2\omega_0 \cos(\omega_0(0))$$

Solving results in $C_1 = u_0$ and $C_2 = \dot{u}_0/\omega_0$. Therefore, the general solution can be given as

$$u = u_0 \cos\omega_0 t + (\dot{u}_0/\omega_0)\sin\omega_0 t \tag{1.12}$$

which describes the motion of the system for all values of $t$. An alternate solution is obtained by writing

$$u = A\cos(\omega_0 t - \varphi) \tag{1.13}$$

where $A$ and $\varphi$ are arbitrary constants of integration determined from initial conditions at $t = 0$. The relationship to $C_1$ and $C_2$ can be found as follows. Expand $u = A\cos(\omega_0 t - \varphi)$ as

$$u = A\cos\omega_0 t\cos\varphi + A\sin\omega_0 t\sin\varphi$$

Comparing with Equation 1.11 gives

$$C_1 = A\cos\varphi \quad \text{and} \quad C_2 = A\sin\varphi$$

so that

$$A = \sqrt{C_1^2 + C_2^2} \quad \text{and} \quad \tan\varphi = \frac{C_2}{C_1}$$

Thus, we have

$$u = \sqrt{C_1^2 + C_2^2}\cos\left(\omega_0 t - \arctan\frac{C_1}{C_2}\right) \tag{1.14}$$

which is an alternative way to write Equation 1.13. In this form, one obtains the maximum displacement, or amplitude of vibration as

$$u_{\max} = A = \sqrt{C_1^2 + C_2^2}$$

The value of $A$ is referred to as the amplitude of vibration and the angle $\varphi$ as the phase angle which lags the motion cos $\omega_0 t$. Equation 1.14 relating to the initial conditions becomes

$$u = \sqrt{u_0^2 + \frac{\dot{u}_0^2}{\omega_0^2}}\cos\left(\omega_0 t - \arctan\frac{\dot{u}_0}{u_0\omega_0}\right) \tag{1.15}$$

Another form of the solution can be obtained by assuming

$$u = A_0\sin(\omega_0 t + \theta) \tag{1.16}$$

where $A = A_0$, giving another alternative solution

$$u = \sqrt{C_1^2 + C_2^2}\sin\left(\omega_0 t + \arctan\frac{C_1}{C_2}\right) \tag{1.17}$$

$$u = \sqrt{u_0^2 + \frac{\dot{u}_0^2}{\omega_0^2}} \sin\left(\omega_0 t + \arctan\frac{u_0\omega_0}{\dot{u}_0}\right) \tag{1.18}$$

Figure 1.7a illustrates the contributions of each term in Equation 1.13 to the overall vibration. It can be seen that the vibration consists of two parts:

a. A vibration that is proportional to cos $\omega_0 t$ and depends on the initial displacement $u_0$ of the weight.
b. A vibration that is proportional to sin $\omega_0 t$ and depends on the initial velocity $\dot{u}_0$.

In this Figure 1.7a, *T* represents the period of free vibration. Figure 1.7b illustrates that the total displacement *u*(*t*) of the oscillation can be obtained by adding projections on the *x*-axis of the two perpendicular vectors $\overline{OA}$ and $\overline{OB}$, rotating with the angular velocity $\omega_0$. The same result will be obtained if, instead of vectors $\overline{OA}$ and $\overline{OB}$, we consider vector $\overline{OC}$, equal to the geometrical sum of the previous two vectors, and take the projection of this vector on the *x*-axis. The magnitude of this vector is $\overline{OC} = \sqrt{u_0^2 + (\dot{u}_0/\omega_0)^2}$ and the angle it makes with the *x*-axis is $\omega_0 t - \varphi$, where $\tan\varphi = (\dot{u}_0/\omega_0)/u_0$, or $\varphi = \tan^{-1}(\dot{u}_0/\omega_0 u_0)$. Therefore, we can express the equation as $u = \overline{OC}\cos(\omega_0 t - \varphi)$ or $u = A\cos(\omega_0 t - \varphi)$. In this expression, *A* is the amplitude of vibration and $\varphi$ is the phase difference.

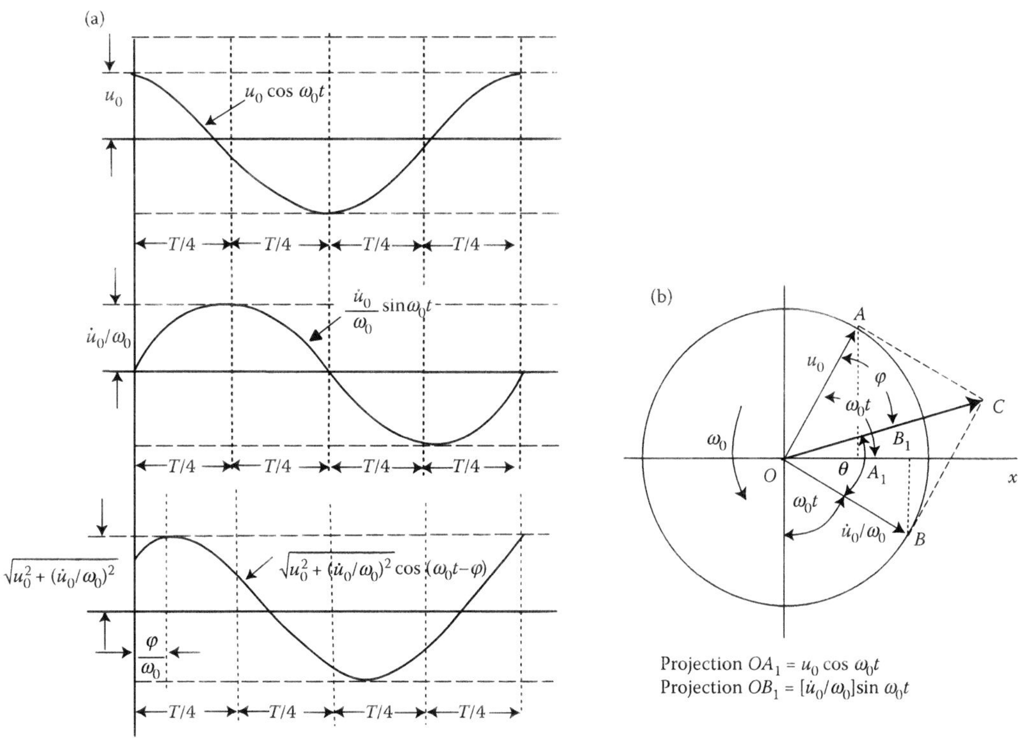


**FIGURE 1.7**
(a) Individual contributions of each term in Equation 1.13 to the overall vibration, (b) contributors to the total displacement of an oscillating system.

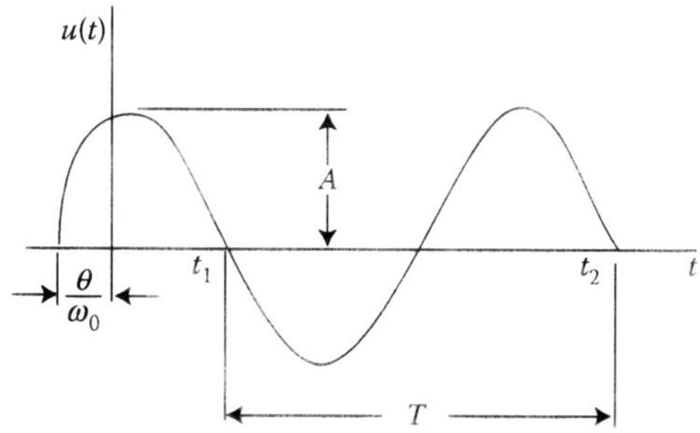


**FIGURE 1.8**
Simple harmonic motion.

Using these solutions, we find that a linear system produces simple harmonic motion with amplitude $A$ and a phase difference $\theta$ as illustrated in Figure 1.8.

We note from Figure 1.8 that at times $t_1$ and $t_2$, $u(t) = 0$. This implies that from Equation 1.13 the argument of the sin functions at $t_1$ and $t_2$ is

$$\omega_0 t_1 + \theta \quad \text{and} \quad \omega_0 t_2 + \theta$$

Since the sine function must be periodic, then

$$\omega_0 t_1 + \theta - (\omega_0 t_2 + \theta) = 2\pi$$

and

$$t_1 - t_2 = \frac{2\pi}{\omega_0}$$

or

$$T = \frac{2\pi}{\omega_0} = 2\pi\sqrt{\frac{m}{k}} \quad \text{(s)} \tag{1.19}$$

The frequency is defined as

$$f = \frac{1}{T} = \frac{\omega_0}{2\pi} = \frac{1}{2\pi}\sqrt{\frac{k}{m}} \quad \text{hertz (Hz) or cycles/s} \tag{1.20}$$

The angular frequency is defined as

$$\omega_0 = 2\pi f = \sqrt{\frac{k}{m}} \quad (\text{s}^{-1}) \text{ or rad/s} \tag{1.21}$$

**EXAMPLE 1.3**

Determine the mass which must be attached to a spring which has a modulus of 2000 N/m such that the resulting frequency of the system would be 15.92 Hz.

**Solution**

The angular frequency of the system is

$$\omega_0 = 2\pi f = 2\pi(15.915) = 100 \text{ rad/s}$$

and the mass is

$$m = \frac{k}{\omega_0^2} = \frac{2000}{(100)^2} = 0.20 \text{ kg}$$

**EXAMPLE 1.4**

A system undergoing vibratory motion has an amplitude of 0.25 in and a period of 0.5 s. Determine the maximum velocity and acceleration that the structure undergoes.

**Solution**

The frequency of the system is

$$f = \frac{1}{\tau} = \frac{1}{0.5} = 2.0 \text{ Hz}$$

The angular frequency of the system is

$$\omega_0 = 2\pi f = 2\pi(2.0) = 12.57 \text{ rad/s}$$

Thus, the velocity and acceleration can be determined as

$$\dot{u} = u\omega_0 = 0.25(12.57) = 3.1425 \text{ in/s}$$
$$\ddot{u} = u\omega_0^2 = 0.25(12.57)^2 = 39.501 \text{ in/s}^2$$

**EXAMPLE 1.5**

An inverted pendulum is supported by two springs and a bracket at $O$ as shown in Figure 1.9. Each spring has a modulus equal to $k$. If the weight of the mass at $A$ is $W$, derive an expression for the frequency of small vibrations, assuming the bar is of negligible weight.

**Solution**

Applying Newton's equation for rotation about a fixed axis gives

$$\sum M_0 = I_0\ddot{\theta}$$

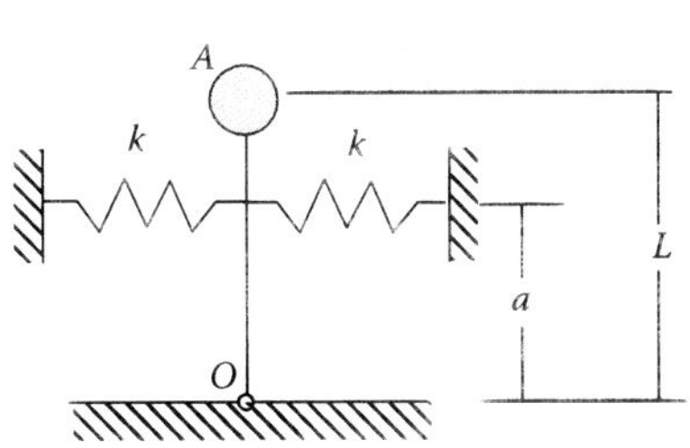


**FIGURE 1.9**
Inverted pendulum.

Thus, we have summing moments about $O$

$$\frac{W}{g}L^2\ddot{\theta} = WL\theta - 2ka^2\theta$$

or

$$\ddot{\theta} + \left(\frac{2gka^2}{WL^2} - \frac{g}{L}\right)\theta = 0$$

Let

$$\omega^2 = \frac{2gka^2}{WL^2} - \frac{g}{L} \quad \text{and} \quad \omega = \sqrt{\frac{2gka^2}{WL^2} - \frac{g}{L}}$$

Hence,

$$f = \frac{1}{2\pi}\omega = \frac{1}{2\pi}\sqrt{\frac{2gka^2}{WL^2} - \frac{g}{L}}$$

**EXAMPLE 1.6**

A car of mass $m_1$ and trailer of mass $m_2$ are inner connected by a spring of stiffness $k$ as shown in Figure 1.10a and the associated free-body diagram is shown in Figure 1.10b. Determine the differential equation for this system in terms of the relative motion between the masses. What is the natural frequency?

**Solution**

The free body diagram is shown in Figure 1.10b, where it has been assumed that $u_1 > u_2$. Applying Newton's second law of motion yields

$$m_1\ddot{u}_1 + ku_1 - ku_2 = 0 \quad m_2\ddot{u}_2 + ku_2 - ku_1 = 0$$

Multiplying the first equation by $m_2$ and the second by $m_1$ and subtracting the second equation from the first gives

$$m_1m_2(\ddot{u}_1 - \ddot{u}_2) + k(m_1 + m_2)(u_1 - u_2) = 0$$

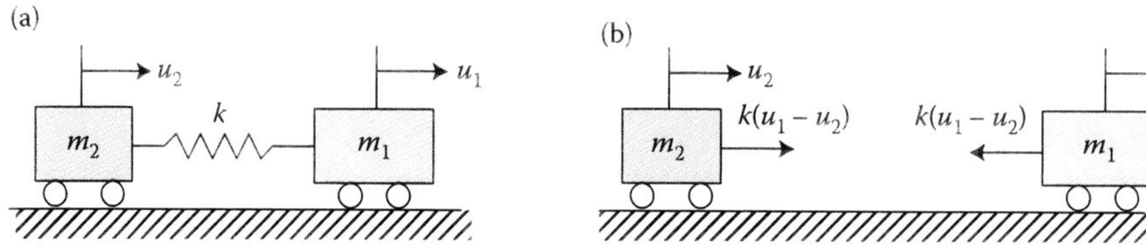


**FIGURE 1.10**
(a) Car, trailer system, (b) free-body diagram of car trailer system.

The relative displacement is taken to be $u = u_1 - u_2$, thus the differential equation of relative motion between the masses is

$$m_1 m_2 \ddot{u} + k(m_1 + m_2)u = 0$$

or

$$\ddot{u} + \frac{k(m_1 + m_2)}{m_1 m_2} u = 0$$

The natural frequency is then given as

$$\omega_0 = \sqrt{\frac{k(m_1 + m_2)}{m_1 m_2}}$$

## 1.5 Damped Free Vibration

Adding a dashpot with (linear damping coefficient $\eta$) in parallel with the existing spring results in a model illustrated in Figure 1.11. Application of Newton's second law to this system results in the familiar expression

$$m\frac{d^2u}{dt^2} = p(t) - ku - \eta\frac{du}{dt} \Rightarrow m\frac{d^2u}{dt^2} + \eta\frac{du}{dt} + ku = p(t)$$

$$m\ddot{u} + \eta\dot{u} + ku = p(t) \tag{1.22}$$

For free vibrations, $p(t) = 0$ and Equation 1.22 reduces to

$$m\ddot{u} + \eta\dot{u} + ku = 0 \tag{1.23}$$

Dividing through by $m$ allows us to write the governing equation for free vibration as

$$\ddot{u} + \frac{\eta}{m}\dot{u} + \frac{k}{m}u = 0$$

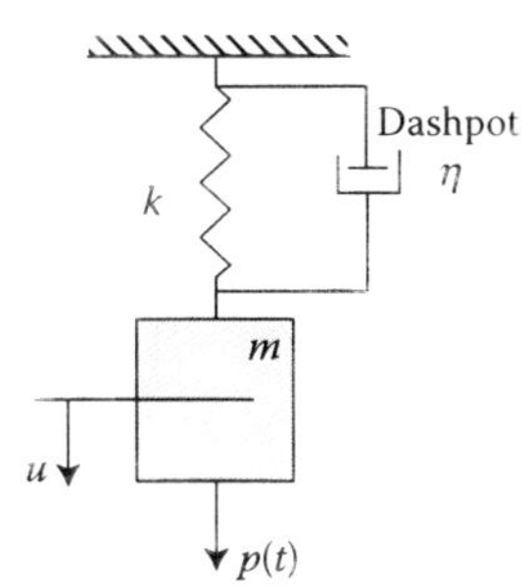


**FIGURE 1.11**
Model of a damped system.

Assuming a solution of the form $u = Ce^{rt}$ (where $r$ is a root to the *characteristic equation*) results in

$$r^2 + \frac{\eta}{m} r + \frac{k}{m} = 0$$

The roots of which are given as

$$r_{1,2} = -\frac{\eta}{2m} \pm \sqrt{\left(\frac{\eta}{2m}\right)^2 - \frac{k}{m}} \tag{1.24}$$

This results in the general solution being given as

$$u(t) = C_1 e^{r_1 t} + C_2 e^{r_2 t} \tag{1.25}$$

where $C_1$ and $C_2$ are constants to be determined from the initial conditions. Note that the roots will be real or complex depending on the value of $\sqrt{(\eta/2m)^2 - (k/m)}$. The solution of this equation contains several distinct cases that arise when $\sqrt{(\eta/2m)^2 - (k/m)} > 0$, $\sqrt{(\eta/2m)^2 - (k/m)} = 0$, and $\sqrt{(\eta/2m)^2 - (k/m)} < 0$. From these cases, it is convenient to define critical damping $\eta_{cr}$, which makes the radical zero as

$$\eta_{cr} = 2m\sqrt{\frac{k}{m}} = 2m\omega_0 = 2\sqrt{km} \tag{1.26}$$

The actual damping in a vibrating system can be specified in terms of $\eta_{cr}$ by introducing the damping ratio $\zeta$. Therefore,

$$\zeta = \frac{\eta}{\eta_{cr}} \tag{1.27}$$

The roots, Equation 1.24, can now be rewritten as

$$\begin{aligned} r_1 &= -\zeta\omega_0 + \omega_0\sqrt{\zeta^2 - 1} \\ r_2 &= -\zeta\omega_0 - \omega_0\sqrt{\zeta^2 - 1} \end{aligned} \tag{1.28}$$

Note that from Equation 1.27, we have

$$\eta = \eta_{cr}\zeta = 2m\omega_0\zeta$$

Thus, we can write the governing equation as

$$\ddot{u} + 2\omega_0\zeta\dot{u} + \omega_0^2 = 0 \tag{1.29}$$

*Case I:* For very large damping (overdamped case), the damping ratio is greater than 1 ($\zeta > 1$). This results in the roots $r_1$ and $r_2$ both being real and negative since $m$, $\eta$, and

$k$ are positive. This means there is no vibration and mass creeps to its equilibrium position. Let $s = \omega_0\sqrt{\zeta^2 - 1}$, then $r_1 = -\zeta\omega_0 + s$ and $r_2 = -\zeta\omega_0 - s$. Thus, the general solution, Equation 1.23, can be written as

$$u(t) = e^{-\zeta\omega_0 t}\left(C_1 e^{st} + C_2 e^{-st}\right) \tag{1.30}$$

The response given by this equation is not periodic. Since both exponent's $r_1$ and $r_2$ are negative, we see that the displacement and the velocity will die out exponentially. We have that the hyperbolic sine and cosine are given as

$$\sinh x = \frac{e^x - e^{-x}}{2} \qquad \cosh x = \frac{e^x + e^{-x}}{2}$$

therefore,

$$e^x = \cosh x + \sinh x \quad e^{-x} = \cosh x - \sinh x \tag{1.31}$$

Substituting Equation 1.31 into Equation 1.30 gives

$$\begin{aligned} u(t) &= e^{-\zeta\omega_0 t}[C_1(\cosh st + \sinh st) + C_2(\cosh st - \sinh st)] \\ &= e^{-\zeta\omega_0 t}[(C_1 + C_2)\cosh st + (C_1 - C_2)\sinh st] \end{aligned}$$

or

$$u(t) = e^{-\zeta\omega_0 t}\left[C_1' \cosh st + C_2' \sinh st\right] \tag{1.32}$$

Using the initial conditions, that at time $t = 0$, $u = u_0$ and $\dot{u} = \dot{u}_0$, the solution becomes

$$\begin{aligned} u(0) &= u_0 = C_1' \\ \dot{u}(0) &= \dot{u}_0 = sC_2' - \zeta\omega_0 C_1' \end{aligned}$$

Solving yields

$$C_1' = u_0 \quad \text{and} \quad C_2' = \frac{\zeta\omega_0 u_0 + \dot{u}_0}{s} \tag{1.33}$$

giving the general solution as

$$u(t) = e^{-\zeta\omega_0 t}\left[u_0 \cosh st + \left(\frac{\zeta\omega_0 u_0 + \dot{u}_0}{s}\right)\sinh st\right]$$

or

$$u(t) = e^{-\zeta\omega_0 t}\left[u_0 \cosh \omega_0\sqrt{\zeta^2 - 1}t + \left(\frac{\zeta\omega_0 u_0 + \dot{u}_0}{\omega_0\sqrt{\zeta^2 - 1}}\right)\sinh \omega_0\sqrt{\zeta^2 - 1}t\right] \tag{1.34}$$

*Case II:* The damping ratio is exactly equal to 1, critically damped case ($\zeta = 1$). Thus, both values of the roots are $r_1 = r_2 = -\omega_0$. However, two constants are required in the solution of Equation 1.30, therefore, the solution takes the form

$$u(t) = (C_1 + C_2 t)e^{-\omega_0 t} \tag{1.35}$$

where again the constants $C_1$ and $C_2$ are determined from the initial conditions. At time $t = 0$, we have

$$u(0) = u_0 = C_1$$
$$\dot{u}(0) = -\omega_0 C_1 + C_2$$

Solving yields

$$C_1 = u_0$$
$$C_2 = \omega_0 u_0 + \dot{u}_0$$

And the response is given as

$$u(t) = [u_0 + (\omega_0 u_0 + \dot{u}_0)t]e^{-\omega_0 t} \tag{1.36}$$

There is no vibration in this case either. The displacement versus time profile for both conditions is illustrated in Figure 1.12 for an initial displacement of 1 and in Figure 1.13 for an initial velocity of 1.

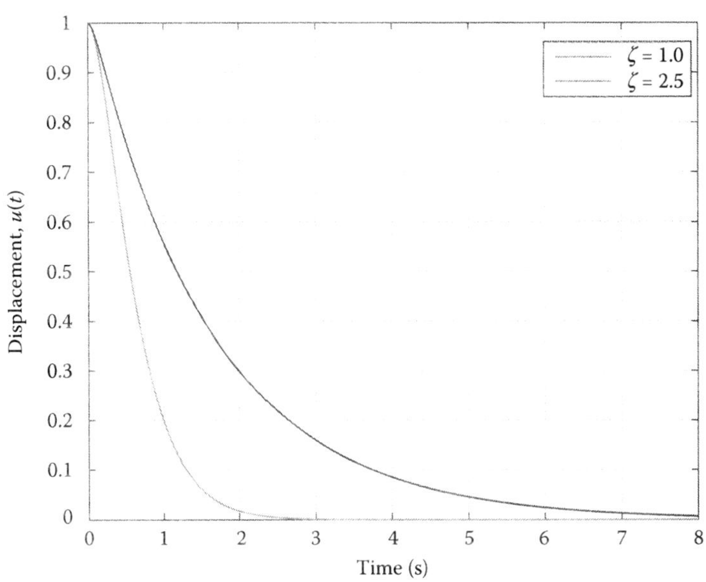


**FIGURE 1.12**
Displacement of critically and overdamped system with an initial displacement.

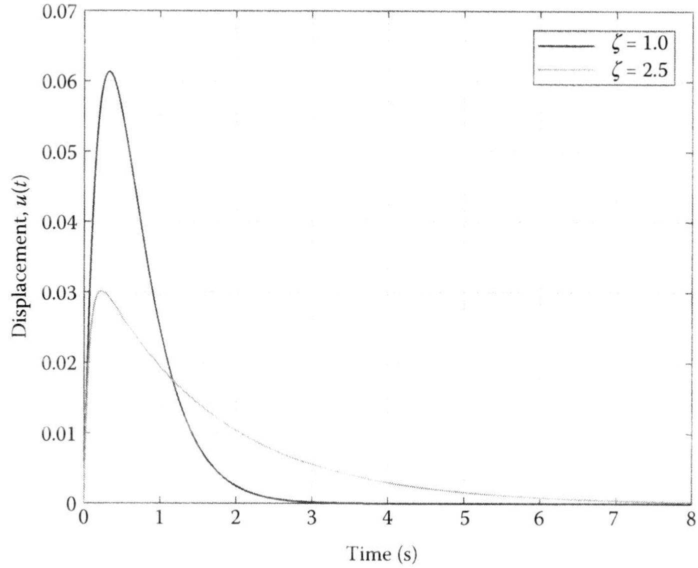


**FIGURE 1.13**
Displacement of critically and overdamped system with an initial velocity.

*Case III:* Small damping, underdamped case ($\zeta < 1$) results in the imaginary roots $r_1 = -\zeta\omega_0 + i\omega_d$ and $r_2 = -\zeta\omega_0 - i\omega_d$, where $\omega_d = \omega_0\sqrt{1-\zeta^2}$ and is termed the damped circular frequency. This results in a solution given as

$$u = e^{-\zeta\omega_0 t}\left(C_1 e^{i\omega_d t} + C_2 e^{-i\omega_d t}\right) \tag{1.37}$$

Using Euler's equations

$$\begin{aligned} e^{i\omega_d t} &= \cos\omega_d t + i\sin\omega_d t \\ e^{-i\omega_d t} &= \cos\omega_d t - i\sin\omega_d t \end{aligned}$$

Equation 1.37 becomes

$$\begin{aligned} u(t) &= e^{-\zeta\omega_0 t}[C_1(\cos\omega_d t + i\sin\omega_d t) + C_2(\cos\omega_d t - i\sin\omega_d t)] \\ &= e^{-\zeta\omega_0 t}[(C_1 + C_2)\cos\omega_d t + i(C_1 - C_2)\sin\omega_d t] \end{aligned}$$

or

$$\begin{aligned} u(t) &= e^{-\zeta\omega_0 t}\left(C_1'\cos\omega_d t + C_2'\sin\omega_d t\right) \\ &= Ae^{-\zeta\omega_0 t}\cos(\omega_d t - \varphi) \\ &= A_0 e^{-\zeta\omega_0 t}\sin(\omega_d t + \theta) \end{aligned} \tag{1.38}$$

where $C_1'$, $C_2'$, $A$, $\varphi$, $A_0$, and $\theta$ are constants that are determined from the initial conditions. Using the initial conditions, that at time $t=0$, $u=u_0$, $\dot{u}=\dot{u}_0$ and $\omega_d t+\theta$, $C_1'$ and $C_2'$ are found as in Equation 1.33 where $s=\omega_d=\omega_0\sqrt{1-\zeta^2}$. Hence,

$$A=A_0=\sqrt{(C_1')^2+(C_2')^2}=\sqrt{u_0^2+\left(\frac{\dot{u}_0}{\omega_d}+\frac{\zeta\omega_0 u_0}{\omega_d}\right)^2} \tag{1.39}$$

and

$$\varphi=\tan^{-1}\left(\frac{C_1'}{C_2'}\right)=\tan^{-1}\left(\frac{u_0\omega_d}{\dot{u}_0+\zeta\omega_0 u_0}\right) \tag{1.40}$$

$$\theta=\tan^{-1}\left(\frac{C_2'}{C_1'}\right)=\tan^{-1}\left(\frac{\dot{u}_0+\zeta\omega_0 u_0}{u_0\omega_d}\right) \tag{1.41}$$

In a system with damping, the motion is neither harmonic nor periodic, as illustrated in Figure 1.14. For this system, the period and frequency are defined as in Figure 1.15.

The period is $T_d=2\pi/\omega_d$ and the frequency is $f_d=\omega_d/2\pi$, where $\omega_d=\omega_0\sqrt{1-\zeta^2}$. Damping increases the period and frequency. From the illustration in Figure 1.15 we note that at time $t'$ we have $u'_{\max}$ for which $\cos(\omega_d t'-\varphi)=1$ and at time $t''$ we have $u''_{\max}$ for which $\cos(\omega_d t''-\varphi)=1$. The cosine term repeats itself if $t''=t'+T_d$. The maximum displacements at times $t'$ and $t''$ are expressed as $u'_{\max}=Ae^{-\zeta\omega_0 t'}$ and $u''_{\max}=Ae^{-\zeta\omega_0(t'+T)}$. Dividing $u'_{\max}$ by $u''_{\max}$ results in

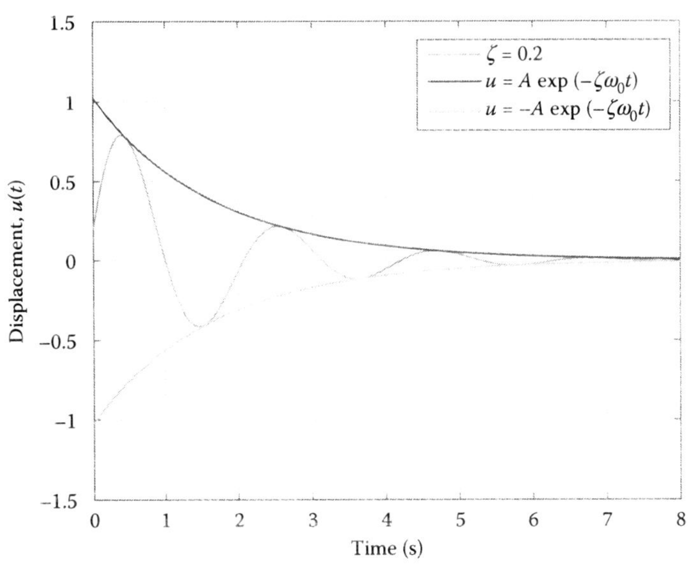


**FIGURE 1.14**
Underdamped case with an initial displacement.

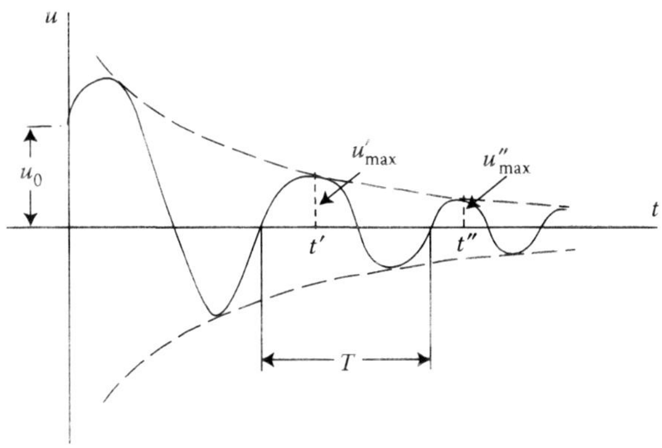


**FIGURE 1.15**
Period and frequency of underdamped case with an initial displacement.

$$\frac{u'_{max}}{u''_{max}} = \frac{Ae^{-\zeta\omega_0 t'}}{Ae^{-\zeta\omega_0 (t'+T_d)}} = e^{\zeta\omega_0 T_d} = e^{\delta}$$

which depends on the damping and period. The quantity $\delta = \zeta\omega_0 T_d$ is called the *logarithmic decrement*. Taking the natural log of $u'_{max}/u''_{max}$ results in a term which can be used to experimentally determine the amount of damping

$$\delta = \ln u'_{max} - \ln u''_{max} = \zeta\omega_0 T = \frac{2\pi\zeta\omega_0}{\omega_d} \tag{1.42}$$

We see that the logarithmic decrement $\delta$ is a measure of the rate of decay of the oscillation. For linear viscous damping $\delta$ = constant, this is not generally true in practice.

**EXAMPLE 1.7**

Consider a mass–spring damped system, as shown in Figure 1.11, having a mass $m = 5$ kg, stiffness coefficient $k = 500$ N/m, and a damping coefficient $\eta = 15$ N s/m. For free vibrations, determine the motion of the system.

**Solution**

The angular frequency of the system is

$$\omega_0 = \sqrt{\frac{k}{m}} = \sqrt{\frac{500}{5}} = 10 \text{ rad/s}$$

The damping ratio is given as $\zeta = \eta/\eta_{cr}$. Since $\eta_{cr} = 2\sqrt{km}$, we find

$$\zeta = \frac{15}{2\sqrt{500(5)}} = 0.15$$

Thus, we have the case of small damping, $\zeta < 1$, and hence

$$\omega_d = \omega_0\sqrt{1-\zeta^2} = 10\sqrt{1-(0.15)^2} = 9.89 \text{ rad/s}$$

Substituting the determined values obtained above into Equation 1.38 gives the motion of the system. Thus,

$$u(t) = Ae^{-9.89t}\cos(9.89t - \varphi) \quad \text{or} \quad u(t) = A_0 e^{-9.89t}\sin(9.89t + \theta)$$

where $A$, $A_0$, $\varphi$, and $\theta$ can be determined from the initial conditions.

**EXAMPLE 1.8**

A plane of mass $m$ is dropping with a vertical velocity "$v_0$." The landing gear consists of a spring of stiffness $k$ and a dashpot with damping coefficient $\eta$ which provides exactly critical damping. The plane is modeled as shown in Figure 1.16. If it is assumed, during landing, that the lift equals the weight, determine what distance, after contact, is required to stop the plane vertically.

**Solution**

For critical damping, we have $\zeta = 1$, and the solution is given by Equation 1.36 with $u_0 = 0$ and $\dot{u}_0 = v_0$ as

$$u(t) = v_0 t e^{-\omega_0 t}$$

To stop the vertical motion of the plane, the velocity $\dot{u}(t)$ must vanish. Therefore,

$$\dot{u}(t) = v_0 e^{-\omega_0 t} - v_0 \omega_0 t e^{-\omega_0 t} = 0, \quad \text{thus} \quad t = \frac{1}{\omega_0}$$

and

$$u = \frac{v_0}{\omega_0 e}$$

**EXAMPLE 1.9**

Consider the system shown in Figure 1.17a with a small mass $m$ fastened to a massless arm of length $l$, with a viscous damper applied at $l/2$. Given the free-body diagram in Figure 1.17b, derive the differential equation of motion. Determine the natural frequency and the critical damping coefficient.

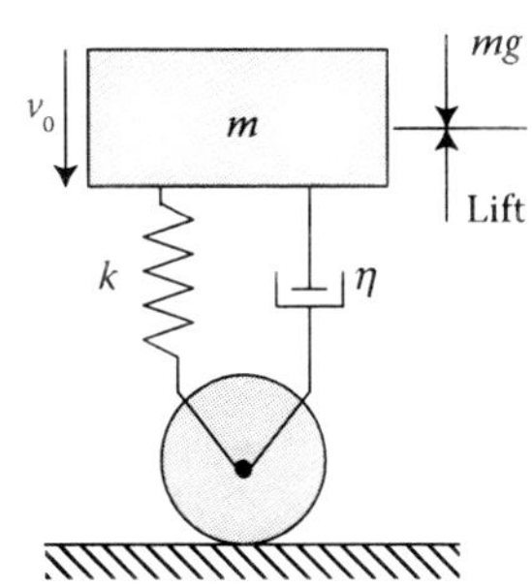


**FIGURE 1.16**
Plane model.

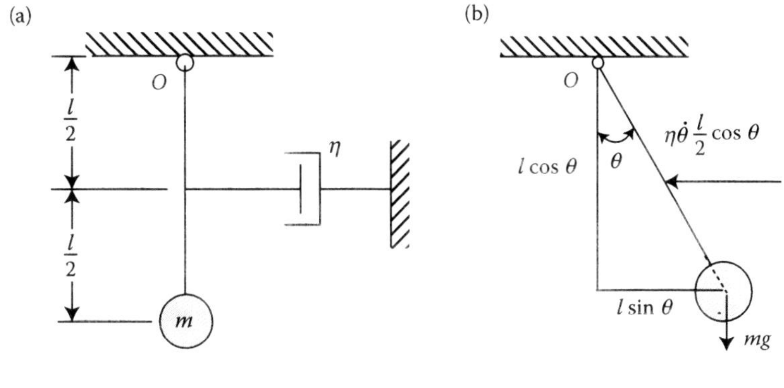


**FIGURE 1.17**
(a) Damped pendulum, (b) free body diagram of damped pendulum.

**Solution**

Applying Newton's second law for rotation about the fixed axis at $O$ gives

$$-mgl\sin\theta - \eta\left(\frac{l}{2}\cos\theta\right)^2\dot{\theta} = mgl^2\ddot{\theta}$$

For small oscillations, $\sin\theta \approx \theta$ and $\cos\theta \approx 1$. Thus, we can write

$$mgl^2\ddot{\theta} + \eta\frac{l^2}{4}\dot{\theta} + mgl\theta = 0$$

or

$$\ddot{\theta} + \frac{\eta}{4mg}\dot{\theta} + \frac{1}{l}\theta = 0$$

The natural frequency of the system is given as $\omega_0 = (1/2)\sqrt{\eta/mg}$. Assuming a solution of the form $\theta = Ce^{rt}$ leads to the characteristic equation $r^2 + (\eta/4mg)r + (1/l) = 0$, which has the roots $r_{1,2} = -(\eta/4mg) \pm \sqrt{(\eta/4mg)^2 - (4/l)}$. For critical damping, we have $\sqrt{(\eta/4mg)^2 - (4/l)} = 0$ or

$$\eta_{cr} = 8mg\sqrt{\frac{1}{l}}$$

## 1.6 Forced Vibration under Harmonic Force

For forced vibrations, the equation for a SDOF system consisting of a mass, spring, and dashpot, subjected to an external force, can be written as

$$m\ddot{u} + \eta\dot{u} + ku = p(t) \tag{1.43}$$

Equation 1.43 is a nonhomogeneous second-order differential equation with constant coefficients. The general solution of this equation consists of two parts. The complementary function, $u_h$, obtained by setting $p(t) = 0$ and solving the homogeneous equation

$$m\ddot{u}_h + \eta\dot{u}_h + ku_h = 0 \tag{1.44}$$

and the particular solution, which depends on the forcing function $p(t)$. The total solution is then given by

$$u(t) = u_h + u_p \tag{1.45}$$

The complementary function is usually called the transient solution since the solution dies out with the presence of damping in the system. The particular solution is referred to as the steady-state solution since the steady-state vibration occurs long after the transient solution has died out.

### 1.6.1 Forced Undamped Harmonic Motion

Let an external harmonic force, which is a sinusoidal function of time, act on the spring–mass system shown in Figure 1.1. The differential equation of motion is given as

$$m\ddot{u} + ku = p_0 \cos\omega_f t$$

or

$$\ddot{u} + \omega_0^2 u = \frac{p_0}{m}\cos\omega_f t \tag{1.46}$$

where $p_0$ is the magnitude of the applied force, $\omega_f$ the frequency of the applied force, and $\omega_0 = \sqrt{k/m}$. The solution of the homogeneous portion of Equation 1.43 is given by Equation 1.9

$$u_h = C_1 \cos\omega_0 t + C_2 \sin\omega_0 t \tag{1.47}$$

For the particular solution we assume, since we only have even derivatives on the left-hand side, that

$$u_p = A\cos\omega_f t \tag{1.48}$$

Substituting Equation 1.47 into Equation 1.45 gives

$$\left(-\omega_f^2 + \omega_0^2\right)A\cos\omega_f t = \frac{p_0}{m}\cos\omega_f t$$

or

$$A = \frac{p_0}{m\left(\omega_0^2 - \omega_f^2\right)} = \frac{p_0}{k\left[1-(\omega_f/\omega_0)^2\right]} = \frac{u_{static}}{1-(\omega_f/\omega_0)^2} \tag{1.49}$$

where $u_{static} = p_0/k$, that is, the deflection of the spring under a static force. The total solution can be expressed in the form

$$u(t) = C_1 \cos \omega_0 t + C_2 \sin \omega_0 t + \frac{p_0}{k - m\omega_f^2} \cos \omega_f t \tag{1.50}$$

Let the initial conditions be $u(0) = u_0$ and $\dot{u}(0) = \dot{u}_0$. Then Equation 1.49 yields

$$C_1 = u_0 - \frac{p_0}{k - \omega_f^2}, \quad C_2 = \frac{\dot{u}_0}{\omega_0} \tag{1.51}$$

Substituting Equation 1.50 into Equation 1.47 yields the total response

$$u(t) = \frac{\dot{u}_0}{\omega_0} \sin \omega_0 t + \left( u_0 - \frac{p_0}{k - m\omega_f^2} \right) \cos \omega_0 t + \frac{p_0}{k - m\omega_f^2} \cos \omega_f t \tag{1.52}$$

The maximum amplitude $A$ in Equation 1.48 can be rewritten as

$$\frac{A}{u_{static}} = \frac{1}{1 - (\omega_f/\omega_0)^2} = \frac{1}{1 - \Omega^2} \tag{1.53}$$

where $\Omega = \omega_f/\omega_0$ represents the frequency ratio. The quantity $A/u_{static}$ represents the ratio of the dynamic amplitude of motion to the static amplitude and is called the *amplification factor, magnification factor,* or the *amplitude ratio.* A plot of Equation 1.53 is shown in Figure 1.18.

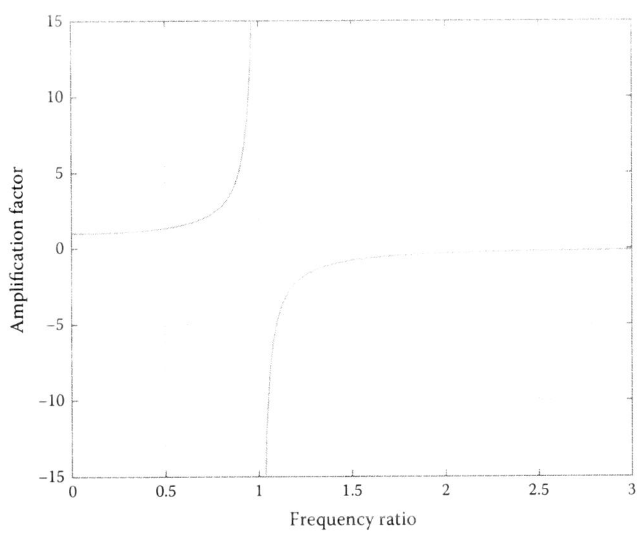


**FIGURE 1.18**
Amplification factor vs frequency ratio.

*Case I:* When $\Omega < 1$ or $\omega_f < \omega_0$, the value is positive, indicates that $u(t)$ and $p(t)$ have the same algebraic sign. The response is given by

$$u_p = A\cos\omega_f t \tag{1.54}$$

The system would move in the same direction as the force and the displacement would be said to be in phase with the excitation as shown in Figure 1.19.

*Case II:* When $\Omega > 1$ or $\omega_f > \omega_0$, the value is negative, which indicates that $u(t)$ and $p(t)$ have opposing algebraic signs. Thus, the system would act to the left if the force acts to the right. The displacement is said to be out of phase with respect to the applied force. The steady-state response can be expressed as

$$u_p = -A\cos\omega_f t \tag{1.55}$$

where the amplitude $A$ is rewritten as a positive quantity. The response is said to be 180° out of phase with the excitation as shown in Figure 1.20.

Therefore, $u_p$ is rewritten as

$$u_p = \frac{u_{static}}{(\Omega^2 - 1)}\cos\omega_f t \tag{1.56}$$

*Case III:* When $\Omega = 1$ or $\omega_f = \omega_0$, the amplitude given by Equation 1.49 or Equation 1.53 becomes infinite. This occurs when the forcing frequency is equal to the natural frequency of the system. This phenomenon is called *resonance*. We need to return to Equation 1.52 which we have rewritten in the following form:

$$u(t) = u_0\cos\omega_0 t + \frac{\dot{u}_0}{\omega_0}\sin\omega_0 t + \frac{p_0}{k}\left[\frac{\cos\omega_f t - \cos\omega_0 t}{1-(\omega_f/\omega_0)^2}\right] \tag{1.57}$$

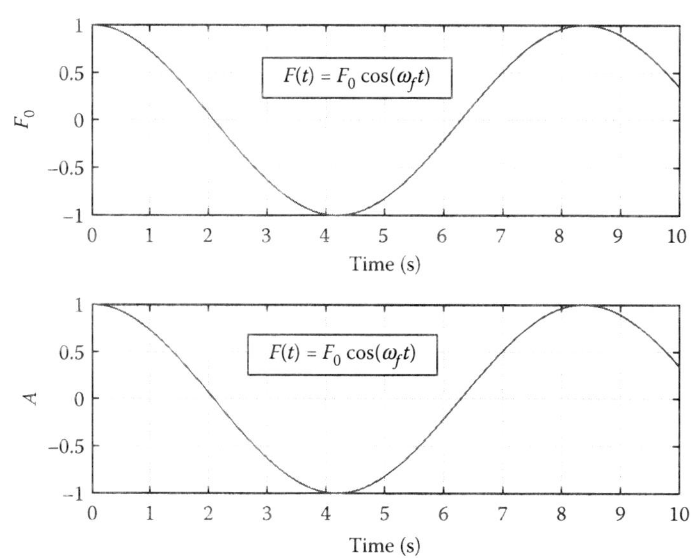


**FIGURE 1.19**
In phase force and displacement.

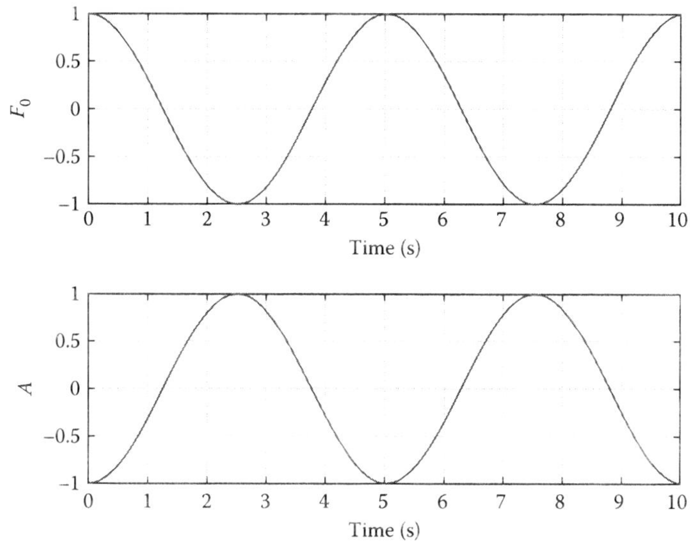


**FIGURE 1.20**
Out of phase force and displacement.

We note that if $\omega_f = \omega_0$, the last term of Equation 1.57 takes an indefinite form 0/0. Because the quotient 0/0 appears, we use L'Hospital's rule and differentiate both numerator and denominator with respect to $\omega_f$ to evaluate the limit of the expression

$$\lim_{\omega_f \to \omega_0}\left[\frac{\cos\omega_f t - \cos\omega_0 t}{1-(\omega_f/\omega_0)^2}\right] = \lim_{\omega_f \to \omega_0}\frac{(d/d\omega_f)(\cos\omega_f t - \cos\omega_0 t)}{(d/d\omega_f)\left[1-(\omega_f/\omega_0)^2\right]} = \lim_{\omega_f \to \omega_0}\frac{t\sin\omega_f t}{2\omega_f/\omega_0^2} = \frac{\omega_0}{2}t\sin\omega_0 t \quad (1.58)$$

Thus the solution corresponding to resonance is given as

$$u(t) = u_0\cos\omega_0 t + \frac{\dot{u}_0}{\omega_0}\sin\omega_0 t + \left(\frac{p_0\omega_0}{2k}\right)t\sin\omega_0 t \quad (1.59)$$

We see from Equation 1.59 that the amplitude grows indefinitely as $t$ increases. The result is shown in Figure 1.21.

When the forcing frequency $\omega_f$ is close, but not equal, to the natural frequency $\omega_0$ of the system, a phenomenon called *beating* can occur. In order to investigate this phenomenon, we will need to look at Equation 1.52. If the initial conditions are taken to be zero, that is, $u_0 = \dot{u}_0 = 0$, then we have

$$u(t) = \frac{p_0}{k}\frac{(\cos\omega_f t - \cos\omega_0 t)}{1-(\omega_f/\omega_0)^2} = \frac{(p_0/m)}{\omega_0^2 - \omega_f^2}(\cos\omega_f t - \cos\omega_0 t) \quad (1.60)$$

A plot of Equation 1.60 is shown in Figure 1.22. To obtain some idea of the phenomenon, we use the identity

$$\cos\alpha - \cos\beta = -2\sin\frac{(\alpha+\beta)t}{2}\sin\frac{(\alpha-\beta)t}{2} = 2\sin\frac{(\alpha+\beta)t}{2}\sin\frac{(\beta-\alpha)t}{2}$$

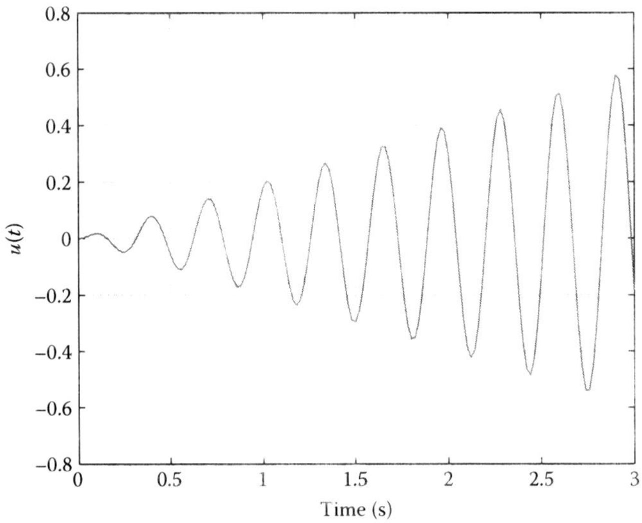


**FIGURE 1.21**
Response as resonance is approached.

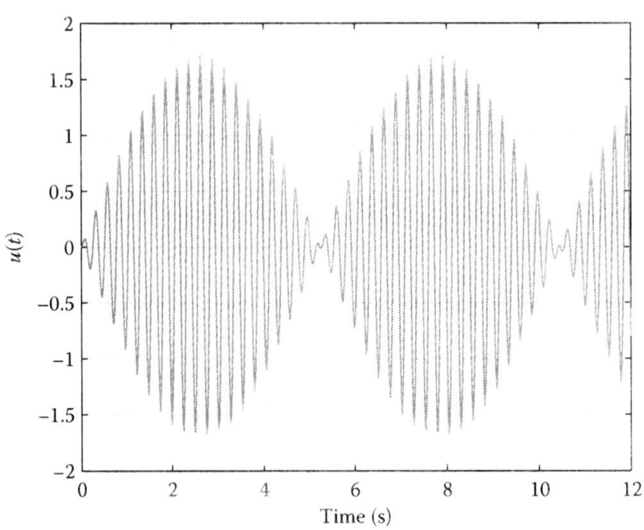


**FIGURE 1.22**
Beating phenomenon.

thus, Equation 1.60 becomes

$$u(t) = \left[\frac{2p_0}{m\left(\omega_0^2 - \omega_f^2\right)} \sin\frac{(\omega_0 - \omega_f)t}{2}\right] \sin\frac{(\omega_0 + \omega_f)t}{2} \tag{1.61}$$

When $(\omega_0 - \omega_f)$ is small, $(\omega_0 + \omega_f)$ is large and $\sin[(\omega_0 + \omega_f)t/2]$ oscillates more rapidly than $\sin[(\omega_0 - \omega_f)t/2]$, therefore, the motion is a rapid oscillation having frequency $(\omega_0 + \omega_f)t/2$ with a slowly varying sinusoidal amplitude

$$\frac{2p_0}{m\left(\omega_0^2-\omega_f^2\right)}\sin\frac{(\omega_0-\omega_f)t}{2}$$

### 1.6.2 Forced Damped Harmonic Motion

For damped harmonic motion, we take the equation in the following form

$$\ddot{u}+2\omega_0\zeta\dot{u}+\omega_0^2u=\left(\frac{p_0}{m}\right)\cos\omega_f t \tag{1.62}$$

Equation 1.61 is a second-order nonhomogeneous differential equation whose solution consists of two parts, the *complementary* solution and the *particular* solution. The complementary solution is the same as that obtained for the free vibration of a damped system that was given above, Equations 1.30, 1.32, or 1.35 depending on the damping ratio $\zeta$. Assuming the damping to be small, that is, less than critical, the complementary solution is given by Equation 1.37. For the particular solution, we can assume a solution of the form

$$u_p=A\cos(\omega_f t-\varphi) \tag{1.63}$$

where $A$ is the amplitude of the oscillation and $\varphi$ is the phase of the displacement with respect to the forcing term. Substituting Equation 1.63 into Equation 1.62 gives

$$A\left[\left(\omega_0^2-\omega_f^2\right)\cos(\omega_f t-\varphi)-2\omega_0\omega_f\zeta\sin(\omega_f t-\varphi)\right]=\left(\frac{p_0}{m}\right)\cos\omega_f t \tag{1.64}$$

Substituting the fundamental trigonometric relations

$$\begin{aligned}\cos(\omega_f t-\varphi)&=\cos\omega_f t\cos\varphi+\sin\omega_f t\sin\varphi\\ \sin(\omega_f t-\varphi)&=\sin\omega_f t\cos\varphi-\cos\omega_f t\sin\varphi\end{aligned} \tag{1.65}$$

into Equation 1.64 yields

$$\begin{aligned}A\left[\left(\omega_0^2-\omega_f^2\right)\cos\varphi+2\omega_0\omega_f\zeta\sin\varphi\right]&=\left(\frac{p_0}{m}\right)\\ A\left[\left(\omega_0^2-\omega_f^2\right)\sin\varphi-2\omega_0\omega_f\zeta\cos\varphi\right]&=0\end{aligned} \tag{1.66}$$

Equation 1.66 can be solved to give the amplitude and phase of the particular solution. This yields

$$A=\frac{p_0}{k}\frac{1}{\sqrt{\left[1-(\omega_f/\omega_0)^2\right]+[2\zeta(\omega_f/\omega_0)]^2}} \tag{1.67}$$

and

$$\varphi = \tan^{-1}\left\{\frac{2\zeta(\omega_f/\omega_0)}{\left[1-(\omega_f/\omega_0)^2\right]}\right\} \tag{1.68}$$

where $\omega_0^2 = k/m$ has been used. Using the frequency ratio $\Omega = \omega_f/\omega_0$, the above equations can be written as

$$A = \frac{p_0}{k}\frac{1}{\sqrt{(1-\Omega^2)^2+(2\zeta\Omega)^2}} \tag{1.69}$$

and

$$\varphi = \tan^{-1}\left(\frac{2\zeta\Omega}{1-\Omega^2}\right) \tag{1.70}$$

The *amplitude ratio, amplification factor,* or the *magnification factor* is

$$X = \frac{A}{u_{static}} = \frac{1}{\sqrt{(1-\Omega^2)^2+(2\zeta\Omega)^2}} \tag{1.71}$$

A plot of the amplification factor as a function of frequency ratio $\Omega$ yields a family of curves which are dependent on $\zeta$ as shown in Figure 1.23. These curves are called *frequency response curves.*

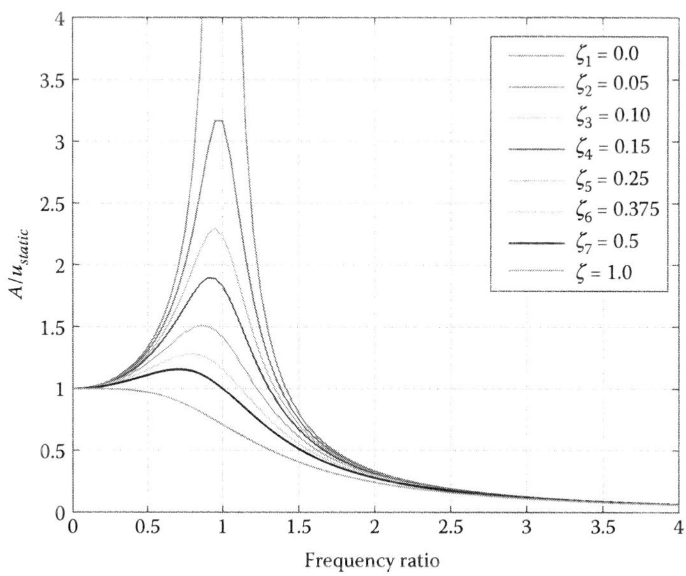


**FIGURE 1.23**
Amplification factor vs frequency ratio for forced damped harmonic motion.

To obtain the maximum and minimum points, Equation 1.71 is differentiated with respect to $\Omega$ and set equal to zero. This leads to

$$\frac{dX}{d\Omega}=\frac{\Omega(1-\Omega^2-2\zeta^2)}{[(1-\Omega^2)^2+(2\zeta\Omega)^2]^{3/2}}=0 \tag{1.72}$$

From Equation 1.71 and Figure 1.23, the following is noted:

1. The initial point on the curves $X=1$ for $\Omega=0$. This will be a minimum point if $\zeta<0.707$.
2. The amplification factor, and therefore the response of the system, is a maximum when the frequency ratio is near $\Omega=1$, depending on which value of damping ratio $\zeta$ is used.
3. The amplification factor approaches zero as $\Omega\to\infty$.
4. From Equation 1.72, the maximum value of the amplification factor occurs for a frequency ratio

$$1-\Omega^2-2\zeta^2=0 \quad \text{or} \quad \Omega=\sqrt{1-2\zeta^2} \tag{1.73}$$

and the corresponding maximum value of the amplification factor $X$ is

$$X_{\text{max}}=\frac{1}{2\zeta\sqrt{1-\zeta^2}} \tag{1.74}$$

The corresponding plot of the phase angle as a function of frequency ratio $\Omega$ is shown in Figure 1.24. As noted, the phase angle depends on the damping factor $\zeta$ and the frequency ratio $\Omega$.

With no damping, $\varphi=0$ from $\Omega=0$ to $\Omega=1$, $\varphi=90°$ for $\Omega=1$, and $\varphi=180°$ for $\Omega>1$. For small values of damping $\zeta\ll 1$, the curve approaches the curve for the zero damping case. For $\Omega=1$, all curves go through the point $\varphi=90°$.

For the case of small damping, the complete solution results in

$$u(t)=A_0e^{-\zeta\omega_0 t}\cos(\omega_d t-\varphi_0)+A\cos(\omega_f t-\varphi) \tag{1.75}$$

where

$$A=\frac{p_0}{k}\frac{1}{\sqrt{(1-\Omega^2)^2+(2\zeta\Omega)^2}},\ \varphi=\tan^{-1}\left(\frac{2\zeta\Omega}{1-\Omega^2}\right),\ \Omega=\frac{\omega_f}{\omega_0}$$

and $A_0$ and $\varphi_0$ can be found from the initial conditions of the problem. From Equation 1.75, and using the initial conditions that at time $t=0$, $u=u_0$ and $\dot{u}=\dot{u}_0$, where $u_0$ and $\dot{u}_0$ are given as

$$\left.\begin{aligned} u_0&=A_0\cos\varphi_0+A\cos\varphi\\ \dot{u}_0&=A_0[-\zeta\omega_0\cos\varphi_0+\omega_d\sin\varphi_0]+\omega_f A\sin\varphi \end{aligned}\right\} \tag{1.76}$$

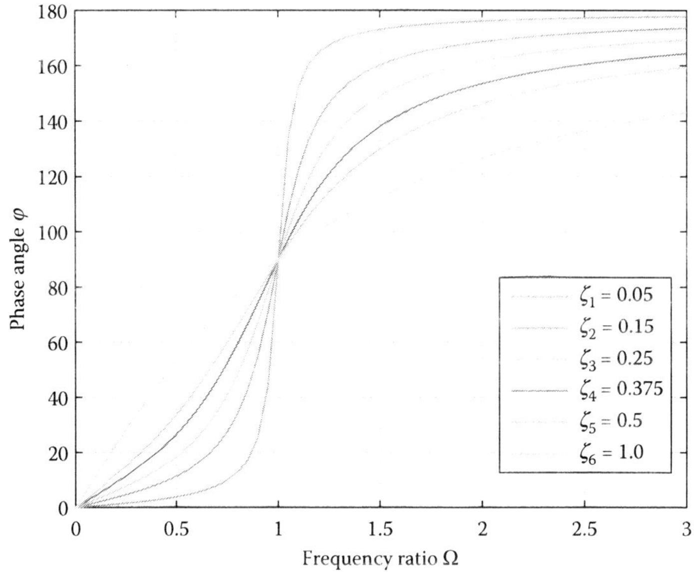


**FIGURE 1.24**
Phase angle vs frequency ratio for forced damped harmonic motion.

From Equations 1.76, we have

$$A_0 \cos\varphi_0 = u_0 - A\cos\varphi$$
$$A_0 \sin\varphi_0 = \frac{1}{\omega_d}[\dot{u}_0 + \zeta\omega_0(u_0 - A\cos\varphi) - \omega_f A\sin\varphi]$$

and using the trigonometric relation $\sin^2\varphi_0 + \cos^2\varphi_0 = 1$ we arrive at the solution for the constants $A_0$ and $\varphi_0$ given as

$$\left.\begin{aligned} A_0 &= \left[(u_0 - A\cos\varphi)^2 + \frac{1}{\omega_d^2}(\zeta\omega_0 u_0 + \dot{u}_0 - \zeta\omega_0 A\cos\varphi_0 - \omega_f A\sin\varphi)^2\right]^{1/2} \\ \varphi_0 &= \tan^{-1}\left(\frac{\sin\varphi_0}{\cos\omega_0}\right) = \tan^{-1}\frac{\zeta\omega_0 u_0 + \dot{u}_0 - \zeta\omega_0 A\cos\varphi_0 - \omega_f A\sin\varphi}{\omega_d(u_0 - A\cos\varphi)} \end{aligned}\right\} \quad (1.77)$$

### 1.6.3 Forced Damped Harmonic Motion Using Complex Format

As before, the total solution will contain both the complementary and particular solutions. The complementary solution, obtained when the right-hand side of the equation is set equal to zero, is the transient solution, since it tends to zero as $t$ tends to infinity. Therefore, to obtain the particular or steady-state solution for force-damped harmonic motion, one can also use complex algebra. Our equation is given by Equation 1.62

$$\ddot{u} + 2\omega_0\zeta\dot{u} + \omega_0^2 u = \left(\frac{p_0}{m}\right)\cos\omega_f t \quad (1.62)$$

where the forcing function is $(p_0/m)\cos\omega_f t$. Let $p(t)$ be a harmonic function and using complex algebra the harmonic function can be taken as

$$p(t) = \text{Re}\left[\left(\frac{p_0}{m}\right)e^{i\omega_f t}\right] = \left(\frac{p_0}{m}\right)\text{Re}(\cos\omega_f t + i\sin\omega_f t) = \left(\frac{p_0}{m}\right)\cos\omega_f t \tag{1.78}$$

where Re denotes the real part, $p_0/m$ is a real constant and $i = \sqrt{-1}$. If we had a sinusoidal forcing function, we would have

$$p(t) = \text{Im}\left[\left(\frac{p_0}{m}\right)e^{i\omega_f t}\right] = \left(\frac{p_0}{m}\right)\text{Im}(\cos\omega_f t + i\sin\omega_f t) = \left(\frac{p_0}{m}\right)\sin\omega_f t \tag{1.79}$$

where Im denotes the imaginary part of the forcing function. The corresponding steady-state response function may be expressed as

$$u(t) = Ue^{i\omega_f t} \tag{1.80}$$

where $U$ is the complex amplitude of the displacement. Substituting Equations 1.80 and 1.62 yields

$$\left(\omega_0^2 - \omega_f^2 + 2i\omega_0\omega_f\right)Ue^{i\omega_f t} = \left(\frac{p_0}{m}\right)e^{i\omega_f t} \tag{1.81}$$

Dividing both sides of Equation 1.81 by $e^{i\omega_f t}$ and solving for $U$ gives

$$U = \frac{p_0/m}{\omega_0^2 - \omega_f^2 + 2i\omega_0\omega_f\zeta} = \frac{p_0/k}{1-\Omega^2 + 2i\zeta\Omega} \tag{1.82}$$

Multiplying the numerator and denominator of Equation 1.82 by its complex conjugate yields

$$U = \frac{p_0/k}{1-\Omega^2 + 2i\zeta\Omega}\left[\frac{1-\Omega^2 - 2i\zeta\Omega}{1-\Omega^2 - 2i\zeta\Omega}\right] = \frac{p_0}{k}\left[\frac{1-\Omega^2}{(1-\Omega^2)^2 + (2\zeta\Omega)^2} - i\frac{2\zeta\Omega}{(1-\Omega^2)^2 + (2\zeta\Omega)^2}\right] \tag{1.83}$$

Recall that

$$\left.\begin{aligned} x + iy &= re^{i\varphi}, \text{ where } r = \sqrt{x^2 + y^2} \\ \cos\varphi &= \frac{x}{r}, \text{ and } \sin\varphi = \frac{y}{r} \\ \varphi &= \tan^{-1}\frac{y}{x} \end{aligned}\right\} \tag{1.84}$$

Therefore, Equation 1.83 can be written as

$$U = \frac{p_0/k}{\sqrt{(1-\Omega^2)^2 + (2\zeta\Omega)^2}}e^{-i\varphi} \tag{1.85}$$

and

$$\varphi = \tan^{-1}\left(\frac{2\zeta\Omega}{1-\Omega^2}\right) \tag{1.86}$$

The particular or steady-state solution Equation 1.80 is

$$u(t) = \frac{p_0/k}{\sqrt{(1-\Omega^2)^2 + (2\zeta\Omega)^2}}e^{-i\varphi}\cdot e^{i\omega_f t} = \frac{p_0/k}{\sqrt{(1-\Omega^2)^2 + (2\zeta\Omega)^2}}e^{i(\omega_f t-\varphi)} \tag{1.87}$$

This is the same as Equation 1.63 if Equations 1.69 and 1.70 are substituted into Equation 1.63. The nondimensional *frequency response function* can be defined as

$$H(i\omega_f) = \frac{U}{p_0/k} = \frac{1}{1-\Omega^2 + 2i\zeta\Omega} \tag{1.88}$$

And its absolute value as

$$|H(i\omega_f)| = \left|\frac{U}{p_0/k}\right| = \frac{1}{\sqrt{(1-\Omega^2)^2 + (2\zeta\Omega)^2}} \tag{1.89}$$

It is seen that

$$u(t) = \frac{p_0}{k}|H(i\omega_f)|\cos(\omega_f t - \varphi) \tag{1.90}$$

Equation 1.89 is the *amplitude ratio, amplification factor,* or the *magnification factor* given in Equation 1.71.

### 1.6.4 Forced Harmonic Motion of the Support

Assume the point to which the spring, viscous damping element, and mass are attached can also move, as illustrated in Figure 1.25.

Since the motion of this point is time dependent, it can have both a velocity and acceleration. The original equation of motion must be modified since the spring and damping forces are no longer the same. Due to the motion of the attachment point, the spring and damping forces are $F_s = -k(u - v)$ and $F_d = -\eta(\dot{u} - \dot{v})$. As a result of this, the equation of motion becomes

$$m\ddot{u} + \eta\dot{u} + ku = \eta\dot{v} + kv \tag{1.91}$$

Defining the right-hand side of Equation 1.91 as the driving force $p(t) = \eta\dot{v} + kv$ allows us to write

$$m\ddot{u} + \eta\dot{u} + ku = p(t) \tag{1.92}$$

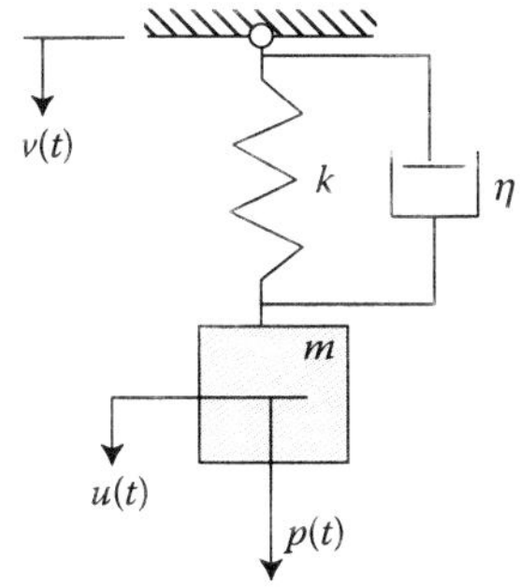


**FIGURE 1.25**
Model of forced harmonic motion of a support.

The solution will have both complementary and particular solutions. For a system with no damping, $\eta = 0$, the solution takes the form

$$m\ddot{u} + ku = kv \tag{1.93}$$

Suppose $v(t) = V_0 \sin \omega_f t$, where $V_0$ and $\omega_f$ are the amplitude and frequency, respectively. The particular solution is a harmonic function $u_p(t) = V \sin \omega_f t$. Substituting this into Equation 1.93 results in

$$\left(k - m\omega_f^2\right)V = kV_0$$

or

$$V = \frac{kV_0}{k - m\omega_f^2} = \frac{V_0}{1-\Omega^2} \tag{1.94}$$

where $\omega_0 = \sqrt{k/m}$ is the natural frequency of the system and $\Omega = \omega_f/\omega_0$ is the frequency ratio. Figure 1.26 shows the variation of $|V|/V_0$ verses $\Omega$. From this we note that if $\Omega \ll 1$, then $V \approx V_0$ (springs do not do anything). If $\Omega = 1$, then $V \to \infty$. Finally, if $\Omega \gg 1$, then $V \to 0$. In order to keep $V$ small, we need a large mass and a soft spring.

For a system with damping $\eta \neq 0$, our equation becomes with $v(t) = V_0 \sin \omega_f t$

$$m\ddot{u} + \eta\dot{u} + ku = kV_0 \sin\omega_f t + \eta\omega_f V_0 \cos\omega_f t \tag{1.95}$$

or

$$\ddot{u} + 2\omega_0\zeta\dot{u} + \omega_0^2 u = V_0\omega_0^2\sqrt{1+(2\zeta\Omega)^2}\,\sin(\omega_f t + \alpha) \tag{1.96}$$

where we have used $\eta = 2m\omega_0\zeta$, $\omega_0^2 = k/m$, and $\Omega = \omega_f/\omega_0$. The steady-state solution can be taken as

$$u_p(t) = A\sin(\omega_f t + \alpha - \theta) \tag{1.97}$$

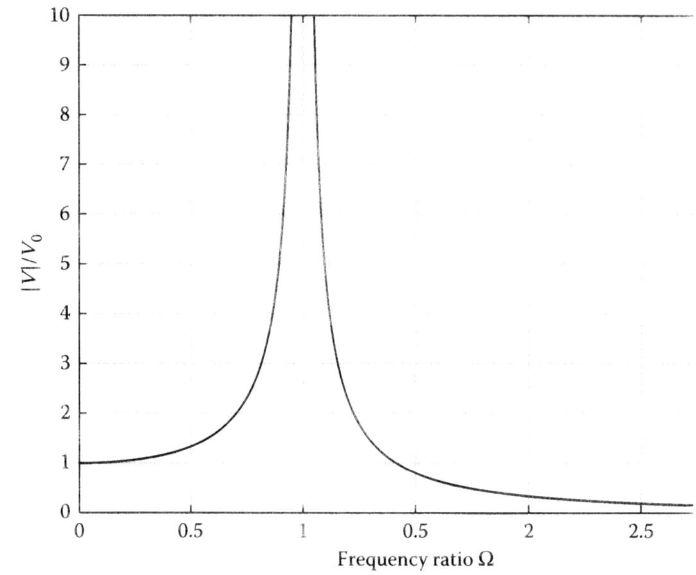


**FIGURE 1.26**
Amplification factor vs frequency ratio.

where the amplitude $A$ and phase angle $\theta$ can be found by substituting Equation 1.97 into Equation 1.96. This yields

$$A = \frac{V_0\sqrt{1+(2\zeta\Omega)^2}}{\sqrt{(1-\Omega^2)^2+(2\zeta\Omega)^2}} \tag{1.98}$$

and

$$\tan\theta = \frac{2\zeta\Omega}{1-\Omega^2} \tag{1.99}$$

The particular or steady-state solution is

$$u_p(t) = \frac{V_0\sqrt{1+(2\zeta\Omega)^2}}{\sqrt{(1-\Omega^2)^2+(2\zeta\Omega)^2}}\sin(\omega_f+\alpha-\theta) \tag{1.100}$$

The relative transmission of the support motion to the support is thus given by

$$\frac{A}{V_0} = \frac{\sqrt{1+(2\zeta\Omega)^2}}{\sqrt{(1-\Omega^2)^2+(2\zeta\Omega)^2}} \tag{1.101}$$

Note that the complete solution requires the complementary solution which depends upon the system being underdamped, critically damped, or overdamped. The ratio of the steady-state amplitude $A$ of the mass and the base amplitude $V_0$ is known as the *displacement transmissibility*, $T_d = A/V_0$, thus

$$T_d = \frac{A}{V_0} = \frac{\sqrt{1+(2\zeta\Omega)^2}}{\sqrt{(1-\Omega^2)^2+(2\zeta\Omega)^2}} \tag{1.102}$$

Equation 1.97 can be rewritten in a different more standard form as

$$u_p(t) = A\sin(\omega_f t - \varphi) \tag{1.103}$$

where

$$\varphi = \theta - \alpha,\ \tan\alpha = \frac{\eta\omega_f}{k} = 2\zeta\Omega \tag{1.104}$$

and

$$A = \frac{V_0\sqrt{1+(2\zeta\Omega)^2}}{\sqrt{(1-\Omega^2)^2+(2\zeta\Omega)^2}} \tag{1.105}$$

Based on Equation 1.103 and the fact that the base displacement is $V_0 \sin \omega_f t$, the phase angle $\varphi$ is the angle by which the steady-state displacement $u_p(t)$ lags behind the displacement of the base. We have that

$$\varphi = \tan^{-1}\left(\frac{\tan\theta - \tan\alpha}{1+\tan\theta\tan\alpha}\right) = \tan^{-1}\left(\frac{2\zeta\Omega/(1-\Omega^2) - 2\zeta\Omega}{1+(2\zeta\Omega)^2/(1-\Omega^2)}\right) = \tan^{-1}\left(\frac{2\zeta\Omega^3}{1-\Omega^2+(2\zeta\Omega)^2}\right) \tag{1.106}$$

A plot of Equations 1.102 and 1.106 are shown for different values of $\zeta$ and $\Omega$ in Figures 1.27 and 1.28, respectively.

It is noted that in Figure 1.27, all curves start at $T_d = 1$ for $\Omega = 1$ and that there is a crossover point at $T_d = 1$, $\Omega = \sqrt{2} = 1.414$. Larger values of $T_d$ occur for $\Omega < 1$ and small damping ratios and smaller values of $T_d$ for $\Omega > 1$ and small damping ratios. The peak value of the curve for $\Omega = \Omega_{T_d} < 1$ can be obtained by differentiating Equation 1.102 with respect to $\Omega$ and set the results to zero. Thus,

$$\frac{dT_d}{d\Omega} = \frac{d}{d\Omega}\sqrt{\frac{1+(2\zeta\Omega)^2}{(1-\Omega^2)^2+(2\zeta\Omega)^2}} = 0 \tag{1.107}$$

This leads to $2\zeta^2\Omega^4 + \Omega^2 - 1 = 0$, which can be solved to give

$$\Omega_{T_d} = \frac{\sqrt{\sqrt{1+8\zeta^2} - 1}}{2\zeta} \tag{1.108}$$

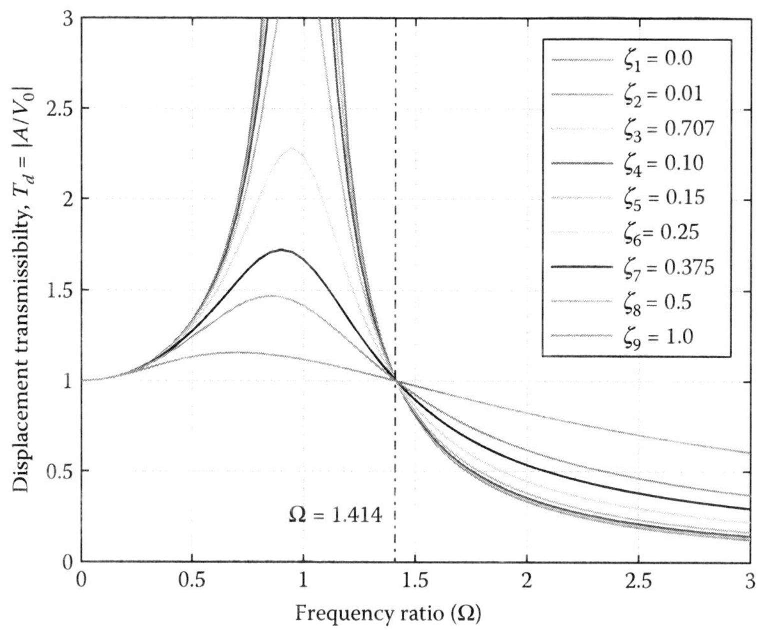


**FIGURE 1.27**
Displacement transmissibility vs frequency ratio.

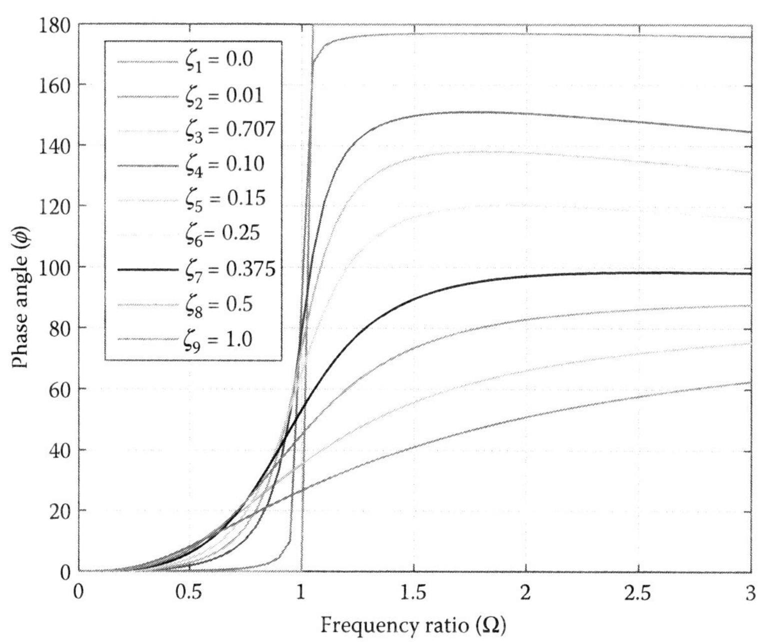


**FIGURE 1.28**
Phase angle vs frequency ratio.

### 1.6.5 Force Transmitted to Base

Consider the force transmitted to the base of Figure 1.25. The total force that is exerted by the spring $k$ and damper $\eta$ is given by

$$f_T = \eta \dot{u} + ku \tag{1.109}$$

For steady-state motion $u(t) = A \sin(\omega_f t - \gamma)$, Equation 1.109 becomes

$$f_T = \eta \omega_f A \cos(\omega_f t - \gamma) + kA \sin(\omega_f t - \gamma) \tag{1.110}$$

where $A$ and $\gamma$ are given by Equations 1.69 and 1.70 as

$$A = \frac{p_0}{k} \frac{1}{\sqrt{(1-\Omega^2)^2 + (2\zeta\Omega)^2}}, \quad \tan\gamma = \frac{2\zeta\Omega}{1-\Omega^2}$$

Equation 1.110 can be written in the form

$$f_T = A\sqrt{k^2 + (\eta\omega_f)^2} \sin(\omega_f t - \gamma + \beta) = kA\sqrt{1+(2\zeta\Omega)^2} \sin(\omega_f t - \varphi) \tag{1.111}$$

where

$$\varphi = \gamma - \beta \text{ and } \tan\beta = 2\zeta\Omega \tag{1.112}$$

The maximum value of the force $f_T$ or the force amplitude that will be transmitted is

$$f_T = Ak\sqrt{1+(2\zeta\Omega)^2} = p_0 \frac{\sqrt{1+(2\zeta\Omega)^2}}{\sqrt{(1-\Omega^2)^2 + (2\zeta\Omega)^2}} \tag{1.113}$$

The transmissibility, TR, or the ratio $f_T/p_0$ is

$$TR = \frac{f_T}{p_0} = \frac{\sqrt{1+(2\zeta\Omega)^2}}{\sqrt{(1-\Omega^2)^2 + (2\zeta\Omega)^2}} \tag{1.114}$$

The right-hand side of Equation 1.114 is the same as the *displacement transmissibility* Equation 1.102 which is plotted in Figure 1.27 along with the phase angle which is plotted in Figure 1.28.

## 1.7 Forced Response to Periodic Loading

In general, many engineering problems exist where the external force is of a more general nature, rather than harmonic. A general forcing function may be periodic or nonperiodic. Periodic forces are those that repeat themselves in time while nonperiodic forces do not repeat themselves in time. A step function would be one example of a nonperiodic forcing function. It is necessary to determine the response of a system to these general forcing functions.

### 1.7.1 Trigonometric Functions and Fourier Series

An external force that repeats itself in equal periods of time $T$ is referred as a periodic force (Greenberg 1998). Thus,

$$p(t) = p(t + nT),\ n = \pm 1, \pm 2, \pm 3, \ldots \tag{1.115}$$

If $p(t)$ is periodic with period $T$, it is necessarily periodic with period $2T$, $3T$, $4T$, etc. An example of a general periodic forcing function, $p(t)$, with period $T$ is shown in Figure 1.29.

The fundamental frequency of a periodic function $\omega_f$ with period $T$ is given by

$$\omega_f = \frac{2\pi}{T} \tag{1.116}$$

A periodic function $p(t)$ can be represented as a Fourier series having the form

$$p(t) = \frac{a_0}{2} + \sum_{n=1}^{\infty} [a_n \cos(n\omega_f t) + b_n \sin(n\omega_f t)] \tag{1.117}$$

where $a_0$, $a_n$, and $b_n$ are constant coefficients. Equation 1.117 holds under the assumption that the right-hand side actually exists, meaning that it converges for all $t$.

We assume that the function $p(t)$ is given but that the constant coefficients $a_0$, $a_n$, and $b_n$ are unknown. To determine the constants, recall the orthogonality conditions for the harmonic functions, $\sin(n\omega_f t)$ and $\cos(n\omega_f t)$, given as

$$\left.\begin{aligned}
&\int_{-T/2}^{T/2} \cos(m\omega_f t)\cos(n\omega_f t)dt = 0 && m \neq n \\
&\int_{-T/2}^{T/2} \cos(m\omega_f t)\sin(n\omega_f t)dt = 0 && m \neq n \\
&\int_{-T/2}^{T/2} \sin(m\omega_f t)\sin(n\omega_f t)dt = 0 && m \neq n \\
&\int_{-T/2}^{T/2} \cos^2(n\omega_f t)dt = \frac{T}{2},\quad \int_{-T/2}^{T/2} \sin^2(n\omega_f t)dt = \frac{T}{2} && n \neq 0
\end{aligned}\right\} \tag{1.118}$$

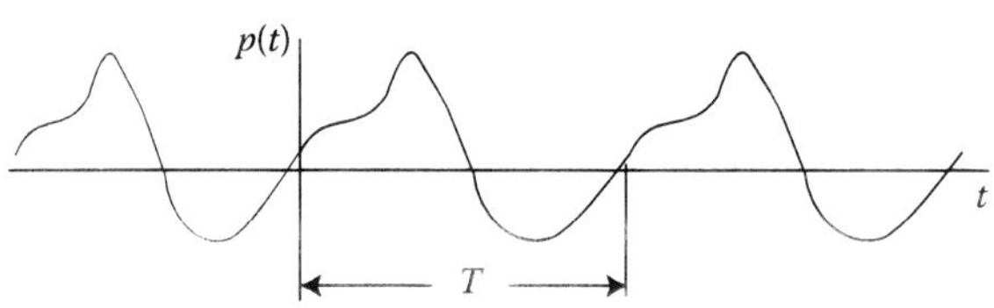


**FIGURE 1.29**
Example of general periodic motion.

and

$$\left.\begin{aligned}&\int_{-T/2}^{T/2} \cos(n\omega_f t)dt = 0 \quad n \neq 0\\&\int_{-T/2}^{T/2} \sin(n\omega_f t)dt = 0\end{aligned}\right\} \tag{1.119}$$

Using these integrals, we can determine $a_0$, $a_n$, and $b_n$. We assume that the series can be integrated term by term. A direct integration can be used for the constant $a_0$, thus,

$$\int_{-T/2}^{T/2} p(t)dt = \int_{-T/2}^{T/2} \left[\frac{a_0}{2} + \sum_{n=1}^{\infty}\{a_n \cos(n\omega_f t) + b_n \sin(n\omega_n t)\}\right]dt = \int_{-T/2}^{T/2} \frac{a_0}{2}dt = T\frac{a_0}{2}$$

and

$$a_0 = \frac{2}{T}\int_{-T/2}^{T/2} p(t)dt \tag{1.120}$$

To determine $a_n$, we would multiply both sides of Equation 1.117 by $\cos(m\omega_f t)$ and integrate over the interval $t = -T/2$ to $t = T/2$ or from $t = 0$ to $t = T$. Hence,

$$\begin{aligned}\int_{-T/2}^{T/2} p(t)\cos(m\omega_f t)dt &= \int_{-T/2}^{T/2} \left[\frac{a_0}{2} + \sum_{n=1}^{\infty}\{a_n \cos(n\omega_f t) + b_n \sin(n\omega_n t)\}\right]\cos(m\omega_f t)dt\\&= \frac{a_0}{2}\int_{-T/2}^{T/2} \cos(m\omega_f t)dt + \sum_{n=1}^{\infty} a_n \int_{-T/2}^{T/2} \cos(n\omega_f t)\cos(m\omega_f t)dt\\&+\sum_{n=1}^{\infty} b_n \int_{-T/2}^{T/2} \sin(n\omega_f t)\cos(m\omega_f t)dt\end{aligned}$$

Based on Equations 1.118 and 1.119, we see that all terms will vanish except for the term $\cos(n\omega_f t)\cos(m\omega_f t)$, where $n = m$. Therefore, we find

$$\int_{-T/2}^{T/2} p(t)\cos(m\omega_f t)dt = a_m\frac{T}{2}$$

or

$$a_m = \frac{2}{T}\int_{-T/2}^{T/2} p(t)\cos(m\omega_f t)dt \quad m = 1, 2, 3, \ldots \tag{1.121}$$

The constant term $b_n$ can be found in a similar manner by multiplying both sides of Equation 1.117 by $\sin(m\omega_f t)$ and integrating over the interval $t = -T/2$ to $t = T/2$. This result gives

$$b_m = \frac{2}{T}\int_{-T/2}^{T/2} p(t)\sin(m\omega_f t)dt \quad m = 1,2,3,\ldots \tag{1.122}$$

Since we have used the factor $a_0/2$ instead of $a_0$ and interchanging the order of summation and integration, the coefficients can be put in the following form:

$$a_n = \frac{2}{T}\int_{-T/2}^{T/2} p(t)\cos(n\omega_f t)dt \quad n = 0,1,2,3,\ldots \tag{1.123}$$

$$b_n = \frac{2}{T}\int_{-T/2}^{T/2} p(t)\sin(n\omega_f t)dt \quad n = 1,2,3,\ldots \tag{1.124}$$

For a finite approximation to a Fourier series, we define an approximation as

$$S_N(t) = \frac{a_0}{2} + \sum_{n=1}^{N}[a_n\cos(n\omega_f t) + b_n\sin(n\omega_f t)] \tag{1.125}$$

called the $N$th partial sum. Not all functions have a Fourier series that involve an infinite number of terms as will be seen in Example 1.10.

**EXAMPLE 1.10**

Develop the Fourier series for the periodic function shown in Figure 1.30. Draw plots showing the partial sums.

**Solution**

Assuming that the frequency of our function $p(t)$ is $\omega_f$, then the Fourier decomposition is given, according to Equation 1.117, as

$$p(t) = \frac{a_0}{2} + \sum_{n=1}^{\infty}[a_n\cos(n\omega_f t) + b_n\sin(n\omega_f t)]$$

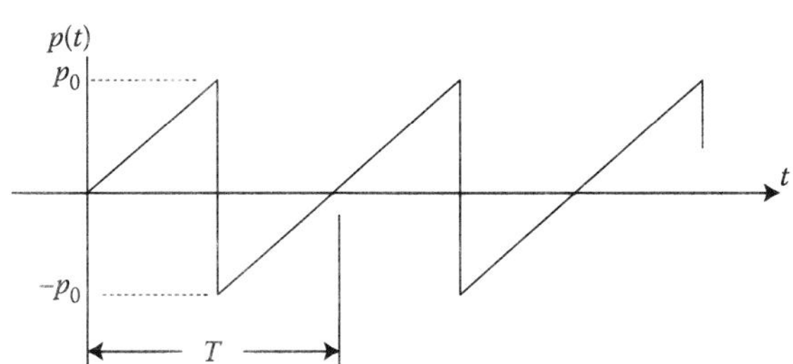


**FIGURE 1.30**
Periodic saw tooth.

The Fourier coefficients are obtained from Equations 1.123 and 1.124. We find

$$a_0 = \frac{2}{T}\int_{-T/2}^{T/2} \frac{2p_0 t}{T}dt = \frac{2p_0}{T}t^2\Big|_{-T/2}^{T/2} = 0$$

$$a_n = \frac{2}{T}\int_{-T/2}^{T/2} \frac{2p_0 t}{T}\cos(n\omega_f t)dt = \frac{4p_0}{T^2}\left[\frac{T^2}{4n^2\pi^2}\cos\left(\frac{2n\pi}{T}t\right) + \frac{T}{2n\pi}t\sin\left(\frac{2n\pi}{T}t\right)\right]_{-T/2}^{T/2} = 0$$

$$b_n = \frac{2}{T}\int_{-T/2}^{T/2} \frac{2p_0 t}{T}\sin(n\omega_f t)dt = \frac{4p_0}{T^2}\left[\frac{T^2}{4n^2\pi^2}\sin\left(\frac{2n\pi}{T}t\right) - \frac{T}{2n\pi}t\cos\left(\frac{2n\pi}{T}t\right)\right]_{-T/2}^{T/2}$$

$$= \begin{cases} \dfrac{2p_0}{n\pi} & \text{if } n \text{ is odd} \\ -\dfrac{2p_0}{n\pi} & \text{if } n \text{ is even} \end{cases}$$

Thus, we have

$$S_n(t) = \frac{2p_0}{\pi}\left[\sin(\omega_f t) - \frac{1}{2}\sin(2\omega_f t) + \frac{1}{3}\sin(3\omega_f t) - \cdots\right]$$

where $\omega_f = 2\pi/T$. A plot is given in Figure 1.31 for $n = 1, 3, 5, 10, 20$, and 50 terms. It is noted that we have a reasonable accurate representation of the function using only the first five terms.

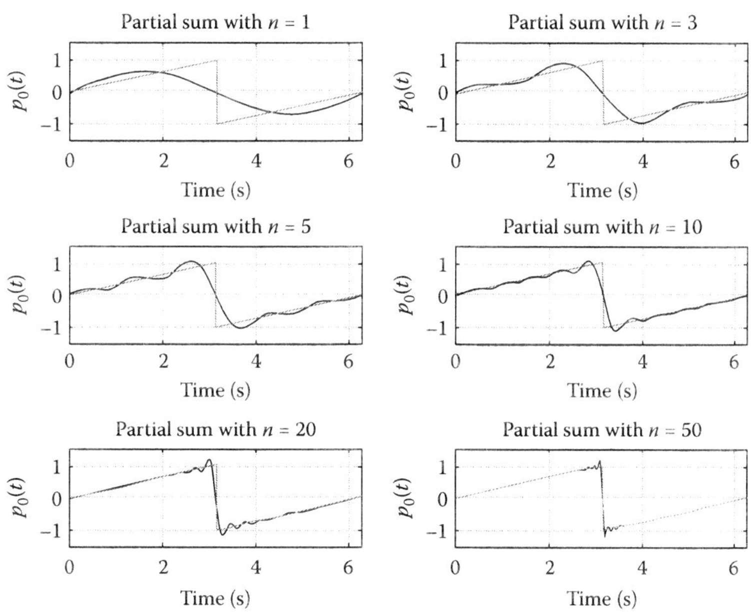


**FIGURE 1.31**
Partial sums of Fourier series.

### 1.7.2 Alternate Forms of Fourier Series

Our periodic function $p(t)$ can be represented as a Fourier series having the form

$$\begin{aligned} p(t) &= \frac{a_0}{2} + \sum_{n=1}^{\infty}\left[a_n \cos(n\omega_f t) + b_n \sin(n\omega_f t)\right] \\ &= \frac{a_0}{2} + \sum_{n=1}^{\infty}\sqrt{a_n^2+b_n^2}\left(\frac{a_n}{\sqrt{a_n^2+b_n^2}}\cos(n\omega_f t) + \frac{b_n}{\sqrt{a_n^2+b_n^2}}\sin(n\omega_f t)\right) \end{aligned} \tag{1.126}$$

Let $A_0 = a_0/2$ and $A_n = \sqrt{a_n^2+b_n^2}$ be the resultant amplitude of the Fourier coefficients having frequency $n\omega_f$ (see Figure 1.32).

$\varphi_n$ and $\beta_n$ are the phase angles which measure the lag or lead of the $n$th harmonic. Thus, Equation 1.125 can be written as

$$\begin{aligned} p(t) &= A_0 + \sum_{n=1}^{\infty} A_n[\cos(n\omega_f t)\cos\varphi_n + \sin(n\omega_f t)\sin\varphi_n] \\ &= A_0 + \sum_{n=1}^{\infty} A_n \cos(n\omega_f t - \varphi_n) \end{aligned} \tag{1.127}$$

or

$$\begin{aligned} p(t) &= A_0 + \sum_{n=1}^{\infty} A_n[\cos(n\omega_f t)\sin\beta_n + \sin(n\omega_f t)\cos\beta_n] \\ &= A_0 + \sum_{n=1}^{\infty} A_n \sin(n\omega_f t + \beta_n) \end{aligned} \tag{1.128}$$

Equations 1.127 and 1.128 can be used in place of Equation 1.117 to calculate the periodic function $p(t)$.

### 1.7.3 Complex Form of Fourier Series

In some instances, in vibration analysis, it is more convenient to use the exponential or complex form of the Fourier series. The Euler identities $e^{i\omega_f t} = \cos\omega_f t + i\sin\omega_f t$ and $e^{-i\omega_f t} = \cos\omega_f t - i\sin\omega_f t$ allow us to write

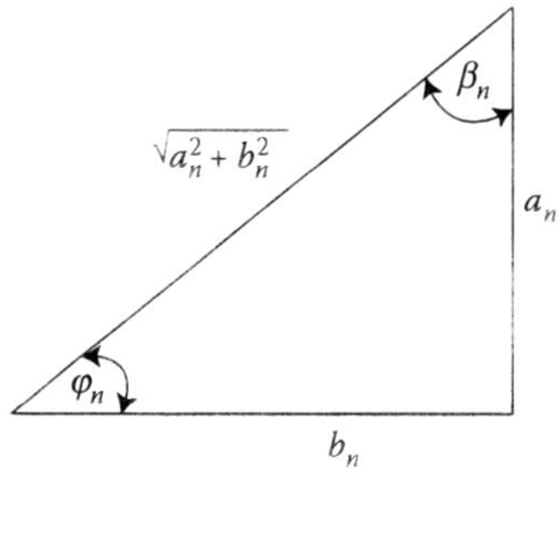


**FIGURE 1.32**
Resultant amplitude of Fourier coefficients.

$$\cos(\omega_f t) = \frac{e^{i\omega_f t} + e^{-i\omega_f t}}{2} \text{ and } \sin(\omega_f t) = \frac{e^{i\omega_f t} - e^{-i\omega_f t}}{2i} \tag{1.129}$$

When Equation 1.129 is substituted into Equation 1.117 for the Fourier series representation, we have

$$p(t) = \frac{a_0}{2} + \sum_{n=1}^{\infty}\left[a_n\left(\frac{e^{in\omega_f t} + e^{-in\omega_f t}}{2}\right) + b_n\left(\frac{e^{in\omega_f t} - e^{-in\omega_f t}}{2i}\right)\right]$$

and after grouping like terms we have

$$p(t) = \frac{a_0}{2} + \sum_{n=1}^{\infty}\left[\left(\frac{a_n - ib_n}{2}\right)e^{in\omega_f t} + \left(\frac{a_n + ib_n}{2i}\right)e^{-in\omega_f t}\right] \tag{1.130}$$

Let

$$c_0 = \frac{a_0}{2},\ c_n = \frac{a_n - ib_n}{2} \text{ and } c_{-n} = \bar{c}_n = \frac{a_n + ib_n}{2} \quad \text{for } n = 1, 2, 3, \ldots \tag{1.131}$$

where $\bar{c}_n$ is the complex conjugate of $c_n$. Then Equation 1.129 for the Fourier series representation of $p(t)$ becomes

$$p(t) = \sum_{n=-\infty}^{\infty} c_n e^{in\omega_f t} \tag{1.132}$$

Equation 1.132 is the *complex* or *exponential* form of the Fourier series representation of $p(t)$. The summation is from $-\infty$ to $\infty$. The Fourier frequencies are negative when $n < 0$. The coefficients are found as

$$c_0 = \frac{a_0}{2} = \frac{1}{T}\int_{-T/2}^{T/2} p(t)dt \tag{1.133}$$

$$c_n = \frac{a_n - ib_n}{2} = \frac{1}{T}\int_{-T/2}^{T/2} p(t)[\cos(n\omega_n t) - i\cos(n\omega_n t)]dt = \frac{1}{T}\int_{-T/2}^{T/2} p(t)e^{-in\omega_f t}dt \quad n = \pm 1, \pm 2, \ldots \tag{1.134}$$

**EXAMPLE 1.11**

Determine $c_n$ for the periodic function $p(t) = 2p_0 t/T$.

**Solution**

$$c_0 = \frac{2p_0}{T}\int_{-T/2}^{T/2} t\,dt \quad c_n = \frac{1}{T}\int_{-T/2}^{T/2} \frac{2p_0 t}{T} e^{-in\omega_f t}\,dt$$

Integrating and substituting $\omega_f = 2\pi/T$, we obtain

$$c_0 = 0 \quad \text{and} \quad c_n = p_0\left\{\frac{-e^{-in\pi}}{i2n\pi} + \frac{e^{-in\pi}}{2(n\pi)^2}\frac{-e^{-in\pi}}{i2n\pi} - \frac{e^{-in\pi}}{2(n\pi)^2}\right\}$$

Now $e^{in\pi} = e^{-in\pi} \cos n\pi = (-1)^n$, so that

$$c_n = p_0(-1)^n\left[-\frac{1}{i2n\pi} + \frac{1}{2n^2\pi^2} - \frac{1}{i2n\pi} - \frac{1}{2n^2\pi^2}\right]$$
$$= -\frac{p_0(-1)^n}{in\pi} = \frac{ip_0(-1)^n}{n\pi}$$

Thus, the complex form of the Fourier expansion of $p(t)$ is

$$p(t) = \sum_{n=-\infty}^{\infty} \frac{ip_0(-1)^n}{n\pi} e^{in\omega_f t}$$

We can convert this back to the trigonometric form, since using the definition of $c_n$ and $c_{-n}$ we have

$$a_0 = 2c_0, \quad a_n = c_n + \bar{c}_n, \quad b_n = i(c_n - \bar{c}_n)$$

So that in our case

$$a_0 = 0, \quad a_n = 0, \quad b_n = i(-1)^n\left[\frac{ip_0}{n\pi} + \frac{ip_0}{n\pi}\right] = -\frac{2p_0(-1)^n}{n\pi}$$

which corresponds to the solution to Example 1.10.

### 1.7.4 Response to General Periodic Forces

A general periodic force can be represented as a sum of sines and cosines. Consider the loading shown in Figure 1.33.

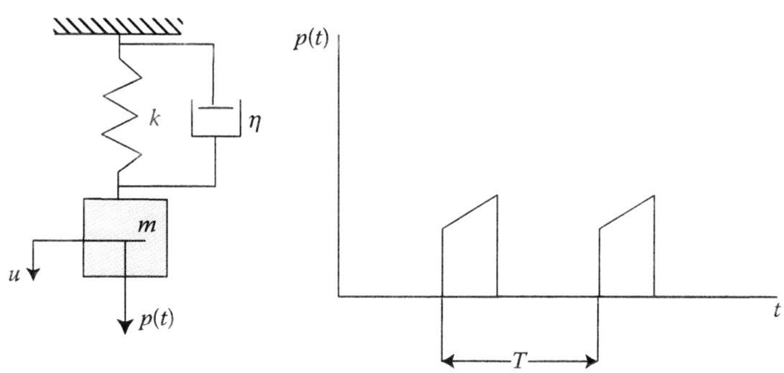


**FIGURE 1.33**
General periodic force.

There are two distinct loads, $p_a(t)$ and $p_b(t)$. These, in turn produce two displacements, $u_a(t)$ and $u_b(t)$. If the load is equal to

$$p(t) = p_a(t) + p_b(t) \tag{1.135}$$

then the corresponding deflection is obtained by

$$u(t) = u_a(t) + u_b(t) \tag{1.136}$$

A proof can be obtained as follows: Consider the equations

$$m\frac{d^2u_a}{dt^2} + \eta\frac{du_a}{dt} + ku_a = p_a \quad \text{and} \quad m\frac{d^2u_b}{dt^2} + \eta\frac{du_b}{dt} + ku_b = p_b$$

Adding these two equations gives

$$m\frac{d^2}{dt^2}(u_a + u_b) + \eta\frac{d}{dt}(u_a + u_b) + k(u_a + u_b) = p_a + p_b$$

or

$$m\frac{d^2u}{dt^2} + \eta\frac{du}{dt} + ku = p(t) \tag{1.137}$$

This result can be applied to an arbitrary periodic force. We have

$$p(t) = \frac{p_0}{2} + \sum_{n=1}^{\infty} p_{cn}\cos n\omega_f t + \sum_{n=1}^{\infty} p_{sn}\sin n\omega_f t \quad \text{where } \omega_f = 2\pi/T \tag{1.138}$$

The corresponding solution is

$$u(t) = \frac{p_0}{2k} + \sum_{n=1}^{\infty} C_{cn}\cos(n\omega_f t - \varphi_n) + \sum_{n=1}^{\infty} C_{sn}\sin(n\omega_f t - \varphi_n) \tag{1.139}$$

where the amplitudes corresponding to the cosine and sine terms are

$$C_{cn} = \left(\frac{p_{cn}}{k}\right)\frac{1}{\sqrt{(1-n^2\Omega^2)^2 + (2\zeta n\Omega)^2}} \quad \text{and} \quad C_{sn} = \left(\frac{p_{sn}}{k}\right)\frac{1}{\sqrt{(1-n^2\Omega^2)^2 + (2\zeta n\Omega)^2}} \tag{1.140}$$

and

$$\varphi_n = \tan^{-1}\left(\frac{2\zeta n\Omega}{1-(n\Omega)^2}\right) \tag{1.141}$$

If $p(t)$ is given, then the Fourier coefficients are determined by

$$p_0 = \frac{2}{T}\int_0^T p(t)dt, \quad p_{cn} = \frac{2}{T}\int_0^T p(t)\cos n\omega_f t\, dt \quad \text{and} \quad p_{sn} = \frac{2}{T}\int_0^T p(t)\sin n\omega_f t\, dt$$

**EXAMPLE 1.12**

A dynamical system, with parameters $m = 1, k = 1$, and $\zeta = 0.6$, is subjected to a square periodic force of period $T = 20$ s as shown in Figure 1.34. Determine the steady-state response.

**Solution**

The forcing function is given as

$$p(t) = \begin{cases} h & \text{for } 0 \le t \le T/2 \\ 0 & \text{for } T/2 \le t \le T \end{cases}$$

The forcing function is given by the Fourier series

$$p(t) = \frac{a_0}{2} + \sum_{n=1}^{\infty}[a_n \cos(n\omega_f t) + b_n \sin(n\omega_f t)]$$

The Fourier constants are found as follows:

$$a_0 = \frac{2}{T}\int_0^T p(t)dt = \frac{2}{T}\int_0^{T/2} h\, dt = \frac{2h}{T}[t]_0^{T/2} = h$$

$$a_n = \frac{2}{T}\int_0^T p(t)\cos(n\omega_f t)dt = \frac{2}{T}\int_0^{T/2} h\cos(n\omega_f t)dt = \frac{2h}{T}\left[\frac{1}{n\omega_f}\sin(n\omega_f t)\right]_0^{T/2} = \frac{2h}{n\omega_f T}\left[\sin\left(\frac{n\omega_f T}{2}\right)\right]$$

$$= 0$$

$$b_n = \frac{2}{T}\int_0^T p(t)\sin(n\omega_f t)dt = \frac{2}{T}\int_0^{T/2} h\sin(n\omega_f t)dt = -\frac{2h}{T}\left[\frac{1}{n\omega_f}\cos(n\omega_f t)\right]_0^{T/2} = \frac{h}{n\pi}(1 - \cos n\pi)$$

$$= \begin{cases} 0 & \text{for } n \text{ even} \\ \dfrac{2h}{n\pi} & \text{for } n \text{ odd} \end{cases}$$

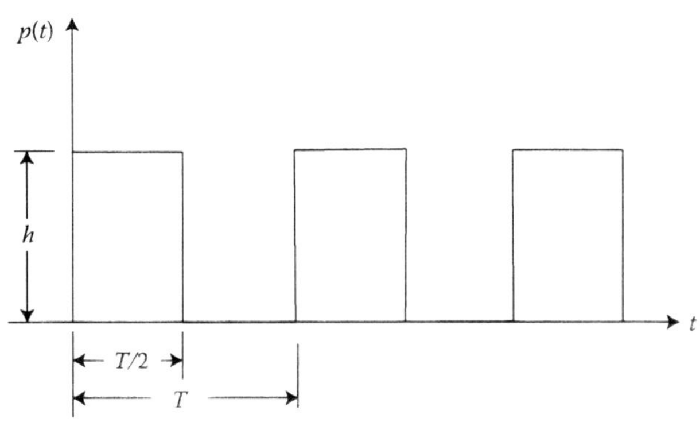


FIGURE 1.34
Square periodic force.

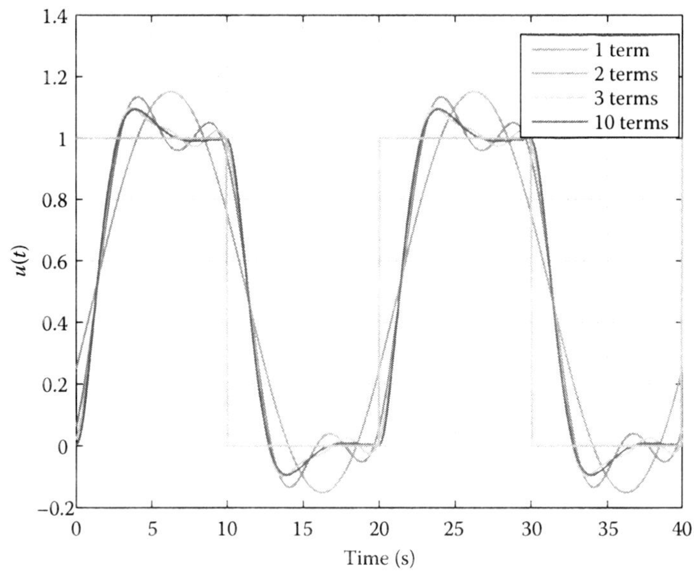


**FIGURE 1.35**
Response to periodic square wave input.

The force can then be written as

$$p(t) = \frac{h}{2} + \sum_{n=1,3,5} \frac{2h}{n\pi} \sin(n\omega_f t)$$

The response of the system is given by the expression

$$u(t) = \frac{p_0}{2k} + \sum_{n=1}^{\infty} \frac{(a_n/k)\cos(n\omega_f t - \varphi_n)}{\sqrt{(1-n^2\Omega^2)^2 + (2\zeta n\Omega)^2}} + \sum_{n=1}^{\infty} \frac{(b_n/k)\sin(n\omega_f t - \varphi_n)}{\sqrt{(1-n^2\Omega^2)^2 + (2\zeta n\Omega)^2}}$$

where

$$\varphi_n = \tan^{-1}\left(\frac{2\zeta n\Omega}{1-(n\Omega)^2}\right) \quad \text{and} \quad \Omega = \frac{\omega_f}{\omega_0}$$

Based on the terms from our forcing function, we have

$$u(t) = \frac{h}{2k} + \sum_{n=1}^{\infty} \frac{(2h/n\pi k)\sin(n\omega_f t - \varphi_n)}{\sqrt{(1-n^2\Omega^2)^2 + (2\zeta n\Omega)^2}}$$

The response of the system is shown in Figure 1.35.

## 1.8 Work Performed by External Forces and Energy Dissipation

The potential and kinetic energies of a linear spring with spring rate $k$ of mass $m$ being displaced through a distance $u$ are expressed as

$$W_p = \frac{1}{2}ku^2, \quad W_k = \frac{1}{2}m\dot{u}^2 \tag{1.142}$$

If an external force $p(t)$ is applied that causes a displacement from $u_1 \to u_2$, the work due to the external force can be written as

$$W_e = \int_{t_1}^{t_2} p(t)du \quad \text{where } du = \frac{du}{dt}dt = \dot{u}\,dt$$

Therefore

$$W_e = \int_{t_1}^{t_2} p(t)\dot{u}(t)dt \tag{1.143}$$

Damping is present in all vibrating systems. Its effect is to remove energy from the system. Dissipated energy is the negative work of the damping force, $R_d = -\eta\dot{u}$, for a displacement from $u_1 \to u_2$.

$$W_d = \int_{u_1}^{u_2} -R_d\,du = -\int_{t_1}^{t_2} R_d\,\dot{u}\,dt = \int_{t_1}^{t_2} \eta\dot{u}^2\,dt \tag{1.144}$$

The energy dissipated is always positive and as time increases, so is $W_d$. Consider harmonic motion with $p(t) = P\sin\omega_f t$ and $u(t) = C\sin(\omega_f t - \varphi)$. We then have that $\dot{u}(t) = C\omega_f\cos(\omega_f t - \varphi)$. We determine the work of the external force $p(t)$ during one cycle to be

$$\Delta W_e = \int_{t_1}^{t_1+T} p(t)\dot{u}(t)dt \tag{1.145}$$

Assuming $t_1 = 0$ and knowing $T = 2\pi/\omega_f$, we find

$$\Delta W_e = \int_{0}^{2\pi/\omega_f} P\sin\omega_f t\, C\omega_f\cos(\omega_f t - \varphi)dt = PC\pi\sin\varphi \tag{1.146}$$

For a case of no damping, $\varphi = 0$ and $\Delta W_e = 0$. For resonance, $\omega_f = \omega_0$, $\varphi = \pi/2$, and $\Delta W_e = PC\pi$, which is the maximum amount of work in one cycle. In order to find the amount of dissipated energy in one cycle, we have

$$\Delta W_d = \int_{0}^{2\pi/\omega_f} \eta C^2\omega_f^2\cos^2(\omega_f t - \varphi)dt = \eta C^2\omega_f\pi \tag{1.147}$$

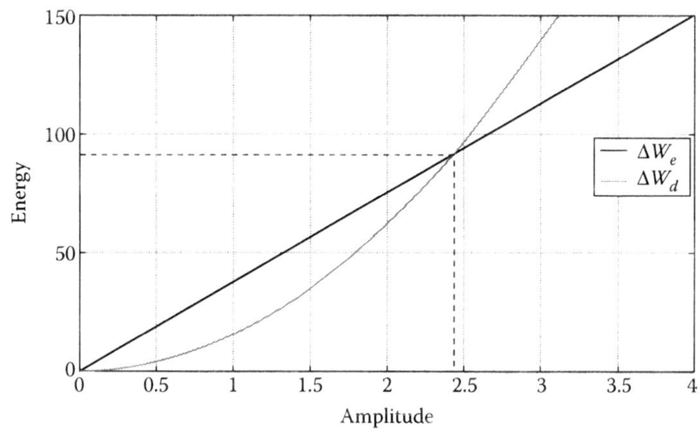


**FIGURE 1.36**
Input energy and dissipated energy.

We see that the input energy, or the work of the external force, increases linearly with amplitude $C$, while the energy dissipated increases as the square of the amplitude $C^2$. The two energies will be equal where they intersect as shown in Figure 1.36.

Therefore, we have from Equations 1.146 and 1.147 that when $\Delta W_e = \Delta W_d$ we find

$$C = \frac{P \sin \varphi}{\eta \omega_f} \tag{1.148}$$

Equation 1.147 expresses the energy dissipated in one cycle due to forced vibration. We can use this expression and equate it to another type of damping and obtain an equivalent viscous damping constant $\eta_{eq}$.

The maximum strain energy is the spring can be found using $W_s = 1/2ku^2$, thus,

$$W_{s_{\max}} = \frac{1}{2} kC^2 \tag{1.149}$$

An interesting relation is the loss (damping) factor $= \Delta W_d / W_{s_{\max}} = 2\pi(\eta/k)\Omega_f$.

### 1.8.1 Material Damping

When considering material damping (Lazan 1968; Bert 1973) we consider a uniaxial tension and a shear test for an elastic material as depicted in Figure 1.37. From these tests, we establish the stress–strain relationship for linear elastic materials as defined in strength of

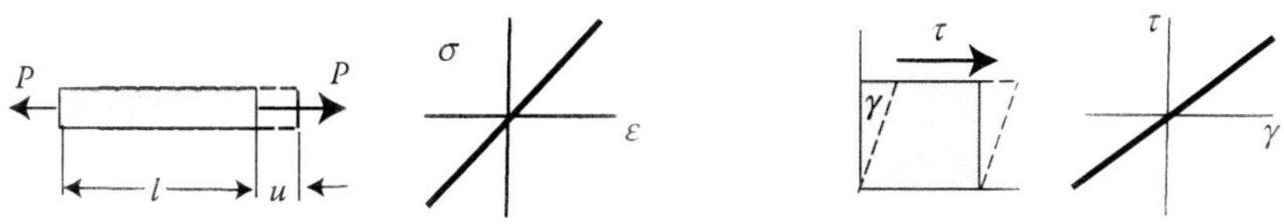


**FIGURE 1.37**
Idealized elastic materials.

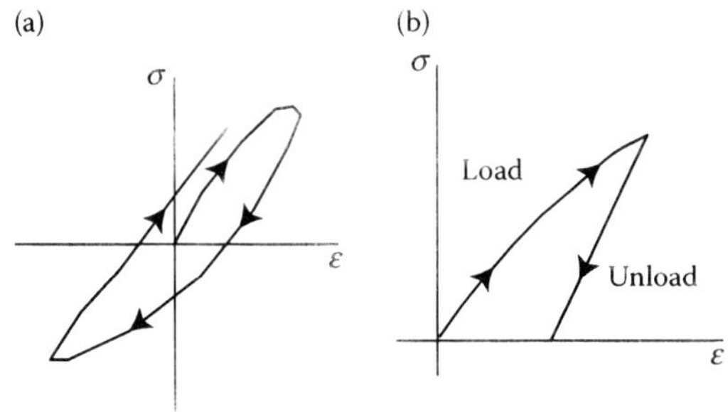


**FIGURE 1.38**
(a) Actual material behavior, (b) idealized elastic–plastic material.

materials. From the tension test, we have $\sigma = P/A$ and $\varepsilon = u/l$. From the $\sigma - \varepsilon$ curve, we establish $\sigma = E\varepsilon$. Similarly, from the shear test, we establish $\tau = G\gamma$. Where $E$ and $G$ are the elastic and shear modulus, respectively.

The $\sigma - \varepsilon$ and $\tau - \gamma$ curves in Figure 1.37 are for an idealized linear elastic material. For an actual material, the behavior is somewhat different when a load–unload cycle is considered. This is illustrated in Figure 1.38a. In order to describe this material behavior, we need to introduce idealized modes that are more realistic than the linear elastic models previously considered. A linear elastic material behaves like a linear spring. For an idealized elastic–plastic material, the load and unload phases are illustrated in Figure 1.38b. From this figure, we note that

$$\sigma = f_1(\varepsilon);\ \text{for}\ \dot{\sigma} > 0 \quad \text{(loading)}$$

$$\sigma = f_2(\varepsilon);\ \text{for}\ \dot{\sigma} < 0 \quad \text{(unloading)}$$

Note that the effect of the stress or strain rates is neglected, models for several types of materials will be developed.

### 1.8.2 Elastic–Plastic Materials

There are several types of elastic–plastic materials, which can be modeled using simple techniques. For example, the behavior of an elastic-perfectly plastic material is modeled as shown in Figure 1.39a and an elastic strain-hardening material in Figure 1.39b. Figure 1.39a represents a low-carbon steel. We see that while loading the stress and strain are related through $\sigma = E\varepsilon$ until the yield stress ($\sigma_{yield} = \sigma_y$) is reached. At this point the strain increases with no increased stress. Upon unloading the stress returns to zero, but since there will be permanent deformation, the strain has a plastic deformation and does not return to zero, it returns to $\varepsilon_{pl}$.

Each segment of the load–unload cycle for this material can be defined as

$$\sigma < \sigma_y;\ \sigma = E\varepsilon \quad \text{(loading)}$$

$$\sigma = \sigma_y;\ \varepsilon\text{-undetermined} \quad \text{(loading)}$$

$$\begin{aligned}\dot{\sigma} < 0;\ \sigma &= \sigma_y - E(\varepsilon_{\max} - \varepsilon) \quad \text{(unloading)}\\ &= E(\varepsilon - \varepsilon_{pl})\end{aligned}$$

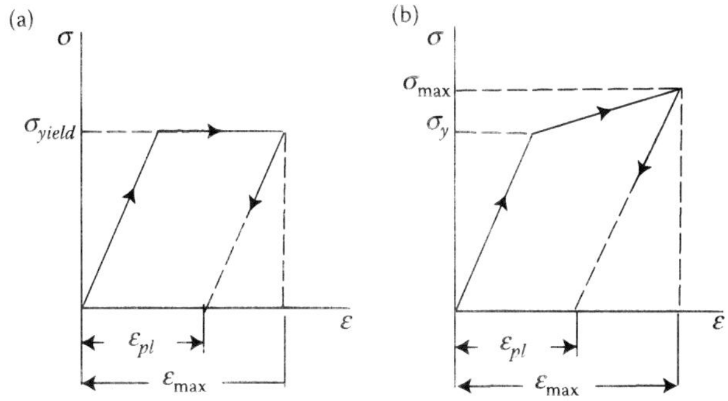


**FIGURE 1.39**
(a) Elastic-perfectly plastic material, (b) elastic strain-hardening material.

In cases where the elastic deformations are small ($\varepsilon_{elastic} \approx 0$) compared to the plastic deformations, we have a rigid-perfectly plastic material. For example, assume the plastic strain is $\varepsilon_{pl} = 10\% = 0.10$ and the yield stress is $\sigma_y = 30{,}000$ psi. The elastic modulus is $E = 30 \times 10^6$ psi and the maximum elastic strain is $\varepsilon_{el} = \sigma_y/E = 10^{-3} \approx 0$. In a similar manner, one can establish a model for a material that exhibits elastic strain-hardening (linear strain hardening) as depicted in Figure 1.39b.

The elastic–plastic models are sufficient for certain problems but they have limitations and disadvantages. For example, they do not model rate effects for which the equations can be nonlinear and require numerical methods for solution. Damping effects can be described using models for viscoelastic materials.

### 1.8.3 Viscoelastic Materials

There are two models that are commonly used when considering viscoelastic material behavior (Lakes 2009). They both use a spring for the linear response and a dashpot for the time-dependent response. The Kelvin–Voigt and Maxwell models are shown in Figure 1.40a and b, respectively. In the Kelvin–Voigt model, we have uniaxial stress and shear stress modeled as

$$\sigma = E\varepsilon + \eta\dot{\varepsilon} \quad \text{or} \quad \tau = G\gamma + \mu\dot{\gamma} \tag{1.150}$$

where $\dot{\varepsilon} = d\varepsilon/dt$ and $\dot{\gamma} = d\gamma/dt$.

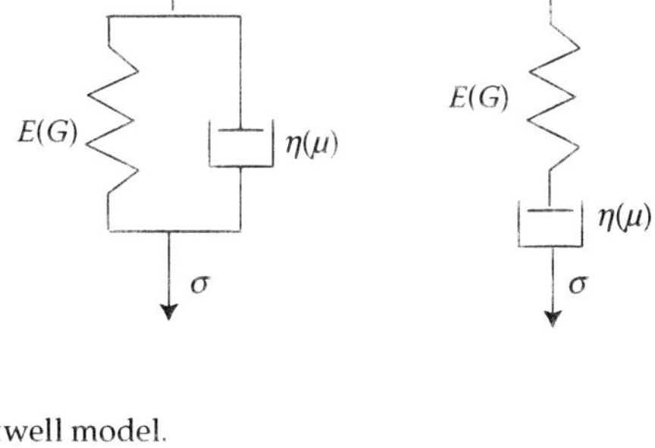


**FIGURE 1.40**
(a) Kelvin–Voigt model, (b) Maxwell model.

The stress at all times is the sum of the stress in the elastic element ($\sigma^e$) and the stress in the viscous element ($\sigma^\eta$), so we can write $\sigma = \sigma^e + \sigma^\eta$. The strains in the two elements are equal ($\varepsilon = \varepsilon^e = \varepsilon^\eta$). The two components of stress can be written as $\sigma^e = E\varepsilon$ and $\sigma^\eta = \eta\dot{\varepsilon}$. The shape of the loading–unloading stress–strain curve depends on the rate of cyclic loading. Although adequate in some cases, very few materials behave according to the relation $\sigma = E\varepsilon + \eta\dot{\varepsilon}$. As a result, we consider the Maxwell model (Figure 1.40b). In this model, the normal and shear strains are modeled as

$$\dot{\varepsilon} = \frac{\dot{\sigma}}{E} + \frac{\sigma}{\eta} \quad \text{or} \quad \dot{\gamma} = \frac{\dot{\tau}}{G} + \frac{\tau}{\mu} \tag{1.151}$$

where $\dot{\varepsilon}_{spring} = \dot{\sigma}/E$, $\dot{\varepsilon}_{daspot} = \sigma/\eta$ and the total strain rate is

$$\dot{\varepsilon}_{total} = \frac{\dot{\sigma}}{E} + \frac{\sigma}{\eta} \tag{1.152}$$

Although the Kelvin–Voigt and Maxwell models are adequate to describe some visco-elastic material behavior, the Kelvin–Voigt model does not describe stress relaxation and the Maxwell model does not describe creep or recovery. There are numerous other solid material models available. They are typically referred to as Zener models (Atanackovic 2002). One such solid model is shown in Figure 1.41. In this model, the material behavior is described by

$$\frac{\eta}{E_1}\dot{\varepsilon} + \varepsilon = \dot{\sigma}\frac{\eta}{E_0 E_1} + \sigma\left(\frac{1}{E_0} + \frac{1}{E_1}\right) \tag{1.153}$$

where $E_0$, $E_1$, and $\eta$ are material constants.

Similarly, for shear deformation,

$$\frac{\mu}{G_1}\dot{\gamma} + \gamma = \dot{\tau}\frac{\mu}{G_0 G_1} + \tau\left(\frac{1}{G_0} + \frac{1}{G_1}\right) \tag{1.154}$$

where $G_0$, $G_1$, and $\mu$ are material constants. For high rates of loading, we find

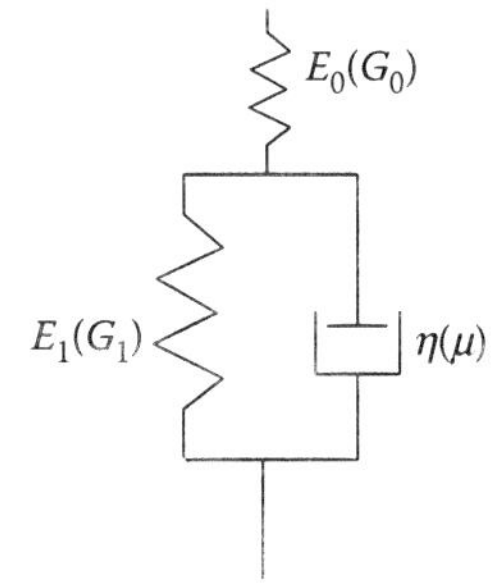


**FIGURE 1.41**
Standard solid model.

$$\frac{\eta}{E_1}\dot{\varepsilon}=\dot{\sigma}\frac{\eta}{E_0E_1} \quad \text{and} \quad \dot{\varepsilon}=\frac{\dot{\sigma}}{E_0} \tag{1.155}$$

For very low-loading rates where the time derivatives are very small, we model the system as two springs in series so that

$$\varepsilon=\sigma\left(\frac{1}{E_0}+\frac{1}{E_1}\right) \tag{1.156}$$

Very few actual materials behave like linear viscoelastic materials. The determination of material constants can be difficult and will not be discussed here.

### 1.8.4 Using Viscoelastic Relations (for Harmonic Functions)

Assume the stress and strain are harmonic in time and can be expressed as

$$\sigma=\sigma_0 e^{i\omega t} \tag{1.157}$$

and we assume $\varepsilon=\varepsilon_0 e^{i\omega t}$ where in general $\sigma_0$ and $\varepsilon_0$ have complex amplitudes. Substituting these into one of our viscoelastic material models yields

1. Kelvin model:

$$\left.\begin{array}{l}\sigma_0 e^{i\omega t}=E\varepsilon_0 e^{i\omega t}+\eta i\omega\varepsilon_0 e^{i\omega t}\\ \text{or}\\ \sigma_0=(E+\eta i\omega)\varepsilon_0\\ \text{or}\\ \sigma_0=E^*(i\omega)\varepsilon_0\end{array}\right\} \tag{1.158}$$

where $(E+\eta i\omega)=E^*(i\omega)$.

2. Maxwell model:

$$\left.\begin{array}{l}\dfrac{1}{E}i\omega\sigma_0 e^{i\omega t}+\dfrac{1}{\eta}i\omega\sigma_0 e^{i\omega t}=Ei\omega\varepsilon_0 e^{i\omega t}\\ \text{or}\\ \sigma_0=\left(\dfrac{Ei\omega}{i\omega+E/\eta}\right)\varepsilon_0\\ \text{or}\\ \sigma_0=E^*(i\omega)\varepsilon_0\end{array}\right\} \tag{1.159}$$

where

$$\frac{Ei\omega}{i\omega+E/\eta}=E^*(i\omega)$$

3. Standard solid model:

$$\left.\begin{aligned}
&\frac{\eta}{E_1} i\omega\varepsilon_0 + \varepsilon_0 = \frac{\eta}{E_0E_1} i\omega\sigma_0 + \left(\frac{1}{E_0} + \frac{1}{E_1}\right)\sigma_0 \\
&\text{or} \\
&\sigma_0 = \left(\frac{(\eta/E_1)i\omega + 1}{(\eta i\omega/E_0E_1) + ((1/E_0) + (1/E_1))}\right)\varepsilon \\
&\text{or} \\
&\sigma_0 = E^*(i\omega)\varepsilon_0
\end{aligned}\right\} \tag{1.160}$$

where

$$E^*(i\omega) = \left(\frac{(\eta/E_1)i\omega + 1}{(\eta i\omega/E_0E_1) + ((1/E_0) + (1/E_1))}\right)$$

Hence, for an elastic material we have $\sigma_0 = E\varepsilon_0$, or in shear $\tau_0 = G\gamma_0$. For any viscoelastic material we have $\sigma_0 = E^*(i\omega)\varepsilon_0$, or in shear $\tau_0 = G^*(i\omega)\gamma_0$.

From these relations, we can form the following conclusion: for harmonic stress and strain ($\sigma_0 e^{i\omega t}$, $\varepsilon_0 e^{i\omega t}$), the viscoelastic relations between $\sigma_0$ and $\varepsilon_0$ are similar to the elastic relations with $E(G)$ replaced by complex $E^*(i\omega)$ and $G^*(i\omega)$. These are summarized for each material by

1. Kelvin model:

$$\left.\begin{aligned}
&E^*(i\omega) = (E + \eta i\omega) \\
&\text{and} \\
&G^*(i\omega) = (G + \eta i\omega)
\end{aligned}\right\} \tag{1.161}$$

2. Maxwell model:

$$\left.\begin{aligned}
&E^*(i\omega) = \frac{Ei\omega}{i\omega + E/\eta} \\
&\text{and} \\
&G^*(i\omega) = \frac{Gi\omega}{i\omega + G/\eta}
\end{aligned}\right\} \tag{1.162}$$

3. Standard solid model:

$$\left.\begin{aligned}
&E^*(i\omega) = \left(\frac{(\eta/E_1)i\omega + 1}{(\eta i\omega/E_0E_1) + ((1/E_0) + (1/E_1))}\right) \\
&\text{and} \\
&G^*(i\omega) = \left(\frac{(\eta/G_1)i\omega + 1}{(\eta i\omega/G_0G_1) + ((1/G_0) + (1/G_1))}\right)
\end{aligned}\right\} \tag{1.163}$$

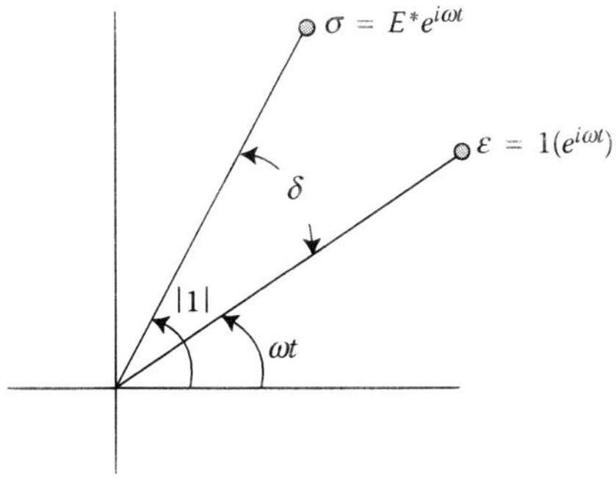


**FIGURE 1.42**
Harmonic stress and strain of unit.

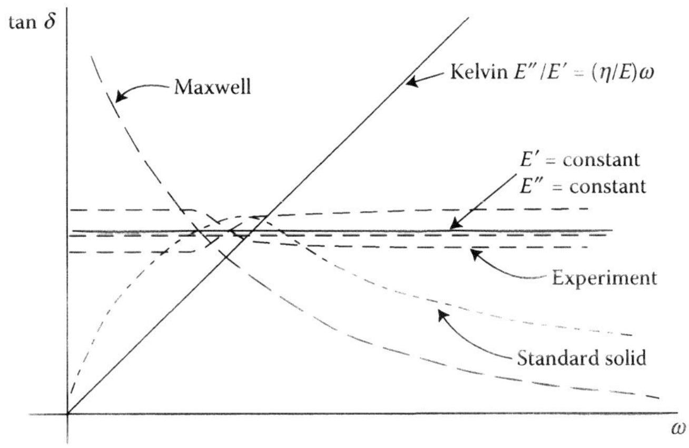


**FIGURE 1.43**
Correlation of frequency vs tan $\delta$ for various material models.

The complex moduli $E^*(i\omega)$ and $G^*(i\omega)$ will also be written as $E^* = E' + iE''$ and $G^* = G' + iG''$ where $E' = E'(\omega)$ and $E'' = E''(\omega)$. For example, in the Kelvin model, we have $E' = \text{constant} = E$ and $E'' = \eta\omega$. It may be useful to visualize the stress and strain rotating in the complex plane at frequency $\omega$. In a viscoelastic material, stress and strain are out of phase. Figure 1.42 illustrates harmonic stress ($\sigma = \sigma_0 e^{i\omega t}$) and the strain ($\varepsilon = \varepsilon_0 e^{i\omega t}$) of a unit magnitude with a phase angle $\delta$ separating them.

Suppose $\varepsilon = 1.0e^{i\omega t}$, then $\sigma = E^* e^{i\omega t}$, but $E^*$ is complex, so $\sigma = |E^*|e^{i\delta}e^{i\omega t} = |E^*|e^{i(\omega t+\delta)t}$. From complex number theory, we know that $\tan\delta = E''/E'$ where $E'$ and $E''$ are known. Figure 1.43 presents a plot of frequency versus tan $\delta$ for the various models we see that the Kelvin and Maxwell models are not very good since their dependency of frequency is too great.

## 1.9 Response to General Forcing Function

In the preceding sections, the steady-state response to harmonic and general periodic forcing functions was considered. It is known that most forcing functions will not be harmonic or periodic and we must determine how to determine the response of a system to an arbitrary forcing function. Before looking into arbitrary forcing functions, let us consider the response to some special types of forcing functions.

### 1.9.1 Dirac Delta or Impulse Function

When a force is applied to a system over a finite time interval and then removed, an impulse loading is applied to the system. The magnitude of the impulse function is on the order of the inverse of the time duration. It is noted that the specific form of the impulse function is not known but for convenience is represented as a rectangular pulse. The time duration of the pulse is typically represented as $\varepsilon$. As $\varepsilon$ approaches zero, the magnitude of the force tends to infinity. However, its time integral exists and is bounded, that is, not infinite. The function described above and depicted in Figure 1.44a is known as the *Dirac delta function* or the *unit impulse function*. The Dirac delta function has the properties

$$\delta(t) = \begin{cases} 0 & \text{for } t \neq 0 \\ \infty & \text{for } t = 0 \end{cases} \quad \text{and} \quad \lim_{\varepsilon \to 0} \int_{-\varepsilon/2}^{\varepsilon/2} \delta(t) = 1 \tag{1.164}$$

or

$$\int_{-\infty}^{\infty} \delta(t)dt = 1 \tag{1.165}$$

The integral in Equation 1.165 is nondimensional, thus $\delta(t)$ has unit as $s^{-1}$. Also, if $p(t)$ is a continuous function that vanishes at infinity, the *Dirac delta filtering property* can be expressed as

$$\int_{-\infty}^{\infty} p(t)\delta(t)dt = p(0) \tag{1.166}$$

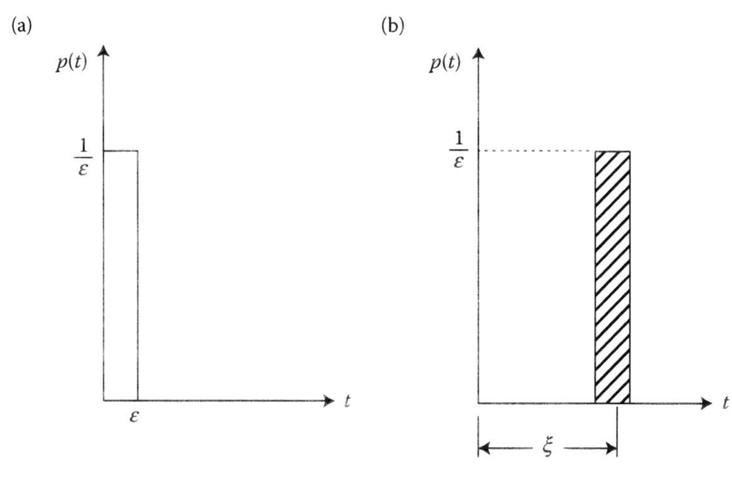


**FIGURE 1.44**
(a) Impulse loading, (b) shifted impulse loading.

This is easy to see since $\delta(t)$ vanishes everywhere except at $t = 0$. Thus, $p(t)\delta(t) = p(0)\delta(t)$ and $p(0)$ can be moved outside the integral since it does not depend on $t$, giving the result of Equation 1.166. The Dirac delta function that results when a time shift is introduced, such that $\delta(t - \xi)$, acting at time $\xi$ as shown in Figure 1.44b has the following properties:

$$\delta(t-\xi) = \begin{cases} 0 & \text{for } t \neq \xi \\ \infty & \text{for } t = \xi \end{cases} \qquad \int_{-\infty}^{\infty} \delta(t-\xi)dt = 1 \tag{1.167}$$

For this case, the filtered integral becomes

$$\int_{-\infty}^{\infty} p(t)\delta(t-\xi)dt = p(\xi) \tag{1.168}$$

### 1.9.2 Response to Dirac Delta Function

Newton's second law of motion states that if a force $p(t)$ acts on a body of mass $m$, the rate of change of momentum of the body is equal to the applied force. Thus,

$$\frac{d}{dt}[m\dot{u}(t)] = p(t) \tag{1.169}$$

Integrating both sides with respect to time gives

$$\int_{t_1}^{t_2} \frac{d}{dt}[m\dot{u}(t)]dt = \int_{t_1}^{t_2} p(t)dt$$

or

$$m[\dot{u}(t_2) - \dot{u}(t_1)] = \int_{t_1}^{t_2} p(t)dt \tag{1.170}$$

We see that the integral on the right-hand side of Equation 1.170 is just the magnitude of the impulse. For a short pulse duration acting on an SDOF system with $t_1 = 0$ and $t_2 = \varepsilon$, the velocity can be determined as

$$\dot{u}(\varepsilon) = \frac{\int_{t_1}^{t_2} p(t)dt}{m} = \frac{1}{m} \tag{1.171}$$

As $\varepsilon$ approaches zero, $\dot{u}(\varepsilon)$ approaches $\dot{u}(0)$. It is noted that the velocity is finite and the change in displacement will be undisturbed, that is the displacement will be zero after the impulse has been applied. The subsequent motion will be a free vibration with initial conditions

$$u(0) = 0, \quad \dot{u}(0) = \frac{1}{m} \tag{1.172}$$

For the case of small viscous damping, the general solution was given by Equation 1.38. Thus,

$$u(t) = e^{-\zeta\omega_0 t}\left(C_1' \cos\omega_d t + C_2' \sin\omega_d t\right)$$

Applying the initial conditions, Equation 1.172, the solution is given as

$$u(t) = \frac{1}{m\omega_d} e^{-\zeta\omega_0 t} \sin\omega_d t \tag{1.173}$$

where $\omega_d = \sqrt{1-\zeta^2}$. The response due to a unit impulse is defined as

$$h(t) = \frac{1}{m\omega_d} e^{-\zeta\omega_0 t} \sin\omega_d t \tag{1.174}$$

For an undamped system, we have

$$h(t) = \frac{1}{m\omega_0} \sin\omega_0 t \tag{1.175}$$

In the same manner, the response to a unit impulsive load applied at $t = \tau$ can be written as

$$h(t-\tau) = \frac{1}{m\omega_d} e^{-\zeta\omega_0 t} \sin\omega_d(t-\tau) \quad \text{for } t > \tau \tag{1.176}$$

The corresponding results for the response to a unit impulsive force for an over-damped system and critically damped system can be obtained from Equations 1.34 and 1.36. Hence, for a load applied at $t = \tau$, we have

$$\begin{aligned} h(t-\tau) &= \frac{e^{-\zeta\omega_0(t-\tau)}}{m\omega_0\sqrt{\zeta^2-1}} \sinh\omega_0\sqrt{\zeta^2-1}\, t \\ h(t-\tau) &= \frac{(t-\tau)}{m} e^{-\zeta\omega_0(t-\tau)} \end{aligned} \tag{1.177}$$

## 1.10 Response to Arbitrary Forcing Function

We can use the previous method of the response of a unit impulse to determine the response to general dynamic loading. Consider a general forcing function $p(t)$ as shown

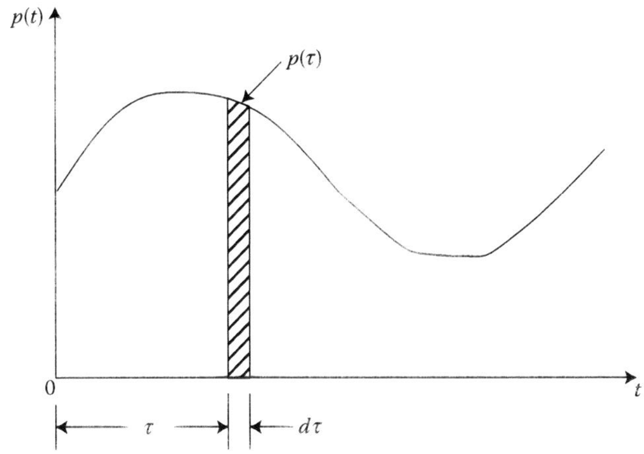


**FIGURE 1.45**
Arbitrary forcing function.

in Figure 1.45. The response to general dynamic loading is obtained by assuming that the force is composed of a series of impulses imposed on the system as shown in Figure 1.45.

We shall consider at time $t = \tau$ the loading $p(\tau)$. This loading acts for a very short duration $d\tau$ and represents the impulse $p(\tau)d\tau$ acting on the structure. Thus, for the time interval $d\tau$, the response due to the impulse $p(\tau)d\tau$ is

$$du(t) = \frac{p(\tau)d\tau}{m\omega_0}\sin\omega_0(t-\tau) \quad t \geq \tau \tag{1.178}$$

It is noted that the forcing function before $t = \tau$ will influence the response at $t = \tau$, however, after $t = \tau$ the forcing function will not have any influence at $t = \tau$. The total displacement is then determined, since the system is considered linear, by summing or integrating all the incremental displacements $du$ over the entire time interval, giving

$$u(t) = \frac{1}{m\omega_0}\int_0^t p(\tau)\sin\omega_0(t-\tau)d\tau \quad t \geq \tau \tag{1.179}$$

Equation 1.179 is known as the *Duhamel integral*. Equation 1.179 can be used to calculate the response of a forced system excited by an arbitrary forcing function. If the system has initial conditions $u_0$ and $\dot{u}_0$ at time $t = 0$, then the total response is the sum of the response caused by the initial conditions and the forced response caused by the forcing function. Hence,

$$u(t) = u_0\cos\omega_0 t + (\dot{u}_0/\omega_0)\sin\omega_0 t + \frac{1}{m\omega_0}\int_0^t p(\tau)\sin\omega_0(t-\tau)d\tau \tag{1.180}$$

For a dissipated system, Equations 1.179 and 1.180 become

$$u(t)=\frac{1}{m\omega_d}\int_0^t p(\tau)e^{-\omega_0\zeta(t-\tau)}\sin\omega_d(t-\tau)d\tau$$

$$u(t)=e^{-\zeta\omega_0 t}[u_0\cos\omega_d t+(\dot{u}_0/\omega_0)\sin\omega_d t]+\frac{1}{m\omega_d}\int_0^t p(\tau)e^{-\zeta\omega_0(t-\tau)}\sin\omega_d(t-\tau)d\tau \tag{1.181}$$

These equations are used to calculate the response of a linear SDOF system to a simple general arbitrary forcing function.

### 1.10.1 Step Response of Undamped System

Consider an undamped spring mass system with zero initial conditions subjected to a step load. A step load is a discontinuous load that is suddenly applied at time $t=0$ and remains constant after application. Figure 1.46 shows a step function applied at time $t=0$. The forcing function is given by

$$p(t)=\begin{cases}p_0 & \text{for } t\geq 0\\ 0 & \text{for } t<0\end{cases} \tag{1.182}$$

Using Equation 1.179, we have

$$u(t)=\frac{1}{m\omega_0}\int_0^t p_0\sin\omega_0(t-\tau)d\tau \tag{1.183}$$

which after integrating becomes

$$u(t)=\frac{p_0}{k}(1-\cos\omega_0 t) \tag{1.184}$$

Here, $\omega_0=\sqrt{k/m}$ has been used and $u_{static}=p_0/k$. A plot of the response given by Equation 1.184 is given in Figure 1.47. The static deflection is the displacement generated by a force $p_0$ as if it was applied to the system statically. The term $(1-\cos\omega_0 t)$ is called the dynamic load factor. Note that peak response to the step force excitation of magnitude $p_0$

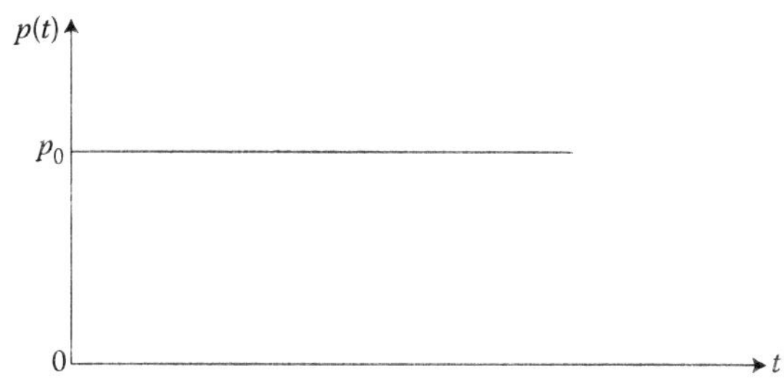


**FIGURE 1.46**
Step loading.

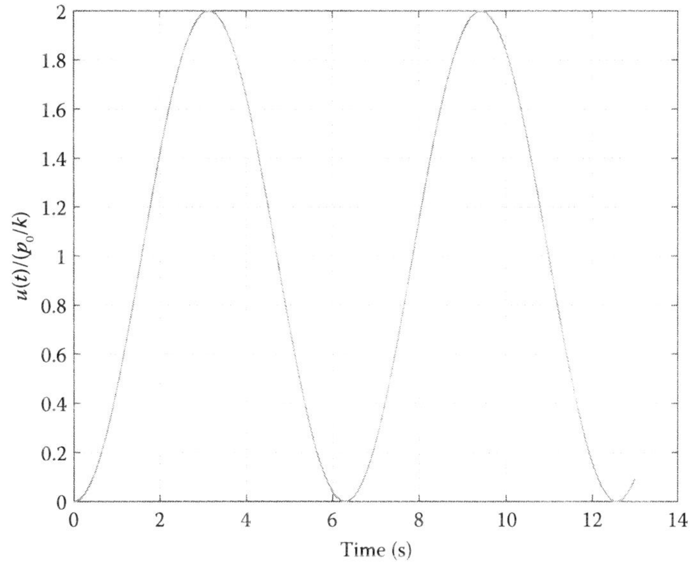


**FIGURE 1.47**
Response to step function.

is twice that of the static deflection which is $u_{\max} = 2u_{static}$. This value can also be found by differentiating Equation 1.184 and setting $\dot{u}(t) = 0$, which gives $\omega_0 \sin \omega_0 t = 0$. The values of $t$ that satisfy this condition are $n\pi/\omega_0$, where $n$ is an odd integer. We see that for $n$ odd that Equation 1.184 gives $u_{\max} = 2u_{static}$.

For a damped system with zero initial conditions subjected to a step load. $p_0$, we have using Equation 1.181

$$u(t) = \frac{p_0}{m\omega_d} \int_0^t e^{-\omega_0\zeta(t-\tau)} \sin \omega_d(t-\tau) d\tau$$

The integration gives, using $z = t - \tau$

$$u(t) = \int_0^t e^{-\zeta\omega_0 z} \sin \omega_d dz = \frac{p_0}{m\omega_d}\left[\frac{e^{-\zeta\omega_0 z}}{(\zeta\omega_0)^2 + \omega_d^2}(-\zeta\omega_0 \sin \omega_d z - \omega_d \cos \omega_d z)\right]_0^t$$

or

$$u(t) = \frac{p_0}{k}\left[1 - e^{-\zeta\omega_0 t}\left\{\cos \omega_d t + \left(\frac{\zeta\omega_0}{\omega_d}\right)\sin \omega_d t\right\}\right] \tag{1.185}$$

where $\omega_0 = \sqrt{k/m}$ and $\omega_d = \sqrt{1-\zeta^2}$ have been used. A plot of Equation 1.185 for various values of $\zeta$ is shown in Figure 1.48. Note that the amplification factor is less than 2 for dissipative systems, since with damping the overshoot beyond the equilibrium is lowered.

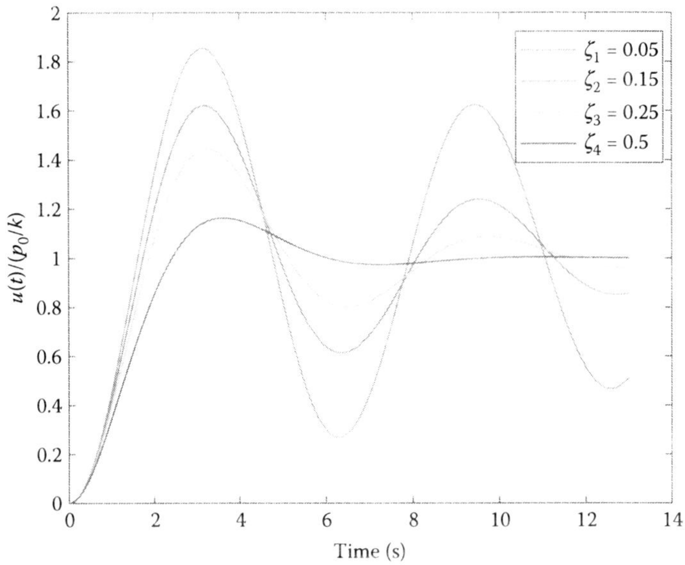


**FIGURE 1.48**
Response of step loading for various values of $\zeta$.

The time of maximum response can again be obtained by differentiating Equation 1.185 and setting the equation to zero. Thus, we find

$$\frac{\dot{u}(t)}{p_0/k} = e^{-\zeta\omega_0 t}\left[\omega_d \sin\omega_d t + \frac{\zeta^2\omega_0^2}{\omega_d}\sin\omega_d t\right] = 0$$

Recall that $\omega_d = \omega_0\sqrt{1-\zeta^2}$, therefore, our above equation can be written as

$$\frac{\dot{u}(t)}{p_0/k} = \frac{\omega_0^2}{\omega_d} e^{-\zeta\omega_0 t}\sin\omega_d t = 0 \tag{1.186}$$

This equation vanishes for

$$t = \frac{n\pi}{\omega_d}, \quad n = 1, 2, 3, \ldots \tag{1.187}$$

For $n = 0$, $t = 0$ and we see that $u(0) = 0$ from Equation 1.185. For $n = 1, 2$, etc., we see that we would obtain the first maximum response peak, the second peak, and so forth. The first maximum displacement occurs for $n = 1$ and is obtained by substituting $t = \pi/\omega_d$ into Equation 1.185 which yields

$$u_{\max} = \frac{p_0}{k}\left(1 + e^{-\zeta\omega_0\pi/\omega_d}\right) = \frac{p_0}{k}\left(1 + e^{-\zeta\pi/\sqrt{1-\zeta^2}}\right), \quad \text{for } n = 1 \tag{1.188}$$

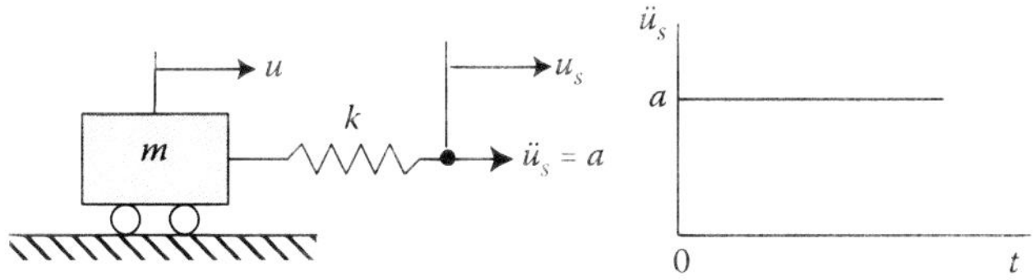


**FIGURE 1.49**
Spring–mass system with suddenly applied acceleration.

**EXAMPLE 1.13**

The free end of the spring of the spring–mass system shown in Figure 1.49 is suddenly given a constant acceleration as depicted. Determine the response of the system.

**Solution**

From Newton's law, we have

$$\sum F_u = m\ddot{u} \quad m\ddot{u} = k(u_s - u) \quad \text{or} \quad m\ddot{u} + ku = ku_s$$

For constant acceleration, we have that $\ddot{u}_s = a$, thus we have by integrating

$$\dot{u}_s = at + C_1 \quad \text{and} \quad u_s = \frac{at^2}{2} + C_1 t + C_2$$

It is noted that at time $t = 0$, $u_s = \dot{u}_s = 0$, and we find that $u_s = at^2/2$. Our differential equation of motion becomes

$$\ddot{u} + \omega_0^2 u = \frac{kat^2}{2m} = \frac{a\omega_0^2 t^2}{2}$$

Applying Equation 1.179 in the form

$$u(t) = \frac{1}{\omega_0}\int_0^t f(\tau)\sin\omega_0(t-\tau)d\tau = \frac{1}{\omega_0}\int_0^t \frac{a\omega_0^2}{2}\tau^2 \sin\omega_0(t-\tau)d\tau = \frac{a\omega_0}{2}\int_0^t (t-z)^2 \sin(\omega_0 z)dz$$

where $z = t - \tau$ has been used. Integrating yields

$$u(t) = \frac{a}{\omega_0^2}[\cos\omega_0 t - 1] + \frac{at^2}{2}$$

A plot of the response is shown in Figure 1.50.

### 1.10.2 Response to External Force That Varies Linearly with Time

Consider a force that linearly increases with time as shown in Figure 1.51. Assume that the force increases up to a finite value $p_0$, which occurs at time $t_0$. The forcing function is then expressed as

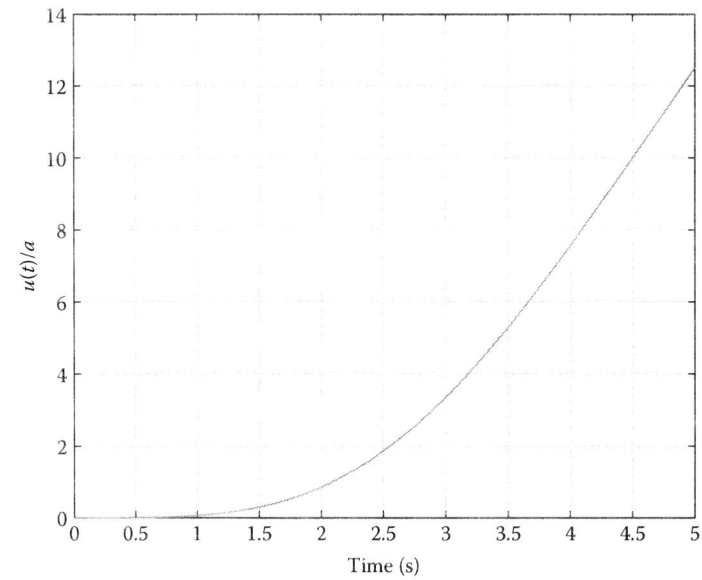


**FIGURE 1.50**
Response due to constant acceleration.

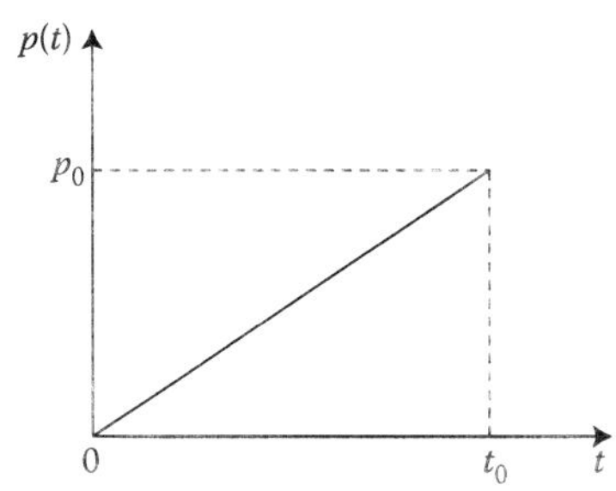


**FIGURE 1.51**
Linearly increasing force.

$$p(t) = p_0 \frac{t}{t_0} \tag{1.189}$$

The response can be determined using Equation 1.179 as

$$u(t) = \frac{1}{m\omega_0} \int_0^t \frac{p_0 \tau}{t_0} \sin \omega_0 (t - \tau) d\tau \tag{1.190}$$

Integrating using the substitution $z = t - \tau$ or by parts yields

$$u(t) = \frac{p_0}{m\omega_0 t_0} \int_0^t (t - z) \sin \omega_0 z \, dz = \frac{p_0}{m\omega_0 t_0} \int_0^t (t \sin \omega_0 z - z \sin \omega_0 z) dz$$

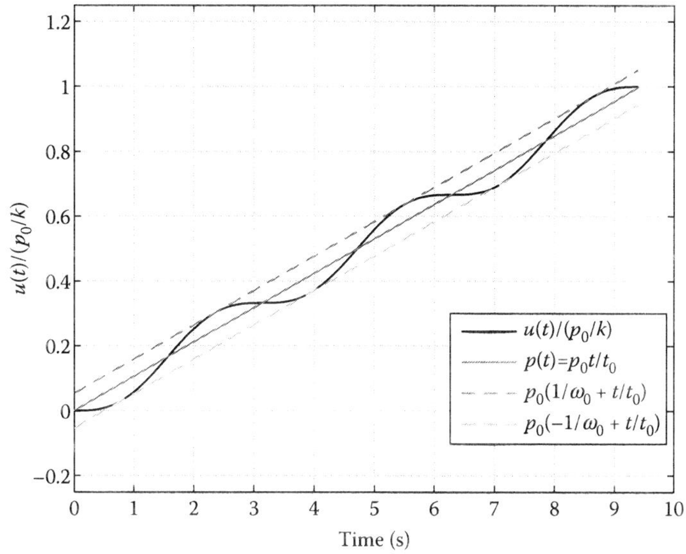


**FIGURE 1.52**
Response to a linearly increasing force.

and

$$u(t) = \frac{p_0}{k}\left[\frac{t}{t_0} - \frac{\sin\omega_0 t}{\omega_0 t_0}\right] \tag{1.191}$$

It is noted that the dynamic load factor is $(t/t_0 - \sin \omega_0 t/\omega_0 t_0)$. A plot of Equation 1.191 is shown in Figure 1.52.

The velocity is obtained by taking the derivative of Equation 1.191 giving

$$\dot{u}(t) = \frac{p_0}{kt_0}(1 - \cos\omega_0 t) \tag{1.192}$$

Equation 1.189 states that the velocity is always positive but vanishes at $t = 0, 2\pi, 4\pi, \ldots$

### 1.10.3 Exponentially Decaying Function

Consider a force that is exponentially decaying with time as shown in Figure 1.53. Assume that the force at $t = 0$ has the value $p_0$. The forcing function is then expressed as

$$p(t) = p_0 e^{-at} \tag{1.193}$$

The response can again be determined using Equation 1.179 as

$$u(t) = \frac{1}{m\omega_0}\int_0^t p_0 e^{-a\tau} \sin\omega_0(t - \tau)d\tau \tag{1.194}$$

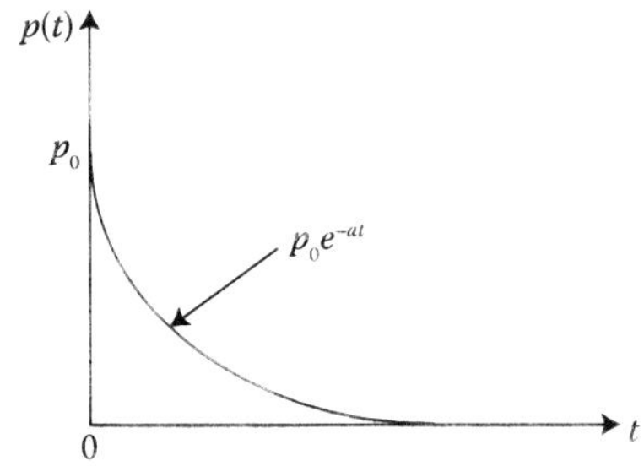


**FIGURE 1.53**
Exponentially decaying force.

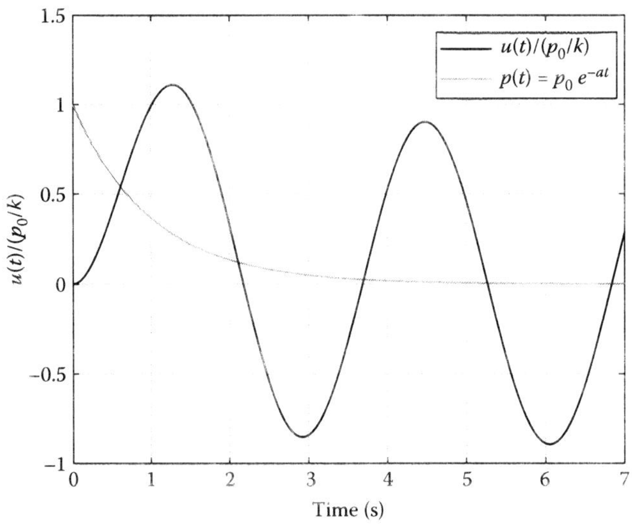


**FIGURE 1.54**
Response due to exponentially decaying force.

Integrating yields

$$u(t) = \left(\frac{p_0}{k}\right)\frac{(a/\omega_0)\sin\omega_0 t - \cos\omega_0 t + e^{-at}}{\left[1+(a/\omega_0)^2\right]} \tag{1.195}$$

A plot of Equation 1.195 is given in Figure 1.54 for $a/\omega_0 = 1/2$.

### 1.10.4 Asymptotic Step Forcing Function

Consider an asymptotic forcing function as shown in Figure 1.55. The forcing function is then expressed as

$$p(t) = p_0(1 - e^{-at}) \tag{1.196}$$

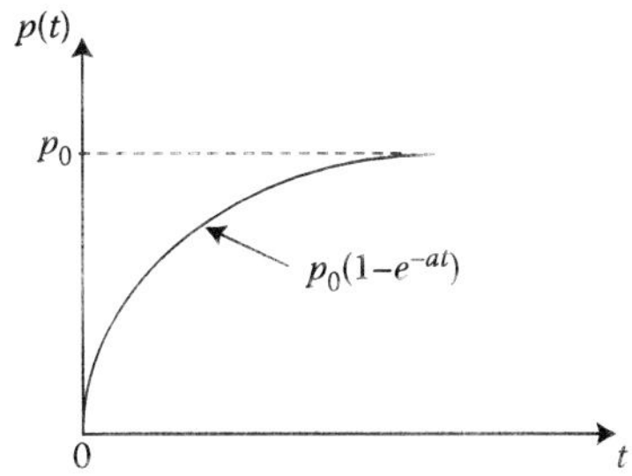


**FIGURE 1.55**
Asymptotic step forcing function.

The response can be determined using Equation 1.179 as

$$u(t) = \frac{1}{m\omega_0}\int_0^t p_0(1 - e^{-a\tau})\sin\omega_0(t-\tau)d\tau \tag{1.197}$$

Integrating using the substitution $z = t - \tau$ yields

$$u(t) = \frac{p_0}{m\omega_0}\int_0^t (1 - e^{-a(t-z)})\sin\omega_0 z\,dz = \frac{p_0}{m\omega_0}\int_0^t \left(\sin\omega_0 z - e^{-a(t-z)}\sin\omega_0 z\right)dz$$

and

$$u(t) = \frac{p_0}{k}\left[1 - \frac{(a/\omega_0)[\sin\omega_0 t + (a/\omega_0)\cos\omega_0 t] + e^{-at}}{1 + (a/\omega_0)^2}\right] \tag{1.198}$$

A plot of Equation 1.198 is shown in Figure 1.56 for the case of $a/\omega_0 = 1/2$.

### 1.10.5 Response to a Ramp-Step Function

Consider the ramp-step force applied as shown in Figure 1.57. The forcing function is given as

$$p(t) = \begin{cases} p_0(t/t_0) & \text{for } t \le t_0 \\ p_0 & \text{for } t > t_0 \end{cases} \tag{1.199}$$

The force response is again determined by using Equation 1.179 given by

$$u(t) = \frac{1}{m\omega_0}\left[\int_0^{t_0} \frac{p_0\tau}{t_0}\sin\omega_0(t-\tau)d\tau + \int_{t_0}^{t} p_0\sin\omega_0(t-\tau)d\tau\right] \tag{1.200}$$

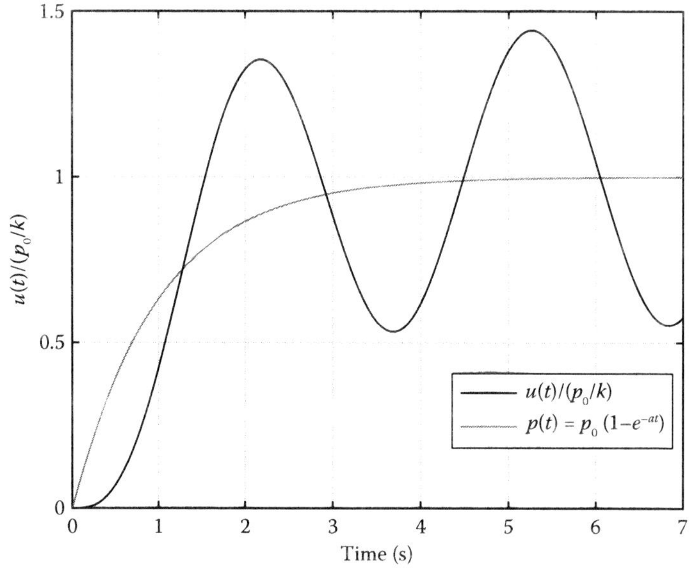


**FIGURE 1.56**
Response to asymptotic-step forcing function.

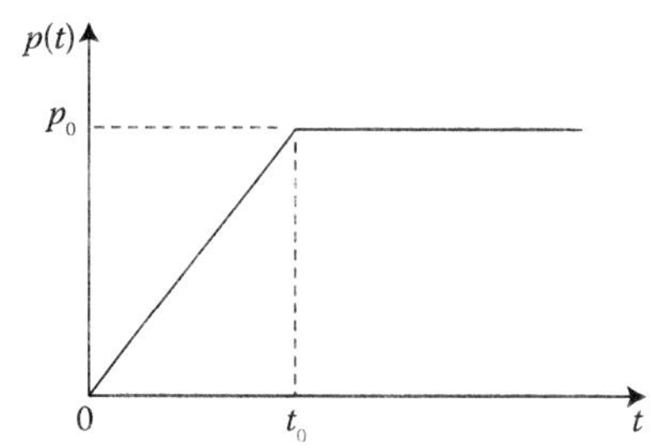


**FIGURE 1.57**
Ramp-step function.

Integrating yields

$$
\begin{aligned}
u(t) &= \frac{1}{m\omega_0}\left[\int_0^{t_0}\frac{p_0\tau}{t_0}\sin\omega_0(t-\tau)d\tau + \int_{t_0}^{t} p_0\sin\omega_0(t-\tau)d\tau\right] \\
&= \frac{p_0}{m\omega_0}\left[\left.\frac{\tau\cos\omega_0(t-\tau)}{\omega_0 t_0}+\frac{\sin\omega_0(t-\tau)}{\omega_0^2 t_0}\right|_0^{t_0}+\left.\frac{\cos\omega_0(t-\tau)}{\omega_0}\right|_{t_0}^{t}\right] \\
&= \frac{p_0}{m\omega_0}\left[\frac{t_0\cos\omega_0(t-t_0)}{\omega_0 t_0}+\frac{\sin\omega_0(t-t_0)}{\omega_0^2 t_0}-\frac{\sin\omega_0 t}{\omega_0^2 t_0}+\frac{1}{\omega_0}-\frac{\cos\omega_0(t-t_0)}{\omega_0 t_0}\right] \\
&= \frac{p_0}{m\omega_0}\left(\frac{1}{\omega_0}+\frac{\sin\omega_0(t-t_0)}{\omega_0^2 t_0}-\frac{\sin\omega_0 t}{\omega_0^2 t_0}\right)=\frac{p_0}{k}\left(1+\frac{\sin\omega_0(t-t_0)}{\omega_0 t_0}-\frac{\sin\omega_0 t}{\omega_0 t_0}\right) \quad t>t_0
\end{aligned}
\tag{1.201}
$$

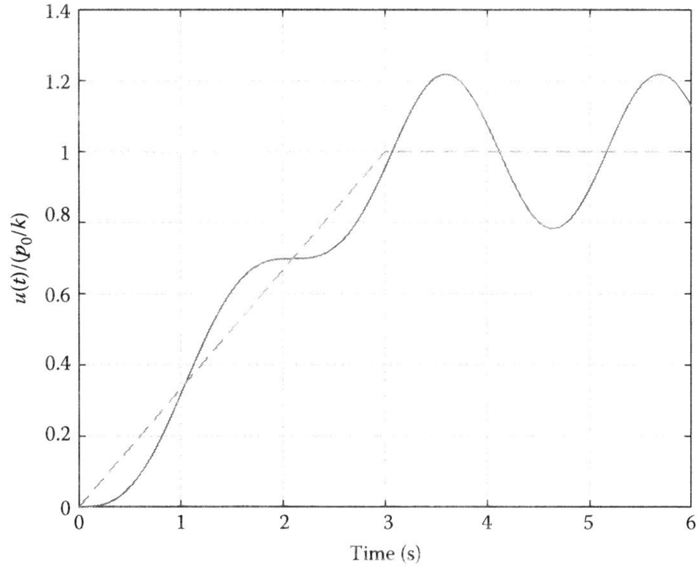


**FIGURE 1.58**
Response to ramp-step function.

For $t < t_0$ we can use the solution obtained from Equation 1.191, thus

$$u(t) = \frac{p_0}{kt_0}\left[t - \frac{\sin\omega_0 t}{\omega_0}\right] \quad t < t_0 \tag{1.202}$$

A plot of Equations 1.201 and 1.202 showing the amplification factor is shown in Figure 1.58.

Here $t_0 = 3$. Note that for $t > t_0$ the system represents harmonic motion around the equilibrium position. Also, the response for $t < t_0$ is similar to what was given for the linearly increasing force given by Equation 1.191 and shown in Figure 1.58. Recall that the velocity given by Equation 1.192 for the linearly increasing force at time $t = t_0$ was either positive or zero. If the displacement increases after $t_0$, then the velocity is different from zero and hence, the displacement will reach a maximum value at $t \geq t_0$. The time can be determined by differentiating Equation 1.201 and setting the expression to zero. Therefore, we have

$$\dot{u}(t) = \frac{p_0}{kt_0}[\cos\omega_0(t - t_0) - \cos\omega_0 t] = 0$$

Using the trigonometric identity $\cos\alpha - \cos\beta = -2\sin(\alpha + \beta)/2 \cdot \sin(\alpha - \beta)/2$, we find

$$\tan\omega_0 t = \tan\frac{\omega_0 t_0}{2}$$

or

$$t = \frac{n\pi}{\omega_0} + \frac{t_0}{2}, \quad n = 0, 1, 2, \ldots \tag{1.203}$$

The maximum displacement can be determined by substituting the value of time back into Equation 1.201. This gives

$$u_{max} = \frac{p_0}{k}\left(1 + \frac{2\sin(n\pi - \omega_0 t_0/2)}{\omega_0 t_0}\right) \tag{1.204}$$

### 1.10.6 Rectangular Impulse

Consider the rectangular impulse shown in Figure 1.59. The rectangular impulse is described by the following forcing function:

$$p(t) = \begin{cases} p_0 & \text{for } t \leq t_0 \\ 0 & \text{for } t \geq t_0 \end{cases} \tag{1.205}$$

The solution for the equation of motion, for $t \leq t_0$, with zero initial conditions is given by Equation 1.184, which is the solution given for a step load of $p_0$. Thus,

$$u(t) = \frac{p_0}{k}(1 - \cos\omega_0 t) \quad \text{for } t \leq t_0 \tag{1.206}$$

For $t \leq t_0$, the rectangular impulse ends and the system is the free vibration mode. The solution is given by Equation 1.12

$$u = u(t_0)\cos\omega_0(t - t_0) + [\dot{u}(t_0)/\omega_0]\sin\omega_0(t - t_0) \tag{1.207}$$

The initial conditions for free vibration portion of the response are obtained from Equation 1.206 evaluated at $t = t_0$. Thus, the displacement and velocity at time $t = t_0$ are given as

$$u(t_0) = \frac{p_0}{k}(1 - \cos\omega_0 t_0) \quad \text{and} \quad \dot{u}(t_0) = \frac{p_0\omega_0}{k}\sin\omega_0 t_0 \tag{1.208}$$

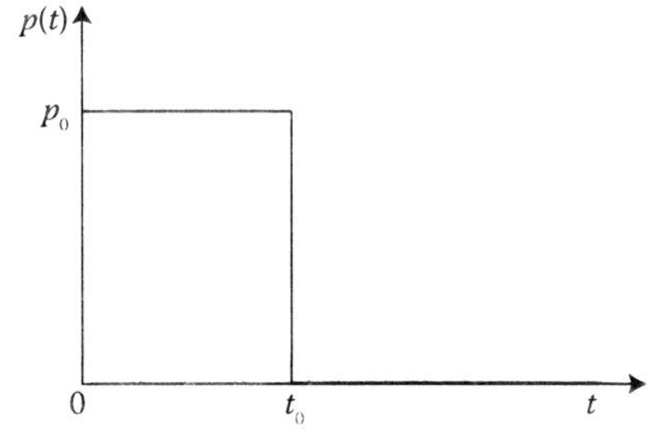


**FIGURE 1.59**
Rectangular impulse.

Substituting Equation 1.208 into Equation 1.207 yields

$$u(t) = \frac{p_0}{k}[(1 - \cos\omega_0 t_0)\cos\omega_0(t - t_0) + \sin\omega_0 t_0 \sin\omega_0(t - t_0)] \quad \text{for } t \geq t_0 \tag{1.209}$$

or, after using the trigonometric identity $\cos(\alpha - \beta) = \cos\alpha\cos\beta + \sin\alpha\sin\beta$

$$u(t) = \frac{p_0}{k}[\cos\omega_0(t - t_0) - \cos\omega_0 t] \quad t > t_0 \tag{1.210}$$

The maximum response will occur in the interval $t < t_0$ or $t > t_0$ depending on the value of the ratio $t_0/T$. A plot of Equations 1.206 and 1.210 is shown in Figure 1.60 for $t_0/T = 1.25$.

For the interval $t < t_0$, we can find $t_{max}$ by setting the derivative of Equation 1.208 to zero. Therefore, we have

$$\dot{u}(t) = \frac{p_0}{k}\omega_0 \sin\omega_0 t = 0$$

which gives

$$\omega t = n\pi \quad n = 1, 3, 5\ldots \tag{1.211}$$

The maximum displacement occurs in the interval $t < t_0$ giving

$$t_{max} = \frac{\pi}{\omega_0} = \frac{T}{2} < t_0 \tag{1.212}$$

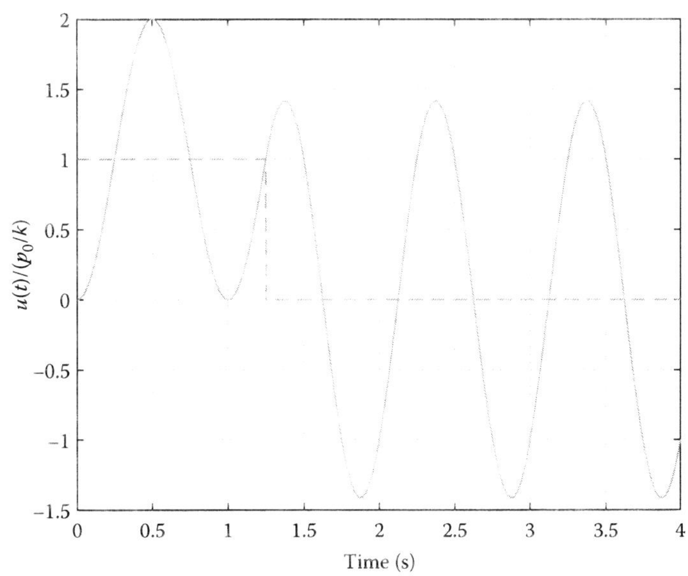


**FIGURE 1.60**
Response to rectangular impulse, $T = 1$ and $t_0/T = 1.25$.

Substituting into Equation 1.206 gives

$$u_{\max} = \frac{p_0}{k}(1-\cos\pi) = 2\frac{p_0}{k} \tag{1.213}$$

as shown in Figure 1.60. For $t_0 < T/2$, the maximum response will occur during the free vibration portion of the response $t > t_0$. This is shown in Figure 1.61 for $t_0/T = 0.25$. The maximum time can be determined by differentiating Equation 1.210, therefore,

$$\dot{u}(t) = \frac{p_0\omega_0}{k}[\sin\omega_0 t - \sin\omega_0(t-t_0)] = 0$$

giving

$$\sin\omega_0 t - \sin\omega_0(t-t_0) = 0$$

$$1-\cos\omega_0 t_0 - \sin\omega_0 t_0 \tan\frac{\omega_0 t_0}{2} = 0$$

or by using trigonometric identities

$$\tan\omega_0 t + \cot\frac{\omega_0 t_0}{2} = 0 \tag{1.214}$$

and $t$ is given as

$$t = (2n+1)\frac{\pi}{2\omega_0} + \frac{t_0}{2}, \quad n = 0,1,2,\ldots \tag{1.215}$$

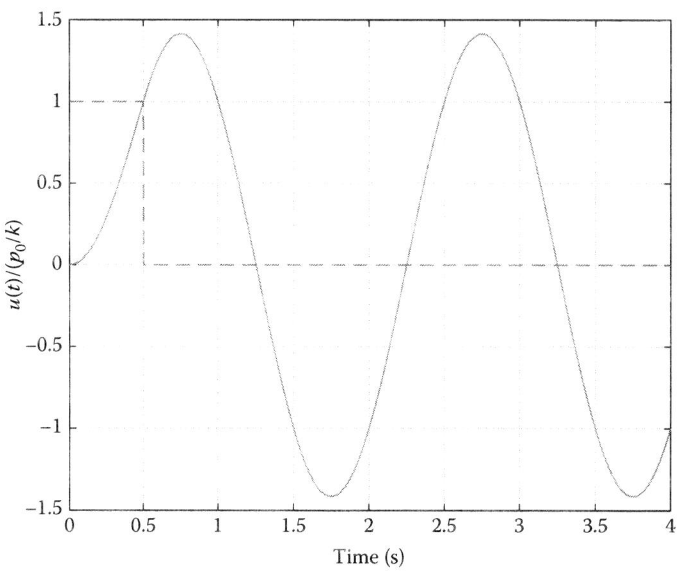


**FIGURE 1.61**
Response of rectangular impulse with $t_0/T = 0.25$ and $T = 2$.

Substituting Equation 1.215 back into Equation 1.210 yields

$$u = \frac{p_0}{k}[\cos\omega_0(t-t_0) - \cos\omega_0 t]$$
$$= -2\frac{p_0}{k}\left[\sin(2n+1)\frac{\pi}{2}\cdot\sin\left(\frac{-\omega_0 t_0}{2}\right)\right]$$

or

$$u_{\max} = 2\frac{p_0}{k}\sin\frac{\omega_0 t}{2} \tag{1.216}$$

Note that for $n = 0$, we have $t = 0.75$ s, which corresponds to our first peak given in Figure 1.61.

### 1.10.7 Triangular Impulse

Consider a triangular pulse as shown in Figure 1.62. The forcing function is given as

$$p(t) = \begin{cases} p_0\left(1 - \dfrac{t}{t_0}\right) & \text{for } t \le t_0 \\ 0 & \text{for } t > t_0 \end{cases} \tag{1.217}$$

Equation 1.179 gives, for an undamped system,

$$\begin{aligned} u(t) &= \frac{1}{m\omega_0}\int_0^t p_0\left(1 - \frac{\tau}{t_0}\right)\sin\omega_0(t-\tau)d\tau \\ &= \frac{p_0}{m\omega_0}\int_0^t \sin\omega_0(t-\tau)d\tau - \frac{p_0}{m\omega_0}\int_0^t \frac{\tau}{t_0}\sin\omega_0(t-\tau)d\tau \end{aligned} \tag{1.218}$$

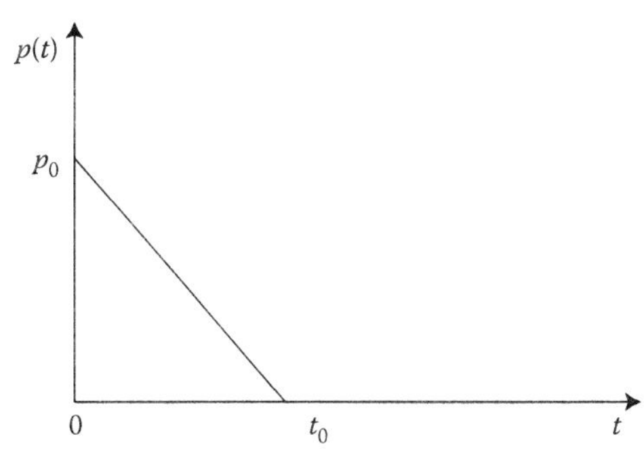


**FIGURE 1.62**
Triangular impulse.

Integrating yields

$$\begin{aligned} u(t) &= \frac{p_0}{m\omega_0}\left[\frac{1}{\omega_0}\cos\omega_0(t-\tau)\right]_0^t - \frac{p_0}{m\omega_0}\left[\frac{\tau\cos\omega_0(t-\tau)}{\omega_0 t_0} + \frac{\sin\omega_0(t-\tau)}{\omega_0^2 t_0}\right]_0^t \\ &= \frac{p_0}{m\omega_0}\left[\frac{1}{\omega_0} - \frac{\cos\omega_0 t}{\omega_0}\right] - \frac{p_0}{m\omega_0 t_0}\left[\frac{t}{\omega_0} - \frac{\sin\omega_0 t}{\omega_0^2}\right] \\ &= \frac{p_0}{k}\left[1 - \frac{t}{t_0} + \frac{\sin\omega_0 t}{\omega_0 t_0} - \cos\omega_0 t\right] \quad \text{for } 0 \le t \le t_0 \end{aligned} \tag{1.219}$$

To investigate the response for $t > t_0$, we use the results obtained for free vibration response given by Equation 1.12 subjected to the initial conditions $u(t_0)$, $\dot{u}(t_0)$ obtained from Equation 1.207 evaluated at $t = t_0$. Thus, we have

$$u(t_0) = \frac{p_0}{k}\left[\frac{\sin\omega_0 t_0}{\omega_0 t_0} - \cos\omega_0 t_0\right] \tag{1.220}$$

and

$$\dot{u}(t_0) = \frac{p_0}{k}\left(\omega_0 \sin\omega_0 t_0 + \frac{\cos\omega_0 t_0}{t_0} - \frac{1}{t_0}\right) \tag{1.221}$$

The result then for $t > t_0$ is

$$u(t) = \frac{p_0}{k}\left[\left(\frac{\sin\omega_0 t_0}{\omega_0 t_0} - \cos\omega_0 t_0\right)\cos\omega_0(t-t_0) + \left(\frac{\cos\omega_0 t_0}{\omega_0 t_0} + \sin\omega_0 t_0 - \frac{1}{\omega_0 t_0}\right)\sin\omega_0(t-t_0)\right] \tag{1.222}$$

The response given by Equations 1.207 and 1.210 is given in Figure 1.63 for $t_0 = 1.5$.

### 1.10.8 Half-Cycle Sine Impulse

Consider a half-cycle sine pulse as shown in Figure 1.64. The forcing function is given as

$$p(t) = \begin{cases} p_0 \sin\omega_f t & \text{for } t \le t_0 \\ 0 & \text{for } t > t_0 \end{cases} \tag{1.223}$$

For the undamped system, we have

$$\begin{aligned} u(t) &= \frac{1}{m\omega_0}\int_0^t p_0 \sin\omega_f \tau \sin\omega_0(t-\tau)d\tau \\ &= \frac{p_0 \sin\omega_0 t}{m\omega_0}\int_0^t \sin\omega_f \tau \cos\omega_0 \tau\, d\tau - \frac{p_0 \cos\omega_0 t}{m\omega_0}\int_0^t \sin\omega_f \tau \sin\omega_0 \tau\, d\tau \end{aligned} \tag{1.224}$$

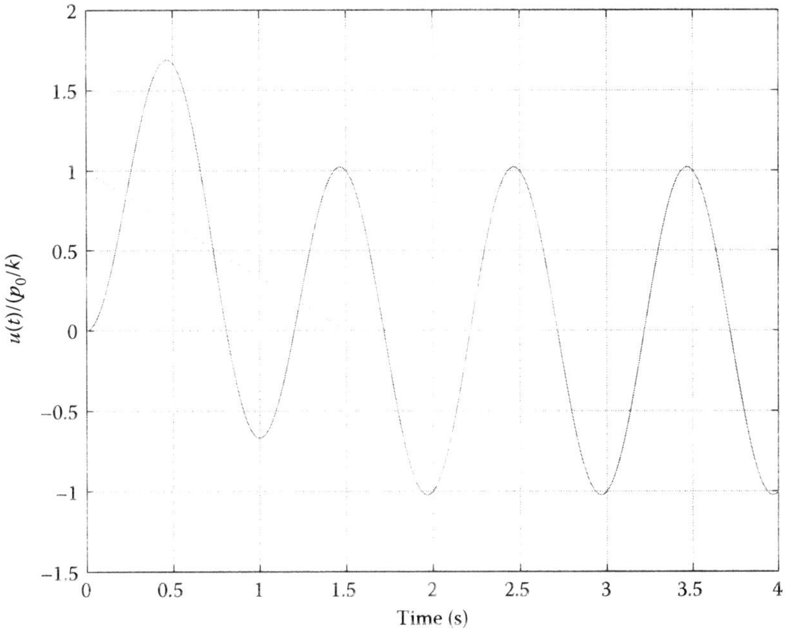


**FIGURE 1.63**
Response to triangular impulse.

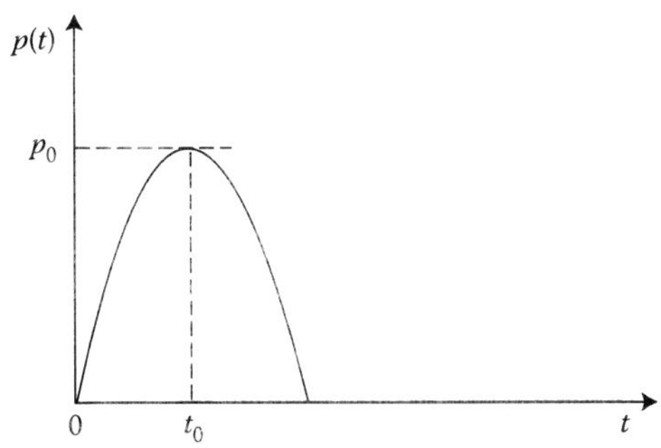


**FIGURE 1.64**
Half-cycle sine impulse.

Integrating yields

$$u(t) = -\frac{p_0 \sin\omega_0 t}{2m\omega_0}\left[\frac{\cos(\omega_f-\omega_0)\tau}{(\omega_f-\omega_0)}+\frac{\cos(\omega_f+\omega_0)\tau}{(\omega_f+\omega_0)}\right]_0^t - \frac{p_0}{2m\omega_0}\left[\frac{\sin(\omega_f-\omega_0)\tau}{(\omega_f-\omega_0)}-\frac{\sin(\omega_f+\omega_0)\tau}{(\omega_f+\omega_0)}\right]_0^t$$

$$= \frac{p_0}{m\omega_0}\left[\frac{\omega_0 \sin\omega_f t + \omega_f \sin\omega_0 t}{\omega_0^2-\omega_f^2}\right] = \frac{p_0/k}{(1-\Omega^2)}\left[\sin\omega_f t - \Omega\sin\omega_0 t\right] \quad \text{for } 0 \le t \le t_0 \tag{1.225}$$

To investigate the response for $t > t_0$, we use the results obtained for free vibration response given by Equation 1.12 subjected to the initial conditions $u(t_0)$, $\dot{u}(t_0)$ obtained from Equation 1.225 evaluated at $t = t_0$. Thus, using

$$u(t) = u_0(t_0)\cos\omega_0(t - t_0) + \frac{\dot{u}_0(t_0)}{\omega_0}\sin\omega_0(t - t_0) \tag{1.226}$$

and

$$\begin{aligned} u(t_0) &= \frac{p_0/k}{(1-\Omega^2)}[\sin\omega_f t_0 - \Omega\sin\omega_0 t_0] \\ \dot{u}(t_0) &= \frac{p_0/k}{(1-\Omega^2)}[\omega_f\cos\omega_f t_0 - \omega_f\cos\omega_0 t_0] \end{aligned} \tag{1.227}$$

the result then for $t > t_0$ is

$$u(t) = \frac{p_0/k}{(1-\Omega^2)}[(\sin\omega_f t_0 - \Omega\sin\omega_0 t_0)\cos\omega_0(t - t_0) + (\omega_f\cos\omega_f t_0 - \omega_f\cos\omega_0 t_0)\sin\omega_0(t - t_0)] \tag{1.228}$$

The response given by Equations 1.227 and 1.228 is given in Figure 1.65 for $t_0 = 0.75$.

The maximum value of the response will occur either in phase I or phase II depending on the value of $t_0$. When $t < t_0$, we take the derivative of Equation 1.228 and equate it to zero. We have that

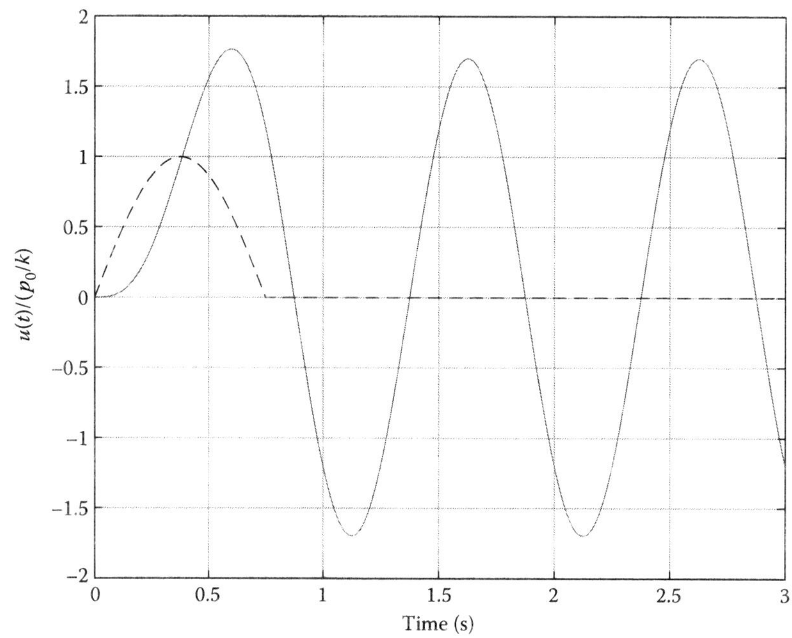


**FIGURE 1.65**
Response to triangular impulse.

$$\frac{du(t)}{dt} = \frac{p_0/k}{(1-\Omega^2)}[\omega_f \cos\omega_f t - \omega_0\Omega\cos\omega_0 t] = 0 \tag{1.229}$$

or

$$\cos\omega_f t = \cos\omega_0 t \quad \text{for } \omega_f t \leq \pi \tag{1.230}$$

Thus, we have

$$\omega_f t_{\max} = 2\pi n \pm \omega_0 t_{\max} \quad n = 0, 1, 2, \ldots \tag{1.231}$$

It is noted that the positive sign corresponds to a local minima and the negative sign to a local maxima. Therefore, the time at which the maximum would occur is given by

$$t_{\max} = \frac{2\pi n}{\omega_f + \omega_0} = \frac{2\pi n}{\pi/t_d + 2\pi/T} = \frac{2n}{1/t_d + 2/T} \quad n = 1, 2, \ldots \tag{1.232}$$

If the maximum occurs the region $t \leq t_d$, then

$$t_{\max} = \frac{2n}{1/t_d + 2/T} \leq t_d \quad \text{or} \quad \frac{t_d}{T} \geq \frac{2n-1}{2} \tag{1.233}$$

The smallest value of $t_d$ occurs for $n = 1$, thus

$$\frac{t_d}{T} \geq \frac{1}{2} \tag{1.234}$$

We have that $t_d/T = 1/2\Omega$ which implies that $\Omega \leq 1$. Thus the maximum response can be determined by substituting $t_{\max}$ as $t_{\max} = 2\pi n/\omega_0(1 + \Omega)$ which gives

$$\frac{u_{\max}}{(p_0/k)} = \frac{1}{(1-\Omega^2)}\left[\sin\left(\frac{2\pi n\Omega}{1+\Omega}\right) - \Omega\sin\left(\frac{2\pi n}{1+\Omega}\right)\right] \tag{1.235}$$

Substituting values for $\Omega$ gives $u_{\max}/(p_0/k) = 1.734$.

## 1.11 Integral Transformations

Integral transforms (Polyanin and Manzhirov 1998) are useful if they allow complicated problems to be turned into simpler problems. The transforms are most useful for solving differential and to a lesser extent, integral equations. The method behind transforms is very simple.

Given a function $y = f(x)$, we introduce a functional $z = I(f)$ where $z$ is a number and $f$ is a function. In equation form, we would write

$$z = I(f) = \int_a^b f(x)dx \tag{1.236}$$

A transformation (or operation) is written as

$$\bar{f}(s) = \Im\{f(t)\} \tag{1.237}$$

where

$$\begin{aligned} f(t) &\rightarrow \text{original function} \\ \bar{f}(s) &\rightarrow \text{its transformation} \\ t &\rightarrow \text{original variable} \\ s &\rightarrow \text{transformation variable (transformation parameter)} \end{aligned}$$

There are many different types of integral transforms, such as Laplace, Fourier, Hankel, Mellin, Hilbert, and just to name a few. Here we will be interested in only two, the Laplace and Fourier transforms.

The Laplace transform of $f(t)$ and the Fourier transform of $f(t)$ are formally defined as

$$\bar{f}(s) = \mathscr{L}\{f(t)\} = \int_0^\infty e^{-st} f(t)dt \tag{1.238}$$

and

$$f^*(i\omega) = \Im\{f(t)\} = \int_{-\infty}^{\infty} f(t)e^{-i\omega t}\, dt \tag{1.239}$$

The inverse transformations for the Laplace and Fourier transformations are

$$f(t) = \mathscr{L}^{-1}\{\bar{f}(s)\} = \frac{1}{2\pi i}\int_{c-i\omega}^{c+i\omega} \bar{f}(s)e^{ts}\, ds \quad c = \text{real constant} \tag{1.240}$$

and

$$f(t) = \Im^{-1}\{f^*(i\omega)\} = \frac{1}{2\pi}\int_{-\infty}^{\infty} f^*(i\omega)e^{i\omega t}\, d\omega \tag{1.241}$$

**TABLE 1.1**

Selected Examples of Laplace and Fourier Transformations

| Laplace | | Fourier | |
|---|---|---|---|
| $f(t)$ | $\bar{f}(s)$ | $f(t)$ | $f^*(i\omega)$ |
| 1 | $\frac{1}{s}$ | $\frac{1}{1+t^2}$ | $\pi e^{-\lvert\omega\rvert}$ |
| $e^{at}$ | $\frac{1}{s-a}$ | $e^{-a\lvert t\rvert}$ | $\frac{2a}{\omega^2+a^2}$ $(a>0)$ |
| $e^{-at}$ | $\frac{1}{s+a}$ | $e^{-a^2t^2}$ | $\frac{\sqrt{\pi}}{a}e^{-\omega^2/4a^2}$ |
| $\sin at$ | $\frac{a}{s^2+a^2}$ | $e^{iat}$ | $2\pi\delta(\omega-a)$ |
| $\cos at$ | $\frac{s}{s^2+a^2}$ | $t^{-1}$ | $-i\pi\,\mathrm{sgn}(\omega)$ |
| $t\sin at$ | $\frac{2as}{(s^2+a^2)^2}$ | $\sin at$ | $i\pi[\delta(\omega+a)-\delta(\omega-a)]$ |
| $t\cos at$ | $\frac{s^2-a^2}{(s^2+a^2)^2}$ | $\cos at$ | $\pi[\delta(\omega+a)+\delta(\omega-a)]$ |
| $t^n$ $(n=0,1,2,3,\ldots)$ | $\frac{n!}{s^{n+1}}$ | $t^n$ | $2\pi i^n\delta^{(n)}(t)$ |

Selected examples of both transformations are given in Table 1.1 and time derivatives of both the Laplace and Fourier transformations can be taken. The resulting properties of each are illustrated in Table 1.2.

Associated with certain kind of integral equations is the concept of convolution theory. The convolution of $f$ and $g$ is written as $f * g$. It is defined as the integral of the product of two functions after one is reversed and shifted. Thus, we define convolution as

$$(f(t)*g(t))=\int_{-\infty}^{\infty} f(\tau)g(t-\tau)d\tau=\int_{-\infty}^{\infty} f(t-\tau)g(\tau)d\tau \tag{1.242}$$

**TABLE 1.2**

Time Derivatives of Laplace and Fourier Transformations

| Laplace ($\bar{f}(s)=\mathscr{L}\{f(t)\}$) | Fourier ($f^*(i\omega)=\Im\{f(t)\}$) |
|---|---|
| $\mathscr{L}\left\{\frac{df}{dt}\right\}=s\bar{f}(s)-f(0)$ | $\Im\left\{\frac{df}{dt}\right\}=i\omega f^*(i\omega)$ |
| $\mathscr{L}\left\{\frac{d^2f}{dt^2}\right\}=s^2\bar{f}(s)-sf(0)-f(0)$ | $\Im\left\{\frac{d^2f}{dt^2}\right\}=(i\omega)^2 f^*(i\omega)$ |
| $\mathscr{L}\left\{\frac{d^nf}{dt^n}\right\}=s^n\bar{f}(s)-s^{n-1}f(0)\cdots-sf^{(n-2)}(0)\cdots-f^{(n-1)}(0)$ | $\Im\left\{\frac{d^nf}{dt^n}\right\}=(i\omega)^n f^*(i\omega)$ |

For the Laplace and Fourier transforms, we have

$$\mathscr{L}^{-1}\{\bar{f}(s)\bar{g}(s)\} = \int_0^t f(t-\tau)g(\tau)d\tau = \int_0^t f(\tau)g(t-\tau)d\tau$$

$$\mathscr{L}\left\{\int_0^t f(t-\tau)g(\tau)d\tau\right\} = \bar{f}(s)\bar{g}(s) \tag{1.243}$$

$$\Im^{-1}\{f^*(i\omega)g^*(i\omega)\} = \int_{-\infty}^t f(t-\tau)g(\tau)d\tau$$

### 1.11.1 Application of the Fourier Transformations to the Viscoelastic Relations

Assume $\sigma(t)$ and $\varepsilon(t)$ arbitrarily with $\sigma^*(i\omega) = \Im\{\sigma(t)\}$ and $\varepsilon^*(i\omega) = \Im\{\varepsilon(t)\}$. For the different models, we would have the following:

1. Kelvin:

$$\sigma^* = (E + i\omega\eta)\varepsilon^* \quad \text{where } \eta = E^*(i\omega) \tag{1.244}$$

2. Maxwell:

$$\left(\frac{1}{E}i\omega + \frac{1}{\eta}\right)\sigma^* = \varepsilon^* i\omega, \quad \sigma^* = \frac{i\omega}{\left(\frac{i\omega}{E} + \frac{1}{\eta}\right)}\varepsilon^*, \quad \sigma^* = \frac{i\omega}{E^*(i\omega)}\varepsilon^* \tag{1.245}$$

3. Standard solid:

$$\left(\frac{\eta}{E_1}i\omega + 1\right)\varepsilon^* = \frac{\eta}{E_o E_1}i\omega\sigma^* + \left(\frac{1}{E_o} + \frac{1}{E_1}\right)\sigma^*$$

or (1.246)

$$\sigma^* = \frac{(\eta/E_1)i\omega + 1}{(\eta/E_o E_1)i\omega + ((1/E_o) + (1/E_1))}\varepsilon^* = E^*(i\omega)\varepsilon^*$$

Thus, for viscoelastic materials, we have the Fourier transform of $\sigma$ and the Fourier transform of $\varepsilon$ yielding a complex modulus ($E^*$) and the relationship $\sigma^* = E^*\varepsilon^*$. In a similar manner, for shear we have $\tau^* = G^*\gamma^*$. For an elastic material, we have $\sigma^* = E\varepsilon^*$ and $\tau^* = G\gamma^*$. If one considers the elastic–viscoelastic analogy, the relationships of interest are $E \to E^*$ and $G \to G^*$.

### 1.11.2 Application of the Laplace Transformations to the Viscoelastic Relations

Assume that at $t = 0$, $\sigma(0) = 0$ and $\varepsilon(0) = 0$, this leads to the following for the viscoelastic models:

1. Kelvin:

$$\bar{\sigma} = (E + \eta s)\bar{\varepsilon} = (E + \bar{E}(s))\bar{\varepsilon} \tag{1.247}$$

2. Maxwell:

$$\left(\frac{s}{E} + \frac{1}{\eta}\right)\bar{\sigma} = s\bar{\varepsilon}, \quad \bar{\sigma} = \frac{s}{((s/E) + (1/\eta))}\bar{\varepsilon} \tag{1.248}$$

3. Standard solid:

$$\left(\frac{\eta}{E_1}s + 1\right)\bar{\varepsilon} = \frac{\eta}{E_0 E_1}s\bar{\sigma} + \left(\frac{1}{E_o} + \frac{1}{E_1}\right)\bar{\sigma}$$

or

$$\bar{\sigma} = \frac{\frac{\eta}{E_1}s + 1}{(\eta/E_0E_1)s + ((1/E_0) + (1/E_1))}\bar{\varepsilon} = \bar{E}(s)\bar{\varepsilon} \tag{1.249}$$

Thus, for viscoelastic materials, we have the relationships $\bar{\sigma} = \bar{E}(s)\bar{\varepsilon}$ and $\bar{\tau} = \bar{G}(s)\bar{\gamma}$. In a similar manner, for an elastic material, we have $\bar{\sigma} = E\bar{\varepsilon}$ and $\bar{\tau} = G\bar{\gamma}$. If one considers the elastic–viscoelastic analogy, the relationships of interest are $E \to \bar{E}(s)$ and $G \to \bar{G}(s)$, where $\bar{E}(s)$ and $\bar{G}(s)$ are called operational parameters.

### 1.11.3 Dirac and Heaviside Functions

The Dirac delta (unit impulse) and Heaviside (unit step) functions are frequently used to define loading conditions. Both are functions of time and are modeled as either starting at time $t = 0$ or at some time $t = t_1$. The Dirac ($\delta(t)$) and Heaviside ($H(t)$) functions are modeled as shown in Figure 1.65. Both functions are subjected to integrations. For the Dirac delta, we have

$$\delta(t) = \begin{cases} +\infty & t = 0 \\ 0 & t \neq 0 \end{cases} \tag{1.250}$$

and it also is constrained to satisfy the identity

$$\int_{-\infty}^{\infty} \delta(t)dt = 1 \tag{1.251}$$

In addition, the delta function satisfies the properties

$$\int_{-\infty}^{\infty} \delta(t)f(t)dt = f(0) \quad \text{and} \quad \int_{-\infty}^{\infty} \delta(t - t_1)f(t)dt = f(t_1) \tag{1.252}$$

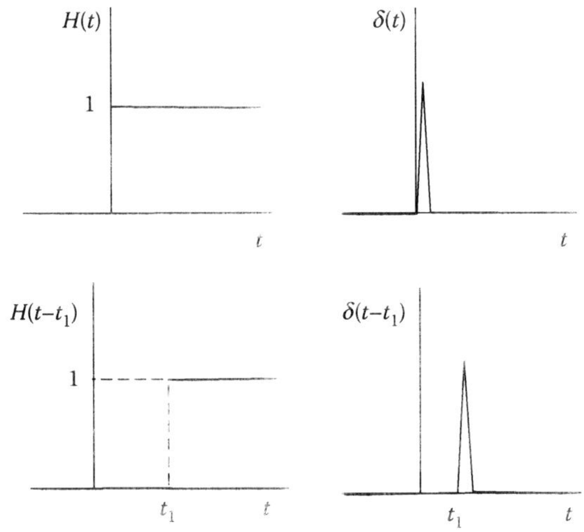


**FIGURE 1.66**
Models of Dirac and Heaviside functions.

For the Heaviside step function, or unit step function, we have

$$H(t)=\begin{cases}0 & t<0\\ 1 & t\geq 0\end{cases} \quad \text{and} \quad H(t-t_1)=\begin{cases}0 & t<t_1\\ 1 & t\geq t_1\end{cases} \tag{1.253}$$

A plot of these functions is depicted in Figure 1.66.

Additional relationships between Dirac and Heaviside functions and Laplace and Fourier transformations are shown below and illustrated in Figure 1.67.

$$\begin{aligned}
&\mathcal{L}\{\delta(t)\}=\int_0^{\infty}\delta(t)e^{-st}dt=e^{-\delta(0)}=1\\
&\mathcal{L}\{\delta(t)\}=\int_0^{\infty}\delta(t)e^{-st}dt=e^{-\delta(0)}=1\\
&\Im\{\delta(t)\}=1\\
&\delta(t)=\frac{dH(t)}{dt}
\end{aligned} \tag{1.254}$$

### 1.11.4 Application of Laplace and Fourier Transforms to ODOF Systems with Material Damping

Consider a one-degree-of-freedom system with damping as modeled in Figure 1.11. The differential equation and initial conditions are given as

$$m\ddot{u}+\eta\dot{u}+ku=P(t) \quad \text{at } t=0,\, u(0)=\dot{u}(0)=0 \tag{1.255}$$

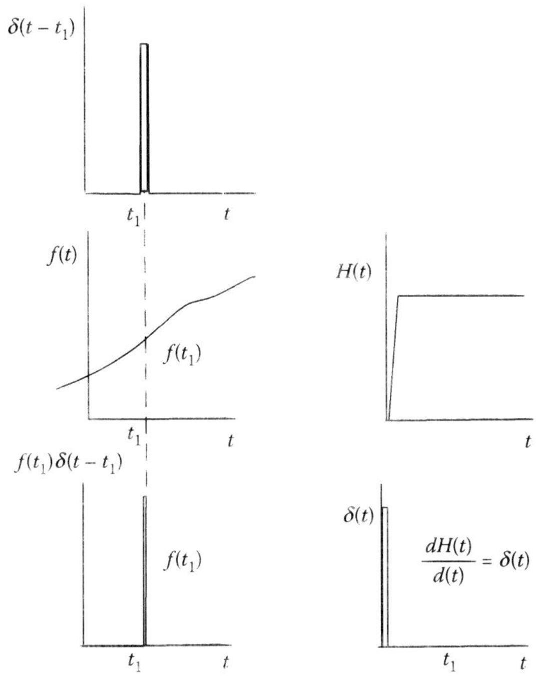


**FIGURE 1.67**
Dirac and Heaviside functions and Laplace and Fourier transformations.

Appling the Laplace transform yields

$$\begin{aligned} &\mathscr{L}\{m\ddot{u}+\eta\dot{u}+ku\}=\mathscr{L}\{P(t)\} \\ &ms^2\bar{u}+\eta s\bar{u}+k\bar{u}=\bar{P}(s) \\ &\bar{u}=\frac{\bar{P}(s)}{ms^2+\eta s+k}=\bar{U}(s)\bar{P}(s) \end{aligned} \tag{1.256}$$

where the transfer function of $u$ is given as

$$\bar{U}(s)=\frac{1}{ms^2+\eta s+k} \tag{1.257}$$

Suppose $P(t)=\delta(t)$, then $\bar{P}(s)=1$ and $\bar{u}=1/(ms^2+\eta s+k)=\bar{U}(s)$, so $\mathscr{L}(\bar{u})=\mathscr{L}(U(t))=\bar{U}(s)$. If $\bar{U}(s)$ is given, then the response to a unit impulse can be obtained by inversion, so that $U(t)=\mathscr{L}^{-1}\{\bar{U}(s)\}$. Thus, $\bar{u}=\bar{U}(s)\bar{P}(s)$. Using the convolution integral theorem, we can write

$$\mathscr{L}^{-1}\{\bar{u}\}=\mathscr{L}^{-1}\{\bar{U}(s)\bar{P}(s)\}$$

and

$$u(t)=\int_0^t U(t-\tau)P(\tau)d\tau \tag{1.258}$$

To find the displacement due to any force, we use the unit impulse response. The transfer function $\bar{U}(s)$ depends on (a) the system, (b) the loading or input, and (c) the investigated quantity.

**EXAMPLE 1.14**

Assume a system as shown in Figure 1.68, where the input is $P(t)$ and the desired output is $\dot{u}(t) = v(t)$. Determine the expression for $\bar{S}(s)$.

**Solution**

Initially assume $a = 0$. The general form of the equation of motion for this system is

$$m\ddot{u} + ku = P(t)$$

We need to find the transfer function of $\nu = du/dt$ or $\bar{v} = s\bar{u}$. If $P(t) = \delta(t)$, then $\bar{P}(s) = 1$, $\bar{v} \rightarrow \bar{V}(s)$, and $\bar{u} \rightarrow \bar{U}(s)$. Then $\bar{V}(s) = s\bar{U}(s)$ and the result will be

$$\bar{V}(s) = \frac{s}{ms^2 + k}$$

As an alternative to this problem, assume $P(t) = 0$ and the input is the displacement $a(t) = \delta(t)$. The desired output is $u(t)$. The governing equation is

$$m\ddot{u} + ku = ka$$

Application of the Laplace transform results in

$$ms^2\bar{U} + k\bar{U} = k \text{ which yields}$$

$$\bar{U} = \frac{k}{k + ms^2}$$

The response to a unit impulse is obtained from

$$U(t) = \mathscr{L}^{-1}\{\bar{U}(s)\} = \frac{k}{m\omega_0} \sin \omega_0 t = \omega_0 \sin \omega_0 t$$

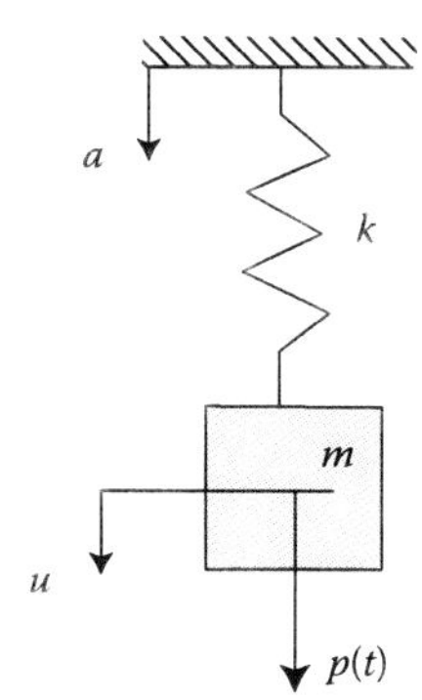


**FIGURE 1.68**
Simple spring–mass system subjected to a time varying input.

As an alternative, assume the desired output is the spring force $\bar{S}(t)$. The solution would be

$$\bar{S}(s) = -\frac{kms^2}{k + ms^2} \quad \text{where we use } s(t) = k(u - a)$$

Next we will consider the spring dashpot model of Figure 1.11. In this example, we explore the frequency response function. The governing equation for this problem is

$$m\ddot{u} + \eta\dot{u} + ku = P(t)$$

Assuming $P(t) = \delta(t)$, then $u(t) = U(t)$. We want to find the Fourier transformation of $U$.

$$\Im\{m\ddot{u} + \eta\dot{u} + ku\} = \Im\{P(t)\}$$

The resulting equation is

$$m(i\omega)^2 u^* + \eta(i\omega)u^* + ku^* = 1$$

and

$$u^* \equiv U^* = \frac{1}{-m\omega^2 + \eta(i\omega) + k}$$

$U^* = U^*(i\omega)$ is the *frequency response function* (F.R.F.) of $u$ for the input $P(t)$, which is also called the *frequency characteristics.* The properties of the frequency response function are such that $U^* = \Im\{U(t)\}$ and for an arbitrary $P(t)$ the solution for $u$ is

$$m(i\omega)^2 u^* + \eta(i\omega)u^* + ku^* = P^* \text{ or } u^* = \frac{P^*}{-m\omega^2 + \eta(i\omega) + k} = U^* P^*$$

In order to obtain $u$ as a function of $t$, $u(t) = \Im^{-1}\{u^*\}$

$$u(t) = \frac{1}{2\pi}\int_{-\infty}^{\infty} U^*(i\omega)P^*(i\omega)e^{i\omega t}\, d\omega$$

This can be solved by (a) tables of Fourier inversions, (b) complex integration, or (c) numerical integration.

The second property of the F.R.R. is if $s$ is replaced by $i\omega$ then $\bar{U}(s)$ becomes $U^*(i\omega)$. The F.R.F. can be defined in another manner. Suppose

$$P(t) = 1e^{i\omega t} \tag{1.259}$$

then the solution for the displacement will be assumed to be

$$u(t) = U^*(i\omega)e^{i\omega t} \tag{1.260}$$

where $U^*(i\omega)$ is the amplitude of $u(t)$ corresponding to $P(t) = 1e^{i\omega t}$. In order to find $U^*(i\omega)$, we substitute (1.259) and (1.260) into the equation of motion resulting in

$$m(i\omega)^2 U^* e^{i\omega t} + \eta(i\omega) U^* e^{i\omega t} + kU^* e^{i\omega t} = 1e^{i\omega t}$$

or

$$U^* = \frac{1}{-m\omega^2 + \eta(i\omega) + k} \tag{1.261}$$

We see that the amplitude is the same as the F.R.F.

### 1.11.5 An Application of $U^*$

If $U^* e^{i\omega t}$ is the solution for $P(t) = 1e^{i\omega t}$, then substituting into the equation of motion yields

$$m\frac{d^2}{dt^2}(U^* e^{i\omega t}) + \eta\frac{d}{dt}(U^* e^{i\omega t}) + k(U^* e^{i\omega t}) = 1e^{i\omega t}$$

Since

$$U^* e^{i\omega t} = \text{Re}[U^* e^{i\omega t}] + i\,\text{Im}[U^* e^{i\omega t}]$$

$$1e^{i\omega t} = \text{Re}\left[1e^{i\omega t}\right] + i\,\text{Im}\left[1e^{i\omega t}\right]$$

then

$$m\frac{d^2}{dt^2}(\text{Re}[U^* e^{i\omega t}] + i\,\text{Im}[U^* e^{i\omega t}]) + \eta\frac{d}{dt}(\text{Re}[U^* e^{i\omega t}] + i\,\text{Im}[U^* e^{i\omega t}])$$
$$+ k(\text{Re}[U^* e^{i\omega t}] + i\,\text{Im}[U^* e^{i\omega t}]) = \text{Re}[1e^{i\omega t}] + i\,\text{Im}[1e^{i\omega t}]$$

Since the real and imaginary part must be equal, we have

$$m\frac{d^2}{dt^2}(\text{Re}[U^* e^{i\omega t}]) + \eta\frac{d}{dt}(\text{Re}[U^* e^{i\omega t}]) + k(\text{Re}[U^* e^{i\omega t}]) = \text{Re}[1e^{i\omega t}] \tag{1.262}$$

$$m\frac{d^2}{dt^2}(\text{Im}[U^* e^{i\omega t}]) + \eta\frac{d}{dt}(\text{Im}[U^* e^{i\omega t}]) + k(\text{Im}[U^* e^{i\omega t}]) = \text{Im}[1e^{i\omega t}] \tag{1.263}$$

There are two possibilities for Equations 1.262 and 1.263:

1. If the loading is $\text{Re}[1e^{i\omega t}]$ the solution is $\text{Re}[U^* e^{i\omega t}]$, or if the loading is $1\cos\omega t$ the solution is $\text{Re}[U^* e^{i\omega t}]$, or if the loading is $P_0 \cos\omega t$ the solution is $P_0(\text{Re}[U^* e^{i\omega t}])$.
2. If the loading is $\text{Im}[1e^{i\omega t}]$ the solution is $\text{Im}[U^* e^{i\omega t}]$, or if the loading is $1\sin\omega t$ the solution is $\text{Im}[U^* e^{i\omega t}]$, or if the loading is $P_0 \sin\omega t$ the solution is $P_0(\text{Im}[U^* e^{i\omega t}])$.

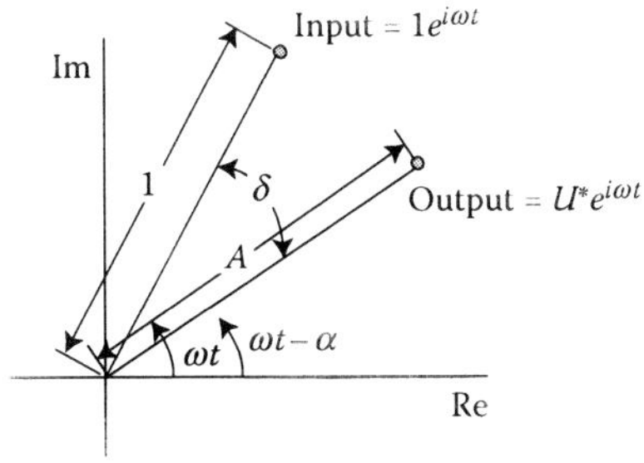


**FIGURE 1.69**
The complex plane representation of $U^*e^{i\omega t}$.

Note that if $U^*(i\omega) = U'(\omega) + iU''(\omega)$, then $U^*e^{i\omega t} = (U' + iU'')(\cos \omega t + i \sin \omega t)$ and

$$\begin{aligned} \text{Re}[U^* e^{i\omega t}] &= U' \cos \omega t - U'' \sin \omega t \\ \text{Im}[U^* e^{i\omega t}] &= U' \sin \omega t + U'' \cos \omega t \end{aligned} \tag{1.264}$$

The interpretation of $U^*e^{i\omega t}$ in the complex plane is as shown in Figure 1.69. The output can be written as

$$\begin{aligned} U^* e^{i\omega t} &= |U^*| e^{-i\alpha} e^{i\omega t} \\ &= A e^{i(\omega t - \alpha)} \end{aligned} \tag{1.265}$$

where

$$A = \sqrt{(U')^2 + (U'')^2} \quad \text{and} \quad \tan \alpha = \frac{-U''}{U'} \tag{1.266}$$

### 1.11.6 Experimental Determination of $U^*$

In order to experimentally determine $U^*$, one needs to measure the displacement and the applied force with displacement and force transducers as illustrated in Figure 1.70.

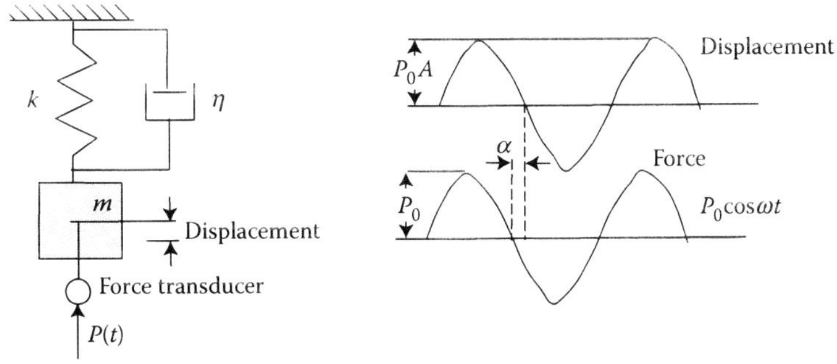


**FIGURE 1.70**
Experimental setup for determining $U^*$.

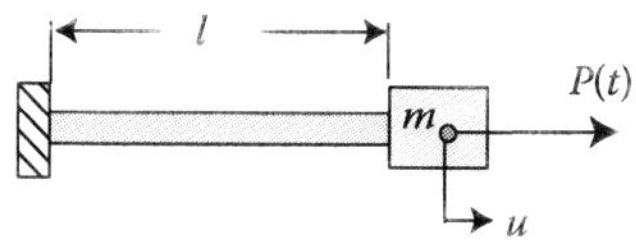


**FIGURE 1.71**
Mass attached to massless rod.

These measured quantities can then be interpreted and will produce both displacement and force plots, where the fore plot is slightly shifted by an amount $\alpha$.

From these plots, we can determine $A$ and $\alpha$, where $A = A(\omega)$ and $\alpha = \alpha(\omega)$. Then, we can determine $U' = A \cos \alpha$ and $U'' = -A \sin \alpha$.

As a practical application, consider a mass $m$ attached to a rod with a cross-sectional area $A$ as shown in Figure 1.71.

If the rod is massless, this becomes a one-degree-of-freedom system. We can model the rod as a spring for which the spring rate is $k = AE/l$. For a displacement $u$, the force in the spring is

$$F_s = ku = \frac{AE}{l} u$$

The equation of motion is

$$m\ddot{u} + ku = P(t) \quad \text{or} \quad m\ddot{u} + \frac{AE}{l} u = P(t)$$

The natural frequency for this system (no damping) is $\omega_o = \sqrt{AE/lm}$. The frequency response function of $u$ is

$$U^* = \frac{1}{(EA/l) - m\omega^2}$$

If $P(t) = P_0 \cos \omega t$, then

$$u(t) = P_o \operatorname{Re}[U^* e^{i\omega t}] = \frac{P_o}{(EA/l) - m\omega^2} \cos \omega t$$

One could take into account the material damping in the bar (linear damping). To find the frequency response function with material damping, replace $E$ with $E^*$ so that

$$U^* = \frac{1}{(E^* A/l) - m\omega^2}$$

where $E^* = E'(\omega) + iE''(\omega)$ or $E^* = E' + iE''$ with $E' =$ constant and $E'' =$ constant. Then find the transfer function $\bar{U}(s)$. For an elastic material

$$\bar{U}(s) = \frac{1}{(EA/l) - ms^2}$$

For a material with linear damping, $E \rightarrow \bar{E}(s)$ and

$$\bar{U}(s) = \frac{1}{(\bar{E}(s)A/l) - ms^2}$$

The response due to a unit impulse (with material damping) is

$$U(t) = \mathscr{L}^{-1}\{\bar{U}(s)\} = \mathscr{L}^{-1}\left\{\frac{1}{(\bar{E}(s)A/l) - ms^2}\right\}$$

| Elastic | Linear-Viscoelastic |
|---|---|
| $m\ddot{u} + S = P(t)$ | $m\ddot{u} + S = P(t)$ assume $u(0) = \dot{u}(0) = 0$ |
| $\bar{u}ms^2 + \bar{S}(s) = \bar{P}(s)$ | $\bar{u}ms^2 + \bar{S}(s) = \bar{P}(s)$ $S =$ spring rate |
| $\bar{S}(s) = A\bar{\sigma}$ | $\bar{S}(s) = A\bar{\sigma}$ |
| $\bar{\sigma} = E\bar{\varepsilon} = \frac{E\bar{u}}{l}$ so $\bar{S}(s) = \frac{AE\bar{u}}{l}$ | $\bar{\sigma} = \bar{E}(s)\bar{\varepsilon} = \frac{\bar{E}(s)\bar{u}}{l}$ so $\bar{S}(s) = \frac{A\bar{E}(s)\bar{u}}{l}$ |
| $\bar{u}ms^2 + \frac{AE\bar{u}}{l} = \bar{P}(s)$ | $\bar{u}ms^2 + \frac{A\bar{E}(s)\bar{u}}{l} = \bar{P}(s)$ |
| $\bar{u} = \frac{\bar{P}(s)}{ms^2 + (AE/l)}$ | $\bar{u} = \frac{\bar{P}(s)}{ms^2 + (A\bar{E}(s)/l)}$ |

As a slightly different application, consider the beam shown in Figure 1.72, where the base is moving an amount $a$.

For the massless bar, we know that due to the type of deflection (a cantilever beam with an end load) we can model the bar as a flexure spring with a spring rate of

$$k = \frac{3EI}{l^3}$$

The spring force is $F_s = k\delta = (3EI/l^3)(u - a)$. The equation of motion is

$$m\ddot{u} + \frac{3EI}{l^3}(u - a) = 0 \quad \text{or} \quad m\ddot{u} + \frac{3EI}{l^3}u = \frac{3EI}{l^3}a$$

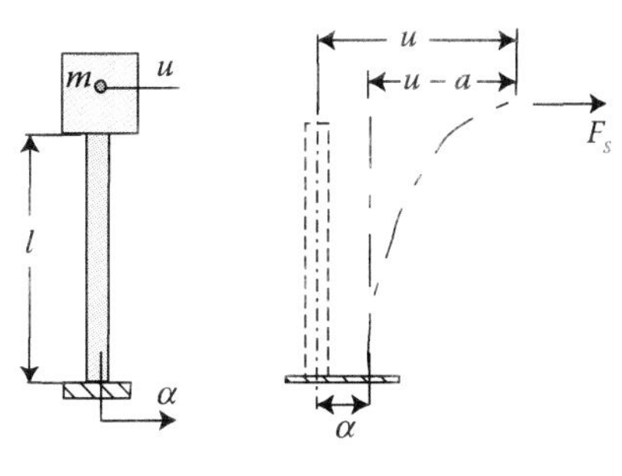


**FIGURE 1.72**
Vertical beam with attached mass.

To find $U^*$ for an input $a(t)$, we can start by assuming $a(t) = 1e^{i\omega t}$. The solution to this is $u = U^* e^{i\omega t}$. Substituting into the equation of motion

$$-m\omega^2 U^* + \frac{3EI}{l^3} U^* = \frac{3EI}{l^3}$$

or

$$U^* = \frac{3EI}{l^3}\left(\frac{1}{(3EI/l^3) - m\omega^2}\right)$$

If we include damping, we have

$$U^* = \frac{3E^* I}{l^3}\left(\frac{1}{(3E^* I/l^3) - m\omega^2}\right)$$

Given $a(t) = a_0 \cos \omega t$, we find that $u(t) = a_0 \,\mathrm{Re}[U^* e^{i\omega t}]$. To determine $U^*$ if the input is $\ddot{a}(t)$ we use

$$m\ddot{u} + \frac{3EI}{l^3} u = \frac{3EI}{l^3} \int \left( \int \ddot{a}\, dt \right) dt$$

Assuming $\ddot{a}(t) = 1e^{i\omega t}$, we are seeking a solution in the form $u(t) = U^* e^{i\omega t}$. Substituting into the equation of motion results in

$$-m\omega^2 U^* + \frac{3EI}{l^3} U^* = \frac{3EI}{l^3}\left(-\frac{1}{\omega^2}\right)$$

or

$$U^* = -\frac{3EI}{l^3 \omega^2}\left(\frac{1}{(3EI/l^3) - m\omega^2}\right)$$

For the one-degree-of-freedom systems, we have used the so-called *deterministic loading.*

## PROBLEMS

1.1 Write the equations of motion for the one-degree-of-freedom systems shown in Figure 1.73a–i. Assume that the loading is in the form of a force $P(t)$, a given displacement $a(t)$, or a given rotation is $\varphi(t)$ as indicated in the figure.

1.2 Find the natural frequency $\omega_0$ of each of the systems in Figure 1.73a–i (write the general expression for $\omega_0$) and calculate values using $E = 30 \times 10^6$ psi, $I = 80$ in$^4$, $A = 10$ in$^2$, $L = 100$ in., and the weight of the mass 500 lbs.

1.3 Find the frequency response functions $U^*(i\omega)$ for the output $u(t)$ and inputs as indicated in Figure 1.73a–i, assuming:

a. An elastic material (without damping)

b. An inelastic material with the complex modulus $E^* = E' + iE''$

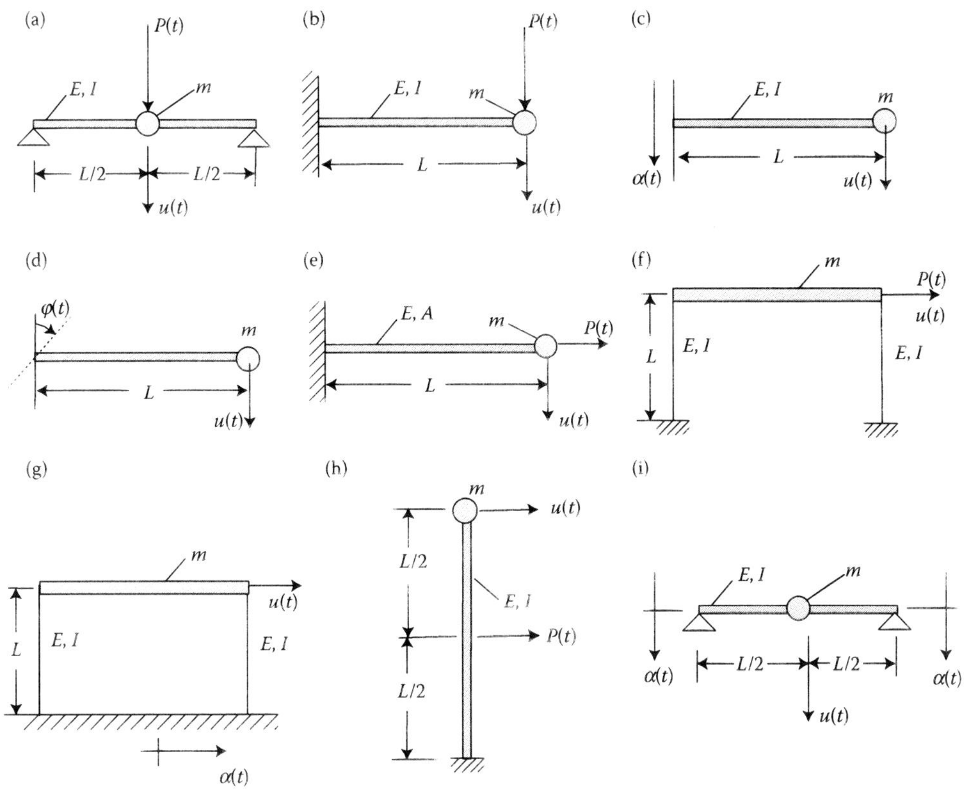


**FIGURE 1.73**
One-degree-of-freedom systems.

For part (A), plot $U^*$ versus the frequency $\omega$ for $0 \le \omega \le 500$, except for case e, where $0 \le \omega \le 2000$. For part (B), plot the absolute values of the frequency functions $|U^*|$ versus the frequency $\omega$ for $0 \le \omega \le 500$. Assume the same numerical data as given in Problem 1.2, with $E'' = 0.03E'$ and $E' = 30 \times 10^6$ psi.

1.4 Find the frequency response functions of the maximum bending moments $M^*(i\omega)$ for systems (a), (b), (c), (d), (f), (g), (h), and (i) of problem 1.1; plot values of $|M^*/P^*|$ or $|M^*/a^*|$ or $|M^*/Q^*|$, as appropriate, versus the frequency $\omega$. Find the frequency response of the axial force, $N^*(i\omega)$, in the system (e); plot $|N^*/P^*|$ versus the frequency $\omega$ for $0 \le \omega \le 500$ being careful to avoid the singularity. Assume $EI = 6.25 \times 10^6$ N m$^2$, $EA = 1.25 \times 10^9$ N, $L = 2.5$ m, and $m = 200$ kg.

1.5 Using Laplace transformations, find the transfer functions for the systems given in Problem 1.1. Determine the transient responses. Assume:

a. Elastic material.

b. Kelvin model, with $E = 30 \times 10^6$ psi, $\eta = 60 \times 10^6$ psi-sec

c. Maxwell model, with $E = 30 \times 10^6$ psi, $\eta = 18 \times 10^8$ psi-sec

d. Standard solid model, with $E_0 = 30 \times 10^6$ psi, $E_1 = 60 \times 10^6$ psi

$$\eta = 60 \times 10^6 \text{ psi-sec}$$

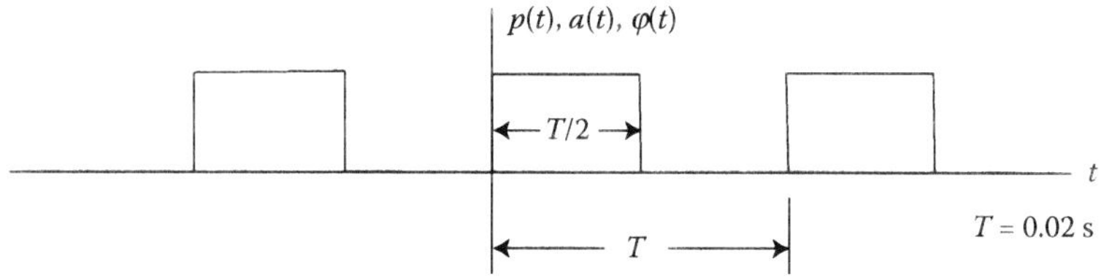


**FIGURE 1.74**
Loading sequence.

1.6 Find the particular solution (steady state) for the displacement if the loading is

$$
\begin{aligned}
P(t) &= P_0 \sin \omega t, \quad P(t) = P_0 \cos \omega t \\
a(t) &= a_0 \sin \omega t, \quad a(t) = a_0 \cos \omega t \\
\varphi(t) &= \varphi_0 \sin \omega, \quad \varphi(t) = \varphi_0 \cos \omega
\end{aligned}
$$

Consider the elastic and inelastic materials as in Problem 1.3, viscoelastic materials as in Problem 1.5.

1.7 Find the solution for the displacement $u(t)$ if the loading is as in Problem 1.6 for $t > 0$, and the initial conditions are

$$u = \dot{u} = 0 \quad \text{for } t = 0$$

1.8 Find the particular solution (steady state) for the displacement for the loading shown in Figure 1.74.

Consider systems as in Problem 1.1, with various types of material. Calculate the first three terms of the Fourier series involved.

## References

Atanackovic, T. M. 2002. A modified Zener model of a viscoelastic body. *Continuum Mech. Thermodyn.*, 14: 137–148.

Bert, C. W. 1973. Material damping: An introductory review of mathematical measures and experimental techniques. *J. Sound Vibr.*, 29: 129–153.

Corradi, M. 2006. A short account of the history of structural dynamics between the nineteenth and twentieth centuries. *Proc. Second Int. Congr. Constr. History*, 1: 837–854.

Greenberg, M. D. 1998. *Advanced Engineering Mathematics*, 2nd ed. Upper Saddle River, NJ: Prentice-Hall.

Lakes, R. 2009. *Viscoelastic Materials*. Cambridge: Cambridge University Press.

Lazan, B. J. 1968. *Damping of Materials and Members in Structural Mechanics*. Oxford: Pergamon Press.

Polyanin A. D. and Manzhirov A. V. 1998. *Handbook of Integral Equations*. Boca Raton: CRC Press.

# 2

# Random Vibrations

## 2.1 Introduction

In Chapter 1, the loadings applied to structures were deterministic, that is, they were known as a function of time. In reality, loading on structures is very difficult to predict or describe in an accurate manner. Typical loadings such as earthquakes, wind loads, landing loads for aircraft, loading by turbulent pressure, etc. are typically random in both time and in the actual magnitude of the load (Li and Chen 2009). For the purpose of analysis, we will consider loads which are random in time only.

A possible time history for a system undergoing a random vibration is shown in Figure 2.1, where the displacement $x$ is plotted as a function of time $t$.

Since the motion is random, the exact value of $x(t)$ at any time $t$ cannot be predicted precisely (Laming and Battin 1956; Solodounikov 1960). However, we can find the probability that $x(t)$ will lie within certain limits at any given time. Therefore, we will consider some fundamental ideas of probability theory.

## 2.2 Probability, Probability Distribution, and Probability Density

The term *event* is used to describe the result of an observation or an experiment. We wish to determine the probability (Hisashi et al. 2012) of some event $A$. We introduce the following definitions:

$n$ = the total number of observations

$n(A)$ = the number of times event $A$ occurs

The probability of event $A$ occurring is defined by $A \equiv P(A) = n(A)/n$, therefore $0 \le P(A) \le 1$. Thus,

If $P(A) = 0$   $A$ is an impossible event

If $P(A) = 1$   $A$ is a sure event

Several rules will be applied:

1. If $A$ and $B$ are two independent events, we define the probability of both $A$ and $B$ as $P(A, B)$. This is written as the product of both probabilities

$$P(A,B) = P(A)P(B) \tag{2.1}$$

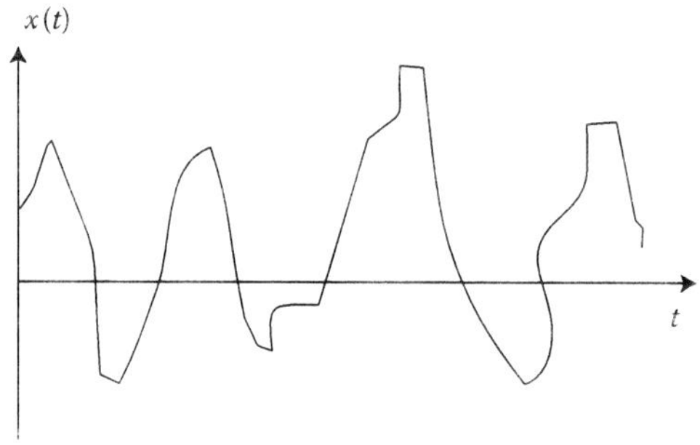


**FIGURE 2.1**
Typical time history for system undergoing random vibration.

2. If $A$ and $B$ are two mutually exclusive events. Either one or the other can occur. We define the probability of either $A$ or $B$ as $P(A \cup B)$. This is written as the sum of both probabilities

$$P(A \cup B) = P(A) + P(B) \tag{2.2}$$

If we assign a real number $X$ to every event $A$, then this $X$ is what we call the random variable. For example, assume event $A$ is a tensile test with the random variable $X$ being the tensile stress at failure. The problem is how to describe the random variable. We can specify the probability that the random variable $X$ is smaller than a given value $x$. This can be written as $P(X \leq x) = F(x)$, where $F(x)$ is the *probability distribution function* of $X$. A random variable described in terms of $F(x)$ is a probability distribution function if

$$F(-\infty) = 0, \quad F(+\infty) = 1;$$
$$F(x) \text{ is a nondecreasing function;}$$
$$F(x) \text{ is continuous to the right.}$$

As a result, we have

$$P(X > x) = 1 - F(x) \tag{2.3}$$

This is proven by

$$\begin{aligned} P(X \leq x \cup X > x) &= P(X \leq x) + P(X > x) = 1 \\ P(X > x) &= 1 - P(X \leq x) \\ &= 1 - F(x) \end{aligned}$$

In addition, we define the *probability density function* of $X$ as

$$f(x) = \frac{dF(x)}{dx} \tag{2.4}$$

As an illustration, consider the results form a tensile test of low-carbon steel. Assume the probability density function is known as shown in Figure 2.2.

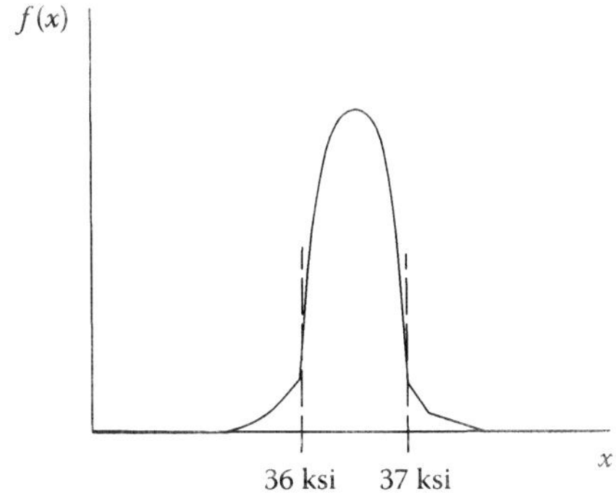


**FIGURE 2.2**
Probability density function from tensile test.

We can determine the probability distribution function $F(x)$ by integration, giving

$$F(x) = \int_{-\infty}^{x} f(x)dx \tag{2.5}$$

where the condition $F(-\infty) = 0$ has been used. Thus, the probability distribution function for the tensile test of low-carbon steel is shown in Figure 2.3.

The probability that $X$ is smaller than $x$ and the probability that $X$ is greater than $x$ come from

$$P(X \le x) + P(X > x) = 1$$
$$P(X > x) = 1 - P(X \le x) = 1 - F(x)$$
$$P(X > x) = \int_{x}^{\infty} f(x)\,dx$$

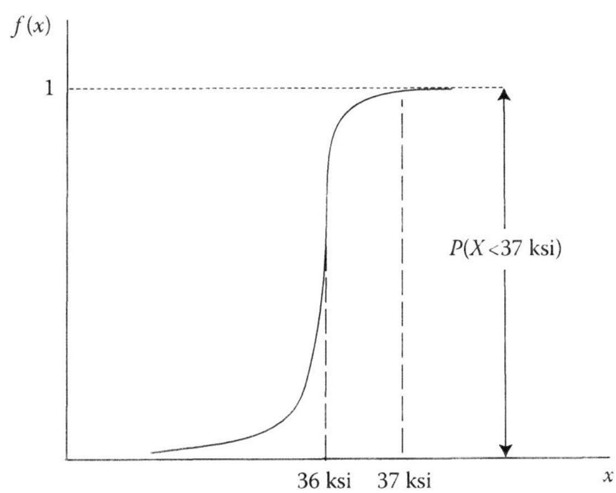


**FIGURE 2.3**
Probability distribution function for tensile test.

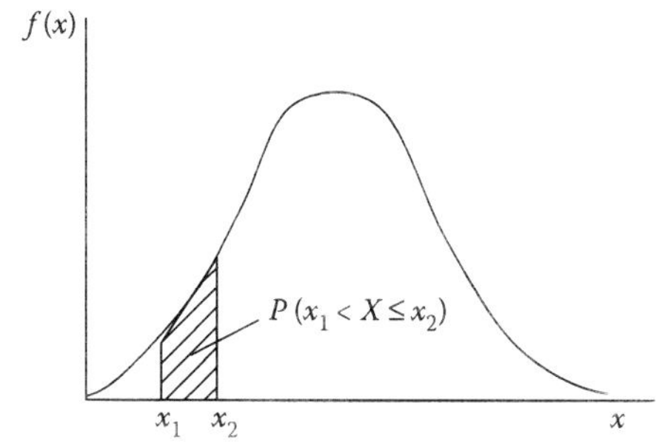


**FIGURE 2.4**
Shape of probability density function.

We can also determine the probability that $X$ is between two possible values of $x$. Assume $f(x)$ has the shape shown in Figure 2.4.

Thus, the proof that the probability that $X$ is between two possible values of $x$ is given as follows:

$$P(x_1 < X \leq x_2) + P(X \leq x_1) + P(X > x_2) = 1$$

$$P(x_1 < X \leq x_2) = 1 - P(X \leq x_1) - P(X > x_2)$$

or

$$\begin{aligned} P(x_1 < X \leq x_2) &= 1 - F(x_1) - [1 - F(x_2)] \\ &= F(x_2) - F(x_1) \\ &= \int_{x_1}^{x_2} f(x)\, dx \end{aligned}$$

## 2.3 Mean, Variance, Standard Deviation, and Distributions

The *mean* value, also called the *average value* or *expected value*, of $X$ is defined as

$$E[X] = \int_{-\infty}^{\infty} x\, f(x)\, dx = \tilde{x} \quad (2.6)$$

Equation 2.6 is interpreted as shown in Figure 2.5.

The mean value only determines the gravity center of the probability distribution function. Full statistical information is possible only if its probability moments are known. In addition, we can define the $n$th moment of $X$ as

$$E[X^n] \equiv \int_{-\infty}^{\infty} x^n f(x)\, dx \quad (2.7)$$

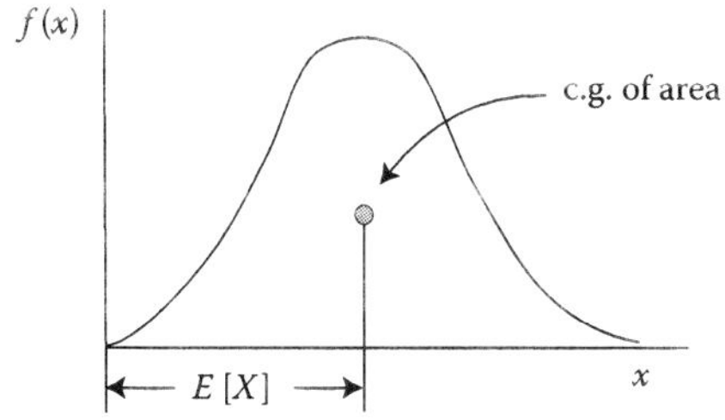


**FIGURE 2.5**
Mean value of X.

The most statistical informative probability moments are defined with respect to the mean value rather than the origin. Thus, we can define the so-called *central probability moment* as

$$E[(X-\tilde{x})^n] \equiv \int_{-\infty}^{\infty} (x-\tilde{x})^n f(x)\,dx \tag{2.8}$$

One of the more useful moments in statistical analysis is the mean square value of X, defined as

$$E[X^2] \equiv \int_{-\infty}^{\infty} x^2 f(x)\,dx = \widetilde{x^2} \tag{2.9}$$

The second moment about the mean is defined as the variance of X and is denoted by

$$E\left[(X-\tilde{x})^2\right] \equiv \mathrm{Var}[X] \equiv \int_{-\infty}^{\infty} (x-\tilde{x})^2 f(x)\,dx \tag{2.10}$$

The standard deviation is denoted by

$$c = \sqrt{\mathrm{Var}[X]} \tag{2.11}$$

or

$$c^2 = \widetilde{x^2} - (\tilde{x})^2 \tag{2.12}$$

and is a measure of the spread of the random variable about the mean as shown in Figure 2.6.

There are different types of distribution, which, in general, can be described by experimentally obtained curves not expressible by any mathematical expressions. However, in many problems the random variables have a probability distribution, which can be expressible in a mathematical form called the *Normal* or *Gaussian* distribution. The normal distribution is useful because of the so-called Central Limit Theorem. In its most general

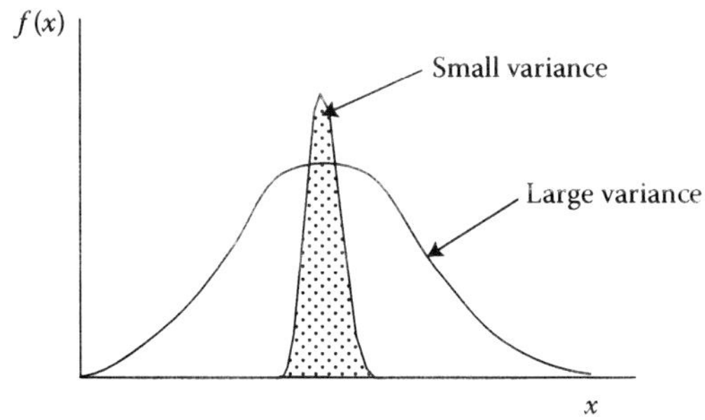


**FIGURE 2.6**
Distributions with small and large variances.

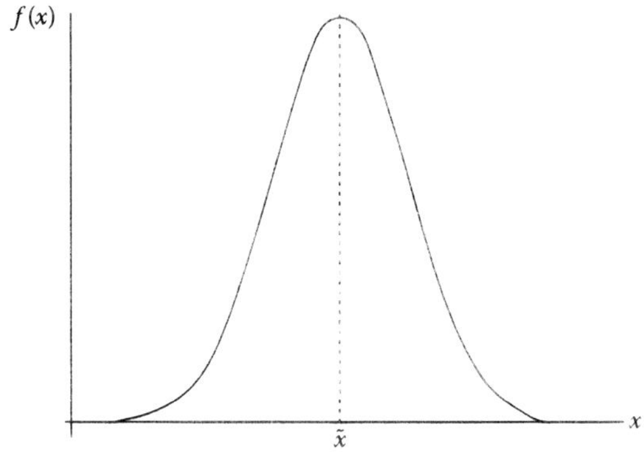


**FIGURE 2.7**
Gaussian probability density function.

form, it states that averages of random variables independently drawn from independent distributions are normally distributed. The Normal or Gaussian distribution is shown in Figure 2.7 and the probability density function $f(x)$ is given as

$$f(x) = \frac{e^{-(x-\tilde{x})^2/2c^2}}{c\sqrt{2\pi}} \tag{2.13}$$

where $\tilde{x}$ and $c$ denote the mean value and the standard deviation of $x$, respectively.

## 2.4 Combined Probabilities

If there are two random variables, $X$ and $Y$, there are different parameters that need to be considered. We could consider $X$ to be the strength of a tensile specimen and $Y$ to be its maximum elongation. In dealing with two random variables, we define the joint distribution function $F(x, y)$, which is related to the probability by

$$F(x, y) = P[(X \leq x, Y \leq y)] \tag{2.14}$$

A joint probability distribution function is nondecreasing and right continuous with respect to each of the multiple dimensions. For two random variables, it satisfies the following relations:

$$
\begin{aligned}
&F_{XY}(-\infty, y) = F_{XY}(x, -\infty) = 0 \\
&F_{XY}(\infty, \infty) = 1 \\
&F_{XY}(x, \infty) = F_X(x) \\
&F_{XY}(\infty, y) = F_Y(y)
\end{aligned}
\tag{2.15}
$$

In addition, the probability density function is defined by

$$
f(x,y) \equiv \frac{\partial^2 F(x,y)}{\partial x \, \partial y} \tag{2.16}
$$

Taking the inverse of Equation 2.16 gives

$$
F_{XY}(x,y) = \int_{-\infty}^{x} \int_{-\infty}^{y} f(x,y) dx dy \tag{2.17}
$$

In the same manner as before it can be shown that

$$
P(a_1 < X \le a_2, b_1 < Y \le b_2) = \int_{a_1}^{a_2} \int_{b_1}^{b_2} f(x,y) dx dy \tag{2.18}
$$

If $X$ and $Y$ are independent variables

$$
P(X \le x, Y \le y) = P(X \le x) P(Y \le y) \tag{2.19}
$$

In addition

$$
F(x,y) = F_X(x) F_Y(y) \tag{2.20}
$$

and

$$
f(x,y) = f_X(x) f_Y(y) \tag{2.21}
$$

Similarly, we can define the general case of the mean values as

$$
\tilde{x} = E[X] \equiv \int_{-\infty}^{\infty} \int_{-\infty}^{\infty} x \, f(x,y) dx \, dy \tag{2.22}
$$

$$
\tilde{y} = E[Y] \equiv \int_{-\infty}^{\infty} \int_{-\infty}^{\infty} y \, f(x,y) dx \, dy \tag{2.23}
$$

and variances by

$$E[(X-\tilde{x})^2] \equiv \int_{-\infty}^{\infty}\int_{-\infty}^{\infty} (x-\tilde{x})^2 f(x,y)dx\,dy = c_x^2 \tag{2.24}$$

$$E[(Y-\tilde{y})^2] \equiv \int_{-\infty}^{\infty}\int_{-\infty}^{\infty} (y-\tilde{y})^2 f(x,y)dx\,dy = c_y^2 \tag{2.25}$$

where $c_x^2$ and $c_y^2$ denote the standard deviation of $X$ and $Y$, respectively.

A new quantity that expresses a relationship between $X$ and $Y$ is the covariance of $X$ and $Y$ defined by

$$E[(X-\tilde{x})(Y-\tilde{y})] \equiv \int_{-\infty}^{\infty}\int_{-\infty}^{\infty} (x-\tilde{x})(y-\tilde{y}) f(x,y)dx\,dy \tag{2.26}$$

The covariance measures that, in a probability sense, how $X$ and $Y$ vary together. A positive covariance indicates that the average of the products $[(X-\tilde{x})][(Y-\tilde{y})]$ is positive, which indicates that $X$ and $Y$ tend to be either both above their means or both below their means simultaneously. Knowing that $X$ and $Y$ are independent, from Equations 2.22 and 2.23, the general cases simplify to

$$E[X] \equiv \int_{-\infty}^{\infty}\int_{-\infty}^{\infty} x\, f_X(x) f_Y(y)dx\,dy = \left(\int_{-\infty}^{\infty} x\, f_X(x)dx\right)\left(\int_{-\infty}^{\infty} f_Y(y)dy\right) = \int_{-\infty}^{\infty} x\, f_X(x)dx \tag{2.27}$$

$$E[Y] \equiv \int_{-\infty}^{\infty}\int_{-\infty}^{\infty} y\, f_X(x) f_Y(y)dx\,dy = \left(\int_{-\infty}^{\infty} f_X(x)dx\right)\left(\int_{-\infty}^{\infty} y\, f_Y(y)dx\right) = \int_{-\infty}^{\infty} y\, f_Y(y)dy \tag{2.28}$$

where

$$\int_{-\infty}^{\infty} f_Y(y)dy = 1 \quad \text{and} \quad \int_{-\infty}^{\infty} f_X(x)dx = 1$$

In addition, for the variance, we have

$$\begin{aligned} E[(X-\tilde{x})^2] &\equiv \int_{-\infty}^{\infty} (x-\tilde{x})^2 f_X(x)dx \\ E[(Y-\tilde{y})^2] &\equiv \int_{-\infty}^{\infty} (y-\tilde{y})^2 f_Y(y)dy \end{aligned} \tag{2.29}$$

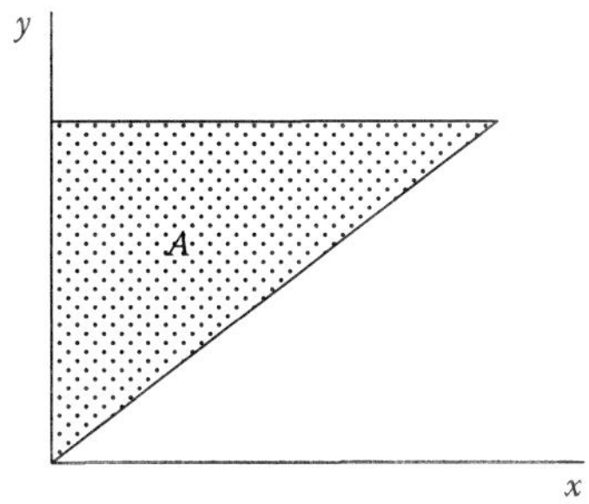


**FIGURE 2.8**
Probability that $X$ and $Y$ are in area $A$.

And similarly, for the covariance, we have

$$E[(X-\tilde{x})(Y-\tilde{y})] \equiv \left(\int_{-\infty}^{\infty}(x-\tilde{x})f_X(x)dx\right)\left(\int_{-\infty}^{\infty}(y-\tilde{y})^2 f_Y(y)dy\right)=0 \tag{2.30}$$

The normal (or Gaussian) distribution of two independent variables is

$$f(x,y)=\frac{e^{-[(x-\tilde{x})^2/2c_x^2+(y-\tilde{y})^2/2c_y^2]}}{c_x c_y\sqrt{2\pi}} \tag{2.31}$$

Consider the tensile test that has been used as a previous illustration. Assume that we define the strength ($X$) and the load ($Y$) as the two independent variables. Defining the probability of failure as

$$P_F = P(Y > X) \quad Y > 0, X > 0$$

Using $f(x, y) = f_X(x)f_Y(y)$, and knowing $f_X(x)$ and $f_Y(y)$, the probability that $X$ and $Y$ are within the area $A$ is illustrated in Figure 2.8 and expressed as

$$P_F = \iint_A f(x,y)dx\,dy \tag{2.32}$$

Since $f(x)$ and $f(y)$ are known, $P_F$ can be determined.

## 2.5 Random Functions

The term used to assess random functions and random sequences is the *stochastic process* (Crandall and Mark 1963; Lin 1966; Bolotin 1984). Random implies a different function each time there is an occurrence. Despite the irregular character of the function, many random phenomena exhibit some measure of statistical regularity. In describing a random function, the mean and mean square values are of great importance. Suppose that the results

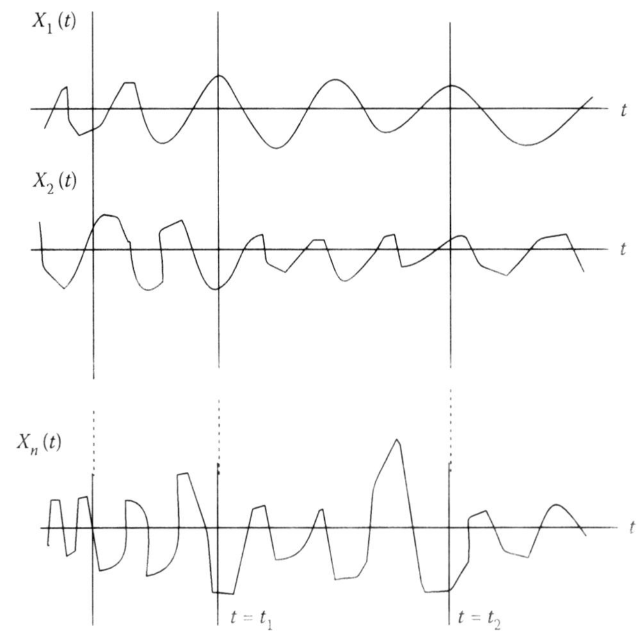


**FIGURE 2.9**
Ensemble of a random process.

of an experiment are recorded continuously in time and given in the form of diagrams (see Figure 2.9).

In order to describe the function, we should know the distribution of $X(t)$ at each time $t$. The distribution function may, in general, depend on time. This, however, is not sufficient; we may have two entirely different random functions with identical distributions at each time $t$, Figure 2.10.

The time-dependence of $X$ will be characterized by the joint distributions

$$
\begin{aligned}
&f[X(t_1), X(t_2)] \\
&f[X(t_1), X(t_2), X(t_3)] \\
&\ldots \\
&f[X(t_1), X(t_2), X(t_3), \ldots, X(t_n)]
\end{aligned}
$$

In a nonstrict sense, a stationary process is one whose statistical properties do not change over time. Thus, a random function $X(t)$ is called *stationary*, in the strict sense, if the joint

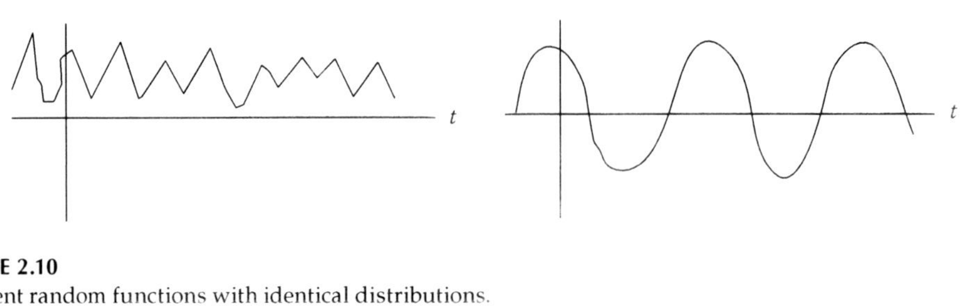


**FIGURE 2.10**
Different random functions with identical distributions.

distribution $f[X(t_1), \ldots, X(t_n)]$ is identical with $f[X(t_1 + \tau), \ldots, X(t_n + \tau)]$ for any $\tau$ and any $n$. This is an extremely strong definition since; in practice it is not possible to determine statistics for all orders of $n$.

### 2.5.1 Correlation Function and Spectral Densities

It was noted that we may have two entirely different random functions with identical distributions at each time $t$. In order to overcome this potential problem, we introduce a correlation function to describe $X(t)$. The *correlation function* of the random process $X(t)$ (sometimes: the *autocorrelation function*) is defined as

$$R_X(t_1, t_2) \equiv E[X(t_1)X(t_2)] \tag{2.33}$$

The prefix *auto* indicates that the two random variables considered, $X(t_1)$ and $X(t_2)$, belong to the same random process. Correlation functions are used to describe the average or mean relation between random variables. Suppose we observe multiple events (see Figure 2.11) and wish to focus on two different times. For each of the $n$ events, we compute $X(t_1)$ and $X(t_2)$. The mean for these events is determined to be

$$\begin{aligned} E[X(t_1)X(t_2)] \cong \frac{1}{n}\Big[&X^{(1)}(t_1)X^{(1)}(t_2) \\ &+X^{(2)}(t_1)X^{(2)}(t_2)+\cdots \\ &+X^{(n)}(t_1)X^{(n)}(t_2)\Big] \end{aligned} \tag{2.34}$$

We note the following properties of the correlation function.

$$\text{If } t_1 = t_2 = t,\ R_X(t,t) = E[X(t)X(t)] = E[X^2(t)] \tag{2.35}$$

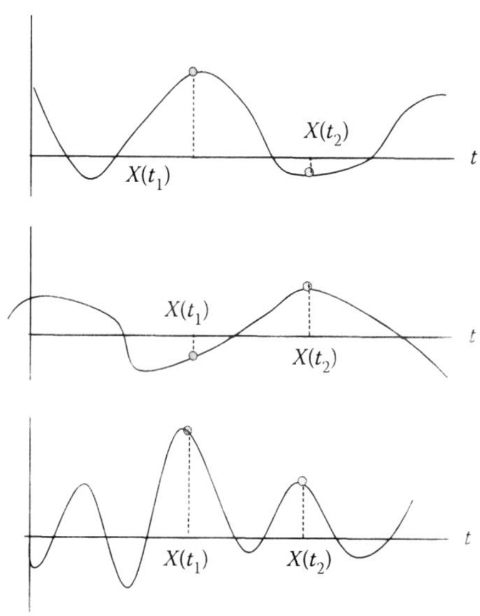


**FIGURE 2.11**
Multiple events.

In other words, $R_X$ is the mean square value of $X(t)$.

$$\text{If } t_2 - t_1 \text{ is large, } R_X(t_1, t_2) \approx E[X(t_1)] \cdot E[X(t_2)] \tag{2.36}$$

For a stationary process

$$R_X(t_1, t_2) = R_X(t_2 - t_1) = R_X(\tau) \tag{2.37}$$

that is, the value of $R_X$ depends on the time interval $\tau = t_2 - t_1$ only and not on the values of $t_1$ and $t_2$. For stationary functions, the mean square value of $X$ is constant in time and $R_X(t, t) = R_X(0)$. Some processes are not stationary in the strict sense; however, their correlation functions depend on $\tau = t_2 - t_1$ only, see Figure 2.12. They are called stationary in the *wide sense.*

A random process is said to be a nonstationary process if its statistics vary with time. Thus, for a nonstationary process, the correlation function $R_X$ depends on $t_1$ and $t_2$, not $\tau$.

For stationary random processes

$$\begin{aligned} R_X(0) &= E[X^2(t)] \equiv \widetilde{x^2} \\ R_X(\infty) &= (E[X(t)])^2 \equiv (\tilde{x})^2 \\ R_X(\tau) &= [X(t_1)X(t_2)] = [X(t_2)X(t_1)] = R_X(-\tau) \end{aligned} \tag{2.38}$$

where $\widetilde{x^2}$ and $(\tilde{x})^2$ do not depend on $t$. Some typical diagrams of $R_X(\tau)$ are shown in Figure 2.13.

When dealing with stationary random processes, we shall assume that they posses the so-called *ergodic property*. It means that the mean value can be obtained by the *time-averaging* procedure.

For a stationary ergodic process, we define the time average of $X$ (see Figure 2.14) as

$$\langle X(t) \rangle = \lim_{T \to \infty} \frac{1}{2T} \int_{-T}^{T} X(t)dt \tag{2.39}$$

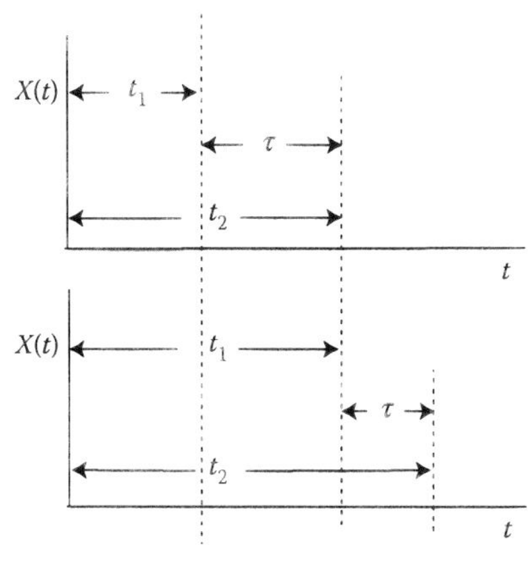


**FIGURE 2.12**
The stationary process correlation function depends only on $\tau$.

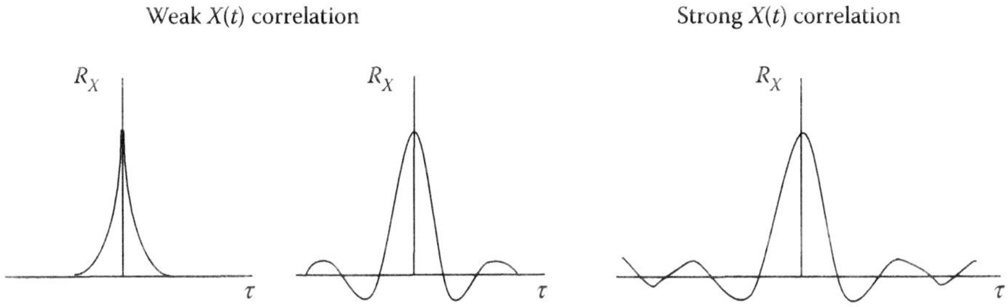


**FIGURE 2.13**
Weak and strong correlation functions.

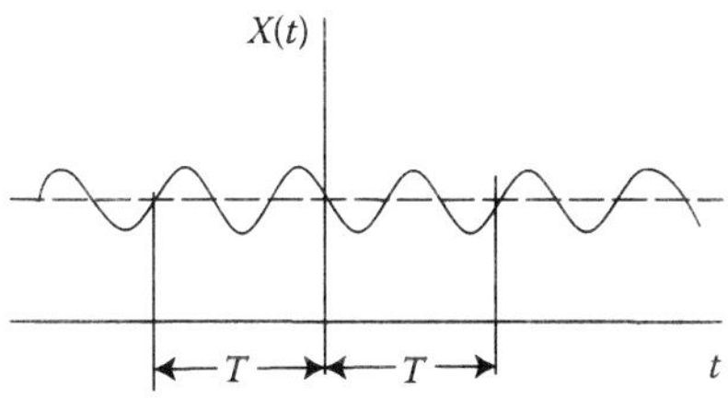


**FIGURE 2.14**
Time average of $X(t)$.

If $T$ is sufficiently large, this can be expressed as

$$\langle X(t)\rangle \cong \frac{1}{2T}\int_{-T}^{T} X(t)dt \tag{2.40}$$

If we assume a function $g(X(t))$, the time average of $g$ is

$$\langle g(X(t))\rangle = \lim_{T\to\infty}\frac{1}{2T}\int_{-T}^{T} g(X(t))dt \tag{2.41}$$

If, for example, we have $g(X(t)) = x^2(t)$, then

$$\langle x^2(t)\rangle = \lim_{T\to\infty}\frac{1}{2T}\int_{-T}^{T} x^2(t)dt$$

Time averaging typically involves one observation for an extended period of time. If multiple observations are used, an approach called ensemble averaging is used (Figure 2.15).

The mean value (or averaging) is called ensemble averaging.

$$\begin{aligned} E[X(t)] &= \int_{-\infty}^{\infty} x(t)f_X(x(t))dx \\ &\approx \frac{1}{n}[x^{(1)}(t) + x^{(2)}(t) + \cdots + x^{(n)}(t)] \end{aligned} \tag{2.42}$$

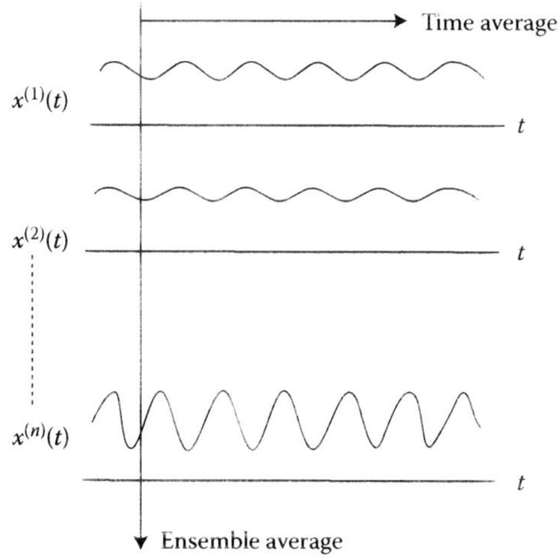


**FIGURE 2.15**
Ensemble average vs time average.

For a stationary process, we assume that the result of time averaging is the same as the result of ensemble averaging, that is, we have a stationary ergodic process. For a stationary ergodic process, we have

$$
\begin{aligned}
E[X(t)] &= \lim_{T\to\infty} \frac{1}{2T} \int_{-T}^{T} x(t)dt \\
E[X^2(t)] &= \lim_{T\to\infty} \frac{1}{2T} \int_{-T}^{T} x(t)dt \\
R_X(\tau) = E[X(t)X(t+\tau)] &= \lim_{T\to\infty} \frac{1}{2T} \int_{-T}^{T} x(t)x(t+\tau)dt
\end{aligned}
\tag{2.43}
$$

If we have a stationary ergodic process, it is always possible to define $X(t)$ in such a way that the mean value $X(t)$ is equal to zero ($E[X(t)] = 0$). This will be assumed in the following.

The concept of the spectral density of a stationary ergodic process can be considered in the following way. Consider a random function $X(t)$. Let $x_T(t)$ (see Figure 2.16) be

$$
\begin{aligned}
x_T(t) &= X(t) \quad \text{for} -T \le t \le T \\
x_T(t) &= 0 \qquad \text{for} \ |t| > T
\end{aligned}
\tag{2.44}
$$

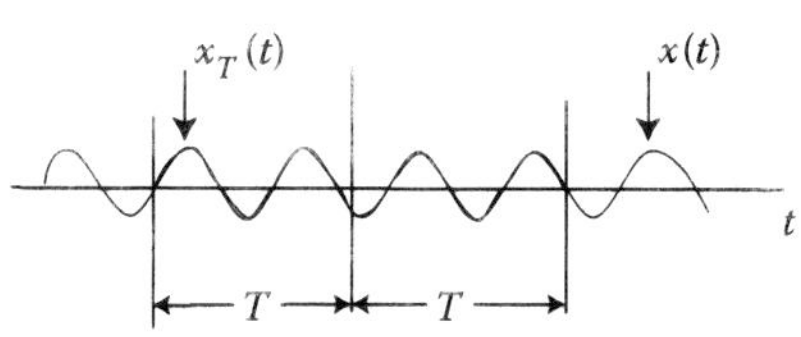


**FIGURE 2.16**
Definition of random function $x_T$.

We now define the Fourier transformation of $x_T$ (see Section 1.10); it is

$$x_T^*(i\omega) = \int_{-\infty}^{\infty} x_T(t)e^{-i\omega t}dt = \int_{-T}^{T} x(t)e^{-i\omega t}dt \tag{2.45}$$

The spectral density $S_x(\omega)$ of the random function $X(t)$ is defined as

$$S_x(\omega) = \lim_{T\to\infty} \frac{1}{2T}\left|x_T^*(i\omega)\right|^2 \tag{2.46}$$

We note that if $T$ is sufficiently large

$$S_x(\omega) \approx \frac{1}{2T}\left|x_T^*(i\omega)\right|^2 \tag{2.47}$$

The spectral density can be used to establish the important relation between $E[X^2]$ and $S_x(\omega)$, that is

$$E[X^2] = R_X(0) = \frac{1}{2\pi}\int_{-\infty}^{\infty} S_x(\omega)d\omega \tag{2.48}$$

The proof is given as follows:
**Proof:** Start with

$$x_T^*(i\omega) = \int_{-\infty}^{\infty} \left|x_T^*(i\omega)\right|^2 d\omega = \int_{-\infty}^{\infty} x_T^*(i\omega)x_T^*(-i\omega)d\omega$$

Note that

$$x_T^*(i\omega) = \int_{-\infty}^{\infty} x_T(t)e^{-i\omega t}dt$$

Therefore, we have

$$\int_{-\infty}^{\infty} x_T^*(i\omega)x_T^*(-i\omega)d\omega = \int_{-\infty}^{\infty}\int_{-\infty}^{\infty} x_T^*(-i\omega)x_T(t)e^{-i\omega t}dt\, d\omega$$

Using the Fourier inversion formula

$$\int_{-\infty}^{\infty}\int_{-\infty}^{\infty} x_T^*(-i\omega)x_T(t)\, e^{-i\omega t}dt\, d\omega = 2\pi\int_{-\infty}^{\infty} x_T^2(t)dt$$

Note that the mean square value of $X$ is given by

$$E[X^2] = \lim_{T\to\infty} \frac{1}{2T} \int_{-T}^{T} x_T^2(t)dt$$

$$= \lim_{T\to\infty} \frac{1}{2T} \frac{1}{2\pi} \int_{-\infty}^{\infty} \left|x_T^*(i\omega)\right|^2 d\omega$$

$$= \frac{1}{2\pi} \int_{-\infty}^{\infty} \lim_{T\to\infty} \frac{1}{2T} \left|x_T^*(i\omega)\right|^2 d\omega$$

$$= \frac{1}{2\pi} \int_{-\infty}^{\infty} S_X(\omega)d\omega$$

Thus, if we know the spectral density, we can calculate the mean square value. The spectral density and the correlation function reflect an apparent periodicity of $X(t)$ in the sense explained in Figure 2.17.

The apparent periodicity of $X(t)$ is reflected in $S_x(\omega)$. Several examples are shown in Figure 2.18. Of particular note are the cases where $S_x(\omega) \approx$ constant and $\Delta\omega$ is very large. This is termed *white noise*. For ideal white noise $S_x(\omega) \to 0$ and $\omega \to \infty$.

Finally note that $S_X(\omega)$ is an even function $S_X(\omega) = S_X(-\omega)$ as shown in Figure 2.19.

We see this from the definition of the spectral density $S_X(\omega)$

$$S_X(\omega) = \lim_{T\to\infty} \frac{1}{2T} x_T^*(i\omega)x_T^*(i\omega) = S_X(-\omega)$$

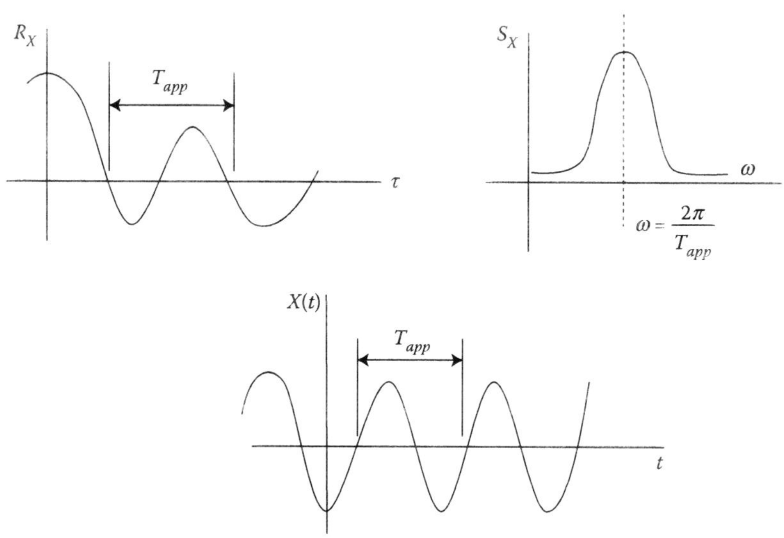


**FIGURE 2.17**
Spectral density and correlation function reflect apparent periodicity of $X(t)$.

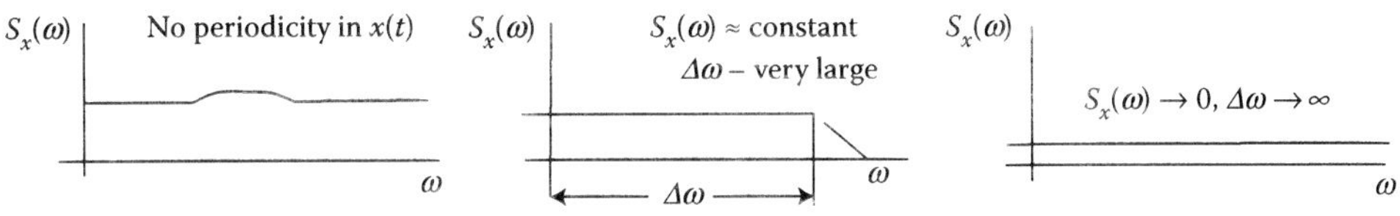


**FIGURE 2.18**
Examples which show the apparent periodicity of $X(t)$ reflected in $S_X(\omega)$.

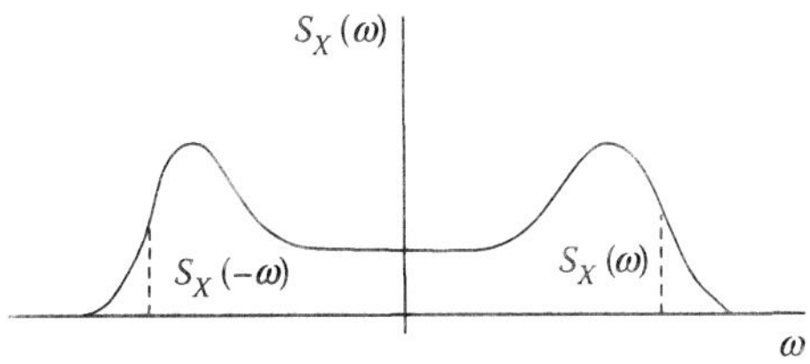


**FIGURE 2.19**
Even function $S_X(\omega)$.

The correlation function $R$ and the spectral density $S$ are connected by

$$\begin{aligned} R(\tau) &= \frac{1}{2\pi}\int_{-\infty}^{\infty} S(\omega)\, e^{i\omega\tau} d\omega = \frac{1}{2\pi}\int_{-\infty}^{\infty} S(\omega)\cos\omega\tau\, d\omega = \frac{1}{\pi}\int_{0}^{\infty} S(\omega)\cos\omega\tau\, d\omega \\ S(\omega) &= \int_{-\infty}^{\infty} R(\tau)\, e^{-i\omega\tau} d\tau = \int_{-\infty}^{\infty} R(\tau)\cos\omega\tau\, d\tau = 2\int_{0}^{\infty} R(\tau)\cos\omega\tau\, d\tau \end{aligned} \tag{2.49}$$

Equations 2.49 are the so-called Wiener–Khinchine relations (Wijker 2009), and except for the factor of 2 they represent a Fourier cosine transform pair. A proof of Equation 2.49 is as follows:

**Proof:** The correlation function $R(\tau)$ is defined by

$$R(\tau) = \frac{1}{2T}\int_{-T}^{T} X_T(t)X_T(t+\tau)dt$$

where

$$X_T(t) = X(t); \quad -T \le t \le T \text{ and } X_T(t) = 0; \quad |t| > T$$

The actual value is given by $R(\tau) = \lim\limits_{T\to\infty} R_T(\tau)$. Thus

$$\begin{aligned} \int_{-\infty}^{\infty} R_T(\tau)e^{-i\omega\tau}d\tau &= \frac{1}{2T}\int_{-\infty}^{\infty} e^{-i\omega\tau}d\tau \int_{-\infty}^{\infty} X_T(t)X_T(t+\tau)dt \\ &= \frac{1}{2T}\int_{-\infty}^{\infty} d\tau \int_{-\infty}^{\infty} X_T(t)X_T(t+\tau)e^{-i\omega(t+\tau)}e^{i\omega t}dt \end{aligned}$$

Noting that $X_T^*(i\omega) = \int_{-\infty}^{\infty} X_T(t)e^{i\omega t}dt$ and $X_T^*(-i\omega) = \int_{-\infty}^{\infty} X_T(t+\tau)e^{-i\omega(t+\tau)}d\tau$ results in

$$\int_{-\infty}^{\infty} R_T(\tau)e^{-i\omega\tau}d\tau = \frac{1}{2T} X_T^*(i\omega)X_T^*(-i\omega)$$

Investigating the limit of both sides as $T \to \infty$ yields

$$\lim_{T\to\infty} \int_{-\infty}^{\infty} R_T(\tau)e^{-i\omega\tau}d\tau = \lim_{T\to\infty} \frac{1}{2T}\left|X_T^*(i\omega)\right|^2$$

Hence

$$R(\tau) = \int_{-\infty}^{\infty} \lim_{T\to\infty} R_T(\tau)e^{-i\omega\tau}d\tau = S(\omega)$$

or

$$S(\omega) = \int_{-\infty}^{\infty} R(\tau)e^{-i\omega\tau}d\tau = 2\int_{0}^{\infty} R(\tau)e^{-i\omega\tau}d\tau$$

A stationary random function $X(t)$ is called an ideal white noise if

$$\begin{aligned} R_X(0) &= E[X^2] \\ R_X(\tau) &= 0 \quad \text{if } \tau \neq 0 \end{aligned} \tag{2.50}$$

and

$$\begin{aligned} S_X(\omega) &\to 0 \\ E[X^2] &= \frac{1}{2\pi}\int_{-\infty}^{\infty} S_X(\omega)d\omega \end{aligned} \tag{2.51}$$

In mechanical systems and in structures, an ideal white noise does not exist. Sometimes it happens that the correlation function $R_X(\tau)$ is very narrow and the spectral density is very flat (Figure 2.20). They may be replaced by

$$R_X(\tau) = A\delta(\tau) \tag{2.52}$$

and

$$S_X(\omega) = \text{constant} = A \tag{2.53}$$

where $A$ is the area enclosed in the $R_X(\tau)$ diagram and $\delta(\tau)$ is the Dirac delta function. These expressions describe a practical white noise.

**FIGURE 2.20**
Practical white noise.

### 2.5.2 Combinations of Random Processes

Suppose now that we have two random functions $X(t)$ and $Y(t)$. The cross-correlation function $R_{XY}$ is defined as

$$R_{XY}(t_1,t_2) = E\left[X(t_1)Y(t_2)\right] \tag{2.54}$$

It has the property

$$R_{XY}\left(t_1,t_2\right) = R_{YX}\left(t_2,t_1\right) \tag{2.55}$$

If $X(t)$ and $Y(t)$ are independent random functions, then

$$R_{XY}(t_1,t_2) = E\left[X(t_1)\right]\cdot E\left[Y(t_2)\right] \tag{2.56}$$

If one of the mean values is equal to zero, then

$$R_{XY}(t_1,t_2) = 0$$

For $n$ random variables $X_1, X_2, \ldots X_i, \ldots X_n$, the correlation matrix $[R]$ can be introduced; its elements are

$$R_{ij}(t_1,t_2) = E\left[X_i(t_1)X_j(t_2)\right] \tag{2.57}$$

The above definition is valid for nonstationary random functions. In case of stationary random functions, we have

$$\begin{aligned} R_{ij}(\tau) &= E[X_i(t)X_j(t+\tau)] \\ &= \lim_{T\to\infty}\frac{1}{2\pi}\int_{-T}^{T} X_i(t)X_j(t+\tau)d\tau \end{aligned} \tag{2.58}$$

The spectral density matrix (for stationary functions) $[S]$ has the elements

$$S_{ij}(\omega) = \lim_{T\to\infty}\frac{1}{2T}x_{iT}^{*}(\omega)x_{jT}^{*}(-\omega) \tag{2.59}$$

or

$$[S(\omega)] = \lim_{T\to\infty} \frac{1}{2T} \left\{x_T^*(\omega)\right\} \left\{x_T^*(-\omega)\right\}^T \tag{2.60}$$

or

$$\begin{aligned} S_{ij}(\omega) &= \Im\{R_{ij}(\tau)\} \\ &= \int_{-\infty}^{\infty} R_{ij}(\tau) e^{-i\omega\tau} d\tau \end{aligned} \tag{2.61}$$

### 2.5.3 Level Crossings

In this section, information will be discussed which may be obtained from the correlation function and the spectral density of a random process. We shall determine the average number of crossings per unit time of a given value $x = x_L$ (Roberts and Spanos 1990; Lutes and Shahram 2004). This problem is illustrated in Figure 2.21.

The problem is to find the expected frequency of crossing of a given value of $x$, that is, we want to establish the number of crossings per unit time. To start, consider our random process $X(t)$, which is stationary, ergodic, and has $E[X] = 0$. Let

$$\dot{X}(t) = \frac{dX}{dt} \tag{2.62}$$

Thus, we have

1. $S_{\dot{X}} = \omega^2 S_X(\omega)$

   **Proof of (1)**:

$$\begin{aligned} S_{\dot{X}}(\omega) &= \lim_{T\to\infty} \frac{1}{2T} \left|X_T^*(i\omega)\right|^2 \\ &= \lim_{T\to\infty} \frac{1}{2T} \left|\dot{X}_T^*(i\omega)\right|^2 \\ &= \lim_{T\to\infty} \frac{1}{2T} \left|i\omega X_T^*(i\omega)\right|^2 \\ &= \omega^2 \lim_{T\to\infty} \frac{1}{2T} \left|X_T^*(i\omega)\right|^2 \\ &= \omega^2 S_X(\omega) \end{aligned}$$

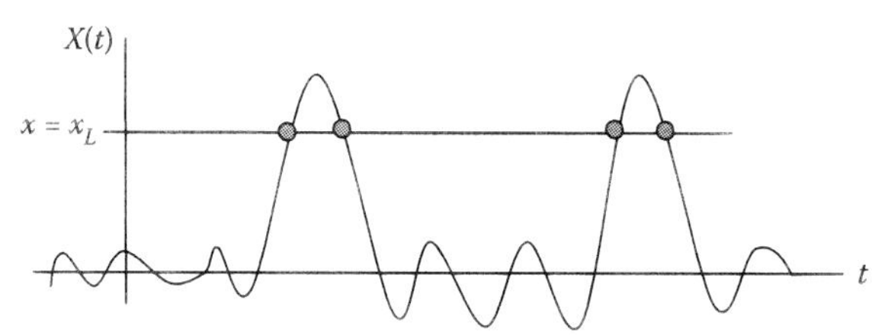


**FIGURE 2.21**
Crossing level with $x = x_L$.

2. $R_{\dot{X}}(\tau) = -\dfrac{d^2R_X(\tau)}{d\tau^2} = -R_X''(\tau)$

   **Proof of (2)**:

$$\begin{aligned}
\Im\{R_X\} &= S_X(\omega) \\
\Im\{R_{\dot{X}}\} &= \omega^2 S_X(\omega)(= S_{\dot{X}}(\omega)) \\
\Im\{R_{\dot{X}}\} &= -(i\omega)^2 S_X(\omega) \\
\Im\{-R_{\dot{X}}\} &= (i\omega)^2 S_X(\omega) \\
\therefore -R_{\dot{X}} &= \frac{d^2R_X}{dt^2}
\end{aligned}$$

3. The covariance of $X$ and $\dot{X}$ is equal to zero

   **Proof of (3)**:

$$\begin{aligned}
\mathrm{cov}(X\dot{X}) &= E[X\dot{X}] \\
&= \lim_{T\to\infty} \frac{1}{2T} \int_{-T}^{T} x(t)\,\dot{x}(t)dt \\
&= \lim_{T\to\infty} \frac{1}{4T}[x^2(t)]_{-T}^{T} \\
&= \lim_{T\to\infty} \frac{1}{4T}[x^2(T) - x^2(-T)] \\
&= 0 \quad \text{Provided } x(T) \text{ and } x(-T) \text{ are finite (bounded)}
\end{aligned}$$

Now find the standard deviation (or mean square value) of $X$ and $\dot{X}$

$$\begin{aligned}
c_1^2 &= [X^2] = \frac{1}{\pi}\int_0^\infty S_X(\omega)d\omega = R_X(0) \\
c_2^2 &= [\dot{X}^2] = \frac{1}{\pi}\int_0^\infty \omega^2 S_X(\omega)d\omega = -R_X''(0)
\end{aligned} \tag{2.63}$$

Assume that $X(t)$ and $\dot{X}(t)$ have Gaussian distributions. Then the joint probability density function of $X(t)$ and $\dot{X}(t)$ is

$$f(\alpha,\beta) = \frac{1}{2\pi c_1 c_2}\exp\left[\frac{-\alpha^2}{2c_1^2} + \frac{-\beta^2}{2c_2^2}\right] \tag{2.64}$$

The probability that $\alpha < X < \alpha + d\alpha$ and $\beta < \dot{X} < \beta + d\beta$ is $f(\alpha, \beta)d\alpha d\beta$ or the probability that $\alpha < X < \alpha + \Delta\alpha$ and $\beta < \dot{X} < \beta + \Delta\beta$ is approximately equal to $f(\alpha, \beta)\Delta\alpha\, \Delta\beta$.

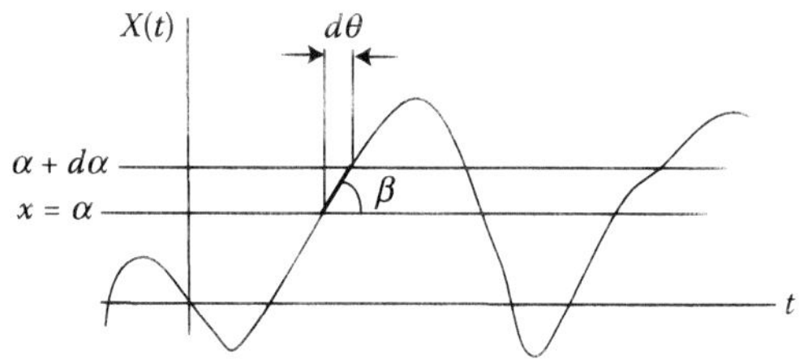


**FIGURE 2.22**
Slope in the interval $\beta$, $\beta + d\beta$.

For the distribution shown, the probability of $X$ in the interval $\alpha$, $\alpha + d\alpha$ with the slope in the interval of $\beta$, $\beta + d\beta$ is $f(\alpha, \beta)d\alpha d\beta$ (see Figure 2.22). The proportion of time spent by $X(t)$ in $(\alpha, \alpha + d\alpha)$ at a slope of $(\beta, \beta + d\beta)$ is

$$\frac{\text{time at } \alpha, \beta}{\text{total time of observation}} = f(\alpha, \beta) d\alpha\, d\beta$$

The duration of one pass through $(\alpha, \alpha + d\alpha)$ is $d\theta = d\alpha / |\beta|$. The number of passes through $(\alpha, \alpha + d\alpha)$ with a slope of $\beta$ is

$$\frac{\text{time at } \alpha, \beta}{\text{total time}} = \frac{(\text{number of passages through } \alpha, \alpha + d\alpha)(\text{duration of one pass})}{\text{total time}}$$

The total number of passages is equal to the time at $\alpha$, $\beta$/duration of one passage or the number of passages per unit time through $\alpha$, $\alpha + d\alpha$ at $\beta$ which is

$$\frac{(\text{time at } \alpha, \beta) / (\text{duration of one pass})}{\text{total time}} = \frac{f(\alpha, \beta)\, d\alpha\, d\beta}{d\alpha / |\beta|} = |\beta|\, f(\alpha, \beta) d\beta$$

In order to determine the number of passages through $(\alpha, \alpha + d\alpha)$ at all values of $\beta$, we add the number of passages corresponding to various values of $\beta$. This gives

$$N_\alpha = \int_{-\infty}^{\infty} |\beta| f(\alpha, \beta) d\beta$$

This answer is not very convenient in this form, hence for $f(\alpha, \beta)$ given by Equation 2.64

$$N_\alpha = \frac{1}{\pi} \frac{c_2}{c_1} e^{-\alpha^2 / 2c_1^2} \tag{2.65}$$

Let $N_0$ be the number of passes through $\alpha = 0$. Then $N_0 = \dfrac{1}{\pi} \dfrac{c_2}{c_1}$ and using Equation 2.63, we have

$$N_0 = \frac{1}{\pi} \left[ \frac{\int_0^\infty \omega^2 S_X(\omega) d\omega}{\int_0^\infty S_X(\omega) d\omega} \right]^{1/2} \tag{2.66}$$

or

$$N_0 = \frac{1}{\pi}\left[\frac{-R''_X(0)}{R_X(0)}\right]^{1/2} \tag{2.67}$$

The average number of crossings through the line $x = \alpha$ is

$$N_\alpha = N_0 e^{-\alpha^2/2c_1^2} \tag{2.68}$$

## 2.6 Dynamic Characteristics of Linear Systems

This section will consider a one-degree-of-freedom system as given in Figure 1.8. The equation of motion is

$$m\ddot{u} + \eta\dot{u} + ku = p(t) \tag{1.22}$$

as given in Section 1.4. The dynamic properties of the system are described in terms of the following functions. Note that for a more detailed account, see Chapter 1.

1. *Response to unit impulse.* If the loading $p(t)$ is assumed as the Dirac delta function $\delta(t)$, then the corresponding solution for $u(t)$ is called the response to unit impulse and denoted by $U(t)$. It can be determined from Equation 1.22 by using the Laplace transform method with $p(t) = \delta(t)$ and $u(0) = \dot{u}(0) = 0$

$$ms^2\bar{u} + \eta s\bar{u} + k\bar{u} = 1$$

Thus

$$\bar{u}(s) = \bar{U}(s) = \frac{1}{ms^2 + cs + k} \tag{2.69}$$

and

$$U(t) = \bar{f}(s) = \mathcal{L}^{-1}\left\{\bar{U}(t)\right\} = \frac{1}{m\omega_d} e^{-\zeta\omega_0 t} \sin\omega_d t \tag{2.70}$$

where $\zeta = \eta / 2m\omega_0$, $\omega_0 = \sqrt{k/m}$, and $\omega_d = \sqrt{1-\zeta^2}$.

Also, the function $U(t)$ may be determined using the Fourier transformation. Again with $p(t) = \delta(t)$, the Fourier transformation of Equation 1.22 is

$$-m\omega^2 u^* + \eta i\omega u^* + ku^* = 1$$

and

$$U^*(\omega) = \frac{1}{-m\omega^2 + \eta i\omega + k} \tag{2.71}$$

By the inverse transformation

$$\begin{aligned} U(t) &= \Im^{-1}\{U^*(\omega)\} \\ &= \frac{1}{2\pi}\int_{-\infty}^{\infty} U^*(\omega)e^{i\omega t}d\omega \\ &= \frac{1}{m\omega_d}e^{-\zeta\omega_n t}\sin\omega_d t \end{aligned} \tag{2.72}$$

2. *Frequency response function.* The frequency function may be defined by the Fourier transform of $U(t)$. Thus, for the system described by Equation 1.22

$$U^*(\omega) = \Im\{U(t)\} = \frac{1}{-m\omega^2 + \eta i\omega + k} \tag{2.73}$$

A different, but equivalent, definition of the frequency response function is as follows: Let $P(t) = 1e^{i\omega t}$ then the solution of Equation 1.2 is $u(t) = U^*(\omega)e^{i\omega t}$. Substituting these two expression into Equation 1.22 gives

$$U^*(\omega) = \frac{1}{-m\omega^2 + \eta i\omega + k} \tag{2.74}$$

which is identical with Equation 2.73.

The following relations are important:

a. If $p(t)$ is an arbitrary function of time, the corresponding displacement $u(t)$ may be expressed as

$$u(t) = \int_{-\infty}^{t} U(t-\tau)p(\tau)d\tau \tag{2.75}$$

b. With $p^*(\omega)$ being the Fourier transform of $p(t)$, the Fourier transform of $u(t)$ is

$$u^*(\omega) = U^*(\omega)p^*(\omega) \tag{2.76}$$

If, instead of the displacement $u(t)$, some other quantity is analyzed, for example, the velocity $v(t)$ or the spring force $s(t)$, the response to unit impulse or the frequency response function can be determined. Thus, we have

For $v(t)$

$$v(t) = \frac{du}{dt}$$
$$v^*(\omega) = i\omega U^*(\omega)$$

For $s(t)$

$$s(t) = kU(t)$$
$$s^*(\omega) = kU^*(\omega)$$

The use of the above functions is similar to what was stated for the displacement. For an arbitrary $p(t)$, we have

$$v(t) = \int_{-\infty}^{t} V(t-\tau)p(\tau)d\tau$$
$$v^*(\omega) = V^*(\omega)p^*(\omega)$$

Finally, the input does not have to be in the form of an external force. In the case of a one-degree-of-freedom system with moving support point, the equation of motion is (see Equation 1.93 with no damping, $\eta = 0$)

$$m\ddot{u} + ku = kv \tag{1.93}$$

To determine the response to unit impulse, we assume the input $v(t) = \delta(t)$. For arbitrary $b(t)$, we have

$$u(t) = \int_{-\infty}^{t} U(t-\tau)b(\tau)d\tau$$
$$= \int_{-\infty}^{t} \frac{k}{m\omega_0} \sin \omega_0(t-\tau)b(\tau)d\tau$$

To find the frequency response function of $u(t)$ for the input $v(t)$, we assume $v(t) = 1e^{i\omega t}$ and $u(t) = U^*(\omega)e^{i\omega t}$. From Equation 1.93, we obtain

$$U^*(\omega) = \frac{k}{-m\omega^2 + k} \tag{2.77}$$

## 2.7 Input–Output Relations for Stationary Random Processes

Consider a one-degree-of-freedom system. Assume the input is given by $x(t)$ for any given external loading and the output is $y(t)$, which could be a displacement, velocity, or stress component, etc. Let

$\bar{\Phi}(s)$ be the transfer function of $y(t)$
$\Phi^*(i\omega)$ be the frequency response function of $y(t)$
$\Phi(t)$ be the unit impulse response

then,

$$\Phi(t) = y(t)\Phi(t) = y(t)x(t) = \delta(t)$$
$$\bar{\Phi}(s) = \mathscr{L}\{\Phi(t)\}$$
$$\Phi^*(i\omega) = \Im\{\Phi(t)\}, \quad \text{also if } x(t) = 1 \cdot e^{i\omega t}, \text{ then } y(t) = \Phi^*(i\omega)e^{i\omega t}$$

We have two types of problems:
**Problem A:** Given $R_x(\tau)$, find $R_y(\tau)$.
For any arbitrary input $x(t)$,

$$y(t) = \int_{-\infty}^{t} \Phi(t-\lambda)x(\lambda)d\lambda$$

Since $\Phi(t - \lambda) = 0$ when $t < \lambda$ or $t > \lambda$, the upper limit can be replaced by infinity so that

$$y(t) = \int_{-\infty}^{\infty} x(t-\lambda)\Phi(\lambda)d\lambda \tag{2.78}$$

By changing the variable from $\lambda$ to $\theta = t - \lambda$, the above equation can be written as

$$\begin{aligned} y(t) &= \int_{-\infty}^{\infty} x(t-\theta)\Phi(\theta)d\theta \\ &= \int_{-\infty}^{\infty} x(t-\lambda)\Phi(\lambda)d\lambda \end{aligned} \tag{2.79}$$

Also, we have

$$y(t+\tau) = \int_{-\infty}^{t+\tau} x(t+\tau-\eta)\Phi(\eta)d\eta = \int_{-\infty}^{\infty} x(t+\tau-\eta)\Phi(\eta)d\eta \tag{2.80}$$

The correlation function of $y$ is

$$R_y(\tau) = \lim_{T\to\infty} \frac{1}{2T} \int_{-T}^{T} y(t)y(t+\tau)dt$$

Substituting Equations 2.79 through 2.80 gives

$$R_y(\tau) = \lim_{T\to\infty} \frac{1}{2T} \int_{-T}^{T} dt \int_{-\infty}^{\infty} x(t-\lambda)\Phi(\lambda)d\lambda \int_{-\infty}^{\infty} x(t+\tau-\eta)\Phi(\eta)d\eta$$

Writing $R_y(\tau)$ as

$$R_Y(\tau) = \lim_{T\to\infty} \int_{-\infty}^{\infty} d\lambda \int_{-\infty}^{\infty} d\eta\ \Phi(\lambda)\Phi(\eta)\frac{1}{2T} \int_{-T}^{T} x(t-\lambda)x(t+\tau-\eta)dt$$

and noting that $t_1 = t - \lambda$ and $t_2 = t + \tau - \eta$, then $t_2 - t_1 = \tau + \lambda - \eta$ and that

$$\lim_{T\to\infty} \frac{1}{2T} \int_{-T}^{T} x(t-\lambda)x(t+\tau-\eta)dt = R_x(t_2 - t_1)$$

We can write

$$R_y(\tau) = \int_{-\infty}^{\infty}\int_{-\infty}^{\infty} \Phi(\lambda)\Phi(\eta)R_x(\tau+\lambda-\eta)d\eta\ d\lambda \tag{2.81}$$

**Problem B:** Given $S_x(\omega)$, find $S_y(\omega)$.

$$\begin{aligned}
S_y(\omega) &= \int_{-\infty}^{\infty} R_y(\tau)e^{-i\omega\tau}d\tau \\
&= \int_{-\infty}^{\infty} d\tau \int_{-\infty}^{\infty} d\lambda \int_{-\infty}^{\infty} d\eta\ e^{-i\omega\tau}\ \Phi(\lambda)\Phi(\eta)R_x(\tau+\lambda-\eta) \\
&= \int_{-\infty}^{\infty} d\tau \int_{-\infty}^{\infty} d\lambda \int_{-\infty}^{\infty} d\eta\ e^{i\omega\lambda}e^{-i\omega\eta}e^{-i\omega(\tau+\lambda-\eta)}\Phi(\lambda)\ \Phi(\eta)R_x(\tau+\lambda-\eta) \\
&= \int_{-\infty}^{\infty} \Phi(\eta)e^{-i\omega\eta}d\eta \int_{-\infty}^{\infty} \Phi(\lambda)e^{i\omega\lambda}d\lambda \int_{-\infty}^{\infty} R_x(\tau+\lambda-\eta)e^{-i\omega(\tau+\lambda-\eta)}d\tau
\end{aligned}$$

Noting that $\tau' = \tau + \lambda - \eta$ and

$$\Phi^*(i\omega) = \int_{-\infty}^{\infty} \Phi(\eta)e^{-i\omega\eta}d\eta$$

$$\Phi^*(-i\omega) = \int_{-\infty}^{\infty} \Phi(\lambda)e^{i\omega\lambda}d\lambda$$

$$\int_{-\infty}^{\infty} R_x(\tau+\lambda-\eta)e^{-i\omega(\tau+\lambda-\eta)}d\tau = \int_{-\infty}^{\infty} R_x(\tau')e^{-i\omega(\tau')}d\tau = S_x(\omega)$$

We find

$$S_y(\omega) = \Phi^*(i\omega)\Phi^*(-i\omega)S_x(\omega) \tag{2.82}$$

or

$$S_y(\omega) = \left|\Phi^*(i\omega)\right|^2 S_x(\omega) \tag{2.83}$$

It is important to note that the relationships between $R_x$ and $R_y$ as well as $S_x$ and $S_y$ are valid if

$$y(t) = \int_{-\infty}^{\infty} \Phi(t-t')x(t')dt'$$

Figure 2.23 shows an interpretation of the input power spectrum, modulus squared of the frequency response function, and the power spectrum of the response.

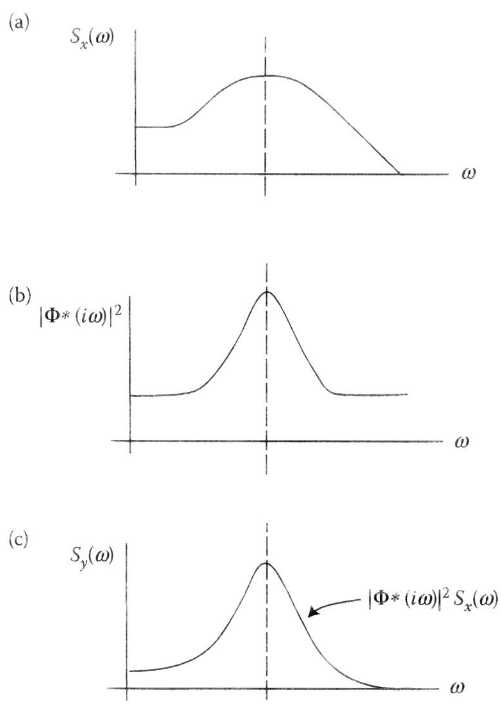


**FIGURE 2.23**
Typical response of oscillator to external excitation. (a) Input power spectrum. (b) Modulus squared of the frequency response function. (c) Power spectrum of the response.

## 2.8 Input–Output Relations for Nonstationary Random Processes

Assume that the input given by $x(t)$ is nonstationary. The output is again $y(t)$. The input correlation function is $R_x(t_1, t_2) = [x(t_1)x(t_2)]$. We want to find the correlation function for the output, $R_y(t_1, t_2) = [y(t_1)y(t_2)]$. We have

$$y(t_1) = \int_0^{t_1} \Phi(t_1 - \lambda_1)x(\lambda_1)d\lambda_1$$

where $x(t)$ is given for $t > 0$. A similar relation can be written for a different time period

$$y(t_2) = \int_0^{t_2} \Phi(t_2 - \lambda_2)x(\lambda_2)d\lambda_2$$

where $\Phi$ is the response to a unit impulse. The mean value and the correlation function of $y$ are

$$E[y(t_1)y(t_2)] = E\left\{\int_0^{t_1}\int_0^{t_2} \Phi(t_1 - \lambda_1)x(\lambda_1)\Phi(t_2 - \lambda_2)x(\lambda_2)d\lambda_1\, d\lambda_2\right\} \tag{2.84}$$

$$R_Y(t_1, t_2) = E[y(t_1)\cdot y(t_2)]\int_0^{t_1}\int_0^{t_2} \Phi(t_1 - \lambda_1)\Phi(t_2 - \lambda_2)[x(\lambda_1)x(\lambda_2)]d\lambda_1\, d\lambda_2$$

$$= \int_0^{t_1}\int_0^{t_2} \Phi(t_1 - \lambda_1)\Phi(t_2 - \lambda_2)E[x(\lambda_1)x(\lambda_2)]d\lambda_1\, d\lambda_2$$

or

$$R_y(t_1, t_2) = \int_0^{t_1}\int_0^{t_2} \Phi(t_1 - \lambda_1)\Phi(t_2 - \lambda_2)R_x(t_1, t_2)d\lambda_1\, d\lambda_2 \tag{2.85}$$

For a special type of nonstationary input, we have $X(t) = A(t)\xi(t)$, where $A(t)$ is a deterministic function and $\xi(t)$ is a stationary random function. Some examples are shown in Figure 2.24.

For $\xi(t)$: $R_\xi(t_1 - t_2) = R_\xi(\tau)$, where $R_\xi(\tau)$ may be found by time averaging or from $S_\xi(\omega)$. By similar arguments as presented before, we find

$$R_y(t_1, t_2) = \int_0^{t_1}\int_0^{t_2} \Phi(t_1 - \lambda_1)\Phi(t_2 - \lambda_2)A(\lambda_1)A(\lambda_2)R_\xi(\lambda_1 - \lambda_2)d\lambda_1\, d\lambda_2 \tag{2.86}$$

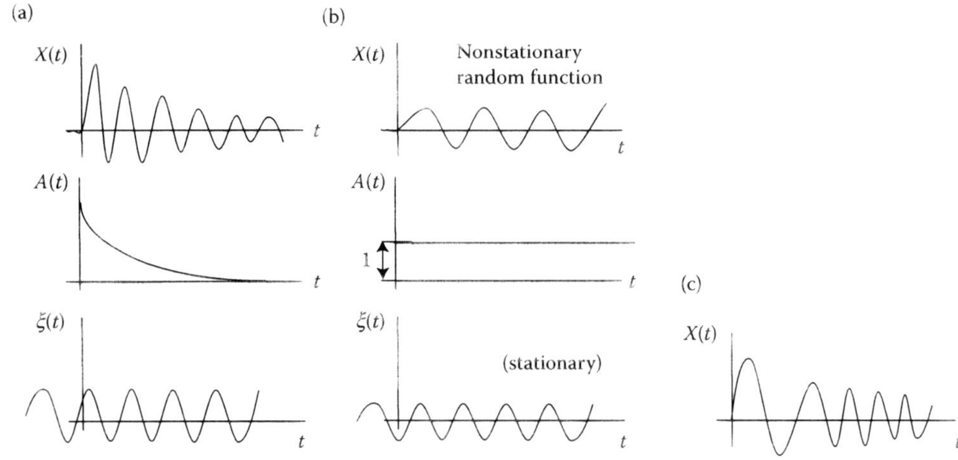


**FIGURE 2.24**
Nonstationary input (a) and (b) typical examples, (c) function that does not belong to the class of functions.

As an example of how to use the above theory, consider a one-degree-of-freedom system whose equation of motion is given by

$$\ddot{y} + 2\eta\dot{y} + \omega_0^2 y = x(t)$$

where $x(t)$ is a nonstationary random function. Determine $\Phi(t)$ for $x(t) = \delta(t)$. This gives

$$\Phi(t) = \begin{cases} e^{-\eta t}\dfrac{1}{\omega_1}\sin\omega_1 t \quad \text{where } \omega_1 = \sqrt{\omega_0^2 - \eta^2} & \text{for } t \geq 0 \\ 0 & \text{for } t < 0 \end{cases}$$

The loading part is written as

$$\begin{aligned} x(t) &= A(t)\xi(t) \\ &= Ae^{-ct}\xi(t) \end{aligned}$$

This is shown in Figure 2.25.

We note that $A$ has dimensions of force/mass and $\xi$ is dimensionless. In addition, we have

$$R_\xi(\tau) = K_0 e^{-\alpha|\tau|}\cos\theta\tau$$

where $K_0$, $\alpha$, and $\theta$ are parameters. The correlation function is depicted in Figure 2.26. And

$$K_0 = R_\xi(0) = \left[\xi^2\right]$$

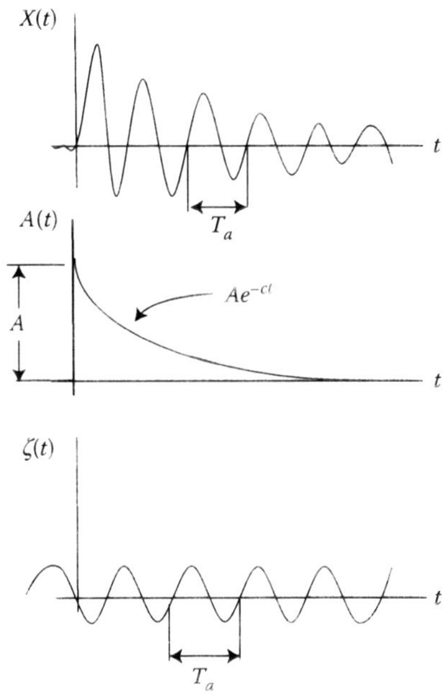


**FIGURE 2.25**
Loading of nonstationary random function.

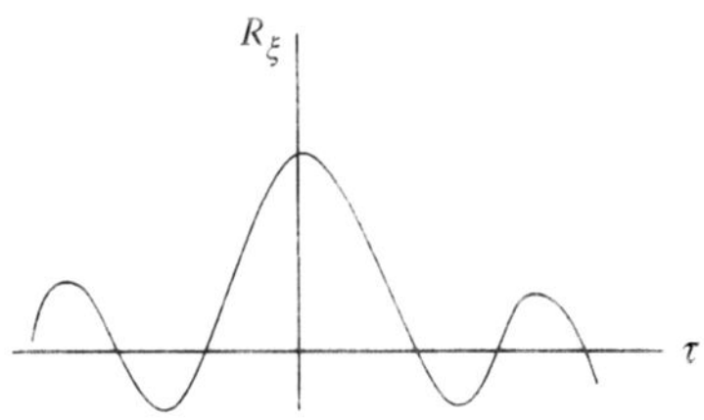


**FIGURE 2.26**
Correlation function $R(\xi)$.

We may take $K_0 = 1$ and we may adjust $A$. The degree of correlation is controlled by $\alpha$. The apparent periodicity of $X(t)$ or $\xi(t)$ is described by $\theta$, where

$$\theta = \frac{2\pi}{T_a}$$

The power spectral density for $S_\xi(\omega)$ is

$$S_\xi(\omega) = \frac{1}{2}\left[\frac{\alpha K_0}{(\omega-\theta)^2+\alpha^2}\right] + \frac{1}{2}\left(\frac{\alpha K_0}{(\omega+\theta)^2+\alpha^2}\right)$$

and shown in Figure 2.27 for different values of $\alpha$ (Bolotin 1984).

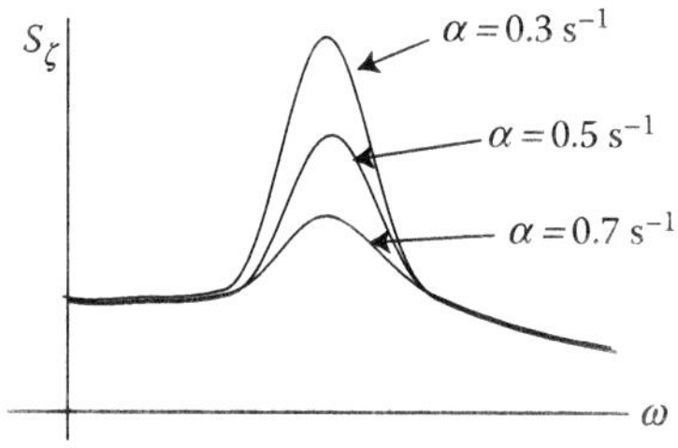


**FIGURE 2.27**
Power spectral densities for various values of $\alpha$.

For $t_1 = t_2 = t$

$$R_y(t,t) = \frac{A^2 e^{-2\eta t}}{\omega_1^2} \int_0^t \int_0^t e^{\beta(\lambda_1+\lambda_2)-\alpha(\lambda_1-\lambda_2)} \sin\omega_1(t-\lambda_1)\sin\omega_1(t-\lambda_2)\cos\theta(\lambda_1-\lambda_2)d\lambda_1\, d\lambda_2$$

Using specific data with the angular frequency of the system being $\omega_1 = 10\ s^{-1}$, the angular frequency $f_1 = \omega_1/2\pi \approx 1.592\ s^{-1}$, the period $T_1 = 1/f_1 \approx 1.6283$ s, $\eta = 0.4\ s^{-1}$, $c = 0.2\ s^{-1}$, and $\alpha = 0.5\ s^{-1}$, we generate the plot shown in Figure 2.28 for various values of $\theta/\omega_1$.

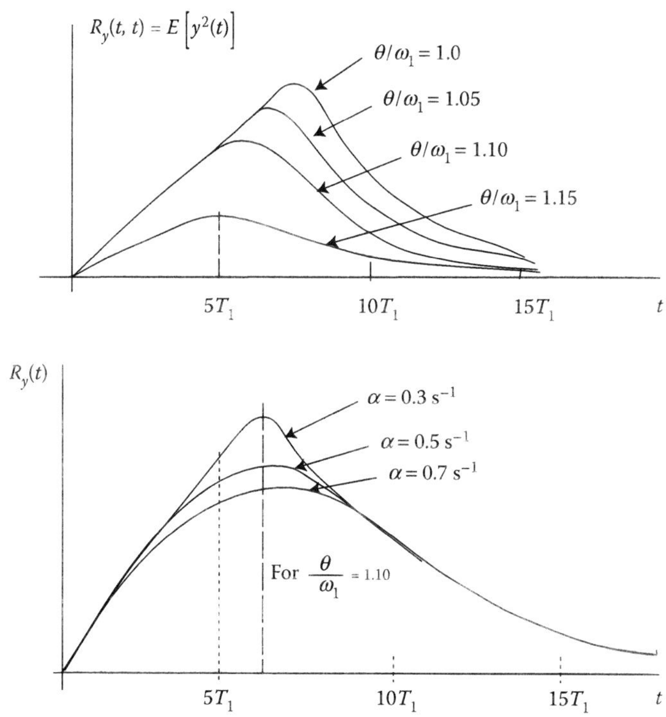


**FIGURE 2.28**
Autocorrelation function.

## PROBLEMS

2.1 What are the dimensions of the correlation function $R_x$ and the spectral density $S_x$ if the random function $X(t)$ is

a. Displacement (in)

b. Acceleration (in/s$^2$)

c. Force (lb)

d. Stress (psi)

2.2 What are the dimensions of the probability function $F_x$ and the probability density function $f_x$ if the random function $X(t)$ is

a. Displacement (in)

b. Acceleration (in/s$^2$)

c. Force (lb)

d. Stress (psi)

2.3 Define a stationary random process.

2.4 Consider a stationary ergodic process $X(t)$; express $E[X]$, $E[X^2]$, and $R_x(\tau)$ as the time averages.

2.5 Prove the relations between the spectral density $S(\omega)$ and the correlation function $R(\tau)$:

$$R(\tau) = \frac{1}{2\pi}\int_{-\infty}^{\infty} S(\omega)e^{i\omega\tau}\, d\omega$$

$$S(\omega) = \int_{-\infty}^{\infty} R(\tau)e^{-i\omega\tau}\, d\tau$$

2.6 Assuming that the input–output relation is $y(t) = \int_{-\infty}^{t} \phi(t-\lambda)x(\lambda)d\lambda$, show that

$$R_y(\tau) = \int_{-\infty}^{\infty}\int_{-\infty}^{\infty} \phi(\lambda)\phi(\eta)R_x(\tau+\lambda-\eta)\,d\lambda\, d\eta \text{ and } S_y(\omega) = |\phi^*(i\omega)|^2\, S_x(\omega)$$

2.7 Derive the expression for the average number of crossings of the value $x = \alpha$, assuming that the joint distribution function $f(\alpha, \beta)$ of $X(t)$ and $\dot{X}(t)$ is

a. Two-dimensional normal

b. Arbitrary

2.8 The spectral density of the loading (input) is given in the form shown in Figure 2.29. Find the spectral density of the outputs (displacement, bending moment, etc.) of the systems described in Problem 1.3 (Chapter 1).

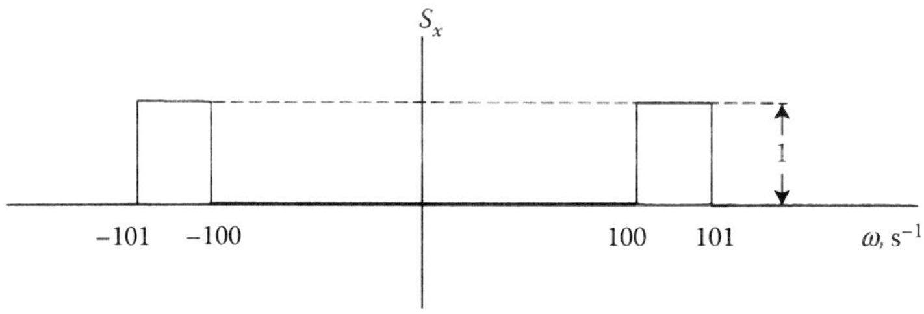


**FIGURE 2.29**
Loading input.

2.9 A random function $X(t)$ has its spectral density as in Problem 2.8. Find the mean square value of $X(t)$. Find the average number of crossings of the values $x = 0$, $x = 1$, and $x = 10$.

## References

Bolotin, V. V. 1984. *Random Vibrations of Elastic Systems*. Dordrecht: Springer Science+Business Media.

Crandall, S. H. and Mark, W. D. 1963. *Vibrations in Mechanical Systems*. Cambridge: Academic Press.

Hisashi, K., Mark, B. L., and Turin, W. 2012. *Probability, Random Processes, and Statistical Analysis*. Cambridge, UK: Cambridge University Press.

Laning, J. H. and Battin, R. H. 1956. *Random Process in Automatic Control*. New York: McGraw-Hill.

Li, J. and Chen, J. 2009. *Stochastic Dynamics of Structures*. Singapore: John Wiley & Sons (Asia).

Lin, Y. K. 1966. *Probability Theory of Structural Dynamics*. New York: McGraw-Hill.

Lutes, L. D. and Shahram, S. 2004. *Random Vibrations—Analysis of Structural and Mechanical Systems*. Oxford, UK: Elsevier Butterworth-Heinemann.

Roberts, J. B. and Spanos, P. D. 1990. *Random Vibrations and Statistical Linearization*. Chichester, England: John Wiley & Sons.

Solodounikov, V. V. 1960. *Introduction to the Statistical Dynamics of Automatic Control Systems*. Mineola: Dover.

Wijker, J. J. 2009. *Random Vibrations in Spacecraft Structures Design*. Dordrecht, Heidelberg, London, New York: Springer.

# 3

# Dynamic Response of SDOF Systems Using Numerical Methods

## 3.1 Introduction

The solution of problems encountered in the previous chapters dealt with excitation functions which were known mathematical functions and rendered what is known as closed-form solutions. These so-called closed-form solutions, for a single-degree-of-freedom (SDOF) system, are usually not possible if the applied external excitation or ground acceleration varies arbitrarily with time or if some portion of the system is nonlinear. For most arbitrary loading, some type of numerical time stepping for integration of the differential equation will be required. The determination of the response of an SDOF system subjected to arbitrary dynamic excitation using numerical methods is usually called *numerical simulation*.

This section considers SDOF systems but can be extended to multi-degree-of-freedom (MDOF) systems. The dynamic response of damped SDOF systems is described by the variations of displacement, velocity, and acceleration of the mass with respect to time. The SDOF system to be solved numerically is

$$m\ddot{u} + \eta\dot{u} + ku = p(t) \tag{3.1}$$

with initial conditions $u(0) = u_0$ and $\dot{u}(0) = \dot{u}_0$. The applied force $p(t)$ is given by a set of discrete values $p_i = p(t_i), i = 0, 1, \ldots, N$ at usually, but not necessarily, constant time intervals $\Delta t_i = t_{i+1} - t_i$. These may be presented in the tabular form or presented graphically as in Figure 3.1.

The response of the system is determined at discrete times $t_i$, giving the displacement, velocity, and acceleration as $u_i$, $\dot{u}_i$, and $\ddot{u}_i$. These values are assumed to be known and satisfying the equation of motion

$$m\ddot{u}_i + \eta\dot{u}_i + ku_i = p_i \tag{3.2}$$

Recurrence formulae for computing $u_{i+1}$, $\dot{u}_{i+1}$, and $\ddot{u}_{i+1}$ can be determined, using numerical methods, at time $t_{i+1}$ that satisfying the equation

$$m\ddot{u}_{i+1} + \eta\dot{u}_{i+1} + ku_{i+1} = p_{i+1} \tag{3.3}$$

where the numerical procedure is applied successively with $i = 0, 1, 2, \ldots$. The time-stepping procedure yields the desired response at all times using the known initial conditions $u(0) = u_0$ and $\dot{u}(0) = \dot{u}_0$ to start the numerical procedure.

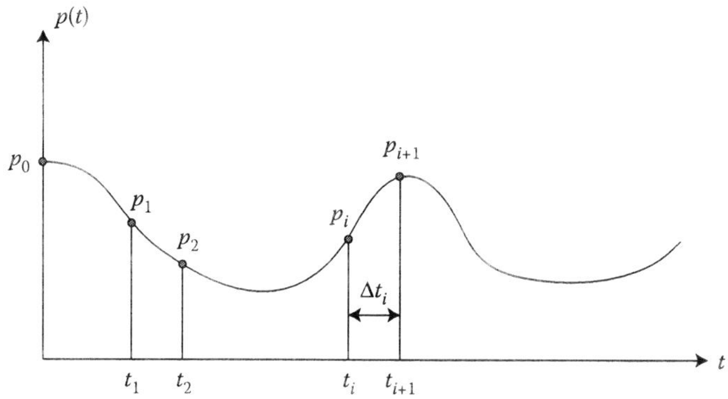


**FIGURE 3.1**
Time-stepping method.

Thus, direct integration of equations of motion involves a stepping along the time dimension. The initial conditions, which have been specified or have been determined, allow the marching scheme to begin. The response can then be computed at as many time points as desired.

In the following section, a brief presentation of several approaches used for the numerical integration of forced SDOF vibration problems will be discussed. Direct integration of the equation of motion provides the response of the SDOF system at discrete interval of time. Response of the system requires the computation of the displacement, velocity, and acceleration. The algorithms described in this section can be applied to many practical problems in structural dynamics.

## 3.2 Interpolating the Excitation Function

Duhamel's integral equation for a damped or undamped SDOF system, for an arbitrary excitation function, can be solved numerically using different numerical integration techniques. Many numerical integration techniques use piecewise-defined excitation functions to interpolate the integrand. Figure 3.2 shows *piecewise-constant* and

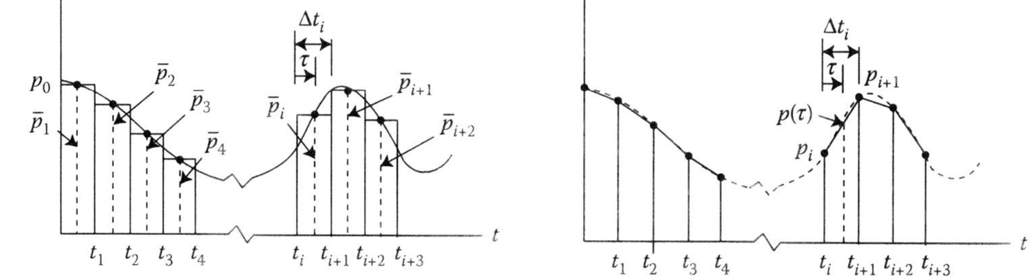

**FIGURE 3.2**
Interpolation of the excitation function. (a) Piecewise-constant interpolation. (b) Piecewise-linear interpolation.

*piecewise-linear* excitation interpolation functions. For piecewise-constant interpolation, $\bar{p}_i$ is the force in the time interval $t_i$ to $t_{i+1}$. This value could be taken at the beginning of the time period to be $p_i$, at the end of the time period to be $p_{i+1}$, or the average value over the time period $\bar{p}_i = (1/2)(p_i + p_{i+1})$.

Consider the piecewise-constant excitation function applied to an undamped SDOF system. The response due to this forcing function is obtained from the super position of Equations 1.12 and 1.183 from Chapter 1. These are

$$u_1(\tau) = u_i \cos\omega_0\tau + \left(\frac{\dot{u}_i}{\omega_0}\right)\sin\omega_0\tau \tag{1.12}$$

and

$$u_2(\tau) = \frac{\bar{p}_i}{k}(1 - \cos\omega_0\tau) \tag{1.183}$$

Here, Equation 1.12 is the free vibration component with initial displacement $u_i$ and initial velocity $\dot{u}_i$ while Equation 1.83 is the response to a constant force $\bar{p}_i$. The displacement at $t_{i+1}$ is obtained by adding $u_1$ and $u_2$ and evaluating these expressions at time $t_{i+1}$. Thus, we have

$$u_{i+1} = u_i \cos\omega_0\Delta t_i + \left(\frac{\dot{u}_i}{\omega_0}\right)\sin\omega_0\Delta t_i + \frac{\bar{p}_i}{k}(1 - \cos\omega_0\Delta t_i) \tag{3.4}$$

The velocity is obtained by differentiating Equation 3.4 with respect to $\Delta t_i$. Hence,

$$\frac{u_{i+1}}{\omega_0} = -u_i \sin\omega_0\Delta t_i + \left(\frac{\dot{u}_i}{\omega_0}\right)\cos\omega_0\Delta t_i + \frac{\bar{p}_i}{k}\sin\omega_0\Delta t_i \tag{3.5}$$

Equations 3.4 and 3.5 are the *recurrence* equations for evaluating the response of the system $(u_{i+1}, \dot{u}_{i+1})$ at time $t_{i+1}$.

In the same manner, we can determine the response of an SDOF system with a piecewise-linear representation of the excitation force *p*(*t*). From Figure 3.2, the piecewise-linear interpolation of the excitation force can be taken as

$$p(\tau) = p_i + \left(\frac{\Delta p_i}{\Delta t_i}\right)\tau \quad \text{where } \Delta p_i = p_{i+1} - p_i \tag{3.6}$$

Again consider an undamped SDOF system. For an undamped system, we have, due to the above piecewise-linear excitation force, three displacement components. These are:

1. Free vibration component with initial displacement $u_i$ and initial velocity $\dot{u}_i$ given again by Equation 1.12.
2. Response to the constant force $p_i$ given by Equation 1.183.
3. Response due to the linearly varying force $(\Delta p_i/\Delta t_i)\,\tau$ given by Equation 1.190.

Superimposing these three solutions gives for the displacement at time $t_{i+1}$ and substituting $\Delta t_i$ for $\tau$ yields

$$u_{i+1} = u_i \cos\omega_0\Delta t_i + \left(\frac{\dot{u}_i}{\omega_0}\right)\sin\omega_0\Delta t_i + \frac{p_i}{k}(1-\cos\omega_0\Delta t_i) + \frac{\Delta p_i}{k\Delta t_i}\left[\Delta t_i - \frac{\sin\omega_0\Delta t_i}{\omega_0}\right] \tag{3.7}$$

and for the velocity

$$\frac{\dot{u}_{i+1}}{\omega_0} = -u_i \sin\omega_0\Delta t_i + \left(\frac{\dot{u}_i}{\omega_0}\right)\cos\omega_0\Delta t_i + \frac{p_i}{k}\sin\omega_0\Delta t_i + \frac{\Delta p_i}{k\Delta t_i}[1-\cos\omega_0\Delta t_i] \tag{3.8}$$

Recurrence formulas for Equations 3.7 and 3.8 may be conveniently expressed as

$$\begin{aligned} u_{i+1} &= au_i + bu_{i+1} + cp_i + dp_{i+1} \\ \dot{u}_{i+1} &= a'u_i + b'u_{i+1} + c'p_i + d'p_{i+1} \end{aligned} \tag{3.9}$$

where the coefficients $a$ through $d'$ are given in Table 3.1. Recurrence Equations 3.9 permit step-by-step integration, when the displacement and velocity are known at time $t = 0$. The acceleration can be obtained for each time step by satisfying the equation of motion for an undamped SDOF system. Thus,

$$\ddot{u}_{i+1} = \frac{1}{m}(p_{i+1} - p_i) \tag{3.10}$$

The same technique used above can be used to solve for the damped system. By using Equations 1.38 and 1.184, we have the response due to free vibrations with the initial displacement $u_i$ and initial velocity $\dot{u}_i$ and the response due to a constant force $p_i$. These expressions are given as

**TABLE 3.1**

Coefficients for Recurrence Formulas for Undamped SDOF System

| |
|---|
| $a = \cos\omega_0\Delta t_i$ |
| $b = \dfrac{1}{\omega_0}\cos\omega_0\Delta t_i$ |
| $c = \dfrac{1}{k\omega_0\Delta t_i}[\sin\omega_0\Delta t_i - \omega_0\Delta t_i\cos\omega_0\Delta t_i]$ |
| $d = \dfrac{1}{k\omega_0\Delta t_i}[\Delta t_i\omega_0 - \sin\omega_0\Delta t_i]$ |
| $a' = -\omega_0\sin\omega_0\Delta t_i$ |
| $b' = \cos\omega_0\Delta t_i$ |
| $c' = \dfrac{1}{k\omega_0\Delta t_i}\left[\omega_0^2\Delta t_i\sin\omega_0\Delta t_i + \omega_0\cos\omega_0\Delta t_i - \omega_0\right]$ |
| $d' = \dfrac{1}{k\omega_0\Delta t_i}[\omega_0 - \omega_0\cos\omega_0\Delta t_i]$ |

$$u_1(\tau) = e^{-\zeta\omega_0\tau}\left[\left(\cos\omega_d\tau + \frac{\zeta\omega_0}{\omega_d}\sin\omega_d\tau\right)u_i + \frac{\dot{u}_i}{\omega_d}\sin\omega_d\tau\right] \tag{1.38}$$

$$u_2(\tau) = \frac{p_i}{k}\left[1 - e^{-\zeta\omega_0\tau}\left\{\cos\omega_d\tau + \left(\frac{\zeta\omega_0}{\omega_d}\right)\sin\omega_d\tau\right\}\right] \tag{1.184}$$

To determine the response to the linearly varying force ($\Delta p_i/\Delta t_i$), we need to solve the following expression given as

$$u_3(\tau) = \frac{1}{m\omega_d}\int_0^t \left(\frac{\Delta p_i}{\Delta t_i}\right)\tau e^{-\omega_0\zeta(t-\tau)}\sin\omega_d(t-\tau)d\tau \tag{3.11}$$

Using the trigonometric identity $\sin\omega_d(t - \tau) = \sin\omega_d t\cos\omega_d\tau - \cos\omega_d t\sin\omega_d\tau$, Equation 3.11 can be written as

$$u_3(\tau) = \frac{\Delta p_i e^{-\omega_0\zeta t}}{m\omega_d\Delta t_i}[A_1(t)\sin\omega_d t - A_2(t)\cos\omega_d t] \tag{3.12}$$

where

$$A_1(t) = \int_0^t \tau e^{\zeta\omega_0\tau}\cos\omega_d\tau\, d\tau$$
$$A_2(t) = \int_0^t \tau e^{\zeta\omega_0\tau}\sin\omega_d\tau\, d\tau \tag{3.13}$$

Integrating Equations 3.13 gives

$$A_1(t) = \frac{e^{at}}{a^2+b^2}\left[\left(at - \frac{a^2-b^2}{a^2+b^2}\right)\cos bt + \left(bt - \frac{2ab}{a^2+b^2}\right)\sin bt\right] + \left(\frac{1}{a^2+b^2}\right)\left(\frac{a^2-b^2}{a^2+b^2}\right)$$
$$A_a(t) = \frac{e^{at}}{a^2+b^2}\left[\left(at - \frac{a^2-b^2}{a^2+b^2}\right)\sin bt - \left(bt - \frac{2ab}{a^2+b^2}\right)\sin bt\right] - \left(\frac{1}{a^2+b^2}\right)\left(\frac{2ab}{a^2+b^2}\right) \tag{3.14}$$

where $a = \zeta\omega_0$, $b = \omega_d = \omega_0\sqrt{1-\zeta^2}$. Equation 3.12 can now be written as

$$u_3(\tau) = \frac{\Delta p_i}{k\Delta t_i}\left[\tau - \frac{2\zeta}{\omega_0} + e^{-\zeta\omega_0\tau}\left(\frac{2\zeta^2-1}{\omega_d}\sin\omega_d\tau + \frac{2\zeta}{\omega_0}\cos\omega_d\tau\right)\right] \tag{3.15}$$

The displacement at $t_{i+1}$ is obtained by adding $u_1$, $u_2$, and $u_3$ and substituting $\Delta t_i$ for $\tau$. Thus,

$$u_{i+1} = Au_i + B\dot{u}_i + Cp_i + Dp_{i+1} \tag{3.16}$$

The velocity at $t_{i+1}$ is obtained by adding the differentials $\dot{u}_1$, $\dot{u}_2$, and $\dot{u}_3$ and substituting $\Delta t_i$ for $\tau$. This gives

$$u_{i+1} = A'u_i + B'\dot{u}_i + C'p_i + D'p_{i+1} \tag{3.17}$$

where the coefficients $A$ through $D'$ are given in Table 3.2.

It is noted that if the time step is constant, the coefficients $A$, $B$, ... , $D'$ need only to be computed once. The time-step size $\Delta t_i$ should be chosen such that there is a close approximation to the excitation function and that the peak responses are not missed. The rule of thumb is to pick $\Delta t < T/10$, where $T$ is the natural period of the structure. The above technique is only applicable to linear systems and usually not applied to MDOF systems. The coefficients given in Table 3.2 depend on $\omega_0$, $\omega_d$, $k$ and $\zeta$, and the time interval $\Delta t_i$.

**EXAMPLE 3.1**

The force exerted on a structure due to shock loading can be represented by $p(t) = p_0(1 - t/t_0)e^{-bt/t_0}$ where $b$ is the waveform parameter. Consider a single-story building shown in Figure 3.3 subjected to the above-described shock loading. Assume that the shock load can be represented by a lateral load applied at the roof level. The total roof mass is 10,000 kg, total lateral stiffness is 4000 kN/m, and the damping ratio is 0.025. Determine the displacement and velocity of the structure with $p_0 = 500$ kN, $b = 4.0$, and $t_0 = 0.75$ s.

**TABLE 3.2**

Coefficients for Recurrence Formulas for Damped SDOF System

$$A = e^{-\zeta\omega_0\Delta t_i}\left(\frac{\zeta}{\sqrt{1-\zeta^2}}\sin\omega_d\Delta t_i + \cos\omega_d\Delta t_i\right)$$

$$B = e^{-\zeta\omega_0\Delta t_i}\left(\frac{1}{\omega_d}\sin\omega_d\Delta t_i\right)$$

$$C = \frac{1}{k}\left[\frac{2\zeta}{\omega_0\Delta t_i} + e^{-\zeta\omega_0\Delta t_i}\left\{\left(\frac{1-2\zeta^2}{\omega_d\Delta t_i} - \frac{\zeta}{\omega_d}\right)\sin\omega_d\Delta t_i - \left(1 + \frac{2\zeta}{\omega_0\Delta t_i}\right)\cos\omega_d\Delta t_i\right\}\right]$$

$$D = \frac{1}{k}\left[1 - \frac{2\zeta}{\omega_0\Delta t_i} + e^{-\zeta\omega_0\Delta t_i}\left\{\left(\frac{2\zeta^2 - 1}{\omega_d\Delta t_i}\right)\sin\omega_d\Delta t_i + \frac{2\zeta}{\omega_0\Delta t_i}\cos\omega_d\Delta t_i\right\}\right]$$

$$A' = -e^{-\zeta\omega_0\Delta t_i}\left(\frac{\omega_0}{\sqrt{1-\zeta^2}}\sin\omega_d\Delta t_i\right)$$

$$B' = e^{-\zeta\omega_0\Delta t_i}\left(\cos\omega_d\Delta t_i - \frac{\zeta}{\sqrt{1-\zeta^2}}\sin\omega_d\Delta t_i\right)$$

$$C' = \frac{1}{k}\left[-\frac{1}{\Delta t_i} + e^{-\zeta\omega_0\Delta t_i}\left\{\left(\frac{\zeta}{\Delta t_i\sqrt{1-\zeta^2}} + \frac{\omega_0}{\sqrt{1-\zeta^2}}\right)\sin\omega_d\Delta t_i + \frac{1}{\Delta t_i}\cos\omega_d\Delta t_i\right\}\right]$$

$$D' = \frac{1}{k\Delta t_i}\left[1 - e^{-\zeta\omega_0\tau}\left(\frac{\zeta}{\sqrt{1-\zeta^2}}\sin\omega_d\Delta t_i + \cos\omega_d\Delta t_i\right)\right]$$

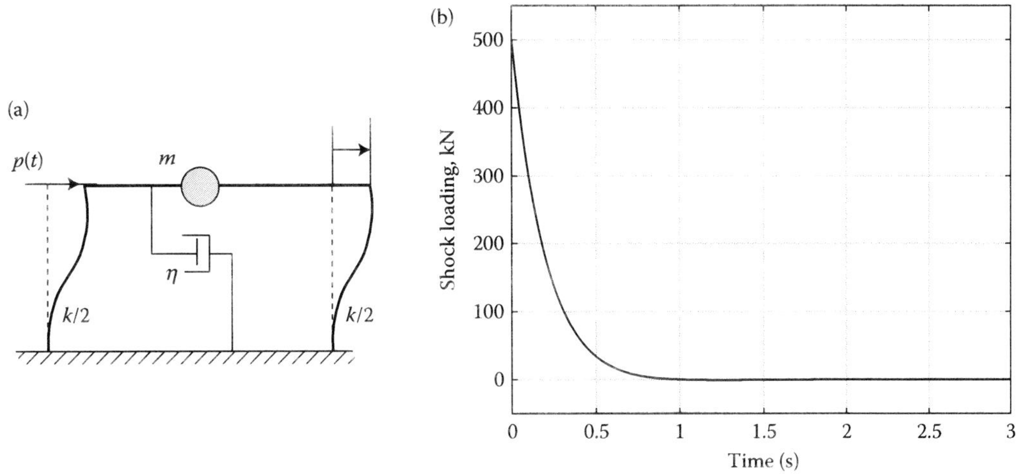


**FIGURE 3.3**
Elementary system: (a) SDOF structure and (b) external shock loading.

**Solution**

We have $k = 4000$ kN/m and the mass $m = 10{,}000$ kg. Thus,

$$\omega_0 = \sqrt{\frac{k}{m}} = \sqrt{\frac{4000 \times 10^3}{10{,}000}} = 20 \text{ rad/s}$$

$$T = \frac{2\pi}{\omega_0} = \frac{2\pi}{20} = 0.314 \text{ s}$$

$$dt = 0.01 < T/10 = 0.0314 \text{ s}$$

This time step should be sufficient since it is less than $T/10$. Using the recurrence equations given in Table 3.2, displacement and velocity time histories for Equations 3.16 and 3.17 are shown in Figure 3.4.

## 3.3 Finite Differences

Finite difference methods (LeVeque 2007) are numerical methods used to approximate solutions to the equations of motion using finite difference equations to approximate the derivatives. The approximations to the derivatives are determined using Taylor's theorem. Consider a function $u(t)$ which is analytic. Being analytic, it can be expanded in a Taylor series in the neighborhood of a grid point $i$ as shown in Figure 3.5. Thus, $u_{i+1}$ and $u_{i-1}$ can be obtained by expanding $u(t)$ in a Taylor series expansion about grid point $i$ as

$$u_{i+1} = u_i + h\dot{u}_i + \frac{h^2}{2!}\ddot{u}_i + \frac{h^3}{3!}\dddot{u}_i + \cdots \tag{3.18}$$

$$u_{i-1} = u_i - h\dot{u}_i + \frac{h^2}{2!}\ddot{u}_i - \frac{h^3}{3!}\dddot{u}_i + \cdots \tag{3.19}$$

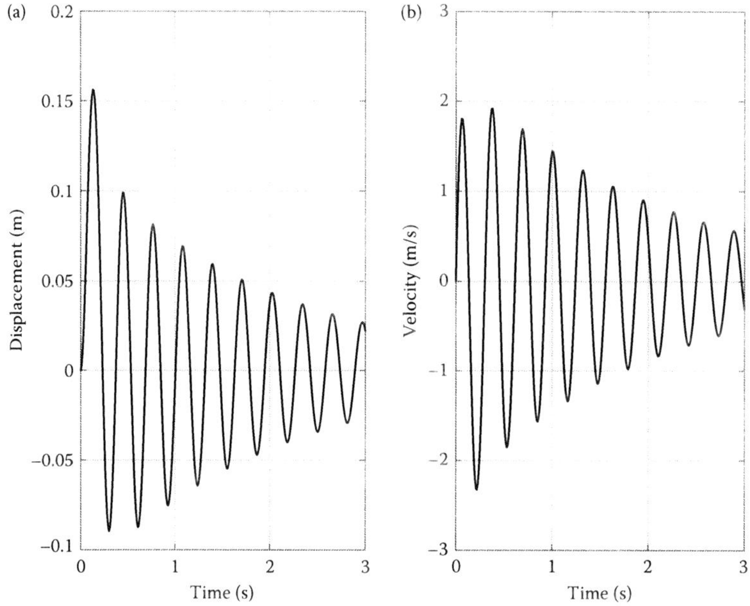


**FIGURE 3.4**
Response history for (a) displacement and (b) velocity of structure with $p_0 = 500$ kN and $t_0 = 0.75$ s.

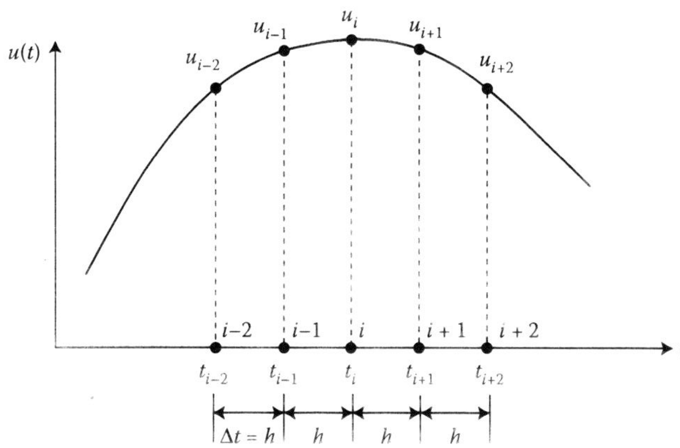


**FIGURE 3.5**
Computational grid for finite difference.

Solving Equation 3.18 and 3.19 for $\dot{u}_i$ gives

$$\dot{u}_i = \frac{u_{i+1} - u_i}{h} - \frac{h}{2}\ddot{u}_i - \frac{h^2}{6}\dddot{u}_i + O(h^3) \tag{3.20}$$

and

$$\dot{u}_i = \frac{u_i - u_{i-1}}{h} - \frac{h}{2}\ddot{u}_i + \frac{h^2}{6}\dddot{u}_i + O(h^3) \tag{3.21}$$

Note that $i$ is a fixed point so that $\ddot{u}_i$, $\dddot{u}_i$, etc. are fixed constants independent of $h$. They depend on $u$ but the function is fixed as we vary $h$. The error will be controlled by the first term $(1/2)h\ddot{u}_i$ and the remaining terms will be small compared to this term, therefore, the error is expected to be approximated by a constant times $h$, where the constant has the value $(1/2)\ddot{u}_i$. Equations 3.20 and 3.21 are known as the *forward* and *backward difference* equations. Equations 3.20 and 3.21 can be written as

$$\dot{u}_i \approx \frac{u_{i+1} - u_i}{h} + O(h) \tag{3.22}$$

and

$$\dot{u}_i \approx \frac{u_i - u_{i-1}}{h} + O(h) \tag{3.23}$$

where $O(h)$ indicates that the error is of the order $h$. If the backward expansion (Equation 3.19) is subtracted from the forward expansion (Equation 3.18), terms of even powers of $h$ cancel giving

$$u_{i+1} - u_{i-1} = 2h\dot{u} + \frac{h^3}{3}\dddot{u}_i + \cdots \tag{3.24}$$

Solving for $\dot{u}_i$ gives

$$\dot{u}_i \approx \frac{u_{i+1} - u_{i-1}}{2h} - \frac{h^2}{6}\dddot{u}_i + \cdots \tag{3.25}$$

or

$$\dot{u}_i \simeq \frac{u_{i+1} - u_{i-1}}{2h} + O(h^2) \tag{3.26}$$

This difference is called central difference and is accurate to $O(h^2)$. If terms up to the second derivative are taken in Equations 3.18 and 3.19 and the two equations added, the central difference for the second derivative is obtained as

$$\ddot{u}_i \approx \frac{u_{i+1} - 2u_i + u_{i-1}}{h^2} + O(h^2) \tag{3.27}$$

This is also accurate to $O(h^2)$. Other higher-order expressions can be obtained for forward and backward differences.

### 3.3.1 Euler Method

In this section, we will investigate the simple numerical integration method of Euler's method (Burden and Faires 2001). It is easy to understand, quick to program, and very helpful to obtain a first impression for the solution. However, it is necessary to take very small step sizes to achieve reasonable precision.

Euler's one step method is the simplest method for approximating the solution to an ordinary differential equation. In this method, the starting point of each time interval is used to determine the slope of the function curve. Thus, the solution would be correct only if the function is linear. The local truncation error using the Euler method is proportional to $h^2$.

The simplest approach to

$$\dot{x} = f(t,x), \quad x(0) = x_0 \tag{3.28}$$

then we have

$$\frac{x_{i+1} - x_i}{h} \approx f(t_i, x_i)$$

Therefore we can write, knowing that $t_i = ih$, for $i = 0, 1, 2, \ldots, N$, that

$$x_{i+1} \approx x_i + hf(t_i, x_i), \quad i = 0,1,2,\ldots,(N-1) \tag{3.29}$$

In the case of SDOF systems, the equation of motion to be solved is

$$m\ddot{u} + \eta\dot{u} + ku = p(t) \tag{3.30}$$

or

$$\ddot{u} + 2\zeta\omega_0\dot{u} + \omega_0^2 u = \frac{p(t)}{m}, \quad u(0) = u_0, \quad \dot{u}(0) = \dot{u}_0 \tag{3.31}$$

where all constants are known along with the complete time history. To use the above numerical Euler solution, we will need to convert our second-order equation into a set of first-order equations. Therefore, if we set $x = \dot{u}$, our second-order system can be written as

$$\begin{aligned} \dot{u} &= x, \quad u(0) = u_0 \\ \dot{x} = \ddot{u} &= \frac{p(t)}{m} - 2\zeta\omega_0 x - \omega_0^2 u, \quad x(0) = \dot{u}_0 \end{aligned} \tag{3.32}$$

Thus, Euler's method uses the equation of motion, Equation 3.30, in the form of the two first-order equations, Equation 3.32. We see that in this method, $u$ and $x$ are updated at each step by

$$\begin{aligned} u_{i+1} &= u_i + hx_i \\ x_{i+1} &= x_i + h\left[\frac{p_i}{m} - 2\zeta\omega_0 x_i - \omega_0^2 u_i\right] \end{aligned} \tag{3.33}$$

The algorithm used in the Euler method is given in Table 3.3.

**TABLE 3.3**

Algorithm Based on Euler's Method

*Initial Computations:*

1. Calculate $m$, $\eta$, and $k$.
2. Form the equation, $\dfrac{d^2u(t)}{dt^2}=\dfrac{1}{m}[p(t)-\eta\dot{u}(t)-ku(t)]=f[\dot{u}(t),u(t),t]$.
3. Write as two coupled first-order differential equations. By a change in variables $\dot{u}=x$, we have

$$\frac{du}{dt}=g(x,u,t)$$
$$\frac{dx}{dt}=f(x,u,t)$$

4. Specify the initial conditions, $u(0)=u_0$ and $x(0)=\dot{u}_0$.
5. Discretize the interval $[t_0, T]$ and select a time step $h=(T-t_0)/N$.

*Calculate: For i = 1, 2, ... , N*

$$u_{i+1}=u_i+h[g(x_i,u_i,t_i)]$$
$$x_{i+1}=x_i+h[f(x_i,u_i,t_i)]$$
$$t_{i+1}=t_i+h$$

**EXAMPLE 3.2**

With the Euler method, calculate the displacement response of the SDOF system of Example 3.1.

**Solution**

The differential equation is given as

$$\ddot{u}+2\zeta\omega_0\dot{u}+\omega_0^2u=\frac{p_0}{m}\left(1-\frac{t}{t_0}\right)e^{-bt/t_0}$$

Making the standard change of variables $\dot{u}=x$, we obtain

$$\dot{u}=x$$
$$\dot{x}=\ddot{u}=\frac{p_0}{m}(1-t/t_0)e^{-bt/t_0}-2\zeta\omega_0x-\omega_0^2u$$

where

$$g(x,u,t)=x \text{ and } f(x,u,t)=\frac{p_0}{m}\left(t-\frac{t}{t_0}\right)e^{-bt/t_0}-2\zeta\omega_0x-\omega_0^2u$$

Using the values $m=10{,}000$ kg, $\omega_0=20$ rad/s, $b=4.0$, $\zeta=0.25$, and $p_0=500$ kN from Example 3.1, we have, using Euler's method, the displacement response shown in Figure 3.6a using a time step of 0.0003 s. Using a larger time step of 0.003 would cause the response to go unstable as shown in Figure 3.6b.

The example demonstrates the generally poor performance of the simple Euler forward method unless the time step chosen is very small. A method called the *modified Euler method* or *Heun's method* is considered an improvement.

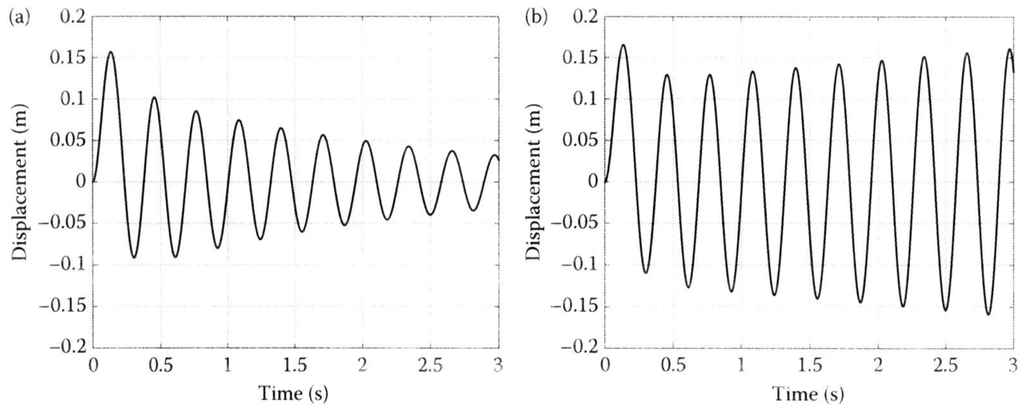


**FIGURE 3.6**
Displacement of Example 3.1 using Euler's method showing (a) stable and (b) unstable response.

### 3.3.2 Modified Euler or Heun's Method

We see that the Euler forward method requires a very small time step to achieve reasonable results. Since the slope was used at the beginning of each time interval, an improvement would be to use the average slope over the end points of each time interval $t_i$ and $t_{i+1}$. For the *modified Euler method*, a better estimate of Equation 3.29 is

$$x_{i+1} \approx x_i + (h/2)\left[f(t_i, x_i) + f\left(t_{i+1}, x_{i+1}^*\right)\right], \quad t_{i+1} = t_i + h \tag{3.34}$$

The method is an example of what is called a *predictor-corrector* technique. The method uses the forward Euler equation to obtain a first approximation to the solution $u(t_{i+1})$, where we have denoted the approximation by $x_{i+1}^*$. Thus, using $x_{i+1}^* = x_i + hf(t_i, x_i)$ gives an estimate of $\dot{x}$ at $t_i$. This is then used to predict a first guess $x_{i+1}^*$, which is used to estimate the average over the interval $h$ of $\dot{x}$. Equation 3.34 is then used to correct the estimate for $x_{i+1}$. Equation 3.34 represents an improvement over Euler method, Equation 3.29, since the local truncation error using Equation 3.34 is proportional to $h^3$.

**EXAMPLE 3.3**

With the modified Euler or Heun's method, calculate the displacement response of the SDOF system of Example 3.1.

**Solution**

Using the same equations and values as used for the Euler method

$$\dot{u} = x$$

$$\dot{x} = \ddot{u} = \frac{p_0}{m}\left(1 - \frac{t}{t_0}\right)e^{-bt/t_0} - 2\zeta\omega_0 x - \omega_0^2 u$$

we can determine the displacement response using the modified Euler method. The results for the displacement response are given in Figure 3.7 for a time step of 0.003 and 0.01.

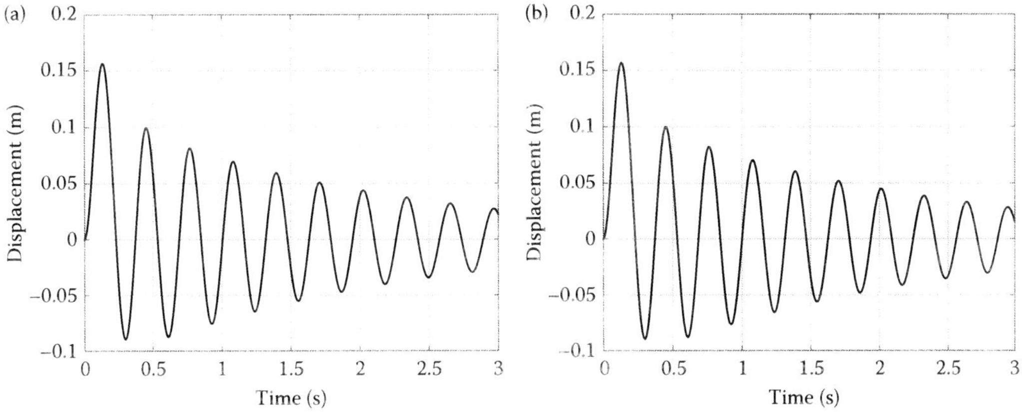


**FIGURE 3.7**
Displacement response using modified Euler method with time step (a) 0.003 and (b) 0.01.

It can be seen that the modified Euler method gives much better results with a much smaller time step and is almost as easy to program and used as the standard Euler method.

### 3.3.3 Runge–Kutta Method

Euler's method was a quick and easy way to determine a first impression for the numerical solution of differential equations. However, it was noted that, unless the time step was chosen very small, it was prone to numerical instabilities. A more robust and intricate numerical technique is the Runge–Kutta method (Butcher 1996). They are easy to program, have good stability characteristics, and are self-starting. However, they require more computer time than other methods of comparable accuracy. Runge–Kutta methods are among the most popular numerical techniques for the solution of ordinary differential equations. The original work was presented in a paper in 1895 by Carl Runge and then modified to handle a system of equations by Wilhelm Kutta in 1901.

Runge–Kutta method is the generalization of the concept used in the modified Euler's method. In the modified Euler's method, the slope of the solution curve was approximated by the slopes of the curve at the end points of the time-step interval in computing the solution. The natural generalization of this concept is computing the slope by taking a weighted average of the slopes taken at a number of points in each subinterval. This method has a local truncation error that is proportional to $h^5$. However, the implementation of the scheme differs from modified Euler's method so that the developed algorithm is explicit in nature.

There are Runge–Kutta methods of all orders, such as second, third, and fourth. However, they can all be written in a generalized form as

$$u_{i+1} = u_i + h\phi(t_i, u_i, h) \tag{3.35}$$

where $\phi(t_i, u_i, h)$ is called an *incremental function*. The $n$th order Runge–Kutta method yields accuracy of order $h^n$. The incremental function can be interpreted as the representative of the slope over the interval $h$. The general form of the incremental function can be written as

$$\phi = w_1k_1 + w_2k_2 + \cdots + w_nk_n \tag{3.36}$$

where the $a$'s are constant and the $k$'s are recurrence relationships given by

$$k_1 = f(t_i, u_i) \tag{3.37}$$

$$k_2 = f(t_i + a_1 h, u_i + b_1 k_1 h) \tag{3.38}$$

$$k_3 = f(t_i + a_2 h, u_i + b_2 k_1 h + b_3 k_2 h) \tag{3.39}$$

$$\vdots$$

$$k_n = f(t_i + a_{n-1} h, u_i + b_{n-1,1} k_1 h + b_{n-1,2} k_2 h + \cdots + b_{n-1,n-1} k_{n-1} h) \tag{3.40}$$

Different types of Runge–Kutta methods can be derived by using different number of terms in the increment function. For example, for $n = 2$, the second-order method (Equation 3.35) is

$$u_{i+1} = u_i + h(w_1 k_1 + w_2 k_2) \tag{3.41}$$

where

$$k_1 = f(t_i, u_i), \quad k_2 = f(t_i + a_1 h, u_i + b_1 k_1) \tag{3.42}$$

The values for the constants $w_1$, $w_2$, $a_1$, and $b_1$ are determined by equating Equations 3.35 and 3.36 to a Taylor series expansion up to the second-order term. This leads to three equations with four unknown constants. The derivation can be found in Butcher (1996). The three equations determined are

$$w_1 + w_2 = 1, \quad w_1 a_1 = w_2 b_1 = \frac{1}{2} \tag{3.43}$$

There are an infinite number of values one can choose for $w_2$, hence an infinite number of second-order Runge–Kutta methods. If the solutions to the ordinary differential equation were quadratic, linear, or constant, every version would give the same result. However, they yield different results when the solution is more complicated. The three values generally picked for $w_2$ are 1/2, 1, and 2/3 which leads to the modified Euler or Heun's method, the midpoint method, and the Ralston method. Thus, using $w_2 = 1/2$, we have $w_1 = 1/2$, and $a_1 = b_1 = 1$ which upon substituting into Equations 3.41 and 3.42 yields for the modified Euler method or second-order Runge–Kutta method

$$u_{i+1} = u_i + \frac{h}{2}(k_1 + k_2) \tag{3.44}$$

where

$$k_1 = f(t_i, u_i) \tag{3.45}$$

$$k_2 = f(t_i + h, u_i + k_1 h) \tag{3.46}$$

For $n = 3$ and $n = 4$, a derivation similar for the second order can be performed leading to the third-order and fourth-order Runge–Kutta method. For the third-order method, there are six equations and eight unknowns. Here, two of the unknowns must be specified. One common versions is given as

$$u_{i+1} = u_i + \frac{h}{6}(k_1 + 4k_2 + k_3) \tag{3.47}$$

where

$$k_1 = f(t_i, u_i) \tag{3.48}$$

$$k_2 = f\left(t_i + \frac{1}{2}h, u_i + \frac{1}{2}k_1 h\right) \tag{3.49}$$

$$k_3 = f(t_i + h, u_i - k_1 h + 2k_2 h) \tag{3.50}$$

The most popular Runge–Kutta method is the fourth-order method. Sometimes, this is called the *classical fourth-order Runge–Kutta method.* This method involves 11 equations and 13 unknowns. Again, two additional conditions must be chosen. The equation is given as

$$u_{i+1} = u_i + \frac{h}{6}(k_1 + 2k_2 + 2k_3 + k_4) \tag{3.51}$$

where

$$k_1 = f(t_i, u_i) \tag{3.52}$$

$$k_2 = f\left(t_i + \frac{1}{2}h, u_i + \frac{1}{2}k_1 h\right) \tag{3.53}$$

$$k_3 = f\left(t_i + \frac{1}{2}h, u_i + \frac{1}{2}k_2 h\right) \tag{3.54}$$

$$k_4 = f(t_i + h, u_i + k_3 h) \tag{3.55}$$

First we note that, just as with the previous two methods, the Runge–Kutta fourth-order method iterates the $t$ values by simply adding a fixed step size of $h$ at each iteration. It is noted that the $u$ iteration equation is a weighted average of four values $k_1$, $k_2$, $k_3$, and $k_4$. That is, multiple estimates of the slope are developed so that an improved average slope for the interval can be made. The algorithm used in the Runge–Kutta fourth-order method is given in Table 3.4.

**EXAMPLE 3.4**

With the Runge–Kutta fourth-order method, calculate the displacement response of the SDOF system of Example 3.1.

**Solution**

Using the same equations and values as used for the Euler method

$$\dot{u} = x$$

$$\dot{x} = \ddot{u} = \frac{p_0}{m}\left(1 - \frac{t}{t_0}\right)e^{-bt/t_0} - 2\zeta\omega_0 x - \omega_0^2 u$$

**TABLE 3.4**

Algorithm Based on Runge–Kutta Fourth-Order Method

*Initial Computations:*

1. Form the equation, $\frac{d^2u(t)}{dt^2} = f[\dot{u}(t), u(t), t]$.
2. Write as two coupled first-order differential equations. By a change in variables $\dot{u} = x$, we have

$$\frac{du}{dt} = g(x,u,t), \quad g(x,u,t) = x$$
$$\frac{dx}{dt} = f(x,u,t), \quad f(x,u,t) = m^{-1}[p(t) - \eta x - ku]$$

3. Specify the initial conditions, $u(0) = u_0$ and $x(0) = \dot{u}_0$.
4. Discretize the interval $[t_0, T]$ and select a time step $h$.

*Calculate for each time step h:*

5. The slope terms

$$k_1 = f(x_i, u_i, t_i) \qquad q_1 = x_i$$
$$k_2 = f\left(x_i + \frac{h}{2}k_1, u_i + \frac{1}{2}q_1, t_i + \frac{1}{2}h\right) \qquad q_2 = x_i + \frac{h}{2}k_1$$
$$k_3 = f\left(x_i + \frac{h}{2}k_2, u_i + \frac{h}{2}q_2, t_i + \frac{1}{2}h\right) \qquad q_3 = x_i + \frac{h}{2}k_2$$
$$k_4 = f\left(x_i + hk_3, u_i + hq_3, t_i + h\right) \qquad q_4 = x_i + hk_3$$

6. The Runge–Kutta equations

$$x_{i+1} = x_i + \frac{h}{6}[k_1 + 2k_2 + 2k_3 + k_4]$$
$$u_{i+1} = u_i + \frac{h}{6}[q_1 + q_2 + q_3 + q_4]$$

we can determine the displacement response using the Runge–Kutta fourth-order method. The results for the displacement response are given in Figure 3.8.

### 3.3.4 Central Difference Method

Our single-degree-of-freedom system is given as

$$m\ddot{u} + \eta\dot{u} + ku = p(t) \tag{3.30}$$

The duration over which the numerical solution of Equation 3.30 is required is divided into constant time steps $h = \Delta t$. The initial conditions again are given as $u(t = 0) = u_0$ and $\dot{u}(t = 0) = \dot{u}_0$. Substituting the central difference expressions for $\dot{u}_i$ and $\ddot{u}_i$, Equations 3.26 through 3.27 into our equation of motion, Equation 3.30 at grid point $i$ gives

$$m\left[\frac{u_{i+1} - 2u_i + u_{i-1}}{h^2}\right] + \eta\left[\frac{u_{i+1} - u_{i-1}}{2h}\right] + ku_i = p_i \tag{3.56}$$

Solving Equation 3.56 for $u_{i+1}$ gives

$$\left[\frac{m}{h^2} + \frac{\eta}{2h}\right]u_{i+1} = \left[\frac{2m}{h^2} - k\right]u_i + \left[\frac{\eta}{2h} - \frac{m}{h^2}\right]u_{i-1} + p_i \tag{3.57}$$

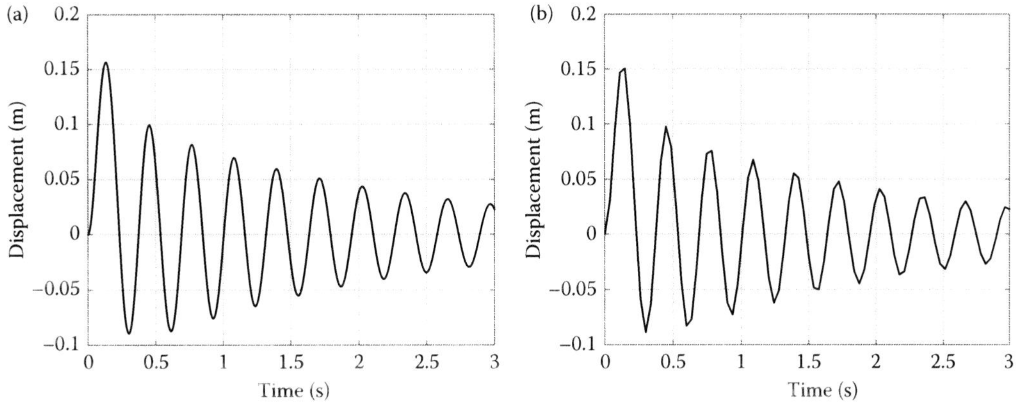


**FIGURE 3.8**
Displacement response using Runge–Kutta fourth-order method with time step (a) 0.003 and (b) 0.0375.

or

$$\left\{\frac{1}{h^2}+\frac{\zeta\omega_0}{h}\right\}u_{i+1}=\frac{p_i}{m}-\left[\omega_0^2-\frac{2}{h^2}\right]u_i-\left[\frac{1}{h^2}-\frac{\zeta\omega_0}{h}\right]u_{i-1} \tag{3.58}$$

where $\omega_0=\sqrt{k/m}$ and $2\zeta\omega_0=\eta/m$. The displacement $u_{i+1}$ can be found using Equation 3.58 provided we know the previous displacements $u_i$ and $u_{i-1}$ along with the current applied external force $p_i$. This type of method is called an *explicit method*. When $i=0$, $u_0$ and $u_{-1}$ are needed to compute $u_1$. Using the initial conditions only gives the values $u_0$ and $\dot{u}_0$, hence we see that the central difference technique is not self-starting. The value for $u_{-1}$ can be determined by using Equations 3.25 and 3.26 evaluated for $i=0$. Thus,

$$\dot{u}_0=\frac{u_1-u_{-1}}{2h}\text{ and }\ddot{u}_0=\frac{u_1-2u_0+u_{-1}}{h^2} \tag{3.59}$$

Using the first expression in Equation 3.59, we can use the value of $u_1$ to eliminate it from the second expression leaving an equation for determining $u_0$. Solving we find

$$u_{-1}=u_0-h\dot{u}_0+\frac{h^2}{2}\ddot{u}_0 \tag{3.60}$$

and from the equation of motion at time $i=0$

$$m\ddot{u}_0+\eta\dot{u}_0+ku_0=p_0$$

or

$$\ddot{u}_0=\frac{p_0-\eta\dot{u}_0-ku_0}{m}=\frac{p_0}{m}-2\zeta\omega_0\dot{u}_0-\omega_0^2u_0 \tag{3.61}$$

**TABLE 3.5**

Algorithm for Central Difference Method

| |
|---|
| *Initial Computations:* |
| 1. Specify $m$, $\eta$, and $k$. |
| 2. Set the initial conditions $u(0) = u_0$ and $\dot{u}(0) = \dot{u}_0$ and calculate $$\ddot{u}_0 = \frac{1}{m}[p_0 - \eta\dot{u}_0 - ku_0]$$ |
| 3. Select a time step $h$ and calculate $$u_{-1} = u_0 - h\dot{u}_0 + \frac{h^2}{2}\ddot{u}_0$$ |
| 4. Form $$\hat{k} = \frac{m}{h^2} + \frac{\eta}{2h}, \quad a = \frac{2m}{h^2} - k, \quad b = \frac{\eta}{2h} - \frac{m}{h^2}$$ |
| *Calculate for each time step:* |
| 5. Form $\hat{p}_i = p_i + a + b$ |
| 6. At time $t_{i+1}$ $u_{i+1} = \frac{\hat{p}_i}{\hat{k}_i}$ |
| 7. At time $t_i$ $\dot{u}_i = \frac{u_{i+1} - u_{i-1}}{2h}$ and $\ddot{u}_i = \frac{u_{i+1} - 2u_i + u_{i-1}}{h^2}$ |

The numerical stability for the central difference technique depends upon the time step chosen. The technique will become *unstable* if the time step is not chosen small enough. To obtain a *stable* solution, the time step $h$ must be picked such that $h < T_i/\pi$, where $T_i$ is the shortest natural period of the system. The algorithm used in the central difference method is given in Table 3.5.

**EXAMPLE 3.5**

With the central difference method, calculate the displacement response of the SDOF system of Example 3.1.

**Solution**

Using the same values used in Example 3.1, the calculation would proceed as follows: first calculate

$$\ddot{u}_0 = \frac{p_0}{m} - 2\zeta\omega_0\dot{u}_0 - \omega_0^2 u_0$$

Then using the initial conditions and the value for $\ddot{u}_0$ calculate

$$u_{-1} = u_0 - h\dot{u}_0 + \frac{h^2}{2}\ddot{u}_0$$

The displacement is then calculated as

$$\left\{\frac{1}{h^2} + \frac{\zeta\omega_0}{h}\right\}u_{i+1} = \frac{p_i}{m} - \left[\omega_0^2 - \frac{2}{h^2}\right]u_i - \left[\frac{1}{h^2} - \frac{\zeta\omega_0}{h}\right]u_{i-1}$$

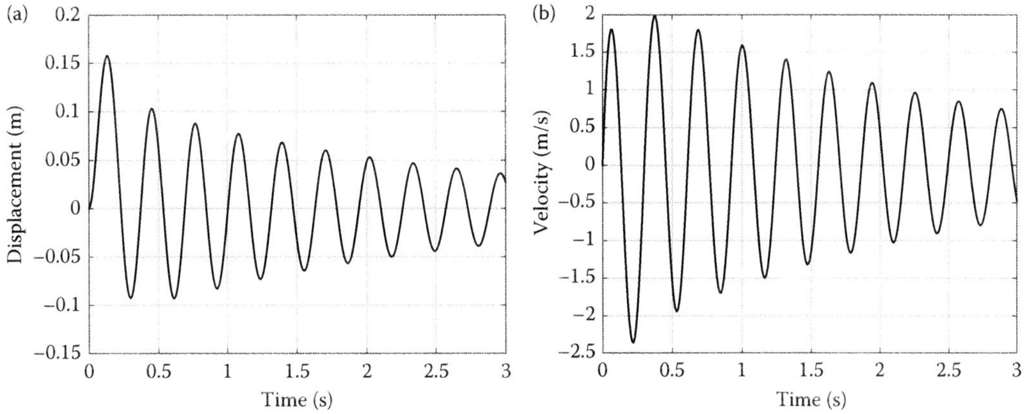


**FIGURE 3.9**
Central difference method using time step 0.03 s. (a) Displacement response and (b) velocity response.

and the velocity and acceleration from

$$\dot{u}_i = \frac{u_{i+1} - u_{i-1}}{2h}, \quad \ddot{u}_i = \frac{u_{i+1} - 2u_i + u_{i-1}}{h^2}$$

The results for the displacement and velocity response are given in Figure 3.9.

## 3.4 Newmark Method

In 1959, Nathan M. Newmark (Newmark 1959) proposed a time integration method that is used for solving the equations for structural dynamics. His method is based on the assumption that the acceleration varies linearly between time intervals. To look at his method, consider a Taylor series expansion with an integral remainder given as

$$f(t_{i+1}) = f(t_i) + hf'(t_i) + \frac{h^2}{2!}f''(t_i) + \cdots + \frac{h^{(n)}}{n!}f^{(n)}(t_i) + \frac{1}{n!}\int_{t_i}^{t_{i+1}} f^{(n+1)}(\tau)(t_{i+1} - \tau)^n d\tau \tag{3.62}$$

where $t_{i+1} = t_i + h$. For $n = 0, f = \dot{u}$ and $n = 1$, $u = 0$ we have for the velocity and displacement

$$\dot{u}_{i+1} = \dot{u}_i + \int_{t_i}^{t_{i+1}} \ddot{u}(\tau)d\tau \tag{3.63}$$

$$u_{i+1} = u_i + h\dot{u}_i + \int_{t_i}^{t_{i+1}} \ddot{u}(\tau)(t_{i+1} - \tau)d\tau \tag{3.64}$$

The integral expressions in Equations 3.63 and 3.64 can be approximated as weighted average of approximate acceleration values. Thus, using

$$\int_{t_i}^{t_{i+1}} \ddot{u}(\tau)d\tau \simeq (1-\gamma)h\ddot{u}_i + \gamma h\ddot{u}_{i+1} \tag{3.65}$$

and

$$\int_{t_i}^{t_{i+1}} \ddot{u}(\tau)(t_{i+1}-\tau)d\tau \simeq [(1-2\beta)\ddot{u}_i + 2\beta\ddot{u}_{i+1}]\frac{h^2}{2} \tag{3.66}$$

The equation for the velocity and displacement can be written as

$$\dot{u}_{i+1} = \dot{u}_i + (1-\gamma)h\ddot{u}_i + \gamma h\ddot{u}_{i+1} \tag{3.67}$$

and

$$u_{i+1} = u_i + h\dot{u}_i + [(1-2\beta)\ddot{u}_i + 2\beta\ddot{u}_{i+1}]\frac{h^2}{2} \tag{3.68}$$

where $0 < \gamma < 1$ and $0 < \beta < 1/2$ are parameters that can be determined to obtain accuracy and stability. These parameters establish how much acceleration is allowed into the velocity and displacement equations at the end of each time interval $h$. Typically, $\gamma$ is set to 1/2, which corresponds to zero artificial damping and $\beta$ is set to a value between 1/6 and 1/4.

The finite difference expressions for the Newmark beta method are obtained using Equations 3.68 and 3.67, thus we have

$$\ddot{u}_{i+1} = \frac{1}{\beta h^2}(u_{i+1}-u_i) - \frac{1}{\beta h}\dot{u}_i - \left(\frac{1}{2\beta}-1\right)\ddot{u}_i \tag{3.69}$$

and

$$\dot{u}_{i+1} = \frac{\gamma}{\beta h}(u_{i+1}-u_i) + \left(1-\frac{\gamma}{\beta}\right)\dot{u}_i + h\left(1-\frac{\gamma}{2\beta}\right)\ddot{u}_i \tag{3.70}$$

The equation of motion at time point $t_{i+1}$ is given as

$$m\ddot{u}_{i+1} + \eta\dot{u}_{i+1} + ku_{i+1} = p_{i+1} \tag{3.71}$$

Substituting Equations 3.69 and 3.70 into Equation 3.71 yields an expression for the displacements as

$$\left[\left(\frac{1}{\beta h^2}\right)m+\left(\frac{\gamma}{\beta h}\right)\eta+k\right]u_{i+1}=\left[\left(\frac{1}{\beta h^2}\right)m+\left(\frac{\gamma}{\beta h}\right)\eta\right]u_i+\left[\left(\frac{1}{\beta h}\right)m+\left(\frac{\gamma}{\beta}-1\right)\eta\right]\dot{u}_i$$
$$+\left[\left(\frac{1}{2\beta}-1\right)m+h\left(\frac{\gamma}{2\beta}-1\right)\eta\right]\ddot{u}_i+p_{i+1} \tag{3.72}$$

Solution of Equation 3.72 gives $u_{i+1}$, which when substituted into Equations 3.69 and 3.70 gives $\dot{u}_{i+1}$ and $\ddot{u}_{i+1}$.

By assigning different values to $\gamma$ and $\beta$, a set of integration equations can be obtained. Some better known versions are given as follows (Hughes 1987):

1. The constant acceleration method, $\gamma = 0$, $\beta = 0$.
2. The average acceleration method, $\gamma = 1/2$, $\beta = 1/4$.
3. The linear acceleration method, $\gamma = 1/2$, $\beta = 1/6$.

The Newmark numerical method is unconditionally stable when

$$2\beta \geq \gamma \geq \frac{1}{2} \tag{3.73}$$

and conditionally stable when

$$\gamma \geq \frac{1}{2},\ \beta < \frac{\gamma}{2} \tag{3.74}$$

$$h \leq h_{cr} = \frac{T}{2\pi\sqrt{\gamma/2-\beta}} \tag{3.75}$$

where $T$ is the period of vibration (Hughes 1987). The algorithm for using the Newmark beta method is given in Table 3.6.

### 3.4.1 Constant Acceleration Method

For the constant acceleration method $\gamma = 0$ and $\beta = 0$, Equations 3.67 and 3.68 become

$$\dot{u}_{i+1} = \dot{u}_i + h\ddot{u}_i \tag{3.76}$$

and

$$u_{i+1} = u_i + h\dot{u}_i + \frac{h^2}{2}\ddot{u}_i \tag{3.77}$$

Equations 3.76 and 3.77 along with the equation of motion at time point $i+1$, Equation 3.71, yield the required set of equations. The acceleration can be determined by substituting Equations 3.76 and 3.77 into Equation 3.71. This gives

**TABLE 3.6**

Algorithm for Newmark Beta Method

| |
|---|
| *Initial Computations:* |
| 1. Specify $m$, $\eta$, and $k$. |
| 2. Set the initial conditions $u(0) = u_0$ and $\dot{u}(0) = \dot{u}_0$ and calculate $$\ddot{u}_0 = \frac{1}{m}[p_0 - \eta\dot{u}_0 - ku_0]$$ |
| 3. Select a time step $h$. |
| 4. Select $\gamma$ and $\beta$. Average acceleration method ($\gamma = 1/2$, $\beta = 1/4$) Linear acceleration method ($\gamma = 1/2$, $\beta = 1/6$) |
| 5. Form $\hat{k} = \left(\frac{1}{\beta h^2}\right)m + \left(\frac{\gamma}{\beta h}\right)\eta + k$ |
| 6. $a = \frac{1}{\beta h^2}m + \frac{\gamma}{\beta h}\eta, b = \frac{1}{\beta h}m + \left(\frac{\gamma}{\beta} - 1\right)\eta$, and $c = \left(\frac{1}{2\beta} - 1\right)m + h\left(\frac{\gamma}{2\beta} - 1\right)\eta$ |
| *Calculate for each time step:* |
| 7. $\hat{p}_{i+1} = p_{i+1} + au_i + b\dot{u}_i + c\ddot{u}_i$ |
| 8. $u_{i+1} = \frac{\hat{p}_{i+1}}{\hat{k}}$ |
| 9. $\ddot{u}_{i+1} = \frac{1}{\beta h^2}(u_{i+1} - u_i) - \left(\frac{1}{\beta h}\right)\dot{u}_i + h\left(1 - \frac{1}{2\beta}\right)\ddot{u}_i$ |
| 10. $\dot{u}_{i+1} = \dot{u}_i + (1-\gamma)h\ddot{u}_i + \gamma h\ddot{u}_{i+1}$ |

$$\ddot{u}_{i+1} = \frac{1}{m}\left[p_{i+1} - ku_i - (\eta + kh)\dot{u}_i - \left(\eta h + \frac{kh^2}{2}\right)\ddot{u}_i\right] \tag{3.78}$$

To start the integration, $u_0$, $\dot{u}_0$, and $\ddot{u}_0$ need to be known. If $u_0$ and $\dot{u}_0$ are known, then $\ddot{u}_0$ can be determined from the equation of motion at time $t = 0$.

### 3.4.2 Average Acceleration Method

For $\gamma = 1/2$ and $\beta = 1/4$, we have the Newmark average acceleration method. The method is based on assuming that the acceleration has a fixed constant value in the time interval $\Delta t$. Due to this assumption, the velocity will vary linearly and the displacement as a quadratic during the time step. This is shown in Figure 3.10a. Equations 3.69, 3.70, and 3.72 now become

$$\ddot{u}_{i+1} = \frac{4}{h^2}(u_{i+1} - u_i - h\dot{u}_i) - \ddot{u}_i \tag{3.79}$$

$$\dot{u}_{i+1} = -\dot{u}_i + \frac{2}{h}(u_{i+1} - u_i) \tag{3.80}$$

$$\left(k + \frac{2\eta}{h} + \frac{4m}{h^2}\right)u_{i+1} = p_{i+1} + \left(\frac{4}{h^2}u_i + \frac{4}{h}\dot{u}_i + \ddot{u}_i\right)m + \left(\frac{2}{h}u_i + \dot{u}_i\right)\eta \tag{3.81}$$

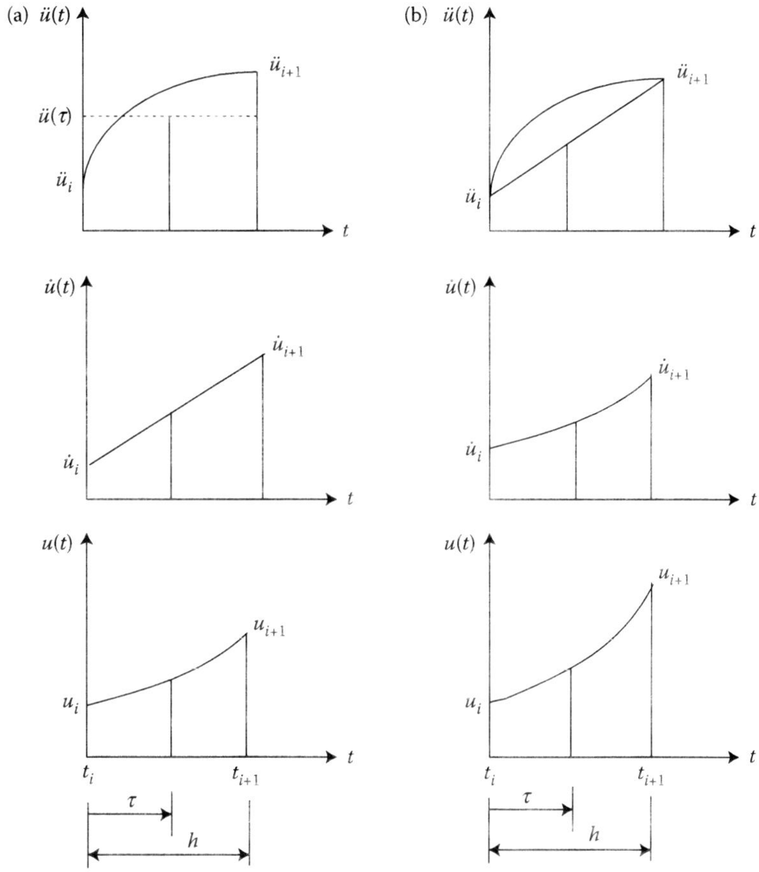


**FIGURE 3.10**
Basic time integration formulas (a) average acceleration method and (b) linear acceleration method.

Knowing $u_{i+1}$, we can calculate $\ddot{u}_{i+1}$ and $\dot{u}_{i+1}$ using Equations 3.79 and 3.80. To start the method, one needs to know the values of $u_0$, $\dot{u}_0$, and $\ddot{u}_0$. Since the initial displacement and velocity are usually given, then one can use the equation of motion at time $t = 0$ to determine $\ddot{u}_0$. For $\gamma = 1/2$ and $\beta = 1/4$, the stability condition (Equation 3.75) becomes

$$\frac{h}{T} < \infty \tag{3.82}$$

which implies that the average acceleration method is unconditionally stable for all values of $h$.

### 3.4.3 Linear Acceleration Method

For the linear acceleration method (Figure 3.10b), $\gamma = 1/2$ and $\beta = 1/6$ and Equations 3.69, 3.70, and 3.72 become

$$\ddot{u}_{i+1} = \frac{6}{h^2}\left(u_{i+1} - u_i + -h\dot{u}_i - \frac{h^2}{3}\ddot{u}_i\right) \tag{3.83}$$

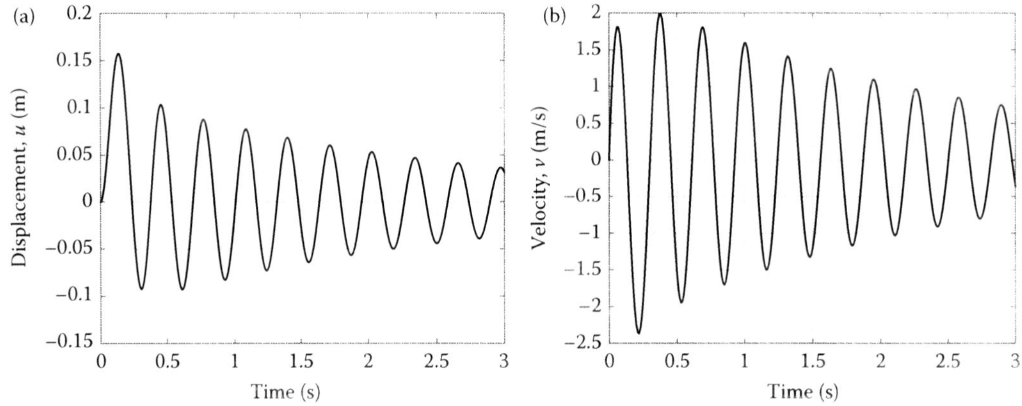


**FIGURE 3.11**
Linear acceleration method using time step 0.03 s. (a) Displacement response and (b) velocity response.

$$\dot{u}_{i+1} = \frac{3}{h}(u_{i+1} - u_i) - 2\dot{u}_i - \frac{h}{2}\ddot{u}_i \tag{3.84}$$

$$\left(k + \frac{6m}{h^2} + \frac{3\eta}{h}\right)u_{i+1} = p_{i+1} + \left(\frac{6}{h^2}u_i + \frac{6}{h}\dot{u}_i + 2\ddot{u}_i\right)m + \left(\frac{3}{h}u_i + 2\dot{u}_i + \frac{h}{2}\ddot{u}_i\right)\eta \tag{3.85}$$

Equation 3.85 gives $u_{i+1}$ and then $\ddot{u}_{i+1}$ and $\dot{u}_{i+1}$ are determined using Equations 3.83 and 3.84. Again, as in the average acceleration method, one needs to know the values of $u_0, \dot{u}_0$, and $\ddot{u}_0$. Since the initial displacement and velocity are usually given, then one can use the equation of motion at time $t = 0$ to determine $\ddot{u}_0$. The linear acceleration method with $\gamma = 1/2$ and $\beta = 1/6$ is conditionally stable with a critical time step, determined from Equation 3.75, as $h/T = 0.5513$.

**EXAMPLE 3.6**

With the linear acceleration method, calculate the displacement response of the SDOF system of Example 3.1.

**Solution**

Using the same values used in Example 3.1, the calculation would proceed as follows: first initialize $u_0$ and $\dot{u}_0$ and then calculate

$$\ddot{u}_0 = \frac{p_0}{m} - 2\zeta\omega_0\dot{u}_0 - \omega_0^2 u_0$$

Using a time step of 0.03 s, the displacement is calculated from Equation 3.87 and then the velocity and acceleration from Equations 3.84 and 3.85, respectively. The results for the displacement and velocity are shown in Figure 3.11.

## 3.5 Wilson-Theta Method

The linear acceleration method when used with the values $\gamma = 1/2$ and $\beta = 1/6$ is only conditional stable where the time step $\Delta t = h$ must be less than the stability limit

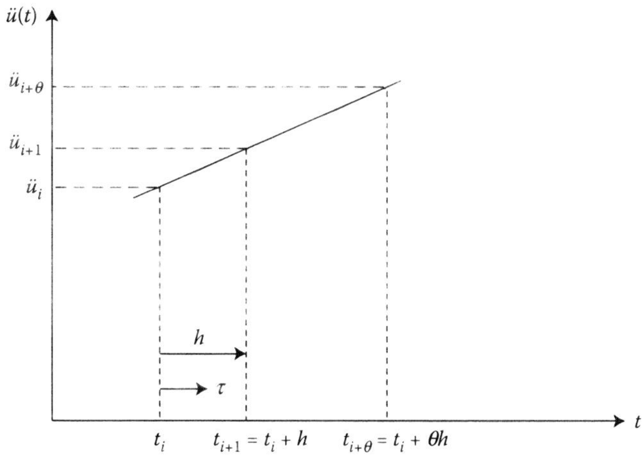


**FIGURE 3.12**
Wilson-theta method.

$h/T = 0.5513$ to ensure that the solutions are bounded. The Wilson-theta method is a modification of the linear acceleration method where the acceleration is assumed to vary in a linear manner, as shown in Figure 3.12, from time $t_i$ to time $t_{i+\theta} = t_i + \theta h$, where $\theta \geq 1.0$. The parameter $\theta$ is selected to yield the desired characteristics of stability and accuracy. This is an implicit scheme and does not require any special starting procedures.

Using dynamic equilibrium equations at time $t_{i+\theta} = t_i + \theta h$, Equation 3.1 gives

$$m\ddot{u}_{i+\theta} + \eta\dot{u}_{i+\theta} + ku_{i+\theta} = p_{i+\theta} \tag{3.86}$$

where

$$p_{i+\theta} = p_i + \theta(p_{i+1} - p_i) \tag{3.87}$$

Since $\ddot{u}(t)$ varies linearly between $t_i$ and $t_{i+\theta}$, as shown in Figure 3.12, the linear change in acceleration $\ddot{u}(t)$, at any time $t_i + \tau$ where $0 \leq \tau \leq \theta h$, can be expressed as

$$\ddot{u}(t_i + \tau) = \ddot{u}_i + \frac{\tau}{\theta h}(\ddot{u}_{i+\theta} - \ddot{u}_i) \tag{3.88}$$

Integrating over time variable $\tau$ gives

$$\dot{u}(t_i + \tau) = \dot{u}_i + \ddot{u}_i\tau + \frac{\tau^2}{2\theta h}(\ddot{u}_{i+\theta} - \ddot{u}_i) \tag{3.89}$$

Integrating again yields the displacement as

$$u(t_i + \tau) = u_i + \dot{u}_i\tau + \frac{\tau^2}{2}\ddot{u}_i + \frac{\tau^3}{6\theta h}(\ddot{u}_{i+\theta} - \ddot{u}_i) \tag{3.90}$$

We now replace the variable $\tau$ with $\theta h$, thus, Equations 3.89 and 3.90 become

$$\dot{u}_{i+\theta} = \dot{u}_i + \frac{\theta h}{2}(\ddot{u}_{i+\theta} + \ddot{u}_i) \tag{3.91}$$

$$u_{i+\theta} = u_i + \theta h \dot{u}_i + \frac{(\theta h)^2}{6}(\ddot{u}_{i+\theta} + 2\ddot{u}_i) \tag{3.92}$$

Equations 3.91 and 3.92 can be solved for $\ddot{u}_{i+\theta}$ and $\dot{u}_{i+\theta}$ in terms of $u_{i+\theta}$. Thus, from Equation 3.92, we have

$$\ddot{u}_{i+\theta} = \frac{6}{(\theta h)^2}(u_{i+\theta} - u_i) - \frac{6}{(\theta h)}\dot{u}_i - 2\ddot{u}_i \tag{3.93}$$

and substituting Equation 3.93 into Equation 3.91 we have for the velocity at $t_{i+\theta}$

$$\dot{u}_{i+\theta} = \frac{3}{\theta h}(u_{i+\theta} - u_i) - 2\dot{u}_i - \frac{\theta h}{2}\ddot{u}_i \tag{3.94}$$

An expression in terms of $u_{i+\theta}$ can be obtained by substituting Equations 3.93 and 3.94 into 3.86. Hence, substituting and collecting terms yields

$$\begin{aligned} \left[\frac{6}{(\theta h)^2}m + \frac{3}{(\theta h)}\eta + k\right]u_{i+\theta} &= p_i + \theta(p_{i+1} - p_i) \\ &+ m\left[\frac{6}{(\theta h)^2}u_i + \frac{6}{(\theta h)}\dot{u}_i + 2\ddot{u}_i\right] \\ &+ \eta\left[\frac{3}{(\theta h)}u_i + 2\dot{u}_i + \frac{\theta h}{2}\ddot{u}_i\right] \end{aligned} \tag{3.95}$$

or

$$\begin{aligned} \left[\frac{6}{(\theta h)^2}m + \frac{3}{(\theta h)}\eta + k\right]u_{i+\theta} &= p_i + \theta(p_{i+1} - p_i) + \left[\frac{6}{(\theta h)^2}m + \frac{3}{(\theta h)}\eta\right]u_i \\ &+ \left[\frac{6}{(\theta h)}m + 2\eta\right]\dot{u}_i + \left[2m + \frac{\theta h}{2}\eta\right]\ddot{u}_i \end{aligned} \tag{3.96}$$

After determining $u_{i+\theta}$ from Equations 3.95 and 3.96, it can be substituted into Equation 3.93 yielding $\ddot{u}_{i+\theta}$, this is then in turn used in Equations 3.90, evaluated at $\tau = h$ to give the acceleration at time $t_{i+1}$.

$$\ddot{u}_{i+1} = \frac{6}{\theta^3 h^2}(u_{i+\theta} - u_i) - \frac{6}{\theta^2 h}\dot{u}_i + \left(1 - \frac{3}{\theta}\right)\ddot{u}_i \tag{3.97}$$

The velocity can be calculated at time $t_{i+1}$ knowing that the acceleration is linear in time or using Equation 3.93 with $\theta = 1$ leads to

$$\dot{u}_{i+1} = \dot{u}_i + \frac{h}{2}(\ddot{u}_{i+1} + \ddot{u}_i) \tag{3.98}$$

Substituting Equation 3.97 yields

$$\dot{u}_{i+1} = \left(1 - \frac{3}{\theta^2}\right)\dot{u}_i + \left(1 - \frac{3}{2\theta}\right)h\ddot{u}_i + \frac{3}{\theta^3 h}(u_{i+\theta} - u_i) \tag{3.99}$$

For the displacement, we have Equation 3.94 with $\theta = 1$

$$u_{i+1} = u_i + h\dot{u}_i + \frac{h^2}{6}(\ddot{u}_{i+1} + 2\ddot{u}_i) \tag{3.100}$$

or again using Equation 3.97

$$u_{i+1} = u_i + \left(1 - \frac{1}{\theta^2}\right)h\dot{u}_i + \left(1 - \frac{1}{\theta}\right)\frac{h^2}{2}\ddot{u}_i + \frac{1}{\theta^3}(u_{i+\theta} - u_i) \tag{3.101}$$

For $\theta = 1$, the Wilson-theta method reduces to the linear acceleration method. In linear problems the method is unconditionally stable for $\theta \geq 1.37$, see Bathe and Wilson (1976). The value of $\theta = 1.4$ is usually used.

**EXAMPLE 3.7**

With the Wilson-theta method, calculate the displacement response of the SDOF system of Example 3.1.

**Solution**

Using the same values used in Example 3.1, the calculation would proceed as follows: first initialize $u_0$ and $\dot{u}_0$ and then calculate

$$\ddot{u}_0 = \frac{p_0}{m} - 2\zeta\omega_0\dot{u}_0 - \omega_0^2 u_0$$

Using a time step of 0.01 s, the displacement is calculated from Equation 3.100 and then the velocity and acceleration from Equations 3.99 and 3.100, respectively. The results for the displacement and velocity are given in Figure 3.13.

The algorithm for using the Wilson-theta method is given in Table 3.7.

## 3.6 HHT-Alpha Method

The HHT method (also called the $\alpha$ method) was introduced by Hilber et al. (1977) as a generalization of the Newmark beta method to control dissipation of the higher

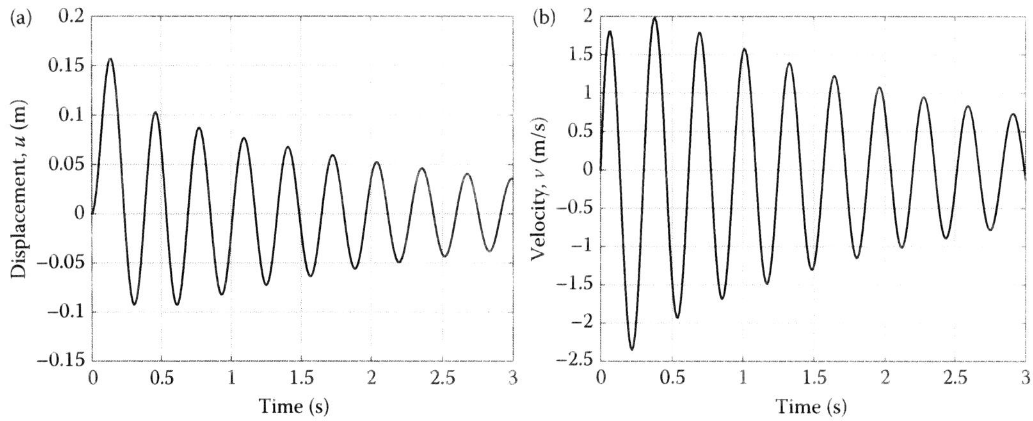


**FIGURE 3.13**
Wilson-theta method using time step 0.01 s. (a) Displacement response and (b) velocity response.

**TABLE 3.7**

Algorithm for Wilson's Theta Method

*Initial Computations:*

1. Specify $m$, $\eta$, and $k$
2. Set the initial conditions $u(0) = u_0$ and $\dot{u}(0) = \dot{u}_0$ and calculate

$$\ddot{u}_0 = \frac{1}{m}[p_0 - \eta\dot{u}_0 - ku_0]$$

3. Specify $\theta$, $\theta \geq 1.0$, usually $\theta = 1.4$
4. Form

$$\hat{k} = \frac{6}{\theta h^2}m + \frac{3}{\theta h}\eta + k$$

5. $a = \frac{6}{(\theta h)^2}m + \frac{3}{\theta h}\eta$, $b = \frac{6}{\theta h}m + 2\eta$ and $c = 2m + \frac{\theta h}{2}\eta$

$$d = \frac{6}{\theta^3 h^2},\ e = -\frac{6}{\theta^2 h} \text{ and } f = \left(1 - \frac{3}{\theta}\right)$$

*Calculate for each time step:*

6. $\hat{p}_{i+\theta} = p_i + \theta(p_{i+1} - p_i) + au_i + b\dot{u}_i + c\ddot{u}_i$
7. $u_{i+\theta} = \frac{\hat{p}_{i+\theta}}{\hat{k}}$
8. $\ddot{u}_{i+1} = d(u_{i+\theta} - u_i) + e\dot{u}_i + f\ddot{u}_i$
9. $\dot{u}_{i+1} = \dot{u}_i + \frac{h}{2}(\ddot{u}_{i+1} + \ddot{u}_i)$
10. $u_{i+1} = u_i + h\dot{u}_i + \frac{h^2}{6}(\ddot{u}_{i+1} + 2\ddot{u}_i)$

frequency modes. It is an implicit technique. This method uses the discrete-time equation of motion in the following form:

$$m\ddot{u}_{i+1} + \eta\dot{u}_{i+1} + ku_{i+1} + \alpha[\eta(\dot{u}_{i+1} - \dot{u}_i) + k(u_{i+1} - u_i)] = (1+\alpha)p_{i+1} - \alpha p_i \tag{3.102}$$

where $\alpha$ is the parameter that controls numerical damping. The method also uses the Newmark equations (3.70) and (3.69) for velocity and acceleration. Selecting $-1/3 \le \alpha \le 0$, $\gamma = 1/2 - \alpha$, and $\beta = (1-\alpha)^2/4$ results in an unconditionally stable second-order accuracy (Hughes 1987). It is noted that $\alpha = 0$ corresponds to the Newmark beta method.

Substituting Equations 3.69 and 3.70 into Equation 3.102 gives after collecting terms

$$\begin{aligned}\left[(1+\alpha)k + (1+\alpha)\frac{\gamma}{\beta h}\eta + \frac{1}{\beta h^2}m\right]u_{i+1} &= (1+\alpha)p_{i+1} - \alpha p_i + m\left[\frac{1}{\beta h^2}u_i + \frac{1}{\beta h}\dot{u}_i + \left(\frac{1}{2\beta} - 1\right)\ddot{u}_i\right]\\ &+ \eta\left[(1+\alpha)\frac{\gamma}{\beta h}u_i + (1+\alpha)\left(\frac{\gamma}{\beta} - 1\right)\dot{u}_i + (1+\alpha)\left(\frac{\gamma}{2\beta} - 1\right)h\ddot{u}_i\right]\\ &+ \alpha\eta\dot{u}_i + \alpha k u_i\end{aligned} \tag{3.103}$$

or

$$\begin{aligned}\left[(1+\alpha)k + (1+\alpha)\frac{\gamma}{\beta h}\eta + \frac{1}{\beta h^2}m\right]u_{i+1} &= (1+\alpha)p_{i+1} - \alpha p_i + \left[\frac{m}{\beta h^2} + (1+\alpha)\frac{\gamma}{\beta h}\eta + \alpha k\right]u_i\\ &+ \left[\frac{1}{\beta h}m + \left\{(1+\alpha)\left(\frac{\gamma}{\beta} - 1\right) + \alpha\right\}\right]\dot{u}_i\\ &+ \left[\left(\frac{1}{2\beta} - 1\right)m + (1+\alpha)\left(\frac{\gamma}{2\beta} - 1\right)h\eta\right]\ddot{u}_i\end{aligned} \tag{3.104}$$

Equations 3.103 and 3.104 are used to calculate the displacement $u_{i+1}$ and Equations 3.69 and 3.70 for $\ddot{u}_{i+1}$ and $\dot{u}_{i+1}$. The algorithm for using the HHT-$\alpha$ method is given in Table 3.8.

**EXAMPLE 3.8**

With the HHT-$\alpha$ method, calculate the displacement response of the SDOF system of Example 3.1.

**Solution**

Using the same values used in Example 3.1, the calculation would proceed as follows: first initialize $u_0$ and $\dot{u}_0$ and then calculate

$$\ddot{u}_0 = \frac{p_0}{m} - 2\zeta\omega_0\dot{u}_0 - \omega_0^2 u_0$$

Using a time step of 0.01 s, the displacement is calculated from Equations 3.103 and 3.104 and then the velocity and acceleration from Equations 3.67 and 3.70, respectively. The results for the displacement and velocity are given in Figure 3.14.

**TABLE 3.8**

Algorithm for HHT-$\alpha$ Method

| |
|---|
| *Initial Computations:* |
| 1. Specify $m$, $\eta$, and $k$. |
| 2. Set the initial conditions $u(0) = u_0$ and $\dot{u}(0) = \dot{u}_0$ and calculate $$\ddot{u}_0 = \frac{1}{m}[p_0 - \eta\dot{u}_0 - ku_0]$$ |
| 3. Specify the integration method:<br>Newmark average acceleration method $\gamma = 1/2$, $\beta = 1/4$, $\alpha = 0$<br>Newmark linear acceleration method $\gamma = 1/2$, $\beta = 1/6$, $\alpha = 0$<br>HHT-$\alpha$ method $\gamma = 1/2 - \alpha$, $\beta = (1 - \alpha)^2/4$, $-1/3 \le \alpha \le 0$ |
| 4. Select time step $h$. |
| 5. Form $$\hat{k} = \frac{1}{\beta h^2}m + (1+\alpha)\frac{\gamma}{\beta h}\eta + (1+\alpha)k$$ |
| 6. $a = \frac{1}{\beta h^2}m + (1+\alpha)\frac{\gamma}{\beta h}\eta + \alpha k, b = \frac{1}{\beta h}m + \left\{(1+\alpha)\left(\frac{\gamma}{\beta} - 1\right) + \alpha\right\}, c = \left(\frac{\gamma}{\beta} - 1\right)m + (1+\alpha)\left(\frac{\gamma}{\beta} - 1\right)h\eta$ |
| 7. $d = \frac{1}{\beta h^2}, e = -\frac{1}{\beta h}, f = -\left(\frac{1}{2\beta} - 1\right)$ |
| *Calculate for each time step:* |
| 8. $\hat{p}_{i+1} = (1+\alpha)p_{i+1} - \alpha p_i + au_i + b\dot{u}_i + c\ddot{u}_i$ |
| 9. $u_{i+1} = \frac{\hat{p}_{i+1}}{\hat{k}}$ |
| 10. $\ddot{u}_{i+1} = d(u_{i+1} - u_i) + e\dot{u}_i + f\ddot{u}_i$ |
| 11. $\dot{u}_{i+1} = \dot{u}_i + (1-\gamma)h\ddot{u}_i + \gamma h\ddot{u}_{i+1}$ |

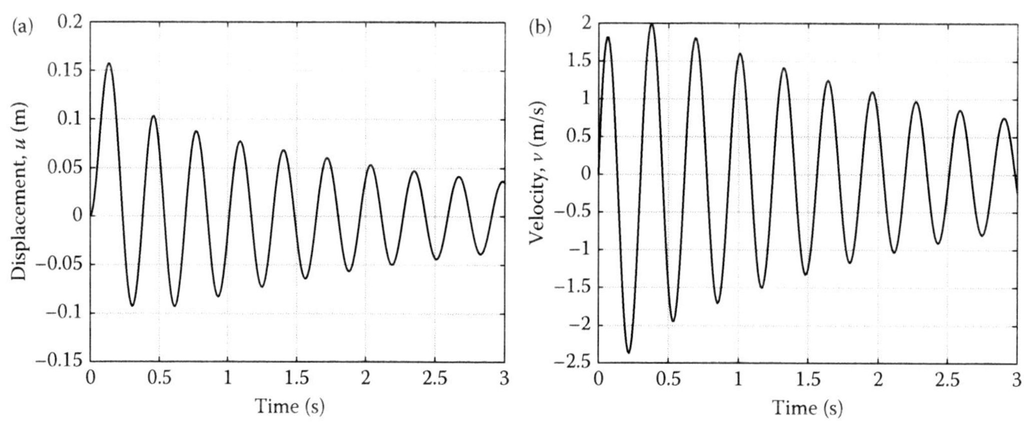


**FIGURE 3.14**
HHT-$\alpha$ method using time step 0.01 s. (a) Displacement response and (b) velocity response.

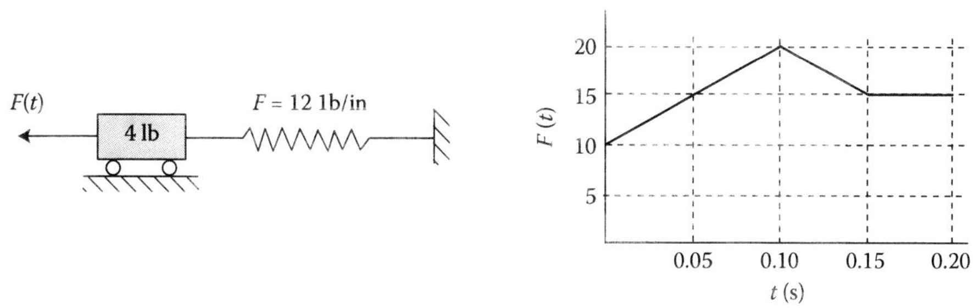


**FIGURE 3.15**
Spring–mass system with an applied force.

## PROBLEMS

3.1 A 4 lb cart is attached to a 12 lb/in spring starts from rest and is subjected to an applied force *F(t)* as illustrated in Figure 3.15. The equation of motion for this system is $m\ddot{x}+kx=F(t)$. Using Euler finite differences, solve the differential equation above and note the position, velocity, and acceleration at various time intervals up to 0.2 s. Assume a time interval of $\Delta t = T/10$.

3.2 Repeat Problem 3.1 using $\Delta t = T = 20 = 0.01$. Compare this solution to that of Problem 3.1 by plotting the displacements, velocities, and accelerations for both cases.

3.3 A spring–mass system is governed by the equation $0.2\ddot{x}+32x-5=100t$, where $0 \leq t \leq 1$ s. Assume the system starts from rest, the time interval is $\Delta t = T/10$, and determine the system response using the Euler finite difference method for $0 \leq t \leq 0.5$ s.

3.4 Repeat Problem 3.3 using $\Delta t = T = 20 = 0.025$. Compare this solution to that of Problem 3.3 by plotting the displacements, velocities, and accelerations for both cases.

3.5 A dynamic system is governed by the equation $\ddot{u}+10\dot{u}+250u=F(t)$, where *F(t)* is shown in Figure 3.16. Assume the system starts from rest, the time interval is $\Delta t = T/10$, and plot the position and velocity using the Euler finite difference method for $0 \leq t \leq 0.5$ s.

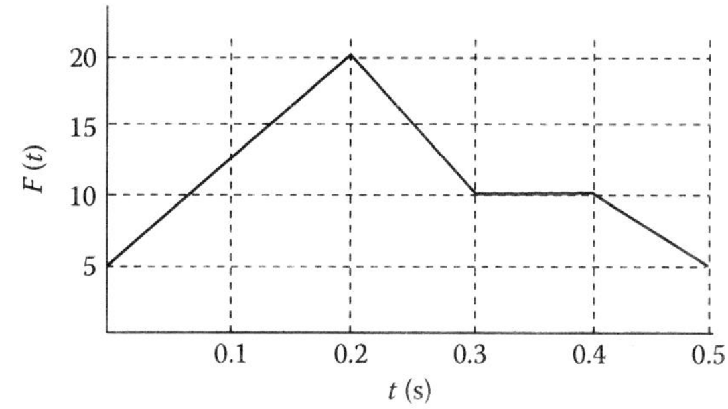


**FIGURE 3.16**
Forcing function for a dynamic system.

3.6 Solve Problem 3.1 using the Runge–Kutta method with $h = 0.02$. Plot position, velocity, and acceleration as a function of time using the Runge–Kutta method and compare it to the solutions obtained in Problem 3.1.

3.7 Solve Problem 3.3 using the Runge–Kutta method with $h = 0.05$. Plot position, velocity, and acceleration as a function of time using the Runge–Kutta method and compare it to the solutions obtained in Problem 3.3.

3.8 A beam of mass $m$ is supported by two identical square steel columns of length $L$. The beam is acted on by a time varying force $P(t)$. The equation of motion for this system is $m\ddot{u} + ((6EI)/L^3)u = P(t)$. The dimension of each column is $a \times a$, and the overall dimensions of the system are such that the equation of motion can be written as $1800\ddot{u} + 3.7\times10^9 a^4 u = 50t$. The application of this system is such that the horizontal displacement, $u$, must not exceed 25 mm. We have available column sizes of $a = 25$ mm, 38 mm, 50 mm and want to approximate the size of the column. Use the Newmark beta method to determine the displacement of this system and approximate the required column size so that the allowable horizontal displacement is not exceeded. Assume $h = 0.2$, $\gamma = 1/2$, $\beta = 1/6$, and that the system starts from rest with the applied load duration being 6 s.

3.9 Assume a single-degree-of-freedom system starts from rest and is defined by $6\ddot{u} + 3.6\dot{u} + 54u = 360t$. Use the Newmark beta method to determine the response (displacement, velocity, and acceleration) of this system. Assume $h = \Delta t = 0.1\,T$, $\gamma = 1/2$, and $\beta = 1/6$. Plot the responses above $0 \geq t \leq 5$ s.

3.10 Work Problem 3.3 using the Newmark beta method with $\gamma = 1/2$, $\beta = 1/6$, and $h = \Delta t = 0.1T$. Compare the results by plotting the displacement for both cases.

## References

Bathe, K. J. and Wilson, E. L. 1976. *Numerical Methods in Finite Element Analysis*. Englewood Cliffs, NJ: Prentice-Hall, Inc.

Burden, R. L. and Faires, J. D. 2001. *Numerical Analysis* 9th Ed. Brooks/Cole, Boston, MA.

Butcher, J. C. 1996. A history of Runge–Kutta methods. *Appl. Numer. Math.*, 20: 247–260.

Hilber, H. M., Hughes, T. J. R., and Talor, R. L. 1977. Improved numerical dissipation for time integration algorithms in structural dynamics. *Earthquake Eng. Struct Dyn.*, 5: 282–292.

Hughes, T. J. R. 1987. *The Finite Element Method-Linear Static and Dynamic Finite Element Analysis.* Englewood Cliffs, NJ: Prentice Hall.

LeVeque, R. J. 2007. *Finite Difference Methods for Ordinary and Partial Differential Equations.* Philadelphia, PA: SIAM.

Newmark, N. M. 1959. A method of computation for structural dynamics. *J. Eng. Mech. Div. ASCE*, 85: 67–94.

# 4

# Systems with Several Degrees of Freedom

## 4.1 Introduction

The previous chapters analyzed simple one-degree-of-freedom systems, where only one coordinate was needed to describe their motion. Multi-degree-of-freedom (MDOF) systems need two or more coordinates to describe their motion (Balachandran and Magrab 2009; Schmitz and Smith 2012). Continuous systems are not included since they require, theoretically, an infinite number of degrees of freedom. However, on many applications continuous systems are approximated as having a finite number of degrees of freedom.

Multiple-degree-of-freedom systems are identified by the number of independent variables required to completely define the motion of the system. A two-degree-of-freedom system involves two variables such as $u_1(t)$ and $u_2(t)$, or $\phi_1(t)$ and $\phi_2(t)$, or $u(t)$ and $v(t)$, or a similar combination. A three-degree-of-freedom system requires three variables such as $u(t)$, $v(t)$, and $\phi(t)$, or $u(t)$, $v(t)$, and $w(t)$, or a similar combination. Examples of two, three, and six-degree-of-freedom (three displacements and three rotations) systems are shown in Figure 4.1a–c.

Solving a multiple-degree-of-freedom system requires:

1. Developing the equations of motion
2. Solving for the free vibrations
3. Solving for the forced vibrations
4. Calculating the frequency response function
5. Calculating the response to a unit impulse
6. Calculating the response for a random loading
7. Taking into account any material damping

## 4.2 Equations of Motion

The equations of motion for a system of particles are developed using Newton's law for each particle. Defining

$$\mathbf{F}^{(i)} = \text{force vector of the } i\text{th particle}$$

$$\mathbf{u}^{(i)} = \text{displacement vector of the } i\text{th particle}$$

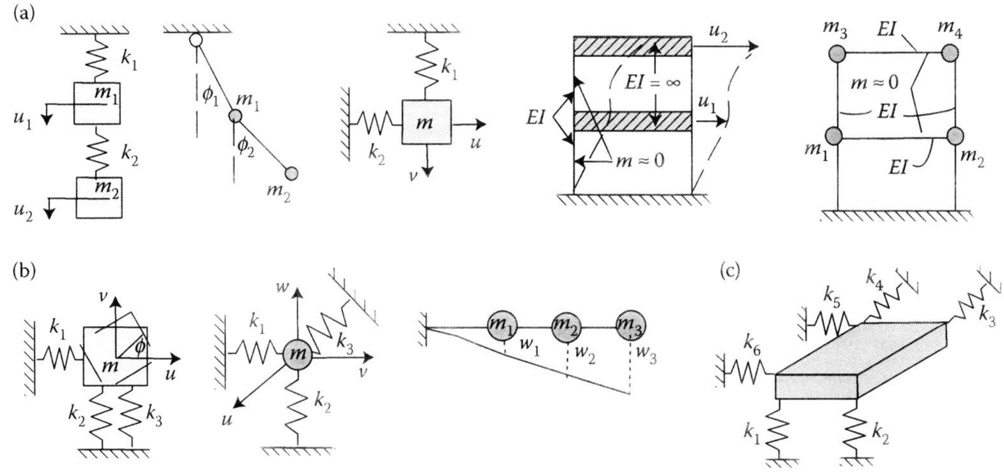


**FIGURE 4.1**
(a) Two-degrees-of-freedom, (b) three-degree-of-freedom systems, and (c) six-degree-of-freedom system.

$$m^{(i)} = \text{mass of the } i\text{th particle}$$

We can write the vector equation

$$\mathbf{F}^{(i)} = m^{(i)} \frac{d^2\mathbf{u}^{(i)}}{dt^2} \tag{4.1}$$

The vector equation represents three scalar equations, which can be written as

$$F_x^{(i)} = m^{(i)} \frac{d^2u_x^{(i)}}{dt^2}, \quad F_y^{(i)} = m^{(i)} \frac{d^2u_y^{(i)}}{dt^2}, \quad F_z^{(i)} = m^{(i)} \frac{d^2u_z^{(i)}}{dt^2} \tag{4.2}$$

If we are dealing with a rigid body or a rigid system of particles, we have two vector equations

$$\mathbf{F} = m\frac{d^2\mathbf{u}}{dt^2} \quad \text{and} \quad \mathbf{M} = \frac{d\mathbf{L}}{dt} \tag{4.3}$$

where **L** is the angular momentum vector whose components are given as

$$\begin{aligned} L_x &= I_{xx}\Psi_x + I_{xy}\Psi_y + I_{xz}\Psi_z \\ L_y &= I_{yx}\Psi_x + I_{yy}\Psi_y + I_{yz}\Psi_z \\ L_z &= I_{zx}\Psi_x + I_{zy}\Psi_y + I_{zz}\Psi_z \end{aligned} \tag{4.4}$$

Here, $\Psi_x$, $\Psi_y$, and $\Psi_z$ are angular velocities about the $x$, $y$, and $z$ axes, respectively, and $I_{xx}$, $I_{xy}$, $I_{xz}$, … are the moments and products of inertia. If $\phi_x$, $\phi_y$, and $\phi_z$ are rotations about the $x$, $y$, and $z$ axes, respectively, then for small rotations

$$\Psi_x = \frac{d\phi_x}{dt}, \quad \Psi_y = \frac{d\phi_y}{dt} \quad \text{and} \quad \Psi_z = \frac{d\phi_z}{dt} \tag{4.5}$$

For a system of $n$ particles, with each of the $i$th particles having coordinates $x_i$, $y_i$, $z_i$, we define the moments and products of inertia as

$$\begin{aligned} I_{xx} &= \sum_{i=1}^{n} m_i\left(y_i^2 + z_i^2\right), \quad I_{yy} = \sum_{i=1}^{n} m_i\left(x_i^2 + z_i^2\right), \quad I_{zz} = \sum_{i=1}^{n} m_i\left(x_i^2 + y_i^2\right) \\ I_{xy} &= \sum_{i=1}^{n} m_i x_i y_i, \quad I_{xz} = \sum_{i=1}^{n} m_i x_i z_i, \quad I_{yz} = \sum_{i=1}^{n} m_i y_i z_i \end{aligned} \tag{4.6}$$

For a rigid body of volume $V$ and density $\rho$, we have for the moments and products of inertia

$$\begin{aligned} I_{xx} &= \int_V \rho(y^2 + z^2)dV, \quad I_{yy} = \int_V \rho(x^2 + z^2)dV, \quad I_{zz} = \int_V \rho(x^2 + y^2)dV \\ I_{xy} &= \int_V \rho xy\, dV, \quad I_{xz} = \int_V \rho xz\, dV, \quad I_{yz} = \int_V \rho yz\, dV \end{aligned} \tag{4.7}$$

Figure 4.2 shows a two-degree-of-freedom spring mass system subjected to forces $P_1$ and $P_2$. The displacement of mass $m_1$ is $u_1$ and $m_2$ is $u_2$.

The equations of motion (one for each mass) will contain four unknowns and are written as

$$P_1 - Z_1 = m_1 \frac{d^2u_1}{dt^2} \quad \text{and} \quad P_2 - Z_2 = m_2 \frac{d^2u_2}{dt^2} \tag{4.8}$$

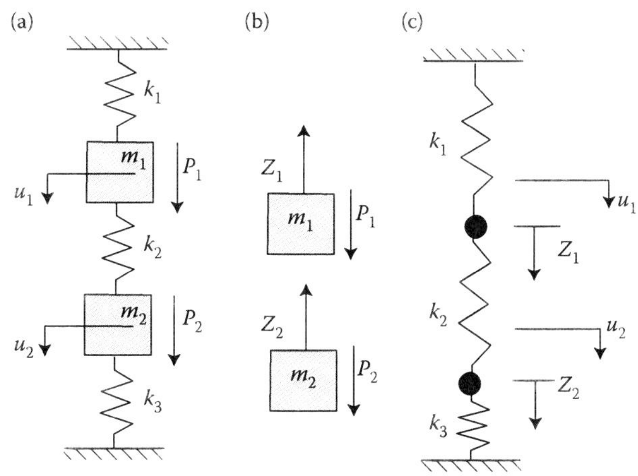


**FIGURE 4.2**
(a) Two-degree-of-freedom spring mass model, (b) free-body of masses, (c) spring deformation.

where $Z_i(i=1,2)$ are the forces exerted by the masses $m_i(i=1,2)$ on the spring arrangement at points $i=1$, 2. Two additional equations that involve the deformation or displacement for the spring arrangement are

$$u_1 = f_{11}Z_1 + f_{12}Z_2, \quad u_2 = f_{21}Z_1 + f_{22}Z_2 \tag{4.9}$$

The coefficients are

$$f_{11} = u_1 \text{ if } Z_1 = 1, Z_2 = 0, \quad f_{12} = u_1 \text{ if } Z_1 = 0, Z_2 = 1$$
$$f_{21} = u_2 \text{ if } Z_1 = 1, Z_2 = 0, \quad f_{22} = u_2 \text{ if } Z_1 = 0, Z_2 = 1$$

where $f_{ij}$ are the *compliance* (or *flexibility*) coefficients. The flexibility coefficients in the matrix form are given as

$$[F] = \begin{bmatrix} f_{11} & f_{12} \\ f_{21} & f_{22} \end{bmatrix} \tag{4.10}$$

Solving Equation 4.8 for $Z_1$ and $Z_2$ substituting into Equation 4.9 results in

$$\begin{aligned} f_{11}m_1\frac{d^2u_1}{dt^2} + f_{12}m_2\frac{d^2u_2}{dt^2} + u_1 &= f_{11}P_1 + f_{12}P_2 \\ f_{21}m_1\frac{d^2u_1}{dt^2} + f_{22}m_2\frac{d^2u_2}{dt^2} + u_1 &= f_{21}P_1 + f_{22}P_2 \end{aligned} \tag{4.11}$$

In matrix format, Equation 4.12 becomes

$$\begin{bmatrix} f_{11} & f_{12} \\ f_{21} & f_{22} \end{bmatrix}\begin{bmatrix} m_1 & 0 \\ 0 & m_2 \end{bmatrix}\begin{Bmatrix} \ddot{u}_1 \\ \ddot{u}_2 \end{Bmatrix} + \begin{Bmatrix} u_1 \\ u_2 \end{Bmatrix} = \begin{bmatrix} f_{11} & f_{12} \\ f_{21} & f_{22} \end{bmatrix}\begin{Bmatrix} P_1 \\ P_2 \end{Bmatrix} \tag{4.12}$$

or

$$[F][M]\ddot{\mathbf{u}} + \mathbf{u} = [F]\mathbf{P} \tag{4.13}$$

where $[F]$ is the flexibility matrix, $[M]$ the mass matrix, $\ddot{\mathbf{u}}$ and $\mathbf{u}$ the acceleration and displacement vector, respectively, and $\mathbf{P}$ the load vector. As an alternate solution, we can keep Equation 4.8 as written and rewrite Equation 4.9 in terms of load as

$$Z_1 = k_{11}u_1 + k_{12}u_2 \quad \text{and} \quad Z_2 = k_{21}u_1 + k_{22}u_2 \tag{4.14}$$

where the coefficients are established as illustrated in Figure 4.3, where axial deflections are shown in the lateral direction for convenience. These coefficients are

$$k_{11} = Z_1 \text{ if } u_1 = 1 \text{ and } u_2 = 0, \quad k_{12} = Z_1 \text{ if } u_1 = 0 \text{ and } u_2 = 1$$
$$k_{21} = Z_2 \text{ if } u_1 = 1 \text{ and } u_2 = 0, \quad k_{22} = Z_2 \text{ if } u_1 = 0 \text{ and } u_2 = 1$$

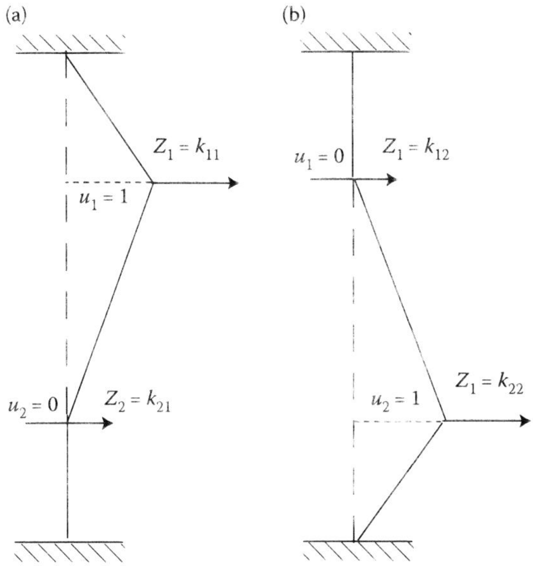


**FIGURE 4.3**
(a) Mass deflection with $u_1 = 1$ and $u_2 = 0$ and (b) mass deflection with $u_1 = 0$ and $u_2 = 1$.

In this approach, $k_{ij}$ are the stiffness coefficients and $[K]$ is the stiffness matrix. Note that the stiffness matrix is the inverse of the flexibility or compliance matrix. That is,

$$[K] = \begin{bmatrix} k_{11} & k_{12} \\ k_{21} & k_{22} \end{bmatrix} = \begin{bmatrix} f_{11} & f_{12} \\ f_{21} & f_{22} \end{bmatrix}^{-1} = [F]^{-1} \tag{4.15}$$

If we substitute Equation 4.14 into Equation 4.8, we obtain

$$m_1 \frac{d^2 u_1}{dt^2} + k_{11}u_1 + k_{12}u_2 = P_1$$
$$m_2 \frac{d^2 u_2}{dt^2} + k_{21}u_1 + k_{22}u_2 = P_2$$

In the matrix format, the above equation becomes

$$\begin{bmatrix} m_1 & 0 \\ 0 & m_2 \end{bmatrix} \begin{Bmatrix} \ddot{u}_1 \\ \ddot{u}_2 \end{Bmatrix} + \begin{bmatrix} k_{11} & k_{12} \\ k_{21} & k_{22} \end{bmatrix} \begin{Bmatrix} u_1 \\ u_2 \end{Bmatrix} = \begin{Bmatrix} P_1 \\ P_2 \end{Bmatrix} \tag{4.16}$$

or

$$[M]\ddot{\mathbf{u}} + [K]\mathbf{u} = \mathbf{P} \tag{4.17}$$

where again $[M]$ is the mass matrix, $\ddot{\mathbf{u}}$ and $\mathbf{u}$ the acceleration and displacement vector, respectively, $\mathbf{P}$ the load vector, and $[K]$ is the stiffness matrix.

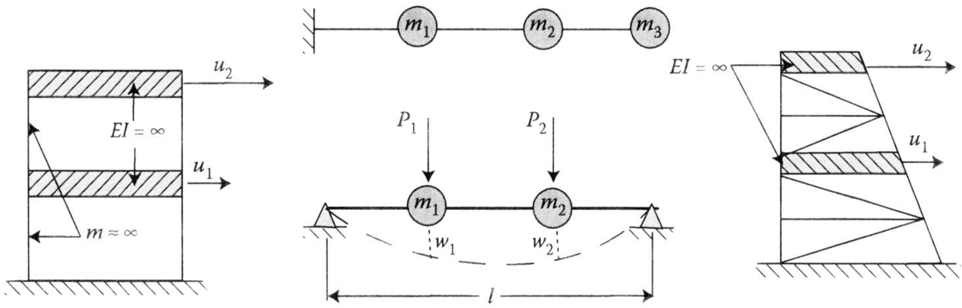


**FIGURE 4.4**
Typical structural systems where the formulation of the problem can be performed using the direct or displacement method.

In both approaches, it is noted that the equation and the motion of the masses are coupled. That is the motion of mass $m_1$ depends on the displacement of mass $m_2$ and vice versa. This method is applicable to systems such as those illustrated in Figure 4.4.

**EXAMPLE 4.1**

Consider a frame with a mass at the end as shown in Figure 4.5. Determine the equations of motion for this system and express them in the matrix form.

**Solution**

For the vertical and horizontal directions, we write

$$-X = m\frac{d^2u}{dt^2}, \quad P - y = m\frac{d^2v}{dt^2}$$

The displacements can be written as $u = f_{xx}X + f_{xy}Y$ and $v = f_{yx}X + f_{yy}Y$, where $f_{ij} = \int_0^l (M_iM_j/EI)dx$. Knowing that $f_{xx} = u$ if $X = 1$, $Y = 0$, $f_{xy} = u$ if $X = 0$, $Y = 1$, $f_{yx} = v$ if $X = 1$, $Y = 0$ and $c_{yy} = v$ if $X = 0$, $Y = 1$, we can substitute and obtain

$$f_{xx}m\frac{d^2u}{dt^2} + f_{xy}m\frac{d^2v}{dt^2} + u = f_{xx}P \quad \text{and} \quad f_{yx}m\frac{d^2u}{dt^2} + f_{yy}m\frac{d^2v}{dt^2} + v = f_{yy}P$$

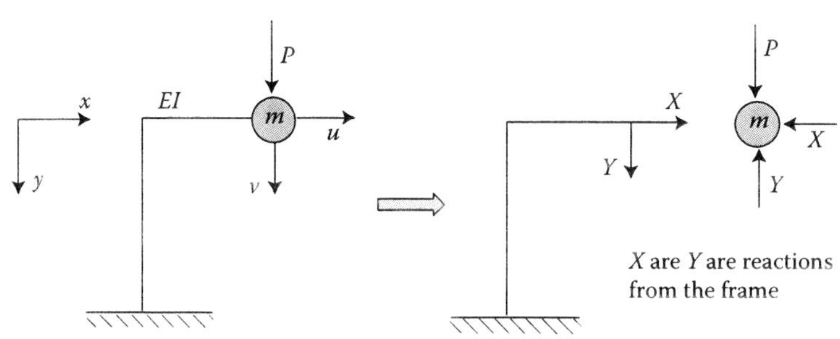


**FIGURE 4.5**
Frame with tip mass $m$ subjected to vertical force $P$.

or in the matrix format

$$\begin{bmatrix} f_{xx} & f_{xy} \\ f_{yx} & f_{yy} \end{bmatrix} \begin{bmatrix} m & 0 \\ 0 & m \end{bmatrix} \begin{Bmatrix} \ddot{u} \\ \ddot{v} \end{Bmatrix} + \begin{Bmatrix} u \\ v \end{Bmatrix} = \begin{Bmatrix} f_{xx} \\ f_{yy} \end{Bmatrix} P$$

**EXAMPLE 4.2**

Consider the three-degree-of-freedom system shown in Figure 4.6. Displacements $u$ and $v$, and rotation $\varphi$ are the three variables required. The springs are assumed to be linear ($F = k\delta$). Determine the equations of motion for this system and express them in the matrix form.

**Solution**

The forces acting on the structure are forces $-X$, $-Y$, and a restoring moment $-M$. It should be noted that the two springs along the bottom are in parallel and can be replaced by an equivalent spring. The fundamental equations of motion are

$$-X = m\frac{d^2u}{dt^2}, \quad P - Y = m\frac{d^2v}{dt^2} \quad \text{and} \quad P\left(\frac{a}{2}\right) - M = I_{zz}\frac{d^2\varphi}{dt^2}$$

Additional equations are needed, so we assume the forces acting on the structure are expressible as linear functions of the displacement. Therefore,

$$X = k_{11}u + k_{12}v + k_{13}\varphi$$

$$Y = k_{21}u + k_{22}v + k_{23}\varphi$$

$$M = k_{31}u + k_{32}v + k_{33}\varphi$$

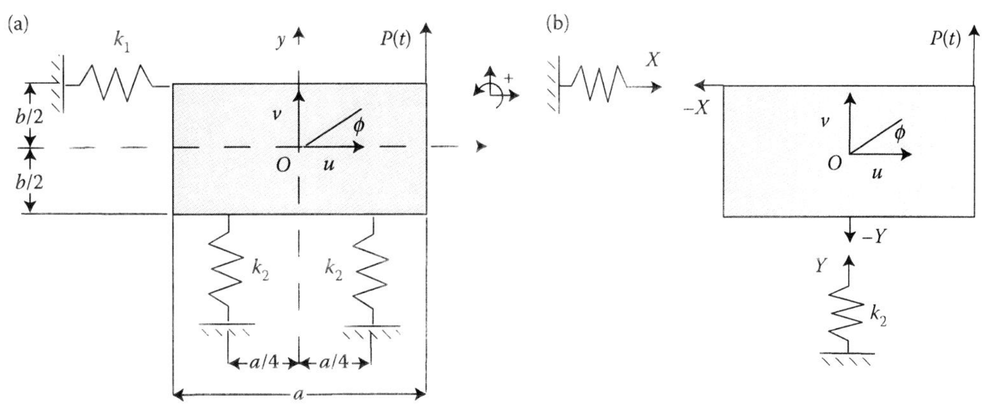


**FIGURE 4.6**
(a) Three-degree-of-freedom system $u$, $v$, $\phi$ and (b) forces on the supporting structure.

where

$k_{11}$ = force in the $x$-direction if $u=1, v=0, \varphi=0$ $\quad k_{11}=k_1(1)$

$k_{12}$ = force in the $x$-direction if $u=0, v=1, \varphi=0$ $\quad k_{12}=0$

$k_{13}$ = force in the $x$-direction if $u=0, v=0, \varphi=1$ $\quad k_{13}=-k_1\left(\frac{b}{2}\right)\varphi=-k_1\left(\frac{b}{2}\right)$

$k_{21}$ = force in the $y$-direction if $u=1, v=0, \varphi=0$ $\quad k_{21}=0$

$k_{22}=2k_2(1)$

$k_{23}$ = force in the $y$-direction if $u=0, v=0, \varphi=1$ $\quad k_{23}=0$

$k_{31}$ = moment if $u=1, v=0, \varphi=0$ $\quad k_{31}=-k_1b/2$

$k_{32}$ = moment if $u=0, v=1, \varphi=0$ $\quad k_{32}=0$

$k_{33}$ = moment if $u=0, v=0, \varphi=1$ $\quad k_{33}=2k_2\left(\frac{a}{4}\right)^2+k_1\left(\frac{b}{2}\right)^2$

In the matrix form, we can express these equations as

$$\begin{Bmatrix} X \\ Y \\ M \end{Bmatrix}=\begin{bmatrix} k_{11} & k_{12} & k_{13} \\ k_{21} & k_{22} & k_{23} \\ k_{31} & k_{32} & k_{33} \end{bmatrix}\begin{Bmatrix} u \\ v \\ \varphi \end{Bmatrix}=\begin{bmatrix} k_1 & 0 & -k_1(b/2) \\ 0 & 2k_2 & 0 \\ -k_1(b/2) & 0 & k_2a^2/8+k_1b^2/4 \end{bmatrix}\begin{Bmatrix} u \\ v \\ \varphi \end{Bmatrix}$$

Substituting into the equations of motion ($-X=m(d^2u/dt^2)$, $P-Y=m(d^2v/dt^2)$, and $P(a/2)-M=I_{zz}(d^2\varphi/dt^2)$) results in

$$m\frac{d^2u}{dt^2}+k_1u-k_1\left(\frac{b}{2}\right)\varphi=0$$

$$m\frac{d^2v}{dt^2}+2k_2v=P$$

$$I_{zz}\frac{d^2\phi}{dt^2}-k_1\left(\frac{b}{2}\right)u+\left[\frac{k_2a^2}{8}+\frac{k_1b^2}{4}\right]\varphi=-P\left(\frac{a}{2}\right)$$

In the matrix format, we have

$$\begin{bmatrix} m & 0 & 0 \\ 0 & m & 0 \\ 0 & 0 & I_z \end{bmatrix}\begin{Bmatrix} \ddot{u} \\ \ddot{v} \\ \ddot{\varphi} \end{Bmatrix}+\begin{bmatrix} k_1 & 0 & -k_1(b/2) \\ 0 & k_2 & 0 \\ -k_1(b/2) & 0 & k_2a^2/8+k_1b^2/4 \end{bmatrix}\begin{Bmatrix} u \\ v \\ \varphi \end{Bmatrix}=\begin{Bmatrix} 0 \\ 1 \\ -a/2 \end{Bmatrix}P$$

or

$$[M]\ddot{\mathbf{x}}+[K]\mathbf{x}=\mathbf{P}$$

## 4.3 Lagrange's Equations

Joseph L. C. Lagrange (1736–1813) showed that the equations of motion for a vibrating system could be developed using a notable simplification by using generalized coordinates and generalized forces (Wells 1967; O'Reilly 2008). Generalized coordinates are defined as the minimum number of independent coordinates necessary to define the configuration of the system that uniquely fix the position of the system in an arbitrary reference frame. Thus, generalized coordinates are equal in number to the degrees of freedom of the system. A single generalized coordinate is thus associated with each degree of freedom; hence a change in any one coordinate does not require a change in any other coordinate. Generalized coordinates do not have to have units of length, or even the same units as each other, they may be angles, area, or any other set of numbers which can uniquely describe the configuration of the system. We use the symbol $q_k$ to describe generalized coordinates which are independent parameters describing the state or configuration of the system. The generalized coordinates can be varied arbitrarily and independently without violating the constraints of the system. Holonomic constraints can be expressed algebraically as

$$\phi_j(q_1, q_2, \ldots, q_n, t) = 0, \quad j = 1, 2, \ldots, m \tag{4.18}$$

and for those which cannot are called *nonholonomic*. A system whose constraint equations, if any, are all of the holonomic form given in Equation 4.18 is called a *holonomic system*. Thus, for holonomic constraints, one can always find a set of independent generalized coordinates by eliminating $m$ coordinates to find $n-m$ independent generalized coordinates.

If the constraints are not dependent on time and are a function of the generalized coordinates alone, then we have *scleronomic constraints*. On the other hand, if the constraints are functions of time, then one has *rheonomic constraints* (Bauchau 2011).

This treatment of dynamical systems is formulated using scalar quantities of kinetic energy $T$, potential energy $V$, and work $W$. Therefore, the potential energy and kinetic energy will be a function of generalized coordinates and velocities. Thus

$$\begin{aligned} V &= V(q_1, q_2, \ldots, q_n) \\ T &= T(q_1, q_2, \ldots, q_n, \dot{q}_1, \dot{q}_2, \ldots, \dot{q}_n) \end{aligned} \tag{4.19}$$

Lagrange's equations for an arbitrary system of forces have the following form:

$$\frac{d}{dt}\left(\frac{\partial T}{\partial \dot{q}_k}\right) - \frac{\partial T}{\partial q_k} = Q_k \tag{4.20}$$

where $Q_k$ are the generalized forces. For potential forces, only our equation becomes

$$\frac{d}{dt}\left(\frac{\partial T}{\partial \dot{q}_k}\right) - \frac{\partial T}{\partial q_k} + \frac{\partial V}{\partial q_k} = 0 \tag{4.21}$$

However, for potential and nonpotential forces, we have

$$\frac{d}{dt}\left(\frac{\partial T}{\partial \dot{q}_k}\right) - \frac{\partial T}{\partial q_k} + \frac{\partial V}{\partial q_k} = Q_k \tag{4.22}$$

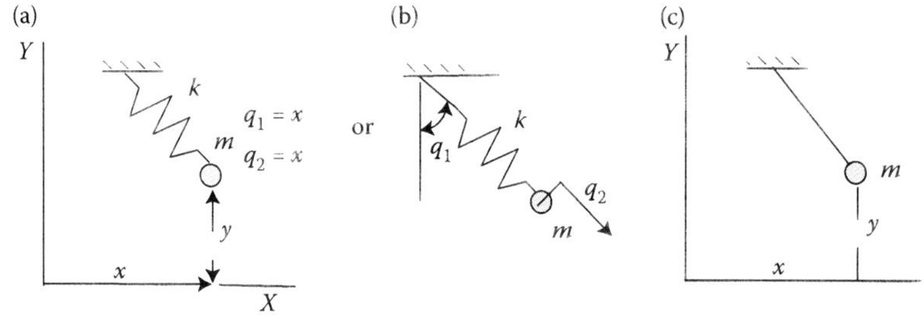


**FIGURE 4.7**
Generalized coordinates coincide with rectangular coordinates (a) and (b). (c) Rectangular coordinates may not be used as generalized coordinates.

Sometimes the generalized coordinates coincide with the rectangular coordinate system as illustrated in Figure 4.7a and b. A counter example would be a mass at the end of an inextensible cable, Figure 4.7c. In that case, there is only one generalized coordinate.

The kinetic energy $T$, of a system, can be defined for three possible systems as follows:

Particle with velocity $v$:

$$T = \frac{1}{2}mv^2 \tag{4.23}$$

System of $n$ particles with masses $m_i$:

$$T = \frac{1}{2}\sum_{i=1}^{n} m_i v_i^2 \tag{4.24}$$

Rigid body:

$$T = \frac{1}{2}\left(mv_{cg}^2 + I\dot{\varphi}_{cg}^2\right) \tag{4.25}$$

where $v_{cg}$ and $I_{cg}$ are the velocity of the center of gravity and moment of inertia about the center of gravity, respectively.

To calculate the kinetic energy in Cartesian coordinates, the velocity is given as

$$v^2 = \left(\frac{dx}{dt}\right)^2 + \left(\frac{dy}{dt}\right)^2 + \left(\frac{dz}{dt}\right)^2 = |\mathbf{v}|^2 \tag{4.26}$$

However, we may need to calculate the kinetic energy, not in $x, y, z$ but in some other system of coordinates. In order to do this, the Cartesian coordinates $x$, $y$, and $z$ can be expressed as functions of generalized coordinates $q_k$, thus

$$\begin{aligned} x &= x(q_1, q_2, \ldots, q_k) \\ y &= y(q_1, q_2, \ldots, q_k) \\ z &= z(q_1, q_2, \ldots, q_k) \end{aligned} \tag{4.27}$$

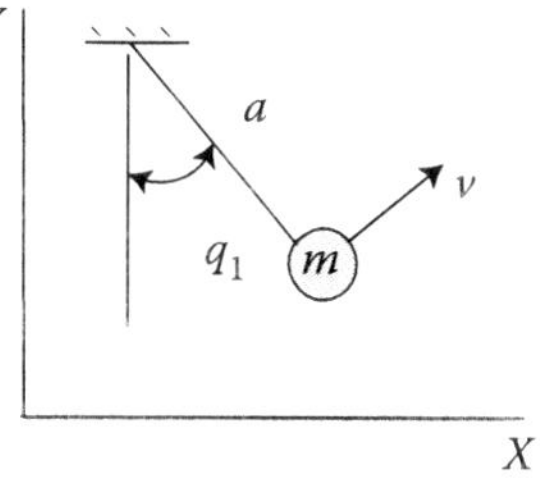


**FIGURE 4.8**
Mass $m$ at the end of a rod of length $a$ subjected to velocity $v$.

where the derivatives are given by

$$\begin{aligned}\frac{dx}{dt} &= \frac{\partial x}{\partial q_1}\dot{q}_1 + \frac{\partial x}{\partial q_2}\dot{q}_2 + \cdots + \frac{\partial x}{\partial q_k}\dot{q}_k \\ \frac{dy}{dt} &= \frac{\partial y}{\partial q_1}\dot{q}_1 + \frac{\partial y}{\partial q_2}\dot{q}_2 + \cdots + \frac{\partial y}{\partial q_k}\dot{q}_k \\ \frac{dz}{dt} &= \frac{\partial z}{\partial q_1}\dot{q}_1 + \frac{\partial z}{\partial q_2}\dot{q}_2 + \cdots + \frac{\partial z}{\partial q_k}\dot{q}_k\end{aligned} \tag{4.28}$$

Alternately, the kinetic energy $T$ may be determined directly as a function of $q_1, q_2, \ldots, q_k$. Consider a mass $m$ at the end of a rod of length $a$ as shown in Figure 4.8. The velocity of the mass and subsequent kinetic energy can be expressed as

$$\begin{aligned}v &= a\dot{q}_1 \\ T &= \frac{1}{2}mv^2 = \frac{1}{2}m(a\dot{q}_1)^2 = \frac{1}{2}ma^2\dot{q}_1^2\end{aligned} \tag{4.29}$$

### 4.3.1 Generalized Forces

In order to compute the generalized forces, $Q_k$, we begin by assuming we have actual force components, $F_i$, acting on the system that has its configuration defined in terms of generalized coordinates. The Cartesian coordinates are given by $x_i = (x_1, x_2, \ldots, x_n)$. Thus, the virtual work of the applied forces is given as

$$\delta W = \sum_{i=1}^{n} F_i \delta x_i \tag{4.30}$$

where $\delta x_i$ is an arbitrary virtual displacement which refers to a change in the system configuration as the result of an arbitrary infinitesimal change in the coordinates, consistent with the forces and constraints imposed on the system. We can write the Cartesian system of coordinates as a function of the generalized coordinates $q_1, q_2, \ldots, q_k$ as

$$x_i = x_i(q_1, q_2, \ldots, q_k) \tag{4.31}$$

Thus, the virtual displacement becomes

$$\begin{aligned}\delta x_i &= \frac{\partial x_i}{\partial q_1}\delta q_1 + \frac{\partial x_i}{\partial q_2}\delta q_2 + \cdots + \frac{\partial x_i}{\partial q_n}\delta q_n \\ &= \sum_{k=1}^{n} \frac{\partial x_i}{\partial q_k}\delta q_k, \quad (i = 1, 2, \ldots, n)\end{aligned} \tag{4.32}$$

Substituting Equation 4.32 into Equation 4.30 gives the virtual work done by the forces as

$$\delta W = \sum_{i=1}^{n} F_i \left( \sum_{k=1}^{n} \frac{\partial x_i}{\partial q_k}\delta q_k \right) = \sum_{k=1}^{n} \left( \sum_{i=1}^{n} F_i \frac{\partial x_i}{\partial q_k} \right) \delta q_k \tag{4.33}$$

Since $W = W(q_k)$, we have

$$\delta W = \sum_{k=1}^{n} \frac{\partial W}{\partial q_k}\delta q_k \tag{4.34}$$

Comparing the coefficients $\delta q_k$ in Equations 4.34 and 4.33 leads to the expression

$$\frac{\partial W}{\partial q_k} = \sum_{i=1}^{n} F_i \frac{\partial x_i}{\partial q_k} \tag{4.35}$$

Define the generalized force $Q_k$ associated with the $k$th system generalized coordinate $q_k$ as

$$Q_k = \sum_{i=1}^{n} F_i \frac{\partial x_i}{\partial q_k}, \quad k = 1, 2, \ldots, n \tag{4.36}$$

It follows that to find, for example $Q_1$, assume $q_1 \neq 0$, all other $q_k' s = 0$. Therefore, the generalized force would be

$$Q_1 = \sum_{i=1}^{n} F_i \frac{\partial x_i}{\partial q_1}$$

**EXAMPLE 4.3**

Assume we have a mass $m$ supported on a wire as shown in Figure 4.9. Determine the generalized force.

**Solution**

Forces $F_1$ and $F_2$ are actual forces and $q_1$ is the generalized coordinate. We have $x_1 = x_1(q_1)$ and $x_2 = x_2(q_1)$. Therefore,

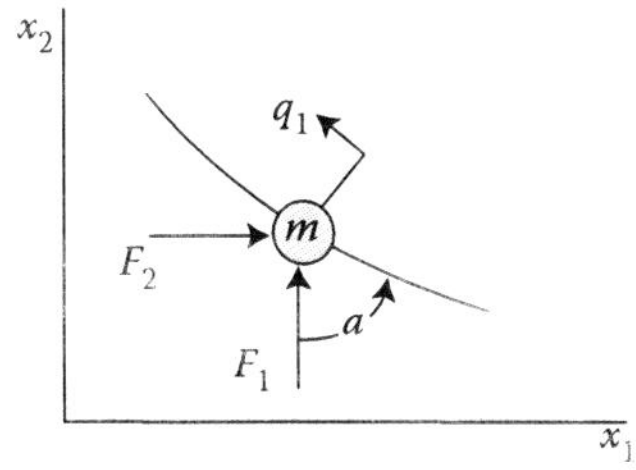


**FIGURE 4.9**
Mass $m$ supported on wire subjected to forces $F_1$ and $F_2$.

$$Q_1 = F_1 \frac{\partial x_1}{\partial q_1} + F_2 \frac{\partial x_1}{\partial q_1}$$

where $\partial x_1/\partial q_1 = -\sin\alpha$ and $\partial x_2/\partial q_1 = \cos\alpha$, so

$$Q_1 = -F_1 \sin\alpha + F_2 \cos\alpha, \quad \alpha = \alpha(q_1)$$

## 4.4 Potential Energy

In a large number of problems, there is a function potential energy function $V = V(q_1, q_2, \ldots, q_n)$ such that the generalized forces acting on the system can be determined using

$$Q_k = -\frac{\partial V}{\partial q_k} \tag{4.37}$$

Here, the $Q_k$ terms are potential forces and $V$ is the potential energy of the system. Potential energy is the energy that results from the configuration or position. A system may have the capacity for doing work as a result of its position in a gravitational field or it may have elastic potential energy as in a stretched spring.

1. *Gravitational forces:* Consider a particle where it is assumed that $Q_k$ is the weight and the particle moves from one position ($z_1$) to another ($z_2$) as illustrated in Figure 4.10.

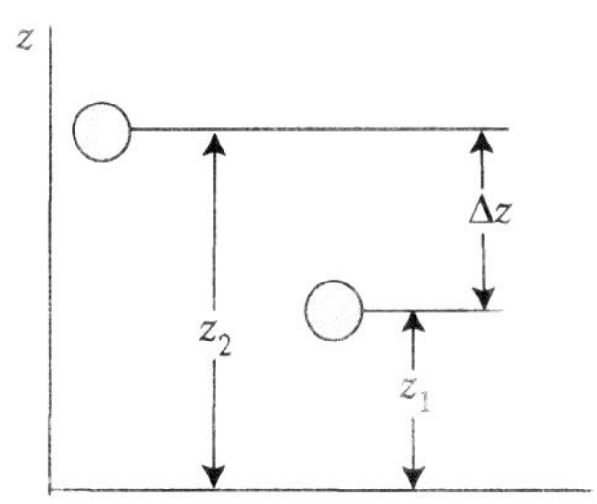


**FIGURE 4.10**
Particle follows an arbitrary path under gravitational force.

Then

$$\Delta V = -Q_k \Delta z = -Q_k(z_2 - z_1)$$

and

$$V = -Qz + \text{constant}$$

Thus,

$$Q = -\frac{\partial V}{\partial z}$$

If $z = q_1$ is the generalize coordinate, then

$$Q = -\frac{\partial V}{\partial q_1}$$

If $Q$ is known, we can find $V$. If $V$ is known, we can find $Q$ since $V = Wz = Wq_1$. Or, $V$ may be used in Lagrange's equations. In the case of gravitational forces, we would have

$$\frac{d}{dt}\left(\frac{\partial T}{\partial \dot{q}_1}\right) - \frac{\partial T}{\partial q_1} + \frac{\partial V}{\partial q_1} = 0$$

2. *Elastic potential energy:* Elastic potential energy is potential energy stored as a result of stretching of a spring. The energy depends on the spring constant $k$ and the stretched distance $q_1$ as shown in Figure 4.11. The elastic potential energy is

$$V = \frac{1}{2}kq_1^2$$

The force acting on the particle is

$$S = Q_1 = -\frac{\partial V}{\partial q_1} = -kq_1$$

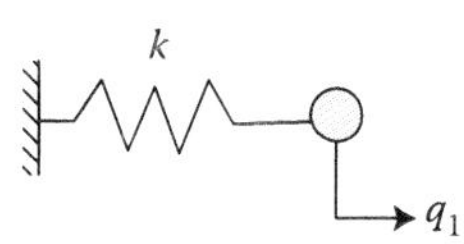


**FIGURE 4.11**
Elastic stretching of an elastic spring.

This means that if spring forces act on the system, the following Lagrange's equation may be used:

$$\frac{d}{dt}\left(\frac{\partial T}{\partial \dot{q}_s}\right)-\frac{\partial T}{\partial q_s}+\frac{\partial V}{\partial q_s}=0$$

If $S$ is the force acting on the particle, the force acting on the spring is

$$N=-S=+\frac{\partial V}{\partial q_1}$$

**EXAMPLE 4.4**

Consider the simply supported beam with a mass $m$ at the midspan location as shown in Figure 4.12. Determine the force acting on the particle ($\partial V/\partial w$).

**Solution**

The deflection $w$, of the mass, is taken as $w = q_1$. The potential energy can be expressed as

$$V=-\sum F_i\delta x_i=\sum F_i'\delta x_i$$

where $F_i'$ is the force acting on the beam. Since $\delta x = \delta q = \delta w$

$$V=\int_0^w F'\delta w$$

From simple beam theory, the deflection (at the midspan location) for a beam with a load ($F'$) applied at the midspan is $w = F'L^3/48EI$. Therefore,

$$V=\int_0^w \frac{48EI}{L^3}w\delta w=\frac{24EI}{L^3}w^2$$

and the force is

$$-\frac{\partial V}{\partial w}=\frac{48EI}{L^3}w$$

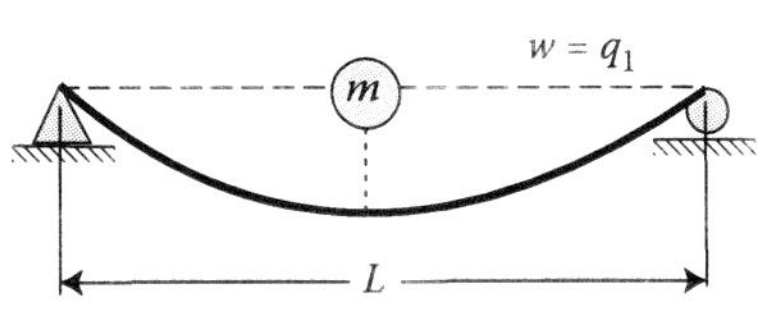


**FIGURE 4.12**
Simply supported beam with mass $m$ located at the midspan $L/2$.

### 4.4.1 Application of Lagrange's Equation

We wish to write Lagrange's equations of motion for the system shown in Figure 4.13.

Begin by noting that the generalized coordinates are $q_1$ and $q_2$. The kinetic energy expression is

$$T = \frac{1}{2} m_1 \dot{q}_1^2 + \frac{1}{2} m_2 \dot{q}_2^2 \tag{4.38}$$

Next, we consider the forces. These are forces due to the potential energy (from the springs) and the nonpotential force ($P_1$ and $P_2$). The potential forces are expressible in terms of the potential energy $V$, where

$$V = \frac{1}{2} k_1 q_1^2 + \frac{1}{2} k_2 q_2^2 \tag{4.39}$$

The nonpotential forces due to $P_1$ and $P_2$ are

$$\begin{aligned} Q_1 &= P_1 \frac{\partial x_1}{\partial q_1} + P_2 \frac{\partial x_2}{\partial q_1} = P_1 \quad (x_1 = q_1) \\ Q_2 &= P_1 \frac{\partial x_1}{\partial q_2} + P_2 \frac{\partial x_2}{\partial q_2} = P_2 \quad (x_2 = q_2) \end{aligned} \tag{4.40}$$

Using the third form of Lagrange's equation, Equation 4.22, gives, for $k = 1$

$$\frac{d}{dt}\left(\frac{\partial T}{\partial \dot{q}_1}\right) - \frac{\partial T}{\partial q_1} + \frac{\partial V}{\partial q_1} = Q_1$$

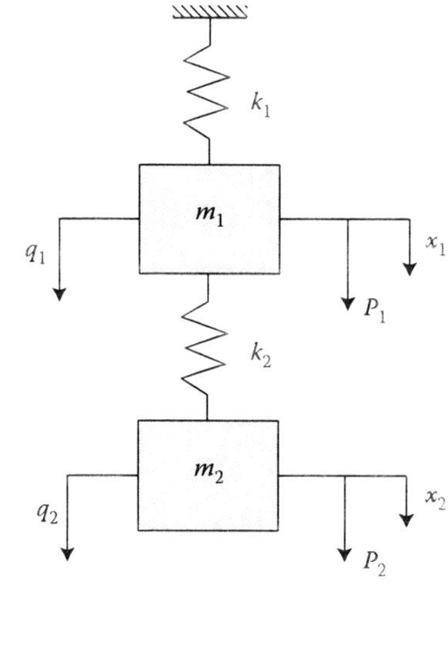


**FIGURE 4.13**
Two-degree-of-freedom system subjected to vertical displacement only.

or

$$m_1\ddot{q}_1 + k_1q_1 - k_2(q_2 - q_1) = P_1 \tag{4.41}$$

In the same manner we find for $k = 2$

$$m_2\ddot{q}_2 + k_2(q_2 - q_1) = P_2 \tag{4.42}$$

## 4.5 Free Vibrations

The simplest dynamic model of a vibrating MDOF system is free undamped vibration. As with single-degree-of-freedom systems, the analysis of free vibrations of MDOF system yields the natural frequencies of the system. Since there is no mechanism for dissipating or adding energy, we have what is known as a *conservative system* (Tong 1960).

Consider the two-degree-of-freedom spring mass system subjected to forces $P_1$ and $P_2$ located as shown in Figure 4.14.

For free vibrations, we have from Equation 4.11 or 4.15 setting $P_1 = P_2 = 0$ results in

$$\begin{aligned} f_{11}m_1\frac{d^2u_1}{dt^2} + f_{12}m_2\frac{d^2u_2}{dt^2} + u_1 = 0 \quad & m_1\frac{d^2u_1}{dt^2} + k_{11}u_1 + k_{12}u_2 = 0 \\ & \text{or} \\ f_{21}m_1\frac{d^2u_1}{dt^2} + f_{22}m_2\frac{d^2u_2}{dt^2} + u_2 = 0 \quad & m_2\frac{d^2u_2}{dt^2} + k_{21}u_1 + k_{22}u_2 = 0 \end{aligned} \tag{4.43}$$

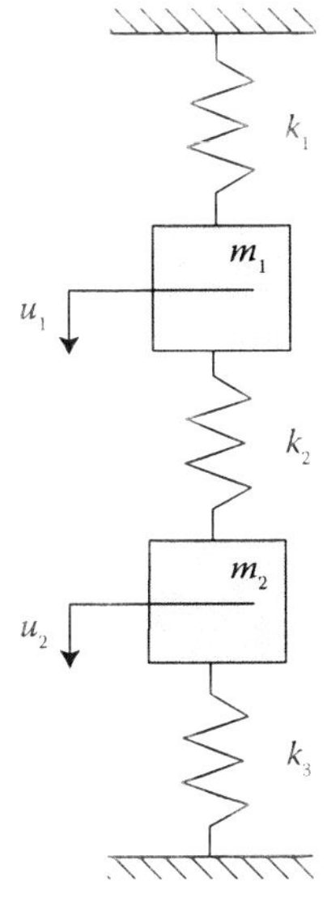


**FIGURE 4.14**
Spring mass system with two attached masses $m_1$ and $m_2$.

The above coupled equations are subject to four initial conditions,

$$\begin{aligned} u_1(t=0) &= u_{10} \quad \dot{u}_1(t=0) = \dot{u}_{10} \\ u_2(t=0) &= u_{20} \quad \dot{u}_2(t=0) = \dot{u}_{20} \end{aligned} \tag{4.44}$$

where the constants $u_{10}$, $u_{20}$ and $\dot{u}_{10}$, $\dot{u}_{20}$ represent the initial displacements and velocities of the two masses $m_1$ and $m_2$, respectively. In a manner analogous to our single degree-of-freedom system, assume solutions of the form

$$\begin{aligned} u_1 &= U_1 \sin(\omega t + \varphi) \\ u_2 &= U_2 \sin(\omega t + \varphi) \end{aligned} \tag{4.45}$$

Substituting Equation 4.45 into Equation 4.43 yields

$$\begin{aligned} (1 - f_{11}m_1\omega^2)U_1 - f_{12}m_2\omega^2 U_2 = 0 \quad (k_{11} - m_1\omega^2)U_1 + k_{12}U_2 = 0 \\ \text{or} \\ -f_{21}m_1\omega^2 U_1 + (1 - f_{22}m_2\omega^2)U_2 = 0 \quad k_{21}U_1 + (k_{22} - m_2\omega^2)U_2 = 0 \end{aligned} \tag{4.46}$$

One possible solution is that $U_1 = U_2 = 0$, which would correspond to the system equilibrium position which gives no information about the response of the system. A non-trivial solution occurs for $U_1$ and $U_2$ only if the determinate of the coefficients is equal to zero. Thus,

$$\begin{vmatrix} (1 - f_{11}m_1\omega^2) & -f_{12}m_2\omega^2 \\ -f_{21}m_1\omega^2 & (1 - f_{22}m_2\omega^2) \end{vmatrix} = 0 \quad \text{or} \quad \begin{vmatrix} (k_{11} - m_1\omega^2) & k_{12} \\ k_{21} & (k_{22} - m_2\omega^2) \end{vmatrix} = 0 \tag{4.47}$$

The frequency equation is obtained by expanding the determinate, remembering that $f_{ij}$ and $k_{ij}$ are symmetric, which results in

$$\begin{aligned} m_1 m_2 \left( f_{11} f_{22} - f_{12}^2 \right) \omega^4 - (m_1 f_{11} + m_2 f_{22})\omega^2 + 1 = 0 \quad \text{(a)} \\ m_1 m_2 (\omega^2)^2 - (m_1 k_{22} + m_2 k_{11})\omega^2 + k_{11} k_{22} - k_{12}^2 = 0 \quad \text{(b)} \end{aligned} \tag{4.48}$$

Equation 4.48 which are quadratic in $\omega^2$ are called the *frequency* or *characteristic equation* of the system. There are two roots that are called *characteristic values* that are determined by applying the quadratic formula to these equations. Thus, we have

$$\begin{aligned} \omega_{1,2}^2 &= \frac{m_1 f_{11} + m_2 f_{22} \mp \sqrt{(m_1 f_{11} + m_2 f_{22})^2 - 4 m_1 m_2 d}}{2 m_1 m_2 \left( f_{11} f_{22} - f_{12}^2 \right)} \quad \text{(a)} \\ \omega_{1,2}^2 &= \frac{m_1 k_{22} + m_2 k_{11} \mp \sqrt{(m_1 k_{22} + m_2 k_{11})^2 - 4 m_1 m_2 d_1}}{2 m_1 m_2} \quad \text{(b)} \end{aligned} \tag{4.49}$$

Where $d = f_{11} f_{22} - f_{12}^2$, $d_1 = k_{11} k_{22} - k_{12}^2$ and $\omega_1$ and $\omega_2$ are *the natural frequencies* of the system. The values $U_1$ and $U_2$ depend on the natural frequencies $\omega_1$ and $\omega_2$. Since we have this dependence let $U_1^{(1)}$ and $U_2^{(1)}$ correspond to $\omega_1$ and $U_1^{(2)}$ and $U_2^{(2)}$ correspond to $\omega_2$. Substituting the characteristic values $\omega_1^2$ and $\omega_2^2$ into Equation 4.46 yields two homogeneous equations, which cannot be solved directly for $U_1$ and $U_2$; however we can obtain the ratios $r_1 = U_2^{(1)}/U_1^{(1)}$ and $r_2 = U_2^{(2)}/U_1^{(2)}$ for our two sets of equations as

$$\begin{aligned} r_1 &= \frac{U_2^{(1)}}{U_1^{(1)}} = \frac{1 - f_{11}m_1\omega_1^2}{f_{12}m_2\omega_1^2} = \frac{f_{21}m_1\omega_1^2}{1 - f_{22}m_2\omega_1^2} \\ r_2 &= \frac{U_2^{(2)}}{U_1^{(2)}} = \frac{1 - f_{11}m_1\omega_2^2}{f_{12}m_2\omega_2^2} = \frac{f_{21}m_1\omega_2^2}{1 - f_{22}m_2\omega_2^2} \end{aligned} \tag{4.50}$$

and

$$\begin{aligned} r_1 &= \frac{U_2^{(1)}}{U_1^{(1)}} = \frac{k_{11} - m_1\omega_1^2}{-k_{12}} = \frac{-k_{21}}{k_{22} - m_2\omega_1^2} \\ r_2 &= \frac{U_2^{(2)}}{U_1^{(2)}} = \frac{k_{11} - m_1\omega_2^2}{-k_{12}} = \frac{-k_{21}}{k_{22} - m_2\omega_2^2} \end{aligned} \tag{4.51}$$

These amplitude ratios represent the mode shapes of vibration. We see that the two ratios given for $r_1$ and $r_2$ in Equation 4.50 or 4.51 are identical. The normal modes of vibration corresponding to $\omega_1$ and $\omega_2$ can be written as

$$\{U\}^{(1)} = \begin{Bmatrix} U_1^{(1)} \\ U_2^{(1)} \end{Bmatrix} = \begin{Bmatrix} U_1^{(1)} \\ r_1 U_1^{(1)} \end{Bmatrix} \quad \text{and} \quad \{U\}^{(2)} = \begin{Bmatrix} U_1^{(2)} \\ U_2^{(2)} \end{Bmatrix} = \begin{Bmatrix} U_1^{(2)} \\ r_2 U_1^{(2)} \end{Bmatrix} \tag{4.52}$$

where $\{U\}^{(1)}$ and $\{U\}^{(2)}$ represent the two modes of vibration and are called *modal vectors* of the two-degree-of-freedom system since they described the mode shapes at natural frequencies $\omega_1$ and $\omega_2$. This means that the following two motions are possible; for frequency $\omega_1$, we have

$$u_1 = U_1^{(1)} \sin(\omega_1 t + \varphi_1) \quad \text{and} \quad u_2 = r_1 U_1^{(1)} \sin(\omega_1 t + \varphi_1) \tag{4.53}$$

Equation 4.52 describes the *first mode* of vibration (also called the *fundamental* mode) and the *second mode* of vibration. The first mode from Equation 4.52 can be written as

$$\{U\}^{(1)} = \begin{Bmatrix} u_1^{(1)}(t) \\ u_2^{(1)}(t) \end{Bmatrix} = \begin{Bmatrix} U_1^{(1)} \sin(\omega_1 t + \varphi_1) \\ r_1 U_1^{(1)} \sin(\omega_1 t + \varphi_1) \end{Bmatrix} \tag{4.54}$$

In the same manner we have for $\omega_2$, the *second mode* of vibration.

$$u_1 = U_1^{(2)} \sin(\omega_2 t + \varphi_2) \quad \text{and} \quad u_2 = r_2 U_1^{(2)} \sin(\omega_2 t + \varphi_2) \tag{4.55}$$

or as

$$\{U\}^{(2)} = \begin{Bmatrix} U_1^{(2)} \\ U_2^{(2)} \end{Bmatrix} = \begin{Bmatrix} U_1^{(2)} \sin(\omega_2 t + \varphi_2) \\ r_2 U_1^{(2)} \sin(\omega_2 t + \varphi_2) \end{Bmatrix} \tag{4.56}$$

For any general type of initial condition, both modes of vibration will get excited. The resulting motion can be obtained by a linear superposition of the two normal modes given by Equations 4.54 and 4.56 as

$$\begin{aligned} u_1(t) &= AU_1^{(1)}\sin(\omega_1 t+\varphi_1)+BU_1^{(2)}\sin(\omega_2 t+\varphi_2) \\ u_2(t) &= Ar_1U_1^{(1)}\sin(\omega_1 t+\varphi_1)+Br_2U_1^{(2)}\sin(\omega_2 t+\varphi_2) \end{aligned} \tag{4.57}$$

where $\omega_1$ and $\omega_2$ are the natural frequencies and $A$, $B$, $\varphi_1$, and $\varphi_2$ are constants. However, since $U_1^{(1)}$ and $U_1^{(2)}$ already involve unknown constants, the constants $A$ and $B$ can be taken as unity without any loss of generality. Therefore, Equation 4.57 becomes

$$\begin{aligned} u_1(t) &= U_1^{(1)}\sin(\omega_1 t+\varphi_1)+U_1^{(2)}\sin(\omega_2 t+\varphi_2) \\ u_2(t) &= r_1U_1^{(1)}\sin(\omega_1 t+\varphi_1)+r_2U_1^{(2)}\sin(\omega_2 t+\varphi_2) \end{aligned} \tag{4.58}$$

The unknown constants $U_1^{(1)}, U_1^{(2)}, \varphi_1$ and $\varphi_2$ can be determined using the initial conditions given in Equation 4.44. Thus, we have

$$\begin{aligned} u_{10} &= U_1^{(1)}\sin\varphi_1+U_1^{(2)}\sin\varphi_2 \quad & u_{20} &= r_1U_1^{(1)}\sin\varphi_1+r_2U_1^{(2)}\sin\varphi_2 \\ \dot{u}_{10} &= \omega_1U_1^{(1)}\cos\varphi_1+\omega_2U_1^{(2)}\cos\varphi_2 \quad & \dot{u}_{20} &= \omega_1r_1U_1^{(1)}\cos\varphi_1+\omega_2r_2U_1^{(2)}\cos\varphi_2 \end{aligned} \tag{4.59}$$

Solving Equation 4.59 for $U_1^{(1)}\sin\varphi_1, U_1^{(2)}\sin\varphi_2, U_1^{(1)}\cos\varphi_1,$ and $U_1^{(2)}\cos\varphi_2$ gives

$$\begin{aligned} U_1^{(1)}\sin\varphi_1 &= \frac{r_2u_{10}-u_{20}}{r_2-r_1} \quad & U_1^{(2)}\sin\varphi_2 &= \frac{u_{20}-r_1u_{10}}{r_2-r_1} \\ U_1^{(1)}\cos\varphi_1 &= \frac{r_2\dot{u}_{10}-\dot{u}_{20}}{\omega_1(r_2-r_1)} \quad & U_1^{(2)}\cos\varphi_2 &= \frac{\dot{u}_{20}-r_1\dot{u}_{10}}{\omega_2(r_2-r_1)} \end{aligned} \tag{4.60}$$

Note that from Equation 4.60 we can form the following expressions:

$$\begin{aligned} U_1^{(1)} &= \left[\left(U_1^{(1)}\sin\varphi_1\right)^2+\left(U_1^{(1)}\cos\varphi_1\right)^2\right]^{1/2} = \frac{1}{(r_2-r_1)}\left[(r_2u_{10}-u_{20})^2+\frac{(r_2\dot{u}_{10}-\dot{u}_{20})^2}{\omega_1^2}\right]^{1/2} \\ U_1^{(1)} &= \left[\left(U_1^{(2)}\sin\varphi_1\right)^2+\left(U_1^{(2)}\cos\varphi_1\right)^2\right]^{1/2} = \frac{1}{(r_2-r_1)}\left[(u_{20}-r_1u_{10})^2+\frac{(\dot{u}_{20}-r_1\dot{u}_{10})^2}{\omega_2^2}\right] \end{aligned} \tag{4.61}$$

and

$$\varphi_1 = \tan^{-1}\left(\frac{\omega_1(r_2u_{10}-u_{20})}{r_2\dot{u}_{10}-\dot{u}_{20}}\right), \quad \varphi_2 = \tan^{-1}\left(\frac{\omega_2(u_{20}-r_1u_{10})}{\dot{u}_{20}-r_1\dot{u}_{10}}\right) \tag{4.62}$$

To illustrate the solution, let us determine the flexibility influence coefficients $c_{ij}$ and stiffness coefficients $k_{ij}$. Recall that the flexibility coefficient is defined as the displacement at coordinate $i$ due to a unit load acting along coordinate $j$. To obtain $f_{11}$ and $f_{21}$, a unit load is applied at coordinate 1 and the deflections are computed due to this load at coordinates

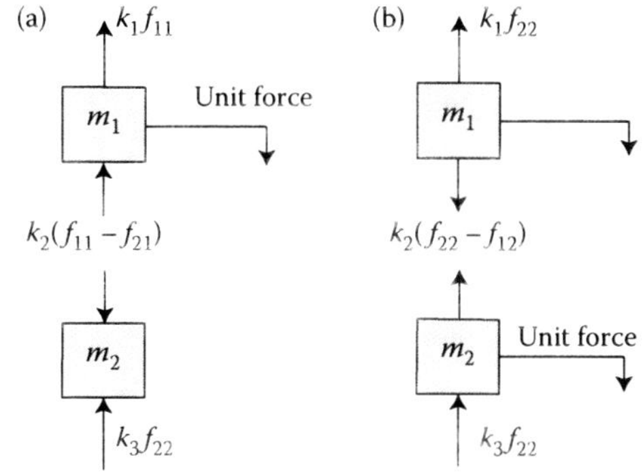


**FIGURE 4.15**
(a) Force diagram for unit force applied to mass $m_1$ and (b) force diagram for unit force applied to mass $m_2$.

1 and 2. In Figure 4.15, a unit force is first applied to mass $m_1$ only resulting in the free-body diagram shown in Figure 4.15a and then a unit force is applied to mass $m_2$ only resulting in the free-body diagram shown in Figure 4.15b.

From Figure 4.15a and b, we have for the forces the following equations:

$$\begin{aligned} &k_1f_{11}+k_2(f_{11}-f_{21})=1 \quad k_1f_{12}=k_2(f_{22}-f_{12}) \\ &k_2(f_{11}-f_{21})=k_3f_{21} \quad k_2(f_{22}-f_{12})+k_3f_{22}=1 \end{aligned} \tag{4.63}$$

Solving Equation 4.63 for $f_{11}$, $f_{22}$, and $f_{12}$ yields

$$f_{11}=\frac{k_2+k_3}{D}, \quad f_{22}=\frac{k_1+k_2}{D}, \quad f_{12}=f_{21}=\frac{k_2}{D} \tag{4.64}$$

where $D = k_1k_2 + k_1k_3 + k_2k_3$. The flexibility coefficients in the matrix format are

$$[F]=\frac{1}{D}\begin{bmatrix} k_2+k_3 & k_2 \\ k_2 & k_1+k_2 \end{bmatrix} \tag{4.65}$$

The stiffness coefficients are determined by taking the inverse of the flexibility matrix, hence

$$[K]=[F]^{-1}=\begin{bmatrix} \dfrac{k_2+k_3}{D} & \dfrac{k_2}{D} \\ \dfrac{k_2}{D} & \dfrac{k_1+k_2}{D} \end{bmatrix}^{-1}=\begin{bmatrix} k_{11} & k_{12} \\ k_{12} & k_{22} \end{bmatrix}=\begin{bmatrix} k_1+k_2 & -k_2 \\ -k_2 & k_2+k_3 \end{bmatrix} \tag{4.66}$$

Consider the spring mass system with masses $m = m_1 = m_2$ and stiffnesses $k = k_1 = k_2 = k_3$, thus, we have $k_{11} = k_{22} = 2k$ and $k_{12} = k_{21} = -k$. Equation 4.48 gives the frequencies as

$$\omega_1^2=\frac{k}{m} \quad \text{and} \quad \omega_2^2=\frac{3k}{m} \tag{4.67}$$

Substituting the smaller frequency $\omega_1^2 = k/m$ into Equation 4.45, 4.49, or 4.50 yields

$$r_1 = \frac{U_2^{(1)}}{U_1^{(1)}} = \frac{-k_{12}}{k_{22} - m_2\omega_1^2} = \frac{k}{2k - m\omega_1^2} = 1$$
$$r_2 = \frac{U_2^{(2)}}{U_1^{(2)}} = \frac{-k_{12}}{k_{22} - m_2\omega_2^2} = \frac{k}{2k - m\omega_2^2} = -1 \quad (4.68)$$

This is the *first mode* of vibration, which is also called the *fundamental mode*, and is represented by substituting Equation 4.68 into Equation 4.53 giving

$$u_1 = U_1^{(1)} \sin(\omega_1 t + \phi_1)$$
$$u_2 = U_1^{(1)} \sin(\omega_1 t + \phi_1) \quad (4.69)$$

where $U_1^{(1)}$ and $\phi_1$ values depend on the initial conditions. Both masses have the same motion and move up and down together and reach their extreme position simultaneously as shown in Figure 4.16a (motion shown in the lateral direction for brevity). If the larger frequency $\omega_2^2 = 3k/m$ is substituted into Equation 4.45, 4.49 or 4.50, we find

$$U_2^{(2)} = -U_1^{(2)} \quad \text{or} \quad r_2 = \frac{U_2^{(2)}}{U_1^{(2)}} = -1 \quad (4.70)$$

and substituting Equation 4.70 into Equation 4.55 yields

$$u_1 = U_1^{(2)} \sin(\omega_2 t + \phi_2)$$
$$u_2 = -U_1^{(2)} \sin(\omega_2 t + \phi_2) \quad (4.71)$$

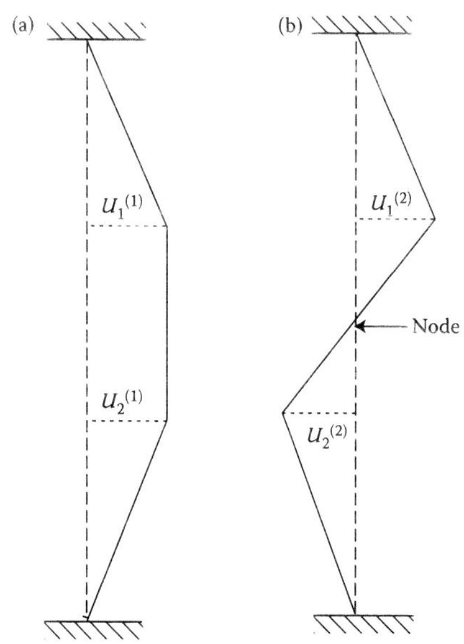


**FIGURE 4.16**
(a) First mode with masses in phase and (b) second mode with masses out of phase.

which describe the *second mode* of vibration. Again $U_1^{(2)}$ and $\phi_2$ depend on the initial conditions. The mode shape is shown in Figure 4.16b. In the second mode, the midpoint of the center spring does not move, such a point is called a node. We see that the two masses move opposite to each other.

The motion of the masses is given by

$$\begin{aligned} u_1(t) &= U_1^{(1)} \sin\left(\sqrt{\left(\frac{k}{m}\right)}t + \varphi_1\right) + U_1^{(2)} \sin\left(\sqrt{\left(\frac{3k}{m}\right)}t + \varphi_2\right) \\ u_2(t) &= U_1^{(1)} \sin\left(\sqrt{\left(\frac{k}{m}\right)}t + \varphi_1\right) - U_1^{(2)} \sin\left(\sqrt{\left(\frac{3k}{m}\right)}t + \varphi_2\right) \end{aligned} \tag{4.72}$$

where the constants are obtained from the initial conditions. Assume that the initial conditions are given as $u_1(0) = 0$, $u_2(0) = 1$, and $\dot{u}_1(0) = 0$, $\dot{u}_2(0) = 0$, thus we have

$$0 = U_1^{(1)} \sin\varphi_1 + U_1^{(2)} \sin\varphi_2 \tag{4.73}$$

$$1 = U_1^{(1)} \sin\varphi_1 - U_1^{(2)} \sin\varphi_2 \tag{4.74}$$

$$0 = \omega_1 U_1^{(1)} \cos\varphi_1 + \omega_2 U_1^{(2)} \cos\varphi_2 \tag{4.75}$$

$$0 = \omega_1 U_1^{(1)} \cos\varphi_1 - \omega_2 U_1^{(2)} \cos\varphi_2 \tag{4.76}$$

Equations 4.73 and 4.74 yield $U_1^{(1)} = 1/(2\sin\varphi_1)$ and $U_1^{(2)} = -1/(2\sin\varphi_2)$ in addition Equations 4.75 and 4.76 yield $\cos\varphi_1 = \cos\varphi_2 = \pi/2$, therefore we have $U_1^{(1)} = 1/2$ and $U_1^2 = -1/2$ and the motion of the masses is given by

$$\begin{aligned} u_1(t) &= \frac{1}{2}\cos\sqrt{\left(\frac{k}{m}\right)}t - \frac{1}{2}\cos\sqrt{\left(\frac{3k}{m}\right)}t \\ u_2(t) &= \frac{1}{2}\cos\sqrt{\left(\frac{k}{m}\right)}t + \frac{1}{2}\cos\sqrt{\left(\frac{3k}{m}\right)}t \end{aligned} \tag{4.77}$$

## 4.6 Frequency Response Function

Assume a loading condition such that $P_2 = 0$. The output displacements are $u_1$ and $u_2$. The equations of motion are

$$\begin{aligned} f_{11}m_1\frac{d^2u_1}{dt^2} + f_{12}m_2\frac{d^2u_2}{dt^2} + u_1 &= f_{11}P_1(t) \\ f_{21}m_1\frac{d^2u_1}{dt^2} + f_{22}m_2\frac{d^2u_2}{dt^2} + u_2 &= f_{21}P_1(t) \end{aligned} \tag{4.78}$$

or we can use

$$\begin{aligned} m_1 \frac{d^2u_1}{dt^2} + k_{11}u_1 + k_{12}u_2 = P_1 \\ m_2 \frac{d^2u_2}{dt^2} + k_{21}u_1 + k_{22}u_2 = 0 \end{aligned} \tag{4.79}$$

Assume $P_1(t) = P_0e^{i\omega_f t}$, $u_1 = U_1^*e^{i\omega_f t}$, and $u_2 = U_2^*e^{i\omega_f t}$, where $U_1^*$ and $U_2^*$ are the frequency response functions of $u_1$, $u_2$ and consider the spring mass system with two attached masses $m_1$ and $m_2$ as shown in Figure 4.14. Substituting these values into Equation 4.78 gives

$$\begin{aligned} -f_{11}m_1\omega_f^2U_1^* - f_{12}m_2\omega_f^2U_2^* + U_1^* = f_{11}P_0 \\ -f_{21}m_1\omega_f^2U_1^* - f_{22}m_2\omega_f^2U_2^* + U_2^* = f_{21}P_0 \end{aligned} \tag{4.80}$$

Solving for $U_1^*$ and $U_2^*$ yields

$$\begin{aligned} U_1^*(i\omega) &= \frac{\begin{vmatrix} f_{11}P_0 & -f_{12}m_2\omega_f^2 \\ f_{21}P_0 & 1 - f_{22}m_2\omega_f^2 \end{vmatrix}}{\Delta} \\ &\text{and} \\ U_2^*(i\omega) &= \frac{\begin{vmatrix} 1 - f_{11}m_1\omega_f^2 & f_{11}P_0 \\ -f_{21}m_1\omega_f^2 & f_{21}P_0 \end{vmatrix}}{\Delta} \end{aligned} \tag{4.81}$$

where

$$\Delta = \begin{vmatrix} 1 - f_{11}m_1\omega_f^2 & -f_{12}m_2\omega_f^2 \\ -f_{21}m_1\omega_f^2 & 1 - f_{22}m_2\omega_f^2 \end{vmatrix} \tag{4.82}$$

Using Equation 4.79 instead of Equation 4.78, we have

$$\begin{aligned} -m_1\omega_f^2U_1^* + k_{11}U_1^* + k_{12}U_2^* = P_1 \\ -m_2\omega_f^2U_2^* + k_{22}U_2^* + k_{12}U_1^* = 0 \end{aligned} \tag{4.83}$$

We arrive at an equivalent set of values for $U_1^*$ and $U_2^*$ given by

$$\begin{aligned} U_1^* &= \frac{\left(k_{22} - m_2\omega_f^2\right)P_0}{\left(k_{11} - m_1\omega_f^2\right)\left(k_{22} - m_2\omega_f^2\right) - k_{12}^2} \\ &\text{and} \\ U_2^* &= \frac{-k_{12}P_0}{\left(k_{11} - m_1\omega_f^2\right)\left(k_{22} - m_2\omega_f^2\right) - k_{12}^2} \end{aligned} \tag{4.84}$$

For the spring mass system shown in Figure 4.14, we have $k_{11} = k_1 + k_2$, $k_{22} = k_2 + k_3$, and $k_{12} = k_{21} = -k_2$, thus Equation 4.84 becomes

$$U_1^* = \frac{\left(k_2 + k_3 - m_2\omega_f^2\right)P_0}{\left(k_1 + k_2 - m_1\omega_f^2\right)\left(k_2 + k_3 - m_2\omega_f^2\right) - k_2^2}$$
$$\text{and} \tag{4.85}$$
$$U_2^* = \frac{k_2 P_0}{\left(k_1 + k_2 - m_1\omega_f^2\right)\left(k_2 + k_3 - m_2\omega_f^2\right) - k_2^2}$$

Let $\omega_1 = \sqrt{k_1/m_1}$, $\omega_2 = \sqrt{k_2/m_2}$, $\Omega_1 = \omega_f/\omega_1$, and $\Omega_2 = \omega_f/\omega_2$, then Equation 4.85 can be written as

$$U_1^* = \frac{\left[1 + (k_3/k_2) - \Omega_2^2\right](P_0/k_1)}{\left[\left\{1 + (k_2/k_1) - \Omega_1^2\right\}\left\{1 + (k_3/k_2) - \Omega_2^2\right\} - (k_2/k_1)\right]}$$
$$\text{and} \tag{4.86}$$
$$U_2^* = \frac{(P_0/k_1)}{\left[\left\{1 + (k_2/k_1) - \Omega_1^2\right\}\left\{1 + (k_3/k_2) - \Omega_2^2\right\} - (k_2/k_1)\right]}$$

Using the values $k_1 = k_3 = k$, $k_2 = 2k$ and $m_1 = m_2 = m$, Equation 4.86 becomes

$$U_1^* = \frac{\left[1.5 - \Omega_2^2\right](P_0/k)}{\left[\left(3 - \Omega_1^2\right)\left(1.5 - \Omega_2^2\right) - 2\right]}$$
$$\text{and} \tag{4.87}$$
$$U_2^* = \frac{(P_0/k)}{\left[\left(3 - \Omega_1^2\right)\left(1.5 - \Omega_2^2\right) - 2\right]}$$

A plot of Equation 4.87 is shown in Figure 4.17, where $\Omega_1 = (\omega_2/\omega_1)\Omega_2$ has been used in Equation 4.87. For $\Omega_2 = \sqrt{1.5}$, we have $\omega_f = \sqrt{1.5}\omega_2$ and the numerator of $U_1^*$ in Equation 4.87 vanishes but not the denominator since it has a finite value equal to –2, thus $U_1^* = 0$. For this same condition, we find that $U_2^* = -P_0/2k$. Also for the case, $\omega_f = 0$, the amplitudes $U_1^*$ and $U_2^*$ become $U_1^* = 0.6P_0/k$ and $U_2^* = -0.4P_0/k$. If the denominator vanishes, both $U_1^*$ and $U_2^*$ become infinite. The value of $\Omega_2$, at which occurs, is obtained by setting the denominator in Equation 4.87 to zero. Therefore,

$$\left(3 - \Omega_1^2\right)\left(1.5 - \Omega_2^2\right) - 2 = 0 \quad \text{or} \quad \left[3 - \left(\frac{\omega_2}{\omega_1}\right)^2 \Omega_2^2\right]\left(1.5 - \Omega_2^2\right) - 2 = 0 \tag{4.88}$$

Expanding Equation 4.88 and using the above values for stiffnesses and frequencies yields two positive roots for $\Omega_2$

$$\Omega_2 = 0.7071 \quad \text{and} \quad \Omega_2 = 1.5811 \tag{4.89}$$

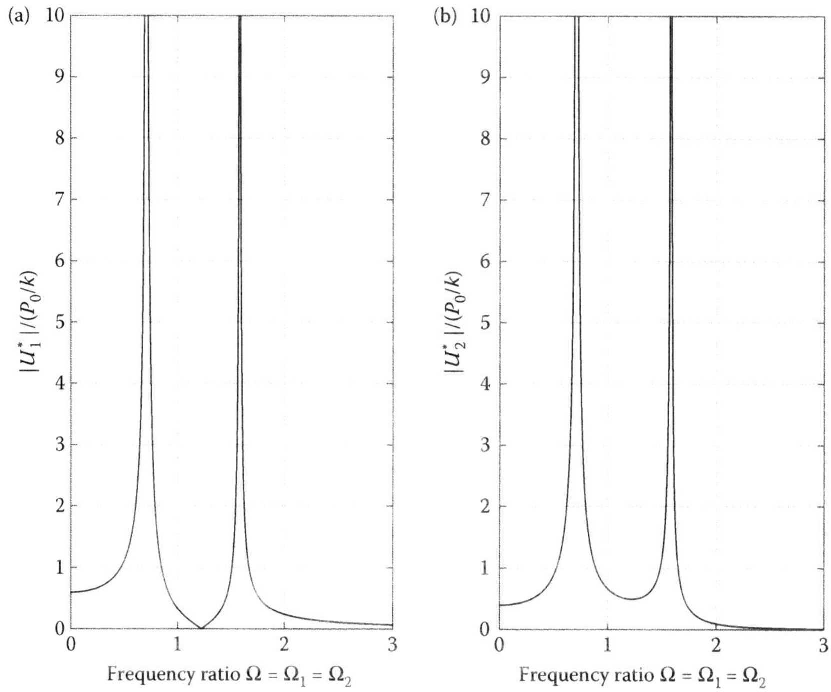


**FIGURE 4.17**
Absolute values of frequency response functions (a) $W_1^*$, and (b) $W_2^*$ for unit load applied at mass $m_1$.

### 4.6.1 Application of Frequency Response Function

Consider the following applications using the above results:

1. If $P_1 = P_{10} \cos \omega_p t$, (i.e., $P_1 = P_{10}\,\mathrm{Re}\{e^{i\omega_p t}\}$ with $P_{10}$ and $\omega_p$ given), then the solution is given by

$$u_1 = P_{10}\,\mathrm{Re}\left\{U_1^* e^{i\omega_p t}\right\} \quad \text{and} \quad u_2 = P_{10}\,\mathrm{Re}\left\{U_2^* e^{i\omega_p t}\right\} \tag{4.90}$$

2. If $P_1 = P_{10} \sin \omega_p t$ (i.e., $P_1 = P_{10}\,\mathrm{Im}\{e^{i\omega_p t}\}$ with $P_{10}$ and $\omega_p$ given), then the solution is given by

$$u_1 = P_{10}\,\mathrm{Im}\left\{U_1^* e^{i\omega_p t}\right\} \quad \text{and} \quad u_2 = P_{10}\,\mathrm{Im}\left\{U_2^* e^{i\omega_p t}\right\} \tag{4.91}$$

3. $P_1(t)$ is a stationary random function. Its spectral density is known and given by $S_{p_1}(\omega)$. For $\omega = \omega_1$ and $\omega = \omega_2$, both denominators go to zero.

### 4.6.2 Transfer Function (Response to Unit Impulse)

Let the input be $P(t) = P_1 = P_2 = 0$ and the output $u_1(t)$ and $u_2(t)$. To find the transfer function, we assume that the loading is given by the Dirac delta function, that is,

$P(t) = \delta(t)$. Assume that the initial conditions are given as $u_1(0) = \dot{u}_1(0) = u_2(0) = \dot{u}_2(0) = 0$. Applying the Laplace transform to the equations of motion and recalling that $L\{\delta(t)\} = 1$, we obtain

$$\begin{gathered} f_{11}m_1s^2\bar{u}_1 + f_{12}m_{12}s^2\bar{u}_2 + \bar{u}_1 = f_{11} \quad m_1s^2\bar{u}_1 + k_{11}\bar{u}_1 + k_{12}\bar{u}_2 = 1 \\ \text{or} \\ f_{21}m_1s^2\bar{u}_1 + f_{22}m_{12}s^2\bar{u}_2 + \bar{u}_2 = f_{21} \quad m_2s^2\bar{u}_2 + k_{21}\bar{u}_1 + k_{22}\bar{u}_2 = 0 \end{gathered} \tag{4.92}$$

where

$$\bar{u}_1 = \mathscr{L}\{u_1\} \quad \text{and} \quad \bar{u}_2 = \mathscr{L}\{u_2\} \tag{4.93}$$

The above equations can be rewritten in the matrix format as

$$\begin{gathered} \begin{bmatrix} 1 + f_{11}m_1s^2 & f_{12}m_{12}s^2 \\ f_{12}m_{12}s^2 & 1 + f_{22}m_{12}s^2 \end{bmatrix} \begin{Bmatrix} \bar{u} \\ \bar{u}_2 \end{Bmatrix} = \begin{Bmatrix} f_{11} \\ f_{21} \end{Bmatrix} \\ \text{or} \\ \begin{bmatrix} k_{11} + m_1s^2 & k_{12} \\ k_{21} & k_{22} + m_2s^2 \end{bmatrix} \begin{Bmatrix} \bar{u} \\ \bar{u}_2 \end{Bmatrix} = \begin{Bmatrix} 1 \\ 0 \end{Bmatrix} \end{gathered} \tag{4.94}$$

Define the determinant as

$$\begin{gathered} \Delta_1 = \begin{vmatrix} 1 + f_{11}m_1s^2 & f_{12}m_{12}s^2 \\ f_{21}m_1s^2 & 1 + f_{22}m_{12}s^2 \end{vmatrix} \\ \text{and} \\ \Delta_2 = \begin{vmatrix} k_{11} + m_1s^2 & k_{12} \\ k_{21} & k_{22} + m_2s^2 \end{vmatrix} \end{gathered} \tag{4.95}$$

Solving the above for $\bar{u}_1(s)$ and $\bar{u}_2(s)$ yields

$$\begin{gathered} \bar{u}_1(s) = \frac{\begin{vmatrix} f_{11} & f_{12}m_{12}s^2 \\ f_{21} & 1 + f_{22}m_{12}s^2 \end{vmatrix}}{\Delta_1} \\ \text{and} \\ \bar{u}_2(s) = \frac{\begin{vmatrix} 1 + f_{11}m_1s^2 & f_{11} \\ f_{21}m_1s^2 & f_{21} \end{vmatrix}}{\Delta_1} \end{gathered} \tag{4.96}$$

or

$$\bar{u}_1(s) = \frac{\begin{vmatrix} 1 & k_{12} \\ 0 & k_{22} + m_2 s^2 \end{vmatrix}}{\Delta_2}$$

and (4.97)

$$\bar{u}_2(s) = \frac{\begin{vmatrix} k_{11} + m_1 s^2 & 1 \\ k_{21} & 0 \end{vmatrix}}{\Delta_2}$$

Thus the response to the unit impulse $P_1 = \delta(t)$ is given as

$$u_1(t) = \mathscr{L}^{-1}\{\bar{u}_1(s)\} \quad \text{and} \quad u_2(t) = \mathscr{L}^{-1}\{\bar{u}_2(s)\} \tag{4.98}$$

### 4.6.3 Arbitrary Forcing Function *P*(*t*)

Assume we have initial conditions $u_1(0) = \dot{u}_1(0) = u_2(0) = \dot{u}_2(0) = 0$. Apply the Laplace transformation to the equations of motion, thus

$$\bar{u}_1(s) = \bar{U}_1(s)\bar{P}_1(s) \quad \text{and} \quad \bar{u}_2(s) = \bar{U}_2(s)\bar{P}_1(s) \tag{4.99}$$

and we have for the displacements

$$\begin{aligned} u_1(t) &= \mathscr{L}^{-1}\{\bar{U}_1(s)\bar{P}_1(s)\} = \int_0^t U_1(t-\tau)P_1(\tau)\,d\tau \\ u_2(t) &= \mathscr{L}^{-1}\{\bar{U}_2(s)\bar{P}_1(s)\} = \int_0^t U_2(t-\tau)P_1(\tau)\,d\tau \end{aligned} \tag{4.100}$$

## 4.7 Damped Free Vibrations

Consider a damped two-degree-of-freedom system as shown in Figure 4.18.

With the forces $P_1 = P_2 = 0$, the equations of motion are given as

$$\begin{aligned} m_1 \frac{d^2u_1}{dt^2} + (\eta_1 + \eta_2)\frac{du_1}{dt} + (k_1 + k_2)u_1 - \eta_2 \frac{du_2}{dt} - k_2 u_2 = 0 \\ m_2 \frac{d^2u_2}{dt^2} + (\eta_2 + \eta_3)\frac{du_2}{dt} + (k_2 + k_3)u_2 - \eta_2 \frac{du_1}{dt} - k_2 u_1 = 0 \end{aligned} \tag{4.101}$$

or in the matrix form as

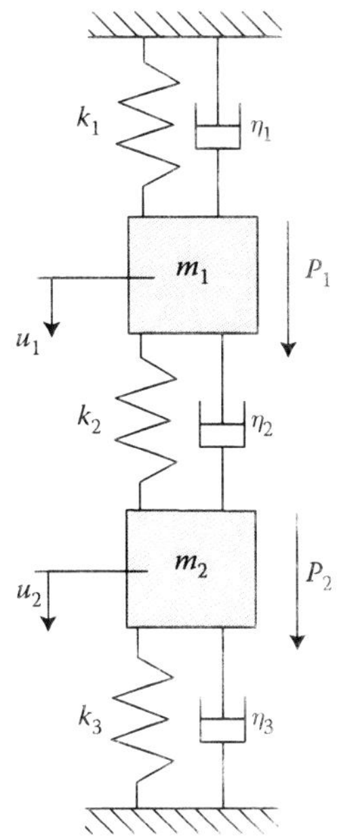


**FIGURE 4.18**
Damped two-degree-of-freedom system.

$$\begin{bmatrix} m_1 & 0 \\ 0 & m_2 \end{bmatrix} \begin{Bmatrix} \ddot{u}_1 \\ \ddot{u}_2 \end{Bmatrix} + \begin{bmatrix} \eta_1+\eta_2 & -\eta_2 \\ -\eta_2 & \eta_2+\eta_3 \end{bmatrix} \begin{Bmatrix} \dot{u}_1 \\ \dot{u}_2 \end{Bmatrix} + \begin{bmatrix} k_1+k_2 & -k_2 \\ -k_2 & k_2+k_3 \end{bmatrix} \begin{Bmatrix} u_1 \\ u_2 \end{Bmatrix} = \begin{Bmatrix} 0 \\ 0 \end{Bmatrix} \tag{4.102}$$

or equivalently as

$$[M]\ddot{\mathbf{u}} + [H]\dot{\mathbf{u}} + [K]\mathbf{u} = 0 \tag{4.103}$$

where $[H]$ is the damping matrix. Using the same method that was used for the single-degree-of-freedom method, we assume a solution as

$$\mathbf{u}(t) = \mathbf{U}e^{st} = \begin{Bmatrix} U_1 \\ U_2 \end{Bmatrix} e^{st} \tag{4.104}$$

Substituting Equation 4.104 into Equation 4.102 yields

$$[s^2[M] + s[H] + [K]]\mathbf{U}e^{st} = 0 \tag{4.105}$$

Since this equation must be satisfied for all time, we have

$$[s^2[M] + s[H] + [K]]\mathbf{U} = 0 \tag{4.106}$$

Equation 4.106 is a homogeneous linear algebraic equation with two unknowns $U_1$ and $U_2$ which can be written in the form

$$\begin{aligned} m_1s^2U_1 + (\eta_1+\eta_2)sU_1 - \eta_2 sU_2 + (k_1+k_2)U_1 - k_2U_2 = 0 \\ m_2s^2U_2 + (\eta_2+\eta_3)sU_2 - \eta_2 sU_1 + (k_2+k_3)U_2 - k_2U_1 = 0 \end{aligned} \tag{4.107}$$

or

$$\begin{bmatrix} m_1s^2 + (\eta_1+\eta_2)s + (k_1+k_2) & -(\eta_2 s + k_2) \\ -(\eta_2 s + k_2) & m_2s^2 + (\eta_2+\eta_3)s + (k_2+k_3) \end{bmatrix} \begin{Bmatrix} U_1 \\ U_2 \end{Bmatrix} = \begin{Bmatrix} 0 \\ 0 \end{Bmatrix} \tag{4.108}$$

Equation 4.108 has a nontrivial solution only if the determinant of the coefficients vanishes. Hence, we have

$$\begin{vmatrix} m_1s^2 + (\eta_1+\eta_2)s + (k_1+k_2) & -(\eta_2 s + k_2) \\ -(\eta_2 s + k_2) & m_2s^2 + (\eta_2+\eta_3)s + (k_2+k_3) \end{vmatrix} = 0 \tag{4.109}$$

or expanding the determinant gives

$$\begin{aligned} s^4 + \left[\frac{\eta_1+\eta_2}{m_1} + \frac{\eta_2+\eta_3}{m_2}\right]s^3 + \left[\frac{k_1+k_2}{m_1} + \frac{k_2+k_3}{m_2} + \frac{\eta_1\eta_2+\eta_1\eta_3+\eta_2\eta_3}{m_1m_2}\right]s^2 \\ + \left[\frac{k_1(\eta_2+\eta_3)+k_2(\eta_1+\eta_3)+k_3(\eta_1+\eta_2)}{m_1m_2}\right]s + \left[\frac{k_1k_2+k_1k_3+k_2k_3}{m_1m_2}\right] = 0 \end{aligned} \tag{4.110}$$

We see that the characteristic equation has four roots. It is noted that there are three possibilities regarding the roots of this equation. These possibilities are (a) all four roots are negative and real (b) two roots real and negative and two which are complex and conjugate and (c) all four roots are complex numbers which would require two pairs of complex conjugates with negative real parts.

Equation 4.110 will represent a vibrating system only when the roots of the characteristic equation are complex. If we assume that the roots are complex, then we know that complex roots of an algebraic equation occur in pairs $a \pm ib$ (complex conjugates). Taking the conjugate pairs in the form

$$\begin{aligned} s_1 = -a_1 + i\omega_{d_1} \quad & s_2 = -a_1 - i\omega_{d_1} \\ s_3 = -a_2 + i\omega_{d_2} \quad & s_4 = -a_2 - i\omega_{d_2} \end{aligned} \tag{4.111}$$

where the damped natural frequencies are $\omega_{d_i}$, $(i = 1,2)$ are $\omega_{d_i} = \omega_{0i}\sqrt{1-\zeta^2}$ and $a_i$, $(i = 1, 2)$ are positive real numbers. Substituting each root from Equation 4.111 into Equation 4.108 gives

$$\begin{bmatrix} m_1s_i^2 + (\eta_1+\eta_2)s_i + (k_1+k_2) & -(\eta_2 s_i + k_2) \\ -(\eta_2 s_i + k_2) & m_2s_i^2 + (\eta_2+\eta_3)s_i + (k_2+k_3) \end{bmatrix} \begin{Bmatrix} U_{1i} \\ U_{2i} \end{Bmatrix} = \begin{Bmatrix} 0 \\ 0 \end{Bmatrix} \tag{4.112}$$

There are four possible ratios that can be obtained from Equation 4.112 corresponding to $i = 1, 2, 3, 4$. Thus, we have using $i = 1, 2, 3, 4$ $r_1$, $r_2$, $r_3$, and $r_4$. This leads to the ratios

$$r_i = \frac{U_{2i}}{U_{1i}} = \frac{\eta_2 s_i + k_2}{m_2 s_i^2 + (\eta_2 + \eta_3)s_i + (k_2 + k_3)} = \frac{m_1 s_i^2 + (\eta_1 + \eta_2)s_i + (k_1 + k_2)}{\eta_2 s_i + k_2} \quad (i = 1,2,3,4) \tag{4.113}$$

The modal amplitudes are related by the amplitude ratios $r_i$, which are given by $U_{2i} = r_i U_{1i}$. The complete solution of Equation 4.104 is

$$\begin{aligned} u_1(t) &= U_{11}e^{s_1 t} + U_{12}e^{s_2 t} + U_{13}e^{s_3 t} + U_{14}e^{s_4 t} \\ u_2(t) &= r_1 U_{11}e^{s_1 t} + r_2 U_{12}e^{s_2 t} + r_3 U_{13}e^{s_3 t} + r_4 U_{14}e^{s_4 t} \end{aligned} \tag{4.114}$$

where the coefficients ($U_{11}$, $U_{12}$, $U_{13}$, and $U_{14}$) are complex conjugate pairs that are determined from the initial conditions for the problem. Substituting the roots $r_i$, ($i$ = 1, 2, 3, 4) into Equation 4.114 gives

$$\begin{aligned} u_1(t) &= U_{11}e^{(-a_1 + i\omega_{d1})t} + U_{12}e^{(-a_1 - i\omega_{d1})t} + U_{13}e^{(-a_2 + i\omega_{d2})t} + U_{14}e^{(-a_2 - i\omega_{d2})t} \\ u_2(t) &= r_1 U_{11}e^{(-a_1 + i\omega_{d1})t} + r_2 U_{12}e^{(-a_1 - i\omega_{d1})t} + r_3 U_{13}e^{(-a_2 + i\omega_{d2})t} + r_4 U_{14}e^{(-a_2 - i\omega_{d2})t} \end{aligned} \tag{4.115}$$

Equation 4.115 can be rewritten as

$$\begin{aligned} u_1(t) &= e^{-a_1 t}\left(U_{11}e^{i\omega_{d1}t} + U_{12}e^{-i\omega_{d1}t}\right) + e^{-a_2 t}\left(U_{13}e^{i\omega_{d2}t} + U_{14}e^{-i\omega_{d2}t}\right) \\ u_2(t) &= e^{-a_1 t}\left(r_1 U_{11}e^{i\omega_{d1}t} + r_2 U_{12}e^{-i\omega_{d1}t}\right) + e^{-a_2 t}\left(r_3 U_{13}e^{i\omega_{d2}t} + r_4 U_{14}e^{-i\omega_{d2}t}\right) \end{aligned} \tag{4.116}$$

Equation 4.116 can be converted into equivalent trigonometric expressions as was done for the single-degree-of-freedom system using the Euler formulas

$$\begin{aligned} e^{i\omega_{d1}t} &= \cos\omega_{d_1}t + i\sin\omega_{d_1}t \quad & e^{-i\omega_{d1}t} &= \cos\omega_{d_1}t - i\sin\omega_{d_1}t \\ e^{i\omega_{d2}t} &= \cos\omega_{d_2}t + i\sin\omega_{d_2}t \quad & e^{-i\omega_{d2}t} &= \cos\omega_{d_2}t - i\sin\omega_{d_2}t \end{aligned} \tag{4.117}$$

Substituting Equation 4.117 into Equation 4.116 yields

$$\begin{aligned} u_1(t) &= e^{-a_1 t}\left[(U_{11} + U_{12})\cos\omega_{d_1}t + i(U_{11} - U_{12})\sin\omega_{d_1}t\right] \\ &\quad + e^{-a_2 t}\left[(U_{13} + U_{14})\cos\omega_{d_2}t + i(U_{13} - U_{14})\sin\omega_{d_2}t\right] \\ u_2(t) &= e^{-a_1 t}\left[(r_1 U_{11} + r_2 U_{12})\cos\omega_{d_1}t + i(r_1 U_{11} - r_2 U_{12})\sin\omega_{d_1}t\right] \\ &\quad + e^{-a_2 t}\left[(r_3 U_{13} + r_4 U_{14})\cos\omega_{d_2}t + i(r_3 U_{13} - r_4 U_{14})\sin\omega_{d_2}t\right] \end{aligned} \tag{4.118}$$

Since $U_{11}$, $U_{12}$, $U_{13}$, and $U_{14}$ are complex conjugate pairs, we can introduce new real constants

$$\begin{aligned} A_1 &= U_{11} + U_{12} \quad & A_2 &= i(U_{11} - U_{12}) \\ A_3 &= U_{13} + U_{14} \quad & A_4 &= i(U_{13} - U_{14}) \end{aligned} \tag{4.119}$$

Also we have

$$\begin{aligned} r_1 &= \alpha_1 + i\beta_1 \quad r_2 = \alpha_1 - i\beta_1 \\ r_3 &= \alpha_2 + i\beta_2 \quad r_3 = \alpha_2 - i\beta_2 \end{aligned} \tag{4.120}$$

which leads to the expressions

$$\begin{aligned} r_1U_{11} + r_2U_{12} &= \alpha_1 A_1 + \beta_1 A_2 \\ i(r_1U_{11} - r_2U_{12}) &= -\beta_1 A_1 + \alpha_1 A_2 \\ r_3U_{13} + r_4U_{14} &= \alpha_2 A_3 + \beta_2 A_4 \\ i(r_3U_{13} - r_4U_{14}) &= -\beta_2 A_3 + \alpha_2 A_4 \end{aligned} \tag{4.121}$$

Substituting Equations 4.119 and 4.121 into Equation 4.118 gives

$$\begin{aligned} u_1(t) &= e^{-\alpha_1 t}\left[A_1 \cos\omega_{d_1}t + A_2 \sin\omega_{d_1}t\right] \\ &\quad + e^{-\alpha_2 t}\left[A_3 \cos\omega_{d_2}t + A_4 \sin\omega_{d_2}t\right] \\ u_2(t) &= e^{-\alpha_1 t}\left[(\alpha_1 A_1 + \beta_1 A_2)\cos\omega_{d_1}t + (-\beta_1 A_1 + \alpha_1 A_2)\sin\omega_{d_1}t\right] \\ &\quad + e^{-\alpha_2 t}\left[(\alpha_2 A_3 + \beta_2 A_4)\cos\omega_{d_2}t + (-\beta_2 A_3 + \alpha_2 A_4)\sin\omega_{d_2}t\right] \end{aligned} \tag{4.122}$$

Rewriting Equation 4.122 as

$$\begin{aligned} u_1(t) &= e^{-\alpha_1 t}\left(A_1 \cos\omega_{d_1}t + A_2 \sin\omega_{d_1}t\right) + e^{-\alpha_2 t}\left(A_3 \cos\omega_{d_2}t + A_4 \sin\omega_{d_2}t\right) \\ u_2(t) &= e^{-\alpha_1 t}\left(r_{11}A_1 \cos\omega_{d_1}t + r'_{11}A_2 \sin\omega_{d_1}t\right) + e^{-\alpha_2 t}\left(r_{22}A_3 \cos\omega_{d_2}t + r'_{22}A_4 \sin\omega_{d_2}t\right) \end{aligned} \tag{4.123}$$

where the real amplitude ratios are

$$\begin{aligned} r_{11} &= \frac{(\alpha_1 A_1 + \beta_1 A_2)}{A_1} \quad r'_{11} = \frac{(-\beta_1 A_1 + \alpha_1 A_2)}{A_2} \\ r_{22} &= \frac{(\alpha_2 A_3 + \beta_2 A_4)}{A_3} \quad r'_{22} = \frac{(-\beta_2 A_3 + \alpha_2 A_4)}{A_4} \end{aligned} \tag{4.124}$$

Equation 4.118 can be rewritten as

$$\begin{aligned} u_1(t) &= X_1\, \mathrm{e}^{-\alpha_1 t} \cos\left(\omega_{d_1}t - \varphi_1\right) + X_2\, \mathrm{e}^{-\alpha_2 t} \cos\left(\omega_{d_2}t - \varphi_2\right) \\ u_2(t) &= X'_1\, \mathrm{e}^{-\alpha_1 t} \cos\left(\omega_{d_1}t - \varphi'_1\right) + X'_2\, \mathrm{e}^{-\alpha_2 t} \cos\left(\omega_{d_2}t - \varphi'_2\right) \end{aligned} \tag{4.125}$$

where

$$\begin{aligned} X_1 &= \sqrt{A_1^2 + A_2^2} \quad X_2 = \sqrt{A_3^2 + A_4^2} \\ X'_1 &= X_1\sqrt{\alpha_1^2 + \beta_1^2} \quad X'_2 = X_2\sqrt{\alpha_2^2 + \beta_2^2} \end{aligned} \tag{4.126}$$

$$
\begin{aligned}
\varphi_1 &= \tan^{-1}\left(\frac{A_2}{A_1}\right) \quad \varphi_2 = \tan^{-1}\left(\frac{A_4}{A_3}\right) \\
\varphi_1' &= \tan^{-1}\left(\frac{r_{11}' A_2}{r_{11} A_1}\right) \quad \varphi_2' = \tan^{-1}\left(\frac{r_{22}' A_4}{r_{22} A_3}\right)
\end{aligned}
\tag{4.127}
$$

The natural modes that exist have an out of phase relationship that complicates the analysis.

## 4.8 Damped Forced Vibration

Consider the two-degree-of-freedom system shown in Figure 4.18 subjected to an external force $P(t)$ applied to mass $m_1$ and whose equations of motion are given as

$$
\begin{bmatrix} m_1 & 0 \\ 0 & m_2 \end{bmatrix} \begin{Bmatrix} \ddot{u}_1 \\ \ddot{u}_2 \end{Bmatrix} + \begin{bmatrix} \eta_{11} & \eta_{12} \\ \eta_{21} & \eta_{22} \end{bmatrix} \begin{Bmatrix} \dot{u}_1 \\ \dot{u}_2 \end{Bmatrix} + \begin{bmatrix} k_{11} & k_{12} \\ k_{21} & k_{22} \end{bmatrix} \begin{Bmatrix} u_1 \\ u_2 \end{Bmatrix} = \begin{Bmatrix} P(t) \\ 0 \end{Bmatrix} \tag{4.128}
$$

where we have set

$$
\begin{aligned}
k_{11} &= k_1 + k_2 \quad k_{22} = k_2 + k_3 \quad k_{12} = k_{21} = -k_2 \\
\eta_{11} &= \eta_1 + \eta_2 \quad \eta_{22} = \eta_2 + \eta_3 \quad \eta_{12} = \eta_{21} = -\eta_2
\end{aligned}
\tag{4.129}
$$

The equations of motions, Equation 4.128, are coupled both elastically and viscously. It is noted that both the mass matrix, damping matrix, and stiffness matrix are symmetric. We shall only consider calculating the particular solution since the homogeneous solution has already been considered in Section 4.5. Although only one of our equations of motion is nonhomogeneous, we must determine particular solutions for both $u_1(t)$ and $u_2(t)$ due to the coupling of the equations. Consider the external force to be harmonic as

$$
P(t) = P_0 e^{i\omega_f t} \tag{4.130}
$$

where the actual applied force is given by the real part of Equation 4.130. Recalling method used for the forced response to harmonic forces of damped systems for one-degree-of-freedom systems leads us to assume that the displacements for a two-degree-of-freedom system are of the form

$$
u_1(t) = \bar{U}_1 e^{i\omega_f t} \quad u_2(t) = \bar{U}_2 e^{i\omega_f t} \tag{4.131}
$$

where $\bar{U}_1$ and $\bar{U}_2$ are complex constants given as

$$
\bar{U}_1 = U_1 e^{-i\psi} \quad \bar{U}_2 = U_2 e^{-i\psi} \tag{4.132}
$$

The actual displacements are given by the real part of Equation 4.131. Substituting Equation 4.131 into Equation 4.128 yields

$$\begin{aligned}&\left[-m_1\omega_f^2+k_{11}+i\eta_{11}\omega_f\right]\bar{U}_1e^{i\omega_f t}+\left[i\eta_{12}\omega_f+k_{12}\right]\bar{U}_2e^{i\omega_f t}=P_0e^{i\omega_f t}\\&\left[-m_2\omega_f^2+k_{22}+i\eta_{22}\omega_f\right]\bar{U}_2e^{i\omega_f t}+\left[i\eta_{12}\omega_f+k_{12}\right]\bar{U}_1e^{i\omega_f t}=0\end{aligned} \tag{4.133}$$

or since the exponential is common to each term we can recast Equation 4.133 into the matrix form as

$$\begin{bmatrix}k_{11}-m_1\omega_f^2+i\eta_{11}\omega_f & i\eta_{12}\omega_f+k_{12}\\ i\eta_{12}\omega_f+k_{12} & k_{22}-m_2\omega_f^2+i\eta_{22}\omega_f\end{bmatrix}\begin{Bmatrix}\bar{U}_1\\ \bar{U}_2\end{Bmatrix}=\begin{Bmatrix}P_0\\ 0\end{Bmatrix} \tag{4.134}$$

The complex amplitudes can be obtained from Equation 4.134 using Cramer's rule as

$$\begin{aligned}\bar{U}_1&=\frac{\left(k_{22}-m_2\omega_f^2+i\eta_{22}\omega_f\right)P_0}{\left(k_{11}-m_1\omega_f^2+i\eta_{11}\omega_f\right)\left(k_{22}-m_2\omega_f^2+i\eta_{22}\omega_f\right)-(i\eta_{12}\omega_f+k_{12})^2}\\ \bar{U}_2&=\frac{(ic_{12}\omega_f+k_{12})P_0}{\left(k_{11}-m_1\omega_f^2+i\eta_{11}\omega_f\right)\left(k_{22}-m_2\omega_f^2+i\eta_{22}\omega_f\right)-(i\eta_{12}\omega_f+k_{12})^2}\end{aligned} \tag{4.135}$$

To simplify the problem let us choose the following values for the mass, damping and stiffness coefficients $m_1=m_2=m$, $\eta_1=\eta_2=\eta_3=\eta$, and $k_1=k_2=k_3=k$. Thus, we have

$$\begin{aligned}&\eta_{11}=\eta_{22}=2\eta \quad \eta_{12}=\eta_{21}=-\eta\\&k_{11}=k_{22}=2k \quad k_{12}=k_{21}=-k\end{aligned} \tag{4.136}$$

Expanding and simplifying Equation 4.135 yields

$$\begin{aligned}\bar{U}_1&=\frac{\left(2k-m\omega_f^2+i2\eta\omega_f\right)P_0}{\left(2k-m\omega_f^2+i2\eta\omega_f\right)\left(2k-m\omega_f^2+i2\eta\omega_f\right)-(i\eta\omega_f+k)^2}\\ \bar{U}_2&=\frac{(k+i\eta\omega_f)P_0}{\left(2k-m\omega_f^2+i2\eta\omega_f\right)\left(2k-m\omega_f^2+i2\eta\omega_f\right)-(i\eta\omega_f+k)^2}\end{aligned} \tag{4.137}$$

or

$$\begin{aligned}\bar{U}_1&=\frac{\left(2k-m\omega_f^2+i2\eta\omega_f\right)P_0}{m^2\omega_f^4-4km\omega_f^2+3k^2-3(\eta\omega_f)^2+i2\eta\omega_f\left(3k-2m\omega_f^2\right)}\\ \bar{U}_2&=\frac{(k+i\eta\omega_f)P_0}{m^2\omega_f^4-4km\omega_f^2+3k^2-3(\eta\omega_f)^2+i2\eta\omega_f\left(3k-2m\omega_f^2\right)}\end{aligned} \tag{4.138}$$

It is noted that if $\eta = 0$, then Equation 4.138 yields the solution for the response of the undamped two-degree-of-freedom system where that portion of the denominator $m^2\omega_f^4 - 4km\omega_f^2 + 3k^2$ corresponds to the frequency equation. Therefore, we can rewrite Equation 4.138 as

$$
\begin{aligned}
\bar{U}_1 &= \frac{\left(2k - m\omega_f^2 + i2\eta\omega_f\right)P_0}{m^2\left(\omega_1^2 - \omega_f^2\right)\left(\omega_2^2 - \omega_f^2\right) - 3(\eta\omega_f)^2 + i2\eta\omega_f\left(3k - 2m\omega_f^2\right)} \\
\bar{U}_2 &= \frac{(k + i\eta\omega_f)P_0}{m^2\left(\omega_1^2 - \omega_f^2\right)\left(\omega_2^2 - \omega_f^2\right) - 3(\eta\omega_f)^2 + i2\eta\omega_f\left(3k - 2m\omega_f^2\right)}
\end{aligned}
\tag{4.139}
$$

To calculate the amplitudes of the forced response given by the real parts of Equation 4.131, some additional relations for complex numbers need to be developed. Recall that for a complex number we have

$$
a + ib = r(\cos\theta + i\sin\theta) = re^{i\theta} = \sqrt{a^2 + b^2}\,e^{i\theta} \quad \text{and} \quad \tan\theta = \frac{b}{a} \tag{4.140}
$$

For the ratio of a complex number, we have

$$
\begin{aligned}
\bar{U} &= \frac{X_1 + iY_1}{X_2 + iY_2} \\
U &= \frac{X_1 + iY_1}{X_2 + iY_2} \cdot \frac{X_2 - iY_2}{X_2 - iY_2} = \frac{(X_1X_2 + Y_1Y_2) + i(Y_1X_2 - X_1Y_2)}{X_2^2 + Y_2^2} \\
\bar{U} &= Ue^{i\beta} = \frac{\sqrt{(X_1X_2 + Y_1Y_2)^2 + (Y_1X_2 - X_1Y_2)^2}}{X_2^2 + Y_2^2} \cdot e^{i\beta} = \sqrt{\frac{X_1^2 + Y_1^2}{X_2^2 + Y_2^2}} \cdot e^{i\beta}
\end{aligned}
\tag{4.141}
$$

where

$$
\tan\beta = \frac{Y_1X_2 - X_1Y_2}{X_1X_2 + Y_1Y_2} \tag{4.142}
$$

Equation 4.139 can now be written as

$$
\begin{aligned}
U_1e^{-i\psi} &= \frac{P_0\sqrt{\left(2k - m\omega_f^2\right)^2 + (2\eta\omega_f)^2} \cdot e^{i\beta}}{\sqrt{\left[m^2\left(\omega_1^2 - \omega_f^2\right)\left(\omega_2^2 - \omega_f^2\right) - 3(\eta\omega_f)^2\right]^2 + (2\eta\omega_f)^2\left(3k - 2m\omega_f^2\right)^2}} \\
U_2e^{-i\psi} &= \frac{P_0\sqrt{k^2 + (\eta\omega_f)^2} \cdot e^{i\beta}}{\sqrt{\left[m^2\left(\omega_1^2 - \omega_f^2\right)\left(\omega_2^2 - \omega_f^2\right) - 3(\eta\omega_f)^2\right]^2 + (2\eta\omega_f)^2\left(3k - 2m\omega_f^2\right)^2}}
\end{aligned}
\tag{4.143}
$$

and $\psi = -\beta$, therefore

$$U_1 = \frac{P_0\sqrt{\left(2k - m\omega_f^2\right)^2 + (2\eta\omega_f)^2}}{\sqrt{\left[m^2\left(\omega_1^2 - \omega_f^2\right)\left(\omega_2^2 - \omega_f^2\right) - 3(\eta\omega_f)^2\right]^2 + (2\eta\omega_f)^2\left(3k - 2m\omega_f^2\right)^2}}$$
$$U_2 = \frac{P_0\sqrt{k^2 + (\eta\omega_f)^2}}{\sqrt{\left[m^2\left(\omega_1^2 - \omega_f^2\right)\left(\omega_2^2 - \omega_f^2\right) - 3(\eta\omega_f)^2\right]^2 + (2\eta\omega_f)^2\left(3k - 2m\omega_f^2\right)^2}} \tag{4.144}$$

as the amplitudes of the forced response. To investigate these amplitudes, they will be reduced to a nondimensional form. For this purpose, we introduce the following values:

$$\omega_1^2 = \frac{\lambda_1 k}{m} = \frac{k}{m} \quad \omega_2^2 = \frac{\lambda_2 k}{m} = \frac{3k}{m}$$
$$\Omega_1 = \frac{\omega_f}{\omega_1} \quad \Omega_2 = \frac{\omega_f}{\omega_2} = \Omega_1\frac{\omega_1}{\omega_2} = \Omega_1\sqrt{\frac{\lambda_1}{\lambda_2}} \tag{4.145}$$
$$U_0 = \frac{P_0}{k} \quad \zeta = \frac{\eta}{m\omega_1}$$

Substituting these variables into Equation 4.144 gives

$$U_1 = \frac{U_0\sqrt{\left(2 - \lambda_1\Omega_1^2\right)^2 + (2\zeta\lambda_1\Omega_1)^2}}{\sqrt{\left[\lambda_1\left(1 - \Omega_1^2\right)\left(\lambda_2 - \lambda_1\Omega_1^2\right) - 3(\zeta\lambda_1\Omega_1)^2\right]^2 + (2\zeta\lambda_1\Omega_1)^2\left(3 - 2\lambda_1\Omega_1^2\right)^2}}$$
$$U_2 = \frac{U_0\sqrt{1 + (\lambda_1\zeta\Omega_1)^2}}{\sqrt{\left[\lambda_1\left(1 - \Omega_1^2\right)\left(\lambda_2 - \lambda_1\Omega_1^2\right) - 3(\zeta\lambda_1\Omega_1)^2\right]^2 + (2\zeta\lambda_1\Omega_1)^2\left(3 - 2\lambda_1\Omega_1^2\right)^2}} \tag{4.146}$$

The steady-state amplitudes given in Equation 4.146 can be plotted versus the frequency ratio $\Omega_1$ for various damping variables $\zeta$. A plot of $U_1/U_0$ and $U_2/U_0$ is shown in Figure 4.19. It is noted that the damping factor $\zeta$ is not the same as what was defined for the SDOF system. This damping factor indicates the amount of damping present in the vibrating system. The curves' behavior is similar to that of an SDOF system except for the large amplitudes that occur at the two resonant frequencies $\omega_f = \omega_1 = \sqrt{k/m}$ and $\omega_f = \omega_2 = \sqrt{3k/m}$. It is noted that the amplitudes decrease as the damping is increased. The two curves are similar since we have an essentially symmetric system.

## 4.9 General Theory of Multi-Degree-of-Freedom Systems

We have looked at various conditions for a two-degree-of-freedom system and now we proceed to investigate a more general treatment of MDOF discrete systems (Humar 1990). It was noted that the two-degree-of-freedom system was computationally more involved than the single-degree-of-freedom system and the mathematical formulation increases as

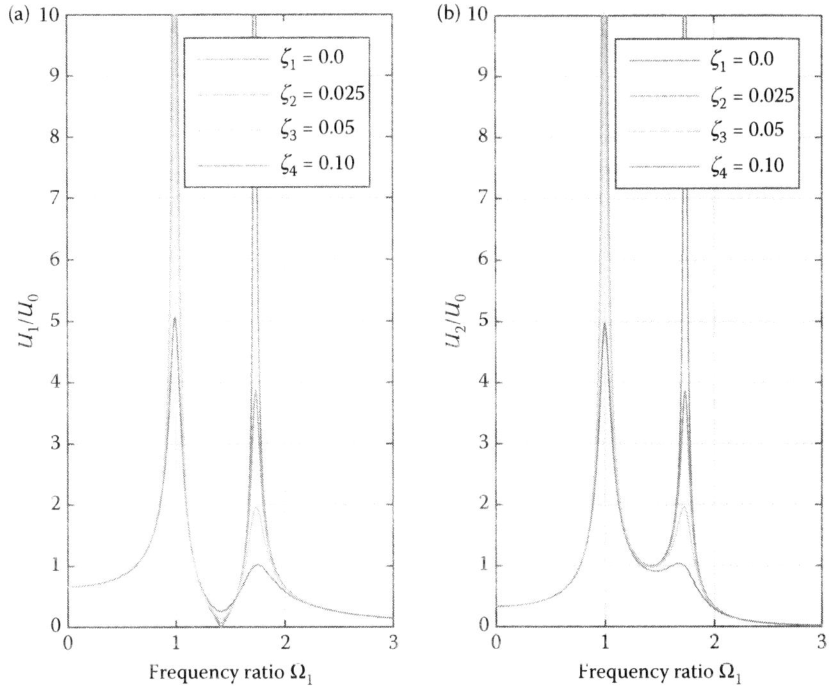


**FIGURE 4.19**
Steady-state amplitudes for two-degree-of-freedom system vs frequency ratio $\Omega_1$.

the number of masses or degrees of freedom increase. As more masses are added, the closer our analysis approaches a continuous system.

In terms of a set of generalized coordinates, the equations of motion are represented using Newton's law as

$$\sum_{j=1}^{n} m_{ij}\ddot{q}_j + \sum_{j=1}^{n} k_{ij}q_j = p_i \quad i = 1,2,\ldots,n \tag{4.147}$$

where $n$ is the number of degrees of freedom, $m_{ij}$ is a mass-type coefficient (mass, moment of inertia, etc.) and $k_{ij}$ is a stiffness coefficient, $q_j$ is a displacement, rotation, or a generalized coordinate and $p_i$ is a force. Using Lagrange's equation, we have for the kinetic and potential energy

$$\begin{aligned} T &= \frac{1}{2}(m_{11}\dot{q}_1\dot{q}_1 + m_{12}\dot{q}_1\dot{q}_2 + \cdots) = \frac{1}{2}\sum_{i=1}^{n}\sum_{j=1}^{n} m_{ij}\dot{q}_i\dot{q}_j \\ V &= \frac{1}{2}(k_{11}q_1q_1 + k_{12}q_1q_2 + \cdots) = \frac{1}{2}\sum_{i=1}^{n}\sum_{j=1}^{n} k_{ij}q_iq_j \end{aligned} \tag{4.148}$$

where $m_{ij}$ are the mass coefficients, $k_{ij}$ the stiffness coefficients of the system. In addition, $k_{ij} = k_{ji}$ and $m_{ij} = m_{ji}$ are symmetric coefficients. Thus, using Lagrange's equation as

$$\frac{d}{dt}\left(\frac{\partial T}{\partial \dot{q}_i}\right)+\frac{\partial V}{\partial q_i}=Q_i \quad i=1,2,\ldots,n \tag{4.149}$$

we have for the $i$th equation of motion

$$\sum_{j=1}^{n} m_{ij}\ddot{q}_j+\sum_{j=1}^{n} k_{ij}q_j=p_i \quad i=1,2,\ldots,n \tag{4.150}$$

Using the flexibility method with flexibility coefficients $f_{ij}$, we have

$$\sum_{k=1}^{n}\sum_{j=1}^{n} f_{ik}m_{kj}\ddot{q}_j+q_i=\sum_{k=1}^{n} f_{ik}p_k \quad i=1,2,\ldots,n \tag{4.151}$$

In special cases we might have $m_{ii}=m_i$, $m_{ij}=0$, $i\neq j$, thus we would then have

$$\sum_{k=1}^{n} f_{ik}m_k\ddot{q}_k+q_i=\sum_{k=1}^{n} f_{ik}p_k \tag{4.152}$$

We shall always have equations of the type shown in Equations 4.150 and 4.151, that is, those with indicial notation. Explicitly, if $i=1$, we could write Equation 4.150 as

$$m_{11}\ddot{q}_1+m_{12}\ddot{q}_2+\cdots+k_{11}q_1+k_{12}q_{12}\cdots=p_1$$

### 4.9.1 Matrix Notation for MDF Systems

The analyses of systems that have a finite number of degrees of freedom require the manipulation of linear algebraic equations. These equations become much more difficult to handle when the number of degrees of freedom is greater than three or four. Many systems are governed by sets of linear algebraic equations, where the use of matrices simplifies the solution (Bishop et al. 1965; Jennings 1977). Let us formulate the vibration equations in matrix notation.

The displacement and force vectors $\mathbf{q}$ and $\mathbf{p}$ are column matrices defined as

$$\mathbf{q}=\{q\}=\begin{Bmatrix} q_1\\ q_2\\ \vdots\\ q_n \end{Bmatrix} \quad \text{and} \quad \mathbf{p}=\{p\}=\begin{Bmatrix} p_1\\ p_2\\ \vdots\\ p_n \end{Bmatrix} \tag{4.153}$$

Similarly, the mass $[M]$, stiffness $[K]$, and flexibility $[F]$ matrices are of the form

$$[M]=\begin{bmatrix} m_{11} & m_{12} & \cdots & m_{1n}\\ m_{21} & m_{22} & \cdots & m_{2n}\\ \vdots & \vdots & \cdots & \vdots\\ m_{n1} & m_{n2} & \cdots & m_{nn} \end{bmatrix},[K]=\begin{bmatrix} k_{11} & k_{12} & \cdots & k_{1n}\\ k_{21} & k_{22} & \cdots & k_{2n}\\ \vdots & \vdots & \cdots & \vdots\\ k_{n1} & k_{n2} & \cdots & k_{nn} \end{bmatrix},[F]=\begin{bmatrix} f_{11} & f_{12} & \cdots & f_{1n}\\ f_{21} & f_{22} & \cdots & f_{2n}\\ \vdots & \vdots & \cdots & \vdots\\ f_{n1} & f_{n2} & \cdots & f_{nn} \end{bmatrix} \tag{4.154}$$

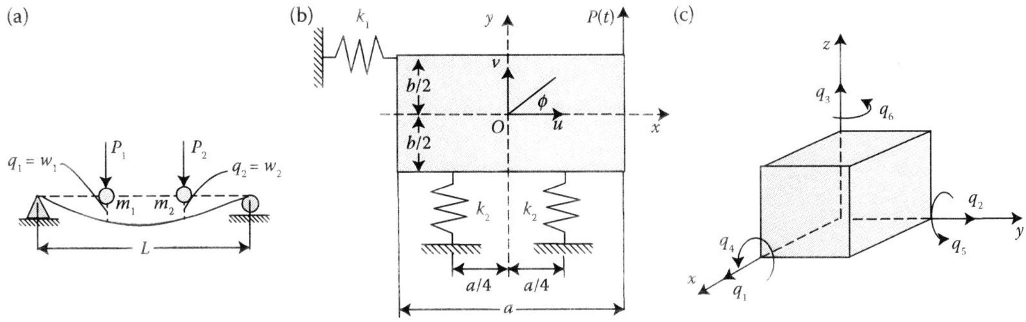


**FIGURE 4.20**
Systems representing (a) two-, (b) three-, (c) six-degree of freedom systems.

It is assumed that these matrices are positive definite. In the matrix format, the equations of motion are written as

$$[M]\ddot{\mathbf{q}} + [K]\mathbf{q} = \mathbf{P} \quad \text{and} \quad [F][M]\ddot{\mathbf{q}} + \mathbf{q} + [F]\mathbf{P} \tag{4.155}$$

or

$$[M]\{\ddot{q}\} + [K]\{q\} = \{P\} \quad \text{and} \quad [F][M]\{\ddot{q}\} + \{q\} = [F]\{P\} \tag{4.156}$$

The size of the matrices will obviously depend on the system being evaluated. Figure 4.20 presents examples of three different systems.

System (a) is a massless beam with two attached masses $m_1$ and $m_2$ subjected to external loads $P_1$ and $P_2$. The $[M]$, $[K]$, $\{q\}$, and $\{P\}$ matrices for this system are

$$[M] = \begin{bmatrix} m_{11} & 0 \\ 0 & m_{22} \end{bmatrix}, \quad [K] = \begin{bmatrix} k_{11} & k_{12} \\ k_{21} & k_{22} \end{bmatrix}, \quad \{q\} = \begin{Bmatrix} w_1 \\ w_2 \end{Bmatrix} \quad \text{and} \quad \{P\} = \begin{Bmatrix} P_1 \\ P_2 \end{Bmatrix}$$

System (b) is a three-degree-of-freedom system with displacements $u$ and $v$, and rotation $\varphi$. The springs are assumed to be linear ($F = k\delta$). The forces acting on the structure are $-X$, $-Y$, and a restoring moment $-M$. The $[M]$, $[K]$, $\{q\}$, and $\{P\}$ matrices for this system are

$$[M] = \begin{bmatrix} m & 0 & 0 \\ 0 & m & 0 \\ 0 & 0 & I_z \end{bmatrix}, [K] = \begin{bmatrix} k_1 & 0 & -k_1(b/2) \\ 0 & k_2 & 0 \\ -k_1(b/2) & 0 & k_2a^2/8 + k_1b^2/4 \end{bmatrix}, \{q\} = \begin{Bmatrix} q_1 \\ q_2 \\ q_3 \end{Bmatrix} = \begin{Bmatrix} u \\ v \\ \phi \end{Bmatrix},$$

$$\{P\} = \begin{Bmatrix} P_1 \\ P_2 \\ P_3 \end{Bmatrix} = \begin{Bmatrix} 0 \\ P(t) \\ M(t) = P(t)(a/2) \end{Bmatrix}$$

System (c) is a three-dimensional six-degree-of-freedom system subjected to three linear displacements and three angular displacements. The $[M]$, $[K]$, $\{q\}$, and $\{P\}$ matrices for this system are

$$[M]=\begin{bmatrix} m & 0 & 0 & 0 & 0 & 0 \\ 0 & m & 0 & 0 & 0 & 0 \\ 0 & 0 & m & 0 & 0 & 0 \\ 0 & 0 & 0 & I_{xx} & I_{xy} & I_{xz} \\ 0 & 0 & 0 & I_{yx} & I_{yy} & I_{yz} \\ 0 & 0 & 0 & I_{zx} & I_{zy} & I_{zz} \end{bmatrix} \quad \text{and} \quad \{q\}=\begin{Bmatrix} q_1 \\ q_2 \\ q_3 \\ q_4 \\ q_5 \\ q_6 \end{Bmatrix}=\begin{Bmatrix} u \\ v \\ w \\ \phi_x \\ \phi_y \\ \phi_z \end{Bmatrix}$$

## 4.10 Free Undamped MDOF Systems

To begin the solution method, we shall restrict the analysis to systems containing no damping elements and free of external forces. In terms of generalized coordinates and these conditions, our equations of freedom are represented with in the matrix form as

$$[M]\{\ddot{q}\}+[K]\{q\}=\{P\} \quad \text{or} \quad [F][M]\{\ddot{q}\}+\{q\}=[F]\{P\} \tag{4.157}$$

which is equivalent to the set of linear simultaneous homogeneous differential equations

$$\sum_{j=1}^{n} m_{ij}\ddot{q}_j+\sum_{j=1}^{n} k_{ij}q_j=0 \quad \text{or} \quad \sum_{k=1}^{n} f_{ik}m_k\ddot{q}_k+q_i=\sum_{k=1}^{n} f_{ik}p_k \quad i=1,2,\ldots,n \tag{4.158}$$

Assume a solution in the following form:

$$\{q(t)\}=\{a\}e^{i\omega t} \quad \{\ddot{q}\}=-\omega^2\{a\}e^{i\omega t} \tag{4.159}$$

where $\{a\}$ represents the possible deformed shapes that are constant in time. Substituting Equation 4.159 into Equations 4.157 or 4.158 yields

$$(-\omega^2[M]\{a\}+[K]\{a\})e^{i\omega t}=\{0\} \quad \text{or} \quad \left(\frac{1}{\omega^2}\{a\}-[F][M]\{a\}\right)e^{i\omega t}=\{0\} \tag{4.160}$$

or

$$([K]-\omega^2[M])\{a\}=\{0\} \quad \text{or} \quad \left(\frac{1}{\omega^2}[I]-[F][M]\right)\{a\}=\{0\} \tag{4.161}$$

where $[I]$ is the unit matrix. It is noted that $\{a\}=\{0\}$ is a solution to Equation 4.161, but is recognized as the trivial solution in that $\{a\}=\{0\}$ implies no motion at all. A nontrivial solution is obtained only if the determinant, known as the frequency determinant, corresponding to $([K]-\omega^2[M])$ or $((1/\omega^2)[I]-[F][M])$ vanishes, that is

$$\left|[K]-\omega^2[M]\right|=0 \quad \text{or} \quad \left|\frac{1}{\omega^2}[I]-[F][M]\right|=0 \tag{4.162}$$

The *frequency equation* or *characteristic equation* obtained from Equation 4.162 when expanded is a polynomial of degree $n$ in $\omega^2$. Thus,

$$(\omega^2)^n + C_1(\omega^2)^{n-1} + C_2(\omega^2)^{n-2} + \cdots + C_{n-1}(\omega^2) + C_n = 0 \tag{4.163}$$

Equation 4.163 possesses, in general, $n$ distinct roots $\omega_r$, which are called *characteristic values* or *eigenvalues*. It can be shown that if the mass matrix $[M]$ is positive definite and the stiffness matrix $[K]$ is either positive definite or positive semidefinite, then the eigenvalues of this polynomial $\omega_1^2, \omega_2^2, \ldots, \omega_n^2$ are real nonnegative numbers. The quantities $\omega_1$, $\omega_2$, …, $\omega_n$ are the *natural circular frequencies* of the system. The natural frequencies can be arranged in the ascending order $\omega_1 < \omega_2 < \cdots < \omega_n$, where all frequencies are assumed distinct. The lowest frequency $\omega_1$ is referred to as the *fundamental frequency*. For each value of $\omega_r$, the matrix $([K] - \omega^2[M])$ or $((1/\omega^2)[I] - [F][M])$ is singular and Equation 4.162 has a solution vector which can be written as $\{a^{(r)}\}$ or

$$\{a^{(r)}\} = \left\{a_1^{(r)} \quad a_2^{(r)} \quad \cdots \quad a_n^{(r)}\right\}^T \tag{4.164}$$

whose components $a_i^{(r)}$ are real numbers and the vector $\{a^{(r)}\}$ is called an *eigenvector, characteristic vector, natural mode*, or *mode shape* (Fu and He 2001) which satisfies

$$\left([K] - \omega_r^2[M]\right)\{a^{(r)}\} = \{0\} \quad \text{or} \quad \left(\frac{1}{\omega_r^2}[I] - [F][M]\right)\{a^{(r)}\} = \{0\} \tag{4.165}$$

If the value of one of the $n$ elements of a natural mode vector $\{a^{(r)}\}$ is assumed arbitrarily, the remaining $n-1$ elements are determined uniquely by solving $n-1$ simultaneous equations. It is noted that the *natural mode* or *mode shape* has been determined but not the mode's amplitude. The set of amplitudes $\{a^{(r)}\}$, $\left\{a_1^{(r)}, a_2^{(r)}, \ldots, a_n^{(r)}\right\}^T$ corresponding to each $\omega_r$ determines the $r$th *natural mode*, and the number of natural modes is equal to the number of frequencies. The resulting modal vectors are called *normal modes* or *principal modes*.

### 4.10.1 Orthogonality of Natural Modes

For an MDOF system, consider $\{a^{(r)}\}$ and $\{a^{(s)}\}$ to be the normal modes corresponding to two natural frequencies $\omega_r$ and $\omega_s$. Then from Equation 4.161, we have

$$\omega_r^2[M]\{a^{(r)}\} = [K]\{a^{(r)}\} \tag{4.166}$$

$$\omega_s^2[M]\{a^{(s)}\} = [K]\{a^{(s)}\} \tag{4.167}$$

Premultiplying Equation 4.166 by $\{a^{(s)}\}^T$ and Equation 4.167 by $\{a^{(r)}\}^T$ where $T$ stands for the transpose yields

$$\omega_r^2\{a^{(s)}\}^T[M]\{a^{(r)}\} = \{a^{(s)}\}^T[K]\{a^{(r)}\} \tag{4.168}$$

$$\omega_s^2\{a^{(r)}\}^T[M]\{a^{(s)}\} = \{a^{(r)}\}^T[K]\{a^{(s)}\} \tag{4.169}$$

Because both $[M]$ and $[K]$ are symmetrical, that is, $[M] = [M]^T$ and $[K] = [K]^T$, we have

$$\{a^{(s)}\}^T[M]\{a^{(r)}\} = \{a^{(r)}\}^T[M]\{a^{(s)}\} \tag{4.170}$$

$$\{a^{(s)}\}^T[K]\{a^{(r)}\} = \{a^{(r)}\}^T[K]\{a^{(s)}\} \tag{4.171}$$

Upon subtracting Equation 4.169 from 4.168 we have

$$\left(\omega_r^2 - \omega_s^2\right)\{a^{(s)}\}^T[M]\{a^{(r)}\} = 0 \tag{4.172}$$

Since $\omega_r \neq \omega_s$, we have the proof

$$\{a^{(s)}\}^T[M]\{a^{(r)}\} = 0 \quad \text{for } \omega_r \neq \omega_s \tag{4.173}$$

Substituting Equation 4.173 into Equation 4.168, we obtain the second orthogonality property

$$\{a^{(s)}\}^T[K]\{a^{(r)}\} = 0 \quad \text{for } \omega_r \neq \omega_s \tag{4.174}$$

Equations 4.173 and 4.174 are known as the orthogonality relationship between the eigenvectors of the modes of free vibration of an undamped system with respect to the mass and stiffness matrices. This relationship is used in solving eigenvalue problems and for solving forced vibration problems. Also, it is a fundamental property of MDOF systems.

### 4.10.2 Normalized Modes

As noted, corresponding to each eigenvalue, $\omega_r^2$, there will be an *eigenvector, characteristic vector, natural mode,* or *mode shape,* given as $\{a^{(r)}\}$, $r = 1, 2, \ldots, n$ and the modes are determined within a multiplicative constant. Therefore, the modes can be scaled or normalized in any convenient manner. When calculating the natural modes $\{a^{(r)}\}$, we assume that the amplitude of some chosen point is unity then the remaining $n - 1$ elements are uniquely determined. This process of scaling a natural mode such that each of its elements has a unique value is called *normalization* and the resulting modal vectors are called *normal modes*. This method may prove to be inconvenient in matrix calculations and another method of standardization which leads to normalized modal vectors or eigenvectors may be adopted. There are many different conventions used by other authors but one that is popular is the following. Instead of using the amplitude $\{a^{(r)}\}$, we use the normalized modal vector $\{\phi^{(r)}\}$, a dimensionless column vector whose $n$ components describe the relative amplitudes of the various points of the system as it vibrates freely in its $r$th mode with circular frequency $\omega_r$. The modal vector is proportional to the mode $\{a^{(r)}\}$, thus

$$\{\phi^{(r)}\} = \alpha_r\{a^{(r)}\} \tag{4.175}$$

where $\alpha_r$ is a constant. The constant $\alpha_r$ is chosen so that

$$\{\phi^{(r)}\}^T[M]\{\phi^{(r)}\} = \bar{m} \tag{4.176}$$

In general we can assume $\bar{m} = 1$, but it can have any value. Usually the value of $\bar{m}$ is chosen for the convenience of the problem under consideration. Note that if $\{\phi^{(r)}\}$ is dimensionless $\bar{m}$ will have the dimensions of mass. When the modal vectors $\{\phi^{(r)}\}$ are collected into a single square matrix $[\Phi]$ of order $n$, the resulting matrix is called the *modal matrix* $[\Phi]$. Thus,

$$[\Phi] = [\{\phi^{(1)}\} \quad \{\phi^{(2)}\} \quad \cdots \quad \{\phi^{(n)}\}] \tag{4.177}$$

or

$$[\Phi] = \begin{bmatrix} \phi_1^{(1)} & \phi_1^{(2)} & \cdots & \phi_1^{(n)} \\ \phi_2^{(1)} & \phi_2^{(2)} & \cdots & \phi_2^{(n)} \\ \vdots & \vdots & \vdots & \vdots \\ \phi_n^{(1)} & \phi_n^{(2)} & \cdots & \phi_n^{(n)} \end{bmatrix} \tag{4.178}$$

Recalling the property of orthogonality (Equation 4.173), we have

$$\{\phi^{(r)}\}^T[M]\{\phi^{(s)}\} = 0 \quad \text{for } r \neq s$$

Using Equation 4.176, we have

$$\begin{aligned} [\Phi]^T[M][\Phi] &= \begin{bmatrix} \{\phi^{(1)}\}^T[M]\{\phi^{(1)}\} & \{\phi^{(1)}\}^T[M]\{\phi^{(2)}\} & \cdots & \{\phi^{(1)}\}^T[M]\{\phi^{(n)}\} \\ \{\phi^{(2)}\}^T[M]\{\phi^{(1)}\} & \{\phi^{(2)}\}^T[M]\{\phi^{(2)}\} & \cdots & \{\phi^{(2)}\}^T[M]\{\phi^{(n)}\} \\ \cdots & \cdots & \cdots & \cdots \\ \{\phi^{(n)}\}^T[M]\{\phi^{(1)}\} & \{\phi^{(n)}\}^T[M]\{\phi^{(2)}\} & \cdots & \{\phi^{(n)}\}^T[M]\{\phi^{(n)}\} \end{bmatrix} \\ &= \begin{bmatrix} \bar{m} & 0 & \cdots & 0 \\ 0 & \bar{m} & \cdots & 0 \\ \cdots & \cdots & \cdots & \cdots \\ 0 & 0 & \cdots & m \end{bmatrix} \\ &= \bar{m}[I] \end{aligned} \tag{4.179}$$

Then,

$$[\Phi]^T[M][\Phi] = \bar{m}[I] \tag{4.180}$$

In addition, another relation that involves the stiffness matrix $[K]$ can be determined since $\{\phi^{(r)}\}$ satisfies Equation 4.165, where

$$[K]\{\phi^{(r)}\} = \omega_r^2[M]\{\phi^{(r)}\} \tag{4.181}$$

Premultiplying both sides of Equation 4.181 by $\{\phi^{(s)}\}^T$ gives

$$\{\phi^{(s)}\}^T[K]\{\phi^{(r)}\} = \omega_r^2\{\phi^{(s)}\}^T[M]\{\phi^{(r)}\} \tag{4.182}$$

but, based on Equation 4.173, this becomes

$$\{\phi^{(s)}\}^T[K]\{\phi^{(r)}\} = \omega_r^2\{\phi^{(s)}\}^T[M]\{\phi^{(r)}\} = 0 \quad \text{for } r \neq s \tag{4.183}$$

and for $r = s$

$$\{\phi^{(r)}\}^T[K]\{\phi^{(r)}\} = \omega_r^2\{\phi^{(r)}\}^T[M]\{\phi^{(r)}\} = \bar{m}\omega_r^2 \tag{4.184}$$

Now as before from the product $[\Phi]^T[K][\Phi]$ gives

$$\begin{aligned}
[\Phi]^T[K][\Phi] &= \begin{bmatrix} \{\phi^{(1)}\}^T[K]\{\phi^{(1)}\} & \{\phi^{(1)}\}^T[K]\{\phi^{(2)}\} & \cdots & \{\phi^{(1)}\}^T[K]\{\phi^{(n)}\} \\ \{\phi^{(2)}\}^T[K]\{\phi^{(1)}\} & \{\phi^{(2)}\}^T[K]\{\phi^{(2)}\} & \cdots & \{\phi^{(2)}\}^T[K]\{\phi^{(n)}\} \\ \cdots & \cdots & \cdots & \cdots \\ \left\{\phi^{(n)}\right\}^T[K]\left\{\phi^{(1)}\right\} & \left\{\phi^{(n)}\right\}^T[K]\left\{\phi^{(2)}\right\} & \cdots & \{\phi^{(n)}\}^T[K]\{\phi^{(n)}\} \end{bmatrix} \\
&= \begin{bmatrix} \bar{m}\omega_1^2 & 0 & \cdots & 0 \\ 0 & \bar{m}\omega_2^2 & \cdots & 0 \\ \cdots & \cdots & \cdots & \cdots \\ 0 & 0 & \cdots & \bar{m}\omega_n^2 \end{bmatrix} = \bar{m}\begin{bmatrix} \omega_1^2 & 0 & \cdots & 0 \\ 0 & \omega_2^2 & \cdots & 0 \\ \cdots & \cdots & \cdots & \cdots \\ 0 & 0 & \cdots & \omega_n^2 \end{bmatrix} \\
&= \bar{m}\left[\omega_i^2\right]_D
\end{aligned} \tag{4.185}$$

where

$$[\omega_i^2]_D = \begin{bmatrix} \omega_1^2 & 0 & \cdots & 0 \\ 0 & \omega_2^2 & \cdots & 0 \\ \cdots & \cdots & \cdots & \cdots \\ 0 & 0 & \cdots & \omega_n^2 \end{bmatrix} = \text{diagonal}\,[\omega_i^2] \tag{4.186}$$

### 4.10.3 Free Vibrations with Give Initial Conditions

The equation governing the free vibration of an undamped system having $n$ degrees of freedom is

$$[M]\{\ddot{q}\} + [K]\{q\} = \{0\} \tag{4.187}$$

where $\{q\}$ is a vector of generalized coordinates $q_i(t)$, $i = 1, 2, \ldots, n$. The general solution of Equation 4.187 can be written in the nonmatrix form as

$$\begin{aligned} q_i(t) &= A_1\phi_i^{(1)}\cos\omega_1 t + B_1\phi_i^{(1)}\sin\omega_1 t + A_2\phi_i^{(2)}\cos\omega_2 t + B_2\phi_i^{(2)}\sin\omega_2 t \\ &\quad + \cdots + A_n\phi_i^{(n)}\cos\omega_n t + B_n\phi_i^{(n)}\sin\omega_n t \end{aligned} \tag{4.188}$$

where $A_1, A_2, \ldots, A_n$ and $B_1, B_2, \ldots, B_n$ are arbitrary constants that can be determined in such a manner that a system with initial conditions is satisfied. Using a better notation, we can rewrite Equation 4.188 as

$$\begin{aligned}\{q\} &= \{\phi^{(1)}\}[A_1 \cos\omega_1 t + B_1 \sin\omega_1 t] + \{\phi^{(2)}\}[A_2 \cos\omega_2 t + B_2 \sin\omega_2 t] \\ &\quad + \cdots + \{\phi^{(n)}\}[A_n \cos\omega_n t + B_n \sin\omega_1 t]\end{aligned} \tag{4.189}$$

or

$$\{q\} = [\Phi]([C]_D\{A\} + [S]_D\{B\}) \tag{4.190}$$

where

$$[C]_D = \begin{bmatrix} \cos\omega_1 t & 0 & 0 & 0 \\ 0 & \cos\omega_2 t & 0 & 0 \\ 0 & 0 & \ddots & 0 \\ 0 & 0 & 0 & \cos\omega_n t \end{bmatrix} \text{ and } [S]_D = \begin{bmatrix} \sin\omega_1 t & 0 & 0 & 0 \\ 0 & \sin\omega_2 t & 0 & 0 \\ 0 & 0 & \ddots & 0 \\ 0 & 0 & 0 & \sin\omega_n t \end{bmatrix} \tag{4.191}$$

are diagonal matrices whose elements are functions cos $\omega_i t$ and sin $\omega_i t$ and $\{A\}$ and $\{B\}$ equal $\{A_1 \quad A_2 \quad \ldots \quad A_n\}^T$ and $\{B_1 \quad B_2 \quad \ldots \quad B_n\}^T$, respectfully. Assume the initial conditions at time $t = 0$ are

$$\{q(t=0)\} = \{q_0\} \quad \text{and} \quad \{\dot{q}(t=0)\} = \{\dot{q}_0\} \tag{4.192}$$

where $\{q_0\}$ and $\{\dot{q}_0\}$ are given initial values. At $t = 0$, we also have that

$$\begin{aligned}&[C]_D = [I], \quad [S]_D = [0] \\ &[\dot{C}]_D = [0], \quad [\dot{S}]_D = [\omega_i]_D\end{aligned} \tag{4.193}$$

where

$$[\omega_i]_D = \begin{bmatrix} \omega_1 & 0 & 0 & 0 \\ 0 & \omega_2 & 0 & 0 \\ 0 & 0 & \ddots & 0 \\ 0 & 0 & 0 & \omega_n \end{bmatrix} \tag{4.194}$$

is a diagonal matrix whose elements are the natural frequencies. Applying the conditions at $t = 0$ to Equation 4.190, we have

$$\begin{aligned}\{q(t=0)\} &= [\Phi]([I]\{A\}) = \{q_0\} \\ \{\dot{q}(t=0)\} &= [\Phi]([\omega_i]_D\{B\}) = \{\dot{q}_0\}\end{aligned} \tag{4.195}$$

from which we obtain

$$[\Phi]\{A\} = \{q_0\} \tag{4.196}$$

$$[\Phi][\omega_i]_D\{B\} = \{\dot{q}_0\} \tag{4.197}$$

so that

$$\{A\} = [\Phi]^{-1}\{q_0\} \tag{4.198}$$

$$\{B\} = \left([\Phi][\omega_i]_D\right)^{-1}\{\dot{q}_0\} \tag{4.199}$$

For large $n$, the inverse of the modal matrix $[\Phi]$ may be time-consuming and difficult. A more convenient expression can be determined by making use of Equation 4.179. Thus, premultiplying both sides of Equations 4.198 and 4.199 by $[\Phi]^T[M]$ gives

$$[\Phi]^T[M]\{q_0\} = [\Phi]^T[M][\Phi]\{A\} = \overline{m}\{A\} \tag{4.200}$$

$$[\Phi]^T[M]\{\dot{q}_0\} = [\Phi]^T[M][\Phi][\omega_i]_D\{B\} = \overline{m}[\omega_i]_D\{B\} \tag{4.201}$$

Therefore,

$$\{A\} = \frac{1}{\overline{m}}[\Phi]^T[M]\{q_0\} \tag{4.202}$$

$$\{B\} = \frac{1}{\overline{m}}[\omega_i]_D^{-1}[\Phi]^T[M]\{\dot{q}_0\} \tag{4.203}$$

Equation 4.190 describes free vibration with initial conditions $\{q_0\}$ and $\{\dot{q}_0\}$ which can be written as

$$\{q\} = \frac{1}{\overline{m}}[\Phi]\left([C]_D[\Phi]^T[M]\{q_0\} + [S]_D[\omega_i]_D^{-1}[\Phi]^T[M]\{\dot{q}_0\}\right) \tag{4.204}$$

Equation 4.204 is much more convenient to use since it only requires matrix multiplications instead of having to take the inverse of the modal matrix.

**EXAMPLE 4.5**

1. Derive the equations of motion for the system shown in Figure 4.21 and determine the natural frequencies and normal modes using $k_1 = k_2 = k_3 = 4000$ N/m, $m_1 = 100$ kg, $m_2 = m_3 = 10$ kg assuming zero damping.
2. Show the orthogonality property of the normal modes.
3. Form the modal matrix for the system.
4. Determine the equation of motion for the system when subjected to initial conditions $\{q_0\} = d\{1 \quad 0 \quad 1\}^T$ and $\{\dot{q}_0\} = v\{0 \quad 0 \quad 1\}$ at time $t_0 = 0$.

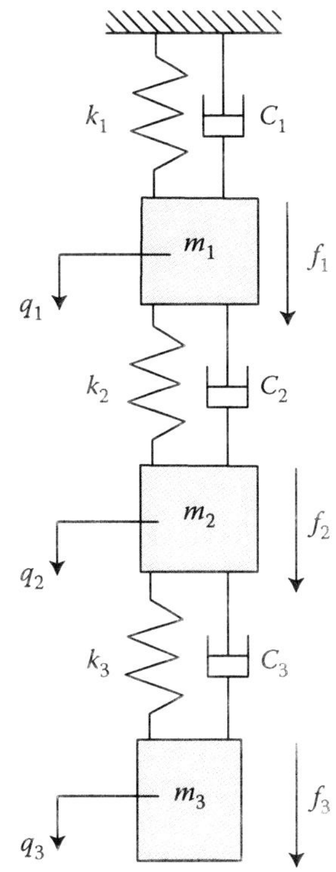


**FIGURE 4.21**
Spring–mass–damping system.

**Solution**

1. The free-body diagram of each of the masses is shown in Figure 4.22, where $q_1$, $q_2$, and $q_3$ are the displacements of masses $m_1$, $m_2$, and $m_3$.
   Summation of forces in the horizontal direction leads to the equilibrium equations

$$m_1\ddot{q}_1 + (c_1 + c_2)\dot{q}_1 - c_2\dot{q}_2 + (k_1 + k_2)q_1 - k_2q_2 = f_1(t)$$
$$m_2\ddot{q}_2 + (c_2 + c_3)\dot{q}_2 - c_2\dot{q}_1 - c_3\dot{q}_3 + (k_2 + k_3)q_2 - k_1q_1 - k_3q_3 = f_2(t)$$
$$m_3\ddot{q}_3 + c_3\dot{q}_3 - c_3\dot{q}_2 + k_3q_3 - k_3q_2 = f_3(t)$$

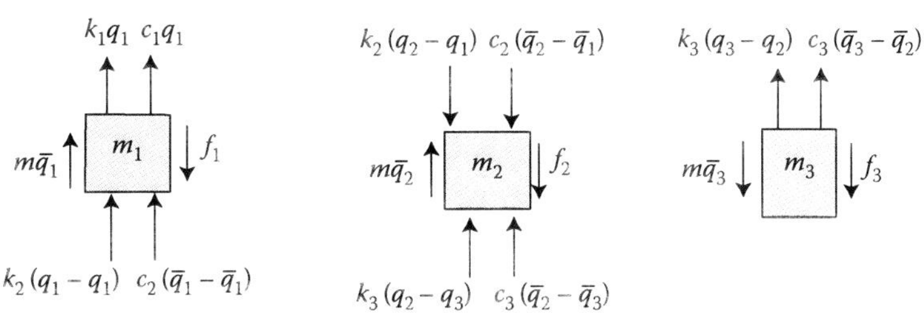


**FIGURE 4.22**
Free-body diagram of forces acting on the masses of the system depicted in Figure 4.21.

In the matrix form, we have

$$\begin{bmatrix} m_1 & 0 & 0 \\ 0 & m_2 & 0 \\ 0 & 0 & m_3 \end{bmatrix} \begin{Bmatrix} \ddot{q}_1 \\ \ddot{q}_2 \\ \ddot{q}_3 \end{Bmatrix} + \begin{bmatrix} c_1 + c_2 & -c_2 & 0 \\ -c_2 & c_2 + c_3 & -c_3 \\ 0 & -c_3 & c_3 \end{bmatrix} \begin{Bmatrix} \dot{q}_1 \\ \dot{q}_2 \\ \dot{q}_3 \end{Bmatrix}$$

$$+ \begin{bmatrix} k_1 + k_2 & -k_2 & 0 \\ -k_2 & k_2 + k_3 & -k_3 \\ 0 & -k_3 & k_3 \end{bmatrix} \begin{Bmatrix} q_1 \\ q_2 \\ q_3 \end{Bmatrix} = \begin{Bmatrix} f_1(t) \\ f_2(t) \\ f_3(t) \end{Bmatrix}$$

or

$$[M]\{\ddot{q}\} + [C]\{\dot{q}\} + [K]\{q\} = \{f(t)\}$$

Thus, we have

$$[M] = \begin{bmatrix} 100 & 0 & 0 \\ 0 & 10 & 0 \\ 0 & 0 & 10 \end{bmatrix} \quad \text{and} \quad [K] = \begin{bmatrix} 8000 & -4000 & 0 \\ -4000 & 8000 & -4000 \\ 0 & -4000 & 4000 \end{bmatrix}$$

And Equation 4.161 becomes

$$\begin{bmatrix} 8000 - 100\omega^2 & -4000 & 0 \\ -4000 & 8000 - 10\omega^2 & -4000 \\ 0 & -4000 & 8000 - 10\omega^2 \end{bmatrix} \begin{Bmatrix} a_1 \\ a_2 \\ a_3 \end{Bmatrix} = \begin{Bmatrix} 0 \\ 0 \\ 0 \end{Bmatrix}$$

or

$$\begin{bmatrix} 80 - \omega^2 & -40 & 0 \\ -400 & 800 - \omega^2 & -400 \\ 0 & -400 & 800 - \omega^2 \end{bmatrix} \begin{Bmatrix} a_1 \\ a_2 \\ a_3 \end{Bmatrix} = \begin{Bmatrix} 0 \\ 0 \\ 0 \end{Bmatrix}$$

The frequency equation is given as

$$\omega^6 - 1280\omega^4 + 240{,}000\omega^2 - 6{,}400{,}000 = 0$$

The roots are

$$\omega_1^2 = 32.0 \quad \omega_2^2 = 188.9 \quad \omega_3^2 = 1059.1$$

and the natural frequencies are

$$\omega_1 = 5.6568\,\text{Hz} \quad \omega_2 = 13.7441\,\text{Hz} \quad \omega_3 = 32.5438\,\text{Hz}$$

The normal modes can be determined by arbitrarily setting $a_1 = 1$ and solving the above equation for $a_2$ and $a_3$. Thus, we find the normal modes as

$$\{a_1\} = \begin{Bmatrix} 0.7665 \\ 0.9200 \\ 1.0000 \end{Bmatrix} \quad \{a_2\} = \begin{Bmatrix} -0.1938 \\ 0.5277 \\ 1.0000 \end{Bmatrix} \quad \{a_3\} = \begin{Bmatrix} -0.0409 \\ 1.0000 \\ -0.6069 \end{Bmatrix}$$

2. The orthogonality property of the normal modes is determined as

$$\{\Phi_1\} = c_1\{a_1\} \quad \{\Phi_2\} = c_2\{a_2\} \quad \{\Phi_3\} = c_3\{a_3\}$$

$$\{a_1\}^T[M]\{a_2\} = 3.5527e^{-15} \; \{a_2\}^T[M]\{a_3\} = 1.7764e^{-15} \; \{a_1\}^T[M]\{a_3\} = -8.8818e^{-16}$$

Notice that the values obtained are not exactly zero. This is due to round off errors.

3. The modal matrix for the system can be obtained as follows:
The normalized modal vector is proportional to the normal mode $\{a_i\}$, so that

$$\{\Phi_i\} = c_i\{a_i\}$$

where the factor $c_i$ is chosen such that

$$\{\Phi_i\}^T[M]\{\Phi_i\} = \bar{m}$$

As mentioned above, the value $\bar{m}$ is selected for convenience of the problem at hand. When the vectors are collected in a single square matrix, we have

$$[\Phi] = [\{\Phi_1\} \quad \{\Phi_2\} \quad \cdots \quad \{\Phi_n\}]$$

Using the modal matrices, we have

$$\{\Phi_1\} = c_1\{a_1\} \quad \{\Phi_2\} = c_2\{a_2\} \quad \{\Phi_3\} = c_3\{a_3\}$$

and forming the matrix product $\{\Phi_i\}^T[M]\{\Phi_i\} = \bar{m}$ we obtain the following:

$$c_1 = 0.1138\sqrt{\bar{m}} \quad c_2 = 0.2459\sqrt{\bar{m}} \quad c_3 = 0.2687\sqrt{\bar{m}}$$

Hence, we find

$$\{\Phi_1\} = \begin{Bmatrix} 0.0872 \\ 0.1047 \\ 0.1138 \end{Bmatrix}\sqrt{\bar{m}} \quad \{\Phi_2\} = \begin{Bmatrix} -0.0477 \\ 0.1298 \\ 0.2459 \end{Bmatrix}\sqrt{\bar{m}} \quad \{\Phi_3\} = \begin{Bmatrix} 0.0110 \\ -0.2687 \\ 0.1631 \end{Bmatrix}\sqrt{\bar{m}}$$

and the modal matrix becomes

$$[\Phi] = \sqrt{\bar{m}}\begin{bmatrix} 0.0872 & -0.0477 & 0.0110 \\ 0.1047 & 0.1298 & -0.2687 \\ 0.1138 & 0.2459 & 0.1631 \end{bmatrix}$$

4. The equation of motion for the system when subjected to initial conditions $\{q_0\} = x_0\{1 \quad 0 \quad 1\}^T$ and $\{\dot{q}_0\} = \dot{x}_0\{0 \quad 0 \quad 1\}$ at time $t = t_0$ we need to calculate the values $\{A\}$ and $\{B\}$ from Equation 4.189. Thus, we find

$$\{A\} = \frac{1}{m}[\Phi]^T[M]\{q_0\} = \frac{d}{\sqrt{m}}\begin{Bmatrix} 9.8580 \\ -2.3110 \\ 2.7310 \end{Bmatrix} \quad \{B\} = \frac{1}{m}[\omega_i]_D^{-1}[\Phi]^T[M]\{\dot{q}_0\} = \frac{v}{\sqrt{m}}\begin{Bmatrix} 0.2012 \\ 0.1789 \\ 0.0501 \end{Bmatrix}$$

and the solution is given as

$$\{q\} = [\Phi]([C]_D\{A\} + [S]_D\{B\})$$

$$\{q\} = \sqrt{m}\begin{bmatrix} 0.0872 & -0.0477 & 0.0110 \\ 0.1047 & 0.1298 & -0.2687 \\ 0.1138 & 0.2459 & 0.1631 \end{bmatrix}\begin{bmatrix} \cos\omega_1(t-t_0) & 0 & 0 \\ 0 & \cos\omega_2(t-t_0) & 0 \\ 0 & 0 & \cos\omega_3(t-t_0) \end{bmatrix}$$

$$\frac{d}{\sqrt{m}}\begin{Bmatrix} 9.8580 \\ -2.3110 \\ 2.7310 \end{Bmatrix} + \sqrt{m}\begin{bmatrix} 0.0872 & -0.0477 & 0.0110 \\ 0.1047 & 0.1298 & -0.2687 \\ 0.1138 & 0.2459 & 0.1631 \end{bmatrix}$$

$$\begin{bmatrix} \sin\omega_1(t-t_0) & 0 & 0 \\ 0 & \sin\omega_2(t-t_0) & 0 \\ 0 & 0 & \sin\omega_3(t-t_0) \end{bmatrix}\frac{v}{\sqrt{m}}\begin{Bmatrix} 0.2012 \\ 0.1789 \\ 0.0501 \end{Bmatrix}$$

$$\{q\} = d\begin{Bmatrix} 1.0000 \\ -2.3110 \\ 1.0000 \end{Bmatrix}\begin{bmatrix} \cos\omega_1(t-t_0) & 0 & 0 \\ 0 & \cos\omega_2(t-t_0) & 0 \\ 0 & 0 & \cos\omega_3(t-t_0) \end{bmatrix}$$

$$+ v\begin{Bmatrix} 0.2012 \\ 0.1789 \\ 0.0501 \end{Bmatrix}\begin{bmatrix} \sin\omega_1(t-t_0) & 0 & 0 \\ 0 & \sin\omega_2(t-t_0) & 0 \\ 0 & 0 & \sin\omega_3(t-t_0) \end{bmatrix}$$

with

$$\omega_1 = 5.6568 \text{ Hz} \quad \omega_2 = 13.7441 \text{ Hz} \quad \omega_3 = 32.5438 \text{ Hz}$$

## 4.11 Forced Undamped Vibrations

We will consider some special cases of forced undamped motion whose equation of motion is given as

$$[M]\{\ddot{q}\} + [K]\{q\} = \{p(t)\} \tag{4.205}$$

### 4.11.1 Steady-State Harmonic Motion

The external forces $\{p(t)\}$ are harmonic with the same frequency thus $\{p(t)\} = \{P\}\mathrm{Re}(e^{i\omega_f t})$ where $\{P\} = \{P_1 \ P_2 \ \cdots \ P_n\}^T$, then the steady-state solution has the form $\{q(t)\} = \mathrm{Re}(\{x\}e^{i\omega_f t})$.

In order to find $\{x\}$, we substitute the previous expressions into the equations of motion, yielding

$$\left(-\omega_f^2[M]+[K]\right)\{x\}=\{P\} \tag{4.206}$$

so that

$$\{x\}=\left([K]-\omega_f^2[M]\right)^{-1}\{P\} \tag{4.207}$$

If the modal matrix $[\Phi]$ is known, then the inverse that is required in Equation 4.207 can be avoided. The column vector $\{x\}$ can be expressed as a linear combination of the modal vectors as

$$\begin{aligned}\{x\}&=\xi_1\left\{\phi_1^{(1)}\right\}+\xi_2\left\{\phi_1^{(2)}\right\}+\cdots+\xi_n\left\{\phi_n^{(n)}\right\}\\&=\left[\left\{\phi_1^{(1)}\right\}\quad\left\{\phi_1^{(2)}\right\}\quad\cdots\quad\left\{\phi_n^{(n)}\right\}\right]\begin{Bmatrix}\xi_1\\\xi_2\\\vdots\\\xi_n\end{Bmatrix}=[\Phi]\{\xi\}\end{aligned} \tag{4.208}$$

Substituting Equation 4.208 into 4.206 gives

$$\left(-\omega_f^2[M]+[K]\right)[\Phi]\{\xi\}=\{P\} \tag{4.209}$$

Premultiplying Equation 4.209 throughout by $[\Phi]^T$ yields

$$\left(-\omega_f^2[\Phi]^T[M][\Phi]+[\Phi]^T[K][\Phi]\right)\{\xi\}=[\Phi]^T\{P\} \tag{4.210}$$

Using Equations 4.179 and 4.185, we obtain

$$\begin{aligned}&\left(-\omega_f^2\bar{m}[I]+\bar{m}\left\lfloor\omega_i^2\right\rfloor_D\right)\{\xi\}=[\Phi]^T\{P\}\\&\bar{m}\left(\left\lfloor\omega_i^2\right\rfloor_D-\omega_f^2[I]\right)\{\xi\}=[\Phi]^T\{P\}\end{aligned} \tag{4.211}$$

and

$$\{\xi\}=\frac{1}{\bar{m}}\left(\left\lfloor\omega_i^2\right\rfloor_D-\omega_f^2[I]\right)^{-1}[\Phi]^T\{P\}=\frac{1}{\bar{m}}\left\lfloor\left(\omega_i^2-\omega_f^2\right)\right\rfloor_D^{-1}[\Phi]^T\{P\} \tag{4.212}$$

Therefore,

$$\{x\}=\frac{1}{\bar{m}}[\Phi]\left(\left\lfloor\omega_i^2\right\rfloor_D-\omega_f^2[I]\right)^{-1}[\Phi]^T\{P\}=\frac{1}{\bar{m}}[\Phi]\left(\left\lfloor\left(\omega_i^2-\omega_f^2\right)\right\rfloor_D\right)^{-1}[\Phi]^T\{P\} \tag{4.213}$$

We notice that if $\omega_f$ is equal to any of the natural frequencies $\omega_r$ resonance takes place. If $\{p(t)\}$ is a periodic force, it can be expanded into a Fourier series and the response to $\{p(t)\}$ can be obtained by the method of superposition similar to that done for the SDOF system in Chapter 1.

### 4.11.2 Arbitrary Forcing Function

The response due to an arbitrary force of varying magnitude can be calculated from the concept of impulse response as was done in Chapter 1 for a single-degree-of-freedom system. Therefore, consider an impulse consisting of a large force acting for a very short duration. It has the effect of giving the mass an initial velocity of

$$\{\dot{q}_0\} = [M]^{-1}\{r\} \tag{4.214}$$

but having an initial displacements

$$\{q_0\} = \{0\} \tag{4.215}$$

Equations 4.214 and 4.215 can be used as the initial conditions for the free vibration problem given by Equation 4.204. Thus, we obtain the equation for the impulse response as

$$\begin{aligned} \{q(t)\} &= \frac{1}{\bar{m}}[\Phi]\left([\sin\omega_i t]_D[\omega_i]_D^{-1}[\Phi]^T[M][M]^{-1}\{r\}\right) \\ &= \frac{1}{\bar{m}}[\Phi][\sin\omega_i t]_D[\omega_i]_D^{-1}[\Phi]^T\{r\} \end{aligned} \tag{4.216}$$

where the column vector $\{r\}$ represents the impulsive forces. When the system is subjected to a time-varying arbitrary force $\{p(t)\}$, we consider the impulse of the force applied at $t=\tau$ over the particular time interval $d\tau$ as $d\{r\} = \{p(\tau)\}d\tau$. Hence, according to Equation 4.216, the response due to the impulse is

$$\{q(t)\} = \frac{1}{\bar{m}}[\Phi][\sin\omega_i(t-\tau)]_D[\omega_i]_D^{-1}[\Phi]^T\{p(\tau)\}d\tau \tag{4.217}$$

where

$$[\sin\omega_i(t-\tau)]_D = \begin{bmatrix} \sin\omega_1(t-\tau) & 0 & 0 & 0 \\ 0 & \sin\omega_2(t-\tau) & 0 & 0 \\ 0 & 0 & \ddots & 0 \\ 0 & 0 & 0 & \sin\omega_n(t-\tau) \end{bmatrix} \tag{4.218}$$

Since the impulses are produced over the time interval $(0-t)$, the total is obtained by summing the differential between 0 and $t$. Therefore,

$$\{q(t)\} = \frac{1}{\bar{m}}[\Phi]\int_0^t [\sin\omega_i(t-\tau)]_D[\omega_i]_D^{-1}[\Phi]^T\{p(\tau)\}d\tau \tag{4.219}$$

For initial conditions at $t = 0$ nonzero, the total response is the sum of the free response caused by nonzero initial conditions and the forced response caused by arbitrary external force. Hence, the complete motion becomes

$$\{q(t)\} = \frac{1}{\bar{m}}[\Phi]\left[[\cos\omega_i t]_D[\Phi]^T[M]\{q_0\} + [\sin\omega_i t]_D[\omega_i]_D^{-1}[\Phi]^T[M]\{\dot{q}_0\}\right.$$
$$\left. + \int_0^t [\sin\omega_i(t-\tau)]_D[\omega_i]_D^{-1}[\Phi]^T\{p(\tau)\}d\tau\right] \tag{4.220}$$

## 4.12 Forced Damped MDOF Systems

Vibration response is always associated with the phenomenon of damping, and it still remains one of the least understood topics. Damping forces depend not only on the vibration of the system but, in addition, on the surrounding medium. Damping is the dissipation of energy from a vibrating structure, which is essentially a conversion of mechanical energy into heat due to the result of the motion. There are several mathematical models that have been used to represent the damping forces which arise due to vibrations. These include simple viscous damping, Coulomb damping, and hysteretic or material damping. Here, only viscous damping, for which the damping force is proportional to the velocity, will be considered since viscous damping leads to linear differential equations with constant coefficients. The motion of an MDOF system is described by our matrix equations as

$$[M]\{\ddot{q}\} + [C]\{\dot{q}\} + [K]\{q\} = \{p(t)\} \tag{4.221}$$

Assuming mode superposition so that physical degrees of freedom are expressed as a linear combination of the modal vectors as

$$\{q(t)\} = [\Phi]\{x(t)\} \tag{4.222}$$

Thus, in the manner above, the MDOF system becomes

$$[\Phi]^T[M][\Phi]\{\ddot{x}\} + [\Phi]^T[C][\Phi]\{\dot{x}\} + [\Phi]^T[K][\Phi]\{x\} = [\Phi]^T\{p(t)\} \tag{4.223}$$

or

$$\bar{m}[I]\{\ddot{x}\} + [C_g]\{\dot{x}\} + \bar{m}\left[\omega_i^2\right]_D\{x\} = \{f\} \tag{4.224}$$

where

$$[\Phi]^T[M][\Phi] = \bar{m}[I], \quad [\Phi]^T[K][\Phi] = \bar{m}\left[\omega_i^2\right]_D, \quad [C_g] = [\Phi]^T[C][\Phi], \quad \{f\} = [\Phi]^T\{p(t)\}$$

As noted above, the modal matrix $[\Phi]$ is able to diagonalize the mass matrix $[M]$ and stiffness matrix $[K]$, however, in general, it is not able to diagonalize the damping matrix $[C]$. However, there is a special case which will be investigated in the next section, where the modal matrix can diagonalize the damping matrix.

### 4.12.1 Proportional Damping

Modal analysis is one of the most popular and efficient method for solving engineering vibration problems. The concept of modal analysis was originated from the linear vibrations of undamped systems. The undamped modes or classical normal modes satisfy an orthogonality relationship as we have seen over the mass and stiffness matrices which uncouple the equations of motion. Recall, if $[\Phi]$ is the modal matrix then $[\Phi]^T[M]$ $[\Phi]$ and $[\Phi]^T[K][\Phi]$ are both diagonal matrices. This significantly simplifies the dynamic analysis because complex multiple degree-of-freedom systems can be essentially treated as a collection of single degree-of-freedom oscillators. Since real systems are not undamped but possess some kind of energy dissipation mechanism or damping, we look at a special system where the damping matrix is linearly related to the mass and stiffness matrices. Such systems are known as *Rayleigh damping* or *proportional damping systems*. Proportional damping is a widely used approach to model dissipative forces in complex engineering structures. Proportional damping expresses the damping matrix as

$$[C]=\alpha[M]+\beta[K] \tag{4.225}$$

where $\alpha$ and $\beta$ are constants. Thus, it is possible to decouple the equations of motion given in Equation 4.221 and the solution will be easily obtainable as a linear combination of natural modes. Modes of classically damped systems preserve the simplicity of the real normal modes as in the undamped case. Premultiplying Equation 4.225 by $[\Phi]^T$ and postmultiplying by $[\Phi]$ gives

$$[\Phi]^T[C][\Phi]=[\Phi]^T(\alpha[M]+\beta[K])[\Phi]=\alpha[\Phi]^T[M][\Phi]+\beta[\Phi]^T[K][\Phi]=\alpha\bar{m}[I]+\beta\bar{m}\left[\omega_i^2\right]_D \tag{4.226}$$

Using Equation 4.222 as the solution to Equation 4.221, substituting Equation 4.225 into 4.221, premultiplying by $[\Phi]^T$, and using Equation 4.226 yields

$$\{\ddot{x}\}+\alpha\{\dot{x}\}+\beta\left[\omega_i^2\right]_D\{\dot{x}\}+\beta\left[\omega_i^2\right]_D\{x\}=\frac{1}{\bar{m}}[\Phi]\{p(t)\} \tag{4.227}$$

or

$$\ddot{x}_i+2\zeta_i\omega_i\dot{x}_i+\omega_i^2x_i=f_i(t)\quad i=1,2,\ldots,n \tag{4.228}$$

where we have made our above expression so that it is analogous to that of a one-degree-of-freedom system. Thus, we used the notation

$$2\zeta_i\omega_i = \alpha + \beta\omega_i^2 \quad \text{and} \quad f_i(t) = \frac{1}{\bar{m}}[\Phi]^T\{p(t)\} \tag{4.229}$$

where

$$\zeta_i = \frac{1}{2}\left(\frac{\alpha}{\omega_i} + \beta\omega_i\right) \tag{4.230}$$

is called the modal damping ratio.

Each of the $n$ equations in Equation 4.228 is uncoupled from all the others. Therefore, we can determine the response of the $i$th mode in the same manner as that for a one-degree of freedom system with viscous damping.

### 4.12.2 Arbitrary Viscous Damping

If the damping matrix [C] is arbitrary, the principal or normal coordinates of the undamped system do not uncouple our equations of motion, Equation 4.221. Therefore, an analytical solution is not possible in what we refer to as the *configuration space*. This worked only when damping was of the proportional type. To analyze arbitrary damping, a more general procedure must be used. Our equations of motion can be reformulated in *state space*. A *state space* is just the set of values which a process can take. Equation 4.221 can be reformulated by first expressing

$$\begin{aligned} \{\dot{q}(t)\} &= \{\dot{q}(t)\} \\ \{\ddot{q}(t)\} &= -[M]^{-1}[C]\{\dot{q}(t)\} - [M]^{-1}[K]\{q(t)\} + [M]^{-1}\{f(t)\} \end{aligned} \tag{4.231}$$

Therefore, using a state vector as $2n$ first-order differential equations, we can write

$$\{x(t)\} = \begin{Bmatrix} \{q(t)\} \\ \{\dot{q}(t)\} \end{Bmatrix} \tag{4.232}$$

Equations 4.231 and 4.232 can be expressed in state-space form as

$$\{\dot{x}(t)\} = [A]\{x(t)\} + [B]\{f(t)\} \tag{4.233}$$

where the $2n \times 2n$ matrices $[A]$ and $[B]$ are given by

$$[A] = \begin{bmatrix} [0] & [I] \\ -[M]^{-1}[K] & -[M]^{-1}[C] \end{bmatrix} \tag{4.234}$$

$$[B] = \begin{bmatrix} [0] \\ [M]^{-1} \end{bmatrix} \tag{4.235}$$

The solution of Equation 4.233 can be determined using modal analysis in state space.

## PROBLEMS

4.1 Write the equations of motion for the systems shown in Figure 4.23a–e. Write the elements of the mass and stiffness (or flexibility) matrices (no damping). Write the frequency equation of each system and solve the equation.

4.2 Find two normalized natural modes for the system in Problem 4.1a. Assume $\bar{m} = 1$. Write the elements $\phi_i^{(r)}$ of the modal matrix $\phi$.

4.3 Write the uncoupled equations of motion for forced vibrations of system of Problem 4.1a.

4.4 Assume that the material of the structure in Problem 4.1a is an arbitrary viscoelastic material (all of the elements of each structure are made from the same material). Write the uncoupled equations of motion for forced vibrations using:

a. Laplace transformations

b. Fourier transformations

4.5 For the system shown in Figure 4.23a, assume a structural rigidity of $EI = 117 \times 10^3$ N m$^3$ and that the masses are related through $m_1 = nm_2$, where $0.2 \leq n \leq 5$. If $m_2 = 225$ kg and the distance between masses is $L = 3$ m, plot the frequencies as a function of $n$.

4.6 For the system shown in Figure 4.23b, assume the beam is made from aluminum ($E = 10 \times 10^6$ psi) and has dimensions such that $I = 2$ in$^4$. The weight of the rigid body is 300 lb. The beam segments are related by $b = na$, where $1 \leq n \leq 5$ and $a = 24''$. Plot the frequencies as a function of $n$.

4.7 For the system shown in Figure 4.21c, assume the rigid bar is a cylinder made from an aluminum alloy ($E = 70$ GPa) and has a weight of 30 N/m. The length of the bar is between 1 and 4 m. The horizontal spring is relatively soft $k_1 = 880$ N/m while springs 2 and 3 are stiffer with $k_2 = k_3 = 3520$ N/m. Plot the frequencies as a function of $L$.

4.8 For the system shown in Figure 4.23d, assume $m = 180$ kg and the three identical supporting rods are aluminum ($E = 70$ GPa). Each rod has a length which is related to its diameter by $L = nd$, where $5 \leq n \leq 30$. Three possible rod diameters are being considered; $d = 12.5$, 25, 37.5 mm. Plot the frequencies as a function of $n$.

4.9 For the system shown in Figure 4.23e, assume the box weighs 200 lb and the three supporting rods are all 1″ in diameter aluminum with $AE \approx 7.9 \times 10^6$ lb. For the crate, we approximate $I \approx 0.863$ in$^4$. The box is a rectangle with dimensions $a = 1$ ft, $b = 2$ ft. The lengths of the rods are $L_1 = 5$ ft and $L_2 = 10$ ft. The box is suspended between 2 and 4 ft above the ground. Plot the frequencies as a function height above the ground.

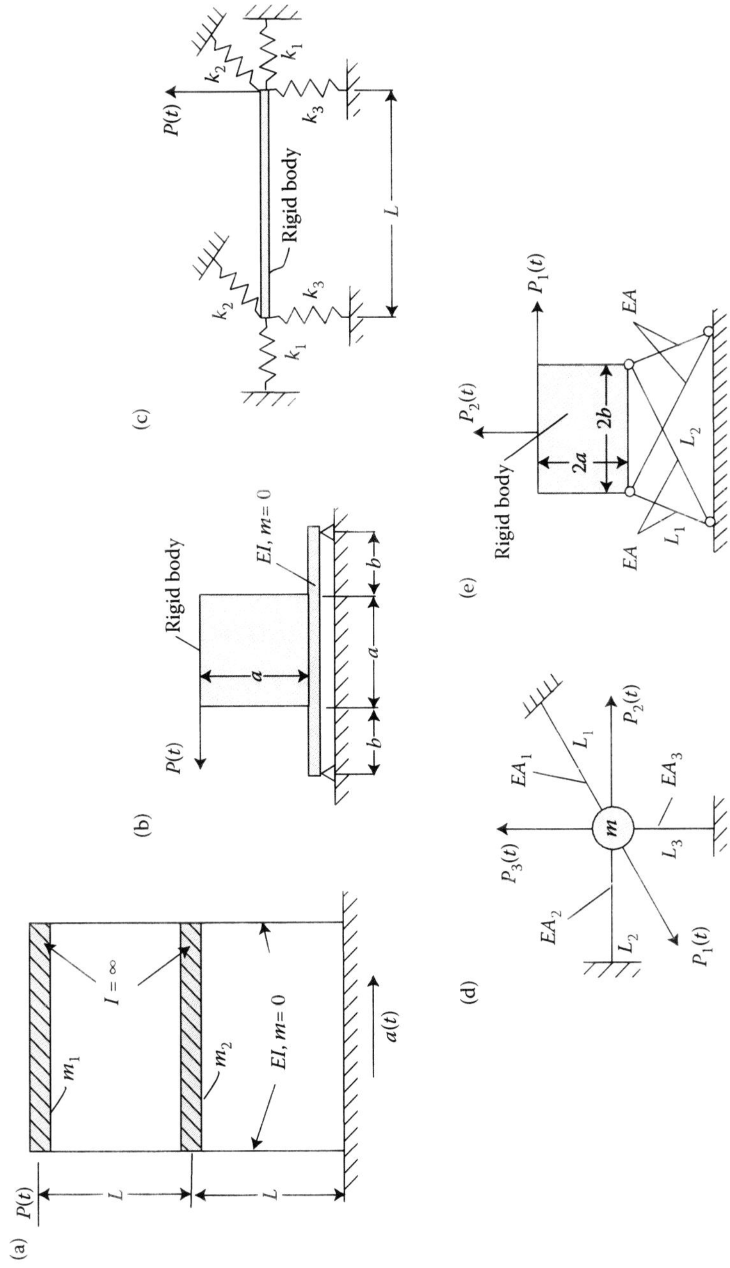


**FIGURE 4.23**
Multiple-degree-of-freedom systems.

## References

Balachandran, B. and Magrab, E.B. 2009. *Vibrations*, 2nd ed. Toronto, ON: Cengage Learning.

Bauchau, O. A. 2011. *Flexible Multibody Dynamics*. New York: Springer.

Bishop, R. E. D., Gladwell, G. M. L., and Michaelson, S. 1965. *The Matrix Analysis of Vibration*. Cambridge, UK: Cambridge University Press.

Fu, Z.-F. and He, J. 2001. *Modal Analysis*. Oxford, Jordan Hill: Butterworth-Heinemann.

Humar, J. L. 1990. *Dynamics of Structures*. Englewood Cliffs, NJ: Prentice-Hall.

Jennings, A. 1977. *Matrix Computation for Engineers and Scientists*. Chichester: John Wiley & Sons.

O'Reilly, O. M. 2008. *Intermediate Dynamics for Engineers*. Cambridge, UK: Cambridge University Press.

Schmitz, T. L. and Smith, K. S. 2012. *Mechanical Vibrations*. New York: Springer.

Tong, K. N. 1960. *Theory of Mechanical Vibration*. New York: John Wiley & Sons, Inc.

Wells, D. A. 1967. *Theory and Problems of Lagrangian Dynamics*. New York: McGraw-Hill, Inc.

# 5

# *Equations of Motion of Continuous Systems*

## 5.1 Introduction

The theory of vibrations of continuous systems is based on the foundations of classical elasticity theory. Therefore, we will give a concise introduction to the theory of elasticity so that the reader will have an understanding of the fundamental relations and equations of the theory of elasticity.

## 5.2 Forces and Stresses

The concepts of stress and strain are the basis of the mechanics of deformable solids (Housner and Vreeland 1965; Barber 2010; Boresi et al. 2011). The external forces which act on an elastic body may be known applied loads or support reactions. External forces may be either *body forces*, which are given in terms of force per unit volume, or *surface forces*, which are forces per unit surface area. Body forces may represent gravitational forces, magnetic forces, or inertia forces if the body is in motion, whereas surface forces could be due to hydrostatic pressure or distributed support reactions. A body responds to the application of external forces and internal forces will be produced between parts of the body. Consider a body that is in equilibrium due to applied loading as shown in Figure 5.1.

To investigate the internal forces at some point $O$ in the interior of the body, we can imagine that the body is separated into two parts by passing a plane through the point $O$ whose normal is $\mathbf{n}$. As the body was in equilibrium so both parts must be in equilibrium. Therefore, forces must be developed on the contact sections of each free body. Recall from strength of materials the concept of force per unit area as shown in Figure 5.2. Thus, a small area $\Delta A$ on the surface of a solid body is subjected to a resultant force $\Delta\mathbf{F}$. The force $\Delta\mathbf{F}$ can be decomposed into two components $\Delta\mathbf{F}_n$ and $\Delta\mathbf{F}_s$, one along the unit normal $\mathbf{n}$ and the other along a unit tangent $\mathbf{s}$ acting in the plane of the area $\Delta A$. The force $\Delta\mathbf{F}_n$ is called the normal force and $\Delta\mathbf{F}_s$ is called the tangential force on $\Delta A$.

It is noted that the forces $\Delta\mathbf{F}$, $\Delta\mathbf{F}_n$, and $\Delta\mathbf{F}_s$ depend on the orientation of the cut and the location of the area $\Delta A$. The *stress* vector or *traction* vector is defined as

$$\boldsymbol{\sigma} = \lim_{\Delta A \to 0} \frac{\Delta\mathbf{F}}{\Delta A} = \frac{d\mathbf{F}}{dA} \tag{5.1}$$

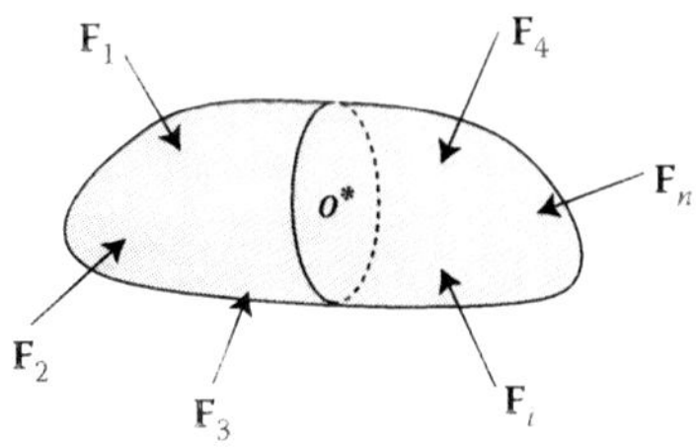


**FIGURE 5.1**
Three-dimensional body acted on by external forces.

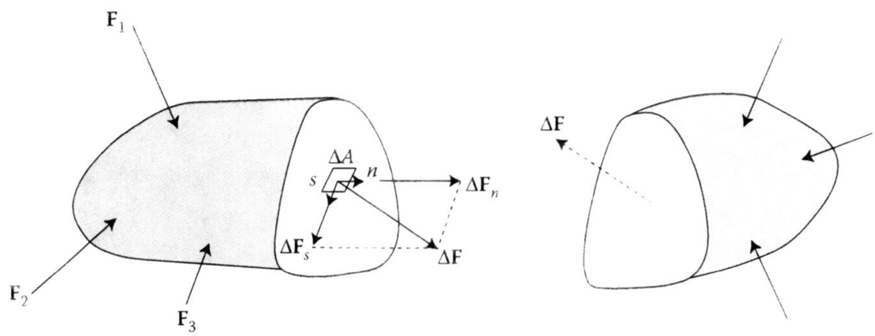


**FIGURE 5.2**
Internal forces acting on a plane with a normal $n$.

Note that from our definition $\boldsymbol{\sigma}$ depends on the position within the body and the direction of the normal $\mathbf{n}$. In the same manner, we can define the normal stress vector $\boldsymbol{\sigma}_n$ and the tangential stress vector $\boldsymbol{\sigma}_s$ that act at a point in the plane. Hence, we have

$$\boldsymbol{\sigma}_n = \lim_{\Delta A \to 0} \frac{\Delta \mathbf{F}_n}{\Delta A} = \frac{d\mathbf{F}_n}{\Delta A} \qquad \boldsymbol{\sigma}_s = \lim_{\Delta A \to 0} \frac{\Delta \mathbf{F}_s}{\Delta A} = \frac{d\mathbf{F}_s}{\Delta A} \tag{5.2}$$

The unit vectors which are associated with $\boldsymbol{\sigma}_n$ and $\boldsymbol{\sigma}_s$ are normal and tangent to the plane cutting the elastic body and exposing the point $O$. At this point, we consider three surface elements mutually perpendicular to the directions of the coordinate axes, where the positive normal directions are in the direction of increasing coordinates. We can consider the elements $dA_1 = dx_2\, dx_3$, $dA_2 = dx_1\, dx_3$, and $dA_3 = dx_1\, dx_2$ perpendicular to the $x_1$, $x_2$, and $x_3$ axes. The components of the stress vectors associated with these elements are

$$\begin{aligned} dA_1 &: \ \sigma_{11}, \sigma_{12}, \sigma_{13} \\ dA_2 &: \ \sigma_{21}, \sigma_{22}, \sigma_{23} \\ dA_3 &: \ \sigma_{31}, \sigma_{32}, \sigma_{33} \end{aligned} \tag{5.3}$$

The components $\sigma_{ij}$ represent the $i$th component of stress on the face whose outer normal points in the $x_j$ direction. This is shown in Figure 5.3. Note that normal stresses are positive in tension and positive shear stresses are in the positive $x_1$, $x_2$, and $x_3$ directions. Two subscripts are used: the first indicates the direction of the normal to the plane and the second one indicates the direction of the component of the stress.

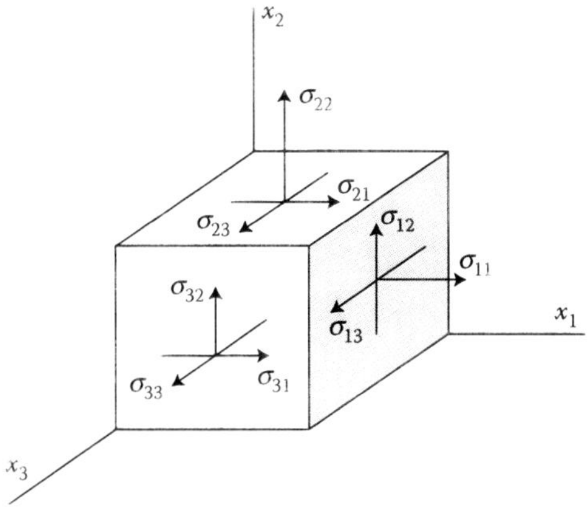


**FIGURE 5.3**
Components of the stress tensor.

The state of stress is determined by these nine components, which constitute what is called the stress tensor. Thus,

$$\sigma_{ij} = \begin{vmatrix} \sigma_{11} & \sigma_{12} & \sigma_{13} \\ \sigma_{21} & \sigma_{22} & \sigma_{23} \\ \sigma_{31} & \sigma_{32} & \sigma_{33} \end{vmatrix} \tag{5.4}$$

In the above equations, we have used an alternative notation called index or indicial notation for stress which is often more convenient for the discussion of elasticity.

## 5.3 Equations of Equilibrium

For a three-dimensional body, recall that Newton's law defines that the summation of force equals the change in linear momentum and the summation of moments equals the change in angular momentum. Using vector notation, the mathematical formulation is given by

$$\begin{aligned} &\sum \text{Forces}: \quad \int_V \mathbf{f}\,dV + \int_A \mathbf{p}\,dA = \int_V \rho\ddot{\mathbf{u}}\,dV \\ &\sum \text{Moments}: \int_V \mathbf{r}\times\mathbf{f}\,dV + \int_A \mathbf{r}\times\mathbf{p}\,dA = \int_V \rho\mathbf{r}\times\ddot{\mathbf{u}}\,dV \end{aligned} \tag{5.5}$$

where

$\mathbf{f}$ = body force vector (force per unit volume);

$\mathbf{p}$ = surface force vector (force per unit area);

$\mathbf{u}$, $\dot{\mathbf{u}}$, and $\ddot{\mathbf{u}}$ = displacement, velocity, and acceleration vectors, respectively;

$\mathbf{r}$ = position vector from a fixed reference point;

$\rho$ = density; $V$ = volume; and $A$ = area

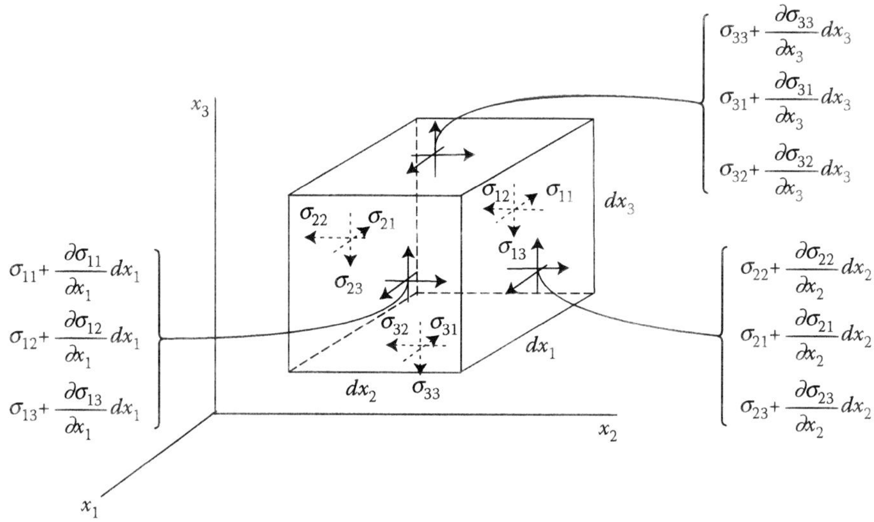


**FIGURE 5.4**
General state of stress and its variation from one point to another.

The system of equations defined by the summation of forces and summation of moments is similar to what is used for rigid bodies. In rigid body dynamics, this system can be integrated. Here, the system cannot be integrated because **u** is an unknown function of position. Therefore, this system is not used in continuous systems. For continuous systems, we use the differential equations of motions. Consider a typical element with infinitesimal dimensions $dx_1$, $dx_2$, and $dx_3$ as shown in Figure 5.4.

We define the stress components to be positive when the components act on a positive face in a positive coordinate direction as mentioned before. Each face of the element has one component, which is a normal stress $\sigma_{ii}$, and two components of shear stresses $\sigma_{ij}$ acting on it. As we move from one point to another, the stress components vary in magnitude. We use the concept of partial differentiation to approximate the amount of change a stress component makes between two points an infinitesimal distance apart. We consider the possibility of having body forces components $f_1$, $f_2$, and $f_3$. Applying Newton's law of motion by summing forces in the $x_1$ direction yields

$$\left(\sigma_{11}+\frac{\partial\sigma_{11}}{\partial x_1}dx_1\right)(dx_2\,dx_3)-\sigma_{11}(dx_2\,dx_3)+\left(\sigma_{21}+\frac{\partial\sigma_{21}}{\partial x_2}dx_2\right)(dx_1\,dx_3)-\sigma_{21}(dx_1\,dx_3)$$
$$+\left(\sigma_{31}+\frac{\partial\sigma_{31}}{\partial x_3}dx_3\right)(dx_1\,dx_2)-\sigma_{31}(dx_1\,dx_2)+f_1(dx_1\,dx_2\,dx_3)=\rho\frac{\partial^2u_1}{\partial t^2}(dx_1\,dx_2\,dx_3) \tag{5.6}$$

Cancelling terms and dividing through by the volume $dx_1\,dx_2\,dx_3$ yield the equation of motion in the $x_1$ direction. In the same manner, summing forces in the $x_2$ and $x_3$ directions yield the equations of motion as

$$\begin{aligned}
\frac{\partial \sigma_{11}}{\partial x_1} + \frac{\partial \sigma_{21}}{\partial x_2} + \frac{\partial \sigma_{31}}{\partial x_3} + f_1 &= \rho \frac{\partial^2 u_1}{\partial t^2} \\
\frac{\partial \sigma_{12}}{\partial x_1} + \frac{\partial \sigma_{22}}{\partial x_2} + \frac{\partial \sigma_{32}}{\partial x_3} + f_2 &= \rho \frac{\partial^2 u_2}{\partial t^2} \\
\frac{\partial \sigma_{13}}{\partial x_1} + \frac{\partial \sigma_{23}}{\partial x_2} + \frac{\partial \sigma_{33}}{\partial x_3} + f_3 &= \rho \frac{\partial^2 u_3}{\partial t^2}
\end{aligned} \tag{5.7}$$

where $u_1$, $u_2$, and $u_3$ are components of the displacement vector $\mathbf{u}$. It is noted that we have used an alternative notation called *index* or *indicial notation* and is often more convenient for general discussions in elasticity. In indicial notation, the coordinate axes $x$, $y$, and $z$ are replaced by $x_1$, $x_2$, and $x_3$, respectively or $x_i$, $i = 1, 2, 3$. The components of stress as mentioned are distinguished by two numerical subscripts $\sigma_{ij}$, the first one indicating the face on which the stress component acts and the second one which specifies its direction. The similarity of the equations suggests replacing any numerical subscript by an alphanumeric subscript which can take on any of the three numerical values 1, 2, and 3. Next consider the summation convention. Repeated indices are always contained within summations or phrased differently; a repeated index *implies* a summation. Therefore, the summation symbol is usually dropped. Consider, for example,

$$\begin{aligned}
\frac{\partial \sigma_{11}}{\partial x_1} + \frac{\partial \sigma_{21}}{\partial x_2} + \frac{\partial \sigma_{31}}{\partial x_3} &= 0 \\
\frac{\partial \sigma_{12}}{\partial x_1} + \frac{\partial \sigma_{22}}{\partial x_2} + \frac{\partial \sigma_{32}}{\partial x_3} &= 0 \quad \text{or} \quad \sum_{i=1}^{3} \frac{\partial \sigma_{ij}}{\partial x_i} \\
\frac{\partial \sigma_{13}}{\partial x_1} + \frac{\partial \sigma_{23}}{\partial x_2} + \frac{\partial \sigma_{33}}{\partial x_3} &= 0
\end{aligned}$$

This repeated index notation is known as Einstein's convention. Any repeated index is called a dummy index. As a repeated index implies a summation over all possible values of the index, we can always relabel a dummy index, for example

$$\begin{aligned}
\mathbf{F} &= F_i \mathbf{e}_i = F_j \mathbf{e}_j = F_k \mathbf{e}_k \\
&= F_1 \mathbf{e}_1 + F_2 \mathbf{e}_2 + F_3 \mathbf{e}_3
\end{aligned}$$

Thus, the equations of motion, Equation 5.7, become

$$\sigma_{ij,i} + f_j = \rho \frac{\partial^2 u_j}{\partial t^2} \tag{5.8}$$

In addition, the stresses are symmetric and $\sigma_{12} = \sigma_{21}$, $\sigma_{13} = \sigma_{31}$, $\sigma_{23} = \sigma_{32}$, or $\sigma_{ij} = \sigma_{ji}$.

## 5.4 Plane Stress

For many problems, the state of stress is simpler than that shown in Figure 5.3. A state of stress in which

$$\sigma_{33} = \sigma_{31} = \sigma_{32} = 0 \tag{5.9}$$

is called a plane state of stress in the $x_1x_2$ plane. In this case, the equations of motion reduce to

$$\begin{aligned} \frac{\partial \sigma_{11}}{\partial x_1} + \frac{\partial \sigma_{21}}{\partial x_2} + f_1 = \rho \frac{\partial^2 u_1}{\partial t^2} \\ \frac{\partial \sigma_{12}}{\partial x_1} + \frac{\partial \sigma_{22}}{\partial x_2} + f_2 = \rho \frac{\partial^2 u_2}{\partial t^2} \end{aligned} \tag{5.10}$$

or

$$\sigma_{ij,i} + f_j = \rho \frac{\partial^2 u_j}{\partial t^2} \quad i = 1,2 \quad j = 1,2 \tag{5.11}$$

The case of plane stress is illustrated in Figure 5.5.

In the case of uniaxial stress, the only nonzero stress components is $\sigma_{11}$. The resulting equation of motion is

$$\frac{\partial \sigma_{11}}{\partial x_1} + f_1 = \rho \frac{\partial^2 u_1}{\partial t^2} \tag{5.12}$$

In many engineering problems, it is sometimes useful to use the so-called stress resultants, defined by

$$\begin{matrix} N_1 = \int\limits_A \sigma_{11}\, dA & M_1 = \int\limits_A z\sigma_{11}\, dA & Q_1 = \int\limits_A \sigma_{13}\, dA \\ \vdots & \vdots & \vdots \end{matrix} \tag{5.13}$$

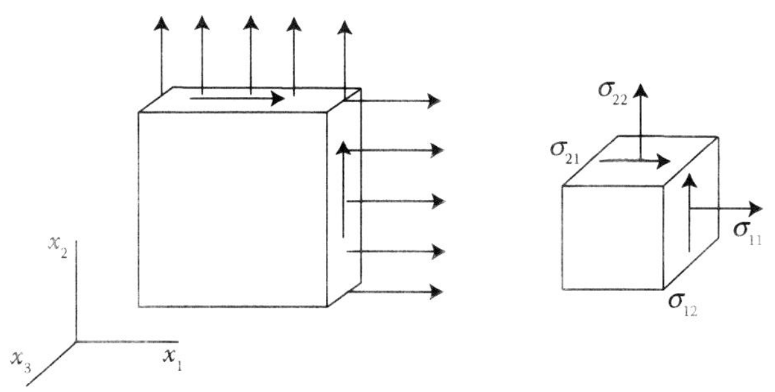


**FIGURE 5.5**
Stress components which define the condition of plane stress.

**EXAMPLE 5.1**

Consider the beam section shown in Figure 5.6, where $b$ is defined as the beam width. Apply a moment and transverse shear to the beam. From the loading, we know $\sigma_{12} = \sigma_{22} = \sigma_{21} = 0$ and we assume $\sigma_{33} \approx 0$. Derive the relation

$$\frac{\partial M}{\partial x} - Q = 0 \tag{5.14}$$

**Solution**

The differential equation of motion in terms of stress components is

$$\frac{\partial \sigma_{11}}{\partial x_1} + \frac{\partial \sigma_{13}}{\partial x_3} = 0 \tag{5.15}$$

Multiply Equation 5.15 by the width $b$ and integrating over the thickness of the beam yields

$$\int_{-c_1}^{c_2} b x_3 \frac{\partial \sigma_{11}}{\partial x_1} dx_3 + \int_{-c_1}^{c_2} b x_3 \frac{\partial \sigma_{31}}{\partial x_3} dx_3 = 0 \tag{5.16}$$

or integrating the second integral by parts yields

$$\frac{\partial}{\partial x_1} \int_{-c_1}^{c_2} b x_3 \sigma_{11}\, dx_3 + (b x_3 \sigma_{31})\Big|_{-c_1}^{c_2} - \int_{-c_1}^{c_2} b \sigma_{31}\, dx_3 = 0 \tag{5.17}$$

Recalling that the stress resultants are $M = \int_{-c_1}^{c_2} b x_3 \sigma_{11}\, dx_3$ and $Q = \int_{-c_1}^{c_2} b \sigma_{13}\, dx_3$, Equation 5.17 becomes

$$\frac{\partial}{\partial x_1} M + (0) - Q = 0$$

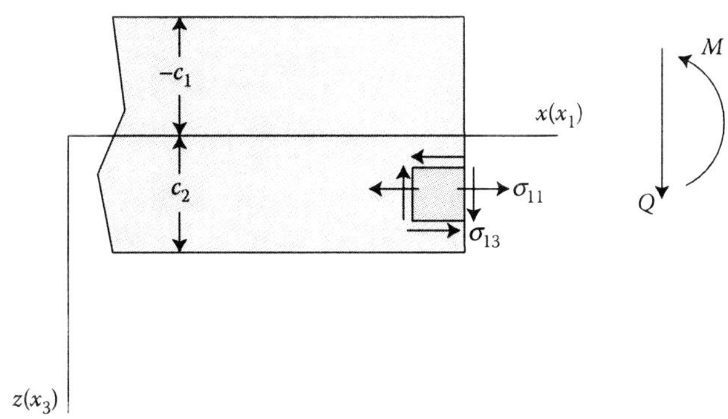


**FIGURE 5.6**
Beam subjected to a bending moment and transverse shear force.

or

$$\frac{\partial M}{\partial x_1} - Q = 0 \tag{5.18}$$

In a three-dimensional body, we have six unknown stresses ($\sigma_{ij}$) and three unknown displacements ($u_i$). Thus, we need additional information. One is the following.

## 5.5 Displacement and Strain Relations

To determine the deformations, a mathematical method that allows us to follow the motion of individual points in the body is needed. Consider the deformation of a three-dimensional body as shown in Figure 5.7. The body under a system of forces is displaced from its initial reference configuration to its deformed configuration.

To analyze the deformation, consider the points $P$ and $Q$ in the initial reference state. Before deformation, the coordinates of points $P$ and $Q$, from Figure 5.7, are ($x_1$, $x_2$, $x_3$) and ($x_1 + dx_1$, $x_2 + dx_2$, $x_3 + dx_3$). The square of the line element connecting $P$ and $Q$ is

$$ds^2 = dx_1^2 + dx_2^2 + dx_3^2 \tag{5.19}$$

Due to the deformation, the particles that were at $P$ and $Q$ in the reference state have been displaced to $\bar{P}$ and $\bar{Q}$ in the deformed state. The coordinates of $\bar{P}$ and $\bar{Q}$ are from Figure 5.7, $(\bar{x}_1, \bar{x}_2, \bar{x}_3)$ and $(\bar{x}_1 + d\bar{x}_1, \bar{x}_2 + d\bar{x}_2, \bar{x}_3 + d\bar{x}_3)$. The square of the line element connecting the $\bar{P}$ and $\bar{Q}$ in the deformed configuration is

$$d\bar{s}^2 = d\bar{x}_1^2 + d\bar{x}_2^2 + d\bar{x}_3^2 \tag{5.20}$$

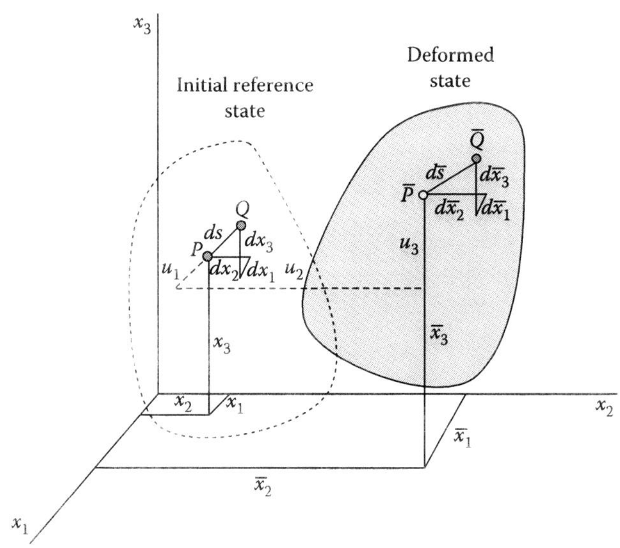


**FIGURE 5.7**
Deformation of a three-dimensional elastic body.

The point $P$ of initial coordinates ($x_1$, $x_2$, $x_3$) is transformed into a point $\bar{P}$ of coordinates (see Figure 5.7)

$$\bar{x}_1 = x_1 + u_1 \quad \bar{x}_2 = x_2 + u_2 \quad \bar{x}_3 = x_3 + u_3 \tag{5.21}$$

The displacement field is represented by the vector of components

$$u_1 = u_1(x_1, x_2, x_3) \quad u_2 = u_2(x_1, x_2, x_3) \quad u_3 = u_3(x_1, x_2, x_3) \tag{5.22}$$

Equation 5.22 defines the displacements in the $x_1$, $x_2$, and $x_3$ directions. They are assumed to be continuous and differentiable functions of the fixed coordinates $x_1$, $x_2$, and $x_3$. It is clear that if the displacement ($u_1$, $u_2$, $u_3$) is given, the displaced state is entirely determined. Substituting Equation 5.21 into Equation 5.20 yields

$$d\bar{s}^2 = dx_1^2 + dx_2^2 + dx_3^2 + 2du_1\,dx_1 + 2du_2\,dx_2 + 2du_3\,dx_3 + du_1^2 + du_2^2 + du_3^2 \tag{5.23}$$

and using Equation 5.19 gives

$$d\bar{s}^2 - ds^2 = 2du_1\,dx_1 + 2du_2\,dx_2 + 2du_3\,dx_3 + du_1^2 + du_2^2 + du_3^2 \tag{5.24}$$

Taking the total differential of each of the expressions in Equation 5.22 yields

$$\begin{aligned} du_1 &= \frac{\partial u_1}{\partial x_1}dx_1 + \frac{\partial u_1}{\partial x_2}dx_2 + \frac{\partial u_1}{\partial x_3}dx_3 \\ du_2 &= \frac{\partial u_2}{\partial x_1}dx_1 + \frac{\partial u_2}{\partial x_2}dx_2 + \frac{\partial u_2}{\partial x_3}dx_3 \\ du_3 &= \frac{\partial u_3}{\partial x_1}dx_1 + \frac{\partial u_3}{\partial x_2}dx_2 + \frac{\partial u_3}{\partial x_3}dx_3 \end{aligned} \tag{5.25}$$

Substituting Equation 5.25 into Equation 5.24 yields

$$\frac{1}{2}(d\bar{s}^2 - ds^2) = \varepsilon_{11}\,dx_1^2 + \varepsilon_{22}\,dx_2^2 + \varepsilon_{33}\,dx_3^2 + 2\varepsilon_{12}\,dx_1\,dx_2 + 2\varepsilon_{23}\,dx_2\,dx_3 + 2\varepsilon_{13}\,dx_1\,dx_3 \tag{5.26}$$

where the coefficients of $dx_1^2, dx_2^2, \ldots$ are given by the following expressions:

$$\begin{aligned} \varepsilon_{11} &= \frac{\partial u_1}{\partial x_1} + \frac{1}{2}\left[\left(\frac{\partial u_1}{\partial x_1}\right)^2 + \left(\frac{\partial u_2}{\partial x_1}\right)^2 + \left(\frac{\partial u_3}{\partial x_1}\right)^2\right] \\ \varepsilon_{22} &= \frac{\partial u_2}{\partial x_2} + \frac{1}{2}\left[\left(\frac{\partial u_1}{\partial x_2}\right)^2 + \left(\frac{\partial u_2}{\partial x_2}\right)^2 + \left(\frac{\partial u_3}{\partial x_2}\right)^2\right] \\ \varepsilon_{33} &= \frac{\partial u_3}{\partial x_3} + \frac{1}{2}\left[\left(\frac{\partial u_1}{\partial x_3}\right)^2 + \left(\frac{\partial u_2}{\partial x_3}\right)^2 + \left(\frac{\partial u_3}{\partial x_3}\right)^2\right] \end{aligned} \tag{5.27}$$

$$2\varepsilon_{12} = \frac{\partial u_2}{\partial x_1} + \frac{\partial u_1}{\partial x_2} + \frac{\partial u_1}{\partial x_1}\frac{\partial u_1}{\partial x_2} + \frac{\partial u_2}{\partial x_1}\frac{\partial u_2}{\partial x_2} + \frac{\partial u_3}{\partial x_1}\frac{\partial u_3}{\partial x_2}$$

$$2\varepsilon_{13} = \frac{\partial u_3}{\partial x_1} + \frac{\partial u_1}{\partial x_3} + \frac{\partial u_1}{\partial x_1}\frac{\partial u_1}{\partial x_3} + \frac{\partial u_2}{\partial x_1}\frac{\partial u_2}{\partial x_3} + \frac{\partial u_3}{\partial x_1}\frac{\partial u_3}{\partial x_3}$$

$$2\varepsilon_{23} = \frac{\partial u_3}{\partial x_2} + \frac{\partial u_2}{\partial x_3} + \frac{\partial u_1}{\partial x_2}\frac{\partial u_1}{\partial x_3} + \frac{\partial u_2}{\partial x_2}\frac{\partial u_2}{\partial x_3} + \frac{\partial u_3}{\partial x_2}\frac{\partial u_3}{\partial x_3}$$

Equation 5.27 is referred to as the *strain–displacement relations*. In mechanics of materials courses, the notation

$$\gamma_{12} = 2\varepsilon_{12} \quad \gamma_{13} = 2\varepsilon_{13} \quad \gamma_{23} = 2\varepsilon_{23} \tag{5.28}$$

is used. It is noted here that the components $\varepsilon_{12}$, $\varepsilon_{13}$, and $\varepsilon_{23}$ can be used in index or indicial notation, whereas the components $\gamma_{12}$, $\gamma_{13}$, and $\gamma_{23}$ cannot. In index or indicial notation, Equation 5.27 becomes

$$2\varepsilon_{ij} = \frac{\partial u_i}{\partial x_j} + \frac{\partial u_j}{\partial x_i} + \frac{\partial u_k}{\partial x_i}\frac{\partial u_k}{\partial x_j} \quad i,j,k = 1,2,3 \tag{5.29}$$

where $\varepsilon_{ij} = \varepsilon_{ji} = \varepsilon_{ij}\,(x_1, x_2, x_3)$.

We shall assume that the strains and derivatives are infinitesimal; therefore, the squares and products of partial derivatives of first order of $u_1$, $u_2$, and $u_3$ are considered negligible compared to the derivatives themselves. Thus, Equations 5.27 and 5.29 are simplified and the strain–displacement relations reduce to

$$\begin{aligned} \varepsilon_{11} &= \frac{\partial u_1}{\partial x_1} & \varepsilon_{22} &= \frac{\partial u_2}{\partial x_2} & \varepsilon_{33} &= \frac{\partial u_3}{\partial x_3} \\ 2\varepsilon_{12} &= \frac{\partial u_2}{\partial x_1} + \frac{\partial u_1}{\partial x_2} & 2\varepsilon_{13} &= \frac{\partial u_3}{\partial x_1} + \frac{\partial u_1}{\partial x_3} & 2\varepsilon_{23} &= \frac{\partial u_3}{\partial x_2} + \frac{\partial u_2}{\partial x_3} \end{aligned} \tag{5.30}$$

or in indicial notation

$$\varepsilon_{ij} = \frac{1}{2}\left(\frac{\partial u_i}{\partial x_j} + \frac{\partial u_j}{\partial x_i}\right) \quad \text{or} \quad \varepsilon_{ij} = \frac{1}{2}(u_{i,j} + u_{j,i}) \quad i,j = 1,2,3 \tag{5.31}$$

where for convenience, instead of the symbol $\partial/\partial x$, we use the subscript comma (,). The component of strain ($\varepsilon_{ij}$) given in the matrix form is

$$\varepsilon_{ij} = \begin{bmatrix} \varepsilon_{11} & \varepsilon_{12} & \varepsilon_{13} \\ \varepsilon_{21} & \varepsilon_{22} & \varepsilon_{23} \\ \varepsilon_{31} & \varepsilon_{32} & \varepsilon_{33} \end{bmatrix} \tag{5.32}$$

To consider the meaning of $\varepsilon_{11}$, $\varepsilon_{22}$, and $\varepsilon_{33}$, consider first the infinitesimal distance $dx_1$ before and after deformation, which is given using Equations 5.24 and 5.25

$$d\bar{s}^2 - ds^2 = 2\,du_1\,dx_1 + du_1^2 = 2\frac{\partial u_1}{\partial x_1}dx_1^2 + \left(\frac{\partial u_1}{\partial x_1}dx_1\right)^2$$

$$\varepsilon_{11}\,dx_1^2 = \frac{1}{2}(d\bar{s}^2 - ds^2) = \frac{\partial u_1}{\partial x_1}dx_1^2 \tag{5.33}$$

$$\varepsilon_{11} = \frac{\partial u_1}{\partial x_1}$$

Note that higher-order terms have been neglected due to being infinitesimal compared to first-order terms. We see that $\varepsilon_{11}$ represents the change in length per unit length of fibers originally parallel to the $x$-axis. The geometrical meaning of $\varepsilon_{22}$, $\varepsilon_{33}$, and their derivation follows in a similar manner. The strain components $\varepsilon_{11}$, $\varepsilon_{22}$, and $\varepsilon_{33}$ are referred to as the *extensional or longitudinal strains*. The remaining strains $\gamma_{12}$, $\gamma_{13}$, and $\gamma_{23}$ are referred to as *shearing strains*. The shearing strains are associated with changes of angles as the material distorts due to the shear stress. To define shear we must look at two directions that form the plane that undergoes shear deformation.

Therefore, consider the cube shown in Figure 5.8 undergoing a pure shear deformation in the $(x_1, x_2)$ plane. The angle between $(x_1, x_2)$ is changed by $\gamma_{12} = 2\varepsilon_{12}$. In the same manner, the angle between $(x_2, x_3)$ is changed by $\gamma_{23} = 2\varepsilon_{23}$ and the angle between $(x_3, x_1)$ is changed by $\gamma_{31} = 2\varepsilon_{31}$.

Consider the linear elements **AB** and **AC**, which in the undeformed state are parallel to axes $x_1$ and $x_2$ and before deformation have lengths $dx_1$ and $dx_2$ as shown in Figure 5.8. Loading deforms the body, which results in changes of length $ds_1'$ and $ds_2'$ and the elements take the positions **A′B′** and **A′C′**. Using Equation 5.23, we have

$$ds_1' = |\mathbf{A'B'}| = dx_1\sqrt{1+\varepsilon_{11}} \quad ds_2' = |\mathbf{A'C'}| = dx_2\sqrt{1+\varepsilon_{22}} \tag{5.34}$$

Also, we can form the expression

$$|\mathbf{B'C'}| = \sqrt{(1+2\varepsilon_{11})dx_1^2 + (1+2\varepsilon_{22})dx_2^2 + 2\gamma_{12}\,dx_1\,dx_2} \tag{5.35}$$

Using the relation

$$(\mathbf{B'C'})^2 = (\mathbf{A'B'})^2 + (\mathbf{A'C'})^2 + 2\cos\left(\frac{\pi}{2} - \gamma_{12}\right)|\mathbf{A'B'}||\mathbf{A'C'}| \tag{5.36}$$

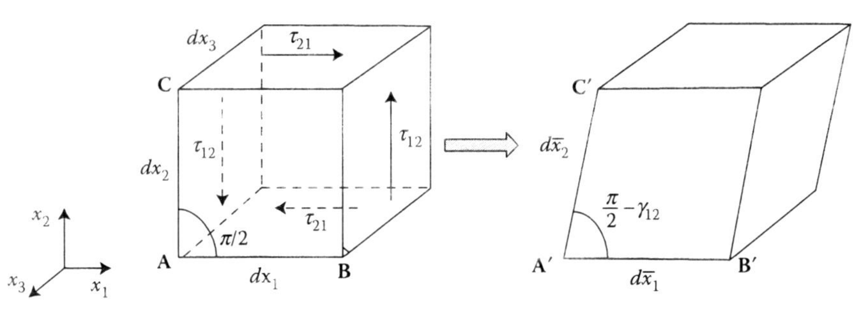


**FIGURE 5.8**
Average shearing strain in the $(x_1, x_2)$ plane.

along with Equations 5.34 and 5.35 yields

$$\sin\gamma_{12} = \frac{2\varepsilon_{12}}{\sqrt{(1+2\varepsilon_{11})(1+2\varepsilon_{22})}} \tag{5.37}$$

The angle $\gamma_{12}$ represents the decrease of the initially right angle between the line segments $dx_1$ and $dx_2$. It defines the distortion of directions $x_1$ and $x_2$ after the deformation. In the same way, the directions $\gamma_{13}$ and $\gamma_{23}$ are found to be

$$\begin{aligned} \sin\gamma_{13} &= \frac{2\varepsilon_{13}}{\sqrt{(1+2\varepsilon_{11})(1+2\varepsilon_{33})}} \\ \sin\gamma_{23} &= \frac{2\varepsilon_{23}}{\sqrt{(1+2\varepsilon_{22})(1+2\varepsilon_{33})}} \end{aligned} \tag{5.38}$$

When the strain components are sufficiently small, Equations 5.37 and 5.38 are approximated by the following expressions:

$$\gamma_{12} = \sin^{-1}(2\varepsilon_{12}) \approx 2\varepsilon_{12} \quad \gamma_{13} = \sin^{-1}(2\varepsilon_{13}) \approx 2\varepsilon_{13} \quad \gamma_{23} = \sin^{-1}(2\varepsilon_{23}) \approx 2\varepsilon_{23} \tag{5.39}$$

## 5.6 Stress–Strain Relations

Recall from strength of materials that, for a one-dimensional case, the stress and strain were related through Hooke's law by the equation $\sigma = E\varepsilon$ and for pure shear by $\tau = G\gamma$. In the most general stress–strain relationship at a point in an elastic material, the stress is related to the strain components in the form

$$\sigma_{ij} = C_{ijkl}\varepsilon_{kl} \tag{5.40}$$

where $C_{ijkl}$ is a *fourth-order stiffness tensor* of material properties. These constitute a sequence of nine equations because each component of stress is a linear combination of strain. For example,

$$\sigma_{12} = C_{1211}\varepsilon_{11} + C_{1212}\varepsilon_{12} + \cdots + C_{1233}\varepsilon_{33} \tag{5.41}$$

A fourth-order tensor has $3^4$ or 81 independent components. However, both $\sigma_{ij}$ and $\varepsilon_{ij}$ are symmetric so that there are only six independent stress components and six independent strain components. Therefore, the elastic constants must be symmetric to the first two subscripts and with the last two subscripts. That is, $C_{ijkl} = C_{jikl}$ and $C_{ijkl} = C_{ijlk}$, where $i, j, k, l = 1, 2, 3$. The number of stiffness coefficients or elastic constants has now been reduced to 36. Therefore, rewriting Equation 5.40 in a pseudo vector stress–strain form yields

$$\begin{Bmatrix} \sigma_{11} \\ \sigma_{22} \\ \sigma_{33} \\ \sigma_{13} \\ \sigma_{23} \\ \sigma_{12} \end{Bmatrix} = \begin{bmatrix} C_{11} & C_{12} & C_{13} & C_{14} & C_{15} & C_{16} \\ C_{21} & C_{22} & C_{23} & C_{24} & C_{25} & C_{26} \\ C_{31} & C_{32} & C_{33} & C_{34} & C_{35} & C_{36} \\ C_{41} & C_{42} & C_{43} & C_{44} & C_{45} & C_{46} \\ C_{51} & C_{52} & C_{53} & C_{54} & C_{55} & C_{56} \\ C_{61} & C_{62} & C_{63} & C_{64} & C_{65} & C_{66} \end{bmatrix} \begin{Bmatrix} \varepsilon_{11} \\ \varepsilon_{22} \\ \varepsilon_{32} \\ \varepsilon_{13} \\ \varepsilon_{23} \\ \varepsilon_{12} \end{Bmatrix} \tag{5.42}$$

Alternatively, the generalized Hooke's law relating strains to stresses can be written as

$$\{\varepsilon\} = [S]\{\sigma\} \tag{5.43}$$

where $[S]$ is the compliance matrix, which is the inverse of the stiffness matrix, $[S] = [C]^{-1}$.

The stiffness matrix or the compliance matrix in this form can be reduced from 36 elastic constants to 21 due to the existence of a strain energy density function, which shows that $C_{ij} = C_{ji}$ or $S_{ij} = S_{ji}$. Thus, Equation 5.42 becomes

$$\begin{Bmatrix} \sigma_{11} \\ \sigma_{22} \\ \sigma_{33} \\ \sigma_{13} \\ \sigma_{23} \\ \sigma_{12} \end{Bmatrix} = \begin{bmatrix} C_{11} & C_{12} & C_{13} & C_{14} & C_{15} & C_{16} \\ & C_{22} & C_{23} & C_{24} & C_{25} & C_{26} \\ & & C_{33} & C_{34} & C_{35} & C_{36} \\ & \text{sym} & & C_{44} & C_{45} & C_{46} \\ & & & & C_{55} & C_{56} \\ & & & & & C_{66} \end{bmatrix} \begin{Bmatrix} \varepsilon_{11} \\ \varepsilon_{22} \\ \varepsilon_{32} \\ \varepsilon_{13} \\ \varepsilon_{23} \\ \varepsilon_{12} \end{Bmatrix} \tag{5.44}$$

A material with 21 independent elastic constants is referred to as an *anisotropic* material. Further simplifications of the stiffness matrix are possible if the material properties have some form of symmetry. If there is one plane of elastic symmetry, then there are 13 independent elastic constants. This material is referred to as a *monoclinic material*. If the material has three mutually perpendicular planes of material symmetry, then we have nine independent elastic constants. This material is referred to as an *orthotropic material*. If there is a plane of material isotropy in one of the planes of an orthotropic body, then the material is called a *transversely isotropic material*. This material has five independent elastic constants. In the simplest case of an *isotropic material*, whose stiffnesses are same in all directions, only two elastic constants are independent. From strength of materials, the equations for the relations between stress and strain for an isotropic material are given as

$$\begin{aligned} \varepsilon_{11} &= \frac{1}{E}(\sigma_{11} - \nu\sigma_{22}) \\ \varepsilon_{22} &= \frac{1}{E}(\sigma_{22} - \nu\sigma_{11}) \\ \varepsilon_{12} &= \frac{\sigma_{12}}{2G} \end{aligned} \tag{5.45}$$

These relations hold for *isotropic* materials, where $E =$ elastic (Young's) modulus, $\nu =$ Poisson's ratio, and $G =$ shear modulus $= E/2(1+\nu)$. Extending to three dimensions, the vector matrix form of Equation 5.45 for isotropic materials would yield

$$\begin{Bmatrix} \varepsilon_{11} \\ \varepsilon_{22} \\ \varepsilon_{33} \\ \varepsilon_{13} \\ \varepsilon_{23} \\ \varepsilon_{12} \end{Bmatrix} = \begin{bmatrix} \frac{1}{E} & \frac{-\nu}{E} & \frac{-\nu}{E} & 0 & 0 & 0 \\ \frac{-\nu}{E} & \frac{1}{E} & \frac{-\nu}{E} & 0 & 0 & 0 \\ \frac{-\nu}{E} & \frac{-\nu}{E} & \frac{1}{E} & 0 & 0 & 0 \\ 0 & 0 & 0 & \frac{1}{2G} & 0 & 0 \\ 0 & 0 & 0 & 0 & \frac{1}{2G} & 0 \\ 0 & 0 & 0 & 0 & 0 & \frac{1}{2G} \end{bmatrix} \begin{Bmatrix} \sigma_{11} \\ \sigma_{22} \\ \sigma_{33} \\ \sigma_{13} \\ \sigma_{23} \\ \sigma_{12} \end{Bmatrix} \tag{5.46}$$

Equation 5.46 in index or indicial notation becomes

$$\varepsilon_{ij} = \frac{1+\nu}{E}\sigma_{ij} - \frac{\nu}{E}\delta_{ij}\sigma_{kk} \quad i,j = 1,2,3 \tag{5.47}$$

where $\delta_{ij} = 1$ if $i = j$ and $\delta_{ij} = 0$ if $i \neq j$. Consider an infinitesimal rectangular parallelepiped of material that encloses a point $P$, Figure 5.9a. If the material at point $P$ undergoes normal strains $\varepsilon_{11}$, $\varepsilon_{22}$, and $\varepsilon_{33}$, then the rectangular parallelepiped new dimensions will be as shown in Figure 5.9b. The initial volume is

$$dV_0 = dx_1\,dx_2\,dx_3 \tag{5.48}$$

The final volume after undergoing normal strains $\varepsilon_{11}$, $\varepsilon_{22}$, and $\varepsilon_{33}$ is

$$\begin{aligned} dV &= (1+\varepsilon_{11})dx_1(1+\varepsilon_{22})dx_2(1+\varepsilon_{33})dx_3 \\ &= [1+(\varepsilon_{11}+\varepsilon_{22}+\varepsilon_{33})+(\varepsilon_{11}\varepsilon_{22}+\varepsilon_{22}\varepsilon_{33}+\varepsilon_{33}\varepsilon_{22})+\varepsilon_{11}\varepsilon_{22}\varepsilon_{33}]dx_1\,dx_2\,dx_3 \end{aligned} \tag{5.49}$$

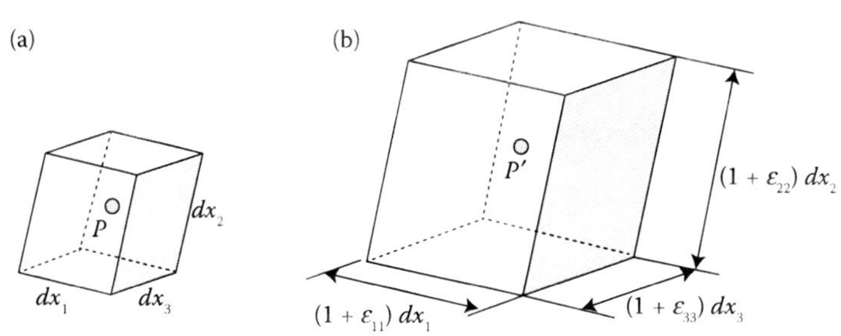


**FIGURE 5.9**
Rectangular parallelepiped, (a) before strain and (b) after strain.

As the strains are assumed to be small, so products of strains may be neglected. The term *volumetric strain* or *cubic dilatation* is defined as

$$\theta = \frac{dV - dV_0}{dV_0} = \varepsilon_{11} + \varepsilon_{22} + \varepsilon_{33} \tag{5.50}$$

To obtain the stresses in terms of strains, Equation 5.47 can be inverted. This leads to the following equation:

$$\begin{Bmatrix} \sigma_{11} \\ \sigma_{22} \\ \sigma_{33} \\ \sigma_{13} \\ \sigma_{23} \\ \sigma_{12} \end{Bmatrix} = \frac{E}{(1-2\nu)(1+\nu)} \begin{bmatrix} (1-\nu) & \nu & \nu & 0 & 0 & 0 \\ \nu & (1-\nu) & \nu & 0 & 0 & 0 \\ \nu & \nu & (1-\nu) & 0 & 0 & 0 \\ 0 & 0 & 0 & (1-2\nu) & 0 & 0 \\ 0 & 0 & 0 & 0 & (1-2\nu) & 0 \\ 0 & 0 & 0 & 0 & 0 & (1-2\nu) \end{bmatrix} \begin{Bmatrix} \varepsilon_{11} \\ \varepsilon_{22} \\ \varepsilon_{33} \\ \varepsilon_{13} \\ \varepsilon_{23} \\ \varepsilon_{12} \end{Bmatrix} \tag{5.51}$$

Equation 5.51 in index or indicial notation becomes

$$\sigma_{ij} = 2G\left[\varepsilon_{ij} + \frac{\nu}{1-2\nu}\varepsilon_{kk}\delta_{ij}\right] \quad i,j = 1,2,3 \tag{5.52}$$

Consider now a case of hydrostatic pressure, which is a case in which no shear stress exists and where all normal stresses are equal to the hydrostatic pressure. Thus, we have

$$\sigma_{11} = \sigma_{22} = \sigma_{33} = -p \tag{5.53}$$

Using our expression for the volumetric strain, Equation 5.50 along with Equations 5.46 and 5.51 yields

$$\theta = \varepsilon_{11} + \varepsilon_{22} + \varepsilon_{33} = \frac{-3}{E}(1-2\nu)p = -\frac{p}{K} \tag{5.54}$$

where

$$K = \frac{E}{3(1-2\nu)} \tag{5.55}$$

is the *bulk modulus*. The bulk modulus is a measure of the pressure required per unit volume change. For an *incompressible* material, we have $K \to \infty$.

An alternate form of the stress–strain relations can be used. Instead of using $E$, $G$, $\nu$, the so-called Lame's constants $\lambda$, and $\mu$ are used. Thus, our stress strain equations become

$$\begin{aligned} &\sigma_{11} = \lambda\theta + 2\mu\varepsilon_{11} \quad && \sigma_{22} = \lambda\theta + 2\mu\varepsilon_{22} \quad && \sigma_{33} = \lambda\theta + 2\mu\varepsilon_{33} \\ &\sigma_{13} = 2\mu\varepsilon_{13} && \sigma_{23} = 2\mu\varepsilon_{23} && \sigma_{12} = 2\mu\varepsilon_{12} \end{aligned} \tag{5.56}$$

Using indicial notation, we would have

$$\sigma_{ij} = \lambda\theta\delta_{ij} + 2\mu\varepsilon_{ij} \tag{5.57}$$

where we have introduced $\lambda$ and $\mu$, which are defined by

$$\lambda = \frac{2G\nu}{1-2\nu} = K - \frac{2}{3}G = \frac{E\nu}{(1+\nu)(1-2\nu)}$$
$$\mu = G = \frac{\lambda(1-2\nu)}{2\nu} = \frac{3}{2}(K-\lambda) = \frac{E}{2(1+\nu)}$$

### 5.6.1 Special Cases

#### *5.6.1.1 Plane Stress and Strain*

Plane stress is a two-dimensional system of stress, where $\sigma_{33} = \sigma_{13} = \sigma_{23} = 0$ and a three-dimensional system of strain. Thus,

$$\sigma_{11} = \frac{E}{1-\nu^2}[\varepsilon_{11} + \nu\varepsilon_{22}] \quad \sigma_{22} = \frac{E}{1-\nu^2}[\varepsilon_{22} + \nu\varepsilon_{11}] \quad \sigma_{12} = 2G\varepsilon_{12} = \frac{E}{1+\nu}\varepsilon_{12} \tag{5.58}$$

or

$$\varepsilon_{11} = \frac{1}{E}[\sigma_{11} - \nu\sigma_{22}] \quad \varepsilon_{22} = \frac{1}{E}[\sigma_{22} - \nu\sigma_{11}] \quad \varepsilon_{33} = \frac{-\nu}{E}[\sigma_{11} + \sigma_{22}] \quad \varepsilon_{12} = \frac{\sigma_{12}}{2G} \tag{5.59}$$

Plane strain is a two-dimensional system of strain, where $\varepsilon_{33} = \varepsilon_{13} = \varepsilon_{23} = 0$ and a three-dimensional system of stress. Thus,

$$\begin{aligned} \sigma_{11} &= \frac{E}{(1-2\nu)(1+v)}[(1-\nu)\varepsilon_{11} + \nu\varepsilon_{22}] \qquad & \sigma_{22} &= \frac{E}{(1-2\nu)(1+v)}[\nu\varepsilon_{11} + (1-\nu)\varepsilon_{22}] \\ \sigma_{33} &= \frac{\nu E}{(1-2\nu)(1+v)}[\varepsilon_{11} + \varepsilon_{22}] & \sigma_{12} &= 2G\varepsilon_{12} = \frac{E}{1+\nu}\varepsilon_{12} \end{aligned} \tag{5.60}$$

or

$$\varepsilon_{11} = \frac{1}{E}[\sigma_{11} - \nu\sigma_{22}] \quad \varepsilon_{22} = \frac{1}{E}[\sigma_{22} - \nu\sigma_{11}] \quad \varepsilon_{12} = \frac{\sigma_{12}}{2G} \tag{5.61}$$

#### *5.6.1.2 Uniaxial Stress*

The only nonzero stress component of stress is $\sigma_{11} \neq 0$. This results in

$$\varepsilon_{11} = \frac{\sigma_{11}}{E} \quad \varepsilon_{22} = \varepsilon_{33} = -\frac{\nu}{E}\sigma_{11} \tag{5.62}$$

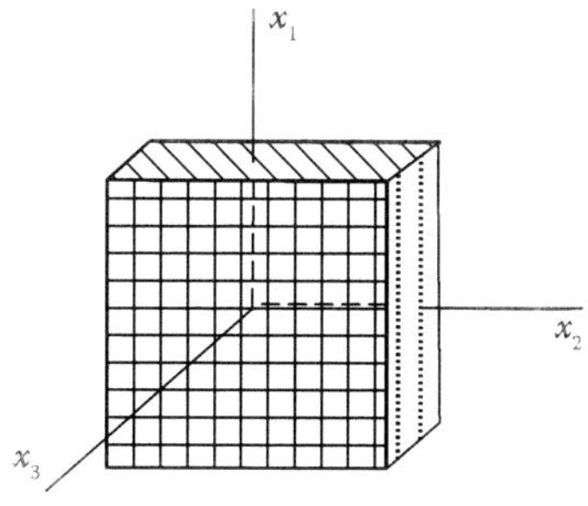


**FIGURE 5.10**
Orthotropic material with three planes of symmetry.

### *5.6.1.3 Anisotropic Materials*

Anisotropic materials are characterized by the general stress–strain relation given by Equation 5.44, where each of the elastic coefficients ($C_{ij}$) must be determined. If there are three planes of elastic moduli symmetry, then the material is characterized as orthotropic. Provided the coordinates ($x_1$, $x_2$, and $x_3$) coincide with the planes of symmetry as shown in Figure 5.10, the stress–strain relations become

$$\begin{aligned} \sigma_{11} &= E_{11}\varepsilon_{11} + E_{12}\varepsilon_{22} + E_{13}\varepsilon_{33} \\ \sigma_{22} &= E_{21}\varepsilon_{11} + E_{22}\varepsilon_{22} + E_{23}\varepsilon_{33} \\ \sigma_{33} &= E_{31}\varepsilon_{11} + E_{32}\varepsilon_{22} + E_{33}\varepsilon_{33} \\ \sigma_{12} &= 2G_{12}\varepsilon_{12} \\ \sigma_{23} &= 2G_{23}\varepsilon_{23} \\ \sigma_{31} &= 2G_{31}\varepsilon_{31} \end{aligned} \tag{5.63}$$

Noting that $E_{12} = E_{21}$, $E_{13} = E_{31}$, and $E_{23} = E_{32}$, we can represent this in a matrix form as

$$\begin{Bmatrix} \sigma_{11} \\ \sigma_{22} \\ \sigma_{33} \\ \sigma_{13} \\ \sigma_{23} \\ \sigma_{12} \end{Bmatrix} = \begin{bmatrix} C_{11} & C_{12} & C_{13} & 0 & 0 & 0 \\ C_{12} & C_{22} & C_{23} & 0 & 0 & 0 \\ C_{13} & C_{23} & C_{33} & 0 & 0 & 0 \\ 0 & 0 & 0 & C_{44} & 0 & 0 \\ 0 & 0 & 0 & 0 & C_{55} & 0 \\ 0 & 0 & 0 & 0 & 0 & C_{66} \end{bmatrix} \begin{Bmatrix} \varepsilon_{11} \\ \varepsilon_{22} \\ \varepsilon_{32} \\ \varepsilon_{13} \\ \varepsilon_{23} \\ \varepsilon_{12} \end{Bmatrix} = \begin{bmatrix} E_{11} & E_{12} & E_{13} & 0 & 0 & 0 \\ E_{12} & E_{22} & E_{23} & 0 & 0 & 0 \\ E_{13} & E_{23} & E_{33} & 0 & 0 & 0 \\ 0 & 0 & 0 & 2G_{13} & 0 & 0 \\ 0 & 0 & 0 & 0 & 2G_{23} & 0 \\ 0 & 0 & 0 & 0 & 0 & 2G_{12} \end{bmatrix} \begin{Bmatrix} \varepsilon_{11} \\ \varepsilon_{22} \\ \varepsilon_{32} \\ \varepsilon_{13} \\ \varepsilon_{23} \\ \varepsilon_{12} \end{Bmatrix}. \tag{5.64}$$

## 5.7 Displacement Equations for Elastic Bodies

When all of the elastic constants are known, we have six unknown stresses and strains ($\sigma_{ij}$ and $\varepsilon_{ij}$) and three unknown displacements ($u_i$) that yield 15 unknowns. We have three equations of motion, six stress–strain relations, and six strain–displacement relations.

We can take the equations of motion and replace the stresses ($\sigma_{ij}$) with the strains ($\varepsilon_{ij}$) using Hooke's law. Then, we can replace the strains with displacements ($u_i$). This results in

$$\begin{aligned}
\mu\left(\frac{\partial^2 u_1}{\partial x_1^2}+\frac{\partial^2 u_1}{\partial x_2^2}+\frac{\partial^2 u_1}{\partial x_3^2}\right)+(\lambda+\mu)\frac{\partial\theta}{\partial x_1}+f_1=\rho\frac{\partial^2 u_1}{\partial t^2}\\
\mu\left(\frac{\partial^2 u_2}{\partial x_1^2}+\frac{\partial^2 u_2}{\partial x_2^2}+\frac{\partial^2 u_2}{\partial x_3^2}\right)+(\lambda+\mu)\frac{\partial\theta}{\partial x_2}+f_2=\rho\frac{\partial^2 u_2}{\partial t^2}\\
\mu\left(\frac{\partial^2 u_3}{\partial x_1^2}+\frac{\partial^2 u_3}{\partial x_2^2}+\frac{\partial^2 u_3}{\partial x_3^2}\right)+(\lambda+\mu)\frac{\partial\theta}{\partial x_3}+f_3=\rho\frac{\partial^2 u_3}{\partial t^2}
\end{aligned} \tag{5.65}$$

The same set of equations using indicial notation is given by

$$\mu\nabla^2 u_i+(\lambda+\mu)\theta_i+f_i=\rho\frac{\partial^2 u_i}{\partial t^2} \quad i=1,2,3 \tag{5.66}$$

where $\nabla^2=\partial^2/\partial x_1^2+\partial^2/\partial x_2^2+\partial^2/\partial x_3^2$. Using symbolic vector notation, we have

$$\mu\nabla^2\mathbf{u}+(\lambda+\mu)\text{grad}(\text{div }\mathbf{u})+\mathbf{f}=\rho\frac{\partial^2\mathbf{u}}{\partial t^2} \tag{5.67}$$

## 5.8 Boundary Conditions

Boundary conditions play an essential role in formulating and solving elasticity problems. Therefore, it is essential to understand their use in formulating a proper solution. Suppose you are given a surface with a load vector **P** applied to an outer surface element as shown in Figure 5.11a. Boundary conditions are usually specified by using the coordinate system

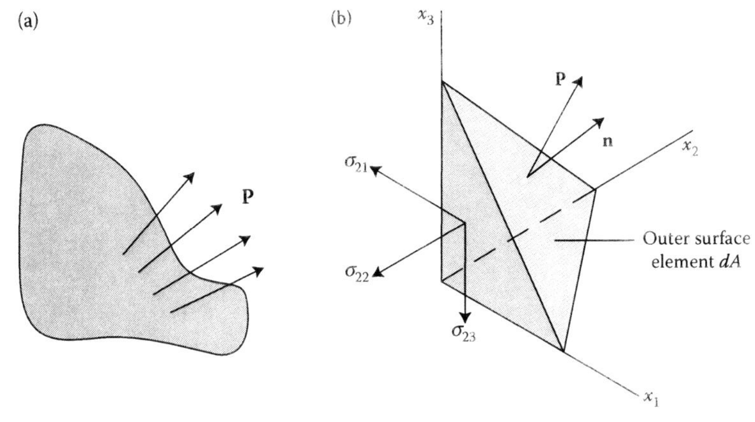


**FIGURE 5.11**
(a) Arbitrary body with a load vector **P** and (b) internal stresses on $x_1$, $x_3$ plane due to a load vector.

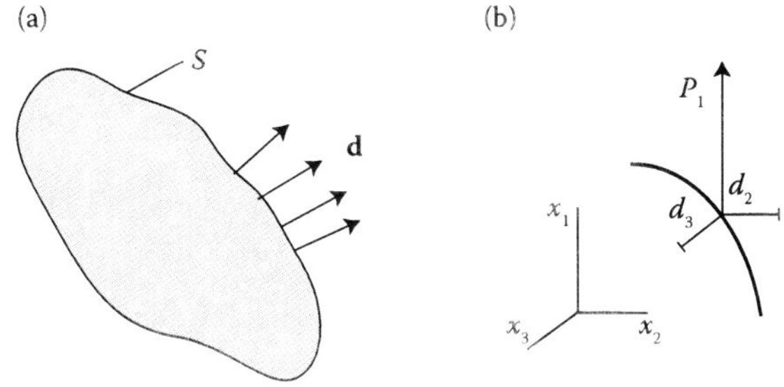


**FIGURE 5.12**
(a) Elastic body with a displacement vector $d$ and (b) example of prescribed force and displacements.

describing the problem at hand. Suppose the load vector has components $P_1$, $P_2$, and $P_3$ along the direction of the coordinate axes. The outer surface element has an outer normal direction **n**; passing a cutting plane through the surface perpendicular to the $x_1$, $x_3$ plane exposes the internal stress component on that plane (see Figure 5.11b). Applying Newton's law to the element produces the stress boundary conditions. These are valid for static and dynamic behavior. For dynamic behavior, the inertia has terms of higher orders and can be neglected. The stress boundary conditions are written as

$$\begin{aligned} P_1 &= \sigma_{11}n_1 + \sigma_{21}n_2 + \sigma_{31}n_3 \\ P_2 &= \sigma_{12}n_1 + \sigma_{22}n_2 + \sigma_{32}n_3 \\ P_3 &= \sigma_{13}n_1 + \sigma_{23}n_2 + \sigma_{33}n_3 \end{aligned} \tag{5.68}$$

Using indicial notation, these are written as

$$P_i = \sigma_{1i}n_1 + \sigma_{2i}n_2 + \sigma_{3i}n_3 = \sigma_{ji}n_j \quad i, j = 1, 2, 3 \tag{5.69}$$

where a repeated subscript indicates summation from 1 to 3. Knowing the stress tensor at an arbitrary point within the body, we can determine from Equation 5.69 the quantities $P_i$ and thus the stress vector.

Now consider a displacement vector **d** given on the outer normal of a surface $S$ as depicted in Figure 5.12a. The displacement boundary conditions are

$$(u_1)_s = d_1 \quad (u_2)_s = d_2 \quad (u_3)_s = d_3 \quad \text{or} \quad (u_i)_s = d_i \tag{5.70}$$

At each point of the outer surface, we have to specify one quantity of the pairs $(P_1, d_1)$, $(P_2, d_2)$, and $(P_3, d_3)$. As an example, let us assume that we are given $P_1$, $d_2$, $d_3$, see Figure 5.12b. In a static application, we have the following formulation: six unknown $\sigma_{ij}$ and six unknown $\varepsilon_{ij}$ terms. We have equations of equilibrium stress–strain relations and compatibility conditions.

## 5.9 Work and Energy

Energy methods (Reddy 2002) are used as an alternative to direct methods to obtain solutions of structural and elasticity problems. Therefore, let us consider the concepts of work and energy. It is noted that *work* is defined as the product of a force and the displacement of its point of application in the direction of the force. On the other hand, energy is the ability to produce or create work.

As loads are applied, the body deforms and work is performed. Let us consider an elastic body that is subjected to an external surface force vector **P**, which can also have body forces represented by a vector **f**. For a body in motion, the displacement vector **u** changes from **u** to **u** + $d$**u**. Thus,

$$dW \equiv \int_S \mathbf{P} \cdot d\mathbf{u}\, dS + \int_V \mathbf{f} \cdot d\mathbf{u}\, dV \tag{5.71}$$

or

$$dW \equiv \int_S P_i\, du_i\, dS + \int_V f_i\, du_i\, dV \tag{5.72}$$

The total work during motion from state $A$ to state $B$ is $W = \int_A^B dW$. This integration is generally impossible except for special cases such as **P** equal to a constant and **f** equal to a constant during the motion from $A$ to $B$. Thus,

$$W = \int_A^B dW = \int_A^B \int_S \mathbf{P} \cdot d\mathbf{u}\, dA + \int_A^B \int_V \mathbf{f} \cdot d\mathbf{u}\, dV \tag{5.73}$$

Changing the order of integration yields

$$W = \int_S \mathbf{P} \cdot \left( \int_A^B d\mathbf{u} \right) dA + \int_V \mathbf{f} \cdot \left( \int_A^B d\mathbf{u} \right) dV = \int_S \mathbf{P} \cdot (\mathbf{u}_B - \mathbf{u}_A) dA + \int_V \mathbf{f} \cdot (\mathbf{u}_B - \mathbf{u}_A) dV \tag{5.74}$$

### 5.9.1 Strain Energy

As a solid deforms, stresses are developed that result in internal forces. These forces perform work in moving through the internal displacements until it reaches its final configuration. If the strained elastic solid were allowed to slowly to return to its unstrained state, the elastic solid would be capable of returning a portion of the work done by the external forces. This capacity of the internal forces to perform work in a strain elastic solid is called *strain energy*. The strained solid due to the deformations is depicted in Figure 5.13, where only a few of the stresses and strains are shown for convenience.

We need to determine the work of the stresses corresponding to changes in the strains $d\varepsilon_{11}$, $d\varepsilon_{22}$, $d\varepsilon_{33}$, $d\varepsilon_{12}$, $d\varepsilon_{23}$, $d\varepsilon_{13}$, and $d\varepsilon_{ij}$. To evaluate the strain energy, we need to compute the differential work done by the internal forces; thus, we have for a three-dimensional elastic solid

$$\begin{aligned} &\sigma_{11}\, dx_2\, dx_3(dx_1\, d\varepsilon_{11}) + \sigma_{22}\, dx_1\, dx_3(dx_2\, d\varepsilon_{22}) + \sigma_{33}\, dx_1\, dx_2(dx_3\, d\varepsilon_{33}) \\ &\quad + 2\sigma_{12}\, dx_2\, dx_3(dx_1\, d\varepsilon_{12}) + 2\sigma_{23}\, dx_3\, dx_1(dx_2\, d\varepsilon_{23}) + 2\sigma_{13}\, dx_2\, dx_3(dx_3\, d\varepsilon_{13}) \\ &\quad = \sigma_{ij}\, d\varepsilon_{ij}\, dx_1\, dx_2\, dx_3 \end{aligned} \tag{5.75}$$

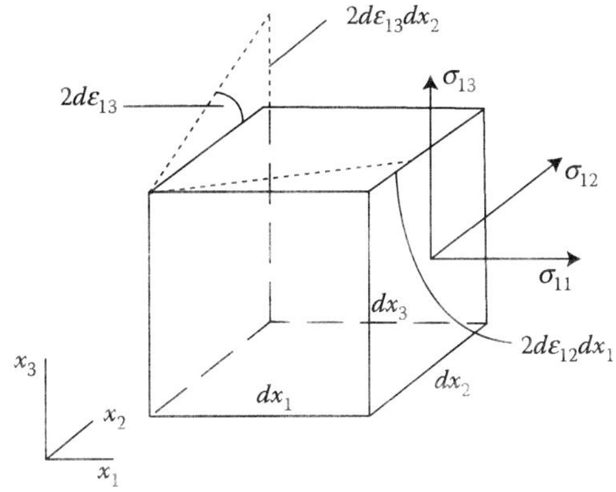


**FIGURE 5.13**
Internal stresses and strains in a three-dimensional solid.

Denoting the *specific strain energy* (or *strain energy density*) as $U_0$, we denote the work of the stresses per unit volume as

$$dU_0 = \sigma_{ij}\, d\varepsilon_{ij} \tag{5.76}$$

Note that $U_0$, the strain energy density, is the mechanical work performed on an element per unit volume at a point during a deformation. The strain energy of the whole body is defined by

$$U = \int_V U_0\, dV \tag{5.77}$$

For a linear elastic body, the stress–strain equation is given as $\sigma_{ij} = \lambda\theta\delta_{ij} + 2\mu\varepsilon_{ij}$, where $\theta = \varepsilon_{11} + \varepsilon_{22} + \varepsilon_{33}$. Therefore,

$$dU_0 = (\lambda\theta\delta_{ij} + 2\mu\varepsilon_{ij})d\varepsilon_{ij} = \lambda\theta\, d\theta + 2\mu\varepsilon_{ij}\, d\varepsilon_{ij} \tag{5.78}$$

If the state of strain changes from 0 to some value $\varepsilon_{ij}$, then we have

$$\begin{aligned} U_0 &= \int_0^{\varepsilon_{ij}} dU_0 = \int_0^{\varepsilon_{ij}} (\lambda\theta\, d\theta + 2\mu\varepsilon_{ij}\, d\varepsilon_{ij}) \\ &= \frac{1}{2}\lambda\theta^2 + \mu\varepsilon_{ij}\varepsilon_{ij} \end{aligned} \tag{5.79}$$

Therefore,

$$U = \int_V U_0\, dV = \int_V \left(\frac{1}{2}\lambda\theta^2 + \mu\varepsilon_{ij}\varepsilon_{ij}\right) dV \tag{5.80}$$

### 5.9.2 Kinetic Energy

Kinetic energy is often defined as the energy of motion. In an equation form, we define the kinetic energy as

$$K \equiv \frac{1}{2}\int_V \rho\,|\mathbf{v}|^2\,dV = \frac{1}{2}\int_V \rho\left(\dot{u}_1^2 + \dot{u}_2^2 + \dot{u}_3^2\right)dV = \frac{1}{2}\int_V \rho\dot{u}_i\dot{u}_i\,dV \tag{5.81}$$

The variation of the kinetic energy when the displacements $u_i$ acquire virtual increments $\delta u_i$ is given as

$$\delta K = K(\dot{\mathbf{u}} + \delta\dot{\mathbf{u}}) - K\dot{\mathbf{u}} = \frac{1}{2}\int_V \rho[(\dot{u}_i + \delta\dot{u}_i)(\dot{u}_i + \delta\dot{u}_i) - \dot{u}_i\dot{u}_i]dV \tag{5.82}$$

and

$$\delta K = \int_V \rho\dot{u}_i\delta\dot{u}_i\,dV = \int_V \rho\frac{\partial}{\partial t}(\dot{u}_i\delta u_i)dV - \int_V \rho\ddot{u}_i\delta u_i\,dV \tag{5.83}$$

where we have assumed that $\delta\dot{u}_i\delta\dot{u}_i \approx 0$. It is understood that $\delta K$ denotes the change of kinetic energy if the velocity changes from $\dot{\mathbf{u}}$ to $\dot{\mathbf{u}} + \delta\dot{\mathbf{u}}$. We note that there is a difference between $\delta\mathbf{u}$ or $\delta u_i$ and $d\mathbf{u}$ or $du_i$. Thus, the actual increment of displacement over the time interval $t$ to $t + dt$ is $du_i = (\partial u_i/\partial t)dt = \dot{u}_i\,dt$ and the virtual displacement $\delta u_i$ represents an arbitrary infinitesimal change of $u_i$ at time $t$ equal to a constant. At a fixed time $t$, we may investigate the value of various functions corresponding to $u_i$ and $u_i + \delta u_i$. It has been assumed that $\delta u_i$ is consistent with the boundary conditions.

## 5.10 Principle of Virtual Work

Virtual work is based on a hypothetical variation of deformation (Reddy 2002). Therefore, consider a body which is in equilibrium under a system of forces. In addition to the actual or true displacements, we assume a system of arbitrary changes of displacement $\delta u_i$ to take place. These displacements are called *virtual displacements* because they do not need actually to take place. The virtual displacements are continuous functions, along with their derivatives, of the coordinates $x_1$, $x_2$, and $x_3$ and satisfy the kinematic boundary conditions. When virtual displacements are imposed on an elastic body, the strains move through *virtual strains* or arbitrary variations in true strains. To determine the virtual strains $\delta\varepsilon_{ij}$, the displacements are replaced by virtual displacements, thus

$$\delta\varepsilon_{ij} = \frac{1}{2}(\delta u_{i,j} + \delta u_{j,i}) \tag{5.84}$$

The corresponding change of $U_0$ is given by

$$\delta U_0 = \sigma_{ij}\delta\varepsilon_{i,j} \tag{5.85}$$

and the change in strain energy $U$ is given by

$$\delta U = \int_V \delta U_0\, dV = \int_V \sigma_{ij}\delta\varepsilon_{i,j}\, dV = \frac{1}{2}\int_V \sigma_{ij}(\delta u_{i,j} + \delta u_{j,i})dV \tag{5.86}$$

Recall that $du_i = (\partial u_i/\partial t)dt = v_i\, dt$ and $\delta u_i$ equals an arbitrary variation of $u_i$. As an example consider $u_1$ as a function of time $t$ as depicted in Figure 5.14.

Let's now apply to the integrals in appearing in Equation 5.86 Green's identity. Let $\phi(x_1, x_2, x_3)$ and $(\psi_1, \psi_2, \psi_3)$ be two continuous functions along with their first and second derivatives; thus, Green's theorem is written as

$$\int_V \left[\frac{\partial\phi}{\partial x_1}\frac{\partial\psi}{\partial x_1} + \frac{\partial\phi}{\partial x_2}\frac{\partial\psi}{\partial x_2} + \frac{\partial\phi}{\partial x_3}\frac{\partial\psi}{\partial x_3}\right] dV = \int_S \phi\left[n_1\frac{\partial\psi}{\partial x_1} + n_2\frac{\partial\psi}{\partial x_2} + n_3\frac{\partial\psi}{\partial x_3}\right] dS$$
$$- \int_V \phi\left[\frac{\partial^2\psi}{\partial x_1^2} + \frac{\partial^2\psi}{\partial x_2^2} + \frac{\partial^2\psi}{\partial x_3^2}\right] dV \tag{5.87}$$

As an example of application, consider the following: let $\sigma_{11} = \phi$, $(\partial\psi/\partial x_1) = \delta u_1$, $\partial\psi/\partial x_2 = \partial\psi/\partial x_3 = 0$, then

$$\int_V \frac{\partial\sigma_{11}}{\partial x_1}\delta u_1\, dV = \int_S \sigma_{11}\delta u_1\eta_1\, dS - \int_V \sigma_{11}\delta\varepsilon_{11}\, dV \tag{5.88}$$

where $\delta\varepsilon_{11} = \delta(\partial u_1/\partial x_1) = (\partial/\partial x_1)\delta u_1$, which is the increment of the strain $\varepsilon_{11}$ corresponding to $\delta u_1$. Similarly, we can apply the other values and Green's identity becomes

$$\int_V \sigma_{ij}\delta u_{i,j}\, dV = \int_S \sigma_{ij}\delta u_i\eta_j\, dS - \int_V \sigma_{ij,j}\delta u_i\, dV \tag{5.89}$$

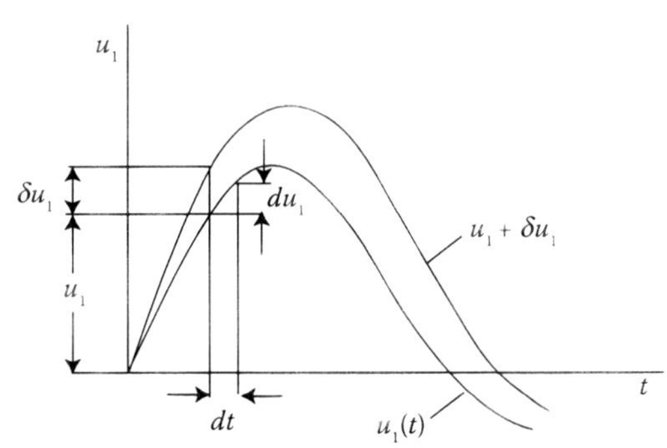


**FIGURE 5.14**
Difference between actual displacement and arbitrary displacement.

therefore,

$$\delta U = \int_S \sigma_{ij}\delta u_i \eta_j \, dS - \int_V \sigma_{ij,j}\delta u_i \, dV \tag{5.90}$$

Taking into account that $p_i = \sigma_{ij}\eta_j$ $(i = 1, 2, 3)$, where $p_i$ are the components of the external loading on the surface $S$ (or we assume $(\delta u_i)_S = 0$, however, if $u_i$ is given on $S$, we have $(u_i)_S = d_i$), and making use of the equation of motion $\sigma_{ij,j} + f_i = \rho\ddot{u}_i$, $i = 1, 2, 3$, Equation 5.90 can be reduced to the form

$$\begin{aligned} \delta U &= \int_S p_i\delta u_i dS + \int_V (f_i - \rho\ddot{u}_i)\delta u_i dV \\ &= \int_S p_i\delta u_i \, dS + \int_V f_i\delta u_i \, dV - \int_V \rho\ddot{u}_i\delta u_i \, dV \end{aligned} \tag{5.91}$$

The total external virtual work done by external forces is $\delta W = \int_S p_i\delta u_i \, dS + \int_V f_i\delta u_i \, dV$. Therefore, we have

$$\delta U = \delta W - \int_V \rho\ddot{u}_i\delta u_i \, dV \tag{5.92}$$

This expression is known as the principle of virtual work or d'Alembert's principle. For a static condition in which $\rho\ddot{u}_i = 0$, this reduces to

$$\delta U = \delta W \tag{5.93}$$

It is noted that the virtual displacements $\delta u_i$ are consistent with the boundary conditions of the body, that is, $(u_i)_S = d_i$. Now suppose that $\delta u_i = du_i$. We would then find that

$$\delta U = dU \rightarrow \delta U = \frac{\partial U}{\partial u_1}\delta u_1 + \frac{\partial U}{\partial u_2}\delta u_2 + \frac{\partial U}{\partial u_3}\delta u_3 + \frac{\partial U}{\partial t}dt$$

and

$$dU = \frac{dU}{dt}dt = \frac{\partial U}{\partial u_1}\frac{\partial u_1}{\partial t}dt + \frac{\partial U}{\partial u_2}\frac{\partial u_2}{\partial t}dt + \frac{\partial U}{\partial u_3}\frac{\partial u_3}{\partial t}dt$$

also,

$$dU = \frac{\partial U}{\partial u_1}du_1 + \frac{\partial U}{\partial u_2}du_2 + \frac{\partial U}{\partial u_3}du_2$$

We also have $\delta W = dW$ for $\delta u_i = du_i$. In an actual static displacement $dU = dW$, therefore, an increase of $U$ corresponds to an increase of $W$, that is, $U = W +$ constant. The principle of virtual work is more powerful than that illustrated before as we can use any virtual displacement (actual or virtual).

## 5.11 Hamilton's Principle

Consider an elastic body whose system varies continuously between two arbitrary instances of time $t_1$ and $t_2$. Start with the principle of virtual work given by $\delta U = \delta W - \int_V \rho \ddot{u}_i \delta u_i \, dV$. Integrating this equation over the time interval $t_1$, $t_2$ yields

$$\int_{t_1}^{t_2} \delta U \, dt = \int_{t_1}^{t_2} \delta W \, dt - \int_{t_1}^{t_2} \int_V \rho \ddot{u}_i \delta u_i \, dV \, dt \tag{5.94}$$

The second integral on the right-hand side of Equation 5.94 is related to our expression for the kinetic energy in Equation 5.81. Thus, from before we had

$$K = \frac{1}{2} \int_V \rho \dot{u}_i \dot{u}_i \, dV \tag{5.95}$$

and

$$\delta K = \int_V \rho \frac{\partial}{\partial t} (\dot{u}_i \delta u_i) dV - \int_V \rho \ddot{u}_i \delta u_i \, dV \tag{5.96}$$

The variation being chosen in such a way that $\delta u_i = 0$ at $t = t_1$ and $t = t_2$, therefore

$$\int_{t_1}^{t_2} \delta K \, dt = \int_V \rho \frac{\partial}{\partial t} (\dot{u}_i \delta u_i) \Big|_{t_1}^{t_2} dV - \int_{t_1}^{t_2} \int_V \rho \ddot{u}_i \delta u_i \, dV \, dt \tag{5.97}$$

As $\int_V \rho (\partial / \partial t) (\dot{u}_i \delta u_i) \big|_{t_1}^{t_2} dV = 0$ because $\delta u_i = 0$ at $t = t_1$ and $t = t_2$, we can write

$$\int_{t_1}^{t_2} \delta U \, dt = \int_{t_1}^{t_2} \delta W \, dt + \int_{t_1}^{t_2} \delta K \, dt \quad \text{or} \quad \delta \int_{t_1}^{t_2} (U - K) dt = \delta \int_{t_1}^{t_2} W \, dt \tag{5.98}$$

If the loading is conservative, that is, if the work done by external loads is independent of the path taken and only dependent on the end points, then Equation 5.98 can be rewritten as

$$\delta \int_{t_1}^{t_2} (U - K - W) dt = 0 \tag{5.99}$$

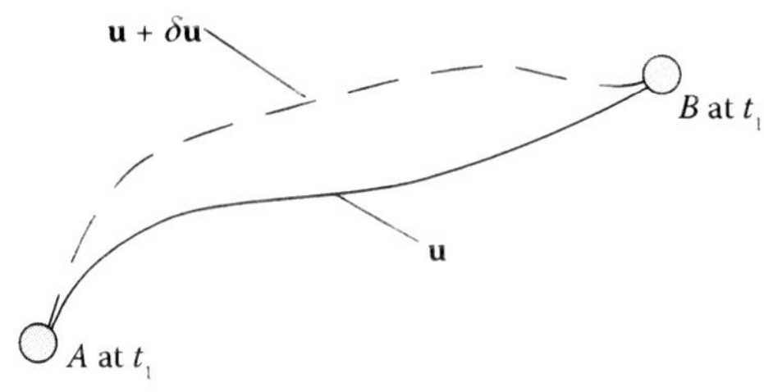


**FIGURE 5.15**
True path and false path between points $A$ and $B$.

The total *potential energy* of the system, Π, is defined by

$$\delta\Pi = \delta U - \delta W \tag{5.100}$$

Integrating gives $\Pi = U - W +$ constant. From Equation 5.99, we obtain

$$\delta \int_{t_1}^{t_2} (\Pi - K)dt = 0 \tag{5.101}$$

which represents *Hamilton's principle* and states that the physical path is that which the integral takes a stationary value.

The integral $\int_{t_1}^{t_2}(\pi - K)dt$ has a stationary (Greenwood 2003; Ardema 2005) or extreme value for $u_i$. Thus, we look at the true path and a false path between $A$ and $B$ that if we calculate the integral, then the value of the false path will have a larger value than the true path as shown in Figure 5.15.

If at some $x$, $df(x)/dx = 0$, then $f(x)$ has a stationary value at $x$. Then $(df/dx)\delta x = 0$ or $\delta f = 0$ as shown in Figure 5.16.

For static problems $K \approx 0$, therefore

$$\delta \int_{t_1}^{t_2} (\Pi - K)dt = \delta \int_{t_1}^{t_2} \Pi\, dt = \delta\Pi(t_2 - t_1) = 0 \tag{5.102}$$

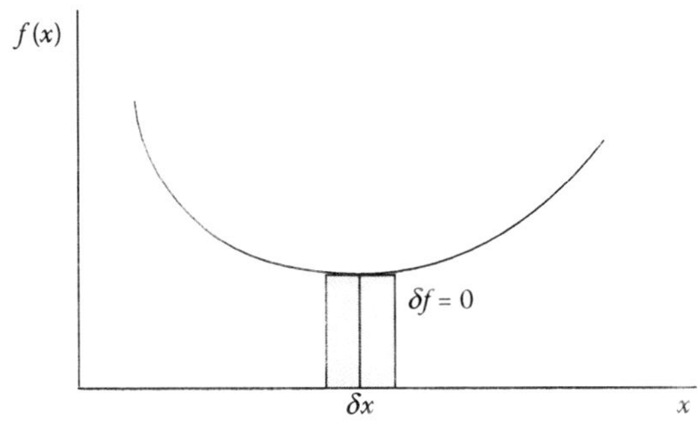


**FIGURE 5.16**
Stationary value of $f(x)$.

or

$$\delta\Pi = 0, \quad \Pi = \text{minimum} \tag{5.103}$$

For a system in equilibrium, $\Pi$ has a stationary value. This is known as the *principle of minimum potential energy*. For different cases of equilibrium, we have the following:

| | |
|---|---|
| In stable equilibrium, | $\delta\Pi = 0$; $\Pi = $ minimum; $\delta^2\Pi > 0$ |
| For unstable equilibrium, | $\delta\Pi = 0$; $\Pi = $ maximum; $\delta^2\Pi < 0$ |
| For neutral equilibrium, | $\delta\Pi = 0$; $\delta^2\Pi = 0$ |

**EXAMPLE 5.2**

Using Hamilton's principle, derive the equation of motion for the beam shown in Figure 5.17. Let us assume that the beam is elastic and obeys Hooke's law ($\sigma = E\varepsilon$) and has a finite length $l$.

**Solution**

The work of deformation or strain energy density $U_0$ is

$$U_0 = \frac{dU}{dV} = \frac{1}{2}\sigma\varepsilon$$

where $\sigma$ is the normal stress in the $x$ direction and $\varepsilon$ is the corresponding deformation. Only strain energy due to bending is considered as the influence of the transverse forces is small. From the beam theory, we have

$$\sigma = E\varepsilon \quad \text{and} \quad \varepsilon = -z\frac{\partial^2 w}{\partial x^2}$$

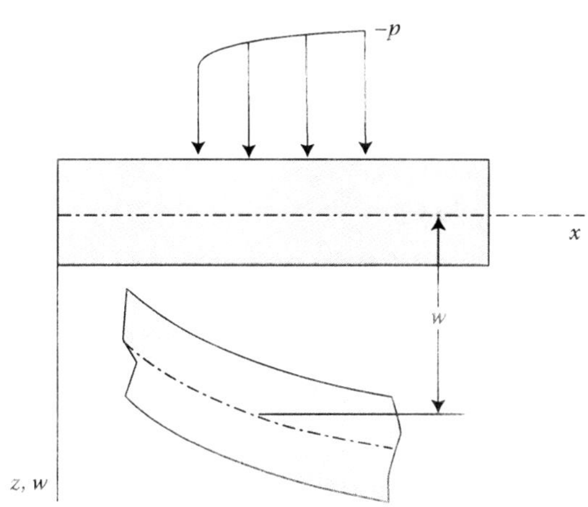


**FIGURE 5.17**
Elastic beam with an external loading $p$ and a deflection $w$.

The strain energy density becomes

$$U_0 = \frac{1}{2}E\left(-z\frac{\partial^2 w}{\partial x^2}\right)^2$$

And the total strain energy of the beam is

$$U = \int_0^l \int_A \frac{1}{2}E\left(-z\frac{\partial^2 w}{\partial x^2}\right)^2 dA\,dx = \int_0^l \frac{1}{2}E\left(z\frac{\partial^2 w}{\partial x^2}\right)^2 dx \int_A z^2\,dA$$

Noting that the second area moment of inertia is defined by $I = \int_A z^2\,dA$, the strain energy becomes

$$U = \frac{EI}{2}\int_0^l \left(\frac{\partial^2 w}{\partial x^2}\right)^2 dx$$

The kinetic energy $K = (1/2)A\rho\int_0^l(\partial w/\partial t)^2 dx$ and the external work $\delta W = \int_0^l \rho\delta w\,dx$. Hamilton's principle has the form

$$\delta\int_{t_1}^{t_2} dt \int_0^l \left[\frac{EI}{2}\left(\frac{\partial^2 w}{\partial x^2}\right)^2 - \frac{A\rho}{2}\left(\frac{\partial w}{\partial t}\right)^2\right] dx = \int_{t_1}^{t_2} dt \int_0^l \rho\delta w\,dx \tag{a}$$

where $\delta w$ satisfies the kinematic boundary conditions at timers $t_1$ and $t_2$. Calculating the variation of the first term gives

$$\delta\int_0^l \left(\frac{\partial^2 w}{\partial x^2}\right)^2 dx = 2\int_0^l \left(\frac{\partial^2 w}{\partial x^2}\right)\frac{\partial^2 \delta w}{\partial x^2} dx$$

Integrating by parts yields

$$\delta\int_0^l \left(\frac{\partial^2 w}{\partial x^2}\right)^2 dx = 2\int_0^l \frac{\partial^4 w}{\partial x^4}\delta w\,dx + 2\left|\frac{\partial^2 w}{\partial x^2}\frac{\partial \delta w}{\partial x} - \frac{\partial^3 w}{\partial x^3}\delta w\right|_{x=0}^{x=l} \tag{b}$$

where the second term

$$2\left|\frac{\partial^2 w}{\partial x^2}\frac{\partial \delta w}{\partial x} - \frac{\partial^3 w}{\partial x^3}\delta w\right|_{x=0}^{x=l}$$

is zero for any set of boundary conditions.

Next compute the second integral, where $\dot{w} = dw/dt$ as

$$\delta\int_{t_1}^{t_2} dt \int_0^l \frac{\rho A}{2}(\dot{w})^2 dx = \int_0^l dx \int_{t_1}^{t_2} A\rho\dot{w}\delta\dot{w}\,dt$$

Integrating by parts

$$\delta \int_{t_1}^{t_2} dt \int_0^l \frac{\rho A}{2} (\dot{w})^2 dx = \int_0^l dx A\rho \left| -\int_{t_1}^{t_2} \ddot{w}\delta w \, dt + (\dot{w}\delta w)_{t_1}^{t_2} \right| \tag{c}$$

where $(\dot{w}\delta w)_{t_1}^{t_2}$ is zero based on initial assumptions on $\delta w$ at times $t_1$ and $t_2$. Combining (b) and (c) results with the initial statement of Hamilton's principle (a) yields

$$\int_{t_1}^{t_2} dt \int_0^l \left[ EI \frac{\partial^4 w}{\partial x^4} + A\rho \ddot{w} \right] \delta w = \int_{t_1}^{t_2} dt \int_0^l \rho \delta w \, dx$$

or

$$\int_{t_1}^{t_2} dt \int_0^l \left[ EI \frac{\partial^4 w}{\partial x^4} + A\rho \ddot{w} - p \right] \delta w \, dx = 0$$

Recall that for $t_1 < t < t_2$, $0 < x < l$, $\delta w$ is arbitrary. Hence, we obtain

$$EI \frac{\partial^4 w}{\partial x^4} + A\rho \ddot{w} = p$$

**EXAMPLE 5.3**

Using Hamilton's principle and the direct methods of the calculus of variations, solve the problem of the forced vibration of a beam. Let us assume that the deflection is of the form $w(x,t) = \sum_n \Phi_n(t) X_n(x)$, where $\Phi_n(t)$ is an unknown, is function of time, and $X_n(x)$ is the $n$th natural mode (or any orthogonal system of functions satisfying the boundary conditions).

**Solution**

We have for the strain energy of the beam

$$U = \frac{EI}{2} \int_0^l \left( \frac{\partial^2 w}{\partial x^2} \right)^2 dx = \frac{EI}{2} \sum_n \Phi_n(t) \int_0^l \left( X_n'' \right)^2 dx$$

Note that

$$\int_0^l \left( \frac{\partial^2 w}{\partial x^2} \right)^2 dx = \int_0^l \left[ \left( \sum_n \Phi_n X_n'' \right) \left( \sum_m \Phi_m X_m'' \right) \right] dx = \int_0^l \sum_m \sum_n \Phi_m \Phi_n X_m'' X_n'' \, dx$$

where we use $\int_0^l X_m'' X_n'' \, dx = 0$ for $m \neq n$. The kinetic energy is

$$K = \frac{A\rho}{2} \sum_n \Phi_n^2 \int_0^l X_n^2 \, dx$$

Let $\delta w = \sum_n \delta\Phi_n X_n$, this results in

$$\delta U = EI \sum_n \Phi_n \delta\Phi_n \int_0^l \left(X_n''\right)^2 dx$$

$$\delta K = A\rho \sum_n \dot{\Phi}_n \delta\dot{\Phi}_n \int_0^l X_n^2 \, dx$$

and

$$\delta W = \sum_n \delta\Phi_n \int_0^l X_n p(x,t) dx$$

Using Hamilton's principle

$$\int_{t_1}^{t_2} dt \, EI \sum_n \Phi_n \delta\Phi_n \int_0^l \left(X_n''\right)^2 dx - \int_{t_1}^{t_2} dt A\rho \sum_n \dot{\Phi}_n \delta\dot{\Phi}_n \int_0^l X_n^2 \, dx = \int_{t_1}^{t_2} dt A\rho \sum_n \delta\Phi_n \int_0^l p(x,t) X_n \, dx$$

Certain integrations can be performed, thus integrating by parts

$$\int_{t_1}^{t_2} \dot{\Phi}_n \delta\dot{\Phi}_n dt = -\int_{t_1}^{t_2} \ddot{\Phi}_n \delta\Phi_n + \left(\dot{\Phi}_n \delta\Phi_n\right)_{t_1}^{t_2} = -\int_{t_1}^{t_2} \ddot{\Phi}_n \delta\Phi_n$$

where the term $(\dot{\Phi}_n \delta\Phi_n)_{t_1}^{t_2}$ vanishes as $\delta\Phi_n = 0$ at $t_1$ and $t_2$. Therefore, we have

$$\int_{t_1}^{t_2} dt EI \sum_n \Phi_n \delta\Phi_n \int_0^l \left(X_n''\right)^2 dx + \int_{t_1}^{t_2} dt A\rho \sum_n \ddot{\Phi}_n \delta\Phi_n \int_0^l X_n^2 \, dx = \int_{t_1}^{t_2} dt A\rho \sum_n \delta\Phi_n \int_0^l p(x,t) X_n \, dx$$

As $\delta\Phi_n$ is arbitrary for $t_1 < t < t_2$, then only if the integrals are equal we have

$$EI \sum_n \Phi_n \int_0^l \left(X_n''\right)^2 dx + A\rho \sum_n \ddot{\Phi}_n \int_0^l X_n^2 \, dx = \sum_n \int_0^l p(x,t) X_n \, dx$$

Thus,

$$EI \, \Phi_n \int_0^l \left(X_n''\right)^2 + A\rho \ddot{\Phi}_n \int_0^l X_n^2 \, dx = \int_0^l p(x,t) X_n \, dx \tag{a}$$

Equation (a) is an ordinary differential equation for $\Phi_n$. Note that

$$\int_0^l (X_n'')^2 \, dx = \int_0^l X_n^{iv} X_n \, dx$$

and

$$EIX_n^{iv} - A\rho\omega^2 X_n = 0$$

if $X_n$ is a natural mode. Thus,

$$\int_0^l (X_n'')^2 \, dx = \frac{A\rho}{EI}\omega_n^2 \int_0^l X_n^2 \, dx$$

Therefore, as $\omega_n$ is the $n$th natural frequency, we can write Equation (a) as

$$\ddot{\Phi}_n + \omega_n^2 \Phi_n = \frac{1}{A\rho} p_n(t)$$

where

$$p_n(t) = \frac{\int_0^l p(x,t) X_n \, dx}{\int_0^l X_n^2 \, dx}$$

## 5.12 General Energy Theorem

In the derivation of the equations of motion, we have, in general, the following: We have Newtonian mechanics when we use Newton's law in deriving d'Alembert's and Hamilton's principle; Lagrangian mechanics when we use d'Alembert's principle to derive Newton's law and Hamilton's principle; and we have Hamiltonian mechanics when we derive Newton's law and d'Alembert's principle (Nowacki 1963).

Now consider a motion, that is a function of time, where $\mathbf{u}_i = \mathbf{u}_i(t)$ and $\delta\mathbf{u}_i = d\mathbf{u}_i = (\partial\mathbf{u}_i/\partial t)dt = \mathbf{v}_i dt$, that is, we have assumed that the virtual displacement $\delta\mathbf{u}_i$ is identical with the actual displacement $d\mathbf{u}_i$. We had the relation

$$\delta U = \int_S p_i \delta u_i \, dS + \int_V (f_i - \rho\ddot{u}_i)\delta u_i \, dV \tag{5.104}$$

which now becomes

$$\delta U = \int_S p_i \dot{\mathbf{u}}_i \, dt \, dS + \int_V (f_i \dot{\mathbf{u}}_i - \rho\ddot{\mathbf{u}}_i\dot{\mathbf{u}}_i) dt \, dV \tag{5.105}$$

From the definition of kinetic energy, we have that $dK/dt = \int_V \rho \ddot{\mathbf{u}}_i \dot{\mathbf{u}}_i \, dV$. Therefore, Equation 5.105 becomes

$$dU = \int_S p_i \dot{\mathbf{u}}_i \, dt \, dS + \int_V f_i \dot{\mathbf{u}}_i \, dt \, dV - dK \tag{5.106}$$

or

$$\frac{dU}{dt} + \frac{dK}{dt} = \int_S p_i \dot{\mathbf{u}}_i \, dS + \int_V f_i \dot{\mathbf{u}}_i \, dV \tag{5.107}$$

The general energy is $U + K$ and the power of external forces is

$$\frac{dw}{dt} = \int_S p_i \dot{\mathbf{u}}_i \, dS + \int_V f_i \dot{\mathbf{u}}_i \, dV \tag{5.108}$$

If there are no external forces, then

$$\begin{aligned} &\frac{dU}{dt} + \frac{dK}{dt} = 0 \\ &\frac{d}{dt}(U + K) = 0 \end{aligned} \tag{5.109}$$

or

$$U + K = \text{constant} \tag{5.110}$$

## 5.13 Rayleigh's Method

Rayleigh's method (Humar 2012) is based on the principle of conservation of energy and therefore applies to conservative systems. This method gives the natural frequencies of elastic systems. Let us assume that the displacement, position, and velocity vectors are $\mathbf{u} = \mathbf{u}_0(\mathbf{x})\cos \omega t$ (or $\sin \omega t$), $\mathbf{x} \equiv (x_1, x_2, x_3)$, and $\dot{\mathbf{u}}_{\max} \equiv |\omega \mathbf{u}_0|$, respectively. Recalling the principle of conservation of energy ($U + K = \text{constant}$), we have

$$U_{\max} = K_{\max} \tag{5.111}$$

**Proof**

Take $t = 0$, then $\dot{\mathbf{u}} = \omega \mathbf{u}_0 \sin \omega t = 0$. Therefore, $K = 0$ and $U_{\max} = C = \text{constant}$. Next, take $\omega t = \pi/2$, so that $U = 0$ and $K_{\max} = C$; hence,

$$U_{\max} = K_{\max}$$

*QED*

For the $n$th mode we have $U_{max} = U_n$ and $K_{max} = K_n$, therefore, we have $U_n = K_n$. Alternately, we could have $U_n = \omega_n^2 K_n'$, where $K_n' = K_n/\omega_n^2$. That is, $K_n$ is proportional to $\omega_n^2$ and $K_n'$ equals the kinetic energy for $\omega_n = 1$. Thus, we have

$$\omega_n^2 = \frac{U_n}{K_n'} \tag{5.112}$$

Suppose that the $n$th mode is approximated by $\Psi(\mathbf{x})$. Then,

$$\omega_n^2 \cong R = \frac{U(\Psi)}{K'(\Psi)} \tag{5.113}$$

The quotient $R$, which was first introduced by Rayleigh, is often called *Rayleigh's quotient*. Equation 5.113 is equivalent to *Rayleigh's method* and $\sqrt{R}$ approximates the $n$th natural frequency. Consider the frequency value $\omega_1$. If $\Psi(x)$ is equal to $X_1(x)$, which is the first natural mode, then we have $\omega_1^2 = R = U(\Psi)/K'(\Psi)$. If $\Psi(x)$ is not identical to $X_1(x)$, then $R > \omega_1^2$.

One must not be led into believing that we have found a way of computing $\omega$, because in forming the energy expressions we have assumed the known solution of motion. The displacement functions $\mathbf{u}_i$ are, however, not known beforehand. In fact, if $\mathbf{u}_i$ were known, the whole problem of free vibration would have been solved and we would know the $\omega^2$.

**Proof**

If $\Psi(x) \neq X_n$, then

$$\Psi(x) = a_1 X_1 + a_2 X_2 + a_3 X_3 + \cdots$$

where $X_1$, $X_2$, $X_3$ are the natural (orthogonal) modes. We also have

$$K'(\Psi) \approx a_1^2 + a_2^2 + a_3^2 + \cdots$$

$$U(\Psi) \approx \omega_1^2 a_1^2 + \omega_2^2 a_2^2 + \omega_3^2 a_3^2 + \cdots$$

Thus,

$$R = \frac{\omega_1^2 a_1^2 + \omega_2^2 a_2^2 + \omega_3^2 a_3^2 + \cdots}{a_1^2 + a_2^2 + a_3^2 + \cdots}$$

As $\omega_1^2 < \omega_2^2 < \omega_3^2 < \cdots$, we have

$$R > \frac{\omega_1^2 a_1^2 + \omega_1^2 a_2^2 + \omega_1^2 a_3^2 + \cdots}{a_1^2 + a_2^2 + a_3^2 + \cdots} = \omega_1^2$$

*QED*

Thus, $R$ always overestimates the actual frequency. To show $K'(\Psi)$, recall for $X_1$

$$K_{\max} = \frac{1}{2}\int_V |\dot{\mathbf{u}}_{\max}|^2 \rho dV$$
$$= \frac{1}{2}\omega_1^2 \int_V |\mathbf{u}_0|^2 \rho dV = \frac{1}{2}\omega_1^2 a_1^2 \int_V X_1^2 \rho dV$$

where $\mathbf{u} = a_1 X_1$.

Suppose we consider a higher frequency, say $\omega_2$. If $\Psi(x)$ is equal to $X_2(x)$, then $R = \omega_2^2$. Suppose that $\Psi(x)$ is not identically equal to $X_2(x)$ and

$$\int_V \Psi(x) X_1(x) dV = 0$$

This condition must be satisfied, but as we do not know $X_1(x)$ exactly, so we do not know if this is satisfied identically. Provided the condition is satisfied, we have

$$R = \frac{U(\Psi)}{K'(\Psi)} > \omega_2^2$$

**Proof**

If $\Psi(x) = a_2 X_2 + a_3 X_3 + a_4 X_4 + \cdots$, then

$$a_1 = \frac{\int_V \Psi(x) X_1 \, dV}{\int_V X_1^2 \, dV} = 0$$

As

$$K'(\Psi) \approx a_2^2 + a_3^2 + a_4^2 + \cdots$$

$$U(\Psi) \approx \omega_2^2 a_2^2 + \omega_3^2 a_3^2 + \omega_4^2 a_4^2 + \cdots$$

we have

$$R = \frac{\omega_2^2 a_2^2 + \omega_3^2 a_3^2 + \omega_4^2 a_4^2 + \cdots}{a_2^2 + a_3^2 + a_4^2 + \cdots}$$

As $\omega_2^2 < \omega_3^2 < \omega_4^2 < \cdots$, we have

$$R > \frac{\omega_2^2 a_2^2 + \omega_3^2 a_3^2 + \omega_4^2 a_4^2 + \cdots}{a_2^2 + a_3^2 + a_4^2 + \cdots} = \omega_2^2$$

*QED*

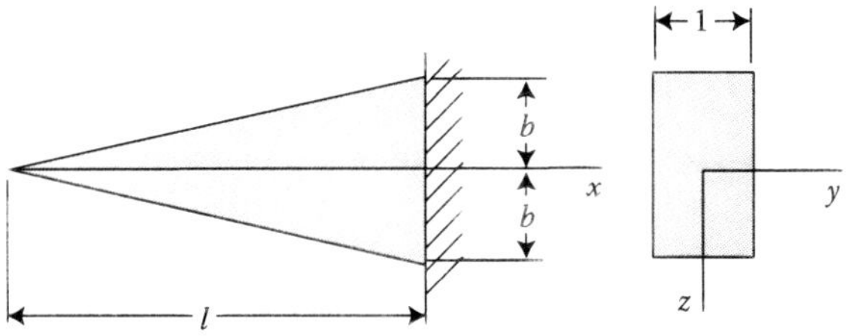


**FIGURE 5.18**
Tapered fixed-free beam of unit width.

**EXAMPLE 5.4**

Consider a tapered beam fixed at one end and free at the other of unit width as shown in Figure 5.18. Using Rayleigh's method, determine the first fundament frequency.

**Solution**

For the beam shown, we define the cross-sectional area and second area moment of inertia in terms of the area and inertia at $x = l$ ($A_0$ and $I_0$) as $A = A_0(x/l)$ and $I = I_0(x/l)^3$. The boundary conditions are

$$\text{At} \quad x = 0: \quad EIw'' = EIw''' = 0$$

$$\text{At} \quad x = l: \quad w(l) = w'(l) = 0$$

The boundary conditions will be satisfied by the function

$$\Psi(x) = a_1\left(1 - \frac{x}{l}\right)^2$$

Compute the following values:

$$2U(\Psi) = \int_0^l EI(\Psi'')^2 dx = \frac{EI_0}{l^3}$$

$$2K'(\Psi) = \int_0^l A\rho(\Psi)^2 dx = \frac{A_0\rho l}{30}$$

Thus, we obtain

$$\omega_1^2 \approx R = \frac{U(\Psi)}{K'(\Psi)} = \frac{30EI_0}{A_0\rho l^4}$$

or

$$\omega_1 \approx \frac{5.48}{l^2}\sqrt{\frac{EI_0}{A_0\rho}}$$

The exact solution is found to be $\omega_1 \approx (5.315\,/\,l^2)\sqrt{EI_0/A_0\rho}$

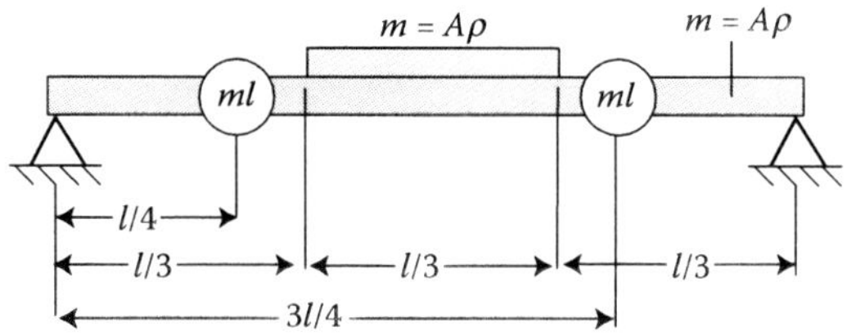


**FIGURE 5.19**
Simply supported beam with mass per unit length along with additional masses.

**EXAMPLE 5.5**

Consider a simply supported beam, as shown in Figure 5.19, that has a mass per unit length, $m$, two concentrated masses $ml$, located at $x = l/4$ and $x = 3\,l/4$, and an additional mass spanning a beam segment of $l/3$ is located at the center of the beam. Using Rayleigh's method, determine the beams' second fundamental frequency $\omega_2$.

**Solution**

It is assumed that $EI$ is constant and the mass per unit length is $m = A\rho$. The first mode of the simply supported beam is assumed as $\Psi_1 = \sin \pi x/l$. Using the first mode leads to the calculation of the first fundamental frequency. Thus, calculating $U(\Psi_1)$ and $K'(\Psi_1)$ yields

$$U(\Psi_1) = \frac{EI\pi^4}{2l^4}\int_0^l \sin^2\frac{\pi x}{l}dx = \frac{EI\pi^4}{4l^3}$$

$$K'(\Psi_1) = \frac{1}{2}\left\{m\int_0^l \sin^2\frac{\pi x}{l}dx + m\int_{l/3}^{2l/3}\sin^2\frac{\pi x}{l}dx + ml\sin^2\frac{\pi(l/4)}{l} + ml\sin^2\frac{\pi(3l/4)}{l}\right\} = \frac{5}{6}ml$$

Thus,

$$\omega_1 = \sqrt{\frac{U(\Psi)}{K'(\Psi)}} = \frac{5.38}{l^2}\sqrt{\frac{EI}{m}}$$

For the second mode let us assume $\Psi_2 = \sin(2\pi x/l)$ as shown in Figure 5.20, where

$$\int_0^l \Psi_2 X_1\,dx = 0$$

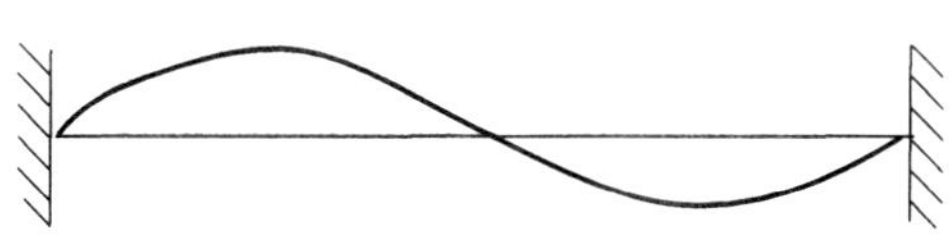

**FIGURE 5.20**
Second mode of a simply supported beam.

Calculating $U(\Psi_2)$ and $K'(\Psi_2)$, we obtain

$$U(\Psi_2) = \frac{4EI\pi^4}{4l^3} \quad \text{and} \quad K'(\Psi_2) = 1.33ml$$

Hence,

$$\omega_2 \approx \sqrt{\frac{U(\Psi_2)}{K'(\Psi_2)}} = \frac{17.1}{l^2}\sqrt{\frac{EI}{m}}$$

## 5.14 Ritz Method

The free vibration case can be derived from Hamilton's principle, Equation 5.101 with $\Pi = U$. Thus, we have

$$\delta \int_{t_1}^{t_2} (U - K)\, dt = 0 \tag{5.114}$$

Let us assume that the displacements are given as

$$\mathbf{u}(\mathbf{x},t) = \mathbf{u}_0(\mathbf{x}) \sin \omega t \tag{5.115}$$

Then the variations in displacements are

$$\mathbf{u}(\mathbf{x},t) + \delta\mathbf{u}(\mathbf{x},t) = [\mathbf{u}_0(\mathbf{x}) + \delta\mathbf{u}_0(\mathbf{x})] \sin \omega t$$

We require the variation in displacement, $\delta\mathbf{u}(\mathbf{x},t)$, to vanish at $t_1 = 0$ and at $t_2 = 2\pi/\omega$. Substituting Equation 5.115 into Equation 5.114 yields

$$\delta \int_{t_1}^{t_2} \left[ U(\mathbf{u}_0) \sin^2 \omega t - \omega^2 K'(\mathbf{u}_0) \cos^2 \omega t \right] dt = 0 \tag{5.116}$$

or

$$\delta \int_{t_1}^{t_2} \left[ \int_V U_0(\mathbf{u}_0) \sin^2 \omega t\, dV - \omega^2 \int_V \frac{1}{2} \rho \,|\, \mathbf{u}_0 \,|^2 \cos^2 \omega t\, dV \right] dt = 0 \tag{5.117}$$

Integrating between $t_1 = 0$ and $t_2 = 2\pi/\omega$ results in

$$\delta\int_V\left[U_0(\mathbf{u}_0)-\frac{\omega^2}{2}\rho\,|\,\mathbf{u}_0\,|^2\right]dV=0 \tag{5.118}$$

If the variation of the integral becomes zero, then $\mathbf{u}_0$ corresponds to a stationary value of the integral. Let us assume that the displacements as $\mathbf{u}_0 = a_1\psi_1 + a_2\psi_2 + \cdots$, where the $a_k$ terms are unknown coefficients and $\psi_k \equiv \mathbf{\Psi}_k(\mathbf{x})$ are known functions satisfying the displacement boundary conditions. Then we have either

$$\frac{\partial}{\partial a_k}\int_V\left[U_0(\mathbf{u}_0)-\frac{\omega^2}{2}\rho\,|\,\mathbf{u}_0\,|^2\right]dV=0 \quad \text{for each} \quad k=1,2,3,\ldots \tag{5.119}$$

$$\begin{aligned} U &= \int_V U_0(\mathbf{u}_0)dV = 0 \\ &= \int_V \{a_1a_1U_0(\psi_1,\psi_1)+a_1a_2U_0(\psi_1,\psi_2)+a_2a_1U_0(\psi_2,\psi_1)+a_2a_2U_0(\psi_2,\psi_2)+\cdots\}dV \end{aligned} \tag{5.120}$$

or

$$U = a_1^2U_{11}+a_1a_2U_{12}+a_2a_1U_{21}+a_2^2U_{22}+\cdots \tag{5.121}$$

where $U_{KL} = \int_V U_0(\psi_K\psi_L)dV$.

In a similar manner,

$$K=\frac{1}{2}\omega^2\int_V \rho\left[a_1^2\psi_1^2+a_1a_2\psi_1\psi_2+a_2a_1\psi_2\psi_1+a_2^2\psi_2^2+a_1a_3\psi_1\psi_3+\cdots\right]dV \tag{5.122}$$

or

$$K=\omega^2\left[a_1^2K_{11}+a_1a_2K_{12}+a_2a_1K_{21}+a_2^2K_{22}+a_1a_3K_{13}+\cdots\right] \tag{5.123}$$

where $K_{KL} = \int_V (1/2)\rho\,\psi_K\psi_L\,dV$.

We note that Equations 5.121 and 5.122 can actually be written as

$$U=\sum_K\sum_L a_Ka_LU_{KL} \tag{5.124}$$

and

$$K=\omega^2\sum_K\sum_L a_Ka_LK_{KL} \tag{5.125}$$

With Equations 5.121 or 5.124 and 5.123 or 5.125, the system (Equation 5.119) becomes

$$\begin{aligned}
&\text{For } K=1: \quad a_1\left(U_{11}-\omega^2K_{11}\right)+a_2\left(U_{12}-\omega^2K_{12}\right)+\cdots+a_N\left(U_{1N}-\omega^2K_{1N}\right)=0\\
&\text{For } K=2: \quad a_1\left(U_{21}-\omega^2K_{21}\right)+a_2\left(U_{22}-\omega^2K_{12}\right)+\cdots+a_N\left(U_{2N}-\omega^2K_{2N}\right)=0\\
&\vdots\\
&\vdots\\
&\text{For } K=N: \quad a_1\left(U_{N1}-\omega^2K_{N1}\right)+a_2\left(U_{N2}-\omega^2K_{N2}\right)+\cdots+a_N\left(U_{NN}-\omega^2K_{NN}\right)=0
\end{aligned} \tag{5.126}$$

This is a system of linear homogeneous equations for $a_1, a_2, \ldots, a_N$. These coefficients determine the shape of the deflected system. As the equations are homogeneous so the only nontrivial solution is if the determinate of the coefficients vanishes. Therefore,

$$\det\begin{vmatrix} U_{11}-\omega^2K_{11} & U_{12}-\omega^2K_{12} & \cdots & U_{1N}-\omega^2K_{1N}\\ U_{21}-\omega^2K_{21} & U_{22}-\omega^2K_{22} & \cdots & U_{2N}-\omega^2K_{2N}\\ \vdots & \vdots & \vdots & \vdots\\ U_{N1}-\omega^2K_{N1} & U_{N2}-\omega^2K_{N2} & \cdots & U_{NN}-\omega^2K_{NN}\end{vmatrix}=0 \tag{5.127}$$

This is the frequency equation. Its roots are $\omega_1, \omega_2, \ldots, \omega_N$. For each root $\omega_i$, we obtain a system of coefficients $a_1, a_2, \ldots, a_N$ and a deflection $\mathbf{u}_0 = a_1\psi_1 + a_2\psi_2 + \cdots + a_N\psi_N$ defining a natural mode.

**EXAMPLE 5.6**

Solve Example 5.5 again, using the Ritz method. The beam is shown in Figure 5.19.

**Solution**

Let us assume the deflection in the form

$$w(x) = a_1\psi_1 + a_2\psi_2 + a_3\psi_3 = a_1\sin\frac{\pi x}{l} + a_2\sin\frac{2\pi x}{l} + a_3\sin\frac{3\pi x}{l}$$

Compute the coefficients $U_{KL}$: As

$$U = a_1^2U_{11} + a_1a_2U_{12} + a_2a_1U_{21} + a_2^2U_{22} + \cdots$$

we find

$$U_{11} = \frac{EI\pi^4}{2l^4}\int_0^l \sin^2\frac{\pi x}{l}dx = \frac{EI\pi^4}{4l^3}$$

$$U_{22} = 4\frac{EI\pi^4}{l^3}$$

$$U_{33} = 81\frac{EI\pi^4}{4l^3}$$

$$U_{12} = U_{21} = 0, U_{13} = U_{31} = U_{32} = U_{23} = 0$$

Next compute the kinetic energy from

$$K = \frac{1}{2}\omega^2\left\{m\int_0^l (w)^2dx + m\int_{l/3}^{2l/3} (w)^2dx + ml(w)^2_{x=l/4} + ml(w)^2_{x=3l/4}\right\}$$

where

$$K_{11} = \frac{5ml}{6}, K_{22} = 1.333ml, K_{33} = \frac{5ml}{6}, K_{13} = K_{31} = \frac{ml}{2}$$

$$K_{12} = K_{21} = 0$$

The frequency equation is then given as

$$\det\begin{vmatrix} \frac{EI\pi^4}{4l^3} - \omega^2\frac{5ml}{6} & 0 & -\frac{\omega^2 ml}{2} \\ 0 & \frac{4EI\pi^4}{l^3} - \omega^2 1.333ml & 0 \\ -\frac{\omega^2 ml}{2} & 0 & \frac{81EI\pi^4}{4l^3} - \omega^2\frac{5ml}{6} \end{vmatrix} = 0$$

The roots are found to be

$$\omega_1 = \frac{5.36}{l^2}\sqrt{\frac{EI}{m}}, \omega_2 = \frac{17.1}{l^2}\sqrt{\frac{EI}{m}}, \omega_3 = \frac{60.2}{l^2}\sqrt{\frac{EI}{m}}$$

We note only a slight difference in the solutions for $\omega_1$ and $\omega_2$ when compared to the solution of Example 5.5 using the Rayleigh method.

### 5.14.1 Property of Ritz Method

Show that the following equation

$$\psi^{(N)}(x) = a_1\psi_1(x) + a_2\psi_2(x) + \cdots + a_N\psi_N(x) \tag{5.128}$$

provides a better approximation to $\omega^2$ than

$$\psi^{(N-1)}(x) = a_1\psi_1(x) + a_2\psi_2(x) + \cdots + a_{N-1}\psi_{N-1}(x) \tag{5.129}$$

From Equation 5.128, if we take $N$ terms, the approximation of the frequency is

$$\omega^{(N)} = \sqrt{\frac{U(\psi^{(N)})}{K'(\psi^{(N)})}} \tag{5.130}$$

$N$–1 term, from Equation 5.129, yields

$$\omega^{(N-1)} = \sqrt{\frac{U(\psi^{(N-1)})}{K'(\psi^{(N-1)})}} \tag{5.131}$$

where $N$ and $(N-1)$ stand for the number of terms not the $N$th and $(N-1)$th frequencies. We must show that $\omega^{(N)} \leq \omega^{(N-1)}$. To show that this is correct we assume the opposite, that is, $\omega^{(N-1)} < \omega^{(N)}$. This leads to a contradiction because $a_N \neq 0$ minimizes $\omega$. For example, $a_1, \ldots,$ $a_N$, $(a_N \neq 0)$ corresponds to the minimum of $\omega^2 = U/K'$, but if $a_N = 0$, this does not hold and a smaller value of $\omega$ is obtained. Thus, $\omega^{(N-1)} < \omega^{(N)}$ is impossible.

### 5.14.2 Another Approach to Ritz's Method

Recall Equation 5.113 where $\omega^2 \approx R = U(\psi(x))/K'(\psi(x))$, where $\psi(x)$ was an approximation to the mode shape. The best approximation for $\psi(x)$ corresponds to the minimum value of $R$. Consider $\psi(x)$ in the following form

$$\psi(x) = a_1\psi_1(x) + a_2\psi_2(x) + \cdots + a_N\psi_N(x) \tag{5.132}$$

We wish to determine the coefficients $a_1, a_2, \ldots, a_N$ to obtain or make $R$ a minimum. This means that $a_1, a_2, \ldots, a_N$ must satisfy the conditions

$$\frac{\partial R}{\partial a_K} = 0 \quad K = 1, 2, \ldots, N \tag{5.133}$$

From the definition of $R$, we have

$$\frac{(\partial U(\psi)/\partial a_K)K'(\psi) - U(\psi)(\partial K'(\psi)/\partial a_K)}{[K'(\psi)]^2} = 0 \tag{5.134}$$

We note that as $U(\psi) = \omega^2 K'(\psi)$, the numerator of this expression becomes

$$\frac{\partial U(\psi)}{\partial a_K}K'(\psi) - \omega^2 K'(\psi)\frac{\partial K'(\psi)}{\partial a_K} = 0 \tag{5.135}$$

With $\omega^2 K'(\psi) = K(\psi)$, this becomes

$$\frac{\partial U(\psi)}{\partial a_K} - \omega^2 \frac{\partial K'(\psi)}{\partial a_K} = 0 \tag{5.136}$$

Which is the same system of equations for $a_K$ as previously obtained.

## 5.15 Galerkin's Method

The method of Galerkin (Washizu 1975; Reddy 2002) belongs to the same general class as those of Rayleigh and Ritz as it seeks to obtain an approximate solution to differential equations with given boundary conditions. The Galerkin method uses functions, which satisfy the boundary conditions exactly, then specializes them in such a way that a satisfactory approximate solution to the differential equation is obtained.

For an elastic beam, Galerkin's method is identical to Ritz's method. For free vibrations, consider a beam for which the equation of motion is

$$EI\frac{\partial^4 w}{\partial x^4}+A\rho\frac{\partial^2 w}{\partial t^2}=0 \tag{5.137}$$

Assuming a solution of the form

$$w(x,t)=W(x)e^{i\omega t} \tag{5.138}$$

results in the equation

$$EI\frac{d^4W(x)}{dx^4}-A\rho\omega^2W(x)=0 \tag{5.139}$$

Assuming $W(x)$ to be a series of the form

$$W(x)=\sum_{i=1}^{n}a_i\psi_i(x) \tag{5.140}$$

where the trial functions $\psi_i(x)$, $i = 1, 2, \ldots, n$ are known independent functions, which form a complete set, and satisfy all boundary conditions. The coefficient term $a_i$, $i = 1, 2, \ldots, n$ is an undetermined coefficient. Equation 5.140 does not satisfy the differential equation exactly so some error will be incurred. Substituting the expression for $W(x)$ into Equation 5.138 yields

$$R_{\text{error}}(W(x),x)=EI\frac{\partial^4}{\partial x^4}\sum_{i=1}^{n}a_i\psi_i(x)-A\rho\omega^2\sum_{i=1}^{n}a_i\psi_i(x) \tag{5.141}$$

where $R_{\text{error}}$ is known as the *residual error*. To determine the coefficient $a_i$, $i = 1, 2, \ldots, n$, we multiply the residual error by $\psi_i(x)$, $i = 1, 2, \ldots, n$, in sequence, integrate the result over the domain of the system, and set the results to zero. Thus, we have

$$\int_0^l \psi_j(x)R_{\text{error}}(W(x),x)dx=0,\quad j=1,2,\ldots,n \tag{5.142}$$

or

$$\int_0^l \left[ EI \frac{\partial^4}{\partial x^4} \sum_{i=1}^{n} a_i \psi_i(x) - A\rho\omega^2 \sum_{i=1}^{n} a_i \psi_i(x) \right] \psi_j(x) dx = 0, \quad j = 1, 2, \ldots, n \tag{5.143}$$

As the equation for $W(x)$ cannot be satisfied for each $x$, we want to satisfy the equation as a *"weighted average"* over the length $l$. Performing the prescribed differentiations and integrations indicated before, we arrive at a system of nonhomogeneous linear algebraic equations

$$\sum_{i=1}^{N} a_i A_{ij} - \omega^2 \sum_{i=1}^{N} a_i B_{ij} = 0 \quad j = 1, 2, \ldots, n \tag{5.144}$$

where

$$A_{ij} = \int_0^l EI \psi_i^{iv} \psi_j \, dx$$

and

$$B_{ij} = \int_0^l A\rho \psi_i \psi_j \, dx$$

For a nontrivial solution, of the $a_i$ terms, its determinant must vanish. Hence,

$$\det \left| A_{ij} - \omega^2 B_{ij} \right| = 0,$$

which in an expanded matrix form becomes

$$\det \begin{vmatrix} A_{11} - \omega^2 B_{11} & A_{12} - \omega^2 B_{12} & \cdots & A_{1n} - \omega^2 B_{1n} \\ A_{21} - \omega^2 B_{21} & A_{22} - \omega^2 B_{22} & \cdots & \vdots \\ \vdots & \vdots & \vdots & \vdots \\ A_{n1} - \omega^2 B_{n1} & A_{n2} - \omega^2 B_{n2} & \cdots & A_{nn} - \omega^2 B_{nn} \end{vmatrix} = 0 \tag{5.145}$$

The equation is an algebraic equation of degree $n$ with respect to $\omega$. Its roots yield the successive values of the frequency $\omega_i$. This is equivalent to the Ritz method, but it does not always have to be the case.

For forced vibrations, we have an equation of motion given as

$$EI \frac{\partial^4 w}{\partial x^4} + A\rho \frac{\partial^2 w}{\partial t^2} = p(x, t) \tag{5.146}$$

Let us assume a solution of the form

$$w(x) = \sum_{i=1}^{n} \phi_i(t)\psi_i(x) \tag{5.147}$$

where $\psi_i(x)$ is the known function of $x$ satisfying the boundary conditions and $\phi_i(t)$ is the unknown function of time. Substituting Equation 5.147 into Equation 5.146 results in the following expression:

$$R_{\text{error}}(w(x),t) = EI\frac{\partial^4}{\partial x^4}\sum_i \phi_i\psi_i + A\rho\frac{\partial^2}{\partial t^2}\sum_i \phi_i\psi_i - p(x,t) \tag{5.148}$$

Imposing the condition

$$\int_0^l R_{\text{error}}(w(x),t)\psi_j(x)dx = 0 \quad j = 1,2,\ldots,n$$

results in

$$\int_0^l \left[EI\frac{\partial^4}{\partial x^4}\sum_i \phi_i\psi_i + A\rho\frac{\partial^2}{\partial t^2}\sum_i \phi_i\psi_i - p(x,t)\right]\psi_j(x)dx = 0, \quad j = 1,2,\ldots,n \tag{5.149}$$

Performing the prescribed operations as before, we arrive at a system of nonhomogeneous linear algebraic equations

$$\sum_i \phi_i A_{ij} + \sum_i \ddot{\phi}_i B_{ij} - f_j(t) = 0, \quad j = 1,2,\ldots,n \tag{5.150}$$

where $A_{ij}$ and $B_{ij}$ are the same as previously defined and

$$f_j(t) = \int_0^l p(x,t)\psi_j(x)dx$$

This is the same as a several degree of freedom system. In the matrix form, we can write

$$[B]\{\ddot{\phi}\} + [A]\{\phi\} = \{f\}$$

or

$$[m]\{\ddot{q}\} + [k]\{q\} = \{p\}$$

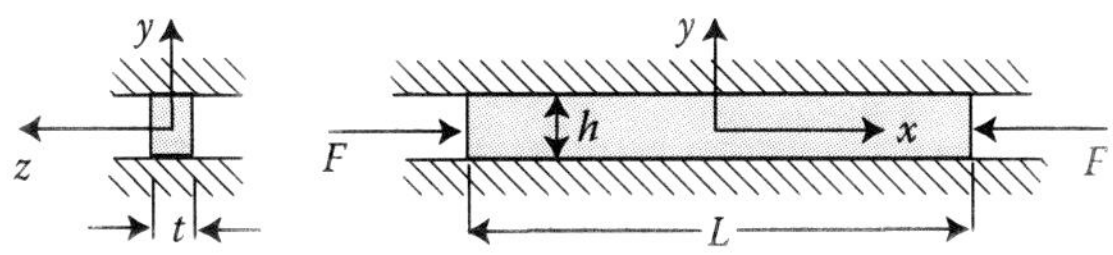


**FIGURE 5.21**
Plate fixed between two rigid walls.

## PROBLEMS

5.1 A plate with the dimension shown in Figure 5.21 is placed between two rigid walls and subjected to an axial compressive force $F$. Determine the displacement equations ($u$, $v$, and $w$) of the plate in the $x$, $y$, and $z$ directions.

5.2 An elastic material ($E = 70$ GPa, $\nu = 0.33$) fills a cavity in a rigid block. The dimensions of the cavity are $a = 75$ mm, $b = 125$ mm, and $L = 300$ mm. A rigid cap is placed on the material and a force $P_0$ is applied to the center of the cap as illustrated in Figure 5.22. The elastic material is compressed an amount $\delta$. Determine the general expressions for the applied force $P_0$ as well as the net forces that exist on the $x$ and $z$ faces of the material ($P_x$, $P_z$). In addition, determine the explicit force in each direction if the axial strain is 0.01%.

5.3 Consider a cantilevered beam of length $L$ subjected to a uniform load $q$ applied over its entire length. Let us assume that $\Psi(x)$ is approximated by the static deflection curve for the beam. In addition, assume the beam dimensions are such that $I = bh^3/12$ and $A = bh$. Use Rayleigh's method to determine the fundamental frequency for this beam. Let us assume that the beam is aluminum and the height can be 25, 50, or 75 mm and the length is between 1 and 3 m. Plot the frequency as a function of beam length for each height.

5.4 The static deflection curve of a uniform beam with a distributed weight $q$ fixed to rigid walls at both ends is $w(x) = (q/24EI)(x^4 - 2Lx^3 + L^2x^2)$. Let us assume that $\Psi(x)$ is approximated by this deflection curve and use Rayleigh's method to determine the fundamental frequency for this beam.

5.5 The tapered circular shaft shown in Figure 5.23 is subjected to a longitudinal vibration, for which the $EI$ term in the strain energy is replaced by $EA$ and $\Psi''$ is replaced by $du/dx$. Using Rayleigh's method, approximate the lowest natural

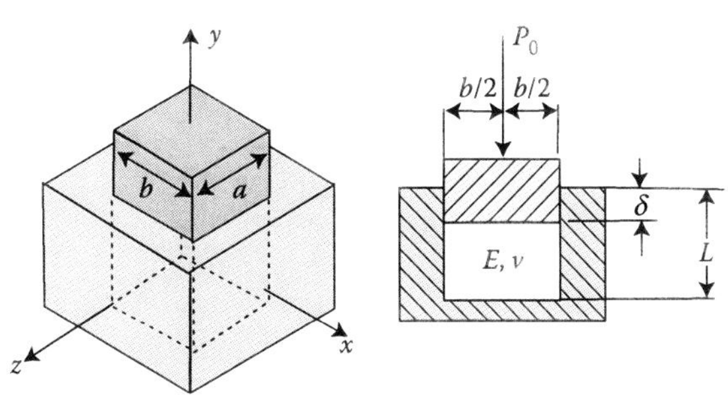


**FIGURE 5.22**
Elastic material compressed by an applied force.

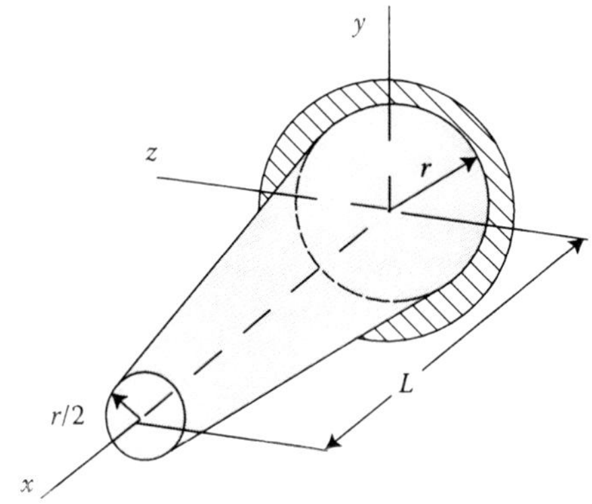


**FIGURE 5.23**
Tapered circular shaft.

frequency of this shaft. Let us assume that the displacement is approximated by $u(x) = A\sin(\pi x/2L)$.

5.6 Let us assume $\Psi(x) = a_1x^2 + a_2x^3$ for a uniform cantilever beam. Determine the first two natural frequencies using the Ritz method.

5.7 For the fixed-fixed beam described in Problem 5.4, we can assume the mode shape to be expressed as $w(x) = a_1\Psi_1(x) + a_2\Psi_2(x)$, where $\Psi_1(x) = (x^4 - 2Lx^3 + L^2x^2)$ and $\Psi_2(x) = (x^5 - 3L^2x^3 + 2L^3x^2)$. Using these expressions for $\Psi_1(x)$ and $\Psi_2(x)$, approximate the lowest two natural frequencies using the Ritz method.

5.8 Use the Ritz method to determine the two lowest frequencies for shaft in Problem 5.5 assuming $\Psi_1 = \sin(\pi x/2L)$ and $\Psi_2 = \sin(3\pi x/2L)$.

5.9 For a torsion bar, $U_{ij} = \int_0^L GJ(\partial\Psi_i/\partial x)(\partial\Psi_j/\partial x)dx$ and $K_{ij} = \int_0^L \rho J(\Psi_i)(\Psi_j)dx$. The 1″ diameter, 48″ long solid aluminum bar in Figure 5.24 is fixed to a wall at one end and has a torsional spring with a spring rate of $k_\tau$ attached at one end. Let us assume that $\Psi_1 = x^3 - 3L^2x$ and $\Psi_2 = x^2 - 2Lx$. Use the Ritz method to determine the two lowest frequency and plot the results as a function of $k_\tau$ for $500 \le k_\tau \le 2000$.

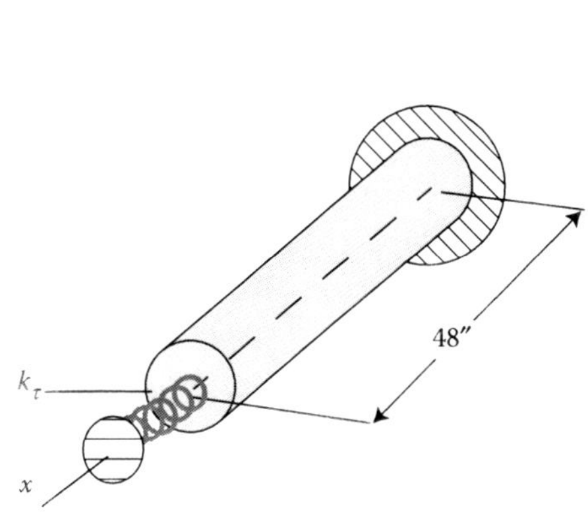


**FIGURE 5.24**
Torsion rod with a torsional spring at the end.

## References

Ardema, M. D. 2005. *Analytical Dynamics—Theory and Applications*. New York: Kluwer Academic/ Plenum Publishers.

Barber, J. R. 2010. *Elasticity*, 3rd revised ed. Dordrecht, Heidelberg, London, New York: Springer Science + Business Media.

Boresi, A. P., Chong, K. P., and Lee, J. D. 2011. *Elasticity in Engineering Mechanics*. Hoboken, NJ: John Wiley & Sons, Inc.

Greenwood, D. T. 2003. *Advanced Dynamics*. Cambridge, UK: Cambridge University Press.

Housner, G. W. and Vreeland, Jr. T. 1965. *The Analysis of Stress and Deformation*. New York: Macmillan.

Humar, J. L. 2012. *Dynamics of Structures*, 3rd ed. London, UK: Taylor & Francis Group.

Nowacki, W. 1963. *Dynamics of Elastic Systems*. New York: John Wiley & Sons, Inc.

Reddy, J. N. 2002. *Energy Principles and Variational Methods in Applied Mechanics*, 2nd ed. Hoboken, NJ: John Wiley & Sons.

Washizu, K. 1975. *Variational Methods in Elasticity and Plasticity*, 2nd ed. Oxford, England: Pergamon Press.

# 6

# Vibration of Strings and Bars

## 6.1 Introduction

Strings are basic structural elements; however, they only support tension. Cables can also be approximated by strings with negligible bending. Strings are used in musical instruments such as a violin or pianos, automotive belts, power transmission lines, suspension bridges, etc. The natural frequencies and mode shapes of a vibrating string are significant especially in the performance characteristics of some musical instruments. Bars (or rods) are also common structural elements found in many engineering applications. A bar supports either a tensile or compressive load along its axis (the $x$-axis by common convention). They are typically slender structural members used to connect (or tie) other structural components together. They are used in bridges, industrial buildings, cranes, airplane fuselages, wings, etc. They can have cross sections of various shapes and sizes. Bars are typically characterized as having a length much greater than their maximum cross-sectional dimension. The vibration of both strings and bars will be considered in this chapter.

## 6.2 Transverse String Vibration

Consider a uniform elastic string or cable of length $l$, having a mass density per unit length $\rho$, stretched to a tension $T$ and supported in some manner at its ends $A$ and $B$ as illustrated in Figure 6.1a. The string is subjected to a distributed transverse force $p(x, t)$, which causes transverse motion of any point on the string which at the coordinate position $x$ is represented by $v(x, t)$. Assuming small displacements and isolating a small segment, $ds$ of the string, the difference in the string tension forces can be modeled as illustrated in Figure 6.1b.

Application of Newton's second law in the vertical direction results in

$$-T\sin\theta + (T + dT)\sin(\theta + d\theta) + p\,dx = \rho(dx)\frac{\partial^2 v}{\partial t^2} \tag{6.1}$$

We note that $dT = (\partial T/\partial x)\partial x$ and since we have assumed small displacements, $\sin\theta \approx \tan\theta \approx \partial v/\partial x$ which allows us to write

$$\sin(\theta + d\theta) \approx \tan(\theta + d\theta) = \frac{\partial v}{\partial x} + \frac{\partial^2 v}{\partial x^2}dx$$

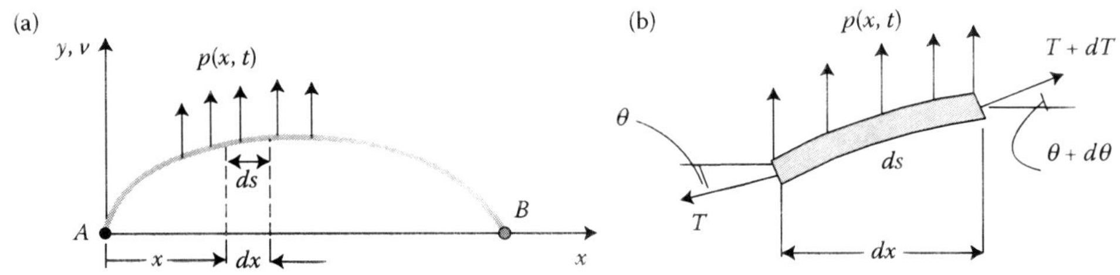


**FIGURE 6.1**
Elastic string (a) displacement and (b) resulting forces on a segment of string.

Using this we can write Equation 6.1 as

$$\frac{\partial}{\partial x}\left[T\frac{\partial v}{\partial x}\right]+p=\rho\frac{\partial^2 v}{\partial t^2}$$

Assuming that the string is uniform with constant tension, the above expression can be written as

$$T\frac{\partial^2 v(x,t)}{\partial x^2}+p(x,t)=\rho\frac{\partial^2 v(x,t)}{\partial t^2} \tag{6.2}$$

If there are no applied loads, $p(x, t) = 0$ and Equation 6.2 reduces to the free vibration case. In addition, if we define $c \equiv \sqrt{T/\rho}$ which is the *wave propagation velocity*, then Equation 6.2 results in the equation governing the free transverse motion of the string and is known as the *one-dimensional wave equation*.

$$\frac{\partial^2 v}{\partial t^2}=c^2\frac{\partial^2 v}{\partial x^2} \tag{6.3}$$

### 6.2.1 Initial and Boundary Conditions

Since Equations 6.2 and 6.3 are second-order partial differential equations in $x$ and $t$, their solution requires both initial and boundary conditions. The initial conditions typically refer to the initial displacement and velocity at time $t = 0$ and are given as

$$\begin{aligned} v(x,t=0)&=v_0(x) \\ \frac{\partial v}{\partial x}(x,t=0)&=\dot{v}_0(x) \end{aligned} \tag{6.4}$$

The boundary conditions for a string or cable can range from the simplest with both ends attached to a rigid support or to a more complicated case, where a mass is connected at each end. As demonstrated by Rao (2007), there are five boundary conditions that prevail, and these are illustrated in Figure 6.2. However, more complicated conditions can be obtained by combining spring, mass, and viscous elements. For all boundary conditions, it is assumed that a constant tension $T$ exists in the string, regardless of whether it appears in the boundary condition or not. The boundary conditions for each case appearing in Figure 6.2 are

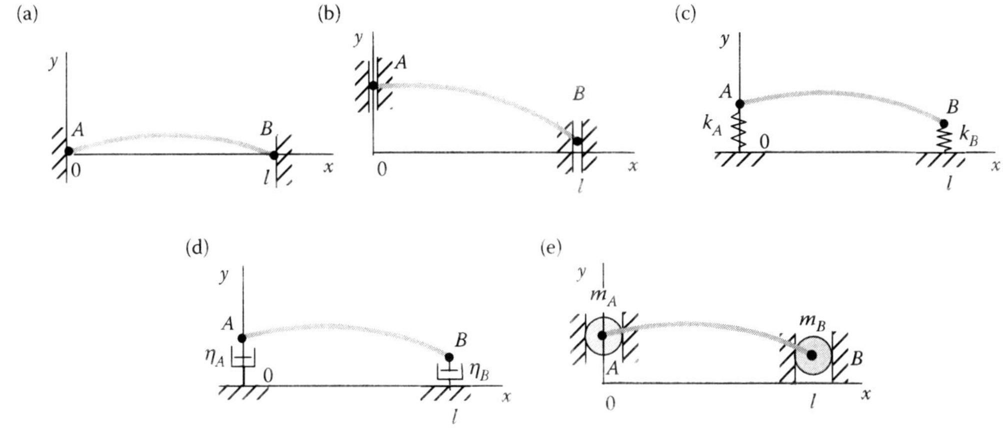


**FIGURE 6.2**
Typical boundary conditions for a sting are (a) both ends fixed, (b) both ends free, (c) both ends attached to springs, (d) both ends attached to dampers, and (e) both ends attached to a mass.

Case (a): Both ends fixed; $v_A(0, t) = v_B(l, t) = 0$

Case (b): Both ends free; $(\partial v_A/\partial x)(0, t) = (\partial v_B/\partial x)(l, t) = 0$

Case (c): Both ends attached to springs; $k_A v_A(0, t) = T(\partial v_A/\partial x)(0, t)$ and $-k_B v_B(l, t) = T(\partial v_B/\partial x)(l, t)$

Case (d): Both ends attached to dampers; $\eta_A(\partial v_A/\partial x)(0, t) = T(\partial v_A/\partial x)(0, t)$ and $-\eta_B(\partial v_B/\partial x)(l, t) = T(\partial v_B/\partial x)(l, t)$

Case (e): Both ends attached to a mass; $m_A(\partial^2 v_A/\partial t^2)(0, t) = T(\partial v_A/\partial x)(0, t)$ and $-m_B(\partial^2 v_B/\partial t^2)(l, t) = T(\partial v_B/\partial x)(l, t)$

## 6.3 General Solution of the Wave Equation

If a taut string is given a small initial displacement and then released, a wave will travel away from the origin of the displacement in both directions at some speed $c$ as illustrated in Figure 6.3. The solution of Equation 6.3 for an infinitely long string can be addressed in various manners.

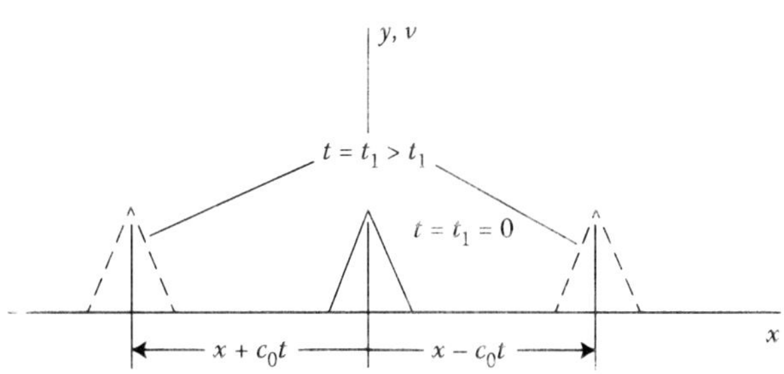


**FIGURE 6.3**
Traveling wave in an infinitely long string.

### 6.3.1 Traveling Wave Solution

The traveling wave solution is commonly referred to as d'Alembert's solution (Nowacki 1963; Rao 2007). The solution consists of two wave functions, $F(x - ct)$ moving in the positive $x$-direction and $G(x + ct)$ moving in the negative $x$-direction. It is assumed that the wave speed $c$ is uniform and there is no distortion of the wave. The solution consists of both waves, so that the displacement of the string can be assumed to be

$$v(x,t) = F(x - ct) + G(x + ct) \tag{6.5}$$

It is not difficult to verify that the solution given by Equation 6.5 is valid by substituting it into Equation 6.3. The nature of the functions $F$ and $G$ is arbitrary, for example, they could be sinusoidal. Assume the initial conditions of displacement and velocities are $v(x, 0) = f(x)$ and $\dot{v}(x, 0) = g(x)$. Using Equation 6.5, we can now write

$$F(x) + G(x) = f(x); \quad -\infty < x < \infty \tag{6.6}$$

$$-cF'(x) + cG'(x) = g(x); \quad -\infty < x < \infty \tag{6.7}$$

Integrating Equation 6.7 from an arbitrary lower limit $\tau = b$ to an arbitrary $x$ yields

$$-F(x) + G(x) = \frac{1}{c}\int_b^x g(\tau)d\tau$$

Combining this with Equation 6.6 yields

$$\begin{aligned} F(x) &= \frac{1}{2}\left[f(x) - \frac{1}{c}\int_b^x g(\tau)d\tau\right] \\ G(x) &= \frac{1}{2}\left[f(x) + \frac{1}{c}\int_b^x g(\tau)d\tau\right] \end{aligned} \tag{6.8}$$

Combining these provides the general solution. Recall that these two expressions are for time $t = 0$. In the general solution, we need to account for the wave traveling in two directions. Therefore, from Equations 6.5 and 6.8, the general solution is

$$v(x,t) = \frac{1}{2}\left[f(x - ct) + f(x + ct)\right] + \frac{1}{2c}\int_{x-ct}^{x+ct} g(\tau)d\tau \tag{6.9}$$

### 6.3.2 Fourier Transformation Solution

d'Alembert's solution is not the only approach to solving the wave equation. An alternate solution can be obtained by using a Fourier transformation (Meirovitch 2001; Rao 2007).

The initial conditions are the same as those presented for d'Alembert's solution, namely, $v(x,0) = f(x)$ and $\dot{v}(x,0) = g(x)$. Multiplying Equation 6.3 by $e^{-i\xi x}$ and integrating from $-\infty$ to $+\infty$ results in

$$\int_{-\infty}^{\infty} \frac{\partial^2 v(x,t)}{\partial x^2} e^{-i\xi x}\, dx = \frac{1}{c^2} \int_{-\infty}^{\infty} \frac{\partial^2 v(x,t)}{\partial t^2} e^{-i\xi x}\, dx \tag{6.10}$$

Define the Fourier transform with respect to $x$, for each fixed $t$, as

$$v^*(\xi,t) = \frac{1}{\sqrt{2\pi}} \int_{-\infty}^{\infty} v(x,t) e^{-i\xi x}\, dx \tag{6.11}$$

If we assume that both $v(x, t)$ and $\partial v(x, t)/\partial x$ tend to zero as $x \to \pm\infty$, the left-hand side can be integrated by parts twice, giving

$$\int_{-\infty}^{\infty} \frac{\partial^2 v(x,t)}{\partial x^2} e^{-i\xi x}\, dx = -\xi^2 \int_{-\infty}^{\infty} v(x,t) e^{-i\xi x}\, dx = -\xi^2 v^*(\xi,t) \tag{6.12}$$

Integrating the right-hand side of Equation 6.10 twice yields

$$\frac{1}{c^2} \int_{-\infty}^{\infty} \frac{\partial^2 v(x,t)}{\partial t^2} e^{-i\xi x}\, dx = \frac{1}{c^2} \frac{\partial}{\partial t} \int_{-\infty}^{\infty} \frac{\partial v(x,t)}{\partial t} e^{-i\xi x}\, dx = \left(\frac{1}{c^2}\right) \frac{\partial^2 v^*(\xi,t)}{\partial t^2} \tag{6.13}$$

Therefore, Equation 6.10 can be written as

$$\frac{d^2 v^*(\xi,t)}{dt^2} + c^2 \xi^2 v^*(\xi,t) = 0 \tag{6.14}$$

which yields, for each fixed $\xi$, a constant coefficient, homogeneous, second-order ordinary differential equation for $v^*(\xi, t)$. The solution to Equation 6.14 can be taken as $v^*(\xi, t) = Ae^{st}$ which yields the solution

$$v^*(\xi,t) = A_1 e^{ic\xi t} + A_2 e^{-ic\xi t} \tag{6.15}$$

where the arbitrary constants $C_1$ and $C_2$ can be determined from the initial conditions of the problem $v(x, 0) = f(x)$ and $\dot{v}(x,0) = g(x)$. To evaluate $v(x, t)$, we need to determine the inverse Fourier transform. Therefore, we first evaluate the arbitrary constant by taking the Fourier transform of the initial conditions. This yields

$$\begin{aligned} v^*(\xi,t=0) &= \frac{1}{\sqrt{2\pi}} \int_{-\infty}^{\infty} f(x) e^{-i\xi x}\, dx = f^*(\xi) \\ \frac{dv^*(\xi,t=0)}{dt} &= \frac{1}{\sqrt{2\pi}} \int_{-\infty}^{\infty} g(x) e^{-i\xi x}\, dx = g^*(\xi) \end{aligned} \tag{6.16}$$

Using Equations 6.15 and 6.16, we obtain $f^*(\xi) = A_1 + A_2$ and $g^*(\xi) = iac(A_1 - A_2)$. Solving these results in

$$A_1 = \frac{f^*(\xi)}{2} + \frac{g^*(\xi)}{2i\xi c}$$

$$A_2 = \frac{f(\xi)}{2} - \frac{g^*(\xi)}{2i\xi c}$$

We can now express Equation 6.15 as

$$v^*(\xi,t) = \frac{f^*(\xi)}{2}(e^{i\xi ct} + e^{-i\xi ct}) + \frac{g^*(\xi)}{2i\xi c}(e^{i\xi ct} - e^{-i\xi ct}) \tag{6.17}$$

Taking the inverse Fourier transform of Equation 6.11 gives

$$v(x,t) = \frac{1}{\sqrt{2\pi}} \int_{-\infty}^{\infty} v^*(\xi,t)e^{i\xi x}\, d\xi$$

Substituting Equation 6.17 into this equation yields

$$\begin{aligned} v(x,t) &= \frac{1}{2}\left\{\frac{1}{\sqrt{2\pi}} \int_{-\infty}^{\infty} f^*(\xi)[e^{i\xi(x+ct)} + e^{i\xi(x-ct)}]d\xi\right\} \\ &+ \frac{1}{2c}\left\{\frac{1}{\sqrt{2\pi}} \int_{-\infty}^{\infty} \frac{g^*}{i\xi}(\xi)[e^{i\xi(x+ct)} - e^{i\xi(x-ct)}]d\xi\right\} \end{aligned} \tag{6.18}$$

In accordance with the rules concerning the exponential Fourier transforms of Equation 6.16, we determined

$$f(x) = \frac{1}{\sqrt{2\pi}} \int_{-\infty}^{\infty} f^*(\xi)e^{i\xi x}\, d\xi \text{ and } g(x) = \frac{1}{\sqrt{2\pi}} \int_{-\infty}^{\infty} g^*(\xi)e^{i\xi x}\, d\xi$$

In view of Equation 6.18, we obtain from the first integral expression

$$f(x \pm ct) = \frac{1}{\sqrt{2\pi}} \int_{-\infty}^{\infty} f^*(\xi)e^{i\xi(x\pm ct)}\, d\xi$$

Integrating the second equation with respect to $\tau$ from $x - ct$ to $x + ct$ produces

$$\int_{x-ct}^{x+ct} g(\tau)d\tau = \frac{1}{\sqrt{2\pi}} \int_{-\infty}^{\infty} \frac{g^*}{i\xi}(\xi)[e^{i\xi(x+ct)} - e^{i\xi(x-ct)}]d\xi$$

Substituting these into Equation 6.18 yields

$$v(x,t) = \frac{1}{2}[f(x-ct)+f(x+ct)] + \frac{1}{2c}\int_{x-ct}^{x+ct} g(\tau)d\tau \tag{6.19}$$

This is the same as Equation 6.9. A third solution technique is to use Laplace transformations as presented by Rao (2007).

## 6.4 Free Vibrations of Finite Length Strings

In Section 6.1, we showed that the vibrations of elastic string are governed by the one-dimensional wave equation

$$\frac{\partial^2 v}{\partial t^2} = c^2 \frac{\partial^2 v}{\partial x^2} \tag{6.3}$$

where $v(x, t)$ is the deflection of the string. One method to solve this partial differential equation is the method of separation of variables. The solution is assumed as

$$v(x,t) = V(x)T(t) \tag{6.20}$$

Differentiating Equation 6.20 and substituting into Equation 6.1 yields

$$\frac{1}{T}\frac{d^2T}{dt^2} = \frac{c^2}{V}\frac{d^2V}{dx^2} \tag{6.21}$$

The expression on the left depends only on $t$ and on the right only on $x$. If they were variable, then changing $t$ or $x$ would affect only the left or the right side where the other side would not be altered. Therefore, their common value must be a constant. Of all possibilities only a negative value can be shown to work. Thus, we choose the constant to be $-\omega^2$ and our equations become

$$\frac{d^2V}{dx^2} + \frac{\omega^2}{c^2}V = 0 \tag{6.22}$$

$$\frac{d^2T}{dt^2} + \omega^2 T = 0 \tag{6.23}$$

The solution of Equations 6.22 and 6.23 can be expressed as

$$V(x) = A\cos\frac{\omega x}{c} + B\sin\frac{\omega x}{c} \tag{6.24}$$

$$T(t) = C\cos\omega t + D\sin\omega t \tag{6.25}$$

where $A$, $B$, $C$, and $D$ are constants that need to be evaluated. The displacements are then given by

$$v(x,t) = \left(A\cos\frac{\omega x}{c} + B\sin\frac{\omega x}{c}\right)(C\cos\omega t + D\sin\omega t) \tag{6.26}$$

Considering a string stretched between two fixed points with a distance $l$ between them, the boundary conditions are $v(0, t) = v(l, t) = 0$. The condition $v(0, t) = 0$ implies that $A = 0$, thus, assuming for convenience that $B = 1$, the solution will be

$$v(x,t) = (C\cos\,\omega t + D\sin\,\omega t)\sin\frac{\omega}{c}x \tag{6.27}$$

The condition that $v(l, t) = 0$ leads to

$$\sin\frac{\omega l}{c} = 0$$

For this to occur $\omega l/c = n\pi$, this means

$$\omega_n = c\frac{n\pi}{l} = \sqrt{\frac{T}{\rho}}\frac{n\pi}{l} \qquad n = 1, 2, 3, \ldots \tag{6.28}$$

As a result, we can write

$$v_n(x,t) = \sin\frac{n\pi x}{l}\left(C_n\cos\frac{cn\pi}{l}t + D_n\sin\frac{cn\pi}{l}t\right)$$

Noting that $v(x,t) = \sum_{n=1}^{\infty} v_n(x,t)$, we can write

$$v(x,t) = \sum_{n=1}^{\infty}\sin\frac{n\pi x}{l}\left(C_n\cos\frac{cn\pi}{l}t + D_n\sin\frac{cn\pi}{l}t\right) \quad \text{for } n = 1, 2, 3, \ldots \tag{6.29}$$

The constants $C_n$ and $D_n$ are determined from the initial conditions $v(x,0)$ and $\dot{v}(x,0)$.

Different results will obviously occur for different boundary conditions. For example, if the string had free-free ends it would be required that $\partial v(0, t)/\partial x = \partial v(l, t)/\partial x = 0$. These conditions imply that $B = 0$ and $-(\omega/c)A\,\sin(\omega l/c) = 0$. This results in the same $\omega_n = c(n\pi/l) = \sqrt{(T/\rho)}(n\pi/l)$ for $n = 1, 2, 3, \ldots$, but the mode shape becomes

$$v = \sum_{n=1}^{\infty}\cos\frac{n\pi x}{l}\left(C_n\cos\frac{cn\pi}{l}t + D_n\sin\frac{cn\pi}{l}t\right) \tag{6.30}$$

**EXAMPLE 6.1**

Assume a string of length $l$ is fixed at end $A$ to a spring with a spring constant $k_A$ and fixed to a wall at end $B$ as shown in Figure 6.4. The string is 1 ft long and has a uniform tension and density so that $T/\rho = 10{,}500\ \text{ft}^2/\text{s}^2$. The spring rate is $k = 50$ lb/ft. Estimate the first three natural frequencies of system.

**Solution**

Assuming the string displacement can be represented as $v = X(x)(C\cos\omega t + D\sin\omega t)$, where we assume $X(x) = A\cos(\omega x/c) + B\sin(\omega x/c)$. The spring at end $A$ and the fixed wall at end $B$ have boundary conditions given as

$$\text{At } A: k_A v(0,t) = T\frac{\partial v(0,t)}{\partial x} \Rightarrow k_A X(0) = T\frac{\partial X(0)}{\partial x}$$

or

$$k_A(A\cos(0) + B\sin(0)) = T\left(\frac{\omega}{c}\right)(-A\sin(0) + B\cos(0)) \Rightarrow TB\frac{\omega}{c} = k_A A$$

$$A = \frac{TB}{k_A}\left(\frac{\omega}{c}\right)$$

$$\text{At } B: v(l,t) = X(l) = 0 = A\cos\left(\frac{\omega l}{c}\right) + B\sin\left(\frac{\omega l}{c}\right)$$

$$\frac{TB}{k_A}\left(\frac{\omega}{c}\right)\cos\frac{\omega l}{c} + B\sin\frac{\omega l}{c} = B\left[\frac{T}{k_A}\left(\frac{\omega}{c}\right)\cos\frac{\omega l}{c} + \sin\frac{\omega l}{c}\right] = 0$$

For a nontrivial solution, we have

$$\frac{T}{k_A}\left(\frac{\omega}{c}\right)\cos\frac{\omega l}{c} + \sin\frac{\omega l}{c} = 0 \Rightarrow \tan\frac{\omega l}{c} = -\frac{T\omega}{k_A c}$$

In order to simplify this equation, we define $\beta \equiv \omega\sqrt{\rho/T} = \omega/c$ which reduces the above equation to

$$\tan\beta l = -\frac{T}{k_A}\beta$$

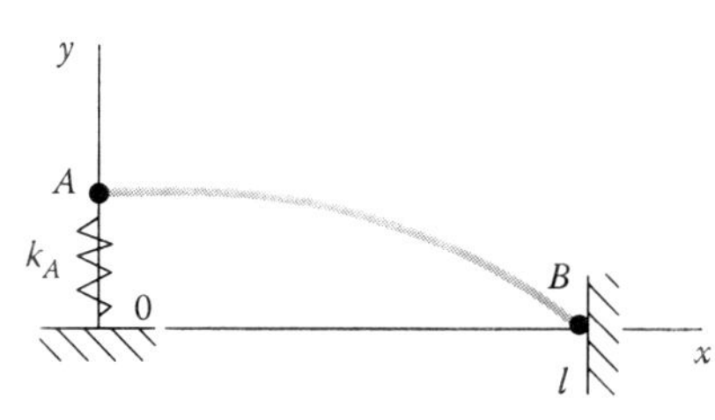


**FIGURE 6.4**
String with a spring at one end and fixed at the other end.

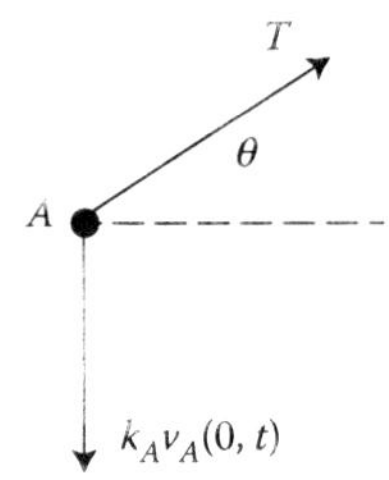


**FIGURE 6.5**
Free-body diagram of point A.

Using $l = 1$ ft, $T = 21$ lb and $k_A = 50$ lb/ft, this becomes

$$\tan \beta l = -0.42\, \beta$$

Solving the above equation yields the following values, $\beta_1 \approx 2.361$, $\beta_2 \approx 5.145$, and $\beta_3 \approx 8.136$. The corresponding frequencies are

$$\omega_1 = \beta_1 \sqrt{\frac{T}{\rho}} = 2.361\sqrt{\frac{21}{0.002}} = 241.9$$

$$\omega_2 = \beta_2 \sqrt{\frac{T}{\rho}} = 5.145\sqrt{\frac{21}{0.002}} = 527.4$$

$$\omega_3 = \beta_3 \sqrt{\frac{T}{\rho}} = 8.136\sqrt{\frac{21}{0.002}} = 833.9$$

Note that as an alternate derivation, we start by developing a free-body diagram of point $A$ as shown in Figure 6.5.

From the free-body diagram, summing forces in the vertical direction yields, $T \sin\theta - k_A v_A(0, t) = 0$. For small angles, we have the approximation $\sin\theta \approx \theta \approx dy/dx$. Since the vertical direction corresponds to a displacement $v_A(0, t)$, which we define in terms of a function of $x$ and a function of $t$ we can write $dy/dx = \partial y/\partial x$ which allows us to write $T(\partial y/\partial x) = k_A v_A(0, t)$, which is the boundary condition for the spring at $A$.

### 6.4.1 Discontinuous Strings

Discontinuities in a string can result from different situations, such as a change in wire diameter along its length, a change in mass density, a mass attached at some location, etc. In order to account for this discontinuity we can identify solutions to Equation 6.2 for each segment of the string. Assume a mass is attached to a fixed-fixed string at some arbitrary location as illustrated in Figure 6.6.

Assume a case of free vibration for a uniform string with a constant density along its length. The string is modeled in two segments defined by

$$T\frac{\partial^2 v_1(x,t)}{\partial x^2} = \rho\frac{\partial^2 v_1(x,t)}{\partial t^2} \quad (0 \le x \le a) \tag{6.31}$$

$$T\frac{\partial^2 v_2(x,t)}{\partial x^2} = \rho\frac{\partial^2 v_2(x,t)}{\partial t^2} \quad (a \le x \le b) \tag{6.32}$$

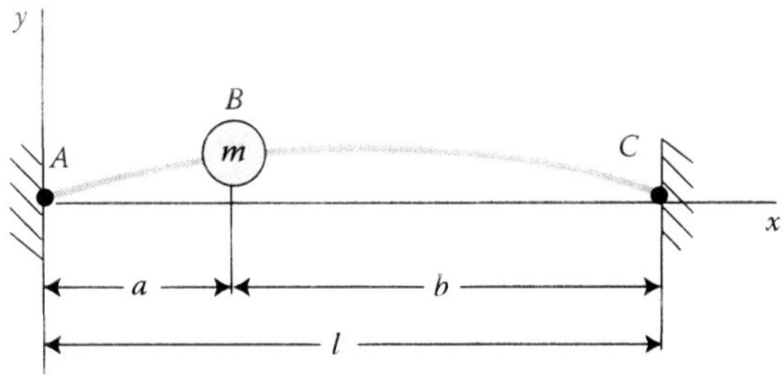


**FIGURE 6.6**
Discontinuous string with an attached mass.

Each segment is modeled independently with an assumed solution in the form

$$v_1(x,t) = V_1(x)(C_1 \cos\omega t + D_1 \sin\omega t),\ v_2(x,t) = V_2(x)(C_2 \cos\omega t + D_2 \sin\omega t)$$

Where we assume

$$V_1(x) = A_1 \cos\frac{\omega x}{c} + B_1 \sin\frac{\omega x}{c} \text{ and } V_2(x) = A_2 \cos\frac{\omega x}{c} + B_2 \sin\frac{\omega x}{c}$$

Defining $\beta$ to be $\omega/c$ and noting that the boundary conditions are $v_1(0, t) = v_2(l, t) = 0$, we have

$$v_1(0,t) = V_1(0) = 0 = A_1 \Rightarrow V_1(x) = B_1 \sin\beta x$$

$$v_2(l,t) = V_2(l) = 0 = A_2 \cos\beta l + B_2 \sin\beta l \Rightarrow A_2 = -B_2\left(\frac{\sin\beta l}{\cos\beta l}\right)$$

Hence, for $V_2(x)$, we have

$$V_2(x) = B_2\left[\sin\beta x - \cos\beta x\left(\frac{\sin\beta l}{\cos\beta l}\right)\right] = \frac{B_2}{\cos\beta l}[\sin\beta x\cos\beta l - \cos\beta x\sin\beta l]$$

$$= A\sin\beta(x-l)$$

where we have replaced $(B_2/\cos\beta l)$ with $A$. At $x = a$, the displacements must be equal, therefore $v_1(a, t) = v_2(a, t)$ and

$$B_1 \sin\beta a(C_1\cos\omega t + D_1\sin\omega t) = A\sin\beta(a-l)(C_2\cos\omega t + D_2\sin\omega t)$$

Therefore, we have

$$(C_1\cos\omega t + D_1\sin\omega t) = (C_2\cos\omega t + D_2\sin\omega t) \Rightarrow C_1 = C_2 = C \text{ and } D_1 = D_2 = D$$

and

$$B_1\sin\beta a = A\sin\beta(x-l)$$

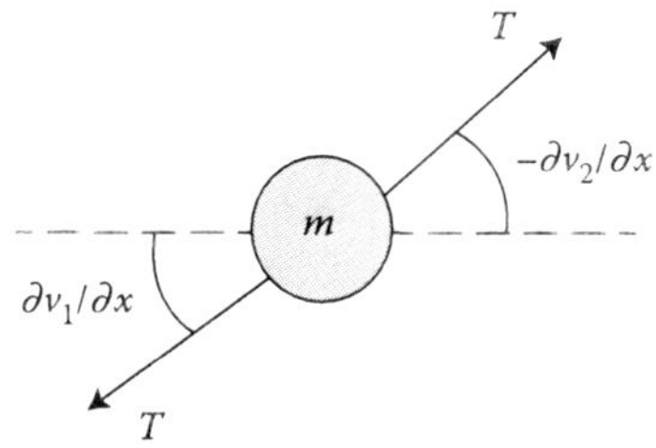


**FIGURE 6.7**
String tensions acting on mass at $x = a$.

$$B_1 = A\left[\frac{\sin\beta(a-l)}{\sin\beta a}\right]$$

The slopes of the two segments of the string are discontinuous at the mass, but they can be related as shown in the free-body diagram of Figure 6.7. Note that the vertical components of the tension $T$ will be in terms of the sine of the angle that each tension makes with the horizontal axis. These can be expressed as $\sin\theta_1 = \partial v_1/\partial x$, $\sin\theta_2 = -\partial v_2/\partial x$. Applying Newton's second law gives

$$T(\sin\theta_2) - T(\sin\theta_1) = m\frac{\partial^2 v_1}{\partial t^2}$$

or

$$T\frac{\partial v_2}{\partial x} - T\frac{\partial v_1}{\partial x} = m\frac{\partial^2 v_1}{\partial t^2} \tag{6.33}$$

The displacements in each part of the string are

$$v_1(x,t) = A\sin\beta x\left(\frac{\sin\beta(a-l)}{\sin\beta a}\right)(C\cos\omega t + D\sin\omega t)$$
$$v_2(x,t) = A\sin\beta(x-l)(C\cos\omega t + D\sin\omega t)$$

Taking the appropriate partial derivatives of $v_1(x, t)$, $v_2(x, t)$ and substituting into Equation 6.33 gives

$$T\beta A\cos\beta(a-l) - T\beta A\left(\frac{\sin\beta(a-l)}{\sin\beta a}\right)\cos(\beta a) = -m\omega^2 A\sin\beta(a-l)$$

$$A\left[T\beta\cos\beta(a-l) - T\beta\left(\frac{\sin\beta(a-l)}{\sin\beta a}\right)\cos(\beta a) + m\omega^2\sin\beta(a-l)\right] = 0$$

$$\frac{D_3}{\sin\beta a}[T\beta\sin\beta l + m\omega^2\sin\beta a\sin\beta(a-l)] = 0$$

The nontrivial solution to this equation is for

$$T\beta \sin \beta l + m\omega^2 \sin \beta a \sin \beta(a-l) = 0 \tag{6.34}$$

Recalling that $\beta = \omega/c$ where $c = \sqrt{T/\rho}$, we find by substituting $\omega^2 = \beta^2 c^2 = \beta^2 T/\rho$ into Equation 6.34 and obtain

$$T\beta\left[\sin \beta l + m\frac{\beta}{\rho}\sin \beta a \sin \beta(a-l)\right] = 0$$

or

$$\sin \beta l + m\frac{\beta}{\rho}\sin \beta a \sin \beta(a-l) = 0 \tag{6.35}$$

## 6.5 Forced Vibrations of Finite Length Strings

In order to solve a forced vibration problem, we start with Equation 6.2. The solution requires both a homogeneous (free vibration) and complementary (nonhomogeneous) solution (Nowacki 1963). The homogeneous and complementary solutions are based on the chosen displacement fields satisfying both boundary and initial conditions. In order to illustrate the solution procedure, we assume a string of length $l$ fixed at both ends with boundary conditions $v(0, t) = v(l, t) = 0$. The complementary solution ($v = v_c$) for this problem has been presented in Equation 6.29 and is repeated below.

$$v_c = v(x,t) = \sum_{n=1}^{\infty} \sin\frac{n\pi x}{l}\left(C_n \cos\frac{cn\pi}{l}t + D_n \sin\frac{cn\pi}{l}t\right) \quad \text{for } n = 1,2,3,\ldots$$

The particular solution can be assumed to be of the form

$$v_p = v(x,t) = \sum_{n=1}^{\infty} \sin\frac{n\pi x}{l}\phi_n(t) \tag{6.36}$$

Substituting Equation 6.36 into 6.2, the equation of motion yields

$$\rho\sum_{n=1}^{\infty} \sin\frac{n\pi x}{l}\left(\frac{d^2\phi_n(t)}{dt^2}\right) + T\sum_{n=1}^{\infty}\left(\frac{n\pi}{l}\right)^2 \sin\frac{n\pi x}{l}\phi_n(t) = p(x,t) \tag{6.37}$$

Multiplying Equation 6.37 by sin $m\pi x/l$, integrating from 0 to $l$ and using the orthogonality condition

$$\int_0^l \sin\frac{n\pi x}{l}\sin\frac{m\pi x}{l}dx = 0 \quad \text{when } n \neq m$$

$$\int_0^l \sin\frac{n\pi x}{l}\sin\frac{m\pi x}{l}dx = \frac{l}{2} \quad \text{when } n = m$$

yields

$$\frac{d^2\phi_n(t)}{\partial t^2} + \left(\frac{n\pi c}{l}\right)^2 \phi_n(t) = \frac{2}{\rho l}P_n(t) \tag{6.38}$$

where $c$ has previously been defined as $c = \sqrt{T/\rho}$ and

$$P_n(t) = \int_0^l p(x,t)\sin\frac{n\pi x}{l}dx \tag{6.39}$$

The solution to Equation 6.38 can be using Duhamel's integral in the form

$$\phi_n(t) = \frac{1}{\omega_n}\left(\frac{2}{\rho l}\right)\int_0^t P_n(\tau)\sin\omega_n(t-\tau)d\tau$$

where $\omega_n = n\pi c/l$. Thus, the solution can be written as

$$\phi_n(t) = \frac{2}{nc\pi\rho}\int_0^t P_n(\tau)\sin\frac{nc\pi(t-\tau)}{l}d\tau \tag{6.40}$$

and the particular solution becomes

$$v_p = \frac{2}{c\pi\rho}\sum_{n=1}^{\infty}\frac{1}{n}\sin\frac{n\pi x}{l}\left(\int_0^t P_n(\tau)\sin\frac{nc\pi}{l}(t-\tau)d\tau\right) \tag{6.41}$$

The solution to the forced vibration of a finite length string fixed at both ends is therefore given as

$$\begin{aligned} v(x,t) = v_c + v_p &= \sum_{n=1}^{\infty}\sin\frac{n\pi x}{l}\left(C_n\cos\frac{cn\pi}{l}t + D_n\sin\frac{cn\pi}{l}t\right) \\ &+ \frac{2}{c\pi\rho}\sum_{n=1}^{\infty}\frac{1}{n}\sin\frac{n\pi x}{l}\left(\int_0^t P_n(\tau)\sin\frac{nc\pi}{l}(t-\tau)d\tau\right) \quad \text{for } n = 1,2,3,\ldots \end{aligned} \tag{6.42}$$

**EXAMPLE 6.2**

A guy wire of length $l$ is stretched between two fixed points to help support a communications tower. The wire is expected to experience wind loads, which produce a uniform load $p_0$ per unit length. The initial displacement and velocity of the cable are zero ($v(x,0) = \dot{v}(x,0) = 0$). Determine the forced vibration response of the cable.

**Solution**

Since there is a uniformly distributed load of $p(x, t) = p_0$

$$P_n(t) = \int_0^l (x,t)\sin\frac{n\pi x}{l}dx\,p = \frac{2lp_0}{n\pi}$$

this gives

$$\int_0^t P_n(\tau)\sin\frac{nc\pi}{l}(t-\tau)d\tau = \frac{2lp_0}{n\pi}\int_0^t \sin\frac{nc\pi}{l}(t-\tau)d\tau = -\frac{2lp_0}{n\pi}\int_t^0 \sin\left(\frac{nc\pi}{l}\right)y\,dy$$

$$= \frac{2l^2p_0}{n\pi^2c}\left(1-\cos\frac{nc\pi t}{l}\right) \quad \text{for } n = 1, 3, 5, \ldots$$

The forced response therefore becomes

$$v(x,t) = \sum_{n=1}^{\infty}\sin\frac{n\pi x}{l}\left(C_n\cos\frac{cn\pi}{l}t + D_n\sin\frac{cn\pi}{l}t\right)$$

$$+\frac{4l^2p_0}{c^2\pi^3\rho}\sum_{n=1,3,5}^{\infty}\frac{1}{n^3}\sin\frac{n\pi x}{l}\left(1-\cos\frac{cn\pi t}{l}\right)$$

Applying the initial conditions $v(x,0) = \dot{v}(x,0) = 0$ results in

$$\text{for } v(x,0) = 0: \sum_{n=1}^{\infty} C_n\sin\frac{n\pi x}{l} = 0$$

$$\text{for } \dot{v}(x,0) = 0: \sum_{n=1}^{\infty}\frac{n\pi c}{l}D_n\sin\frac{n\pi x}{l} = 0$$

Therefore, $A_n = B_n = 0$ and the forced vibration response of the finite length string is

$$v(x,t) = \frac{4l^2p_0}{c^2\pi^3\rho}\sum_{n=1,3,5}^{\infty}\frac{1}{n^3}\sin\frac{n\pi x}{l}\left(1-\cos\frac{cn\pi t}{l}\right)$$

## 6.6 Longitudinal Vibrations of Bars

The longitudinal vibrations of a bar (Weaver et al. 1990) are assessed by considering a bar of finite length $l$. A representative volume element of length $dx$ is isolated for analysis as shown in Figure 6.8. We shall assume that the cross sections of the bar remain plane and that all particles in every cross section only move in the axial direction. Along with longitudinal extensions and compressions, some lateral deformation will take place. In our discussion of longitudinal motion, we will consider cases such that the length of longitudinal waves is large in comparison to the bar's cross section. Also the force at each end of

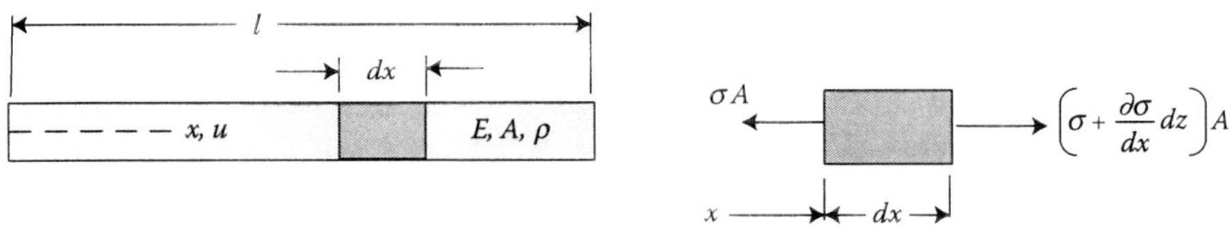


**FIGURE 6.8**
Representative volume element of a bar.

the bar is expressed in terms of stress and accommodations are made for a change in force over the length $dx$. The bar is assumed to have a cross-sectional area $A$, an elastic modulus $E$, and a density $\rho = \gamma/g$ (where $\gamma$ is the weight per volume or specific weight). The rod is assumed to undergo a time-dependent axial displacement $u = u(x, t)$.

In the cross section of the bar at a value $x$ equals a constant there acts a force $\sigma A$, while in the cross section $x + dx$ the force is $(\sigma + (\partial\sigma/\partial x)dx)A$. The equations of motion are developed by applying Newton's law to the representative volume element shown in Figure 6.8. The equation of motion has the form

$$\left(\sigma + \frac{\partial \sigma}{\partial x} dx\right) A - \sigma A + pA\,dx = \rho A \frac{\partial^2 u}{\partial t^2} dx$$

or

$$\rho \frac{\partial^2 u(x,t)}{\partial t^2} = \frac{\partial \sigma}{\partial x} dx + p(x,t) \tag{6.43}$$

Assuming the material behaves elastically, we apply Hook's law $(\sigma = E\varepsilon = E(\partial u/\partial x))$ which results in

$$\rho \frac{\partial^2 u}{\partial t^2} - E \frac{\partial^2 u}{\partial x^2} = p(x,t) \tag{6.44}$$

If there are no applied loads, $p(x, t) = 0$ and defining $a^2 = E/\rho = Eg/\gamma$ results in a linear partial differential equation with constant coefficients, thus

$$\frac{\partial^2 u}{\partial t^2} = a^2 \frac{\partial^2 u}{\partial x^2} \tag{6.45}$$

Equation 6.45 has the same form as Equation 6.3 is often called the *one-dimensional wave equation*. This is to indicate the displacement patterns propagate in the axial direction at the velocity $a$.

## 6.7 Free Vibrations of Bars

If there are no applied external forces acting on the bar and the ends are free, we can assume a solution of the form

$$u = X(x)(A\cos\omega t + B\sin\omega t) \tag{6.46}$$

where $X(x)$ is a function of $x$, $\omega$ is an unknown frequency, and $A$ and $B$ are unknown constants. Substituting (6.46) into (6.45) yields

$$a^2 \frac{d^2X}{dx^2} + \omega^2 X = 0 \tag{6.47}$$

The solution to this equation is

$$X = C\cos\frac{\omega x}{a} + D\sin\frac{\omega x}{a} \tag{6.48}$$

where $C$ and $D$ are constants that are determined form the boundary conditions. For a bar, there are three typical boundary conditions: both ends free; one end fixed; and both ends fixed. These three conditions are illustrated in Figure 6.9.

*Free ends*: If the bar is free at both ends as in Figure 6.9a, we note that no displacements occur at $x = 0$ and $x = l$ so the strain is zero and subsequently $\sigma = 0$. Therefore, at each end of the bar, we have

$$\left(\frac{\partial u}{\partial x}\right) = 0 \Rightarrow \left(\frac{\partial X}{\partial x}\right) = 0$$

Based on this, we can differentiate Equation 6.48 with respect to $x$ and obtain

$$\frac{\partial X}{\partial x} = -\frac{\omega}{a} C\sin\frac{\omega x}{a} + \frac{\omega}{a} D\cos\frac{\omega x}{a}$$

and using the boundary conditions at $x = 0$ and $x = l$ results in two equations

$$C(0) + D\frac{\omega}{a} = 0 \text{ and } -\frac{\omega}{a} C\sin\frac{\omega l}{a} + \frac{\omega}{a} D\cos\frac{\omega l}{a} = 0$$

From the above equation, we find $D = 0$ and $-(\omega/a)C\sin(\omega l/a) = 0$. For a nontrivial solution, $\sin(\omega l/a) = 0$ implying $(\omega l/a) = n\pi$ or $\omega_n = (n\pi a/l)$ for $n = 1, 2, 3$. Recalling that we defined $a^2 = E/\rho = Eg/\gamma$, we can write

$$\omega_1 = \frac{\pi a}{l} = \frac{\pi}{l}\sqrt{\frac{Eg}{\gamma}}, \quad \omega_2 = \frac{2\pi}{l}\sqrt{\frac{Eg}{\gamma}}, \ldots, \quad \omega_n = \frac{n\pi}{l}\sqrt{\frac{Eg}{\gamma}} \tag{6.49}$$

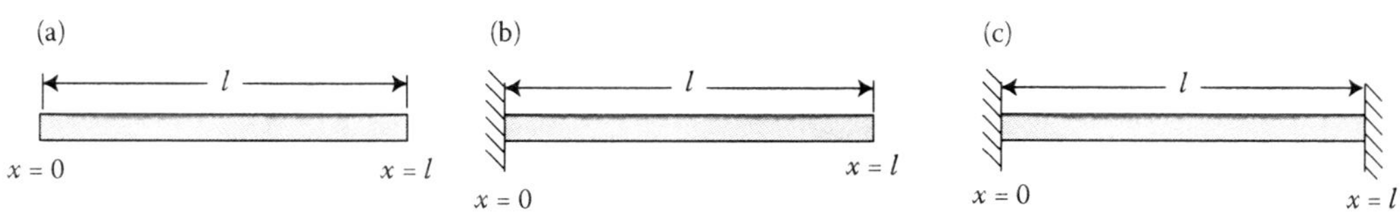


**FIGURE 6.9**
Support conditions for (a) free-free, (b) fixed-free, (c) fixed-fixed rod.

The constant C is arbitrary since the boundary conditions are satisfied for any C. Thus, for each $\omega_n$

$$X_n(x) = C_n \cos\frac{n\pi x}{l} \tag{6.50}$$

and

$$u_n = X_n(A_n \cos\omega_n t + B_n \sin\omega_n t) = C_n \cos\frac{n\pi x}{l}\left(A_n \cos\frac{n\pi a}{l}t + B_n \sin\frac{n\pi a}{l}t\right) \tag{6.51}$$

Assuming, for convenience, that $C_n = 1$ and noting that $u(x,t) = \Sigma_{n=1}^{\infty} u_n$, we can write

$$u(x,t) = \sum_{n=1}^{\infty} \cos\frac{n\pi x}{l}\left(A_n \cos\frac{n\pi a}{l}t + B_n \sin\frac{n\pi a}{l}t\right) \tag{6.52}$$

For the case of pure rigid body motion ($\omega = 0$), we have from Equation 6.47 that $d^2X/dx^2 = 0$. Integrating this equation yields the linear relation $X_0 = C_0 + D_0x$. Using boundary conditions $\partial X/\partial x = 0 \Rightarrow D_0 = 0$ and $X_0 = C_0 = 1$ (arbitrarily set to 1), we have

$$u_0 = X_0(A_0 + B_0t) = (A_0 + B_0t) \tag{6.53}$$

The physical meaning of $u_0$ is that it represents rigid body motion along the $x$-axis. From this expression, we note that $\partial u_0/\partial x = 0$ and $\partial u_0/\partial t = B_0 =$ constant. The general form of the solution is given by

$$u = \sum_{n=1}^{\infty} \cos\frac{n\pi x}{l}(A_n \cos\omega_n t + B_n \sin\omega_n t) + (A_0 + B_0t) \tag{6.54}$$

The constants $A_n$ and $B_n$ are determined from the initial conditions while $A_0$ and $B_0$ are typically considered negligible.

*Fixed-free ends*: For a fixed-free condition shown in Figure 6.9b the boundary conditions are $u(0, t) = X(0) = 0$ and $\partial u(l, t)/\partial x = \partial x(l)/\partial x = 0$. Using these boundary conditions, we determine

$$C = 0, \quad D\cos\frac{\omega l}{a} = 0$$

This implies that $D$ is arbitrary if $\cos(\omega l/a)$. Therefore, we have

$$\frac{\omega l}{a} = \frac{n\pi}{2} \quad \text{for } n = 1,3,5,\ldots \quad \text{and} \quad \omega_n = \frac{n\pi a}{2l} \quad \text{for } n = 1,3,5,\ldots$$

Following the same procedure as previously presented, we obtain the solution

$$u(x,t) = \sum_{n=1}^{\infty} \sin\frac{n\pi x}{2l}\left(A_n \cos\frac{n\pi a}{2l}t + B_n \sin\frac{n\pi a}{2l}t\right) \quad \text{for } n = 1,3,5,\ldots \tag{6.55}$$

The constants $A_n$ and $B_n$ are determined from the initial conditions.

*Fixed-fixed ends*: For a fixed-fixed condition shown in Figure 6.9c, the boundary conditions are $u(0, t) = X(0) = 0$ and $u(l, t) = X(l) = 0$. Using these boundary conditions in Equation 6.48 gives

$$C = 0, \quad D \sin \frac{\omega l}{a} = 0$$

This implies that $D$ is arbitrary if $\sin(\omega l/a) = 0$. Therefore, we have

$$\frac{\omega l}{a} = \frac{n\pi}{2} \quad \text{for } n = 1,2,3,4,\ldots \quad \text{and} \quad \omega_n = \frac{n\pi a}{2l} \quad \text{for } n = 1,2,3,4,\ldots$$

Following the same procedure as before, we find

$$u = \sum_{n=1}^{\infty} \sin \frac{n\pi x}{l} \left( A_n \cos \frac{n\pi a}{l} t + B_n \sin \frac{n\pi a}{l} t \right) \quad \text{for } n = 1,2,3,4,\ldots \tag{6.56}$$

As before, we determine the constants $A_n$ and $B_n$ from the initial conditions.

The first two natural modes, $X_n(x)$, each of these conditions is illustrated in Figure 6.10.

### 6.7.1 Orthogonality of Natural Modes

When a bar vibrates in its $n$th natural mode, it has harmonic motion as

$$u_n = X_n(A_n \cos \omega_n t + B_n \sin \omega_n t)$$

By substituting into the equation of motion, a solution is obtained which has the form

$$X_n = C_n \cos \frac{\omega_n x}{a} + D_n \sin \frac{\omega_n x}{a}$$

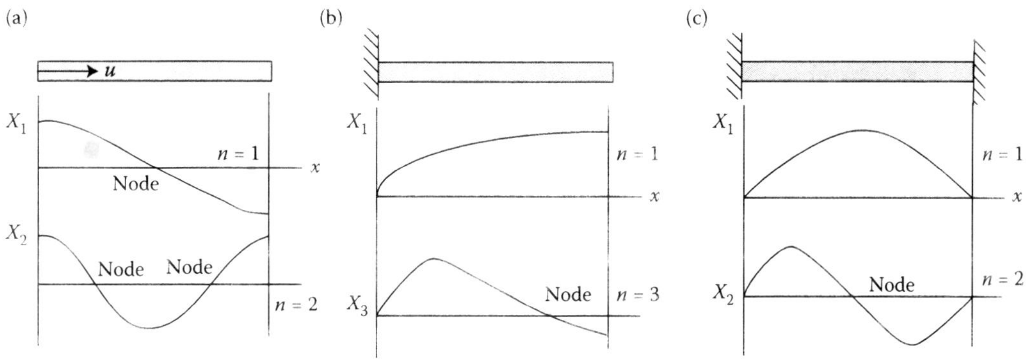


**FIGURE 6.10**
First two free vibration natural mode shapes for (a) free-free, (b) fixed-free, and (c) fixed-fixed rods.

Each natural frequency or eigenvalue $\omega_n$ is related to a specific mode shape or eigenfunction $X_n$. Equation 6.45 can be expressed in a somewhat different form by replacing $d^2X/dx^2$ with $X''$. Using this substitution, we can relate the $m$th and $n$th modes or eigenfunctions by using the two equations

$$\omega_m\text{:}\ \frac{\omega_m^2}{a^2}X_m = -X_m'' \quad \text{(a)}$$

$$\omega_n\text{:}\ \frac{\omega_n^2}{a^2}X_n = -X_n'' \quad \text{(b)}$$

Multiplying (a) by $X_n$ and (b) by $X_m$ and integrating them from 0 to $l$ in $x$ yields

$$\frac{\omega_m^2}{a^2}\int_0^l X_m X_n\,dx = -\int_0^l X_m'' X_n\,dx \qquad \text{(c)}$$

$$\frac{\omega_n^2}{a^2}\int_0^l X_n X_m\,dx = -\int_0^l X_n'' X_m\,dx \qquad \text{(d)}$$

Integrating the term on the right-hand side of (c) and (d) by parts gives

$$\frac{\omega_m^2}{a^2}\int_0^l X_m X_n\,dx = -\left[X_n X_m'\right]_0^l + \int_0^l X_m' X_n'\,dx \qquad \text{(e)}$$

$$\frac{\omega_n^2}{a^2}\int_0^l X_n X_m\,dx = -\left[X_m X_n'\right]_0^l + \int_0^l X_n' X_m'\,dx \qquad \text{(f)}$$

For fixed or free ends, the boundary conditions are zero at $x = 0$ or $l$. Therefore, subtracting equation (f) from (e) yields

$$\frac{\omega_m^2 - \omega_n^2}{a^2}\int_0^l X_m X_n\,dx = 0 \tag{6.57}$$

For distinct eigenvalues $\omega_m \neq \omega_n$, we have

$$\int_0^l X_m X_n\,dx = 0 \quad m \neq n \tag{6.58}$$

Using Equation 6.58 in Equation (e) gives

$$\int_0^l X_m' X_n'\,dx = 0 \quad m \neq n \tag{6.59}$$

From Equation (c), again using Equation 6.58, it is noted that

$$\int_0^l X_m'' X_n \, dx = 0 \quad m \neq n \tag{6.60}$$

Equation 6.58 is the orthogonality condition for natural modes or eigenfunctions. In addition, Equations 6.59 and 6.60 show that the orthogonality condition also exists among their derivatives.

### 6.7.2 Free Vibrations with Given Initial Conditions

Initial conditions are required to determine an explicit solution for a particular situation. The general solution for free vibrations can be written as

$$u(x,t) = \sum_n X_n(x)(A_n \cos \omega_n t + B_n \sin \omega_n t) \tag{6.61}$$

The initial conditions are defined by first letting $u(x, 0) = f_1(x)$ and $\dot{u}(x,0) = f_2(x)$. The functions $f_1(x)$ and $f_2(x)$ are known functions of $x$. The constants $A_n$ and $B_n$ are determined in order to satisfy the initial conditions of the problem. Since $u(x,0) = f_1(x)$, Equation 6.61 becomes

$$\sum_n X_n A_n = f_1(x)$$

Multiplying both sides by $X_m$ and integrating from 0 to $l$ results in

$$\sum_n \int_0^l X_n X_m A_m \, dx = \int_0^l f_1(x) X_m \, dx$$

The integral on the left is zero unless $n = m$. Therefore,

$$\sum_n \int_0^l X_m^2 A_m \, dx = \int_0^l f_1(x) X_m \, dx$$

Solving for $A_m$ yields

$$A_m = \frac{\int_0^l f_1(x) X_m \, dx}{\int_0^l X_m^2(x) dx} \quad m = 1, 2, 3, \ldots, \infty \tag{6.62}$$

Next we consider the second boundary condition, $\dot{u}(x,0) = f_2(x)$, which results in Equation 6.61 being written as

$$\sum_n \omega_n X_n B_n = f_2(x)$$

Multiplying both sides by $X_m$ and integrating from 0 to $l$ results in

$$\sum_n \omega_n B_n \int_0^l X_n X_m \, dx = \int_0^l f_2(x) X_m \, dx$$

Following the same procedures as before, $B_m$ becomes

$$B_m = \frac{\int_0^l f_2(x) X_m \, dx}{\omega_m \int_0^l X_m^2(x) dx} \qquad m = 1,2,3,\ldots,\infty \tag{6.63}$$

**EXAMPLE 6.3**

Consider the fixed-free bar shown in Figure 6.11 that is initially stretched an amount $\varepsilon$ at its end. The initial conditions are $u(x, 0) = \varepsilon x$, and $\dot{u}(x,0) = 0$. Determine the appropriate expression for the axial displacement $u(x, t)$ as a function of time.

**Solution**

Since this is a fixed-free beam, so the general form of the solution is given by Equation 6.55. In addition, we note that $X_n = \sin(n\pi x/2l)$ for $n = 1,3,5,\ldots$. From the initial condition for the velocity $\dot{u}(x,0) = 0 = f_2$, we determine $B_n = 0$. Since $u(x, 0) = \varepsilon x = f_1$, we find that

$$A_n = \frac{\int_0^l \varepsilon x \sin \frac{n\pi x}{2l} dx}{\int_0^l (\sin(n\pi x/2l))^2 dx} = \frac{\frac{4\varepsilon}{n^2\pi^2}(-1)^{(n-1)/2}}{l/2} = \frac{8\varepsilon}{ln^2\pi^2}(-1)^{(n-1)/2}$$

Therefore, the displacement is

$$u(x,t) = \sum_n \sin \frac{n\pi x}{2l} A_n \cos \omega_n t$$

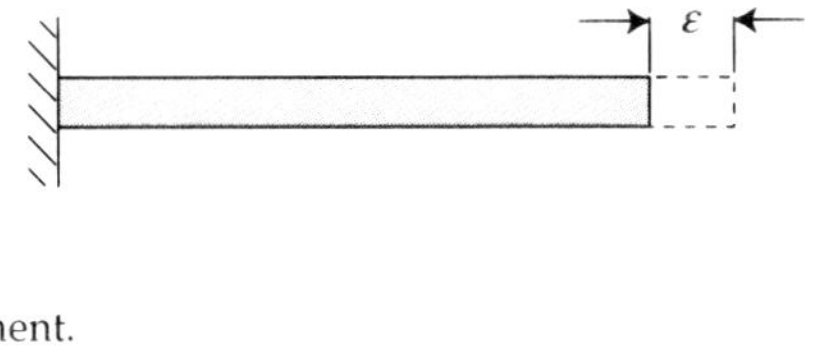


FIGURE 6.11
Bar with an initial end displacement.

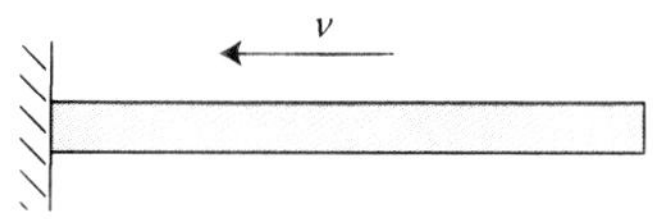


**FIGURE 6.12**
Fixed-free rod with a constant velocity.

or

$$u(x,t) = \sum_n \sin\frac{n\pi x}{2l}\left(\frac{8\varepsilon}{ln^2\pi^2}(-1)^{(n-1)/2}\right)\cos\omega_n t$$

**EXAMPLE 6.4**

Consider a fixed-free bar with a constant negative velocity and initial conditions $u(x,0) = f_1 = 0$ and $\dot{u}(x,0) = f_2 = -v$ as illustrated in Figure 6.12. Determine the expression for the displacement.

**Solution**

As in Example 6.3, we have $X_n = \sin(n\pi x/2l)$ for $n = 1,3,5,\ldots$ and since $u(x,0) = f_1 = 0$ we have $A_n = 0$. Using the initial condition $\dot{u}(x,0) = f_2 = -v$ results in

$$B_n = \frac{\int_0^l -v\sin(n\pi x/2l)dx}{\omega_n\int_0^l (\sin(n\pi x/2l))^2 dx} = \frac{-2v}{\omega_n}\int_0^l \sin\frac{n\pi x}{2l} = \frac{-4v}{\omega_n n\pi}$$

Therefore, the displacement is

$$u(x,t) = \sum_n \sin\frac{n\pi x}{2l}B_n\sin\omega_n t$$

or

$$u(x,t) = \sum_n \sin\frac{n\pi x}{2l}\left(\frac{-4v}{\omega_n n\pi}\right)\sin\omega_n t$$

## 6.8 Forced Vibrations of Bars

A forced vibration occurs when there is a driving force (generally a function of time) applied to the bar. Figure 6.13 shows a bar fixed at one end and subjected to a time-dependent end load at the free end. For the special case of an end load, we consider the frequency response function. In other words, we want to determine $U^*(x, i\omega)$.

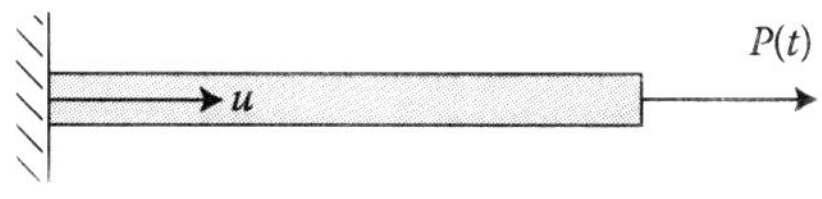


**FIGURE 6.13**
Tine-dependent end loaded bar.

We start by assuming $P(t) = 1 \cdot e^{i\omega t}$ and $u(x,t) = U^*(x, i\omega)e^{i\omega t}$. Substituting into the equation of motion, Equation 6.45, results in

$$(i\omega)^2 U^*(x, i\omega) = a^2 \frac{\partial^2 U^*(x, i\omega)}{\partial x^2}$$

where

$$U^*(x, i\omega) = C \sin\frac{\omega x}{a} + D\cos\frac{\omega x}{a}$$

At the free end of the bar there is an applied load $P(t)$, resulting in a stress $\sigma = P(t)/A$ as well as an accompanying strain $\varepsilon = \partial u/\partial x$. The boundary conditions are $u(0, t)$ which implies $U^*(0, i\omega) = 0$, and since $E\varepsilon = \sigma$ we have that $EA(\partial u(l, t)/\partial x) = P(t)$ implies that $(\partial U^*(l, i\omega)/\partial x) = 1/AE$. Therefore, we can establish that $D = 0$ and

$$C\frac{\omega}{a}\cos\frac{\omega l}{a} = \frac{1}{AE} \Rightarrow C = \frac{1}{AE}\left(\frac{a}{\omega}\right)\left(\frac{1}{\cos(\omega l/a)}\right)$$

This results in the solution being given as

$$U^*(x, i\omega) = \frac{1}{AE}\left(\frac{a}{\omega}\right)\left(\frac{\sin(\omega/a)x}{\cos(\omega/a)l}\right) \tag{6.64}$$

The displacement is therefore given as

$$u(x,t) = U^*(x, i\omega)e^{i\omega t} = \frac{1}{AE}\left(\frac{a}{\omega}\right)\left(\frac{\sin(\omega/a)x}{\cos(\omega/a)l}\right)e^{i\omega t} \tag{6.65}$$

At some arbitrary constant value of $x$, the value of $U^*(x, i\omega)$ asymptotically approaches infinity for certain values of $\omega$. This is illustrated in Figure 6.14.

Similar procedures apply for imposed displacements as illustrated in Figure 6.15. For example, consider determining $U^*$ if the input is a displacement of the form $b(t)$.

Similarly, if the input is $P(t) = P_0 \cos \omega t$, then one would find $u(x, t) = P_0\text{Re}\{U^*(x, i\omega)e^{i\omega t}\}$. Alternately, if the input is $P(t) = P_0 \sin \omega t$, then we would have $u(t) = P_0\,\text{Im}\{U^*(x, i\omega)e^{i\omega t}\}$.

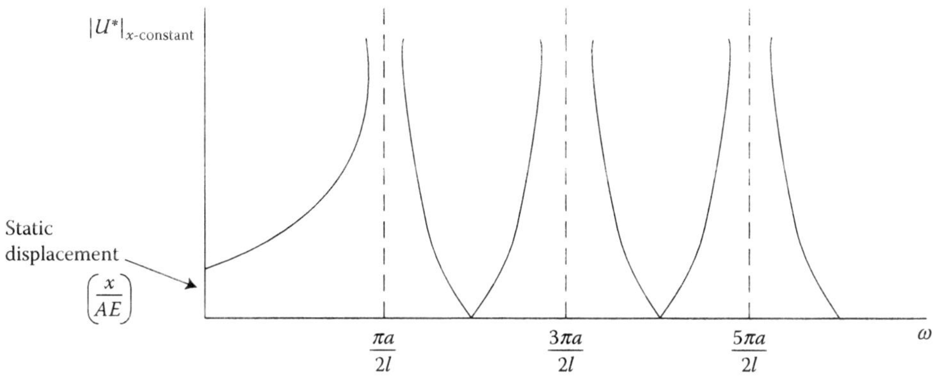


**FIGURE 6.14**
Variation of $U^*(x, i\omega)$ with $\omega$ for a constant $x$.

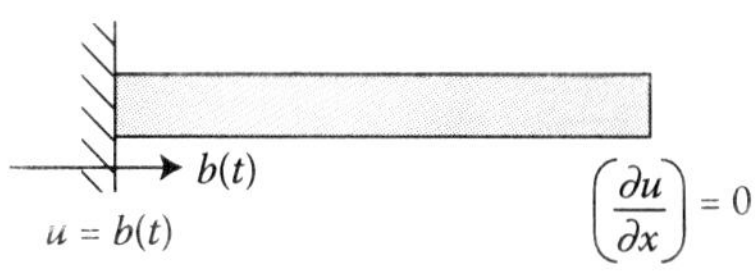


**FIGURE 6.15**
Imposed displacements at the fixed end of a rod.

## 6.9 Material with Damping

If material damping exists, Equation 6.65 is altered by replacing $E$ with $E^*(i\omega)$ and $a$ with $a^*(i\omega) = \sqrt{E^*(i\omega)/\rho}$ as initially discussed in Section 1.8.1. This results in Equation 6.65 being written as

$$U^*(x, i\omega) = \frac{1}{E^*(i\omega)A}\left(\frac{a^*(i\omega)}{\omega}\right)\left(\frac{\sin(\omega/(a^*(i\omega)))x}{\cos(\omega/(a^*(i\omega)))l}\right) \tag{6.66}$$

The effect of material damping is to reduce the magnitude of $U^*(x, i\omega)$ at those frequencies corresponding to infinity in Figure 6.14. This results in a much smoother curve shape as illustrated in Figure 6.16.

## 6.10 Forced Vibrations and Natural Mode Expansion Method

Consider a bar that instead of an applied end load, we assume an arbitrary load per unit length $p(x, t)$ as illustrated in Figure 6.17. This model is similar to that previously established, except that the arbitrary loading per unit length is included as shown.

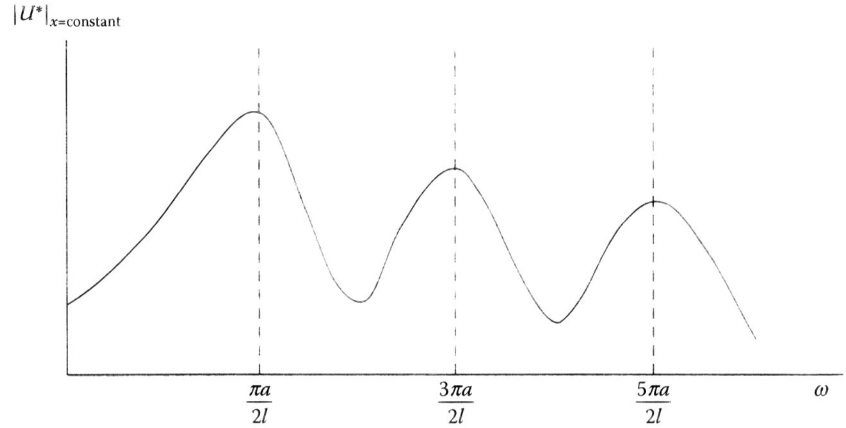


**FIGURE 6.16**
Variation of $U^*(x, i\omega)$ with $\omega$ for a constant $x$ with material damping.

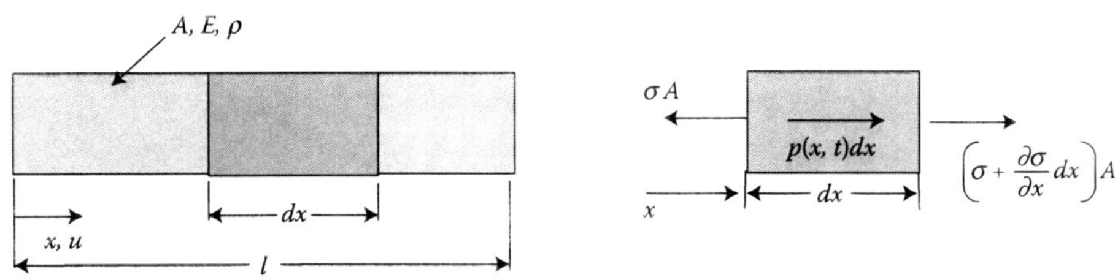


**FIGURE 6.17**
Forced vibration model for natural mode (expansion method).

Applying Newton's law to a typical element of the bar is

$$\rho A \frac{\partial^2 u}{\partial t^2} dx = A \frac{\partial \sigma}{\partial x} dx + p(x,t)dx$$

or

$$\rho A \frac{\partial^2 u}{\partial t^2} = A \frac{\partial \sigma}{\partial x} + p(x,t)$$

Using Hooke's law, we have $\sigma = E\varepsilon = E(\partial u/\partial x)$ and $\partial\sigma/\partial x = E(\partial^2 u/\partial x^2)$ which yield

$$\rho A \frac{\partial^2 u}{\partial t^2} - AE \frac{\partial^2 u}{\partial x^2} = p(x,t)$$

Again, we define $a^2$ to be $E/\rho$ or $Eg/\gamma$, which results in

$$\frac{\partial^2 u}{\partial t^2} - a^2 \frac{\partial^2 u}{\partial x^2} = \frac{1}{\rho A} p(x,t) \tag{6.67}$$

The natural frequencies and modes are designated as $\omega_n$ and $X_n$, respectively. Assume that the loading is given by

$$p(x,t)=\sum_{n=1}^{\infty}P_n(t)X_n(x) \tag{6.68}$$

For any given $p(x, t)$, we can determine $P_n(t)$ by following the previously established procedures. Multiplying both sides of Equation 6.68 by $X_m(t)$ and integrating from 0 to $l$ in $x$ yields

$$\int_0^l p(x,t)X_m\,dx=\int_0^l\sum_{n=1}^{\infty}P_nX_nX_m\,dx=\sum_{n=1}^{\infty}P_n\int_0^l X_nX_m\,dx$$

and $\int_0^l X_nX_m\,dx=0$ if $n\neq m$. Therefore,

$$\int_0^l p(x,t)X_m\,dx=P_m\int_0^l X_m^2\,dx$$

or

$$P_m=\frac{\int_0^l p(x,t)X_m\,dx}{\int_0^l X_m^2\,dx} \tag{6.69}$$

**EXAMPLE 6.5**

Consider a fixed-free (cantilevered) bar subjected to a load expressed as $p(x, t)=\sum_n P_n(t)$ $\sin(n\pi x/2l)$ and where $X_n(x)=\sin(n\pi x/2l)$ $n=1,3,5,\ldots$. Determine an expression for $P_n(t)$.

**Solution**

Using Equation 6.69 directly results in

$$P_n(t)=\frac{\int_0^l p(x,t)\sin(n\pi x/2l)dx}{\int_0^l(\sin(n\pi x/2l))^2dx}$$
$$=\frac{2}{l}\int_0^l p(x,t)\sin\frac{n\pi x}{2l}dx$$

In addition to the above, we also assume a solution in the following form

$$u(x,t)=\sum_{n=1}^{\infty}f_n(t)X_n(x) \tag{6.70}$$

where the function $f_n(t)$ is unknown. To determine $f_n(t)$, we substitute Equations 6.70 and 6.68 into the equation of motion, Equation 6.67, which results in

$$\sum_n X_n(x)\ddot{f}_n(t) - a^2 \sum_n f_n(t) X_n''(x) = \frac{1}{A\rho} \sum_n P_n(t) X_n(x) \tag{6.71}$$

From free vibrations we have $-\omega_n^2 X_n = a^2 X_n''$ or $X_n'' = (-\omega_n^2 X_n / a^2)$. Therefore,

$$\sum_n X_n(x)\ddot{f}_n(t) + \sum_n f_n(t)\omega_n^2 X_n(x) = \frac{1}{A\rho} \sum_n P_n(t) X_n(x)$$

Since the series on the left-hand side equals the series on the right-hand side, we can write

$$\ddot{f}_n(t) + \omega_n^2 f_n(t) = \frac{1}{A\rho} P_n(t) \tag{6.72}$$

or

$$f_n(t) = \int_0^t F_n(t-\tau) P_n(\tau) d\tau \tag{6.73}$$

where $F_n(t)$ is the solution for $P_n(t) = \delta(t)$. In this way, we have the particular solution for Equation 6.67 being expressed as

$$u(x,t) = \sum_n f_n(t) X_n(x) \tag{6.74}$$

If a set of initial conditions is given, we can write the general solution for $u(x, t)$ in the form

$$u(x,t) = \sum_n X_n (A_n \cos \omega_n t + B_n \sin \omega_n t) + \sum_n f_n(t) X_n(x) \tag{6.75}$$

where the constants $A_n$ and $B_n$ are determined from initial conditions. If $f_n$ follows from the convolution integral, Equation 6.73, then Equation 6.74 satisfies the initial conditions given as $u(x,0) = \dot{u}(x,0) = 0$. For homogeneous initial conditions, $f_n$ can be found by Laplace transformations.

The solution can also be determined using a different method. Equation 6.67 can also be written as

$$\sum_n \ddot{f}_n X_n - a^2 \sum_n f_n X_n'' = \frac{1}{A\rho} p(x,t) \tag{6.76}$$

where it has been assumed that $u(x, t) = \sum_n f_n X_n$. Using $X_n'' = -(\omega_n^2/a^2)X_n$, Equation 6.76 results in

$$\sum_n \left(\ddot{f}_n + \omega_n^2 f_n\right) X_n = \frac{1}{A\rho} p(x,t)$$

Multiplying by $X_m$ and integrating over the length of the bar from 0 to $l$ gives

$$\int_0^l \sum_n \left(\ddot{f}_n + \omega_n^2 f_n\right) X_m X_n \, dx = \int_0^l \frac{1}{A\rho} X_m p(x,t) dx$$

or

$$\left(\ddot{f}_m + f_m \omega_m^2\right) \int_0^l X_m^2 \, dx = \frac{1}{A\rho} \int_0^l X_m p(x,t) dx$$

This equation can be written as

$$\ddot{f}_m + \omega_m^2 f_m = \frac{P_m(t)}{A\rho} \tag{6.77}$$

where

$$P_m(t) = \frac{\int_0^l p(x,t) X_m \, dx}{\int_0^l X_m^2 \, dx}$$

## 6.11 Concentrated Force

A concentrated force does not need to be applied at the ends of a bar. Consider a concentrated axial load applied at some arbitrary location along the span of the bar. Consider a bar of length $l$ with a concentrated force applied at $x = l_1$ as shown in Figure 6.18.

Since the applied load is concentrated at a specific point, we can define it using the Dirac delta function, which is zero everywhere except at specific locations as was defined in Chapter 1. The load is therefore defined by

$$p(x,t) = P(t)\delta(x - l_1)$$

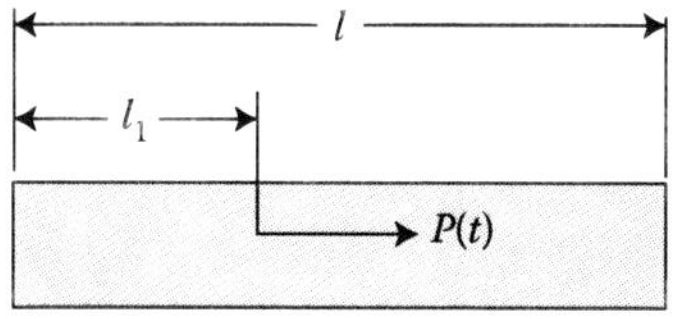


**FIGURE 6.18**
Bar subjected to an intermediate concentrated load.

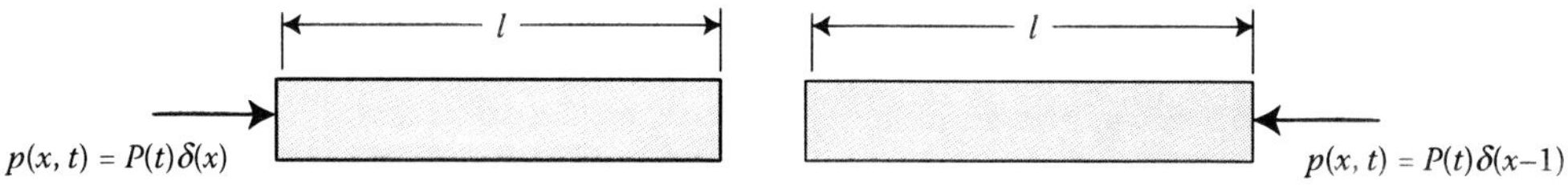


**FIGURE 6.19**
Bar with concentrated end load and associated expressions for $p(x, t)$.

This translates to the conditions that $p(x, t) = 0$ for $0 \le x < l_1$ and $p(x, t) = 0$ for $l_1 < x \le l$. These conditions lead to in

$$\int_0^l p(x,t)dx = P(t)\int_0^l \delta(x - l_1)dx = P(t)$$

Using Equation 6.69, we have

$$P_m(t) = \frac{\int_0^l p(x,t)X_m\,dx}{\int_0^l X_m^2\,dx} = \frac{P(t)\int_0^l \delta(x - l_1)X_m(x)dx}{\int_0^l X_m^2\,dx} = \frac{P(t)X_m(l_1)}{\int_0^l X_m^2\,dx}$$

The same approach can be applied to a bar with and end load. Figure 6.19 shows the two possible end loads (left or right ends) and the associated expression for $p(x, t)$.

## 6.12 Bar with Concentrated End Load and Associated Expressions for $p(x, t)$

Along similar lines, we can determine the displacement $u(x, t)$ for a bar that has its base displaced by some amount. Assume a model such as that shown in Figure 6.20, where the base of the bar is subjected to a specified displacement $b(t)$.

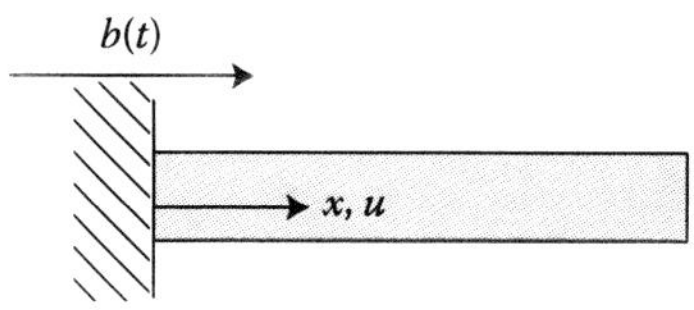


**FIGURE 6.20**
Bar with a specified base displacement.

We start with the equation of motion given by Equation 6.45

$$\frac{\partial^2 u}{\partial t^2} = a^2 \frac{\partial^2 u}{\partial x^2} \tag{6.45}$$

The boundary conditions are $u = b(t)$ at $x = 0$ and $\partial u/\partial x = 0$ at $x = l$. Introducing a new variable $\xi$, we have

$$\xi(x,t) = u(x,t) - b(t)$$

Therefore, we can define the displacement as

$$u(x,t) = \xi(x,t) + b(t) \tag{6.78}$$

Substituting Equation 6.78 into Equation 6.45 yields

$$\frac{\partial^2 \xi(x,t)}{\partial t^2} = a^2 \frac{\partial^2 \xi(x,t)}{\partial x^2} - \frac{\partial^2 b(t)}{\partial t^2} \tag{6.79}$$

The boundary conditions are: at $x = 0$, $u = b(t)$, so $\xi = 0$ and at $x = l$, $\partial u/\partial x = 0$, so $\partial \xi/\partial x = 0$. The variable $\xi$ is the solution of a bar fixed at $x = 0$ and free at $x = l$, thus

$$\frac{1}{A\rho} p(x,t) \equiv -\ddot{b}(t)$$

For a solution, we assume

$$\xi(x,t) = \sum_n X_n(x) f_n(t)$$

where $X_n(x)$ corresponds to the beam shown and $f_n(t)$ is unknown function of time. Letting

$$\ddot{b}(t) = \sum_n X_n(x) B_n(t)$$

where

$$B_n(t) = \frac{\int_0^l \ddot{b}(t) X_n \, dx}{\int_0^l X_n^2 \, dx} = \frac{2}{l} \int_0^l \ddot{b}(t) \sin \frac{n\pi x}{2l} \, dx$$

$$= \frac{2}{l} \ddot{b}(t) \frac{2l}{n\pi} \left( -\cos \frac{n\pi x}{2l} \right) \Bigg|_0^l = \frac{4\ddot{b}(t)}{n\pi} \qquad n = 1, 3, 5, \ldots$$

Equation 6.79 is similar to Equation 6.67, thus the $n$th generalized coordinate $\ddot{f}_n(t)$ we have

$$\ddot{f}_n(t)+\omega_n^2 f_n(t)=-B_n(t)$$

or

$$\ddot{f}_n(t)+\omega_n^2 f_n(t)=-\frac{4\ddot{b}(t)}{n\pi} \tag{6.80}$$

The solution to Equation 6.80 can be expressed, using a Duhamel integral, as

$$f_n(t)=-\frac{1}{\omega_n}\int_0^l X_n(x)dx\int_0^t \frac{4\ddot{b}(\tau)}{n\pi}\sin\omega_n(t-\tau)d\tau \tag{6.81}$$

The total solution for $\zeta(x, t)$ is obtained by superposing all the natural mode responses as

$$\zeta(x,t)=-\sum_n \frac{X_n(x)}{\omega_n}\int_0^l X_n(x)dx\int_0^t \frac{4\ddot{b}(\tau)}{n\pi}\sin\omega_n(t-\tau)d\tau \tag{6.82}$$

The longitudinal vibration of the bar can be determined from Equation 6.78 as

$$u(x,t)=b(t)-\sum_n \frac{X_n(x)}{\omega_n}\int_0^l X_n(x)dx\int_0^t \frac{4\ddot{b}(\tau)}{n\pi}\sin\omega_n(t-\tau)d\tau \tag{6.83}$$

## PROBLEMS

6.1 A string is fixed to a mass and a spring at point $A$ and fixed to a wall at point $B$ as shown in Figure 6.21a and a free-body diagram of the mass is shown in Figure 6.21b. Use a free-body diagram approach to develop an expression which allows you to determine the natural frequencies. Approximate the first two

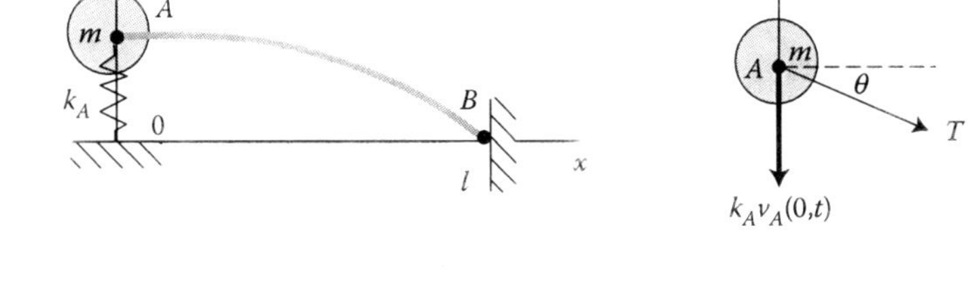


**FIGURE 6.21**
String with a spring and mass at one end.

natural frequencies by assuming the cable length is $l = 1$, the spring rate is $k_A = 25$, the mass is $m = 0.005$, and $T/\rho = 10{,}000$.

6.2 Assume a fixed-fixed string of length $l$ is subjected to an applied load. Determine an expression for the steady-state forced vibration response if the applied load is:

a. A concentrated force $p(x, t) = P(t)\delta(x - x_0) = P_0\delta(x - x_0)$

b. A constant velocity ($v$) moving load

$$p(x,t) = \begin{cases} P(t)\delta(x - vt) & 0 \le vt \le l \\ 0 & vt > l \end{cases}$$

6.3 For the bars shown in Figure 6.22, find the natural frequencies and modes of longitudinal vibrations and show the shapes of the first five modes. All bars have a length $L$.

6.4 For the bars in Figure 6.23, find the general expressions for the displacement $u(x, t)$ caused by arbitrary forces $P(t)$ acting on the bars. Use normal mode expansions.

6.5 Find the frequency response functions of the displacement $u$ for the bar and loadings shown in Problem 6.4. Use natural mode expansions. Develop expressions for $U^*(x, \omega)$ if the material is elastic, inelastic with $E^* = E' + iE''$, a Kelvin-type material, and a Maxwell-type material.

6.6 A viscoelastic bar as shown in Figure 6.24 whose properties are specified by the complex modulus $E^* = E + i\omega\eta$, where $E$ and $\eta$ are constants, is subjected to an end load $P(t) = P_0 \cos \omega t$. Determine a particular solution for the axial displacement $u(x, t)$ and the axial stress $\sigma(x, t)$ using:

a. The natural mode method

b. The closed form solution

**FIGURE 6.22**
Bars with different end conditions.

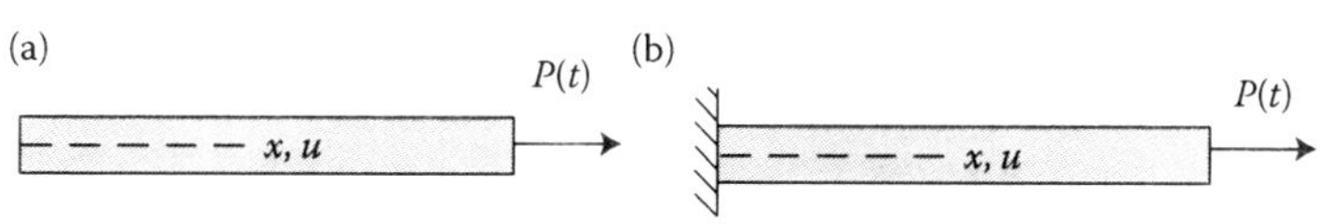


**FIGURE 6.23**
End-loaded bars with different end conditions.

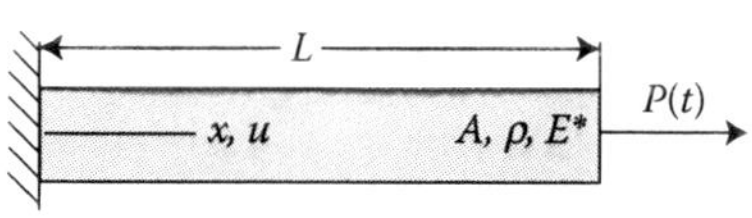


**FIGURE 6.24**
Viscoelastic bar with an end load.

## References

Meirovitch, L. 2001. *Fundamentals of Vibrations*. New York: McGraw-Hill Book Co.
Nowacki, W. 1963. *Dynamics of Elastic Systems*. New York: John Wiley & Sons, Inc.
Rao, S. S. 2007. *Vibrations of Continuous Systems*. Hoboken: John Wiley and Sons.
Weaver, W., Jr., Timoshenko, S. P., and Young, D. H. 1990. *Vibration Problems in Engineering*, 5th ed. New York: John Wiley & Sons.

# 7

# Beam Vibrations

## 7.1 Introduction

Beams are a common structural elements subjected to transverse loads applied perpendicular to the longitudinal axis ($x$-axis by convention) of the beam. Transverse loads typically occur in the $x$–$z$ plane and are in the $z$ direction. The transverse loads can be concentrated or distributed on the top, bottom, or on both surfaces. The assumed deformation that results from the loading defines the complexity of the governing equations. The simplest set of governing equations is for the so-called shear beam, in which only shear deformation is assumed. A more rigorous set of equations is developed using the Euler–Bernoulli beam theory sometimes called the classical beam theory, Euler beam theory, or Bernoulli beam theory, because it is simple and provides reasonable engineering approximations for many problems. Both of these theories are typically introduced in undergraduate strength of materials courses and can be found in any number of strength of materials texts (Beer, Johnston, and Dewolf 2002; Hibbeler 2014). There are, however limiting assumptions made when developing the Euler–Bernoulli theory. The most rigorous beam deflection theory is the Timoshenko beam theory, which is an elasticity solution to the problem of beam bending and can be found in various texts, including Timoshenko (1955) and Timoshenko and Gere (1972). All three of these theories are presented in this chapter.

## 7.2 Shear Beam

A shear beam is one that undergoes only a shearing type of deformation with no bending deformation. A displacement $w(x, t)$ exists due to the shear deformation in the $x$–$z$ plane. The beam is loaded in such a manner that it may be assumed to deform in such a manner as to produce only a shear deformation exists. Examples of the types of structures to which this type of deformation is applicable are illustrated in Figure 7.1.

The equations of motion of this type of beam are developed using the model shown in Figure 7.2. An applied force $p(x, t)$ is assumed to act as shown and we assume the representative volume element has a mass per unit length of $m$. Using Newton's second law, we determine

$$p(x,t)dx + \frac{\partial Q}{\partial x}dx = mdx\frac{\partial^2 w}{\partial t^2}$$

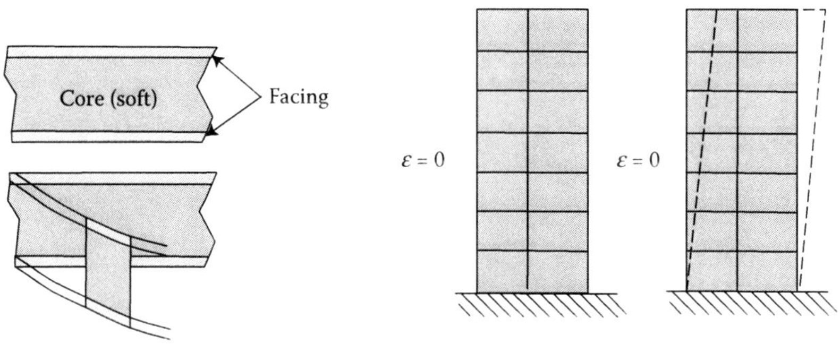


**FIGURE 7.1**
Typical structures in which only shear deformation exists.

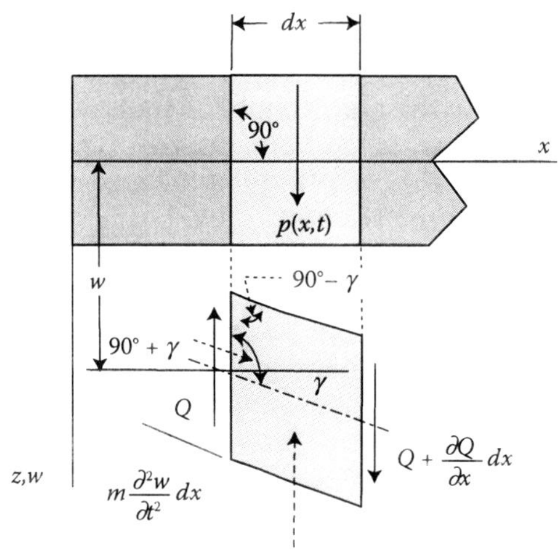


**FIGURE 7.2**
Model shear beam deformation.

or

$$m\frac{\partial^2 w}{\partial t^2} = p(x,t) + \frac{\partial Q}{\partial x} \tag{7.1}$$

The unknown shear force $Q$ and displacement $w$ are related by the shear strain $\gamma$ given by the expression

$$\gamma = \frac{\partial w}{\partial x} = \frac{Q}{K} \tag{7.2}$$

where $K$ is the shear rigidity of the beam (the shear force corresponds to $\gamma = 1$), as defined in most undergraduate strength of material texts (Timoshenko and Gere 1972). Therefore, the shear force and displacement are related by $Q = K(\partial w/\partial x)$. Using this relationship to eliminate $Q$ from Equation 7.1 yields

$$m\frac{\partial^2 w}{\partial t^2} - K\frac{\partial^2 w}{\partial x^2} = p(x,t) \tag{7.3}$$

**TABLE 7.1**

Boundary Conditions for Free-Free, Fixed-Free, and Fixed-Fixed Beams

| End Conditions | $x = 0$ | $x = l$ |
|---|---|---|
| Free-free | $Q = 0, \frac{\partial w}{\partial x} = 0$ | $Q = 0, \frac{\partial w}{\partial x} = 0$ |
| Fixed-free | $w = 0$ | $\frac{\partial w}{\partial x} = 0$ |
| Fixed-fixed | $w = 0$ | $w = 0$ |

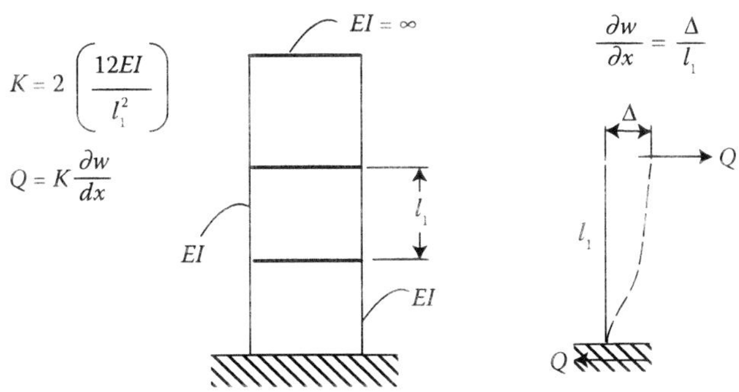


**FIGURE 7.3**
Relationship between shear rigidity, shear force, and lateral displacement.

For free vibrations, the loading $p(x, t)$ is set to zero. Defining $a^2 = (K/m)$, then Equation 7.3 can be written as

$$\frac{\partial^2 w}{\partial t^2} = a^2 \frac{\partial^2 w}{\partial x^2} \tag{7.4}$$

This is the same equation that was obtained for the longitudinal vibrations of bars. Boundary conditions are identified for different conditions as shown in Table 7.1.

The method of solution for free vibrations and forced vibrations is the same as that obtained for the longitudinal vibrations of bars, except that $u$ has been replaced by $w$. Figure 7.3 shows a frame with rigidity $EI$ except on the top member, which has an infinite rigidity ($EI = \infty$). The relationships between the shear rigidity $K$, shear force $Q$, and the lateral displacement $w$ are depicted in Figure 7.3.

## 7.3 Euler–Bernoulli Theory

Bending vibration of beams differs from the transverse vibration of strings or the longitudinal vibration of bars. Using the solid mechanics theory of beams commonly referred to as *classical or simple beam* theory, we shall consider the transverse vibrations of a prismatic beam with constant cross section and mass which lies in the $x$–$z$ plane, which is assumed

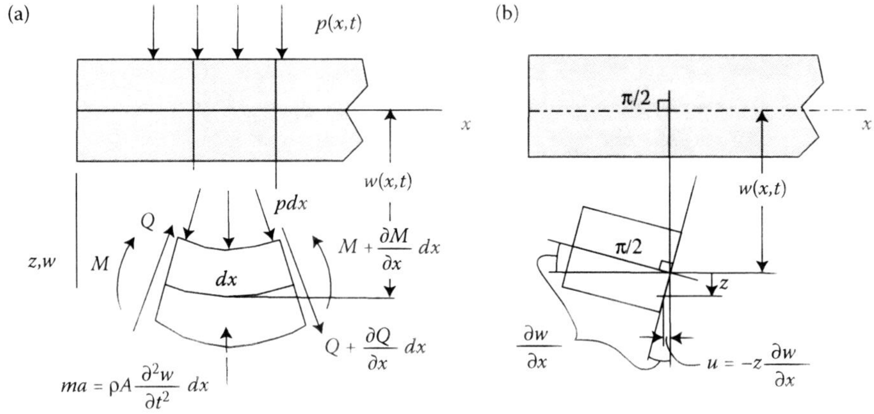


**FIGURE 7.4**
Model defining (a) loads and (b) deflections used to determine beam equation of motion.

to be a plane of symmetry for any cross section. The beam undergoes small displacements and the material is elastic. It is noted that for *in-plane displacements,* the cross-section of the beam is considered infinitely rigid, that is, no deformations occur in the plane of the cross section. Additional assumptions which deal with *out of plane displacements* are as follows: the cross section of the beam remains plane after deformation and remains normal to the deformed axis of the beam. This is considered the classical Euler–Bernoulli beam theory, which was developed in the mid-1700s. A more comprehensive theory was developed by S.P. Timoshenko (1921). In the Timoshenko theory, the assumption that plane sections remain plane and perpendicular to the deformed axis is relaxed to allow sections to undergo a shear strain. This will be presented in a subsequent section. The model used to develop the equation of motion is shown in Figure 7.4a. The volume element in this figure has a mass density per unit volume $\rho$ and a cross-sectional area $A$. Applying Newton's second law to the deformed element and neglecting the rotational inertia effects results in

$$\rho A dx \frac{\partial^2 w}{\partial t^2} = \frac{\partial Q}{\partial x} dx + p dx$$

or

$$\rho A \frac{\partial^2 w}{\partial t^2} = \frac{\partial Q}{\partial x} + p \tag{7.5}$$

As we are neglecting rotational inertia effects, the moment equation is effectively

$$\frac{\partial M}{\partial x} dx - Q dx = 0$$

or

$$Q = \frac{\partial M}{\partial x} \tag{7.6}$$

Substituting Equation 7.6 into Equation 7.5 gives

$$\rho A\frac{\partial^2 w}{\partial t^2}=\frac{\partial^2 M}{\partial x^2}+p(x,t) \tag{7.7}$$

Upon loading, the beam deforms and undergoes some rotation as indicated in Figure 7.4b. The strain in the $x$ direction depends on the distance a particular point is from the deformed neutral bending axis. At some distance $z$ from the deformed neutral bending axis, the $x$ displacement is given by $u=-z(\partial w/\partial x)$, such that the stress $\sigma_x$ and strain $\varepsilon_x$ are expressed as

$$\varepsilon_x=\frac{\partial u}{\partial x}=-z\frac{\partial^2 w}{\partial x^2}\quad \sigma_x=E\varepsilon_x-zE\frac{\partial^2 w}{\partial x^2}$$

As the bending moment is related to stress, using the above stress value, we can write

$$M=\int_A z\sigma_x dA=-\int_A z^2E\frac{\partial^2 w}{\partial x^2}dA=-EI\frac{\partial^2 w}{\partial x^2} \tag{7.8}$$

As a result, we have

$$Q=-\frac{\partial}{\partial x}\left(EI\frac{\partial^2 w}{\partial x^2}\right) \tag{7.9}$$

Using Equations 7.6 and 7.9 and substituting into the equation of motion, Equation 7.7, results in

$$\frac{\partial^2}{\partial x^2}\left(EI\frac{\partial^2 w}{\partial x^2}\right)+\rho A\frac{\partial^2 w}{\partial t^2}=p(x,t) \tag{7.10}$$

Both initial and boundary conditions must be specified before this equation can be solved. If the material of the beam is homogeneous, then $E$ is constant and if the cross section is constant then $I$ is constant, so Equation 7.10 can be written as

$$EI\frac{\partial^4 w}{\partial x^4}+\rho A\frac{\partial^2 w}{\partial t^2}=p(x,t) \tag{7.11}$$

### 7.3.1 Free Vibrations

When the load is absent $p(x, t)=0$, then Equation 7.11 reduces to the free vibration condition. In addition, if we define $a^2=EI/\rho A$, the equation of motion for free vibration of a beam becomes

$$\frac{\partial^2 w}{\partial t^2}+a^2\frac{\partial^4 w}{\partial x^4}=0 \tag{7.12}$$

It is immediately noted that this equation is not the form of the wave equation and does not have the dimension of velocity. When an elastic beam performs a normal mode of vibration, the deflection at any location varies harmonically with time and can be found using separation of variables and represented as

$$w(x,t) = X(x)(A\cos\omega t + B\sin\omega t)$$

where $X(x) = f(x)$ defines the shape of the normal mode of vibration. Substituting our expression for the beam deflection $w(x, t)$ into Equation 7.12 results in

$$\frac{d^4X}{dx^4} - \frac{\omega^2}{a^2}X = 0$$

or

$$\frac{d^4X}{dx^4} - \beta^4X = 0 \tag{7.13}$$

where we have used

$$\beta^4 = \frac{\omega^2}{a^2} = \frac{\rho A\omega^2}{EI} \tag{7.14}$$

For the solution of Equation 7.13, the exponential form

$$X(x) = Ce^{sx} \tag{7.15}$$

is assumed where $C$ and $s$ are constants and $s$ is to be determined by satisfying the differential equation. Substituting Equation 7.15 into Equation 7.13 gives the relation

$$s^4 - \beta^4 = 0 \tag{7.16}$$

or

$$(s^2 - \beta^2)(s^2 + \beta^2) = 0 \tag{7.17}$$

which has the roots

$$s = \pm\beta, \quad s = \pm i\beta \tag{7.18}$$

so that the solution for Equation 7.13 is

$$X(x) = \bar{C}_1e^{\beta x} + \bar{C}_2e^{-\beta x} + \bar{C}_3e^{i\beta x} + \bar{C}_4e^{-i\beta x} \tag{7.19}$$

**TABLE 7.2**

Boundary Conditions for Simple Supports, Fixed, and Free Ends of Beams

| Support Condition | Boundary Conditions |
|---|---|
| Simple | $X = 0, \dfrac{d^2X}{dx^2} = 0$ |
| Fixed | $X = 0, \dfrac{dX}{dx} = 0$ |
| Free | $\dfrac{d^2X}{dx^2} = 0, \dfrac{d^3X}{dx^3} = 0$ |

which can be written in the equivalent form

$$X(x) = C_1 \sin \beta x + C_2 \cos \beta x + C_3 \sinh \beta x + C_4 \cosh \beta x \tag{7.20}$$

The constants $C_1$, $C_2$, $C_3$, and $C_4$ are determined from the boundary and initial conditions of the problem. The boundary conditions for simple supports, fixed and free ends, are given in Table 7.2.

Vibrations and mode shapes are determined by evaluating Equation 7.20 using the four boundary conditions, two at each end. We then obtain a set of four homogeneous equations in terms of the four unknown constants $C_1$, $C_2$, $C_3$, and $C_4$. For a nontrivial solution of these equations, the determinant of the coefficients must vanish. This leads to the characteristic or frequency equation of the problem, which has an infinite number of solutions for $\beta$. For each value of $\beta$, there is a value of $\omega$ given by Equation 7.14. The roots of the frequency equation or eigenvalues are $\omega_1, \omega_2, \ldots, \omega_n$. For each root, we find the constants $C_1$, $C_2$, $C_3$, and $C_4$. One of these values is arbitrary and the rest are related by some functions. In other words, the mode shapes can be determined but not the amplitudes. In other words, for each $\omega_n$, we find the corresponding mode shape $X_n$. The normal or natural modes can be superimposed to produce the total response of the beam as

$$w = \sum_n X_n(x)(A_n \cos \omega_n t + B_n \sin \omega_n t) \tag{7.21}$$

where $\omega_n$ is the $n$th natural frequency and $X_n$ is the $n$th natural mode. The orthogonality of the normal or natural modes provides a general solution for the mode shape $X(x)$. Consider $\omega_n$, $X_n$ and $\omega_m$, $X_m$ so that we have the two equations

$$\frac{d^4X_m}{dx^4} = \frac{\omega_m^2}{a^2} X_m \tag{7.22}$$

$$\frac{d^4X_n}{dx^4} = \frac{\omega_n^2}{a^2} X_n \tag{7.23}$$

Multiplying Equation 7.22 by $X_n$ and the second Equation 7.23 by $X_m$ and integrating from $x = 0$ to $x = l$ yields

$$\int_0^l \frac{d^4X_m}{dx^4} X_n\,dx = \frac{\omega_m^2}{a^2}\int_0^l X_mX_n\,dx \tag{7.24}$$

$$\int_0^l \frac{d^4X_n}{dx^4} X_m\,dx = \frac{\omega_n^2}{a^2}\int_0^l X_nX_m\,dx \tag{7.25}$$

Integrating the left-hand side of Equations 7.24 and 7.25 by parts twice yields

$$\left[\frac{d^3X_m}{dx^3}X_n\right]_0^l - \left[\frac{d^2X_m}{dx^2}\frac{dX_n}{dx}\right]_0^l + \int_0^l \frac{d^2X_m}{dx^2}\frac{d^2X_n}{dx^2}\,dx = \frac{\omega_m^2}{a^2}\int_0^l X_mX_n\,dx \tag{7.26}$$

$$\left[\frac{d^3X_n}{dx^3}X_m\right]_0^l - \left[\frac{d^2X_n}{dx^2}\frac{dX_m}{dx}\right]_0^l + \int_0^l \frac{d^2X_n}{dx^2}\frac{d^2X_m}{dx^2}\,dx = \frac{\omega_n^2}{a^2}\int_0^l X_nX_m\,dx \tag{7.27}$$

The integrated values evaluated at $x = 0$ and $x = l$ vanish for any boundary condition. Hence, subtracting Equation 7.27 from Equation 7.26 yields

$$\frac{\omega_m^2 - \omega_n^2}{a^2}\int_0^l X_mX_n dx = 0 \tag{7.28}$$

If $m \neq n$, then the eigenvalues are distinct, $\omega_n \neq \omega_m$, then the natural modes form an orthogonal set of functions and

$$\int_0^l X_mX_n dx = 0 \quad (m \neq n) \tag{7.29}$$

Substituting Equation 7.29 into Equation 7.26 yields

$$\int_0^l \frac{d^2X_m}{dx^2}\frac{d^2X_n}{dx^2}\,dx = 0 \quad (m \neq n) \tag{7.30}$$

and using Equation 7.24 we have

$$\int_0^l \frac{d^4X_m}{dx^4} X_n\,dx = 0 \quad (m \neq n) \tag{7.31}$$

Equations 7.29 through 7.31 comprise the orthogonality conditions for the transverse vibration of a prismatic beam.

The general solution for $X(x)$ is

$$X(x) = C_1 \sin \beta x + C_2 \cos \beta x + C_3 \sinh \beta x + C_4 \cosh \beta x \tag{7.32}$$

which can also be written in an equivalent form as

$$X(x) = D_1(\cos\beta x + \cosh\beta x) + D_2(\cos\beta x - \cosh\beta x) + D_3(\sin\beta x + \sinh\beta x) + D_4(\sin\beta x - \sinh\beta x) \tag{7.33}$$

**EXAMPLE 7.1**

Consider an 18 in long free-free steel beam with cross-sectional dimensions resulting in a value for $a = 86.5 \times 10^3$ in$^2$/s. Determine the first three natural frequencies.

**Solution**

We begin by focusing on the boundary conditions, which are

$$x = 0: \frac{d^2X}{dx^2} = 0, \quad \frac{d^3X}{dx^3} = 0$$

and

$$x = l: \frac{d^2X}{dx^2} = 0, \quad \frac{d^3X}{dx^3} = 0$$

Substituting these into Equation 7.32 yields

$$x = 0: \frac{d^2X}{dx^2} = 0 = D_1(0) + D_2(2) + D_3(0) + D_4(0) \Rightarrow D_2 = 0$$

$$\frac{d^3X}{dx^3} = 0 = D_1(0) + D_2(0) + D_3(0) + D_4(2) \Rightarrow D_4 = 0$$

$$x = l: \frac{d^2X}{dx^2} = 0 = D_1(-\cos \beta l + \cosh \beta l) + D_3(-\sin \beta l + \sinh \beta l)$$

$$\frac{d^3X}{dx^3} = 0 = D_1(\sin \beta l + \sinh \beta l) + D_3(-\cos \beta l + \cosh \beta l)$$

The frequency equation is given by the determinant

$$\begin{vmatrix} -\cos \beta l + \cosh \beta l & -\sin \beta l + \sinh \beta l \\ \sin \beta l + \sinh \beta l & -\cos \beta l + \cosh \beta l \end{vmatrix} = 0$$

and

$$(-\cos\beta l + \cosh\beta l)^2 - (\sin\beta l + \sinh\beta l)(-\sin\beta l + \sinh\beta l) = 0$$

Expanding this expression results in

$$\cos^2\beta l - 2\cos\beta l\cosh\beta l + \cosh^2 Kl\beta l + \sin^2\beta l - \sinh^2\beta l = 0$$

Using the trigonometric identities $\cos^2\beta l + \sin^2\beta l = 1$ and $\cosh^2\beta l - \sinh^2\beta l = 1$, this reduces to

$$\cos\beta l\cosh\beta l = 1$$

The first three are $\beta_1 l = 0$, $\beta_2 l = 4.730$, and $\beta_3 l = 7.853$. As the length of the beam is 18 in, we determine $\beta_1 = 0$, $\beta_2 = 0.263$, and $\beta_3 = 0.436$. As $\omega = a\beta^2$, we can determine

$$\omega_1 = 0, \quad \omega_2 \approx 5985, \quad \omega_3 \approx 16,448$$

For each $\omega_n$, $n \geq 2$ we can assume, for example, $D_1 = 1$ and determine $D_3$ from our boundary conditions at $x = l$. For the condition $\omega = \omega_1 = 0$, we can write $X_1(x)$ as $a + bx$, which satisfies the equation for the mode $X_1(x)$. The mode $X_1(x)$ represents the rigid body motion part of $w(x, t)$.

**EXAMPLE 7.2**

Consider a beam of length $l$ which is pinned at the left end and supported by a roller at the right end (simply supported). Determine a general expression for $X_1, X_2, \ldots, X_n$.

**Solution**

We start with the general expression for $X(x)$ and apply the boundary conditions from Table 7.2 to Equation 7.33. Thus, we find that at $x = 0$: $X(0) = 0 = D_1(2) + D_2(0) + D_3(0) + D_4(0) \Rightarrow D_1 = 0$

$$\frac{d^2X}{dx^2} = 0 = D_2(2) + D_3(0) + D_4(0) \Rightarrow D_2 = 0$$

and at $x = l$: $X(l) = 0 = D_3(\sin\beta l + \sinh\beta l) + D_4(\sin\beta l - \sinh\beta l)$

$$\frac{d^2X}{dx^2} = 0 = D_3(-\sin\beta l + \sinh\beta l) + D_4(-\sin\beta l - \sinh\beta l)$$

These two equations yield $D_3 = D_4$ and if $\sin\beta l = 0$, then $D_3$ and $D_4$ are not identically zero. From this, we obtain

$$\beta_n l = n\pi \quad \text{or} \quad \beta_n = \frac{n\pi}{l}$$

The natural frequencies of a simply supported beam are

$$\omega_1 = aK_1^2 = a\left(\frac{\pi}{l}\right)^2$$

$$\omega_2 = aK_2^2 = a\left(\frac{2\pi}{l}\right)^2$$

$$\vdots$$

$$\omega_n = aK_n^2 = a\left(\frac{n\pi}{l}\right)^2$$

The mode shapes are defined by $X(x) = 2D_3 \sin \beta x$, where $D_3$ is arbitrary. For example, if $D_3 = 1/2$, then we have $X(x) = \sin \beta x$. If

$$\beta = \beta_1 = \frac{\pi}{l} \qquad X = X_1 = \sin\frac{\pi x}{l}$$

$$\beta = \beta_2 = \frac{\pi}{l} \qquad X = X_2 = \sin\frac{2\pi x}{l}$$

$$\vdots \qquad\qquad \vdots$$

$$\beta = \beta_n = \frac{n\pi}{l} \qquad X = X_n = \sin\frac{n\pi x}{l}$$

### 7.3.2 Free Vibration with Given Initial Conditions

The general solution for free vibrations is given by $w(x,t) = \sum_{n=1}^{\infty} X_n(x)(A_n \cos\omega_n t + B_n \sin\omega_n t)$. The natural frequencies are $\omega_n$ and mode shapes are $X_n(x)$. The constants $A_n$ and $B_n$ are determined from the initial conditions. The initial conditions are usually specified at time $t = 0$ as $w(x, 0) = w_0(x)$ and $\dot{w}(x, 0) = \dot{w}_0(x)$, where $w_0(x)$ and $\dot{w}_0(x)$ are given functions of $x$. Using the initial conditions generates two equations:

$$w(x,0) = \sum_{n=1}^{\infty} X_n(x)A_n = w_0(x) \tag{7.34}$$

and

$$\dot{w}(x,0) = \sum_{n=1}^{\infty} \omega_n X_n(x)B_n = \dot{w}_0(x) \tag{7.35}$$

Multiplying Equation 7.34 by $X_m$ and integrating from 0 to $l$ gives

$$\sum_{n=1}^{\infty} A_n \int_0^l X_n X_m \, dx = \int_0^l w_0(x)X_m \, dx$$

If $n = m$, then we obtain the only nonzero term resulting in

$$A_m \int_0^l X_m^2 \, dx = \int_0^l w_0(x)X_m \, dx$$

Therefore,

$$A_m = \frac{\int_0^l w_0(x) X_m \, dx}{\int_0^l X_m^2 \, dx} \quad m = 1, 2, 3, \ldots \tag{7.36}$$

Following a similar procedure and multiplying Equation 7.35 by $X_m$ and integrating from 0 to $l$ yields

$$\sum_{n=1}^{\infty} B_n \omega_n \int_0^l X_n X_m \, dx = \int_0^l \dot{w}_0(x) X_m \, dx$$

Again, if $n = m$, we obtain the only nonzero term which results in

$$B_m \omega_m \int_0^l X_m^2 \, dx = \int_0^l \dot{w}_0(x) X_m \, dx$$

Therefore,

$$B_m = \frac{\int_0^l \dot{w}_0(x) X_m \, dx}{\omega_m \int_0^l X_m^2 \, dx} \quad m = 1, 2, 3, \ldots \tag{7.37}$$

An alternate derivation of the expressions for $A_m$ and $B_m$ is obtained by expanding $w_o(x)$ in a series. Let

$$w_0(x) = \sum_{n=1}^{\infty} K_n X_n \quad \text{where} \quad K_n = \frac{\int_0^l w_0(x) X_n \, dx}{\int_0^l X_n^2 \, dx}$$

With this result we see that Equation 7.34 becomes

$$\sum_{n=1}^{\infty} A_n X_n = \sum_{n=1}^{\infty} K_n X_n$$

or $A_n = K_n$. In the same manner for $B_n$, we expand $\dot{w}_0(x)$ as

$$\dot{w}_0(x) = \sum_{n=1}^{\infty} L_n X_n \quad \text{where} \quad L_n = \frac{\int_0^l \dot{w}_0(x) X_n \, dx}{\int_0^l X_n^2 \, dx}$$

Thus, Equation 7.35 becomes

$$\sum_{n=1}^{\infty} B_n \omega_n X_n = \sum_{n=1}^{\infty} L_n X_n$$

which results in $B_n = L_n/\omega_n$.

**EXAMPLE 7.3**

Assume a beam of length $l$ is dropped onto simple supports with a constant velocity $\dot{w}_0$ = constant at time $t = 0$. Determine the expression for $w(x, t)$.

**Solution**

The boundary conditions are $w(x,0) = 0$ and $\dot{w}(x,0) = \dot{w}_0$. For the simply supported beam, we have

$$X_n = \sin\frac{n\pi x}{l} \quad \text{and} \quad \omega_n = a\left(\frac{n\pi}{l}\right)^2 \quad \text{for} \quad n = 1,2,3,\ldots$$

For $t \geq 0$, the general expression for the response $w(x, t)$ is

$$w(x,t) = \sum_{n=1}^{\infty} X_n(x)(A_n \cos\omega_n t + B_n \sin\omega_n t)$$

$$A_n = \frac{\int_0^l w_0(x) X_n dx}{\int_0^l X_n^2 dx} = \frac{\int_0^l (0)(\sin(n\pi x/l))dx}{\int_0^l (\sin(n\pi x/l))^2 dx} = 0$$

$$B_n = \frac{1}{\omega_n}\frac{\int_0^l \dot{w}_0(x) X_n\, dx}{\int_0^l X_n^2\, dx} = \frac{1}{\omega_n}\frac{\int_0^l \dot{w}_0(x)\sin(n\pi x\,/\,l)dx}{\int_0^l (\sin(n\pi x\,/\,l))^2 dx} = \frac{2}{\omega_n l}\int_0^l \dot{w}_0(x)\sin\frac{n\pi x}{l}dx$$

$$= \frac{2\dot{w}_0}{\omega_n l}\left[-\frac{l}{n\pi}\cos\frac{n\pi x}{l}\right]_0^l = \frac{2\dot{w}_0}{\omega_n l}\left[-\frac{l}{n\pi}\cos n\pi + \frac{l}{n\pi}\right]$$

$$= \frac{4\dot{w}_0}{\omega_n n\pi} \quad \text{for} \quad n = 1,3,5,\ldots$$

and $B_n = 0$ for $n = 2, 4, 6, \ldots$. Thus, we have for the response of the beam

$$w(x,t) = \sum_{n=1,3,5,\ldots}^{\infty} \frac{4}{n\pi}\left(\frac{\dot{w}_0}{\omega_n}\right)\sin\frac{n\pi x}{l}\sin\omega_n t \quad \text{for} \quad t \geq 0$$

### 7.3.3 Forced Vibrations

For forced vibrations, we solve Equation 7.11 with the external forcing term $p(x, t) \neq 0$. The solution of this equation consists of the complementary (homogeneous) solution with $p(x, t) = 0$ and the particular (nonhomogeneous) solution.

One approach to solving this problem is using the natural mode expansion method, where it is assumed that

$$p(x,t) = \sum f_n(t)X_n(x) \tag{7.38}$$

As above $X_n$ is the $n$th natural mode and $f_n(t)$ is a function of time. Multiplying both sides of Equation 7.38 by $X_m$ and integrating from 0 to $l$ yields

$$\int_0^l p(x,t)X_m\,dx = \sum_n f_n(t)\int_0^l X_nX_m\,dx = f_m(t)\int_0^l X_m^2\,dx$$

or

$$f_n(t) = \frac{\int_0^l p(x,t)X_n\,dx}{\int_0^l X_n^2\,dx} \tag{7.39}$$

The solution for $w(x, t)$ is assumed as

$$w(x,t) = \sum_n \phi_n(t)X_n(x) \tag{7.40}$$

where $\phi(t)$ is an unknown function of time. Substituting Equations 7.38 and 7.40 into Equation 7.11, results in the equation of motion as

$$\sum_n \left( \frac{d^4X_n}{dx^4}\phi_n + \frac{1}{a^2}X_n\frac{d^2\phi_n}{dt^2} \right) = \frac{1}{EI}\sum_n f_n(t)X_n \tag{7.41}$$

Recalling that $(d^4X_n)/(dx^4) = \left(\omega_n^2/a^2\right)X_n$ allows us to write

$$\sum_n \left( \frac{\omega_n^2}{a^2}X_n\phi_n + \frac{1}{a^2}X_n\frac{d^2\phi_n}{dt^2} \right) = \frac{1}{EI}\sum_n f_n(t)X_n$$

or

$$\frac{1}{a^2}\frac{d^2\phi_n}{dt^2} + \frac{\omega_n^2}{a^2}\phi_n = \frac{1}{EI}f_n(t) \tag{7.42}$$

This is also written in the form

$$\ddot{\phi}_n + \omega_n^2\phi_n = \frac{a^2}{EI}f_n(t) \tag{7.43}$$

Recalling that we defined $a^2 = EI/\rho A$, this can also be written as

$$\ddot{\phi}_n + \omega_n^2 \phi_n = \frac{1}{\rho A} f_n(t)$$

The solution for $\phi_n$ is the same as that obtained for a single-degree-of-freedom system. Namely

$$\phi_n(t) = \frac{1}{\omega_n}\left(\frac{1}{\rho A}\right)\int_0^t f_n(\tau)\sin\omega_n(t-\tau)d\tau \tag{7.44}$$

An alternate approach to solving Equation 7.11 begins with using $a^2 = EI/\rho A$ to rewrite the equation of motion as

$$\frac{\partial^4 w}{\partial x^4} + \frac{1}{a^2}\frac{\partial^2 w}{\partial t^2} = \frac{1}{EI}p(x,t) \tag{7.45}$$

We then assume a displacement $w(x, t) = \sum_n \phi_n(t) X_n(x)$ and substitute it into Equation 7.45 giving

$$\sum_n X_n'''' \phi_n + \frac{1}{a^2}\sum_n X_n \ddot{\phi}_n = \frac{1}{EI}p(x,t) \tag{7.46}$$

Noting that $X_n'''' = (\partial^4 X_n/\partial x^4) = (\omega_n^2/a^2) X_n$ allows us to write Equation 7.46 as

$$\sum_n \left(\omega_n^2 X_n \phi_n + X_n \ddot{\phi}\right) = \frac{a^2}{EI}p(x,t)$$

Multiplying both sides by $X_m$ and integrating from 0 to $l$ yields

$$\omega_n^2 \phi_n \int_0^l X_n^2\,dx + \ddot{\phi}_n \int_0^l X_n^2\,dx = \frac{a^2}{EI}\int_0^l p(x,t) X_n\,dx$$

or

$$\left(\ddot{\phi}_n + \omega_n^2 \phi_n\right)\int_0^l X_n^2\,dx = \frac{a^2}{EI}\int_0^l p(x,t) X_n\,dx \tag{7.47}$$

Equation 7.47 can be written in the following form:

$$\ddot{\phi}_n + \omega_n^2 \phi_n = \frac{a^2}{EI} f_n(t) \tag{7.48}$$

where

$$f_n(t) = \frac{\int_0^l p(x,t)X_n\,dx}{\int_0^l X_n^2\,dx} \tag{7.49}$$

We note that Equation 7.49 is identical to Equation 7.43.

**EXAMPLE 7.4**

Consider the simply supported beam in Figure 7.5. A concentrated force as a function of time, $P(t)$, is applied at the location $x = c$. Determine a general expression for the particular solution and a representative form of the general solution assuming the initial conditions are $w(x, 0) = w_0(x)$ and $\dot{w}(x,0) = \dot{w}_0(x)$.

**Solution**

Knowing that the beam is simply supported with boundary conditions $w(x, 0) = 0$, $w(l, 0) = 0$ and initial conditions $\dot{w}(x,0) = \dot{w}_0$, we have from Example 7.2 that

$$X_n = \sin\frac{n\pi x}{l} \quad \text{and} \quad \omega_n = a\left(\frac{n\pi}{l}\right)^2 \quad \text{for} \quad n = 1,2,3,\ldots$$

Recalling from Chapter 5 that concentrated loads can be defined using a Dirac delta function, we can assume the loading as

$$p(x,t) = P(t)\delta(x-c)$$

and

$$\int_0^l p(x,t)dx = P(t)\int_0^l \delta(x-c)dx = P(t)$$

We have that

$$p(x,t) = \sum_n f_n X_n$$

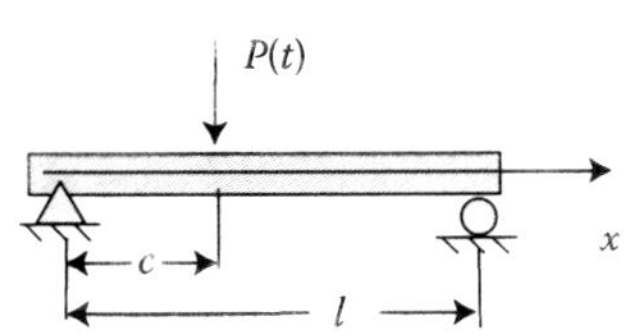


FIGURE 7.5
Simply supported beam with an applied concentrated force.

hence, from Equation 7.39 or 7.49 we can form the following:

$$f_n = \frac{\int_0^l p(x,t)X_n dx}{\int_0^l X_n^2 dx} = \frac{P(t)\int_0^l \delta(x-c)\sin(n\pi x/l)dx}{\int_0^l \sin^2(n\pi x/l)dx}$$

$$= \frac{2P(t)}{l}\sin\frac{n\pi c}{l}$$

Using Equation 7.44, we establish the solution for $\phi_n$ to be

$$\phi_n(t) = \left(\frac{1}{\omega_n l}\right)\left(\frac{2}{\rho A}\right)\sin\frac{n\pi c}{l}\int_0^t P(\tau)\sin\omega_n(t-\tau)d\tau$$

The deflection for $w(x, t)$ is assumed as

$$w(x,t) = \sum_n \phi_n(t)X_n(x)$$

Making the simple assumption that the forcing term is $P(t) = P_0 \sin \Omega t$, the expression for $\phi_n(t)$ becomes

$$\phi_n(t) = \left(\frac{1}{\omega_n l}\right)\left(\frac{2}{\rho A}\right)P_0\sin\frac{n\pi c}{l}\int_0^t \sin\Omega\tau\,\sin\omega_n(t-\tau)d\tau$$

Integrating we find

$$\phi_n(t) = \frac{2}{\rho A}P_0\sin\frac{n\pi c}{l}\left[\frac{1}{l\left(\omega_n^2-\Omega^2\right)}\sin\Omega t - \frac{\Omega}{\omega_n l\left(\omega_n^2-\Omega^2\right)}\sin\omega_n t\right]$$

With this evaluated, we determine the particular solution to be

$$w(x,t) = w_p(x,t) = \frac{2P_0}{\rho A}\left(\sum_{n=1}^{\infty}\frac{\sin(n\pi c/l)}{l\left(\omega_n^2-\Omega^2\right)}\sin\frac{n\pi x}{l}\right)\sin\Omega t - \frac{2\Omega P_0}{\rho A}\sum_{n=1}^{\infty}\frac{\sin(n\pi c/l)}{l\omega_n\left(\omega_n^2-\Omega^2\right)}\sin\frac{n\pi x}{l}\sin\omega_n t$$

This is the particular solution. It has the properties that $w(x,0) = \dot{w}(x,0) = 0$. As the initial conditions were $w(x, 0) = w_0(x)$ and $\dot{w}(x,0) = \dot{w}_0(x)$, the form of the general solution is

$$w(x,t) = w_p(x,t) + \sum_n X_n(A_n\cos\omega_n t + B_n\sin\omega_n t)$$

For the general solution given in Example 7.4, we note that the constants $A_n$ and $B_n$ are determined in the usual manner, however, care must be taken. In the general solution given above, $w_p(x, t)$ may not satisfy the homogeneous initial conditions. In that case, the determination of $A_n$ and $B_n$ must be modified. We could, for example, say that $w(x, 0) = w_0(x)$, where $w_0(x)$ is given. Then,

$$w_0(x) = w_p(x,0) + \sum_n X_n A_n$$

where $w_p(x, 0)$ is known as $\sum_n X_n \phi_n(0)$ and $\sum_n X_n A_n$ is to be determined. The particular solution for displacement at time $t = 0$ is

$$w_p(x,0) = \sum_n X_n(\phi_n(0) + A_n) = \sum_n K_n X_n$$

Therefore,

$$A_n = K_n - \phi_n(0)$$

In addition, we have that $\dot{w}(x,0) = \dot{w}_0(x)$ which leads to

$$\dot{w}_0(x) = \dot{w}_p(x,0) + \sum_n X_n B_n$$

and

$$\dot{w}_p(x,0) = \sum_n X_n(\dot{\phi}_n(0) + \omega_n B_n) = \sum_n L_n X_n$$

leading to

$$B_n = \frac{1}{\omega_n}(L_n - \dot{\phi}_n(0))$$

A forced vibration does not have to be in the form of an applied load. It could, for example, be in the form of a moving base support as in an earthquake. For such cases, an approach similar to that for an applied force is used. Consider the column of length $l$ shown in Figure 7.6, where the base is subjected to a time-dependent horizontal displacement $b(t)$ causing the deflection shown.

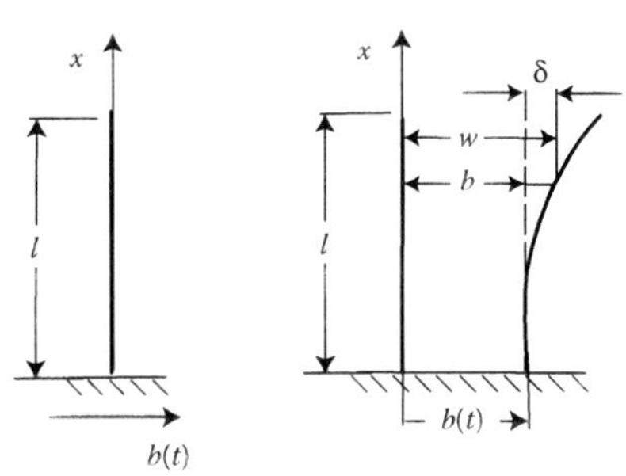


**FIGURE 7.6**
Vertical column subjected to a lateral base displacement.

As there is no directly applied load, we can treat this as a free vibration problem where the equation of motion is given by Equation 7.12

$$\frac{\partial^2 w}{\partial t^2} + a^2 \frac{\partial^4 w}{\partial x^4} = 0 \tag{7.12}$$

The boundary conditions at the base and at the free end of the column are given in Table 7.2 and are as follows:

$$\text{At } x = 0: w(0,t) = b(t), \quad \frac{\partial w(0,t)}{\partial x} = 0 \tag{7.50}$$

$$\text{At } x = l: \frac{\partial^2 w(l,t)}{\partial x^2} = 0, \quad \frac{\partial^3 w(l,t)}{\partial x^3} = 0 \tag{7.51}$$

The displacement of the column $w(x, t)$ is related to the displacement of the base $b(t)$ and the problem can be changed by introducing a dummy variable $\varsigma(x, t) = w(x, t) - b(t)$ (or $w(x, t) = \varsigma(x, t) + b(t)$). Using this relation, the equation of motion, Equation 7.12, becomes

$$\frac{\partial^4 \varsigma}{\partial x^4} + \frac{1}{a^2}\frac{\partial^2 \varsigma}{\partial t^2} = -\frac{1}{a^2}\frac{\partial^2 b}{\partial t^2} \tag{7.52}$$

This resembles the traditional equation of motion for forced vibration, that is, Equation 7.11. In addition to the equation of motion changing, the boundary conditions change to

$$\text{At } x = 0: \varsigma(0,t) = 0, \quad \frac{\partial \varsigma(0,t)}{\partial x} = 0 \tag{7.53}$$

$$\text{At } x = l: \frac{\partial^2 \varsigma(l,t)}{\partial x^2} = 0, \quad \frac{\partial^3 \varsigma(l,t)}{\partial x^3} = 0 \tag{7.54}$$

As previously established, the solution for $\varsigma(x, t)$ is assumed to be

$$\varsigma(x,t) = \sum_n \phi_n(t) X_n(x) \tag{7.55}$$

where $X_n(x)$ is the $n$th natural mode of a cantilever beam and $\phi_n(t)$ is an unknown function. The right-hand side of Equation 7.52 requires an expression for the second partial derivative of the base displacement with respect to time. As the right-hand side of Equation 7.52 represents $p(x, t)$ for the traditional forced vibration problem, we expand $\ddot{b}(t)$ in the series

$$\ddot{b}(t) = \sum_n f_n(t) X_n(x) \tag{7.56}$$

Following previously established procedures, we have

$$f_n = \frac{\int_0^l \ddot{b}(t) X_n dx}{\int_0^l X_n^2 dx} = \frac{\ddot{b}(t) \int_0^l X_n dx}{\int_0^l X_n^2 dx} = \ddot{b}(t) C_n \tag{7.57}$$

Substituting into the revised equation of motion, Equation 7.52, results in

$$\ddot{\phi}_n + \omega_n^2 \phi_n = -\ddot{b}(t) C_n \tag{7.58}$$

This is essentially Equation 7.43, with exception that $(a^2/EI)f_n(t)$ in Equation 7.43 is replaced by $-\ddot{b}(t)C_n$. The solution for $\phi_n$ proceeds as before and the end result is that we can express the displacement as

$$w(x,t) = b(t) + \sum_n \phi_n(t) X_n(x) \tag{7.59}$$

### 7.3.4 Application of the Laplace Transformations

The Laplace transformation (Thompson and Dahleh 1998) can be applied to Equation 7.43 in order to obtain the solution for $\phi_n(t)$. We can also apply the Laplace transformation to the equation of motion (Equation 7.11) shown below:

$$\frac{\partial^4 w}{\partial x^4} + \frac{A\rho}{EI} \frac{\partial^2 w}{\partial t^2} = \frac{1}{EI} p(x,t)$$

The transformed equation of motion becomes

$$\frac{\partial^4 \bar{w}}{\partial x^4} + \frac{A\rho}{EI} s^2 \bar{w} + \frac{A\rho}{EI}[-s w_0(x) - \dot{w}_0(x)] = \frac{1}{EI} \bar{p}(x,s) \tag{7.60}$$

where

$$\bar{w}(x,s) = \mathscr{L}\{w(x,t)\} \quad \text{and} \quad \bar{p}(x,s) = \mathscr{L}\{p(x,t)\}$$

In solving for $\bar{w}(x,s)$, we assume the following:

$$\bar{p}(x,s) = \sum_n \bar{f}_n(s) X_n(x) \tag{7.61}$$

$$w_0(x) = \sum_n K_n X_n(x) \tag{7.62}$$

$$\dot{w}_0(x) = \sum_n L_n X_n(x) \tag{7.63}$$

$$\bar{w}(x,s) = \sum_n \bar{\phi}_n(s)X_n(x) \tag{7.64}$$

Substituting Equations 7.61–7.64 into Equation 7.60 and using $X_n'''' = (\omega_n/a)^2 X_n$ yields

$$\left(\frac{\omega_n}{a}\right)^2 \bar{\phi}_n(s) + \frac{A\rho}{EI} s^2 \bar{\phi}_n(s) + \frac{A\rho}{EI}[-sK_n - L_n] = \frac{1}{EI}\bar{f}_n(s)$$

which is written in the form

$$\left[\left(\frac{\omega_n}{a}\right)^2 + \frac{A\rho}{EI}s^2\right]\bar{\phi}_n(s) + \frac{A\rho}{EI}[-sK_n - L_n] = \frac{1}{EI}\bar{f}_n(s) \tag{7.65}$$

In order to solve for the inverse Laplace transform, $\phi_n(t) = \mathcal{L}^{-1}\{\bar{\phi}_n(s)\}$, the result must be the same as through the standard application of the natural mode expansion. Here, the initial condition and the loading are taken into account.

For a linear viscoelastic beam for which $E \rightarrow \bar{E}(s)$, the equation of motion, Equation (7.60), with $w_0(x) = \dot{w}_0(x) = 0$ becomes

$$\frac{\partial^4 \bar{w}}{\partial x^4} + \frac{A\rho}{\bar{E}(s)I} s^2 \bar{w} = \frac{1}{\bar{E}(s)I}\bar{p}(x,s) \tag{7.66}$$

The equation $X_n'''' = (\omega_n/a)^2 X_n$ remains unchanged which means that we use the modes of the elastic beam, that is $a = \sqrt{EI/\rho A}$. We assume that

$$\bar{p}(x,s) = \sum_n \bar{f}_n(s)X_n(x) \quad \text{and} \quad \bar{w}(x,s) = \sum_n \bar{\phi}_n(s)X_n(x)$$

and substituting into the equation of motion yields

$$\left(\frac{\omega_n}{a}\right)^2 \bar{\phi}_n(s) + \frac{A\rho}{\bar{E}(s)I} s^2 \bar{\phi}_n(s) = \frac{1}{\bar{E}(s)I}\bar{f}_n(s) \tag{7.67}$$

and

$$\phi_n(t) = \mathcal{L}^{-1}\{\bar{\phi}_n(s)\}$$

### 7.3.5 Frequency Response Function

The applied loads typically considered for a beam are as follows: uniform loads, linearly distributed loads, and concentrated loads as illustrated in Figure 7.7. Consider a loading condition defined by equation

$$p(x,t) = q(x)f(t) \tag{7.68}$$

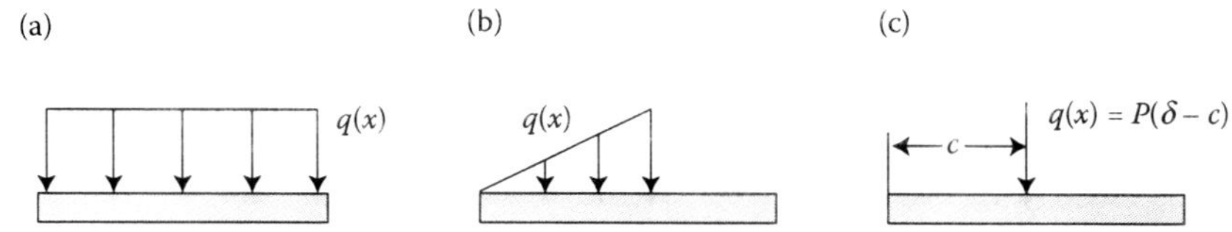


**FIGURE 7.7**
Typical beam loads: (a) uniform, (b) linear, and (c) concentrated.

where $q(x)$ defines the shape of the loading and $f(x)$ defines the time dependence of loading. In order to determine the frequency response function of the loading, we begin by assuming $f(t) = e^{i\omega t}$ so that

$$p(x,t) = q(x)e^{i\omega t} \tag{7.69}$$

$$w(x,t) = W^*(x,i\omega)e^{i\omega t} \tag{7.70}$$

Substituting Equations 7.69 and 7.70 into the beam equation of motion Equation 7.11 results in

$$\frac{\partial^4 W^*}{\partial x^4} - \frac{\omega^2}{a^2}\frac{\partial^2 W^*}{\partial t^2} = \frac{1}{EI}q(x) \tag{7.71}$$

Equation 7.71 contains material properties $E$ and $a$, which represent traditional linear elastic materials. For materials with damping, we replace $E$ by $E^* = E' + iE''$ and $a \to a^* = \sqrt{(E^*I)/(A\rho)}$. In order to solve this equation, we assume the normal mode expansion

$$q(x) = \sum_n q_n X_n \tag{7.72}$$

where

$$q_n = \frac{\int_0^l q(x)X_n dx}{\int_0^l X_n^2 dx} \tag{7.73}$$

and

$$W^*(x,i\omega) = \sum_n W_n^*(i\omega)X_n \tag{7.74}$$

We note that $W_n^*(i\omega)$ is an unknown function. Substituting Equations 7.72 and 7.74 into Equation 7.71 yields

$$W_n^* X_n'''' - \frac{\omega^2}{a^{*2}} W_n^* X_n = \frac{1}{E^* I} q_n X_n \tag{7.75}$$

For the natural modes, we have $X_n'''' = (\omega_n/a)^2 X_n$, where we have used the elastic material modes for which $a$ is real and equal to $\sqrt{(EI)/(A\rho)}$. Thus,

$$\left[\frac{\omega_n^2}{a^2} - \frac{\omega^2}{a^{*2}}\right] W_n^* = \frac{1}{E^* I} q_n$$

or

$$W_n^* = \left(\frac{1}{\left(\omega_n^2/a^2\right) - \left(\omega^2/a^{*2}\right)}\right)\left(\frac{q_n}{E^* I}\right)$$

The final result for $W_n^*$ can be written as

$$W_n^*(x, i\omega) = \frac{1}{E^* I} \sum_{n=1}^{\infty} \frac{q_n}{\left(\omega_n^2/a^2\right) - \left(\omega^2/a^{*2}\right)} X_n(x) \tag{7.76}$$

The damping causes the magnitude of $W_n^*(x, i\omega)$ to diminish with increasing frequency. At a constant $x$ location along the span of the beam, this decrease in magnitude would resemble that illustrated in Figure 7.8.

Applications of the frequency response function are related to the type of applied load. For example, assume a harmonic loading condition with

$$p(x,t) = q(x)\cos \omega t$$

The solution for this is $w(x, t) = \text{Re}[W^*(x, i\omega)e^{i\omega t}]$. Similarly, if the loading is

$$p(x,t) = q(x)\sin \omega t$$

the solution is $w(x, t) = \text{Im}[W^*(x, i\omega)e^{i\omega t}]$. Suppose $f(t)$ is a stationary random function and its spectral density is $S_f(\omega)$. The spectral density of $w$ (output) is given by

$$S_w(x, \omega) = |W^*(x, i\omega)|^2 S_f(\omega) \tag{7.77}$$

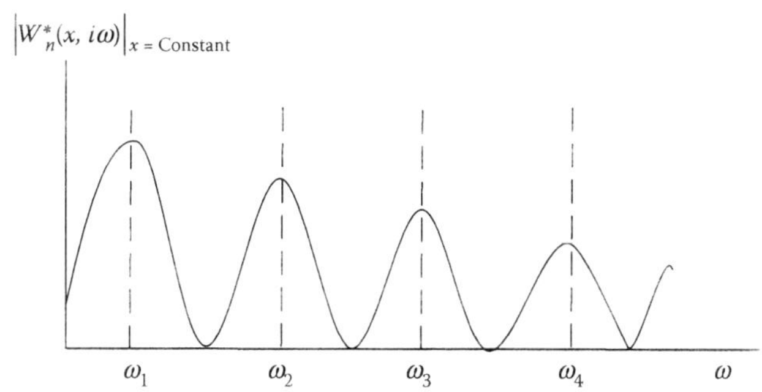


**FIGURE 7.8**
Variation of $W_n^*(x, i\omega)$ with increased frequency.

Assume that a bending moment $M(x, t)$ is applied at some location $x$ along the beam at time $t$. To determine the frequency response function of the bending moment, we would begin with

$$M(x,t) = -EI\frac{\partial^2 w}{\partial x^2}$$

Subsequently, we express this as

$$M^*(x,i\omega)e^{i\omega t} = -E^*I\frac{\partial^2 W^*(x,i\omega)}{\partial x^2}e^{i\omega t} \tag{7.78}$$

or

$$M^*(x,i\omega) = -E^*I\frac{\partial^2 W^*(x,i\omega)}{\partial x^2} \tag{7.79}$$

The spectral density of $M(x, t)$ is then given by $S_M(x, \omega) = |M^*(x, i\omega)|^2 S_f(\omega)$. Following the procedures of Hurty and Rubinstein (1964) we would write

$$S_W = \left|\sum_n W_n^*(i\omega)X_n\right|^2 S_f(\omega) \tag{7.80}$$

As an approximation to this expression, others would write

$$S_W = \left\{\sum_n \left|M_n^*(i\omega)X_n\right|^2\right\} S_f(\omega) \tag{7.81}$$

This, however, is an approximation and should not be used when the system has frequencies that are close together.

For cases in which the boundary is moving as illustrated in Figure 7.9, there are two methods for determining the frequency response function.

*Method (a):* One method involves a change of variables where we define $w(x, t) = b(t) + \xi(x, t)$. Note that this is the same approach taken for a moving boundary in Section 7.2.3. Following this procedure, we assume $b(t) = e^{i\omega t}$ and use normal mode expansion for $\xi(x, t)$.

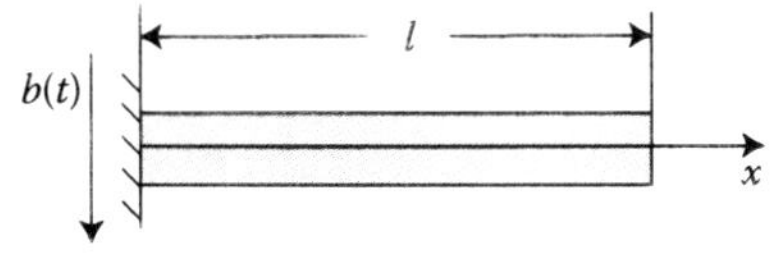


**FIGURE 7.9**
Cantilevered beam with a moving boundary.

*Method (b)*: The second method is a closed form solution. In the absence of a driving force, the equation of motion is

$$\frac{\partial^4 w}{\partial x^4} + \frac{1}{a^2}\frac{\partial^2 w}{\partial t^2} = 0 \tag{7.82}$$

The boundary conditions for the beam in Figure 7.9 are

$$\text{At } x = 0: w(0,t) = 0 \quad \text{and} \quad \frac{\partial w(0,t)}{\partial x} = 0 \tag{7.83}$$

$$\text{At } x = l: \frac{\partial^2 w(l,t)}{\partial x^2} = 0 \quad \text{and} \quad \frac{\partial^3 w(l,t)}{\partial x^3} = 0 \tag{7.84}$$

We assume that $b(t) = e^{i\omega t}$ and $w(x, t) = W^*(x, i\omega)e^{i\omega t}$. The equation for $W^*(x, i\omega)$ is then given by substituting the expression for $w(x, t)$ into Equation 7.82. Hence,

$$\frac{d^4 W^*(x,i\omega)}{dx^4} - \left(\frac{\omega^2}{a^2}\right)\frac{d^2 W^*(x,i\omega)}{dt^2} = 0 \tag{7.85}$$

where $\omega$ is a parameter, not the frequency. The general solution is given by

$$W^*(x,i\omega) = C_1 \sin\beta x + C_2 \cos\beta x + C_3 \sinh\beta x + C_4 \cosh\beta x \tag{7.86}$$

where $\beta = \sqrt{\omega/a} = \sqrt[4]{(\omega^2 A\rho)/EI}$. The constants $C_1$, $C_2$, $C_3$, and $C_4$ are determined from the boundary conditions, Equations 7.83 and 7.84, thus, we form

$$\text{At } x = 0: W^*(0,i\omega) = 1 \quad \text{and} \quad \frac{\partial W^*(0,i\omega)}{\partial x} = 0 \tag{7.87}$$

$$\text{At } x = l: \frac{\partial^2 W^*(l,i\omega)}{\partial x^2} = 0 \quad \text{and} \quad \frac{\partial^3 W^*(l,i\omega)}{\partial x^3} = 0 \tag{7.88}$$

Therefore, we have four linear equations for the constants $C_1$, $C_2$, $C_3$, and $C_4$. When $\omega \neq \omega_n$, there exist unique solutions for $C_1$, $C_2$, $C_3$, and $C_4$. Similarly, closed form solutions are possible for frequency response functions corresponding to inputs that are end loads such as an applied force or moment as shown in Figure 7.10. Closed form solutions are only possible with end loadings, homogeneous differential equation, and nonhomogeneous boundary conditions.

Method (b) for a material with damping has the equation for $W^*(x, i\omega)$ as

$$\frac{\partial^4 W^*(x,i\omega)}{\partial x^4} - \frac{\omega^2}{a^{*2}} W^*(x,i\omega) = 0 \quad \text{where} \quad a \to a^* \quad \text{and} \quad E \to E^*$$

The general solution for $W^*(x, i\omega)$ is the same as before, except $k \to k^* = \sqrt{\omega/a^*}$. The four equations $C_1$, $C_2$, $C_3$, and $C_4$ have complex coefficients. Therefore, $C_1$, $C_2$, $C_3$, and $C_4$ in general must be complex.

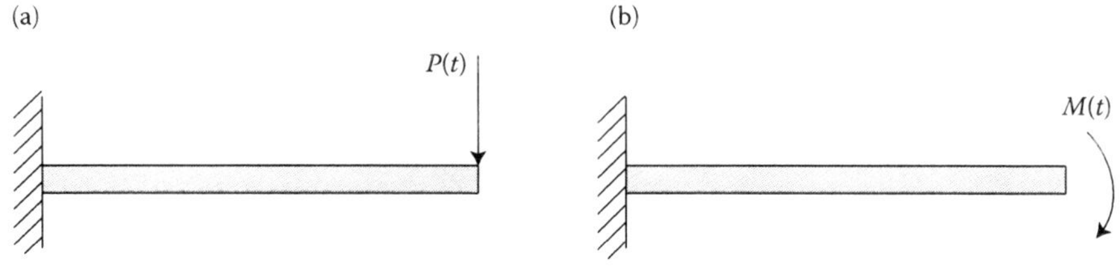


**FIGURE 7.10**
End load inputs: (a) end load $P(t)$ and (b) end moment load $M(t)$.

### 7.3.6 Effect of Axial Force

Assuming an axial force $T$ is applied to the beam shown in Figure 7.11a. We can make the same small displacement assumptions applicable to classical beam theory and investigate the effects of the applied load $T$. We start with the model of a representative volume element shown in Figure 7.11b.

Applying Newton's second law to the representative volume element in Figure 7.11b, we obtain

$$pdx + \frac{\partial Q}{\partial x}dx - T\frac{\partial^2 w}{\partial x^2}dx = A\rho\frac{\partial^2 w}{\partial t^2}dx$$

As previously determined, $M = -EI(\partial^2 w/\partial x^2)$ and $Q = -EI(\partial^3 w/\partial x^3)$, resulting in the governing equation

$$A\rho\frac{\partial^2 w}{\partial t^2} + T\frac{\partial^2 w}{\partial x^2} + EI\frac{\partial^4 w}{\partial x^4} = p(x,t) \tag{7.89}$$

The free vibration case results when we assume $p(x, t) = 0$. Assume

$$w(x,t) = X(x)(A\cos\omega t + B\sin\omega t) \tag{7.90}$$

where $\omega$ and $X(x)$ are the natural frequency and normal mode function, respectively. Substituting Equation 7.90 into Equation 7.89 yields

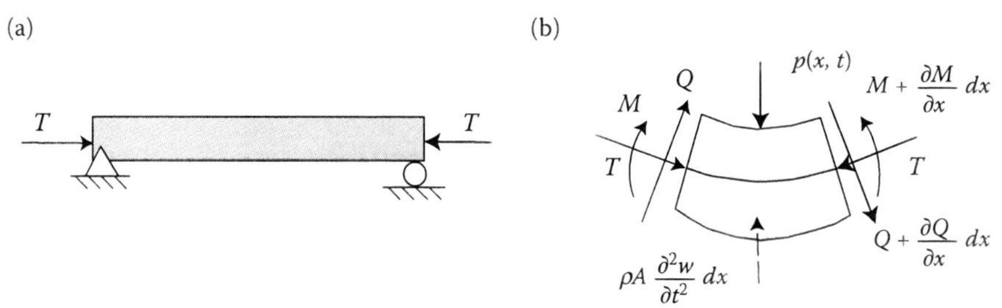


**FIGURE 7.11**
Simply supported beam (a) with an axial compressive load applied and (b) resulting representative volume element.

$$EI\frac{d^4X}{dx^4}+T\frac{d^2w}{dx^2}-A\rho\omega^2\frac{\partial^2 w}{\partial t^2}-0 \tag{7.91}$$

The solution of Equation 7.91 satisfying the prescribed end conditions yields the normal mode functions $X(x)$. $X(x)$ will satisfy the case of a simply supported beam using boundary conditions previously defined by taking

$$X_n=\sin\frac{n\pi x}{l} \quad n=1,2,3,\ldots \tag{7.92}$$

Substituting Equation 7.92 into 7.91 gives the $n$th frequency of vibration as

$$\omega=\omega_n=\frac{an^2\pi^2}{l^2}\sqrt{1-\frac{Tl^2}{n^2\pi^2EI}} \tag{7.93}$$

If $0<T<T_{cr}=(EI\pi^2/l^2)$, then $\omega>0$ and real for every $n$. Thus, the free vibration case of a simply supported beam is just harmonic motion ($\cos\omega_n t$, $\sin\omega_n t$).

If $T>T_{cr}=(EI\pi^2/l^2)$, then some frequencies ($\omega_1,\ldots$) become imaginary. In this case, $\cos\omega_1 t$, $\sin\omega_1 t$ increases exponentially and $w\to\infty$. These results lead to the dynamic criterion of stability. A system is stable if all natural frequencies are real.

When the axial force is tensile, the argument presented is the same as before and only the sign of $T$ changes. Equation 7.93 will change to the following:

$$\omega_n=\frac{an^2\pi^2}{l^2}\sqrt{1+\frac{Tl^2}{n^2\pi^2EI}} \tag{7.94}$$

The conclusion from this equation is that an axial tensile force $T$ will increase the frequency $\omega_n$. If the moment of inertia $I$ is very small, for example if $Tl^2/(n^2\pi^2EI)\gg 1$, then

$$\omega_n\approx\frac{an^2\pi^2}{l^2}\sqrt{\frac{Tl^2}{n^2\pi^2EI}}\approx\frac{n\pi}{l}\sqrt{\frac{T}{A\rho}} \tag{7.95}$$

This represents the $n$th natural frequency of a string with a tensile force $T$ applied.

Substituting Equation 7.91 into Equation 7.90, we obtain the natural mode of vibration $n$ sinusoidal half waves and summing all modes yields the general solution of a simply supported beam with an axial force as

$$w(x,t)=\sum_{n=1}^{\infty}\sin\frac{n\pi x}{l}(A_n\cos\omega_n t+B_n\sin\omega_n t) \tag{7.96}$$

If the axial force is compressive, then $\omega_n$ is given by Equation 7.93 and if there is tension, then $\omega_n$ is given by Equation 7.94.

## 7.4 Timoshenko Beam Theory (Effects of Shear Deformation and Rotary Inertia)

The Euler–Bernoulli theory or as stated the *classical or simple beam* theory presented earlier was based on two major assumptions: first the effects of rotary had been neglected and secondly, that straight lines normal to the midplane of the beam before deformation remain straight and normal to the midplane of the beam after deformation. These assumptions led to a simple beam equation for determining the deflection of the beam, which did not include any shear deformation. This leads to a somewhat inconsistent beam theory. A more consistent beam theory is the *Timoshenko beam theory* or *thick beam theory*, (Weaver et al. 1990) which was developed early in the twentieth century. This theory relaxes the aforementioned normality condition and takes into account shear deformation and rotational inertia effects. This leads to the assumption of having a constant shear strain through the thickness, which then requires a uniform shear stress at every point of the beam. This is not true in reality. To compensate for this effect, Timoshenko proposed a *shear correction factor* for the shear stress, which depends on the shape of the cross section. These effects are typically taken into account when considering deep beams (large height to length ratios), high frequencies, sandwich beams, and tall frames. The influence of the shear correction factor has been studied by various researches, including Rosinger and Ritchi (1977) and Cowper (1966). The resulting equation is fourth order but, unlike the Euler–Bernoulli beam theory, there is also a second-order partial derivative present. Taking into account the additional types of deformation essentially lowers the stiffness of the beam, resulting in larger deflection under a static load. A simple depiction of the difference in deformations between the Timoshenko and Euler–Bernoulli beams is illustrated in Figure 7.12.

The assumptions made when including rotary inertia and shear deformation are consistent with those associated with a Timoshenko beam: (a) elastic beam, (b) small deflections, and (c) plane sections remain plane.

### 7.4.1 Equations of Motion

Consider the Timoshenko beam model shown in Figure 7.13, which undergoes shear deformation in addition to pure bending. The slope of the center line $\partial w/\partial x$ is affected by both the bending moment and shear force. The bending moment rotates the face of the cross section through an angle $\psi$, and the shear force turns the center line to assume the

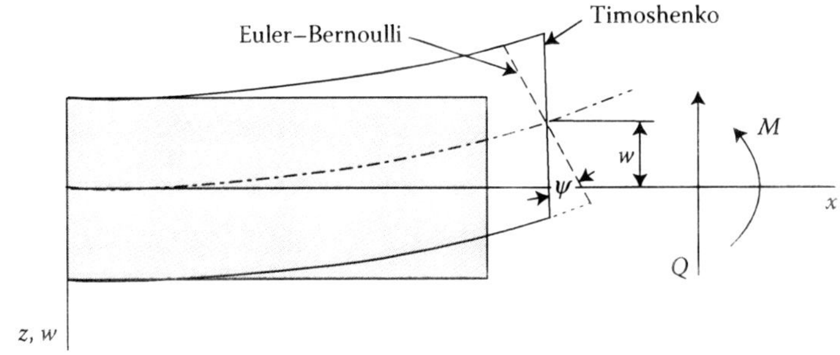


**FIGURE 7.12**
Comparison of assumed deformation for Euler–Bernoulli and Timoshenko beam theories.

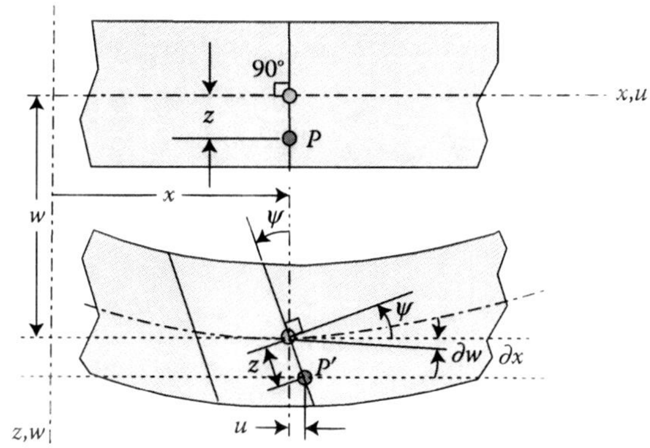


**FIGURE 7.13**
Timoshenko beam deformations.

slope $\partial w/\partial x$, where the angle of the face of the beam remains unchanged as depicted in Figure 7.13.

For small deformations, the displacements of point $P$ are

$$u(x,z,t) = z\psi(x,t) \tag{7.97}$$

$$w(x,z,t) = w(x,t) \tag{7.98}$$

In view of the displacements given in Equations 7.97 and 7.98, the strain–displacement relations are

$$\varepsilon_{xx} = \frac{\partial u}{\partial x} = \frac{\partial (z\psi)}{\partial x} = z\frac{\partial \psi}{\partial x} \tag{7.99}$$

$$\gamma_{xz} = \frac{\partial u}{\partial z} + \frac{\partial w}{\partial x} = \psi + \frac{\partial w}{\partial x} \tag{7.100}$$

Using the one-dimensional Hooke's law, the equations for the normal and shear stress become in terms of displacements

$$\sigma_{xx} = E\varepsilon_{xx} = Ez\frac{\partial \psi}{\partial x} \quad \sigma_{xz} = G\gamma_{xz} = Ez\frac{\partial \psi}{\partial x} \tag{7.101}$$

The bending moment and shear force are determined in terms of displacements by integrating the normal and shear stresses, Equation 7.101, over the area of the cross section. Thus, we have

$$M = \int_A z\sigma_{xx} dA = \int_A Ez^2 \frac{\partial \psi}{\partial x} dA = EI\frac{\partial \psi}{\partial x} \tag{7.102}$$

$$Q = k_s \int_A \sigma_{xz} dA = k_s GA\left(\psi + \frac{\partial w}{\partial x}\right) \tag{7.103}$$

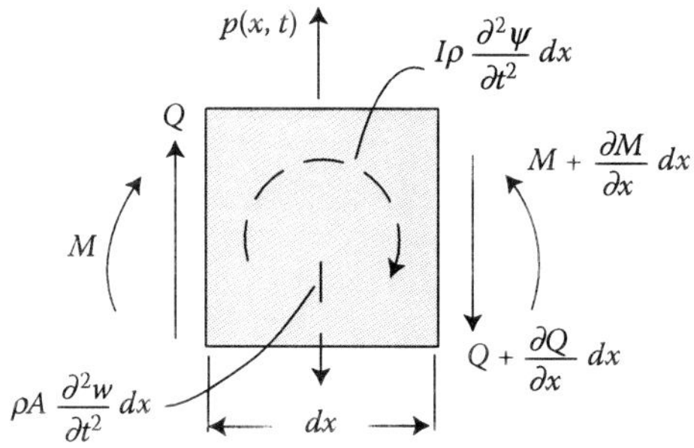


**FIGURE 7.14**
Timoshenko beam volume element.

where $k_s$ is a shear correction factor associated with the beams cross-sectional geometry, which is associated with the shear strain to compensate for the error caused by assuming a constant transverse shear stress distribution through the beam thickness.

Considering a representative volume element of the beam, shown in Figure 7.14, and applying Newton's second law of motion, we generate two equations. One for translation in the lateral direction and another for the rotational motion about point $A$, which yields

$$\rho A\frac{\partial^2 w}{\partial t^2}dx = \frac{\partial Q}{\partial x}dx + pdx \tag{7.104}$$

$$I\rho\frac{\partial^2 \psi}{\partial t^2}dx = \frac{\partial M}{\partial x}dx - Qdx \tag{7.105}$$

Substituting the expressions for $M$ and $Q$ into the equations of motion, Equations 7.104 and 7.105 result in

$$\rho A\frac{\partial^2 w}{\partial t^2} - k_s GA\left(\frac{\partial^2 w}{\partial x^2} + \frac{\partial \psi}{\partial x}\right) = p(x,t) \tag{7.106}$$

$$\rho I\frac{\partial^2 \psi}{\partial t^2} = EI\frac{\partial^2 \psi}{\partial x^2} - k_s AG\left(\frac{\partial w}{\partial x} + \psi\right) \tag{7.107}$$

The boundary conditions are slightly different than before as we have introduced new parameters into the model. The biggest difference in the boundary conditions is for the fixed end. As there is shear deformation, the slope at the fixed end does not have to be zero because shear deformations may be large. The boundary conditions are as follows:

$$\text{Simple supported ends}: w = 0 \quad \text{and} \quad M = 0 \rightarrow \frac{\partial \psi}{\partial x} = 0 \tag{7.108}$$

$$\text{Free ends}: M = 0 \rightarrow \frac{\partial \psi}{\partial x} = 0 \quad \text{and} \quad Q = 0 \rightarrow \frac{\partial w}{\partial x} - \psi = 0 \tag{7.109}$$

$$\text{Fixed ends}: w = 0 \quad \text{and} \quad \psi = 0 \tag{7.110}$$

### 7.4.2 Free Vibrations

For the free vibration problem, we set $p(x, t) = 0$. In addition, we assume

$$\begin{aligned} w &= W(x)e^{i\omega t} \\ \psi &= \Psi(x)e^{i\omega t} \end{aligned} \tag{7.111}$$

Note that we could also have used $\cos \omega t$ or $\sin \omega t$. Substituting Equation 7.111 into Equations 7.106 and 7.107 with $k_s = 1$ yields

$$AGW''(x) + A\rho\omega^2 W(x) + AG\Psi'(x) = 0 \tag{7.112}$$

$$AGW'(x) - EI\Psi''(x) + AG\Psi(x) - I\rho\omega^2\Psi = 0 \tag{7.113}$$

Next assume that $W(x)$ and $\Psi(x)$ can be expressed as

$$W(x) = Ce^{\lambda x} \tag{7.114}$$

$$\Psi(x) = \hat{C}e^{\lambda x} \tag{7.115}$$

Substituting Equation 7.111 into Equations 7.112 and 7.113 yields

$$A\rho\omega^2 C + AG\lambda^2 C + AG\lambda\hat{C} = (A\rho\omega^2 + AG\lambda^2)C + AG\lambda\hat{C} = 0 \tag{7.116}$$

$$AG\lambda C - EI\lambda^2\hat{C} + AG\hat{C} - I\rho\omega^2\hat{C} = AG\lambda C + (-I\rho\omega^2 + AG - EI\lambda^2)\hat{C} + = 0 \tag{7.117}$$

The unknowns are $C, \hat{C}$, and $\lambda$. We have two equations and three unknowns. An additional equation is that the determinant of the coefficients must be equal to zero. Thus,

$$\begin{vmatrix} A\rho\omega^2 - AG\lambda^2 & AG\lambda \\ AG\lambda & -I\rho\omega^2 - AG - EI\lambda^2 \end{vmatrix} = 0 \tag{7.118}$$

From the above-mentioned equations, we determine the roots for $\lambda$, which depend on $\omega$. These can be expressed as $\lambda_1 - \lambda_1(\omega)$, $\lambda_2 - \lambda_2(\omega)$, $\lambda_3 = \lambda_3(\omega)$, and $\lambda_4 = \lambda_4(\omega)$. For each $\lambda$, we find $C$ and $\hat{C}$ from two algebraic equations, Equations 7.116 and 7.117, or exactly speaking the ratio between $C$ and $\hat{C}$. Thus, $\hat{C} = \alpha C$, which yields

$$\text{For } \lambda_1: \hat{C}_1 = \alpha_1 C_1 = -\left(\frac{AG\lambda_1^2 + A\rho\omega^2}{AG\lambda_1}\right)C_1$$

$$\text{For } \lambda_2: \hat{C}_2 = \alpha_2 C_2 = -\left(\frac{AG\lambda_2^2 + A\rho\omega^2}{AG\lambda_2}\right)C_2$$

$$\text{For } \lambda_3: \hat{C}_3 = \alpha_3 C_3 = -\left(\frac{AG\lambda_3^2 + A\rho\omega^2}{AG\lambda_3}\right)C_3$$

$$\text{For } \lambda_4: \hat{C}_4 = \alpha_4 C_4 = -\left(\frac{AG\lambda_4^2 + A\rho\omega^2}{AG\lambda_4}\right)C_4$$

and

$$W(x) = C_1 e^{\lambda_1 x} + C_2 e^{\lambda_2 x} + C_3 e^{\lambda_3 x} + C_4 e^{\lambda_4 x} \tag{7.119}$$

$$\Psi(x) = \alpha_1 C_1 e^{\lambda_1 x} + \alpha_2 C_2 e^{\lambda_2 x} + \alpha_3 C_3 e^{\lambda_3 x} + \alpha_4 C_4 e^{\lambda_4 x} \tag{7.120}$$

Now we use the boundary conditions to determine $C_1$, $C_2$, $C_3$, $C_4$, and $\omega$. Consider, for example, a built in beam as illustrated in Figure 7.15.

For this beam, the boundary conditions are from Equation 7.110, $W(0) = W(l) = \Psi(0) = \Psi(l) = 0$. Using these boundary conditions yields

$$\begin{aligned}
C_1 + C_2 + C_3 + C_4 &= 0 \\
\alpha_1 C_1 + \alpha_2 C_2 + \alpha_3 C_3 + \alpha_4 C_4 &= 0 \\
C_1 e^{\lambda_1 l} + C_2 e^{\lambda_2 l} + C_3 e^{\lambda_3 l} + C_4 e^{\lambda_4 l} &= 0 \\
\alpha_1 C_1 e^{\lambda_1 l} + \alpha_2 C_2 e^{\lambda_2 l} + \alpha_3 C_3 e^{\lambda_3 l} + \alpha_4 C_4 e^{\lambda l} &= 0
\end{aligned}$$

These four equations can be expressed in the matrix form and the only nontrivial solution will result if the determinant vanishes, thus,

$$\begin{vmatrix}
1 & 1 & 1 & 1 \\
\alpha_1 & \alpha_2 & \alpha_3 & \alpha_4 \\
e^{\lambda_1 l} & e^{\lambda_2 l} & e^{\lambda_3 l} & e^{\lambda_4 l} \\
\alpha_1 e^{\lambda_1 l} & \alpha_2 e^{\lambda_2 l} & \alpha_3 e^{\lambda_3 l} & \alpha_4 e^{\lambda l}
\end{vmatrix} = 0$$

This is the frequency equation for $\omega_1, \omega_2, \ldots$, so for each $\omega$ we will have $C_1^{(i)}, C_2^{(i)}, C_3^{(i)}, C_4^{(i)}$. One of these is arbitrary and the rest are functions of it.

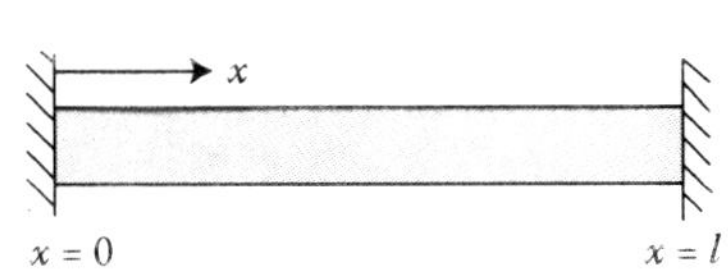


**FIGURE 7.15**
Built in beam.

For a simply supported beam, we assume

$$W(x) = C_m \sin \frac{m\pi x}{l}$$

$$\Psi(x) = \hat{C}_m \cos \frac{m\pi}{l}$$

The boundary conditions from Equation 7.108 are

$$\text{At } x = 0: W(0) = 0, \quad \frac{d\Psi(0)}{dx} = 0 \quad \text{and} \quad \text{At } x = l: W(l) = 0, \quad \frac{d\Psi(l)}{dx} = 0$$

Substituting the expressions for $W(x)$ and $\Psi(x)$ into Equations 7.112 and 7.113 results in

$$A\rho\omega^2 C_m - AG\left(\frac{m\pi}{l}\right)^2 C_m - AG\left(\frac{m\pi}{l}\right) C_m' = 0$$

$$-I\rho\omega^2 \hat{C}_m + AG\left(\frac{m\pi}{l}\right) C_m + AG\hat{C}_m + EI\left(\frac{m\pi}{l}\right)^2 \hat{C}_m = 0$$

For a nontrivial solution

$$\begin{vmatrix} AG\left(\frac{m\pi}{l}\right)^2 - A\rho\omega^2 & AG\left(\frac{m\pi}{l}\right) \\ AG\left(\frac{m\pi}{l}\right) & EI\left(\frac{m\pi}{l}\right)^2 + AG - I\rho\omega^2 \end{vmatrix} = 0$$

This is the frequency equation. For each $m$, we find $\omega_{mB}^2$ (for bending) and $\omega_{mS}^2$ (for shear) or $\omega_{mB}$ and $\omega_{mS}$. For $\omega_{mB}$, we obtain $C_{mB}$ and $\hat{C}_{mB}$ from the equations of motion, where one of them can be arbitrary. For $\omega_{mS}$, we obtain $C_{mS}$ and $\hat{C}_{mS}$, where again one of them can be arbitrary. Too see what we have determined, let us consider the $\omega_{mB}$ term. We have

$$W(x) = C_{mB} \sin \frac{m\pi x}{l}$$

$$\Psi(x) = \hat{C}_{mB} \cos \frac{m\pi x}{l}$$

Let us assume $m = 1$. If $\omega_{mB}$ is close to the $m$th natural bending frequency of a classical beam, then we obtain a deformation similar to that shown in Figure 7.16. The shear deformation is small as $\psi$ is close to $-\partial w/\partial x$.

**FIGURE 7.16**
Beam deformations for $\omega_{mB}$ close to $m$th natural bending frequency.

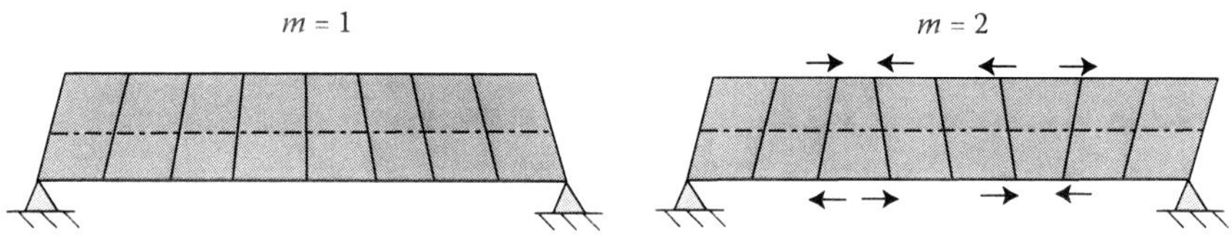


**FIGURE 7.17**
Beam deformations for $m = 1$ and $m = 2$ when shear deformations are large.

For the $\omega_{mS}$ term, we have

$$W(x) = C_{mS} \sin \frac{m\pi x}{l}$$

$$\Psi(x) = \hat{C}_{mS} \cos \frac{m\pi}{l}$$

When $\hat{C}_{mS} \gg C_{mS}$, the bending deformations are small and the shear deformations are large, as shown by the sketches in Figure 7.17 for $m = 1$ and $m = 2$.

### *7.4.2.1 Effect of Shear Deformation and Rotary Inertia*

As previously noted, one of the reasons for including rotary inertia is when considering deep beams. The implication being that cross-sectional properties $I$ and $A$ of the beam can provide a relative measure of the effect of rotary inertia. Equations 7.106 and 7.107 can be combined to yield a single equation by eliminating $\psi$ or $w$. As demonstrated by Rao (2011), eliminating $\psi$ yields

$$EI\frac{\partial^4 w}{\partial x^4} + \rho A\frac{\partial^2 w}{\partial t^2} - I\rho\left(1 + \frac{E}{k_s G}\right)\frac{\partial^4 w}{\partial x^2 \partial t^2} + \frac{I\rho^2}{k_s G}\frac{\partial^4 w}{\partial t^4} = p + \frac{I\rho}{k_s AG}\frac{\partial^2 p}{\partial t^2} - \frac{EI}{k_s AG}\frac{\partial^2 p}{\partial x^2} \tag{7.121}$$

For free vibrations, this reduces to

$$EI\frac{\partial^4 w}{\partial x^4} + \rho A\frac{\partial^2 w}{\partial t^2} - I\rho\left(1 + \frac{E}{k_s G}\right)\frac{\partial^4 w}{\partial x^2 \partial t^2} + \frac{I\rho^2}{k_s G}\frac{\partial^4 w}{\partial t^4} = 0 \tag{7.122}$$

The effects of shear deformation and rotary inertia can be studied independently. If we consider rotary inertia alone, the terms containing the shear correction factor $k_s$ in Equation 7.122 are neglected and we obtain

$$EI\frac{\partial^4 w}{\partial x^4} + \rho A\frac{\partial^2 w}{\partial t^2} - I\rho\frac{\partial^4 w}{\partial x^2 \partial t^2} = 0 \tag{7.123}$$

Similarly, if we consider only shear deformation, the terms originating from $\rho I(\partial^2\psi/\partial t^2)$ in Equation 7.107 are neglected and the equation of motion becomes

$$EI\frac{\partial^4 w}{\partial x^4} + \rho A\frac{\partial^2 w}{\partial t^2} - \frac{EI\rho}{k_s G}\frac{\partial^4 w}{\partial x^2 \partial t^2} = 0 \tag{7.124}$$

In order to explore the effects of shear deformation and rotary inertia, we define $\beta^2 = EI/\rho A$ and $r^2 = I/A$ where $r$ is the radius of gyration. In addition, let us assume a simply supported beam of length $l$ for which a solution is taken in the form

$$w(x,t) = C \sin \frac{n\pi x}{l} \cos \omega_n t \tag{7.125}$$

which satisfies the boundary conditions at $x = 0, l$. Using Equation 7.123, Equation 7.122 can be written as

$$\omega_n^4 \left( \frac{\rho r^2}{k_s G} \right) - \omega_n^2 \left( 1 + \frac{n^2 \pi^2 r^2}{l^2} + \frac{n^2 \pi^2 r^2}{l^2} \frac{E}{k_s G} \right) + \left( \frac{\beta^2 n^4 \pi^4}{l^4} \right) = 0 \tag{7.126}$$

The solution of this equation is a quadratic equation in $\omega_n^2$ resulting in two possible values for $\omega_n$. The resulting frequency equation for each condition (either rotary inertia or shear deformation), along with two forms of the equation of motion, one using the values $\beta^2$ and $r^2$, is given in Table 7.3.

Neglecting both shear deformation and rotary inertia results in the Euler–Bernoulli theory, for which the frequency equation is $\omega_n^2 = ((\beta^2 n^4 \pi^4)/l^4)$.

In order to explore the influence of shear deformation and rotary inertia, consider a beam of rectangular cross-section having a width $b$ and height $h$. The cross-sectional area is $A = bh$ and the second area moment of inertia $I = bh^3/12$, which result in $r^2 = h^2/12$. Comparing the frequency equation for the Euler–Bernoulli theory to that of the frequency equation when rotary inertia $\left(\omega_n^2\right)_{inert}$ and shear deformation $\left(\omega_n^2\right)_{shear}$ are considered independently, we obtain

$$\frac{\omega_n^2}{\left(\omega_n^2\right)_{inert}} = \left(1 + \frac{n^2 \pi^2 r^2}{l^2}\right) = \left(1 + \frac{n^2 \pi^2}{2} \left(\frac{h}{l}\right)^2\right)$$

$$\frac{\omega_n^2}{\left(\omega_n^2\right)_{shear}} = \left(1 + \frac{n^2 \pi^2 r^2}{l^2} \frac{E}{KG}\right) = \left(1 + \frac{n^2 \pi^2}{12} \left(\frac{h}{l}\right)^2 \frac{E}{KG}\right)$$

**TABLE 7.3**

Equations of Motion and Frequency Equation for Rotary Inertia and Shear Deformation Effects

| Term Considered | Equation of Motion | Frequency Equation |
|---|---|---|
| Rotary inertia | $EI \frac{\partial^4 w}{\partial x^4} + \rho A \frac{\partial^2 w}{\partial t^2} - I\rho \frac{\partial^4 w}{\partial x^2 \partial t^2} = 0$ | $\omega_n^2 = \frac{\beta^2 n^4 \pi^4}{l^4(1 + ((n^2\pi^2 r^2)/l^2))}$ |
| | $\beta^2 \frac{\partial^4 w}{\partial x^4} + \frac{\partial^2 w}{\partial t^2} - r^2 \frac{\partial^4 w}{\partial x^2 \partial t^2} = 0$ | |
| Shear deformation | $EI \frac{\partial^4 w}{\partial x^4} + \rho A \frac{\partial^2 w}{\partial t^2} - \frac{EI\rho}{k_s G} \frac{\partial^4 w}{\partial x^2 \partial t^2} = 0$ | $\omega_n^2 = \frac{\beta^2 n^4 \pi^4}{l^4(1 + ((n^2\pi^2 r^2)/l^2)(E/KG))}$ |
| | $\beta^2 \frac{\partial^4 w}{\partial x^4} + \frac{\partial^2 w}{\partial t^2} - \frac{Er^2}{k_s G} \frac{\partial^4 w}{\partial x^2 \partial t^2} = 0$ | |

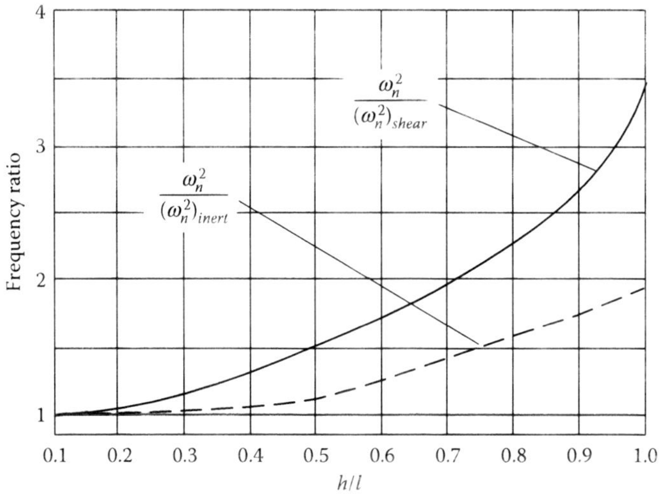


**FIGURE 7.18**
Ratio of Euler–Bernoulli to Timoshenko beam natural frequencies as a function of a simply supported beam height to length.

As the ratio of beam height to length increases, the effects of rotary inertia and shear deformation become more prominent. For $n = 1$ and a value of $E/KG$ consistent with a rectangular steel beam, the frequency ratios of the Euler–Bernoulli beam compared to those for the Timoshenko beam with rotary inertia and shear deformation considered independently are shown in Figure 7.18 as a function of $h/l$. As seen, the effect of both rotary inertia and shear deformation increases as the ratio of beam height to length increases. In other words, for short tall beams, one should include rotary inertia and shear deformation.

### 7.4.3 Forced Vibration

A closed form solution of Equations 7.38 and 7.39 or 7.41 is not available (Schmitz and Smith 2012) for forced vibrations. Any solution to the forced vibration problem consists of the complementary (homogeneous) solution with $p(x, t) = 0$ and the particular (nonhomogeneous) solution. One approach to solving this problem is using the natural mode expansion method as discussed in Section 7.2.3, where we begin by assuming $p(x, t) = \sum f_n(t)X_n(x)$. As in previous cases, $X_n$ is the $n$th natural mode and $f_n(t)$ is a function of time. Then, we assume $w(x, t) = \sum_n \phi_n(t)X_n(x)$. Substituting $p(x, t)$ and $w(x, t)$ into the equation(s) of motion follows the procedures previously established for forced vibration problems to define the frequencies, mode shapes, etc. Alternately, one could use finite elements to establish a solution to a specific problem.

## PROBLEMS

7.1 Each beam shown in Figure 7.19 has a length $L$, a uniform cross-sectional area $A$, and a constant $EI$ along its span. Determine the natural frequencies and modes for these beams.

7.2 Each beam shown in Figure 7.20 has a length $L$, a uniform cross-sectional area $A$, and a constant $EI$ along its span. Determine the particular solution for forced vibrations of the beams.

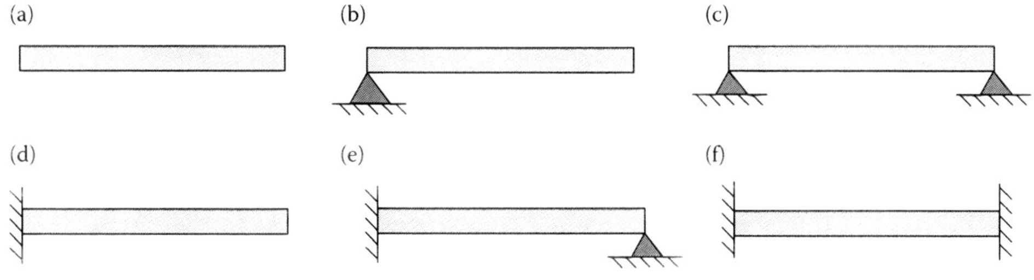


**FIGURE 7.19**
Unloaded beams with various end supports.

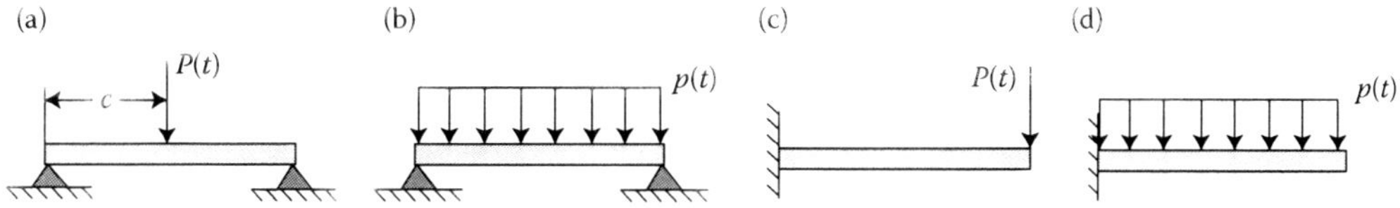


**FIGURE 7.20**
Beams with various loading conditions.

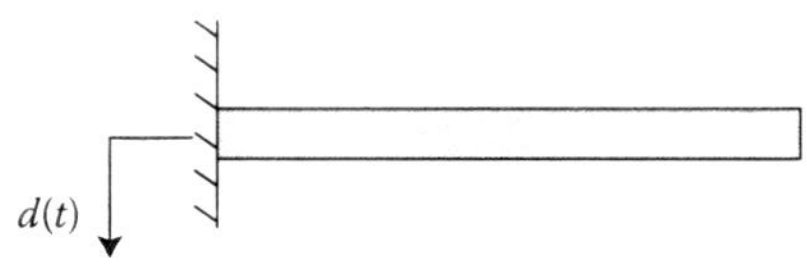


**FIGURE 7.21**
Cantilevered beam with time-dependent support motion.

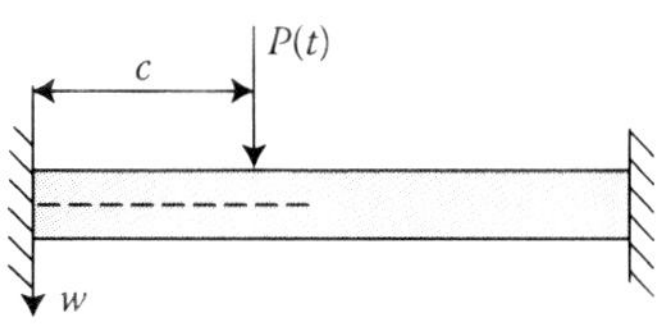


**FIGURE 7.22**
Fixed-fixed beam with a concentrated load.

7.3 Find the frequency response functions for the beams and loadings in Problem 7.2. Let us assume that the material is

a. An elastic material without damping.

b. An inelastic material with $E^* = E' + iE''$, where $E' =$ constant, $E'' =$ constant.

7.4 Find the frequency response function and the response to a unit input for the beams shown in Figure 7.21.

7.5 Describe the method of analysis for the loadings in Problems 7.2 and 7.4 defined as stationary and nonstationary random functions of time.

7.6 Find the displacement $w$ for the shear beam shown in Figure 7.22, knowing that $P(t) = P_0H(t)$. Use natural mode expansion.

7.7 Determine the deflection $w$ for the cantilevered shear beam shown in Figure 7.23, knowing that $p(x, t) = pH(t)$.

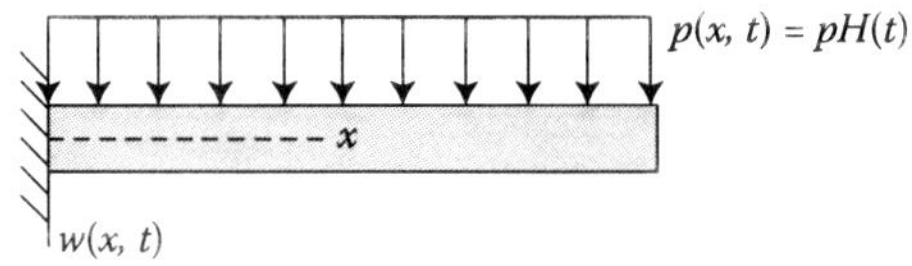


**FIGURE 7.23**
Cantilevered beam with a uniform load.

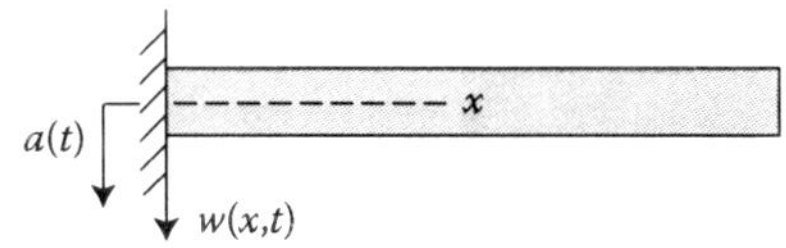


**FIGURE 7.24**
Cantilever beam with accelerating support.

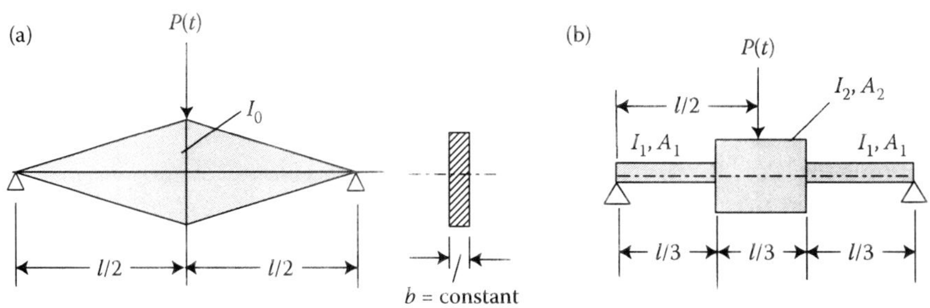


**FIGURE 7.25**
Simply supported beams of various shapes.

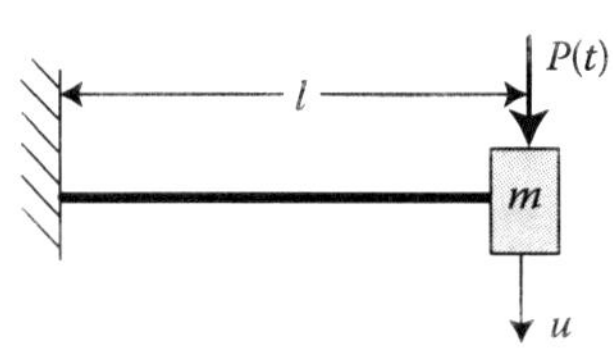


**FIGURE 7.26**
Cantilevered beam with an end load.

7.8 The support of a cantilever shear beam performs a stationary random motion as illustrated in Figure 7.24. The spectral density of $a(t)$ is $S_a(\omega)$. Determine the spectral density of the deflection $w$.

7.9 Using Rayleigh's method, find the lowest natural frequencies of the beams shown in Figure 7.25.

7.10 Using Ritz's method, find the first three natural frequencies of the beams given in Problem 7.9.

7.11 Using Galerkin's method, solve the problem of forced vibrations of the beams given in Problem 7.9. Find the frequency response functions for the following inputs.

7.12 Consider the one-degree of freedom system shown in Figure 7.26. The material for the beam has the complex modulus $E^* = E' + \eta E''$, where $E'$ and $E''$ are constants.

a. Find the response to $P(t) = \delta(t)$.

b. Find a particular solution for $u(t)$ if $P(t) = P_0 \cos \omega t$

7.13 You are considering the design of a 4 ft long simply supported steel beam with a rectangular cross section of width $b$ and height $h$. For space limitations, the beam width may not exceed 3 in, but the height may be varied. You are asked to select a preliminary limiting beam height based on the frequency of free vibration for the first three modes. The requirement is that the frequency ratio based on the Timoshenko beam theory be slightly less than that based on the Euler–Bernoulli theory.

## References

Beer, F. P., Johnston, R. E., and Dewolf, J. T. 2002. *Mechanics of Materials*. New York: McGraw-Hill.

Cowper, G. R. 1966. On the correction for shear of the differential equation for transverse vibration of prismatic bars. *J. Appl. Mech.*, 33: 335–340.

Hibbeler, R. C. 2014. *Mechanics of Materials*. Upper Saddle River, NJ: Pearson Education.

Hurty, W. C. and Rubinstein, M. F. 1964. *Dynamics of Structures*. Upper Saddle River, NJ: Prentice Hall.

Rao, S. S. 2011. *Mechanical Vibrations*, 5th ed. Upper Saddle River, NJ: Prentice Hall.

Rosinger, H. E. and Ritchi, I. G. 1977. On Timoshenko's correction for shear in vibrating isotropic beams. *J. Phys. D: Appl. Phys.*, 10: 1461–1466.

Schmitz, T. L. and Smith, K. S. 2012. *Mechanical Vibrations Modeling and Measurement*. New York: Springer.

Thompson, W. T. and Dahleh, M. D. 1998. *Theory of Vibration with Applications*, 5th ed. Upper Saddle River, NJ: Prentice Hall.

Timoshenko, S. P. 1955. *Strength of Materials*. New York: Van Nostrand.

Timoshenko, S. P. 1921. On the correction for shear of the differential equation for transverse vibration of prismatic bars. *London Phil. Mag.* 41: 744–746.

Timoshenko, S. P. and Gere, J. M. 1972. *Mechanics of Materials*. New York: Van Nostrand.

Weaver, W., Timoshenko, S. P., and Young, D. H. 1990. *Vibration Problems in Engineering*, 5th ed. Hoboken, NJ: John Wiley and Sons.

# 8

# *Continuous Beams and Frames*

## 8.1 Introduction

Beams that are supported at three or more supports and thus having two or more spans are defined as *continuous beams*. These types of beams are common to civil engineering bridges and building frame structures. Continuous beams and frames are considered to be moment resisting. The connection between the beams and columns is classified as rigid and is designed to transmit the beam-end moments and shear forces into the column. No bracing system is required to resist lateral loads and frame stability is provided by the rigidity of the connections and the stiffness of the members only. Continuous frames for structures are used where vertical bracing is not acceptable. A key advantage of continuous frames is the ability to minimize the depths of the beams. In this chapter, we consider several methods for the analysis of continuous beams and frames.

## 8.2 Slope–Deflection Method

The slope–deflection method was introduced in 1914 by George A. Manley (Manley 1915) as an analysis method for beams and frames. In this method, the relationship between moments at the end of a member and the corresponding rotations and displacements are established. The basic assumption is that a typical member can flex but the shear and axial deformations are negligible. By forming slope–deflection equations and applying joint and shear equilibrium conditions, the rotation angles (or the slope angles) are determined. Substituting them back into the slope–deflection equations, member-end moments are readily determined. Frames can be constructed in various configurations as illustrated in Figure 8.1. The intersections of connecting members are the joints and are labeled $K$ and $L$ for reference. We assume the mass of each element is uniformly distributed along the member and there is no axial deformation (axil strain is zero), but there can be flexure. In assessing the behavior, we focus on joints $K$ and $L$ as illustrated in Figure 8.2. Associated with each joint there is a deflection $w$, a moment $M$, shear force $Q$, and a rotation $\phi$ (Nowacki 1963; McCormac 2007).

Assuming harmonic motion, we can define the displacement, rotation, end forces, and moments at each joint as

$$w_K = W_K e^{i\omega t}, \quad \phi_K = \Phi_K e^{i\omega t}, \quad w_L = W_L e^{i\omega t}, \quad \phi_L = \Phi_L e^{i\omega t} \tag{8.1}$$

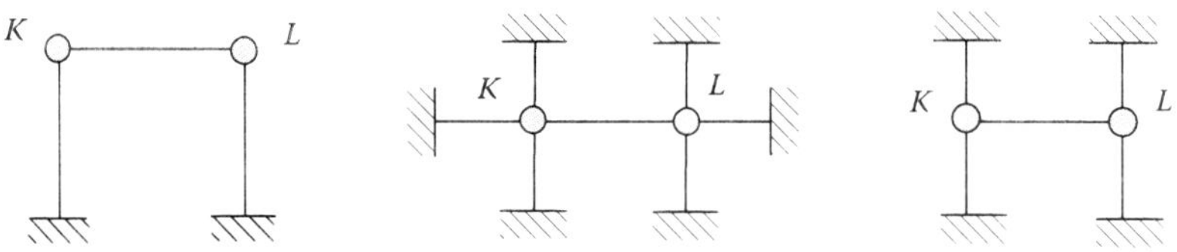


**FIGURE 8.1**
Various possible frame configurations.

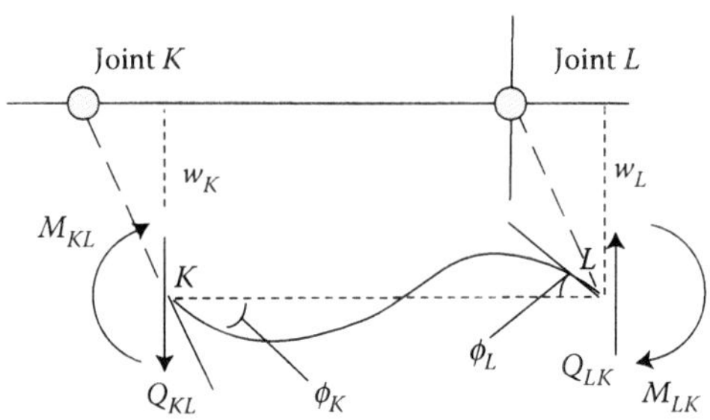


**FIGURE 8.2**
Deflection, rotation, and loads at two joints.

and

$$M_{KL}e^{i\omega t},\quad M_{LK}e^{i\omega t},\quad Q_{KL}e^{i\omega t},\quad Q_{LK}e^{i\omega t} \tag{8.2}$$

For free vibrations of bar $KL$, we can write Equation 7.11 with $p(x,t) = 0$ as

$$EI\frac{\partial^4 w}{\partial x^4} + A\rho\frac{\partial^2 w}{\partial t^2} = 0 \tag{8.3}$$

Using the assumption that $w(x, t) = W(x)e^{i\omega t}$, Equation 8.3 becomes

$$EI\left(\frac{\partial^4 W}{\partial x^4}\right) - A\rho\omega^2 W = 0 \tag{8.4}$$

The solution of this equation has been previously shown to be

$$W(x) = C_1 \sin\beta x + C_2 \cos\beta x + C_3 \sinh\beta x + C_4 \cosh\beta x \tag{8.5}$$

where $\beta^4 = (\rho A\omega^2/EI)$.

We determine the constants $C_1$, $C_2$, $C_3$, and $C_4$ from the boundary conditions at each end of member $K$–$L$. For convenience, we assume joint $K$ is associated with the coordinate $x = 0$ and joint $L$ with $x = l$. The boundary conditions at each joint are

$$W(0)=W_K; \quad \frac{dW(0)}{dx}=\Phi_K$$
$$W(l)=W_L; \quad \frac{dW(l)}{dx}=\Phi_L \tag{8.6}$$

Therefore, we can write

$$W_K=C_1\sin\beta(0)+C_2\cos\beta(0)+C_3\sinh\beta(0)+C_4\cosh\beta(0) \tag{8.7}$$

$$\frac{\Phi_K}{\beta}=C_1\cos\beta(0)-C_2\sin\beta(0)+C_3\cosh\beta(0)+C_4\sinh\beta(0) \tag{8.8}$$

$$W_L=C_1\sin\beta l+C_2\cos\beta l+C_3\sinh\beta l+C_4\cosh\beta l \tag{8.9}$$

$$\frac{\Phi_K}{\beta}=C_1\cos\beta l-C_2\sin\beta l+C_3\cosh\beta l+C_4\sinh\beta l \tag{8.10}$$

In addition to the deflection and slope at each joint, we can define the bending moment and shear force at each joint. We can express the moment amplitudes $M_{KL}$, $M_{LK}$ and shear force amplitudes $Q_{KL}$, $Q_{LK}$ as linear functions of the angles $\phi_K$, $\phi_L$ and the displacements $W_k$, $W_L$. The amplitude of the bending moment and shear force are determined from

$$M_{KL}=-EI\frac{d^2W_K}{dx^2}, \quad M_{LK}=-EI\frac{d^2W_L}{dx^2}$$
$$Q_{KL}=-EI\frac{d^3W_K}{dx^3}, \quad Q_{LK}=-EI\frac{d^3W_L}{dx^3} \tag{8.11}$$

respectively. The limits on $x$ are $x=0$ and $x=l$, which results in

$$M_{KL}=\frac{EI}{l}\left[c(\beta l)\Phi_K+s(\beta l)\Phi_L-r(\beta l)\frac{W_L}{l}+t(\beta l)\frac{W_K}{l}\right] \tag{8.12}$$

$$M_{LK}=\frac{EI}{l}\left[s(\beta l)\Phi_K+c(\beta l)\Phi_L-t(\beta l)\frac{W_L}{l}+r(\beta l)\frac{W_K}{l}\right] \tag{8.13}$$

$$Q_{KL}=\frac{EI}{l^2}\left[t(\beta l)\Phi_K+r(\beta l)\Phi_L-n(\beta l)\frac{W_L}{l}+m(\beta l)\frac{W_K}{l}\right] \tag{8.14}$$

$$Q_{LK}=-\frac{EI}{l^2}\left[r(\beta l)\Phi_K+t(\beta l)\Phi_L-m(\beta l)\frac{W_L}{l}+n(\beta l)\frac{W_K}{l}\right] \tag{8.15}$$

where

$$c(\beta l)=(\beta l)\frac{\cosh\beta l\sin\beta l-\sinh\beta l\cos\beta l}{1-\cosh\beta l\cos\beta l} \tag{8.16}$$

$$s(\beta l) = (\beta l)\frac{\sinh\beta l - \sin\beta l}{1 - \cosh\beta l\cos\beta l} \tag{8.17}$$

$$r(\beta l) = (\beta l)^2\frac{\cosh\beta l - \cos\beta l}{1 - \cosh\beta l\cos\beta l} \tag{8.18}$$

$$t(\beta l) = (\beta l)^2\frac{\sinh\beta l\sin\beta l}{1 - \cosh\beta l\cos\beta l} \tag{8.19}$$

$$m(\beta l) = (\beta l)^3\frac{\sinh\beta l\cos\beta l + \cosh\beta l\sin\beta l}{1 - \cosh\beta l\cos\beta l} \tag{8.20}$$

$$n(\beta l) = (\beta l)^3\frac{\sinh\beta l + \sin\beta l}{1 - \cosh\beta l\cos\beta l} \tag{8.21}$$

The unknowns for the problem are $\Phi_K$, $\Phi_L$ and $W_K$, $W_L$. In order to solve for these unknowns, we need to write the equations of equilibrium for the joints and bars. The procedure for solving the unknown values depends on the type of modal deformation one expects. In order to illustrate this, we can consider a simple three-element frame, which can experience either symmetric or antisymmetric modes as illustrated in Figure 8.3.

Both three-membered frames are fixed at points $A$ and $B$ for either mode and have no deflection or rotation. Each member has identical properties $E$, $I$, $\rho$, and $A$ as well as the same length $l$. For the frame with symmetric modes as shown in Figure 8.3a, there are no joint rotations, and we have that $\Phi_1 = -\Phi_2$ or $\Phi_K = -\Phi_L$ with $U = W = 0$. Assuming one unknown rotation, say $\Phi_1$, we examine joint 1 first. For joint 1, there is no angular rotation so we have $M_{12} + M_{1A} = 0$. Using Equations 8.12 and 8.13 with $W_K = W_L = 0$

$$M_{12} = M_{KL} = \frac{EI}{l}\left[c(\beta l)\Phi_K + s(\beta l)\Phi_L - r(\beta l)\frac{W_L}{l} + t(\beta l)\frac{W_K}{l}\right] = \frac{EI}{l}\left[c(\beta l)\Phi_1 + s(\beta l)\Phi_2\right]$$

$$M_{1A} = M_{LK} = \frac{EI}{l}\left[s(\beta l)\Phi_K + c(\beta l)\Phi_L - t(\beta l)\frac{W_L}{l} + r(\beta L)\frac{W_K}{l}\right] = \frac{EI}{l}\left[c(\beta l)\Phi_1\right]$$

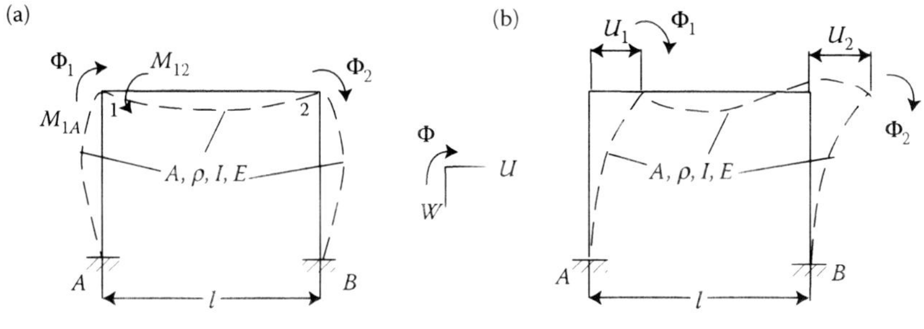


**FIGURE 8.3**
Simple three element frame with either (a) symmetric or (b) antisymmetric modes.

Since $M_{12} + M_{1A} = 0$ and $\Phi_1 = -\Phi_2$, we find

$$M_{12} + M_{1A} = 0 = \frac{EI}{l}\left[c(\beta l)\Phi_1 + s(\beta l)\Phi_2\right] + \frac{EI}{l}\left[c(\beta l)\Phi_1\right] \Rightarrow \Phi_1\left[2c(\beta l) - s(\beta l)\right] = 0 \quad (8.22)$$

The frequency equation is

$$2c(\beta l) - s(\beta l) = 0 \quad (8.23)$$

from which we need to find $\beta_1 l, \beta_2 l, \ldots, \beta_n l$ and subsequently $\omega_1, \omega_2, \ldots, \omega_n$. Solving Equation 8.23 for the first three values yields

$$\beta_1 l = 3.5564 \quad \text{or} \quad \omega_1 = \frac{12.6480}{l^2}\sqrt{\frac{EI}{\rho A}}$$

$$\beta_2 l = 7.4295 \quad \text{or} \quad \omega_2 = \frac{55.1981}{l^2}\sqrt{\frac{EI}{\rho A}}$$

$$\beta_3 l = 9.8488 \quad \text{or} \quad \omega_3 = \frac{96.9987}{l^2}\sqrt{\frac{EI}{\rho A}}$$

$$\vdots \qquad\qquad \vdots$$

If, on the other hand, we have antisymmetric modes as in Figure 8.3b, we have $U_1 = U_2$, $\Phi_1 = \Phi_2$, and $W_1 = -W_2 = 0$. Due to the displacements associated with this mode, the $W$'s in Equations 8.12 through 8.15 are replaced with $U$'s. As with the case of symmetric modes, we have the condition that the moments must sum to zero, that is, $M_{12} + M_{1A} = 0$. Neglecting the longitudinal vibration of the beam girder, we still have to take into account the inertia force of its weight in horizontal displacements. The shear-force equation becomes

$$Q_{1A} + \omega^2 A\rho l U_1 = 0 \quad (8.24)$$

From Equations 8.12 and 8.13, we have

$$M_{12} = \frac{EI}{l}\Phi_1[c(\beta l) + s(\beta l)] \quad \text{and} \quad M_{1A} = \frac{EI}{l}\left[c(\beta l)\Phi_1 - t(\beta l)\frac{U_1}{l}\right]$$

Since we have that $M_{12} + M_{1A} = 0$, we can find

$$\frac{EI}{l}\left[c(\beta l)\Phi_1 + s(\beta l)\Phi_1 + c(\beta l)\Phi_1 - t(\beta l)\frac{U_1}{l}\right] = \frac{EI}{l}\left[2c(\beta l)\Phi_1 + s(\beta l)\Phi_1 - t(\beta l)\frac{U_1}{l}\right] = 0$$

or

$$[2c(\beta l) + s(\beta l)]\Phi_1 - t(\beta l)\frac{U_1}{l} = 0 \quad (8.25)$$

For the shear force (noting that the $W$'s are replaced by $U$'s) in Equations 8.14 and 8.15, we have

$$Q_{1A} = Q_{LK} = -\frac{EI}{l^2}\left[r(\beta l)\Phi_K + t(\beta l)\Phi_L - m(\beta l)\frac{U_L}{l} + n(\beta l)\frac{U_K}{l}\right] = -\frac{EI}{l^2}\left[t(\beta l)\Phi_1 - m(\beta l)\frac{U_1}{l}\right] \quad (8.26)$$

Substituting Equation 8.26 into Equation 8.24 yields

$$-t(kl)\Phi_1 + \left[m(kl) - \omega^2 A\rho\frac{l^3}{EI}\right]\frac{U_1}{l} = 0 \quad (8.27)$$

A nontrivial solution exists if the determinate of the coefficients is zero, or

$$\begin{vmatrix} 2c(\beta l) + s(\beta l) & -t(\beta l) \\ -t(\beta l) & m(\beta l) - A\rho\dfrac{\omega^2 l^3}{EI} \end{vmatrix} = 0$$

Thus,

$$[2c(\beta l) + s(\beta l)]\left[m(\beta l) - \omega^2 A\rho\frac{l^3}{EI}\right] - [t(\beta l)]^2 = 0 \quad (8.28)$$

This is the frequency equation which is needed to determine the eigenfrequencies $\omega_1$, $\omega_2$, … , $\omega_n$. For each value of $\omega$, we must determine $\Phi_1$ and $U_1$, which will give us the shape of the deformed frame. One may be picked arbitrarily, that is, $\Phi_1$ or $U_1$.

### 8.2.1 Forced Vibrations

Consider a single bar with an applied distributed load $q(x)e^{i\omega t}$ as shown in Figure 8.4. Since the vibrations are forced, the moments and shear forces shown in Figure 8.4 are those resulting from the applied forces and are denoted with a superscript "$p$" indicating that they are related to the particular solution. For this model, we wish to determine $M_{KL}^p, Q_{KL}^p, M_{LK}^p, Q_{LK}^p$.

Thus, in Equations 8.12 through 8.15 for moments and forces, we include $M_{KL}^p, Q_{KL}^p, M_{LK}^p$, and $Q_{LK}^p$. Then, we solve for $\Phi_K$, $\Phi_L$, $W_K$, and $W_L$. It is to be noted that we may only use this method for harmonic motion.

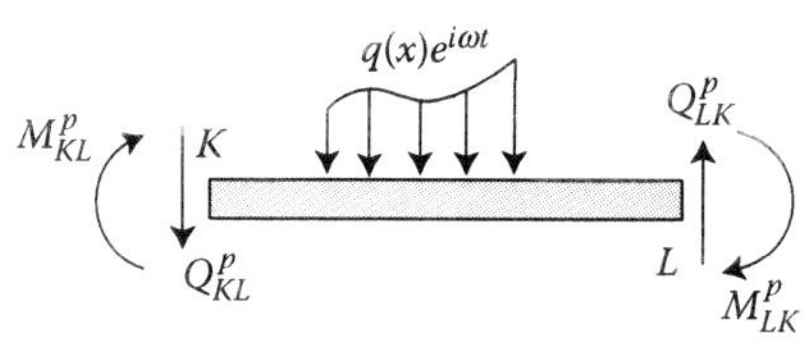


**FIGURE 8.4**
Beam segment with an applied load.

## 8.3 Vibrations of Frames with Axial Forces

Consider a frame member $K$–$L$ subjected to an axial compressive force $T$. When the frame member is subjected to an axial compressive load, the deflection, shear, and moment forces are represented as shown in Figure 8.5. As before, we must find relations between $M_{KL}$, $Q_{KL}$, $M_{LK}$, $Q_{LK}$ and $\Phi_K$, $\Phi_L$, $W_K$, $W_L$.

In order to solve the problem, we assume a displacement

$$w(x,t) = W(x)e^{i\omega t} \tag{8.29}$$

Substituting Equation 8.29 into the equation of motion of a beam with an axial compressive load, Equation 7.31 with the external load $p(x, t) = 0$ yields

$$EI\frac{d^4W}{dx^4} + T\frac{d^2W}{dx^2} - \omega^2\rho AW = 0 \tag{8.30}$$

The solution to this ordinary fourth-order differential equation with constant coefficients is given by

$$W(x) = C_1\sin\frac{\eta_1 x}{l} + C_2\cos\frac{\eta_1 x}{l} + C_3\sinh\frac{\eta_2 x}{l} + C_4\cosh\frac{\eta_2 x}{l} \tag{8.31}$$

where

$$\eta_1 = \sqrt{\frac{\alpha^2}{2} + \sqrt{\frac{\alpha^4}{4} + \beta^4}}, \quad \eta_2 = \sqrt{-\frac{\alpha^2}{2} - \sqrt{\frac{\alpha^4}{4} + \beta^4}}$$

$$\alpha^2 = \frac{Tl^2}{EI}, \quad \beta^4 = \frac{A\rho l^4\omega^2}{EI}$$

We determine the constants of integration $C_1$, $C_2$, $C_3$, and $C_4$ from the boundary conditions $W(0) = W_K$, $W(l) = W_L$, $(dW(0))/dx = \Phi_K$ and $(dW(l))/dx = \Phi_L$. Then we determine the end moments and forces. The results are expressed in a manner similar to Equations 8.12 through 8.21 as

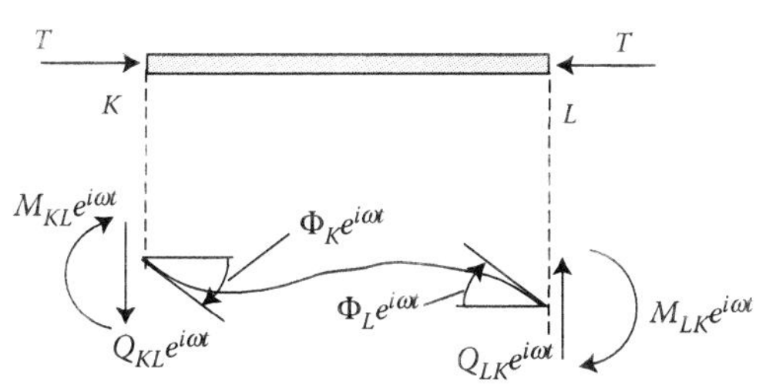


**FIGURE 8.5**
Frame member subjected to an axial compressive force.

$$M_{KL} = \frac{EI}{l}\left[c(\alpha,\beta)\Phi_K + s(\alpha,\beta)\Phi_L - r(\alpha,\beta)\frac{W_L}{l} + t(\alpha,\beta)\frac{W_K}{l}\right] \tag{8.32}$$

$$M_{LK} = \frac{EI}{l}\left[s(\alpha,\beta)\Phi_K + c(\alpha,\beta)\Phi_L - t(\alpha,\beta)\frac{W_L}{l} + r(\alpha,\beta)\frac{W_K}{l}\right] \tag{8.33}$$

$$Q_{KL} = \frac{EI}{l^2}\left[t(\alpha,\beta)\Phi_K + r(\alpha,\beta)\Phi_L - n(\alpha,\beta)\frac{W_L}{l} + m(\alpha,\beta)\frac{W_K}{l}\right] \tag{8.34}$$

$$Q_{LK} = -\frac{EI}{l^2}\left[r(\alpha,\beta)\Phi_K + t(\alpha,\beta)\Phi_L - m(\alpha,\beta)\frac{W_L}{l} + n(\alpha,\beta)\frac{W_K}{l}\right] \tag{8.35}$$

where

$$c(\alpha,\beta) = \left(\frac{\eta_1^2+\eta_2^2}{\Delta}\right)(\eta_2\cosh\eta_2\sin\eta_1 - \eta_1\cos\eta_1\sinh\eta_2) \tag{8.36}$$

$$s(\alpha,\beta) = \left(\frac{\eta_1^2+\eta_2^2}{\Delta}\right)(\eta_1\sinh\eta_2 - \eta_2\sin\eta_1) \tag{8.37}$$

$$r(\alpha,\beta) = \left(\frac{\eta_1^2+\eta_2^2}{\Delta}\right)(\eta_1\eta_2)(\cosh\eta_2 - \cos\eta_1) \tag{8.38}$$

$$t(\alpha,\beta) = \frac{\eta_1\eta_2}{\Delta}\left[2\eta_1\eta_2\sinh\eta_2\sin\eta_1 + \left(\eta_2^2-\eta_1^2\right)(\cosh\eta_2\cos\eta_1 - 1)\right] \tag{8.39}$$

$$n(\alpha,\beta) = \left(\frac{\eta_1^2+\eta_2^2}{\Delta}\right)(\eta_1\eta_2)(\eta_2\sinh\eta_2 + \eta_1\sin\eta_1) \tag{8.40}$$

$$m(\alpha,\beta) = \left(\frac{\eta_1^2+\eta_2^2}{\Delta}\right)(\eta_1\eta_2)(\eta_2\sinh\eta_2\cos\eta_1 + \eta_1\sin\eta_1\cosh\eta_2) \tag{8.41}$$

$$\Delta = 2\eta_1\eta_2(1-\cosh\eta_2\cos\eta_1) + \left(\eta_2^2-\eta_1^2\right)\sinh\eta_2\sin\eta_1 \tag{8.42}$$

If there is no axial load $\alpha = 0$ which results in $\eta_1 = \eta_1 = \beta$. If there are no vibrations (stability problem), we have that $w = \eta_2 = 0$ and $\eta_1 = \alpha$.

**EXAMPLE 8.1**

Consider the forced harmonic vibration of the frame shown in Figure 8.6. Symmetric motion for the frame will occur. Determine an expression for $\Phi_1$ along member 1–2.

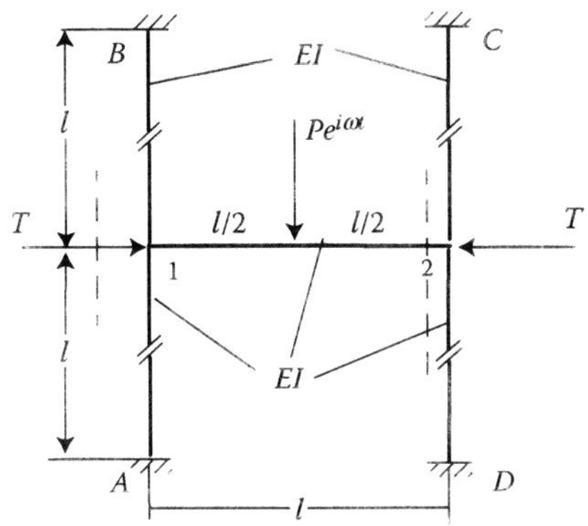


**FIGURE 8.6**
Frame subjected to an applied load $P$ and a compressive force $T$.

**Solution**

The only unknowns are $\Phi_1$ and $\Phi_2$, but because of symmetric modes we have $\Phi_1 = -\Phi_2$. In order to determine $\Phi_1$, we need to determine the moment at joint 1. The internal moment as joint 1 comes from vertical members $A$-1, $B$-1, and horizontal member 1–2 must combine so that for equilibrium

$$M_{1A} + M_{1B} + M_{12} = 0$$

Using Equations 8.32 and 8.33 and noting that we have $W_L = W_K = 0$, and that the rotation at location $A$ ($K$ in the governing equations) is $\Phi_K = 0$. Thus, we arrive at

$$M_{LK} = M_{1A} = \frac{EI}{l} c(0, \beta)\Phi_1$$

and

$$M_{KL} = M_{1B} = \frac{EI}{l} c(0, \beta)\Phi_1$$

where $\beta$ depends on the forcing frequency $\omega$. For horizontal member 1–2, we augment Equation 8.33 by adding the moment $M_{12}^P$ generated by the applied force so that we have

$$M_{KL} = M_{12} = \frac{EI}{l}[c(\alpha, \beta)\Phi_1 + s(\alpha, \beta)\Phi_2] + M_{12}^P$$

where $\alpha$ depends on $T$ and $\alpha \neq 0$. In order to find $M_{12}^P$, we use the slope–deflection method again by referring to the model of element 1–2 as illustrated in Figure 8.7. In this model, we react the external applied load $Pe^{i\omega t}$ with internal shear forces $Q_{E1}$ and $Q_{E2}$. At joint $E$, we have that $Q_{E1} + Q_{E2} = P$. At point $E$, the rotation is zero 0 but there is a vertical deflection $\Phi_E = 0$ and ($W_E \neq 0$).

We express $Q_{E1}$ and $Q_{E2}$ in terms of $W_E$. These two terms are defined from Equations 8.34 and 8.35 noting that $\Phi_K = \Phi_L = W_L = 0$ and using $l_{1E} = l/2$, $\alpha_{1E} = \alpha/2$, and $\beta_{1E} = \beta/2$. Thus, we can write

$$\frac{2EI}{l_{1E}^3} m(\alpha_{1E}, \beta_{1E})W_E = P \Rightarrow W_E = \frac{Pl_{1E}^3}{2EIm(\alpha_{1E}, \beta_{1E})}$$

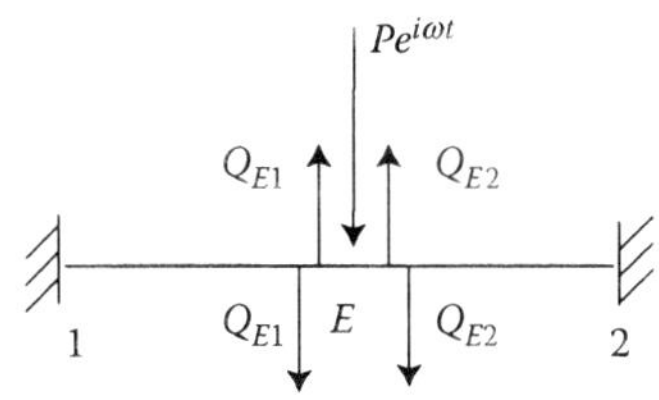


**FIGURE 8.7**
Model of beam segment 1–2.

Then we can solve for $M_{12}^P$ by noting that $M_{1E} = M_{12}^P$ (or $M_{KL}$ from Equation 8.32).

$$M_{12}^P = M_{1E} = -\frac{EI}{l_{1E}} r(\alpha_{1E}, \beta_{1E}) \frac{W_E}{l_{1E}} = -\frac{Pl}{4} \frac{r(\alpha_{1E}, \beta_{1E})}{m(\alpha_{1E}, \beta_{1E})}$$

Substituting back into the moment equilibrium equation results in

$$[c(\alpha,\beta) + 2c(0,\beta) - s(\alpha,\beta)]\Phi_1 + \frac{l}{EI} M_{12}^P = 0$$

Solving for $\Phi_1$ gives

$$\Phi_1 = -\Phi_2 = \left(\frac{Pl^2}{4EI}\right) \frac{r(\alpha/2, \beta/2)}{m(\alpha/2, \beta/2)\left[c(\alpha,\beta) + 2c(0,\beta) - s(\alpha,\beta)\right]}$$

If we want to determine the natural frequency, we set $P = M_{12}^P = 0$, resulting in

$$[c(\alpha,\beta) + 2c(0,\beta) - s(\alpha,\beta)]\Phi_1 = 0$$

The frequency equation is given by

$$[c(\alpha,\beta) + 2c(0,\beta) - s(\alpha,\beta)] = 0$$

For a given $T$ (or $\alpha$), we find $\omega$ which satisfies this equation. If there are imaginary values of $\omega$ which satisfy the frequency equation, the frame is unstable. Recall that

$$w(x,t) = W(x)e^{\bar{\omega}t}; \quad \text{where } \bar{\omega} \text{ is real}$$

## PROBLEMS

8.1 Assume a beam segment $KL$ of length $l$ may be subjected to the end loads and moments shown in Figure 8.8. Using the slope–deflection method, develop the governing relations between the loads $Q_{KL}$, $Q_{LK}$ and moments $M_{KL}$, $M_{LK}$ and the slopes and deflections ($W_K$, $\phi_K$, $W_L$, and $\phi_L$) for free vibrations.

8.2 Using the slope–deflection method, determine the frequency equation for the continuous beams and frames shown in Figure 8.9. Assume each segment of these structures is made from the same material.

8.3 Find the frequency equation and the frequency response function of the beam shown in Figure 8.10. Use the slope–deflection method.

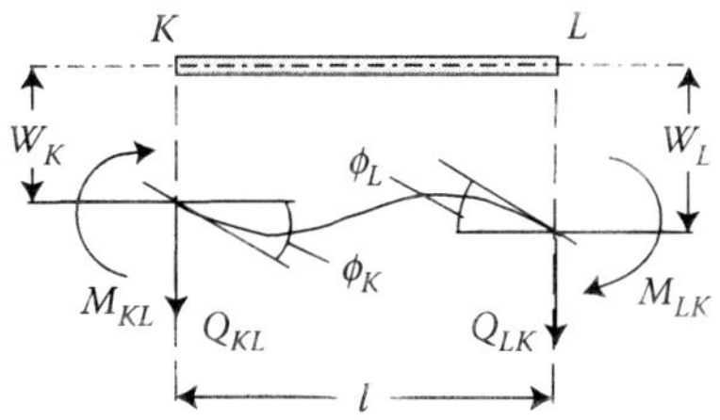


**FIGURE 8.8**
Beam segment with end loads and moments.

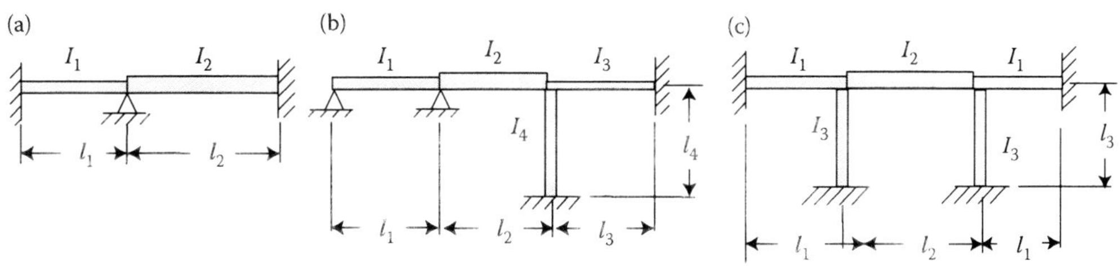


**FIGURE 8.9**
Continuous beams and frames.

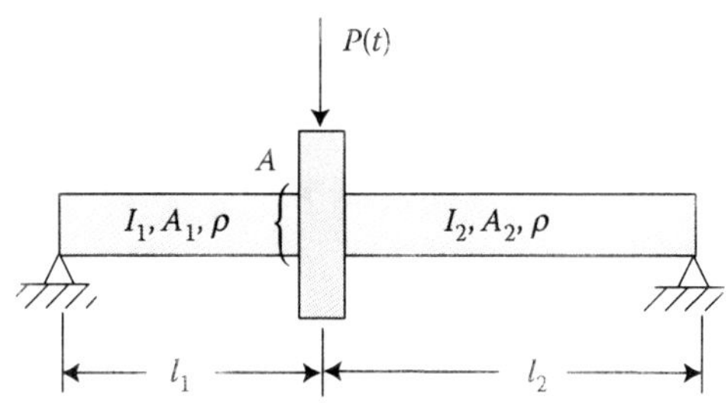


**FIGURE 8.10**
Simply supported beam with a concentrated load.

## References

Manley, G. A. 1915. *Studies in Engineering*. Minneapolis: University of Minnesota.
McCormac, J. C. 2007. *Structural Analysis: A Classical and Matrix Approach*, Boston: Addison-Wesley.
Nowacki, W. 1963. *Dynamic of Elastic Systems*. New York: John Wiley & Sons.

# 9

# *Vibrations of Plates*

## 9.1 Introduction

Plate vibrations are a special case of the more general topic of mechanical vibrations. Their analysis is more complex than that of beams and bars. Particularly, thin flat plates of various shapes are found in many structures, such as vehicles, aircraft, ships, missiles, etc. and the natural frequencies and mode shapes of the plates are indispensable to the prediction of their dynamic response. The equations of motion for plate vibrations are simpler than those for general three-dimensional vibration problems because one of the plate dimensions, the thickness, is much smaller than the other two dimensions. Plates are among the most common structural elements and two-dimensional plate theory gives excellent approximation to the actual motion of the plate. A plate can be defined as a load-carrying structural member bounded by two parallel surfaces whose thickness $h$ between the two surfaces is small compared with the other dimensions of the body (such as length, or width of a rectangle, diameter of a circle, etc.). The plane that is parallel to and equidistant from the two surfaces is called the *midplane* of the plate. A plate is considered thin if the plate's thickness is small compared to the length and width dimension. That is, if $h < 20L$ where $L$ is the length of the smallest side of the plate. A plate that has $h > 20L$ is considered a thick plate and requires a different set of equations than those used for the thin plate. Thin plate theory is referred to as *classical plate theory* (CPT) and does not take into account shear effects. Relations between the load and deflection of plates are developed following the procedures similar to those for beams. Numerous references exist regarding the general plate analysis, including Timoshenko and Woinowsky-Krieger (1959); Panc (1975); Ugural (1981); Lowe (1982); Reddy (2006); and Chakraverty (2009).

## 9.2 Equations of Motion

A rectangular plate is assumed to have width and length dimensions of $a$, $b$ and a thickness $h$ as depicted in Figure 9.1. For rectangular plates, it is convenient to use a rectangular right-handed coordinate system as shown in the figure, where the $x$, $y$, and $z$ coordinates are defined from the middle surface of the plate ($h/2$). Specifically, with reference to the figure we assume: (a) plate is thin, (b) plane sections remain plane, (c) there are no transverse

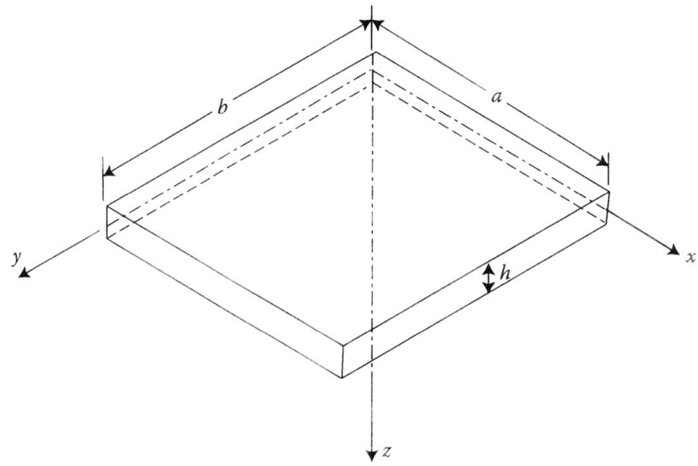


**FIGURE 9.1**
Basic plate geometry.

shear deformations, (d) there is no rotary inertia, and (e) the plate is elastic, isotropic, and homogeneous.

Upon application of a load in the $z$-direction, plate deformation in the $x$–$z$ and $y$–$z$ planes occurs as illustrated in Figure 9.2. At an arbitrary location $z$ below the midplane of the plate, the displacements in the $x$ ($u$) and $y$ ($v$) directions are defined in terms of the $z(w)$ deflection of the plate. From Figure 9.2, we see that the displacements $u$ and $v$ are proportional to the angles $\partial w/\partial x$ and $\partial w/\partial y$, thus

$$u = -z\frac{\partial w}{\partial x}, \quad v = -z\frac{\partial w}{\partial y} \tag{9.1}$$

The deformation results in strains, which are related to the stresses shown in Figure 9.3a through the constitutive relations. Associated with the stresses are internal shear forces and moments. These forces and moments are illustrated in Figure 9.3b. The strains in the $x$-, $y$-plane associated with these deformations are defined as

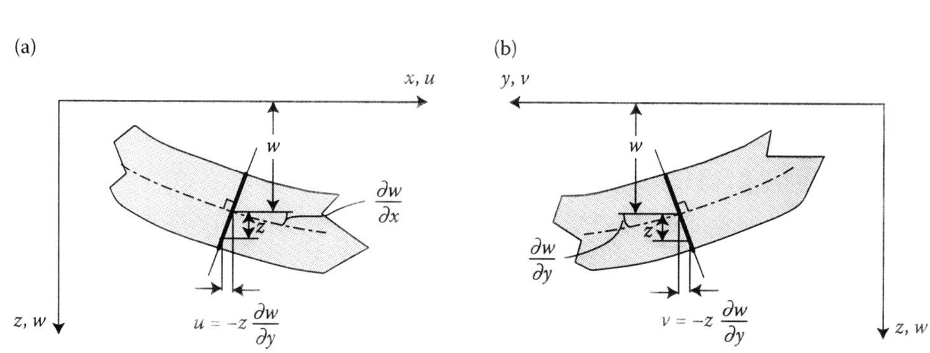


**FIGURE 9.2**
Plate deformations in the (a) $x$–$z$, and (b) $y$–$z$ planes under an applied load in the $z$-direction.

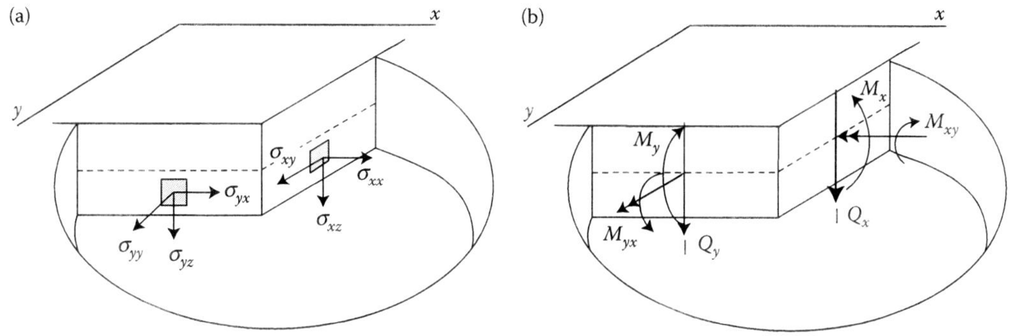


**FIGURE 9.3**
Plate (a) stresses and (b) forces and moments associated to plate deformations.

$$\begin{aligned} \varepsilon_{xx} &= \frac{\partial u}{\partial x} = -z\frac{\partial^2 w}{\partial x^2} \\ \varepsilon_{yy} &= \frac{\partial v}{\partial y} = -z\frac{\partial^2 w}{\partial y^2} \\ \gamma_{xy} &= \frac{1}{2}\left(\frac{\partial u}{\partial y} + \frac{\partial v}{\partial x}\right) = -z\frac{\partial^2 w}{\partial x \partial y} \end{aligned} \tag{9.2}$$

The bending moments $M_x$, $M_y$, twisting moments $M_{xy} = M_{yx}$, and shear forces $Q_x$, $Q_y$ are defined in the traditional manner by the relations

$$M_x = \int_{-h/2}^{h/2} \sigma_{xx} z\, dz, \quad M_y = \int_{-h/2}^{h/2} \sigma_{yy} z\, dz \tag{9.3}$$

$$M_{xy} = \int_{-h/2}^{h/2} \sigma_{xy} z\, dz = M_{yx} = \int_{-h/2}^{h/2} \sigma_{yx} z\, dz \tag{9.4}$$

$$Q_x = \int_{-h/2}^{h/2} \sigma_{xz} z\, dz, \quad Q_y = \int_{-h/2}^{h/2} \sigma_{yz} z\, dz \tag{9.5}$$

Assuming the normal out of plane stress is approximately zero ($\sigma_{zz} \approx 0$), a state of plane stress exists in the $x$-, $y$-plane so that the strains $\varepsilon_x$, $\varepsilon_y$, and $\varepsilon_{xy}$ are related to the stresses $\sigma_{xx}$, $\sigma_{yy}$, and $\sigma_{xy}$ by the relations

$$\varepsilon_{xx} = \frac{1}{E}(\sigma_{xx} - \nu\sigma_{yy}), \quad \varepsilon_{yy} = \frac{1}{E}(\sigma_{yy} - \nu\sigma_{xx}), \quad \varepsilon_{xy} = \frac{1}{G}\sigma_{xy} = \frac{2(1+\nu)}{E}\sigma_{xy} \tag{9.6}$$

Equations 9.2 and 9.6 can be combined to give

$$\sigma_{xx} = -\frac{Ez}{1-\nu^2}\left(\frac{\partial^2 w}{\partial x^2} + \nu\frac{\partial^2 w}{\partial y^2}\right), \quad \sigma_{yy} = -\frac{Ez}{1-\nu^2}\left(\frac{\partial^2 w}{\partial y^2} + \nu\frac{\partial^2 w}{\partial x^2}\right), \quad \sigma_{xy} = -\frac{Ez}{1+\nu}\frac{\partial^2 w}{\partial x \partial y} \tag{9.7}$$

Integration of Equations 9.3 and 9.4 using Equation 9.7 yields

$$M_x = \int_{-h/2}^{h/2} \sigma_{xx} z\,dz = -\int_{-h/2}^{h/2} \frac{E}{1-\nu^2}\left(\frac{\partial^2 w}{\partial x^2} + \nu\frac{\partial^2 w}{\partial y^2}\right) z^2\,dz = -D\left(\frac{\partial^2 w}{\partial x^2} + \nu\frac{\partial^2 w}{\partial y^2}\right) \tag{9.8}$$

where $D$ is the *flexural rigidity* of the plate and is defined by

$$D = \int_{-h/2}^{h/2} \frac{E}{1-\nu^2} z^2\,dz = \frac{Eh^3}{12(1-\nu^2)} \tag{9.9}$$

Similarly,

$$M_y = \int_{-h/2}^{h/2} \sigma_{xx} z\,dz = -D\left(\frac{\partial^2 w}{\partial y^2} + \nu\frac{\partial^2 w}{\partial x^2}\right) \tag{9.10}$$

$$M_{xy} = M_{yx} = \int_{-h/2}^{h/2} \sigma_{xy} z\,dz = -D(1-\nu)\frac{\partial^2 w}{\partial x\,\partial y} \tag{9.11}$$

Note that because of the equality of the shear stresses $\sigma_{xy} = \sigma_{yx}$ it follows that $M_{xy} = M_{yx}$. The units for the bending and torsional moments are (in lb/in) or N m/m. This is one of the primary differences between the moments defined in undergraduate strength of materials and the moments used in plate theory. The equations of motion are developed using the free-body diagram in Figure 9.4. The representative volume element shown in this figure has a thickness $h$ and planform dimensions $dx$ and $dy$.

The mass of the element is expressed as $\rho h(dx\,dy)$. Starting with the forces in the $z$-direction and applying Newton's second law results in

$$\frac{\partial Q_x}{\partial x} dx\,dy + \frac{\partial Q_y}{\partial y} dx\,dy + p(x,y,t)dx\,dy - h\,dx\,dy\,\rho\frac{\partial^2 w}{\partial t^2} = 0$$

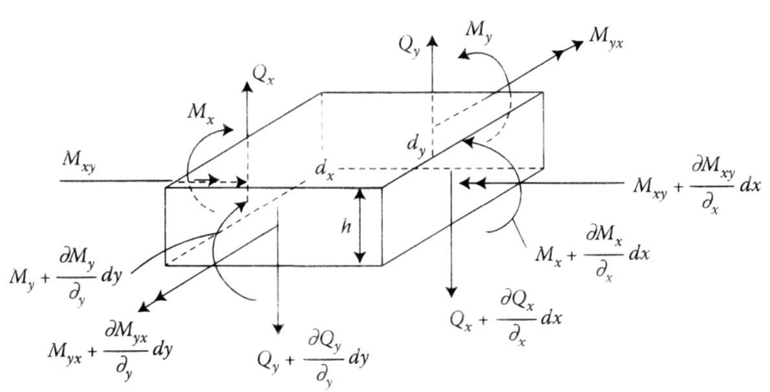


**FIGURE 9.4**
Free-body diagram of a representative volume element of a plate.

which reduces to

$$\frac{\partial Q_x}{\partial x}+\frac{\partial Q_y}{\partial y}-h\rho\frac{\partial^2 w}{\partial t^2}=-p(x,y,t) \tag{9.12}$$

Taking moments about $y$ yields

$$\frac{\partial M_x}{\partial x}dx\,dy-Q_x\,dx\,dy+\frac{\partial M_{yx}}{\partial y}dx\,dy=0$$

or

$$Q_x=\frac{\partial M_x}{\partial x}+\frac{\partial M_{yx}}{\partial y} \tag{9.13}$$

Taking moments about $x$ yields

$$\frac{\partial M_y}{\partial y}dx\,dy-Q_y\,dx\,dy+\frac{\partial M_{xy}}{\partial x}dx\,dy=0$$

or

$$Q_y=\frac{\partial M_y}{\partial y}+\frac{\partial M_{xy}}{\partial x} \tag{9.14}$$

An equation of equilibrium can be expressed in terms of the derivatives of the moments by eliminating $Q_x$, $Q_y$ from Equations 9.12 through 9.14, giving

$$\frac{\partial^2 M_x}{\partial x^2}+2\frac{\partial^2 M_{xy}}{\partial x\partial y}+\frac{\partial^2 M_y}{\partial y^2}-h\rho\frac{\partial^2 w}{\partial t^2}=-p(x,y,t) \tag{9.15}$$

The differential equation for the deflection of the plate is obtained by substituting the moment–curvature relations, Equations 9.8 through 9.11 into Equation 9.15, giving

$$\frac{\partial^2}{\partial x^2}\left\{D\left(\frac{\partial^2 w}{\partial x^2}+\nu\frac{\partial^2 w}{\partial y^2}\right)\right\}+2(1-\nu)\frac{\partial^2}{\partial x\partial y}\left(D\frac{\partial^2 w}{\partial x\partial y}\right)+\frac{\partial^2}{\partial y^2}\left\{D\left(\frac{\partial^2 w}{\partial y^2}+\nu\frac{\partial^2 w}{\partial x^2}\right)\right\}+\rho h\frac{\partial^2 w}{\partial t^2}=p(x,y,t) \tag{9.16}$$

or

$$\nabla^2(D\nabla^2 w)-(1-\nu)\left[\frac{\partial^2 D}{\partial x^2}\frac{\partial^2 w}{\partial y^2}+\frac{\partial^2 D}{\partial y^2}\frac{\partial^2 w}{\partial x^2}-2\frac{\partial^2 D}{\partial x\partial y}\frac{\partial^2 w}{\partial x\partial y}\right]+\rho h\frac{\partial^2 w}{\partial t^2}=p(x,y,t) \tag{9.17}$$

if $D$ is a function of $x$ and $y$. Here

$$\nabla^2 \equiv \frac{\partial^2}{\partial x^2} + \frac{\partial^2}{\partial y^2}$$

and is called the *Del* operator or the two-dimensional Laplacian operator. If the plate has a constant bending rigidity $D$, Equation 9.17 simplifies to

$$D\nabla^4 w(x,y,t) + \rho h \frac{\partial^2 w(x,y,t)}{\partial t^2} = p(x,y,t) \tag{9.18}$$

where $\nabla^4(\bullet) = \nabla^2\nabla^2(\bullet)$, or we can write Equation 9.18 as

$$\frac{\partial^4 w}{\partial x^4} + 2\frac{\partial^4 w}{\partial x^2\,\partial y^2} + \frac{\partial^4 w}{\partial y^4} + \frac{h\rho}{D}\frac{\partial^2 w}{\partial t^2} = \frac{p(x,y,t)}{D} \tag{9.19}$$

For comparative purposes, we note that for a beam we have an equation of the form

$$\frac{\partial^4 w}{\partial x^4} + \frac{A\rho}{EI}\frac{\partial^2 w}{\partial t^2} = \frac{p}{EI}$$

In order to solve the equation of motion, we must know the boundary and initial conditions. Since Equation 9.19 is a fourth-order differential equation, two boundary conditions, either for the displacements or for the internal forces, are required at each boundary. The boundary conditions depend on the constraints at the ends of the plate. The boundary conditions are applied along lines of $x$ or $y$ being some constant value. Typical support conditions are simple supports, built-in edges, and free edges. The appropriate boundary conditions for each of these are as follows:

Simply supported:

$$x = \text{constant: } w = 0,\ M_x = 0 \rightarrow \frac{\partial^2 w}{\partial x^2} = 0 \tag{9.20}$$

$$y = \text{constant: } w = 0,\ M_y = 0 \rightarrow \frac{\partial^2 w}{\partial y^2} = 0 \tag{9.21}$$

Built-in edges:

$$x = \text{constant: } w = 0, \frac{\partial w}{\partial x} = 0 \tag{9.22}$$

$$y = \text{constant: } w = 0, \frac{\partial w}{\partial y} = 0 \tag{9.23}$$

Free edges: A free-edge condition is more complicated to define and requires additional information. In this case, there are no external restrictions on the displacement of the edge. Thus, there are no forces or moments applied to the edge.

That is for the edge $x$ = constant,

$$M_x = 0, \quad Q_x = 0, \quad M_{xy} = 0 \tag{9.24}$$

To be consistent with the assumptions of plate theory, the conditions in Equation 9.24 need to be combined reducing the number of boundary conditions from three to two. Starting with $M_x = 0$, we note that Equation 9.8 can be written as

$$M_x = 0 \Rightarrow \frac{\partial^2 w}{\partial x^2} + \nu \frac{\partial^2 w}{\partial y^2} = 0$$

The shearing force at the edge of the plate consists of two terms, the transverse shear and the twisting or torsional moment. The vertical edge force $\bar{Q}_x$ per unit length can be written at point $A$, as illustrated in Figure 9.5, as

$$\bar{Q}_x = Q_x + \frac{\partial M_{xy}}{\partial y}$$

This method of reducing the number of boundary conditions is attributed to Kirchhoff. Thus, if $x$ = constant is a free edge, $\bar{Q}_x = 0$ so that $Q_x + (\partial M_{xy}/\partial y) = 0$. As a result, the second boundary condition becomes

$$\frac{\partial^3 w}{\partial x^3} + (2-\nu)\frac{\partial^3 w}{\partial x \, \partial y^2} = 0$$

In summary, for $x$ = constant the two boundary conditions are

$$\frac{\partial^2 w}{\partial x^2} + \nu \frac{\partial^2 w}{\partial y^2} = 0, \quad \frac{\partial^3 w}{\partial x^3} + (2-\nu)\frac{\partial^3 w}{\partial x \, \partial y^2} = 0 \tag{9.25}$$

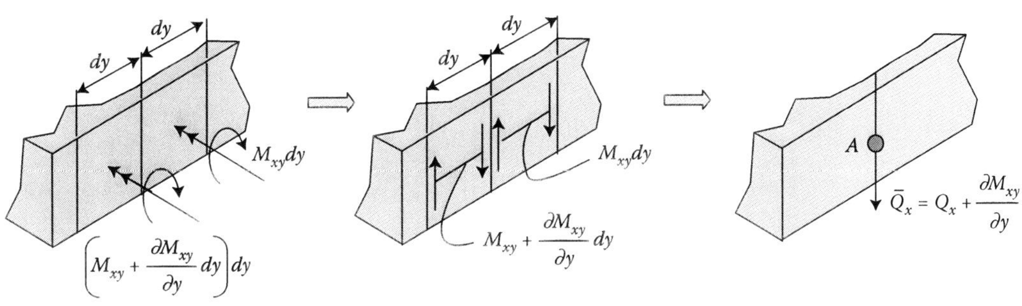


**FIGURE 9.5**
Model for replacing $M_{xy}$ with a series of shear forces and defining $\bar{Q}_x$.

Following similar reasoning for $y$ = constant, we begin knowing $M_y = 0$ and we define $\bar{Q}_y = 0$. The result is the following two boundary conditions:

$$\frac{\partial^2 w}{\partial y^2} + \nu \frac{\partial^2 w}{\partial x^2} = 0 \quad \frac{\partial^3 w}{\partial y^3} + (2-\nu)\frac{\partial^3 w}{\partial y\, \partial x^2} = 0 \tag{9.26}$$

The final information needed to solve the equation of motion is the initial conditions. These conditions are typically, at $t = 0$;

$$w(x, y, 0) = w_0(x, y)$$
$$\dot{w}(x, y, 0) = \dot{w}_0(x, y)$$

where $w_0(x, y)$ and $\dot{w}_0(x, y)$ are given.

### 9.2.1 Solution to Plate Equations

In order to illustrate the procedure for solving the plate equations, consider the free vibration of a simply supported rectangular plate, of thickness $h$ with edge lengths $a$ and $b$ as shown in Figure 9.6. Since the plate is simply supported on all edges, the boundary conditions are

$$\begin{aligned} &\text{At } x = 0, a : w = 0, \frac{\partial^2 w}{\partial x^2} = 0 \\ &\text{At } y = 0, b : w = 0, \frac{\partial^2 w}{\partial y^2} = 0 \end{aligned} \tag{9.27}$$

For free vibrations $p(x, y, t) = 0$ and the equation of motion (Equation 9.19) becomes

$$\frac{\partial^4 w}{\partial x^4} + 2\frac{\partial^4 w}{\partial x^2\, \partial y^2} + \frac{\partial^4 w}{\partial y^4} = -\frac{h\rho}{D}\frac{\partial^2 w}{\partial t^2} \tag{9.28}$$

Assume a solution of the form

$$w(x, y, t) = W(x, y)(A\cos\omega t + B\sin\omega t)$$

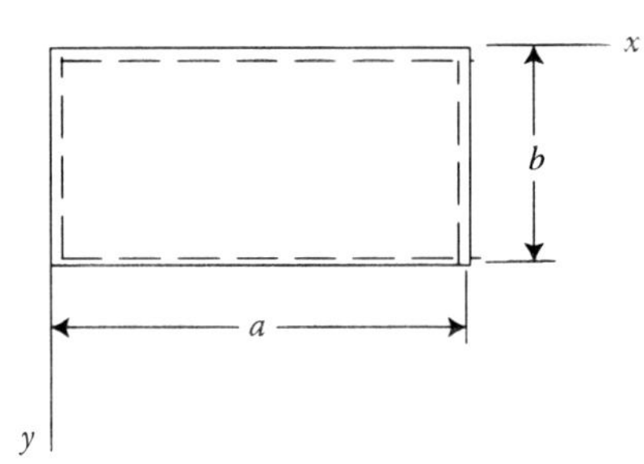


**FIGURE 9.6**
Simply supported rectangular plate.

where $A$ and $B$ are arbitrary constants. Substituting into the equation of motion, Equation 9.28 yields

$$\frac{\partial^4 W}{\partial x^4}+2\frac{\partial^4 W}{\partial x^2\,\partial y^2}+\frac{\partial^4 W}{\partial y^4}=\frac{h\rho}{D}\omega^2 W \tag{9.29}$$

or

$$\nabla^4 W-\lambda^4 W=0 \tag{9.30}$$

where

$$\lambda^4=\frac{\rho h\omega^2}{D} \tag{9.31}$$

Assume

$$W(x,y)=\sin\frac{m\pi x}{a}\sin\frac{n\pi y}{b},\quad (m,n=1,2,3,\ldots) \tag{9.32}$$

The boundary conditions are satisfied by the above assumed value of $W(x, y)$. Substituting Equation 9.32 into Equation 9.29 or Equation 9.30 gives

$$\left(\frac{m\pi}{a}\right)^4+2\left(\frac{m\pi}{a}\right)^2\left(\frac{n\pi}{b}\right)^2+\left(\frac{n\pi}{b}\right)^4=\lambda^4\omega^2$$

or

$$\pi^4\left(\frac{m^2}{a^2}+\frac{n^2}{b^2}\right)^2=\lambda^4$$

This equation will be satisfied if the frequency satisfies

$$\omega_{mn}\equiv\lambda_{mn}^2=\pi^2\sqrt{\frac{D}{h\rho}}\left(\frac{m^2}{a^2}+\frac{n^2}{b^2}\right) \tag{9.33}$$

Depending on the values of $m$ and $n$, Equation 9.33 will obviously yield different values of $\omega_{mn}$. Additionally, the mode shapes differ. A summary of the expressions for $\omega_{mn}$ and $W_{mn}$ for selected values of $m$ and $n$ is given in Table 9.1. An illustration of the mode shapes for each $m$ and $n$ combination in Table 9.1 is illustrated in Figure 9.7.

Knowing $\omega_{mn}$, consider the initial conditions at time $t = 0$ as

$$w(x,y,0)=w_0(x,y)$$
$$\dot{w}(x,y,0)=\dot{w}_0(x,y)$$

**TABLE 9.1**

Frequency and Mode Shape for Selected Values of $m$ and $n$ for a Simply Supported Rectangular Plate

| $m$ | $n$ | $\omega_{mn}$ | $W_{mn}$ |
|---|---|---|---|
| 1 | 1 | $\omega_{11}=\pi^2\sqrt{\frac{D}{h\rho}}\left(\frac{1}{a^2}+\frac{1}{b^2}\right)$ | $W_{11}=\sin\frac{\pi x}{a}\sin\frac{\pi y}{b}$ |
| 2 | 1 | $\omega_{21}=\pi^2\sqrt{\frac{D}{h\rho}}\left(\frac{4}{a^2}+\frac{1}{b^2}\right)$ | $W_{21}=\sin\frac{2\pi x}{a}\sin\frac{\pi y}{b}$ |
| 1 | 2 | $\omega_{12}=\pi^2\sqrt{\frac{D}{h\rho}}\left(\frac{1}{a^2}+\frac{4}{b^2}\right)$ | $W_{12}=\sin\frac{\pi x}{a}\sin\frac{2\pi y}{b}$ |
| 2 | 2 | $\omega_{22}=\pi^2\sqrt{\frac{D}{h\rho}}\left(\frac{4}{a^2}+\frac{4}{b^2}\right)$ | $W_{22}=\sin\frac{2\pi x}{a}\sin\frac{2\pi y}{b}$ |

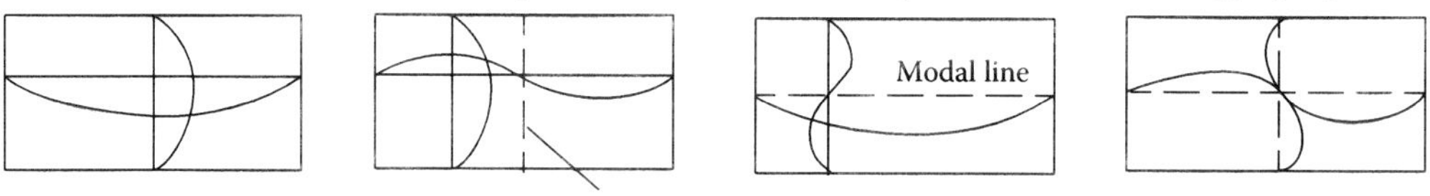


**FIGURE 9.7**
Mode shapes for selected values of $m$ and $n$ for a simply supported rectangular plate.

where $w_0(x, y)$ and $\dot{w}_0(x,y)$ are given known functions. The general solution of the free vibration problem is given by

$$w(x,y)=\sum_m\sum_n \sin\frac{m\pi x}{a}\sin\frac{n\pi y}{b}(A_{mn}\cos\omega_{mn}t+B_{mn}\sin\omega_{mn}t) \tag{9.34}$$

The terms $A_{mn}$ and $B_{mn}$ are determined so that they satisfy the initial conditions, thus we have

$$\begin{aligned} w_0(x,y)&=\sum_m\sum_n \sin\frac{m\pi x}{a}\sin\frac{n\pi y}{b}A_{mn} && \text{(a)}\\ \dot{w}_0(x,y)&=\sum_m\sum_n \sin\frac{m\pi x}{a}\sin\frac{n\pi y}{b}\omega_{mn}B_{mn} && \text{(b)}\end{aligned} \tag{9.35}$$

Multiplying both sides of Equation 9.35 by $\sin(i\pi x/a)\sin(j\pi y/b)$ and integrating from 0 to $a$ in $x$ and 0 to $b$ in $y$ yields

$$\begin{aligned} \int_0^a\int_0^b w_0(x,y)\sin\frac{i\pi x}{a}\sin\frac{j\pi y}{b}dxdy&=\sum_m\sum_n\int_0^a\int_0^b \sin\frac{m\pi x}{a}\sin\frac{n\pi y}{b}A_{mn}\sin\frac{i\pi x}{a}\sin\frac{j\pi y}{b}dxdy && \text{(a)}\\ \int_0^a\int_0^b \dot{w}_0(x,y)\sin\frac{i\pi x}{a}\sin\frac{j\pi y}{b}dxdy&=\sum_m\sum_n\int_0^a\int_0^b \sin\frac{m\pi x}{a}\sin\frac{n\pi y}{b}\omega_{mn}B_{mn}\sin\frac{i\pi x}{a}\sin\frac{j\pi y}{b}dxdy && \text{(b)}\end{aligned} \tag{9.36}$$

The integrals are nonzero only, if and only if, $m = i$ and $n = j$. Therefore, Equation 9.36 can be written as

$$A_{ij}\int_0^a\int_0^b \sin^2\frac{i\pi x}{a}\sin^2\frac{j\pi y}{b}\,dx\,dy = \int_0^a\int_0^b w_0(x,y)\sin\frac{i\pi x}{a}\sin\frac{j\pi y}{b}\,dx\,dy \quad \text{(a)}$$

$$B_{ij}\omega_{ij}\int_0^a\int_0^b \sin^2\frac{i\pi x}{a}\sin^2\frac{j\pi y}{b}\,dx\,dy = \int_0^a\int_0^b \dot{w}_0(x,y)\sin\frac{i\pi x}{a}\sin\frac{j\pi y}{b}\,dx\,dy \quad \text{(b)} \tag{9.37}$$

From Equations 9.37 we can determine $A_{mn}$ and $B_{mn}$ as

$$A_{mn} = \frac{4}{ab}\int_0^a\int_0^b w_0(x,y)\sin\frac{m\pi x}{a}\sin\frac{n\pi y}{b}\,dx\,dy \tag{9.38}$$

and

$$B_{mn} = \frac{4}{ab\omega_{mn}}\int_0^a\int_0^b \dot{w}_0(x,y)\sin\frac{m\pi x}{a}\sin\frac{n\pi y}{b}\,dx\,dy \tag{9.39}$$

where we have substituted the subscripts $mn$ for $ij$. If explicit plate dimensions and material properties are known, we can define $\omega_{mn}$ from Equation 9.33, $A_{mn}$ and $B_{mn}$ from Equations 9.38 and 9.39, respectively. Then, we can determine $w(x, y)$ using Equation 9.34.

### 9.2.2 Solutions for Other Boundary Conditions

For a free vibration problem, we assume a solution in the form $w(x, y, t) = W(x, y)(A\cos\omega t + B\sin\omega t)$. From the equation of motion, we obtain an expression for $W$

$$\nabla^2\nabla^2 W - \lambda^4 W = 0$$

where

$$\lambda^4 = \frac{\rho h\omega^2}{D}$$

And for simply supported boundaries, we assumed that

$$W(x,y) = \sin\frac{m\pi x}{a}\sin\frac{n\pi y}{b}$$

To solve for a plate that has arbitrary boundary conditions, we use the technique of separation of variables, and assume that

$$W(x,y) = X(x)Y(y) \tag{9.40}$$

Substituting Equation 9.40 into Equation 9.29 gives

$$X''''Y + 2X''Y'' + XY'''' - \lambda^4 XY = 0 \tag{9.41}$$

or

$$\frac{X''''}{X} + 2\frac{X''}{X}\frac{Y''}{Y} + \frac{Y''''}{Y} - \lambda^4 = 0 \tag{9.42}$$

where, in Equation 9.41 or 9.42 , a prime denotes a derivative with respect to its argument. It is generally assumed that the variables in Equation 9.42 cannot be separated. However, Equation 9.42 can easily be separated by simply differentiating Equation 9.42 with respect to $x$ and $y$. Thus, differentiating with respect to $x$ and $y$, we have

$$\frac{d}{dx}\left(\frac{X''''}{X}\right) + 2\frac{Y''}{Y}\frac{d}{dx}\left(\frac{X''}{X}\right) = 0 \quad \text{and} \quad 2\frac{X''}{X}\frac{d}{dy}\left(\frac{Y''}{Y}\right) + \frac{d}{dy}\left(\frac{Y''''}{Y}\right) = 0$$

Therefore, in order for the separation of variables to occur, it is required that

$$Y''(y) = -\gamma^2 Y(y), \quad Y''''(y) = -\gamma^2 Y''(y) \tag{9.43}$$

or

$$X''(x) = -\alpha^2 X(x), \quad X''''(x) = -\alpha^2 X''(x) \tag{9.44}$$

or both are satisfied. Equation 9.43 can be satisfied by

$$Y_n(y) = \bar{A}\sin\gamma_n y + \bar{B}\cos\gamma_n y \tag{9.45}$$

or if we use Equation 9.44 we have

$$X_m(x) = \hat{A}\sin\alpha_m x + \hat{B}\cos\alpha_m x \tag{9.46}$$

where $\bar{A}$, $\bar{B}$, $\hat{A}$, and $\hat{B}$ are constants and

$$\gamma_n = \frac{n\pi}{b}, \quad n = 1, 2, \ldots, \quad \alpha_m = \frac{m\pi}{a}, \quad m = 1, 2, \ldots \tag{9.47}$$

Assume that two edges of the plate are simply supported and the remaining two edges have arbitrary boundary conditions as shown in Figure 9.8, which has simple supports at $y = 0$ and $y = b$, and arbitrary boundary conditions at $x = 0$ and $x = a$.

For the plate simply supported along the edges $y = 0$ and $y = b$, we have from Equation 9.45 that

$$Y_n(y) = \bar{A}\sin\gamma_n y, \qquad n = 1, 2, \ldots \tag{9.48}$$

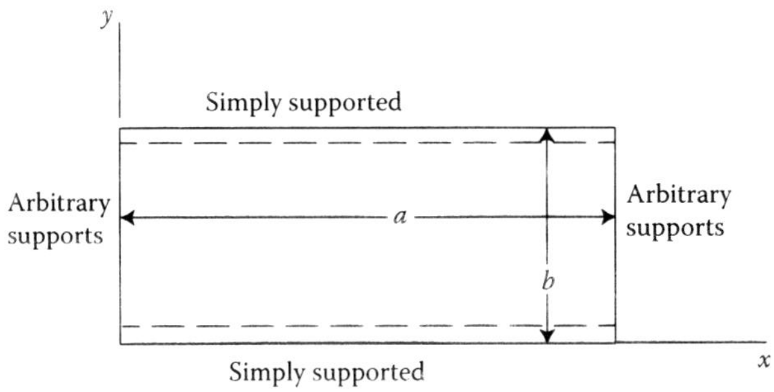


**FIGURE 9.8**
Plate with arbitrary boundary conditions along two edges and simple supports along the others.

where $\bar{A}$ is a constant. The boundary conditions

$$Y_n(0) = Y_n(b) = Y''(0) = Y''(b) = 0 \tag{9.49}$$

are satisfied by Equation 9.48 for any integer $n$ and therefore, the boundary conditions

$$w(x,0,t) = w(x,b,t) = \nabla^2 w(x,0,t) = \nabla^2 w(x,b,t) = 0 \tag{9.50}$$

are also satisfied. Since we have assumed that the plate is simply supported at $y = 0$ and $y = b$, we can substitute Equation 9.48 into Equation 9.41 yielding the expression

$$\frac{d^4X(x)}{dX^4} - 2\gamma_n^2 \frac{d^2X(x)}{dX^2} - \left(\lambda^4 - \gamma_n^4\right)X(x) = 0 \tag{9.51}$$

For the homogeneous Equation 9.51, its particular solution has, assuming $\lambda^4 > \gamma_n^4$, the form

$$X(x) = e^{sx} \tag{9.52}$$

Upon substituting Equation 9.52 into Equation 9.51 yields the characteristic equation of Equation 9.51 as

$$s^4 - 2\gamma_n^2 s^2 - \left(\lambda^4 - \gamma_n^4\right) = 0 \tag{9.53}$$

By solving Equation 9.53, one can obtain the characteristic roots

$$s_{1,2} = \pm\sqrt{\lambda^2 + \gamma_n^2}, \quad s_{3,4} = \pm i\sqrt{\lambda^2 - \gamma_n^2} \tag{9.54}$$

where $i = \sqrt{-1}$. Letting $\varepsilon^2 = \lambda^2 - \gamma_n^2$ and $\delta^2 = \lambda^2 + \gamma_n^2$, the solution of Equation 9.51 can be written as

$$X(x) = C_1 \sin\delta x + C_2 \cos\delta x + C_3 \sinh\varepsilon x + C_4 \cosh\varepsilon x \tag{9.55}$$

where $C_1$, $C_2$, $C_3$, and $C_4$ are arbitrary constants. Now we apply the boundary conditions at $x = 0$ and $x = a$ to Equation 9.55. Assume, for example, that at $x = 0$ and $x = a$ both ends are built-in. We have

$$\begin{aligned} X(0) &= C_1(0)+C_2(1)+C_3(0)+C_4(1)=0 \\ X'(0) &= \delta C_1(1)-C_2(0)+\varepsilon C_3(1)+C_4(0)=0 \\ X(a) &= C_1 \sin\delta a + C_2\cos\delta a + C_3\sinh\varepsilon a + C_4\cosh\varepsilon a = 0 \\ X'(x) &= \delta C_1\cos\delta a - \delta C_2\sin\delta a + \varepsilon C_3\cosh\varepsilon a + \varepsilon C_4\sinh\varepsilon a = 0 \end{aligned} \tag{9.56}$$

The system of equations, Equation 9.56, has nontrivial solution when the determinant of the coefficient matrix for the unknown quantities for $C_i = 0, i = 1, 2, 3, 4$ is equal to zero. In this case, the determinant has the following form:

$$\begin{vmatrix} 0 & 1 & 0 & 1 \\ \delta & 0 & \varepsilon & 0 \\ \sin\delta a & \cos\delta a & \sinh\varepsilon a & \cosh\varepsilon a \\ \delta\cos\delta a & -\delta\sin\delta a & \varepsilon\cosh\varepsilon a & \varepsilon\sinh\varepsilon a \end{vmatrix} = 0 \tag{9.57}$$

Expanding the determinate gives the frequency equation

$$(\delta^2-\varepsilon^2)(\sinh\varepsilon a\sin\delta a)+2\delta\varepsilon(\cosh\varepsilon a\cos\delta a-1)=0$$

or

$$\gamma_n^2(\sinh\varepsilon a\sin\delta a)+\delta\varepsilon(\cosh\varepsilon a\cos\delta a-1)=0 \tag{9.58}$$

For any specific value of $n$, there will be successive values $\lambda$, that is $\lambda_{1n}$, $\lambda_{2n}$, …, $\lambda_{mn}$. For each $\lambda_{mn}$ we find the frequency $\omega$ that satisfies the frequency (Equation 9.58). Thus, we have

$$\omega_{mn} = \lambda_{mn}^2\sqrt{\frac{D}{h\rho}} \tag{9.59}$$

where the corresponding natural mode is given by

$$W_{mn}(x,y) = X_{mn}(x)\sin\frac{n\pi y}{b} \tag{9.60}$$

Other boundary conditions can be analyzed using the same procedure as above. It is noted that there are six possible combinations of boundary conditions when the edges $y = 0$ and $y = b$ are simply supported. Using SS, C, and F for edges being simply supported, clamped or fixed, and free we have that the remaining five boundary conditions would, respectively, be SS–SS–SS–SS, SS–F–SS–F, SS–C–SS–SS, SS–F–SS–SS, and SS–F–SS–C.

Once again the orthogonality of natural modes can be used if we consider $\omega_{mn}$, $W_{mn}$, and $\omega_{KL}$, $W_{KL}$ for the simply supported plate as shown in Figure 9.7. We have

$$\begin{aligned} \nabla^4 W_{mn} &= \frac{h\rho}{D}\omega_{mn}^2 W_{mn} \quad \text{(a)} \\ \nabla^4 W_{KL} &= \frac{h\rho}{D}\omega_{KL}^2 W_{KL} \quad \text{(b)} \end{aligned} \tag{9.61}$$

Multiply the first Equation 9.61a by $W_{KL}$, the second (9.61b) by $W_{mn}$, and integrate over the plate area and subtract the results. This gives

$$\int_0^a\int_0^b \left\{\left(\nabla^4 W_{mn}\right)W_{KL} - \left(\nabla^4 W_{KL}\right)W_{mn}\right\}dx\,dy = \frac{h\rho}{D}\left(\omega_{mn}^2 - \omega_{KL}^2\right)\int_0^a\int_0^b W_{mn}W_{KL}\,dx\,dy$$

Integrating the left-hand side by parts, and using the end conditions we find

$$\left(\omega_{mn}^2 - \omega_{KL}^2\right)\int_0^a\int_0^b W_{mn}W_{KL}\,dx\,dy = 0 \tag{9.62}$$

If $\omega_{mn}^2 \neq \omega_{KL}^2$, we have

$$\int_0^a\int_0^b W_{mn}W_{KL}\,dx\,dy = 0 \tag{9.63}$$

### 9.2.3 Forced Vibrations

A major problem in the analysis of plates is the determination of the deformations in a plate subjected to time-dependent transverse surface loads. Consider, therefore, a simply supported rectangular plate with edge dimensions $a$, $b$ subjected to an applied load $p = p(x, y, t)$. The equation of motion, Equation 9.18, is

$$\nabla^4 w(x,y,t) = -\frac{h\rho}{D}\frac{\partial^2 w(x,y,t)}{\partial t^2} + \frac{p(x,y,t)}{D}$$

The boundary conditions for a simply supported plate are known, Equations 9.20 and 9.21. We need to determine a solution that satisfies both the equation of motion and the boundary conditions. Using the *modal expansion method,* we assume a displacement field as

$$w(x,y,t) = \sum_{m=1}^{\infty}\sum_{n=1}^{\infty}\Phi_{mn}(t)\sin\frac{m\pi x}{a}\sin\frac{n\pi y}{b} \tag{9.64}$$

where $\Phi_{mn}(t)$ is an unknown function of time. The boundary conditions are satisfied by the assumed displacement field given in Equation 9.64. Substituting Equation 9.64 into Equation 9.18 yields

$$\sum_{m=1}^{\infty}\sum_{n=1}^{\infty}\left\{\left[\left(\frac{m\pi}{a}\right)^2+\left(\frac{n\pi}{b}\right)^2\right]^2\Phi_{mn}+\frac{h\rho}{D}\ddot{\Phi}_{mn}\right\}\sin\frac{m\pi x}{a}\sin\frac{n\pi y}{b}=\frac{p(x,y,t)}{D}$$

Multiplying both sides of the above expression by $\sin(i\pi x/a)\sin(j\pi y/b)$ and integrating over the plate area gives

$$\sum_{m=1}^{\infty}\sum_{n=1}^{\infty}\left\{\left[\left(\frac{m\pi}{a}\right)^2+\left(\frac{n\pi}{b}\right)^2\right]^2\Phi_{mn}+\frac{h\rho}{D}\ddot{\Phi}_{mn}\right\}\int_0^a\int_0^b\sin\frac{m\pi x}{a}\sin\frac{n\pi y}{b}\sin\frac{i\pi x}{a}\sin\frac{j\pi y}{b}\,dxdy$$
$$=\int_0^a\int_0^b\frac{p(x,y,t)}{D}\sin\frac{i\pi x}{a}\sin\frac{j\pi y}{b}\,dx\,dy$$

This yields the governing equation for $\Phi_{mn}$ as

$$\ddot{\Phi}_{mn}+\omega_{mn}^2\Phi_{mn}=\frac{1}{h\rho}P_{mn}(t) \tag{9.65}$$

where

$$P_{mn}(t)=\frac{4}{ab}\int_0^a\int_0^b p(x,y,t)\sin\frac{m\pi x}{a}\sin\frac{n\pi y}{b}\,dx\,dy \tag{9.66}$$

From Equation 9.65, we have

$$\Phi_{mn}(t)=C_1e^{r_1t}+C_2e^{r_2t}+\Phi_{mn}^p(t) \tag{9.67}$$

where $C_1$ and $C_2$ are constants that are determined from the initial conditions and $\Phi_{mn}^p(t)$ is the particular solution. We find from the homogeneous solution that $r_1$ and $r_2$ are the roots of the equation

$$r^2+\omega_{mn}^2=0,\quad r_1=-i\omega_{mn},\quad r_2=i\omega_{mn} \tag{9.68}$$

and the particular solution is given as

$$\Phi_{mn}^p(t)=\frac{1}{\rho h\omega_{mn}}\int_0^t P_{mn}(\tau)\sin\omega_{mn}(t-\tau)d\tau \tag{9.69}$$

The solution, Equation 9.67, can be written as

$$\Phi_{mn}(t)=A_{mn}\cos\omega_{mn}t+B_{mn}\sin\omega_{mn}t+\Phi_{mn}^p(t) \tag{9.70}$$

giving the total solution as

$$w(x,y,t)=\sum_{m=1}^{\infty}\sum_{n=1}^{\infty}\left[\Phi_{mn}^{p}(t)+A_{mn}\cos\omega_{mn}t+B_{mn}\sin\omega_{mn}t\right]\sin\frac{m\pi x}{a}\sin\frac{n\pi y}{b} \tag{9.71}$$

We determine $A_{mn}$ and $B_{mn}$ from the initial conditions, which are typically expressed at $t=0$ as

$$w(x,y,0)=w_0(x,y)$$

$$\dot{w}(x,y,0)=\dot{w}_0(x,y)$$

Thus, we have for $A_{mn}$, $B_{nm}$ the expressions

$$A_{mn}=\frac{4}{ab}\int_0^a\int_0^b w_0(x,y)\sin\frac{m\pi x}{a}\sin\frac{n\pi y}{b}\,dxdy-\Phi_{mn}^{p}(0) \tag{9.72}$$

and

$$B_{mn}=\frac{1}{\omega_{mn}}\left\{\frac{4}{ab}\int_0^a\int_0^b \dot{w}_0(x,y)\sin\frac{m\pi x}{a}\sin\frac{n\pi y}{b}\,dx\,dy-\dot{\Phi}_{mn}^{p}(0)\right\} \tag{9.73}$$

**EXAMPLE 9.1**

A simply supported plate with edge dimensions $a$, $b$, and thickness $h$ is subjected to the loading condition given as $p(x,y,t)=P(x,y)e^{i\omega_f t}$. Assume $\Phi_{mn}=C_{mn}e^{i\omega_f t}$ and determine an expression for the displacement $w(x, y, t)$.

**Solution**

Given $p(x,y,t)=P(x,y)e^{i\omega_f t}$ and $\Phi_{mn}=C_{mn}e^{i\omega_f t}$, we use Equation 9.60 to write

$$-\omega_f^2C_{mn}+\omega_{mn}^2C_{mn}=\frac{1}{h\rho}P_{mn}$$

We define $P_{mn}(t)$ using Equation 9.61 as

$$P_{mn}(t)=\frac{4}{ab}\int_0^a\int_0^b P(x,y)\sin\frac{m\pi x}{a}\sin\frac{n\pi y}{b}\,dx\,dy$$

We now solve the governing equation to determine $C_{mn}$, thus

$$C_{mn}=\frac{P_{mn}}{h\rho}\left(\frac{1}{\omega_{mn}^2-\omega_f^2}\right)$$

and with zero initial displacement and velocity the displacement is then given by

$$w(x,y,t)=\left\{\sum_{m=1}^{\infty}\sum_{n=1}^{\infty}\frac{P_{mn}}{h\rho}\left(\frac{1}{\omega_{mn}^2-\omega_f^2}\right)\sin\frac{m\pi x}{a}\sin\frac{n\pi y}{b}\right\}e^{i\omega_f t}$$

Note that if $\omega_f \to \omega_{mn}$ (resonance) for any $m$, $n$ the displacement approaches infinity ($w \to \infty$).

**EXAMPLE 9.2**

A simply supported plate with edge dimensions $a$, $b$, and thickness $h$ is subjected to the loading condition

$$\begin{aligned} p(x,y,t)&=0 \quad \text{for } t\le 0\\ p(x,y,t)&=p_0 \quad \text{for } t>0 \end{aligned}$$

Determine the response assuming the initial conditions to be zero.

**Solution**

Since the initial conditions are zero, the response is given by the steady-state solution

$$w(x,y,t)=\sum_{m=1}^{\infty}\sum_{n=1}^{\infty}\Phi_{mn}^{p}(t)\sin\frac{m\pi x}{a}\sin\frac{n\pi y}{b}$$

where

$$\Phi_{mn}^{p}(t)=\frac{1}{\rho h\omega_{mn}}\int_0^t P_{mn}(\tau)\sin\omega_{mn}(t-\tau)d\tau$$

and

$$P_{mn}(t)=\frac{4}{ab}\int_0^a\int_0^b p(x,y,t)\sin\frac{m\pi x}{a}\sin\frac{n\pi y}{b}dx\,dy$$

Substituting for $p(x, y, t) = p_0$ yields

$$\begin{aligned} P_{mn}(t)&=\frac{4}{ab}\int_0^a\int_0^b p_0\sin\frac{m\pi x}{a}\sin\frac{n\pi y}{b}dx\,dy\\ &=\frac{4p_0}{ab}\frac{\left(\cos(m\pi x/a)\right)_0^a}{(m\pi/a)}\frac{\left(\cos(m\pi y/b)\right)_0^b}{(m\pi/b)}\\ &=\begin{cases}\dfrac{16p_0}{\pi^2 mn} & \text{for } m,n \text{ both odd}\\ 0 & m \text{ and/or } n \text{ even}\end{cases} \end{aligned}$$

Therefore,

$$\Phi^{p}_{mn}(t) = \frac{16p_0}{\pi^2 mn\rho h\omega_{mn}} \int_0^t \sin\omega_{mn}(t-\tau)\,d\tau$$
$$= \frac{16p_0}{\pi^6 mnD((m^2/a^2)+(n^2/b^2))^2}(1-\cos\omega_{mn}t)$$

and the solution is given as

$$w(x,y,t) = \sum_{m=1,3,5,\ldots}^{\infty} \sum_{n=1,3,5,\ldots}^{\infty} \frac{16p_0}{\pi^6 mnD((m^2/a^2)+(n^2/b^2))^2}(1-\cos\omega_{mn}t)\sin\frac{m\pi x}{a}\sin\frac{n\pi y}{b}$$

### 9.2.4 Composite Plates

The equations of motion previously presented applied to plates made from isotropic homogeneous materials. When considering composite materials, we no longer have an isotropic material, but one in which the elastic properties depend on direction (fiber orientation). This complicates the analysis of the equations of motion because the bending rigidity ($D = Eh^3/[12(1-\nu^2)]$) is no longer a single term. A complete presentation on the structural behavior of composite materials is beyond the scope of this text. Those unfamiliar with laminate analysis are referred to Appendix A, which provides a brief introduction to the governing equations for the analysis of laminated composites materials.

The nomenclature used for laminate analysis differs from that used for homogeneous materials. In classical lamination theory (CLT), the displacements ($u$, $v$, and $w$) are defined with respect to the midsurface of the laminate. As a result, the displacements of the midsurface of the laminate in each direction are defined by $U_0$, $V_0$, and $W_0$. Using the traditional CLT nomenclature of $U_{0,x} = \partial U_0/\partial x$, $U_{0,xx} = \partial^2 U_0/\partial x^2$, etc., the relationship between loads and displacements for a laminated plate is

$$\begin{Bmatrix} N_x \\ N_y \\ N_{xy} \\ M_x \\ M_y \\ M_{xy} \end{Bmatrix} = \begin{bmatrix} A_{11} & A_{12} & A_{16} & B_{11} & B_{12} & B_{16} \\ A_{12} & A_{22} & A_{26} & B_{12} & B_{22} & B_{26} \\ A_{16} & A_{26} & A_{66} & B_{16} & B_{26} & B_{66} \\ B_{11} & B_{12} & B_{16} & D_{11} & D_{12} & D_{16} \\ B_{12} & B_{22} & B_{26} & D_{12} & D_{22} & D_{26} \\ B_{16} & B_{26} & B_{66} & D_{16} & D_{26} & D_{66} \end{bmatrix} \begin{Bmatrix} U_{0,x} \\ V_{0,y} \\ U_{0,y}+V_{0,x} \\ -W_{0,xx} \\ -W_{0,yy} \\ -2W_{0,xy} \end{Bmatrix} \tag{9.74}$$

Following the same procedures used to define the equations of motion for an isotropic material, it is obvious from Equation 9.74 that the equations of motion for a laminated composite are somewhat more complicated. Extended derivations of this equation can be found in various reference texts, including Reddy (1997). The equation most applicable to the discussions herein is

$$\begin{aligned} &D_{11}W_{0,xxxx} + 4D_{16}W_{0,xxxy} + 2(D_{12}+2D_{66})W_{0,xxyy} + 4D_{26}W_{0,xyyy} + D_{22}W_{0,yyyy} - B_{11}U_{0,xxx} \\ &- 3B_{16}U_{0,xxy} - (B_{12}+2B_{66})U_{0,xyy} - B_{26}U_{0,yyy} - B_{16}V_{0,xxx} - (B_{12}+2B_{66})V_{0,xxy} - 3B_{26}V_{0,xyy} \\ &- B_{22}V_{0,yyy} + \rho\dddot{W}_0 = q \end{aligned} \tag{9.75}$$

As a result of the relative complexity of Equation 9.43, the discussions herein are limited to specially orthotropic plates for which $[B] = 0$ and the only nonzero terms of $[A]$ and $[D]$ are $A_{11}$, $A_{12}$, $A_{22}$, $A_{66}$, $D_{11}$, $D_{12}$, $D_{22}$, and $D_{66}$. This presentation is intended to illustrate approaches and procedures involved in exploring plate-vibration problems. Solutions to more complex problems are best obtained through appropriate numerical techniques.

#### 9.2.4.1 Free Vibrations of a Simply Supported Plate

Fore a specially orthotropic ($[B] = 0$) simply supported rectangular plate with no loads ($q = N_x = N_y = N_{xy} = 0$), the equation of motion reduces to

$$D_{11}\frac{d^4W_0}{dx^4} + 2(D_{12} + 2D_{66})\frac{d^4W_0}{dx^2\,dy^2} + D_{22}\frac{d^4W_0}{dy^4} + \rho\frac{d^2W_0}{dt^2} = 0 \tag{9.76}$$

In order to satisfy boundary and initial conditions, we assume a solution of the form $W_0 = W(x, y)e^{i\omega t}$. The equation of motion reduces to

$$D_{11}W_{,xxxx} + 2(D_{12} + 2D_{66})W_{,xxyy} + D_{22}W_{,yyyy} - \rho\omega^2 W = 0$$

The boundary conditions at $x = 0$ and $x = a$ are

$$W(x,y) = M_x = -D_{11}W_{,xx} - D_{12}W_{,yy} = 0$$

and at $y = 0, b$:

$$W(x,y) = M_x = -D_{12}W_{,xx} - D_{22}W_{,yy} = 0$$

Assuming a displacement field of $W(x, y) = A_{mn}\sin(m\pi x/a)\sin(n\pi y/b)$, where $m$ and $n$ are integers, we get

$$D_{11}\left(\frac{m\pi}{a}\right)^4 + 2(D_{12} + 2D_{66})\left(\frac{m\pi}{a}\right)^2\left(\frac{n\pi}{b}\right)^2 + D_{22}\left(\frac{n\pi}{b}\right)^4 - \rho\omega_{mn}^2 = 0$$

This is obviously different from the corresponding solution for an isotropic plate in which the bending rigidity is simply defined by $D$ as opposed to $D_{11}$, $D_{12}$, etc. Defining the aspect ratio of the plate as $R = a/b$, we can solve the equation above for $\omega_{mn}$

$$\omega_{mn} = \frac{\pi^2}{R^2b^2}\frac{1}{\sqrt{\rho}}\sqrt{D_{11}m^4 + 2(D_{12} + 2D_{66})m^2n^2R^2 + D_{22}n^4R^4} \tag{9.77}$$

The simplest solution is for $m = n = 1$, in which case

$$\omega_{11} = \frac{\pi^2}{R^2b^2}\frac{1}{\sqrt{\rho}}\sqrt{D_{11} + 2(D_{12} + 2D_{66})R^2 + D_{22}R^4} \tag{9.78}$$

For comparative purposes, we note that for an isotropic plate, we have $\omega_{11} = (\pi^2/b^2)\sqrt{(D/\rho)}[(b/a)^2 + 1]$. Assuming the plate is square ($R = 1$) and further assuming $D_{11}/D_{22} = 10$ and $(D_{12} + 2D_{66})/D_{22} = 1$, we get $\omega_{11}$ for the composite and isotropic plates to be

$$\omega_{11\text{-composite}} = 3.61\frac{\pi^2}{b^2}\sqrt{\frac{D_{22}}{\rho}} \qquad \omega_{11\text{-isotropic}} = 2\frac{\pi^2}{b^2}\sqrt{\frac{D}{\rho}}$$

A one-to-one comparison of frequencies would obviously require explicit values for $D$, $D_{22}$, $\rho$, etc., which is not the intent of this presentation. Suffice it to say that the difference in frequencies results from a difference in materials.

Along similar lines, assume the laminate had its lamina arranged in what is termed an angle-ply arrangement. In this case $D_{16} \neq 0$ and $D_{26} \neq 0$, which results in the equation of motion being

$$D_{11}W_{0,xxxx} + 4D_{16}W_{0,xxxy} + 2(D_{12} + 2D_{66})W_{0,xxyy} + D_{26}W_{0,xyyy} + D_{22}W_{0,yyyy} + \rho W_{0,tt} = 0 \quad (9.79)$$

Adding the terms $D_{16}$ and $D_{26}$ into the equation of motion results in boundary conditions which are somewhat different from those of the previous case. In this, the boundary conditions are

$$\begin{aligned} &\text{at } x = 0, a: \quad W(x,y) = M_x = -D_{11}W_{,xx} - D_{12}W_{,yy} - 2D_{16}W_{xy} = 0 \\ &\text{at } y = 0, b: \quad W(x,y) = M_y = -D_{12}W_{,xx} - D_{22}W_{,yy} - 2D_{26}W_{xy} = 0 \end{aligned} \quad (9.80)$$

If we once again assume $W(x, y) = A_{mn}\sin(m\pi x/a)\sin(n\pi y/b)$, the equation of motion becomes

$$\begin{aligned} &D_{11}\left(\frac{m\pi}{a}\right)^4 + 4D_{16}\left(\frac{m\pi}{a}\right)^3\left(\frac{n\pi}{b}\right) + 2(D_{12} + 2D_{66})\left(\frac{m\pi}{a}\right)^2\left(\frac{n\pi}{b}\right)^2 \\ &+ 4D_{26}\left(\frac{m\pi}{a}\right)\left(\frac{n\pi}{b}\right)^3 + D_{22}\left(\frac{n\pi}{b}\right)^4 - \rho\omega_{mn}^2 = 0 \end{aligned} \quad (9.81)$$

Introducing the aspect ratio ($R = a/b$), we can solve for $\omega_{mn}$ and get

$$\omega_{mn} = \frac{\pi^2}{R^2b^2}\frac{1}{\sqrt{\rho}}\sqrt{D_{11}m^4 + 4D_{16}m^3nR + 2(D_{12} + 2D_{66})m^2n^2R^2 + 4D_{26}mn^3R^3 + D_{22}n^4R^4} \quad (9.82)$$

As earlier, we consider a square orthotropic plate ($R = 1$) with the appropriate terms of $[D]$ approximated to be $D_{11}/D_{22} = 10$, $(D_{12} + 2D_{66})/D_{22} = 1$, $4D_{16}/D_{22} = -0.3$, and $4D_{26}/D_{22} = -0.03$. The frequency equation above therefore becomes

$$\omega_{mn} = \frac{\pi^2}{b^2}\frac{1}{\sqrt{\rho}}\sqrt{D_{22}\left(10m^4 - 0.3m^3n + 2m^2n^2 - 0.03mn^3 + n^4\right)} \quad (9.83)$$

For the case of $m = n = 1$, we get

$$\omega_{11} = \frac{3.56\pi^2}{b^2}\sqrt{\frac{D_{22}}{\rho}} \tag{9.84}$$

For the specialty orthotropic laminate, we obtained $\omega_{11} = (3.62\pi^2/b^2)\sqrt{D_{22}/\rho}$. Although the difference between these two solutions does not appear significant, it does indicate the influence of material behavior (in particular the $[D]$ matrix) on the solution.

**EXAMPLE 9.3**

Assume two rectangular cross-ply laminated plate is made from AS/3501 graphite/epoxy. One laminate has the stacking arrangement $[0/90_5]_s$ and the other $[90_5/0]_s$. The $[\bar{Q}]$ for each lamina is

$$[Q] = [\bar{Q}]_0 = \begin{bmatrix} 20.14 & 0.392 & 0 \\ 0.390 & 1.307 & 0 \\ 0 & 0 & 1.03 \end{bmatrix} \times 10^6 \text{ psi and } [\bar{Q}]_{90} = \begin{bmatrix} 1.307 & 0.392 & 0 \\ 0.390 & 20.14 & 0 \\ 0 & 0 & 1.03 \end{bmatrix} \times 10^6 \text{ psi}$$

The thickness of each lamina is $t_{ply} = 0.005$ in and the total plate thickness is $h = 0.06$ in. Consider the $[0/90_5]_s$ laminate to be Case I and the $[90_5/0]_s$ laminate to be Case II. Assume each plate is simply supported and has an aspect ratio of $R = a/b = 2$. For the lowest natural frequency ($\omega_{11}$), determine the ratio $(\omega_{11\text{-Case I}})/(\omega_{11\text{-Case II}})$.

**Solution**

For the $[0/90_5]_s$ (Case I) and the $[90_5/0]_s$ (Case II) laminates, we use Equation A.11 to determine

$$[D]_{\text{Case I}} = \sum [\bar{Q}]_k \left( t_k\bar{z}_k^2 + \frac{t_k^3}{12} \right) = 2[\bar{Q}]_0 \left\{ t_{ply}\left(\frac{11t_{ply}}{2}\right)^2 + \frac{t_{ply}^3}{12} \right\} + [\bar{Q}]_{90}\frac{(10t_{ply})^3}{12}$$

$$= \begin{bmatrix} 1330.7 & 56.45 & 0 \\ 56.45 & 1757.6 & 0 \\ 0 & 0 & 117.1 \end{bmatrix} \times 10^6(0.005)^3 = \begin{bmatrix} 166.3 & 7.06 & 0 \\ 7.06 & 219.7 & 0 \\ 0 & 0 & 14.6 \end{bmatrix}$$

$$[D]_{\text{Case II}} = \sum [\bar{Q}]_k \left( t_k\bar{z}_k^2 + \frac{t_k^3}{12} \right) = [\bar{Q}]_0 \left(\frac{2t_{ply}}{12}\right)^3 + 2[\bar{Q}]_{90} \left\{ t_{ply}\frac{(7t_{ply})^2}{2} + \frac{(5t_{ply})^3}{12} \right\}$$

$$= \begin{bmatrix} 72.67 & 18.03 & 0 \\ 18.03 & 913.9 & 0 \\ 0 & 0 & 47.3 \end{bmatrix} \times 10^6(0.005)^3 = \begin{bmatrix} 9.08 & 2.25 & 0 \\ 2.25 & 114.2 & 0 \\ 0 & 0 & 5.91 \end{bmatrix}$$

From Equation 9.77 with $R = 2$ and $m = n = 1$, we have

$$\omega_{11} = \frac{\pi^2}{4b^2}\frac{1}{\sqrt{\rho}}\sqrt{D_{11} + 8(D_{12} + 2D_{66}) + 16D_{22}}$$

Using the information above, we find

$$\omega_{11\text{-Case I}} = \frac{\pi^2}{4b^2}\frac{1}{\sqrt{\rho}}\sqrt{166.3 + 8(7.06 + 2(14.6)) + 16(219.7)} = 15.76\frac{\pi^2}{b^2}\frac{1}{\sqrt{\rho}}$$

$$\omega_{11\text{-Case II}} = \frac{\pi^2}{4b^2}\frac{1}{\sqrt{\rho}}\sqrt{9.08 + 8(2.25 + 2(5.91)) + 16(114.2)} = 11.03\frac{\pi^2}{b^2}\frac{1}{\sqrt{\rho}}$$

Therefore,

$$\frac{\omega_{11\text{-Case I}}}{\omega_{11\text{-Case II}}} = \frac{15.76}{11.03} \approx 1.43$$

#### 9.2.4.2 *Forced Vibrations*

It should be apparent that the principal difference between the evaluation of isotropic and orthotropic plates is the behavior of the material, which is evident in the equations of motion. The solution procedures for forced vibration problems involving laminate composite plates are identical to those for isotropic plates. A more comprehensive discussion of forced vibrations for laminated composite plates can be obtained from various references specifically dedicated to composite materials, such as Reddy (2004).

## 9.3 Circular Plates

We have been using Cartesian ($x$, $y$, $z$) coordinates, but for circular plates we write the plate equation in cylindrical coordinates ($r$, $\phi$, $z$), where the relationship between $x$, $y$, and $\phi$ is shown in Figure 9.9. The two systems are easily related by the simple relations $x = r\cos\phi$, $y = r\sin\phi$, $r = \sqrt{x^2 + y^2}$, and $\phi = \tan^{-1} y/x$.

The partial derivatives needed to transform the plate equations in Cartesian coordinates into cylindrical coordinates are

$$\frac{\partial r}{\partial x} = \frac{x}{r} = \cos\phi, \quad \frac{\partial r}{\partial y} = \frac{y}{r} = \sin\phi \tag{9.85}$$

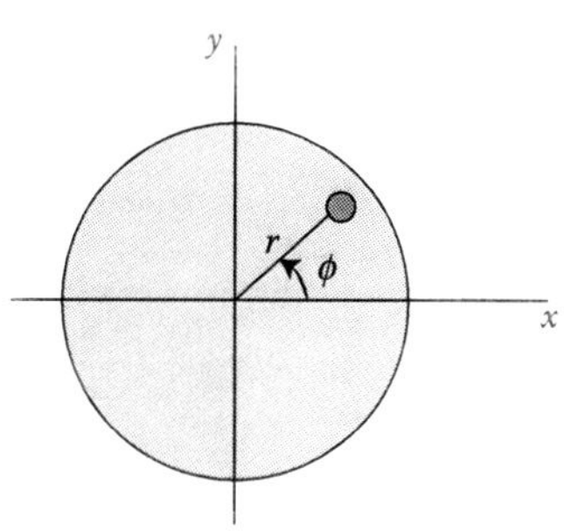


**FIGURE 9.9**
Coordinate system used for circular plates.

and

$$\frac{\partial \phi}{\partial x} = -\frac{y}{r^2} = -\frac{\sin\phi}{r}, \quad \frac{\partial \phi}{\partial y} = \frac{x}{r^2} = \frac{\cos\phi}{r} \tag{9.86}$$

Since the plate deflection $w$ is a function of $r$ and $\phi$, we can use the chain rule of differentiation to give

$$\frac{\partial w}{\partial x} = \frac{\partial w}{\partial r}\frac{\partial r}{\partial x} + \frac{\partial w}{\partial \phi}\frac{\partial \phi}{\partial x} = \frac{\partial w}{\partial r}\cos\phi - \frac{\sin\phi}{r}\frac{\partial w}{\partial \phi} \tag{9.87}$$

$$\frac{\partial w}{\partial y} = \frac{\partial w}{\partial r}\frac{\partial r}{\partial y} + \frac{\partial w}{\partial \phi}\frac{\partial \phi}{\partial y} = \frac{\partial w}{\partial r}\sin\phi + \frac{\cos\phi}{r}\frac{\partial w}{\partial \phi} \tag{9.88}$$

For the expressions for $\partial^2 w/\partial x^2$ and $\partial^2 w/\partial y^2$, we use the operations $\partial/\partial x$ and $\partial/\partial y$ on Equations 9.87 and 9.88 which yields

$$\frac{\partial^2 w}{\partial x^2} = \left(\cos\phi\frac{\partial}{\partial r} - \frac{\sin\phi}{r}\frac{\partial}{\partial \phi}\right)\left(\cos\phi\frac{\partial w}{\partial r} - \frac{\sin\phi}{r}\frac{\partial w}{\partial \phi}\right) \tag{9.89}$$

$$\frac{\partial^2 w}{\partial y^2} = \left(\sin\phi\frac{\partial}{\partial r} + \frac{\cos\phi}{r}\frac{\partial}{\partial \phi}\right)\left(\sin\phi\frac{\partial w}{\partial r} + \frac{\cos\phi}{r}\frac{\partial w}{\partial \phi}\right) \tag{9.90}$$

### 9.3.1 Equations of Motion

The equation of motion for a rectangular plate given in Equation 9.18 is expressed in terms of the Del operator $\nabla^2 \equiv \partial^2/\partial x^2 + \partial^2/\partial y^2$. In cylindrical coordinates, we use the expression

$$\nabla^2 w = \frac{\partial^2 w}{\partial x^2} + \frac{\partial^2 w}{\partial y^2} = \frac{\partial^2 w}{\partial r^2} + \frac{1}{r}\frac{\partial w}{\partial r} + \frac{1}{r^2}\frac{\partial^2 w}{\partial \phi^2} \tag{9.91}$$

The plate equation of motion in cylindrical coordinates becomes

$$\left(\frac{\partial^2}{\partial r^2} + \frac{1}{r}\frac{\partial}{\partial r} + \frac{1}{r^2}\frac{\partial^2}{\partial \phi^2}\right)\left(\frac{\partial^2 w}{\partial r^2} + \frac{1}{r}\frac{\partial w}{\partial r} + \frac{1}{r^2}\frac{\partial^2 w}{\partial \phi^2}\right) + \frac{h\rho}{D}\frac{\partial^2 w}{\partial t^2} = \frac{p(x,y,t)}{D} \tag{9.92}$$

For free vibrations, we assume $p = 0$ and $w(r, \phi, t) = W(r, \phi)e^{i\omega t}$. Substituting these into the equation of motion results in

$$\nabla^2\nabla^2 W - \lambda^4 W = 0 \tag{9.93}$$

or

$$(\nabla^2 + \lambda^2)(\nabla^2 - \lambda^2)W = 0 \tag{9.94}$$

where $\lambda^4 = (h\rho/D)\omega^2$. Equation 9.93 or 9.94 may be replaced by two equations

$$\frac{\partial^2 W}{\partial r^2} + \frac{1}{r}\frac{\partial W}{\partial r} + \frac{1}{r^2}\frac{\partial^2 W}{\partial \phi^2} + \lambda^2 W = 0 \tag{9.95}$$

$$\frac{\partial^2 W}{\partial r^2} + \frac{1}{r}\frac{\partial W}{\partial r} + \frac{1}{r^2}\frac{\partial^2 W}{\partial \phi^2} - \lambda^2 W = 0 \tag{9.96}$$

Let $\tilde{W}$ satisfy Equation 9.95, which means

$$\nabla^2 \tilde{W} + \lambda^2 \tilde{W} = 0$$

Then,

$$\nabla^2 \nabla^2 \tilde{W} + \lambda^2 \nabla^2 \tilde{W} = 0$$

However, since $\nabla^2 \tilde{W} = -\lambda^2 \tilde{W}$, we have that

$$\nabla^2 \nabla^2 \tilde{W} - \lambda^2 \nabla^2 \tilde{W} = 0 \tag{9.97}$$

Thus, we see that $\tilde{W}$ satisfies the equation of motion. In a similar manner, let $\bar{W}$ satisfy Equation 9.96, which means

$$\nabla^2 \bar{W} - \lambda^2 \bar{W} = 0$$

Then,

$$\nabla^2 \nabla^2 \bar{W} - \lambda^2 \nabla^2 \bar{W} = 0$$

However, since $\nabla^2 \bar{W} = \lambda^2 \bar{W}$, we have that

$$\nabla^2 \nabla^2 \bar{W} - \lambda^2 \nabla^2 \bar{W} = 0 \tag{9.98}$$

Thus, $\bar{W}$ satisfies the equation of motion. If we assume $W(r, \phi) = R(r)\cos n\phi$ (or $W(R, \phi) = R(r)$ $\sin n\phi$), then Equations 9.95 and 9.96 become

$$\frac{d^2 R}{dr^2} + \frac{1}{r}\frac{dR}{dr} + \left(\lambda^2 - \frac{n^2}{r^2}\right) R = 0 \tag{9.99}$$

$$\frac{d^2 R}{dr^2} + \frac{1}{r}\frac{dR}{dr} + \left(-\lambda^2 - \frac{n^2}{r^2}\right) R = 0 \tag{9.100}$$

These two equations are of the Bessel type. From Equation 9.99, we have

$$R = C_n J_n(\lambda r) + E_n Y_n(\lambda r) \tag{9.101}$$

where

$$J_n = \text{ Bessel's function of the first kind, of order } n$$

$$Y_n = \text{ Bessel's function of the second kind, of order } n$$

From Equation 9.100

$$R = D_n J_n(i\lambda r) + F_n Y_n(i\lambda r) \tag{9.102}$$

where

$$J_n(i\lambda r) \equiv I_n(\lambda r) = \text{modified Bessel's function of the first kind, of order } n$$

$$Y_n(i\lambda r) \equiv K_n(\lambda r) = \text{modified Bessel's function of the second kind, of order } n$$

In Equations 9.101 and 9.102, the terms $C_n$, $D_n$, $E_n$, and $F_n$ are arbitrary constants. We can write that the general solution of the Equation 9.93 or 9.94 can be expressed as a combination of both solutions above. That is,

$$W(r,\phi) = \left[C_n J_n(\lambda r) + D_n J_n(i\lambda r) + E_n Y_n(\lambda r) + F_n Y_n(i\lambda r)\right]\cos n\phi \tag{9.103}$$

where $C_n$, $D_n$, $E_n$, and $F_n$ follow from the boundary conditions. The boundary conditions for a circular plate depend on whether the plate is annular or solid as depicted in Figure 9.10. For the annular plate of Figure 9.10a, there are two boundary conditions. One is at the inner radius $r = a$ and the other at the outer radius $r = b$. In order to satisfy the boundary conditions for a solid circular plate of Figure 9.10b at the inner radius ($r = a$), only two constants are needed from the Equation 9.103. However, the two are not picked at random. We see that $Y_n(\lambda r)$ and $Y_n(i\lambda r)$ must be rejected, so we take $E_n = F_n = 0$ and the solution becomes

$$W(r,\phi) = \left[C_n J_n(\lambda r) + D_n J_n(i\lambda r)\right]\cos n\phi \tag{9.104}$$

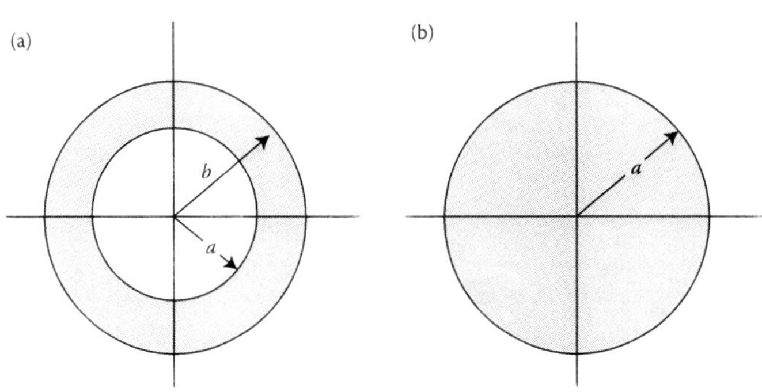


**FIGURE 9.10**
Circular plates which are (a) annular, and (b) solid.

For a clamped edge at $r = a$, we have

$$W(r,\phi) = \frac{\partial W}{\partial r} = 0$$

so that the boundary conditions are

$$\begin{aligned} C_n J_n(\lambda a) + D_n J_n(i\lambda a) &= 0 \\ C_n J_n'(\lambda a) + iD_n J_n'(i\lambda a) &= 0 \end{aligned} \tag{9.105}$$

A nontrivial solution for Equation 9.105 for $C_n$ and $D_n$ is obtained if the determinant of coefficients vanishes. Thus,

$$\begin{vmatrix} J_n(\lambda a) & J_n(i\lambda a) \\ J_n'(\lambda a) & J_n'(i\lambda a) \end{vmatrix} = 0 \tag{9.106}$$

This gives the frequency equation. For each $n$ we obtain an infinite number of $\lambda$'s, which satisfy the equation, or an infinite number of frequencies, that is

$$\lambda_{1n}, \lambda_{2n}, \lambda_{3n}, \ldots$$

$$\omega_{1n}, \omega_{2n}, \omega_{3n}, \ldots, \omega_{mn} \quad m = 1,2,3,\ldots \quad n = 1,2,3,\ldots$$

Four modes of circular plates are illustrated in Figure 9.11 for various combinations of $m$ and $n$. Figure 9.11a is associated with the lowest frequency with $n = 0$, $m = 1$. Similarly, Figure 9.11b with $n = 0$, $m = 2$, Figure 9.11c with $n = 1$, $m = 1$ and $W = R(r)$ cos $\phi$, and Figure 9.11d with $n = 1$, $m = 2$ illustrate the variety in mode shapes. We have, for $\omega_{mn}$, that $W_{mn}^{(c)}(r,\phi) = R_{mn}(r)\cos n\phi$. Also, for the same frequency we have $W_{mn}^{(s)}(r,\phi) = R_{mn}(r)\sin n\phi$. The shape of $W_{mn}^{(s)}$ is the same as the shape of $W_{mn}^{(c)}$. The only difference is that $W_{mn}^{(c)}$ are symmetric with respect to $\phi = 0$, while $W_{mn}^{(s)}$ are antisymmetric with respect to $\phi = 0$.

Suppose that the initial conditions are given at $t = 0$ as

$$\begin{aligned} w(r,\phi) &= w_0(r,\phi) \\ \dot{w}(r,\phi) &= \dot{w}_0(r,\phi) \end{aligned} \tag{9.107}$$

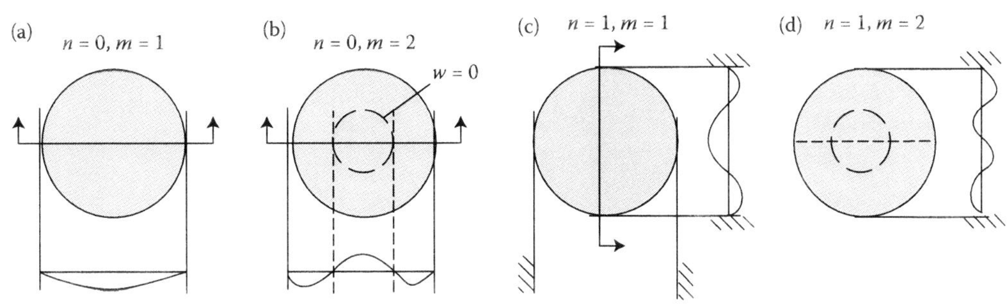


**FIGURE 9.11**
Modes of vibration for a circular plate with selected values of $m$ and $n$.

If $w_0(r, \phi)$ and $\dot{w}_0(r,\phi)$ are symmetric with respect to $\phi = 0$, we can take the general solution of free vibration in the form

$$w(r,\phi,t) = \sum_{m=1}\sum_{n=1} R_{mn}(r)\cos n\phi(A_{mn}\cos\omega_{mn}t + B_{mn}\sin\omega_{mn}t) \tag{9.108}$$

The constants are determined from the following two equations

$$\sum_{m=1}\sum_{n=1} R_{mn}(r)\cos n\phi A_{mn} = w_0(r,\phi) \tag{9.109}$$

$$\sum_{m=1}\sum_{n=1} R_{mn}(r)\cos n\phi B_{mn}\omega_{mn} = \dot{w}_0(r,\phi) \tag{9.110}$$

Proceeding as before, multiply Equations 9.109 and 9.110 by $R_{ij}(r)\cos j\phi$ and integrate from 0 to $a$ and 0 to $2\pi$, which yields

$$A_{ij} = \frac{\int_0^a\int_0^{2\pi} w_0 R_{ij}(r)\cos j\phi\, dr\, d\phi}{\int_0^a\int_0^{2\pi} (R_{ij}(r)\cos j\phi)^2 dr\, d\phi} \tag{9.111}$$

$$B_{ij} = \frac{1}{\omega_{ij}}\frac{\int_0^a\int_0^{2\pi} \dot{w}_0 R_{ij}(r)\cos j\phi\, dr\, d\phi}{\int_0^a\int_0^{2\pi} (R_{ij}(r)\cos j\phi)^2 dr\, d\phi} \tag{9.112}$$

For antisymmetric initial conditions, we start with

$$w(r,\phi,t) = \sum_{m}\sum_{n} R_{mn}(r)\sin n\phi(A_{mn}\cos\omega_{mn}t + B_{mn}\sin\omega_{mn}t) \tag{9.113}$$

and follow the same procedures to determine $A_{ij}$ and $B_{ij}$.

### 9.3.2 Forced Vibrations

For forced vibrations, we solve the complete equation of motion with the driving force included, as given by Equation 9.14

$$\nabla^2\nabla^2 w + \frac{\rho h}{D}\frac{\partial^2 w}{\partial t^2} = \frac{p(r,\phi,t)}{D}$$

Use normal mode expansion with the following conditions: If $p(r, \phi, t)$ is symmetric in $\phi$, use $W_{mn}^{(c)}$; If $p(r, \phi, t)$ is anti-symmetric in $\phi$, use $W_{mn}^{(s)}$. In order to solve a forced vibration problem, one follows the procedures used for forced vibrations of rectangular plates.

## 9.4 Approximate Solutions

Approximations to the solution of vibration problems typically require using the strain energy of the deformable body. A brief review is presented below. Recall our definitions for strain–displacement and the stress–strain relation presented in Chapter 5 and repeated below for convenience

$$\varepsilon_x = -z\frac{\partial^2 w}{\partial x^2}, \varepsilon_y = -z\frac{\partial^2 w}{\partial y^2}, \varepsilon_{xy} = -z\frac{\partial^2 w}{\partial x\,\partial y}$$

$$\sigma_x = \frac{E}{1-\nu^2}(\varepsilon_x + \nu\varepsilon_y), \sigma_y = \frac{E}{1-\nu^2}(\varepsilon_y + \nu\varepsilon_x), \sigma_{xy} = \frac{E}{1+\nu}\varepsilon_{xy}$$

The specific strain energy per unit volume is defined by $\delta U_0 = \sigma_x\delta\varepsilon_x + \sigma_y\delta\varepsilon_y + 2\sigma_{xy}\delta\varepsilon_{xy}$. Under the assumption that the stress and strain are proportional (Hooke's law applies), then the expression above can be integrated resulting in

$$U_0 = \frac{1}{2}(\sigma_x\varepsilon_x + \sigma_y\varepsilon_y + 2\sigma_{xy}\varepsilon_{xy})$$

$$U_0 = \frac{1}{2}\left[\frac{Ez^2}{1-\nu^2}\frac{\partial^2 w}{\partial x^2}\left(\frac{\partial^2 w}{\partial x^2} + \nu\frac{\partial^2 w}{\partial y^2}\right) + \frac{Ez^2}{1-\nu^2}\frac{\partial^2 w}{\partial y^2}\left(\frac{\partial^2 w}{\partial y^2} + \nu\frac{\partial^2 w}{\partial x^2}\right) + \frac{Ez^2}{1+\nu}\left(\frac{\partial^2 w}{\partial x\,\partial y}\right)^2\right] \tag{9.114}$$

The strain energy for a plate is given by

$$U = \iint_A dx\,dy \int_{-h/2}^{h/2} U_0\,dz \tag{9.115}$$

Integrating Equation 9.115 with respect to $z$ yields

$$U = \frac{1}{2}\iint_A D\left\{\left(\frac{\partial^2 w}{\partial x^2} + \frac{\partial^2 w}{\partial y^2}\right)^2 - 2(1-\nu)\left[\frac{\partial^2 w}{\partial x^2}\frac{\partial^2 w}{\partial y^2} - \left(\frac{\partial^2 w}{\partial x\,\partial y}\right)^2\right]\right\}dx\,dy \tag{9.116}$$

Where, as before, the plates flexural rigidity is given as $D = Eh^3/[12(1-\nu^2)]$, the kinetic energy is defined by

$$K = \frac{1}{2}\iint_A h\rho\left(\frac{\partial w}{\partial t}\right)^2 dx\,dy \tag{9.117}$$

and the work due to external forces is defined by

$$\delta W = \iint_A p(x,y)\,\delta w\,dx\,dy \tag{9.118}$$

where the increment in work $\delta W$ corresponds to the increment in displacement $\delta w$. For free vibrations, we assume $w(x, y, t) = \Phi(x, y)e^{i\omega t}$, which yields the maximum strain energy as

$$U_{\max} = \frac{1}{2}\iint_A D\left\{\left(\frac{\partial^2\Phi}{\partial x^2} + \frac{\partial^2\Phi}{\partial y^2}\right)^2 - 2(1-\nu)\left[\frac{\partial^2\Phi}{\partial x^2}\frac{\partial^2\Phi}{\partial y^2} - \left(\frac{\partial^2\Phi}{\partial x\,\partial y}\right)^2\right]\right\} dx\,dy \tag{9.119}$$

To obtain the maximum kinetic energy, we substitute in the maximum deflection $w(x, y, t) = \Phi(x, y)e^{i\omega t}$ to obtain

$$K_{\max} = \frac{\omega^2}{2}\iint_A h\rho W^2\,dx\,dy = \omega^2 K'_{\max} \tag{9.120}$$

where

$$K'_{\max} = \frac{1}{2}\iint_A h\rho W^2\,dx\,dy \tag{9.121}$$

### 9.4.1 Rayleigh Method

Rayleigh's method is based on the principle of conservation of energy. The Rayleigh method provides an approximation for the $m$, $n$th mode as given by $\Psi_{mn}(x, y) = W_{mn}(x, y)$, where $W_{mn}(x, y)$ is the $m$, $n$th mode of the plate of constant stiffness and mass. Using the kinetic and strain energy expressions above, we estimate the frequency as

$$\omega^2 = \frac{U_{\max}(\Psi_{mn})}{K'_{\max}(\Psi_{mn})} \tag{9.122}$$

Another approach to defining modes is illustrated in Figure 9.12 and is given by $\Psi_{mn}(x, y) = X_m(x)Y_n(y)$, where

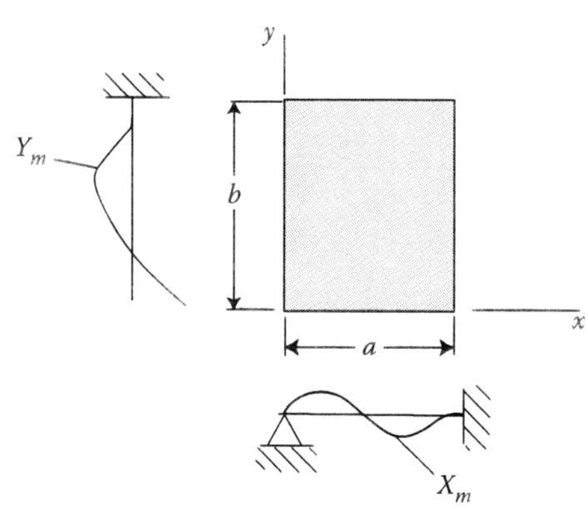


**FIGURE 9.12**
Approximate modes.

$X_m(x)$ = the $m$th natural mode of a beam of length $a$ and the boundary conditions at $x=0$, $x=a$ are the same as for the plate.

$Y_n(y)$ = the $n$th natural mode of a beam of length $b$ and the boundary conditions at $y=0$, $y=b$ are the same as for the plate.

**EXAMPLE 9.4**

Assume an isotropic circular plate with a diameter $D=2a$ is completely clamped around its circumference. Determine the fundamental frequency using Rayleigh's method.

**Solution**

The boundary conditions around the circumference of the plate are $w(a,\ \phi)=\partial w(a,\ \phi)/\partial r=0$ and must be satisfied by the assumed mode shape. For free vibrations assume $w(r,\phi,t)=W(r,\phi)e^{i\omega t}$, where we can select a function for $W(r,\phi)$ which satisfies the boundary conditions. A function that satisfies these conditions is

$$W(r,\phi)=(a^2-r^2)^2$$

Since we are considering a circular plate, Equations 9.111 and 9.112 must be considered in polar coordinates. Referring to Equation 9.86, we can write $U_{max}$ and $K_{max}$ as

$$U_{max}=\frac{1}{2}\int_0^a\int_0^{2\pi}D\left\{\left(\frac{\partial^2 w}{\partial r^2}+\frac{1}{r}\frac{\partial w}{\partial r}+\frac{1}{r^2}\frac{\partial^2 w}{\partial \phi^2}\right)^2 - 2(1-\nu)\left[\frac{\partial^2 w}{\partial r^2}\left(\frac{1}{r}\frac{\partial w}{\partial r}+\frac{1}{r^2}\frac{\partial^2 w}{\partial \phi^2}\right)-\left[\frac{\partial}{\partial r}\left(\frac{1}{r}\frac{\partial w}{\partial \phi}\right)\right]^2\right]\right\}r\,dr\,d\phi$$

$$K_{max}=\frac{\rho h\omega^2}{2}\int_0^a\int_0^{2\pi}\left[\left(a^2-r^2\right)^2\right]^2 r\,dr\,d\phi$$

$$=\frac{\rho h\omega^2}{2}\int_0^a\int_0^{2\pi}(a^8-4a^6r^2+6a^4r^4-4a^2r^6+r^8)r\,dr\,d\phi$$

Performing the integrations results in

$$U_{max}=\frac{32}{3}D\pi a^6 \text{ and } K_{max}=\frac{\pi\rho h\omega^2 a^{10}}{10}$$

Applying Equation 9.122 gives

$$\omega^2=\frac{(32/3)D\pi a^6}{\pi\rho h a^{10}/10}=\frac{320}{3}\frac{D}{\rho h a^4}$$

or

$$\omega\approx\frac{10.33}{a^2}\sqrt{\frac{D}{\rho h}}$$

The function $W(r, \phi) = (a - r)^2$ also satisfies the boundary conditions, but a moment singularity defined by the function $(1/r)(\partial W/\partial r) = 2(1 - (a/r))$ exists at $r = a$ which makes it an unacceptable function.

### 9.4.2 Ritz Method

Using the Ritz method, we approximate the natural modes in the form

$$\Psi_{mn}(x,y) = \sum_{m=1}^{\infty}\sum_{n=1}^{\infty} A_{mn}W_{mn}$$

where $W_{mn}(x, y)$ is the same as in the Rayleigh method and $A_{mn}$ is a set of parameters as presented in Chapter 5. Another approach is to use

$$\Psi_{mn}(x,y) = \sum_{m}\sum_{n} A_{mn}X_m(x)Y_n(y)$$

Using this approach, $A_{mn}$ and $\omega^2$ follow from the relation

$$\frac{\partial(U_{\max} - K_{\max})}{\partial A_{mn}} = 0 \tag{9.123}$$

Although similar in appearance to Rayleigh's method, the Ritz method does produce somewhat different results for the same problem as was illustrated by Examples 5.5 and 5.6.

### 9.4.3 Galerkin's Method

As discussed in Chapter 5, Galerkin's method requires knowing the differential equations of motion. As before, we assume the solution in the form of an approximating function. The approximating function can be taken as

$$\Psi_{mn}(x,y) = \sum_{m=1}^{\infty}\sum_{n=1}^{\infty} A_{mn}W_{mn}$$

or

$$\Psi_{mn}(x,y) = \sum_{m=1}^{\infty}\sum_{n=1}^{\infty} X_m(x)Y_n(y)$$

## 9.5 Sandwich Plates

A sandwich plate consists of a core of thickness $h$ which is sandwiched between two cover plates of thickness $f$, as illustrated in Figure 9.13. We assume $f \ll h$. The plates are assumed to be rectangular with dimensions $a \times b$. The primary assumptions are: (a) small

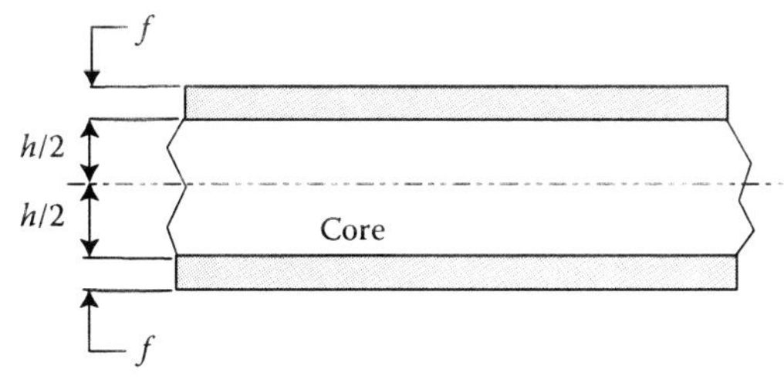


**FIGURE 9.13**
Sandwich beam geometry.

deflections, (b) elastic materials (both core and cover plates), and (c) transverse shear deformation and rotary inertia are taken into account. Since these plates typically consist of different materials for the core and cover plates, they generally behave as orthotropic materials. Although orthotropic in nature, they are not treated the same as continuous fiber-laminated composites are in CLT.

### 9.5.1 Equations of Motion

The equations of motion are formulated using the Reissner–Mindlin theory. They are developed based on the plate deformations in the $x$–$z$ and $y$–$z$ planes as illustrated in Figure 9.14. The vertical displacement is given by $w(z) = w$. At an arbitrary distance $z$ from the midplane of the plate, the $x$ and $y$ displacements ($u(z)$ and $v(z)$, respectively) are expressed based on a small angle assumption ($\tan\phi \approx \phi$ and $\tan\psi \approx \psi$) as $\tan\phi = u(z)/z \Rightarrow u(z) = z\tan\phi \approx z\phi$ and $\tan\psi = v(z)/z \Rightarrow v(z) = z\tan\psi \approx z\psi$. If transverse shear deformation is neglected, everything is expressible in terms of $w$, with $\phi = -\partial w/\partial x$ and $\psi = -\partial w/\partial y$.

Based on the deformations shown in Figure 9.14, we note that three functions that need to be determined are $w = w(x, y)$, $\phi = \phi(x, y)$, and $\psi = \psi(x, y)$.

The plate consists of two distinct components, the core and the cover plates. Each of these can be different materials, which requires the formulation of two different stress–strain relations based on the elastic moduli of the constituent components (core and cover

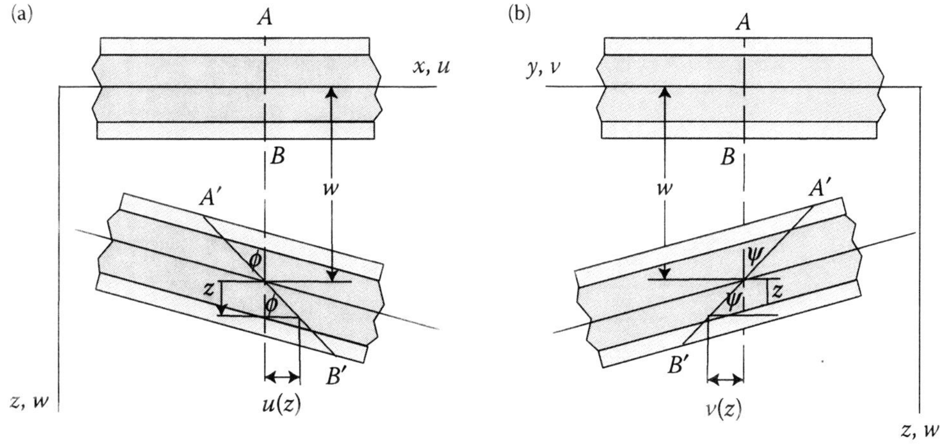


**FIGURE 9.14**
Assumed plate deformations in the (a) $x$–$z$ and (b) $y$–$z$ planes.

plates). In order to distinguish between the core and the cover plates, we use $E$, $G$, etc. for the core and $E'$, $G'$, etc. for the cover plates. In addition, we have no guarantee that the constituent materials will be isotropic. Therefore, the elastic moduli will have subscripts such as $E_{xx}$ and $E_{xy}$, etc. Using this notation, the first subscript refers to the direction in which the modulus is measured and the second subscript refers to the direction of load application. For example, $E_{xy}$ is determined to be the elastic modulus in the $x$-direction when a specimen is loaded in the $y$-direction. This is the same as the elastic moduli $E_{11}$, $E_{22}$ etc. used in CLT. In sandwich construction, we are dealing with three layers of variable thickness as opposed to $N$ layers of variable thickness, so the formulation of constitutive relations is less stringent.

The normal stress–strain relations for the core are

$$\sigma_{xx} = E_{xx}\varepsilon_{xx} + E_{xy}\varepsilon_{yy}, \quad \sigma_{yy} = E_{yy}\varepsilon_{yy} + E_{yx}\varepsilon_{xx} \tag{9.124}$$

We assume that shear stress and strain are expressed as $\sigma_{yz} = 2G_y\varepsilon_{yz}$, $\sigma_{xz} = 2G_x\varepsilon_{xz}$, and $\sigma_{xy} = 2G_z\varepsilon_{xy}$, where the positive direction of stress and associated shear deformation are depicted in Figure 9.15. We note that $E_{xy} = E_{yx}$. In order to define stresses based on deformation of the core, we must know $E_{xx}$, $E_{yy}$, $E_{xy} = E_{yx}$, $G_x$, $G_y$, and $G_z$. If the core material is isotropic

$$E_{xx} = E_{yy} = \frac{E}{1-\nu^2}, \quad E_{xy} = E_{yx} = \frac{\nu E}{1-\nu^2}, \quad G_x = G_y = G_z = \frac{E}{2(1+\nu)} \tag{9.125}$$

For the cover plates (face plates), we neglect transverse shear deformation and have

$$\sigma'_{xx} = E'_{xx}\varepsilon'_{xx} + E'_{xy}\varepsilon'_{yy}, \quad \sigma'_{yy} = E'_{yy}\varepsilon'_{yy} + E'_{yx}\varepsilon'_{xx}, \quad \sigma'_{xy} = 2G'_z\varepsilon'_{xy} \tag{9.126}$$

The equations of motion are derived based on CPT. Therefore, as has been previously demonstrated, we need to consider

$$\begin{aligned}
&\text{Moment about } y: \quad \frac{\partial M_x}{\partial x} + \frac{\partial M_{xy}}{\partial y} - Q_x = I\frac{\partial^2 \phi}{\partial t^2} \\
&\text{Moment about } x: \quad \frac{\partial M_{xy}}{\partial x} + \frac{\partial M_y}{\partial y} - Q_y = I\frac{\partial^2 \psi}{\partial t^2} \\
&\text{Forces along } z: \quad \frac{\partial Q_x}{\partial x} + \frac{\partial Q_y}{\partial y} + p = M\frac{\partial^2 w}{\partial t^2}
\end{aligned} \tag{9.127}$$

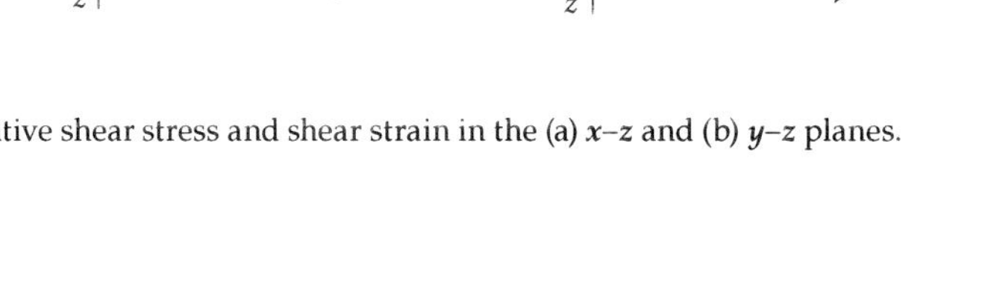


**FIGURE 9.15**
Definitions of positive shear stress and shear strain in the (a) $x$–$z$ and (b) $y$–$z$ planes.

where

$$M = \text{mass per unit area} = \rho h + 2\rho' f$$

$$I = \text{moment of inertia about } x \text{ or } y \text{ per unit area} = \rho\frac{h^3}{12} + 2\rho' f\left(\frac{h}{2}\right)^2$$

We want to express the components of strain in terms of $\phi$, $\psi$, and $w$ for the cover plates (primed) and core (unprimed). We use the traditional definitions of $\varepsilon_{xx} = \partial u/\partial x$, $\varepsilon_{yy} = \partial v/\partial y$, and $\varepsilon_{xy} = (1/2)(\partial u/\partial y + \partial v/\partial x)$ with $u = z\phi$ and $v = z\psi$. Therefore, noting that for the cover plates $z \approx h/2$, so that

$$\begin{aligned}
&\varepsilon_{xx} = z\frac{\partial\phi}{\partial x}, \quad \varepsilon_{xx} = \pm\frac{h}{2}\frac{\partial\phi}{\partial x}; \quad \varepsilon_{yy} = z\frac{\partial\psi}{\partial x}, \quad \varepsilon'_{yy} = \pm\frac{h}{2}\frac{\partial\psi}{\partial x} \\
&\varepsilon_{xy} = \frac{z}{2}\left(\frac{\partial\phi}{\partial y} + \frac{\partial\psi}{\partial x}\right); \quad \varepsilon'_{xy} = \pm\frac{h}{4}\left(\frac{\partial\phi}{\partial y} + \frac{\partial\psi}{\partial x}\right) \\
&\varepsilon_{xz} = \frac{1}{2}\left(\frac{\partial w}{\partial x} + \frac{\partial u}{\partial z}\right) = \frac{1}{2}\left(\phi + \frac{\partial w}{\partial x}\right) \quad \text{or} \quad \gamma_{xz} = 2\varepsilon_{xz} = \phi + \frac{\partial w}{\partial x} \\
&\varepsilon_{yz} = \frac{1}{2}\left(\frac{\partial w}{\partial y} + \frac{\partial v}{\partial z}\right) = \frac{1}{2}\left(\psi + \frac{\partial w}{\partial y}\right) \quad \text{or} \quad \gamma_{yz} = 2\varepsilon_{yz} = \psi + \frac{\partial w}{\partial y}
\end{aligned} \tag{9.128}$$

Next we express the stresses in terms of $\phi$, $\psi$, and $w$. For the core, we have

$$\begin{aligned}
&\sigma_{xx} = E_{xx}z\frac{\partial\phi}{\partial x} + E_{xy}z\frac{\partial\psi}{\partial y} \quad \sigma_{xz} = 2G_x\varepsilon_{xz} = G_x\left(\phi + \frac{\partial w}{\partial x}\right) \\
&\sigma_{yy} = E_{yy}z\frac{\partial\psi}{\partial y} + E_{yx}z\frac{\partial\phi}{\partial x} \quad \sigma_{yz} = 2G_y\varepsilon_{yz} = G_y\left(\psi + \frac{\partial w}{\partial y}\right) \\
&\sigma_{xy} = 2G_z\frac{z}{2}\left(\frac{\partial\phi}{\partial y} + \frac{\partial\psi}{\partial x}\right)
\end{aligned} \tag{9.129}$$

and for the cover plates

$$\begin{aligned}
&\pm\sigma'_{xx} = E'_{xx}\frac{h}{2}\frac{\partial\phi}{\partial x} + E'_{xy}\frac{h}{2}\frac{\partial\psi}{\partial y} \quad \pm\sigma'_{yy} = E'_{yy}\frac{h}{2}\frac{\partial\psi}{\partial y} + E'_{xy}\frac{h}{2}\frac{\partial\phi}{\partial y} \\
&\pm\sigma'_{xy} = 2G'_x\frac{h}{4}\left(\frac{\partial\phi}{\partial y} + \frac{\partial\psi}{\partial x}\right)
\end{aligned} \tag{9.130}$$

The stress resultants (primed for the facing and unprimed for the core) are now used to define the applied loads in terms of $\phi$, $\psi$, and $w$. We use a model similar to that used for defining moment–stress relations in undergraduate strength of materials. One of the big difference is that for plate theory, the subscript associated with the bending moment refers to the direction in which a stress will result as opposed to the axis about which bending

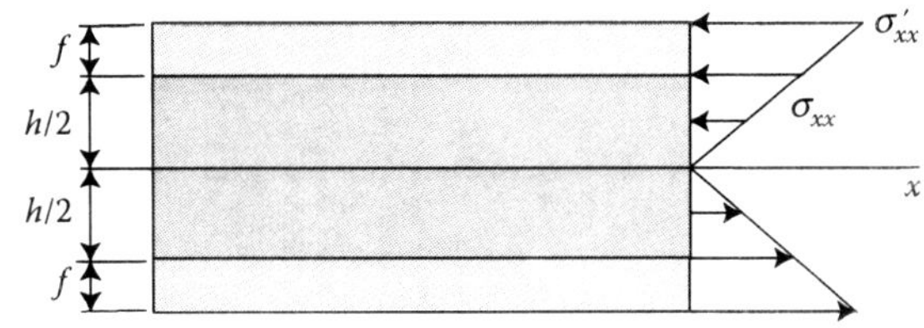


**FIGURE 9.16**
Model for defining moment–stress relations for a moment in the $x$-direction.

occurs (as presented in undergraduate strength of materials). Figure 9.16 is the model used for defining the moment–stresses relation in the $x$-direction.

From Figure 9.16, we see that the moment producing a normal stress in the $x$-direction can be expressed in terms of the stress in the $x$-direction as

$$M_x = \int_{-h/2}^{h/2} \sigma_{xx} z\,dz + \left(2\sigma'_{xx} f\right)\left(\frac{h}{2}\right) \tag{9.131}$$

Integrating Equation 9.131 gives

$$M_x = D_x \frac{\partial \varphi}{\partial x} + D_{xy} \frac{\partial \psi}{\partial y} \tag{9.132}$$

where

$$\begin{aligned} D_x &= E_{xx} \frac{h^3}{12} + E'_{xx} \frac{h^2 f}{2} \\ D_{xy} &= E_{xy} \frac{h^3}{12} + E'_{xy} \frac{h^2 f}{2} \end{aligned} \tag{9.133}$$

In a similar manner, we have for $M_y$ and $M_{xy}$

$$M_y = \int_{-h/2}^{h/2} \sigma_{yy} z\,dz + 2\sigma'_{yy} f \left(\frac{h}{2}\right) = D_y \frac{\partial \psi}{\partial y} + D_{xy} \frac{\partial \varphi}{\partial x} \tag{9.134}$$

$$M_{xy} = H_{xy} \left(\frac{\partial \psi}{\partial x} + \frac{\partial \varphi}{\partial y}\right) \tag{9.135}$$

where

$$\begin{aligned} D_y &= E_{yy} \frac{h^3}{12} + E'_{yy} \frac{h^2 f}{2} \\ H_{xy} &= G_z \frac{h^3}{12} + G'_z \frac{h^2 f}{2} \end{aligned} \tag{9.136}$$

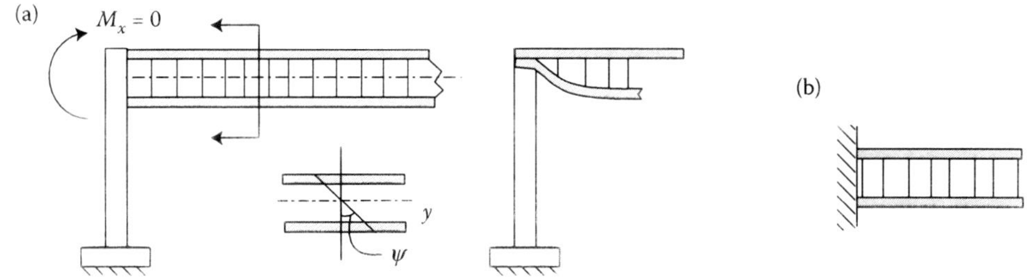


**FIGURE 9.17**
Illustrations of (a) simply supported and (b) built-in boundaries for sandwich plates.

For the core alone, we have

$$Q_x = \int_{-h/2}^{h/2} \sigma_{xz}\, dz = K_x \left( \varphi + \frac{\partial w}{\partial x} \right), \quad \text{where } K_x = G_x h \tag{9.137}$$

$$Q_y = K_y \left( \psi + \frac{\partial w}{\partial y} \right), \quad \text{where } K_y = G_y h \tag{9.138}$$

Substituting these expressions for $M_x$, $M_y$, …, $Q_y$ into the equations of motion, Equation 9.127 yields

$$\begin{aligned}
&D_x \frac{\partial^2 \phi}{\partial x^2} + D_{xy} \frac{\partial^2 \psi}{\partial x \partial y} + H_{xy} \frac{\partial^2 \psi}{\partial x \partial y} + H_{xy} \frac{\partial^2 \phi}{\partial y^2} - K_x \left( \phi + \frac{\partial w}{\partial x} \right) = I \frac{\partial^2 \varphi}{\partial t^2} \\
&D_y \frac{\partial^2 \psi}{\partial y^2} + D_{xy} \frac{\partial^2 \phi}{\partial x \partial y} + H_{xy} \frac{\partial^2 \phi}{\partial x \partial y} + H_{xy} \frac{\partial^2 \psi}{\partial y^2} - K_y \left( \psi + \frac{\partial w}{\partial y} \right) = I \frac{\partial^2 \psi}{\partial t^2} \\
&K_x \frac{\partial \phi}{\partial x} + K_y \frac{\partial \psi}{\partial y} + K_x \frac{\partial^2 w}{\partial x^2} + K_y \frac{\partial^2 w}{\partial y^2} + p = M \frac{\partial^2 w}{\partial t^2}
\end{aligned} \tag{9.139}$$

As with previous cases, the solution of these equations requires knowledge of both the loading and boundary conditions. The boundary conditions for simply supported and built-in edges of a sandwich beam are illustrated in Figure 9.17. A free-edge condition is not illustrated. The boundary conditions are summarized in Table 9.2. The typical

**TABLE 9.2**
Typical Boundary Conditions for Sandwich Plates

| Type of Support | Boundary Condition |
|---|---|
| Simple supported edge at $x$ = constant | $M_x = 0 \Rightarrow \frac{\partial \phi}{\partial x} = 0,\ w = 0,\ \psi = 0$ |
| Built-in edge at $x$ = constant | $w = 0,\ \phi = 0,\ \psi = 0$ |
| Free edge at $x$ = constant | $M_x = 0 \Rightarrow D_x \frac{\partial \phi}{\partial x} + D_{xy} \frac{\partial \psi}{\partial y} = 0$ |
| | $Q_x = 0 \Rightarrow \phi + \frac{\partial w}{\partial x} = 0 \quad M_{xy} = 0 \Rightarrow \frac{\partial \phi}{\partial y} + \frac{\partial \psi}{\partial x} = 0$ |

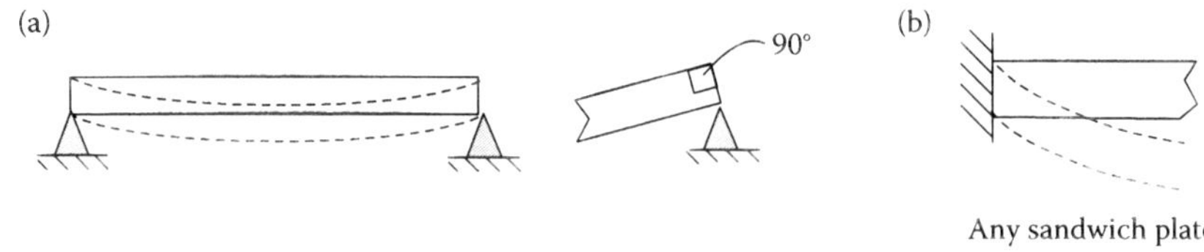


**FIGURE 9.18**
Deflection for (a) simply supported and (b) built-in soft core sandwich plates.

objective for a sandwich plate is to determine the frequency response function in terms of the deformations $\varphi$, $\psi$, and $w$.

For a simply supported homogeneous plate with a soft core, the deflection would resemble the shape shown in Figure 9.18a, where the original perpendicularity of adjacent corners of the plate would remain unchanged (as indicated). The above deflection of a soft core simply supported plate is very similar to the deflection pattern of a built-in plate as shown in Figure 9.18c. Thus, we can use the boundary conditions of the simply supported sandwich plate for other types of boundary conditions, such as a built-in plate.

### 9.5.2 Free Vibrations

For free vibrations, there is no applied load and we can assume the deformations to be

$$\begin{aligned} w(x,y,t) &= C_{mn} \sin \alpha_m x \sin \beta_n y e^{i\omega t} \\ \phi(x,y,t) &= A_{mn} \cos \alpha_m x \sin \beta_n y e^{i\omega t} \\ \psi(x,y,t) &= B_{mn} \sin \alpha_m x \cos \beta_n y e^{i\omega t} \end{aligned} \tag{9.140}$$

where $\alpha_m = m\pi/a$ and $\beta_n = n\pi/b$. In order to determine $\omega$, $A_{mn}$, $B_{mn}$, and $C_{mn}$, we substitute the above relations into the equations of motion, which results in

$$\begin{aligned} &\left(D_x\alpha_m^2 + H_{xy}\beta_n^2 + K_x - I\omega^2\right)A_{mn} + \left(D_{xy}\alpha_m\beta_n + H_{xy}\alpha_m\beta_n\right)B_{mn} + K_x\alpha_m C_{mn} = 0 \\ &\left(D_{xy}\alpha_m\beta_n + H_{xy}\alpha_m\beta_n\right)A_{mn} + \left(D_y\beta_n^2 + H_{xy}\alpha_m^2 + K_y - I\omega^2\right)B_{mn} + K_y\beta_m C_{mn} = 0 \\ &K_x\alpha_m A_{mn} + K_y\beta_n B_{mn} + \left(K_x\alpha_m^2 + K_y\beta_n^2 - M\omega^2\right)C_{mn} = 0 \end{aligned} \tag{9.141}$$

These three liner equations in terms of $A_{mn}$, $B_{mn}$, and $C_{mn}$ have a nontrivial solution only if the determinant of the coefficients is zero. Thus, the frequency equation is

$$\begin{vmatrix} D_x\alpha_m^2 + H_{xy}\beta_n^2 + K_x - I\omega^2 & D_{xy}\alpha_m\beta_n + H_{xy}\alpha_m\beta_n & K_x\alpha_m \\ D_{xy}\alpha_m\beta_n + H_{xy}\alpha_m\beta_n & D_y\beta_n^2 + H_{xy}\alpha_m^2 + K_y - I\omega^2 & K_y\beta_m \\ K_x\alpha_m & K_y\beta_n & K_x\alpha_m^2 + K_y\beta_n^2 - M\omega^2 \end{vmatrix} = 0 \tag{9.142}$$

Since this results in a cubic equation in $\omega^2$, there are three frequencies for each $m$ and $n$, which would be $\omega_{mn}^{(1)}$, $\omega_{mn}^{(2)}$, and $\omega_{mn}^{(3)}$ (e.g., $m = 1, n = 1 \rightarrow \omega_{11}^{(1)}, \omega_{11}^{(2)}, \omega_{11}^{(3)}$). In addition, there are three solutions for $A_{mn}$, $B_{mn}$, and $C_{mn}$. The three modes can be characterized as

$$\text{For } \omega_{mn}^{(1)}, \text{ we have the solutions } A_{mn}^{(1)}, B_{mn}^{(1)}, C_{mn}^{(1)}$$

$$\text{For } \omega_{mn}^{(2)}, \text{ we have the solutions } A_{mn}^{(2)}, B_{mn}^{(2)}, C_{mn}^{(2)}$$

$$\text{For } \omega_{mn}^{(3)}, \text{ we have the solutions } A_{mn}^{(3)}, B_{mn}^{(3)}, C_{mn}^{(3)}$$

### 9.5.3 Forced Vibrations

For forced vibrations, we no longer have $p(x, y, t) = 0$. In order to illustrate as solution procedure, assume a simply supported plate under the uniform loading $p(x, y, t) = 1 \cdot e^{i\omega t}$. We can define $W^*(x, y)$, $\Phi^*(x, y)$, and $\Psi^*(x, y)$ as the frequency response functions. Therefore, we write the deformations as

$$w(x,y,t) = W^*(x,y)e^{i\omega t}, \quad \phi(x,y,t) = \Phi^*(x,y)e^{i\omega t}, \quad \psi(x,y,t) = \Psi^*(x,y)e^{i\omega t} \tag{9.143}$$

Assuming a Fourier expansion for $p(x, y)$, $W^*(x, y)$, $\Phi^*(x, y)$, and $\Psi^*(x, y)$ yields

$$p(x,y,t) = \left( \sum_{m=1}^{\infty} \sum_{n=1}^{\infty} P_{mn}(x,y) \sin \alpha_m x \sin \beta_n y \right) e^{i\omega t} \tag{9.144}$$

where $\alpha_m = m\pi/a$, $\beta_n = n\pi/b$, and

$$P_{mn}(x,y) = \frac{\displaystyle\int_0^a \int_0^b \sin \alpha_m x \sin \beta_n y \, dx \, dy}{\displaystyle\int_0^a \int_0^b (\sin \alpha_m x \sin \beta_n y)^2 dx \, dy} \tag{9.145}$$

Integrating Equation 9.145 gives

$$P_{mn}(x,y) = \begin{cases} \dfrac{16}{mn\pi^2} & m,n = 1,3,5,\ldots \\ 0 & m \text{ and/or } n \text{ even} \end{cases} \tag{9.146}$$

The remaining three terms are

$$\begin{aligned} w(x,y,t) &= \left( \sum_m \sum_n W_{mn} \sin \alpha_m x \sin \beta_n y \right) e^{i\omega t} \\ \phi(x,y,t) &= \left( \sum_m \sum_n \Phi_{mn} \cos \alpha_m x \sin \beta_n y \right) e^{i\omega t} \\ \psi(x,y,t) &= \left( \sum_m \sum_n \Psi_{mn} \sin \alpha_m x \cos \beta_n y \right) e^{i\omega t} \end{aligned} \tag{9.147}$$

where

$$
\begin{aligned}
W^*(x,y) &= \sum_m \sum_n W_{mn} \sin\alpha_m x \sin\beta_n y \\
\Phi^*(x,y) &= \sum_m \sum_n \Phi_{mn} \cos\alpha_m x \sin\beta_n y \\
\Psi^*(x,y) &= \sum_m \sum_n \Psi_{mn} \sin\alpha_m x \cos\beta_n y
\end{aligned} \tag{9.148}
$$

Note that $W_{mn}$, $\Phi_{mn}$, and $\Psi_{mn}$ are unknowns that must be determined from the boundary conditions. For the simply supported plate, the boundary conditions are

$$x = 0 : M_x = 0 \Rightarrow \frac{\partial \phi}{\partial x} = 0, \quad x = a : M_x = 0 \Rightarrow \frac{\partial \phi}{\partial x} = 0, w = 0, \psi = 0$$

all boundary conditions are satisfied

$$y = 0 : M_y = 0 \Rightarrow \frac{\partial \psi}{\partial x} = 0, y = b : M_y = 0 \Rightarrow \frac{\partial \psi}{\partial x} = 0, w = 0, \phi = 0$$

all boundary conditions are satisfied

Substituting into the equations of motion results in

$$
\begin{aligned}
&\left(D_x\alpha_m^2 + H_{xy}\beta_n^2 + K_x - I\omega^2\right)\Phi_{mn} + \left(D_{xy}\alpha_m\beta_n + H_{xy}\alpha_m\beta_n\right)\Psi_{mn} + K_x\alpha_m W_{mn} = 0 \\
&\left(D_{xy}\alpha_m\beta_n + H_{xy}\alpha_m\beta_n\right)\Phi_{mn} + \left(D_y\beta_n^2 + H_{xy}\alpha_m^2 + K_y - I\omega^2\right)\Psi_{mn} + K_y\beta_m W_{mn} = 0 \\
&K_x\alpha_m\Phi_{mn} + K_y\beta_n\Psi_{mn} + \left(K_x\alpha_m^2 + K_y\beta_n^2 - M\omega^2\right)W_{mn} = P_{mn}
\end{aligned} \tag{9.149}
$$

where for each $m = 1, 3, 5, \ldots, n = 1, 3, 5, \ldots$.

For a simply supported sandwich plate, the natural modes of $\varphi$, $\psi$, and $w$ are $W_{mn}\sin\alpha_m x \sin\beta_n y$, $\Phi_{mn}\cos\alpha_m x \sin\beta_n y$, and $\Psi_{mn}\sin\alpha_m x \cos\beta_n y$. For an arbitrary loading, we express the load in a power series as $p(x, y, t) = P_{mn}(t)\sin\alpha_m x \sin\beta_n y$. We assume solutions of the form

$$
\begin{aligned}
w(x,y,t) &= \sum_m \sum_n W_{mn} \sin\alpha_m x \sin\beta_n y \\
\phi(x,y,t) &= \sum_m \sum_n \Phi_{mn} \cos\alpha_m x \sin\beta_n y \\
\psi(x,y,t) &= \sum_m \sum_n \Psi_{mn} \sin\alpha_m x \cos\beta_n y
\end{aligned} \tag{9.150}
$$

We determine $P_{mn}(t)$ from the relation

$$P_{mn}(t) = \frac{\int_0^a \int_0^b p(x,y,t) \sin\alpha_m x \sin\beta_n y \, dx\, dy}{\int_0^a \int_0^b (\sin\alpha_m x \sin\beta_n y)^2 dx\, dy} \tag{9.151}$$

The functions $W_{mn}(t)$, $\Phi_{mn}(t)$, and $\Psi_{mn}(t)$ are unknown. If we substitute the above equations into the equations of motion in terms of $\varphi$, $\psi$, and $w$, we will find three ordinary differential equations in $W_{mn}(t)$, $\Phi_{mn}(t)$, and $\Psi_{mn}(t)$. If $P_{mn}(t)$ is a harmonic function, we assume harmonic functions for $W_{mn}(t)$, $\Phi_{mn}(t)$, and $\Psi_{mn}(t)$.

## 9.6 Equations for Plates of Variable Thickness

If we assume that a plate having variable thickness has no abrupt changes in thickness, then the corresponding expressions for plates of constant thickness can be used for plates having variable thickness. The moment equations remain unchanged and are

$$\begin{aligned} M_x &= -D\left(\frac{\partial^2 w}{\partial x^2} + \nu\frac{\partial^2 w}{\partial y^2}\right) \\ M_y &= -D\left(\frac{\partial^2 w}{\partial y^2} + \nu\frac{\partial^2 w}{\partial x^2}\right) \\ M_{xy} &= M_{yx} = -D(1-\nu)\frac{\partial^2 w}{\partial x\,\partial y} \end{aligned} \tag{9.152}$$

where $D$ is the plate flexural rigidity, given as $D = Eh^3/12(1 - \nu^2)$, are valid for plates of variable thickness. Equations of motion in terms of $M_x$, $M_y$, $M_{xy}$, $Q_x$, and $Q_y$ remain unchanged. Recall that

$$\begin{aligned} Q_x &= \frac{\partial M_x}{\partial x} + \frac{\partial M_{xy}}{\partial y} \\ &= \frac{\partial}{\partial x}\left[-D\left(\frac{\partial^2 w}{\partial x^2} + \nu\frac{\partial^2 w}{\partial y^2}\right)\right] + \frac{\partial}{\partial y}\left[-D(1-\nu)\frac{\partial^2 w}{\partial x\,\partial y}\right] \\ Q_y &= \frac{\partial M_y}{\partial y} + \frac{\partial M_{xy}}{\partial x} \\ &= \frac{\partial}{\partial y}\left[-D\left(\frac{\partial^2 w}{\partial y^2} + \nu\frac{\partial^2 w}{\partial x^2}\right)\right] + \frac{\partial}{\partial x}\left[-D(1-\nu)\frac{\partial^2 w}{\partial x\,\partial y}\right] \end{aligned} \tag{9.153}$$

The differential equation for the deflection $w(x, y, t)$ of a vibrating plate having flexural rigidity $D(x, y)$ may be obtained by substituting Equation 9.75 into Equation 9.12, which yields

$$\nabla^2(D\nabla^2 w) - (1-\nu)\left[\frac{\partial^2 D}{\partial x^2}\frac{\partial^2 w}{\partial y^2} - 2\frac{\partial^2 D}{\partial x\,\partial y}\frac{\partial^2 w}{\partial x\,\partial y} + \frac{\partial^2 D}{\partial y^2}\frac{\partial^2 w}{\partial x^2}\right] + \rho h(x,y)\frac{\partial^2 w}{\partial t^2} = p(x,y,t) \tag{9.154}$$

Closed-form solutions of Equation 9.154 are possible for only very simple variations of the plate thickness $h(x, y)$. However, numerical methods such as finite difference or finite element can be used to obtain a solution or approximate method such as the Galerkin or Rayleigh–Ritz method can be used.

## PROBLEMS

9.1 Determine the natural frequency or write the frequency equation for the rectangular plates supported as shown in Figure 9.19.

9.2 An aluminum plate is supported as shown in Figure 9.20. Using Ritz's method, determine the frequency equations for the plate. Assume $m = n = 1$ and $b = pa$, where $p = 1, 2, 3$ and $12'' \le a \le 48''$. Plot the frequency versus $a$.

9.3 The plates shown in Figure 9.21 are subjected to a uniform normal pressure $p = p_0 H(t)$. Determine the deflection assuming that at time $t = 0$, $w(0) = \dot{w}(0) = 0$.

9.4 Given a thin elastic plate with arbitrary boundary conditions, assume the natural frequencies $\omega_{mn}$ and natural modes $W_{mn}$ are known ($W_{mn}$ form an orthogonal set). For a uniform loading $p(x, y, t) = p(t)$, determine:

a. The frequency response functions of the deflection.

b. The deflection caused by $p(t) = P_0 \cos \omega t$ ($P_0$ = constant)

c. The deflection caused by $p(t) = P_0 \delta(t)$ ($P_0$ = constant) and $w(x, y, 0) = \dot{w}(x, y, 0) = 0$

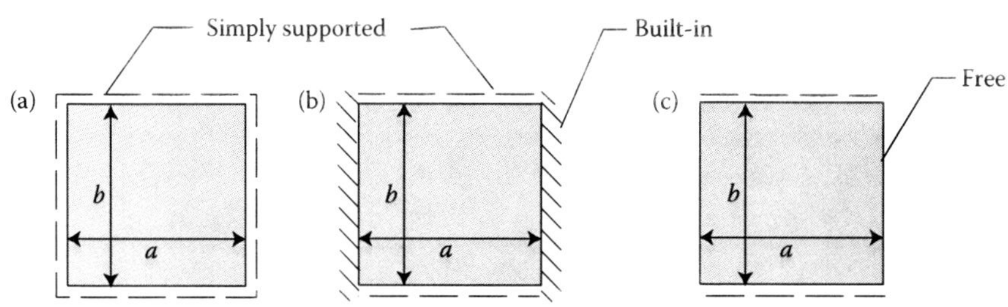


**FIGURE 9.19**
Rectangular plates with various supports.

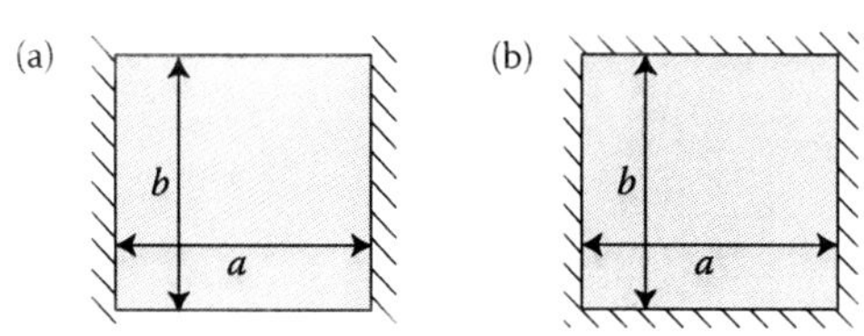


**FIGURE 9.20**
Rectangular plates with fixed-free and fixed-fixed boundaries.

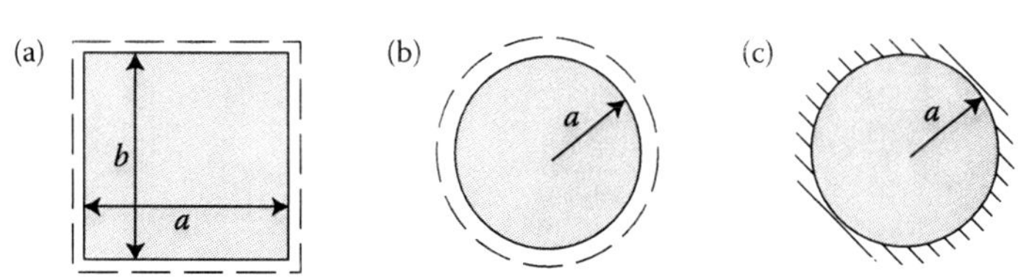


**FIGURE 9.21**
Rectangular and circular plates.

9.5 For the plate in Problem 9.4, assume a loading in the form of a concentrated force $P(t)$ applied at $(x = \xi, y = \eta)$. Find:

a. The frequency response function of the deflection

b. The deflection caused by $P(t) = P_0 \cos \omega t$ ($P_0$ = constant)

c. The deflection caused by $P(t) = P_0 \delta(t)$ ( $P_0$ = constant)

9.6 Assume a 0.05 in thick $[15/-15_4]_s$ simply supported angle-ply laminated plate made from a material with a ply thickness of $t_{ply} = 0.005$ in and a density of $\rho = 0.05$ lb/in$^3$. For this material, we know

$$[Q] = \begin{bmatrix} 22.35 & 0.493 & 0 \\ 0.493 & 1.59 & 0 \\ 0 & 0 & 0.81 \end{bmatrix} \times 10^6 \text{ psi}$$

Determine the frequency of free vibration $\omega_{11}$ and plot it along with the $\omega_{11}$ for a steel plate ($E = 30 \times 10^6$ psi, $\nu = 0.33$ $\rho = 0.28$ lb/in$^3$), assuming the planform dimensions of the plate are within the range $0.5 \leq a/b \leq 5$, where $b = 12$.

## References

Chakraverty, S. 2009. *Vibration of Plates*. Boca Raton: CRC Press.

Lowe, P. G. 1982. *Basic Principles of Plate Theory*. High Holborn, London: Surrey University Press.

Panc, V. 1975. *Theories of Elastic Plates*. The Netherlands: Noordhoff.

Reddy, J. N. 1997. *Mechanics of Laminated Composite Plates*. Boca Raton: CRC Press.

Reddy, J. N. 2004. *Mechanics of Laminated Composite Plates and Shells*. Boca Raton: CRC Press.

Reddy, J. N. 2006. *Theory and Analysis of Elastic Plates and Shells*. New York: CRC Press.

Timoshenko, S. P. and S. Woinowsky-Krieger. 1959. *Theory of Plates and Shells*. New York: McGraw-Hill.

Ugural, A. C. 1981. *Stresses in Plates and Shells*. New York: McGraw-Hill.

# 10

# Vibration of Shells

## 10.1 Introduction

The most commonly used structural components in industry applications are beams, plates, and shells. They are widely used with applications in airplane fuselages, building domes, pipes, etc. In comparison to beams and plates, the dynamic behavior of shells is more complicated due to the curvatures, which couple the bending and in-plane deformations. This adds a degree of complexity and yields systems of coupled equations in some models. Modeling of shells is more complicated than modeling plates due to the fact that shells have initial curvature. Plates can be considered shells without having initial curvature. Curvilinear coordinates are used to describe shells due to the initial curvatures. Despite the large amount of work performed on shell theories, which can be found in the literature (Timoshenko and Woinowsky-Krieger 1959; Gol'Denveizer 1961; Kil'chevskiy 1965; Calladine 1983; Huang 1989; Wempner and Talaslidis 2003; Soedel 2005; Ugural 2010), analytical solutions are available only for simple shapes and fixed, simply supported and free boundary conditions, which may be unrealistic in practice.

Shells are three-dimensional bodies, where one dimension (thickness) is much smaller than the two dimensions. Almost all shell theories reduce the three-dimensional problem into a two-dimensional one.

## 10.2 Cylindrical Shells

A cylindrical shell of radius $R$, thickness $h$, and length $L$ is modeled in Figure 10.1. Let $u$, $v$, and $w$ denote the displacements in the axial $x$, circumferential $\theta$, and radial $r$ directions, respectively. The shell may be subjected to forces in the $x$, $y$, and $z$ directions. Due to the curvature of the shell, we use $y = R\theta$ and $p_y = p_\theta$. For cylindrical shells, we impose the following assumptions:

1. The shells are thin with $h/R < 0.1$ and $h/L < 0.1$.
2. Small deflections.
3. Use classical bending theory (Donnell's type theory).
4. Waves in the circumferential directions must not be too few in numbers.

When subjected to applied loads, the shell will experience internal forces and moments as shown in Figure 10.2.

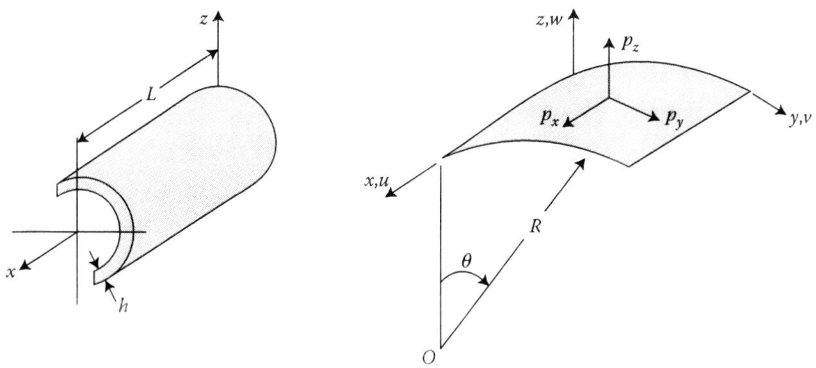


**FIGURE 10.1**
Cylindrical shell coordinates and applied loads.

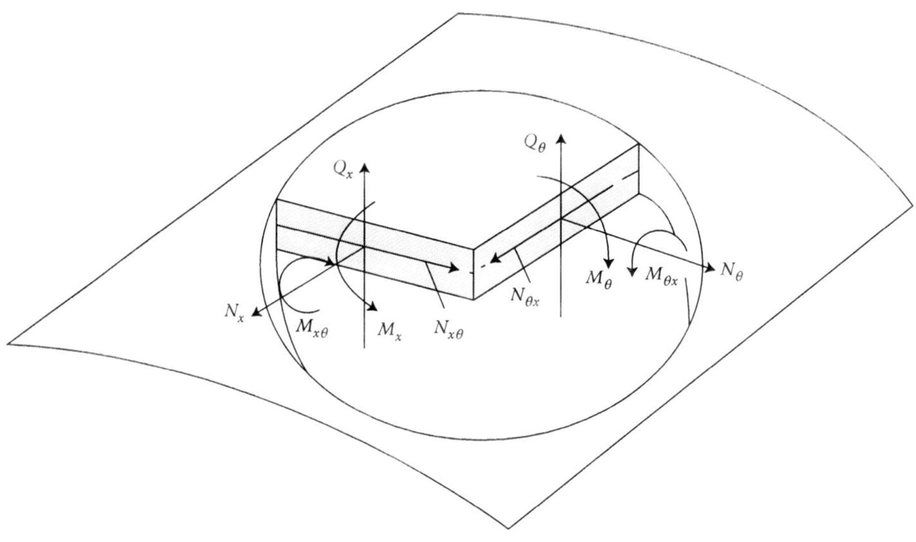


**FIGURE 10.2**
Internal forces and moments resulting from external loads.

### 10.2.1 Equations of Motion

The equations of motion are developed using a series of models for forces and moments acting on representative volume elements as shown in Figure 10.3. In these models, the forces $N_x$, $N_\theta$, $N_{x\theta}$, $N_{\theta x}$, $Q_x$, and $Q_\theta$ and moments $M_x$, $M_\theta$, $M_{x\theta}$, and $M_{\theta x}$ are per unit length of the section. Therefore, the units for forces are lb/in and for moments in-lb/in as they were for plates.

Summing forces in the $x$ direction is

$$\frac{\partial N_x}{\partial x} R\,dx\,d\theta + \frac{\partial N_{\theta x}}{\partial \theta} dx\,d\theta + p_x R\,dx\,d\theta = \rho h R\,dx\,d\theta \frac{\partial^2 u}{\partial t^2}$$

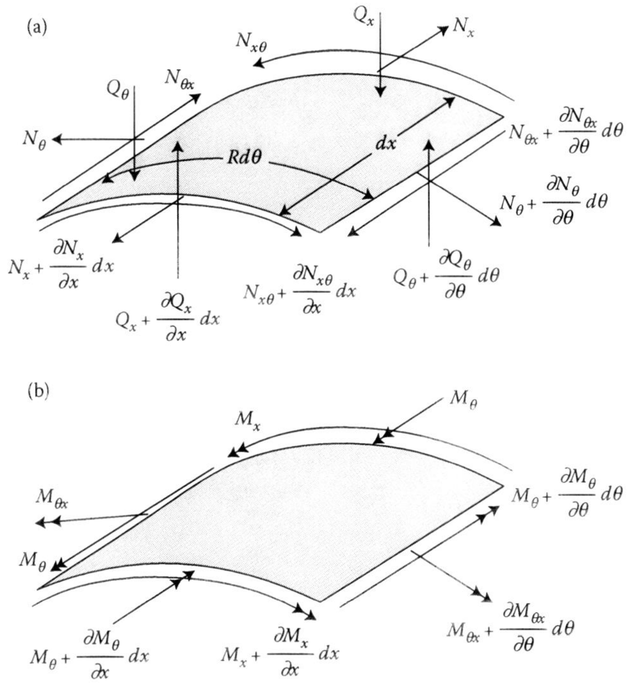


**FIGURE 10.3**
Representative volume element showing (a) forces and (b) moments acting on a shell.

or

$$\frac{\partial N_x}{\partial x} + \frac{1}{R}\frac{\partial N_{\theta x}}{\partial \theta} + p_x = \rho h \frac{\partial^2 u}{\partial t^2} \tag{10.1}$$

Similarly, summing forces in the $\theta$ direction results in

$$\frac{1}{R}\frac{\partial N_\theta}{\partial \theta} + \frac{\partial N_{x\theta}}{\partial x} + p_\theta = \rho h \frac{\partial^2 v}{\partial t^2} \tag{10.2}$$

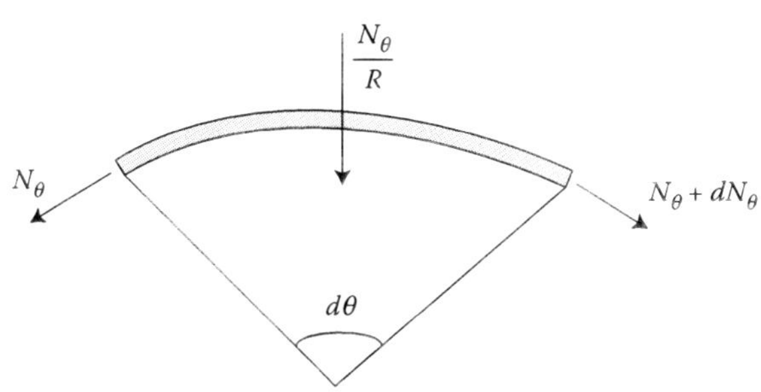


**FIGURE 10.4**
Forces in the z direction.

Summing forces in the $z$ direction, we refer to Figure 10.4, which results in

$$\frac{\partial Q_x}{\partial x} + \frac{1}{R}\frac{\partial Q_\theta}{\partial \theta} - \frac{N_\theta}{R} + p_z = \rho h \frac{\partial^2 w}{\partial t^2} \tag{10.3}$$

Summing moments about the $x$-axis yields

$$\frac{\partial M_{x\theta}}{\partial x} + \frac{1}{R}\frac{\partial M_\theta}{\partial \theta} - Q_\theta = 0 \tag{10.4}$$

Summing moments about the $y$-axis yields

$$\frac{1}{R}\frac{\partial M_{\theta x}}{\partial \theta} + \frac{\partial M_x}{\partial x} - Q_x = 0 \tag{10.5}$$

Summing moments about the $z$-axis yields

$$N_{x\theta} - N_{\theta x} - \frac{M_{\theta x}}{R} = 0 \tag{10.6}$$

We note that Equation 10.6 will not be used as we assume that $N_{\theta x} \approx N_{x\theta}$; $(M_{\theta x}/R) \cong 0$. The relevant equations of motion Equations 10.1 through 10.5 are not sufficient to solve for the number of unknowns. Therefore, we must consider the deformation of the shell in both the $x$–$z$ and $y$–$z$ planes as well as in the $z$ direction as illustrated in Figure 10.5a–c, respectively. For the $x$–$z$ plane, we use $w(z) \approx w$ and $u(z) \approx u - z(\partial w/\partial x)$. In a similar manner, for the $y$–$z$ plane, we use $v(z) \approx v - (z/R)(\partial w/\partial \theta)$. In these relations, $u$, $v$, and $w$ are the displacements of the midsurface of the shell. As we know the displacements, we can calculate the components of strain from the equations of elasticity.

$$\varepsilon_x(z) = \frac{\partial u(z)}{\partial x} = \frac{\partial u}{\partial x} - z\frac{\partial^2 w}{\partial x^2} \tag{10.7}$$

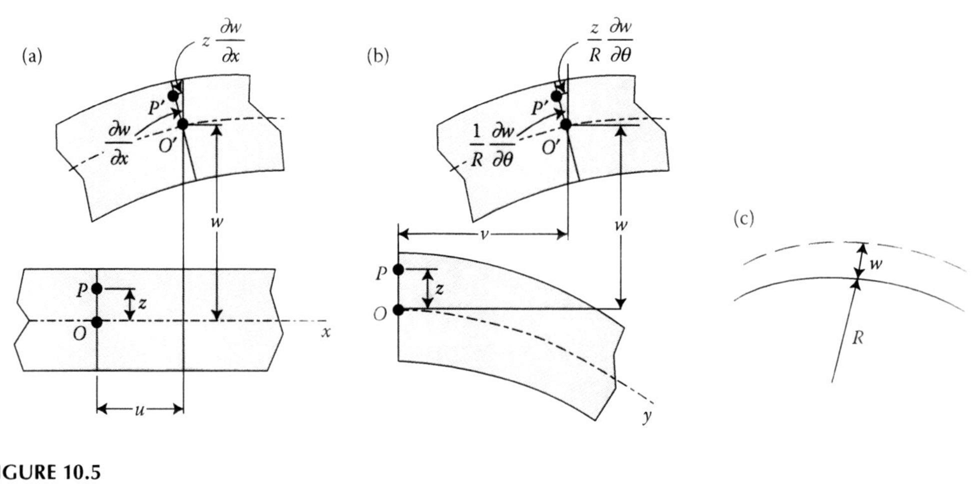


**FIGURE 10.5**
Shell deflections in the (a) $x$–$z$ and (b) $y$–$z$ planes and (c) $z$ direction.

$$\varepsilon_\theta(z) = \frac{1}{R}\frac{\partial v(z)}{\partial \theta} + \frac{w(z)}{R} = \frac{1}{R}\frac{\partial v}{\partial \theta} - \frac{z}{R^2}\frac{\partial^2 w}{\partial \theta^2} + \frac{w}{R} \tag{10.8}$$

$$2\varepsilon_{x\theta}(z) = \frac{1}{R}\frac{\partial u(z)}{\partial \theta} + \frac{\partial v(z)}{\partial x} = \frac{1}{R}\frac{\partial u}{\partial \theta} + \frac{\partial v}{\partial x} - 2\frac{z}{R}\frac{\partial^2 w}{\partial x \partial \theta} \tag{10.9}$$

We note that when $v = 0$

$$\varepsilon_\theta = \frac{R + w - R}{R} = \frac{w}{R} \tag{10.10}$$

With the components of strain determined, we can now form the stress–strain relations (for plane stress) as

$$\begin{aligned}\sigma_x &= \frac{E}{1-\nu^2}(\varepsilon_x + \nu\varepsilon_\theta) \\ &= \frac{E}{1-\nu^2}\left[\frac{\partial u}{\partial x} - z\frac{\partial^2 w}{\partial x^2} + \nu\left(\frac{1}{R}\frac{\partial v}{\partial \theta} - \frac{z}{R^2}\frac{\partial^2 w}{\partial \theta^2} + \frac{w}{R}\right)\right]\end{aligned} \tag{10.11}$$

$$\begin{aligned}\sigma_\theta &= \frac{E}{1-\nu^2}(\varepsilon_\theta + \nu\varepsilon_x) \\ &= \frac{E}{1-\nu^2}\left[\frac{1}{R}\frac{\partial v}{\partial \theta} - \frac{z}{R^2}\frac{\partial^2 w}{\partial \theta^2} + \frac{w}{R} + \nu\left(\frac{\partial u}{\partial x} - z\frac{\partial^2 w}{\partial x^2}\right)\right]\end{aligned} \tag{10.12}$$

$$\begin{aligned}\sigma_{x\theta} = \sigma_{\theta x} &= \frac{E}{2(1+\nu)}(2\varepsilon_{x\theta}) \\ &= \frac{E}{2(1+\nu)}\left[\frac{1}{R}\frac{\partial u}{\partial \theta} + \frac{\partial v}{\partial x} - 2\frac{z}{R}\frac{\partial^2 w}{\partial x \partial \theta}\right]\end{aligned} \tag{10.13}$$

With these three relations, we can express the forces in terms of the displacements $u$, $v$, and $w$ as

$$N_x = \int_{-h/2}^{h/2} \sigma_x \, dz = \frac{Eh}{1-\nu^2}\left[\frac{\partial u}{\partial x} + \nu\left(\frac{1}{R}\frac{\partial v}{\partial \theta} + \frac{w}{R}\right)\right] \tag{10.14}$$

$$N_\theta = \int_{-h/2}^{h/2} \sigma_\theta \, dz = \frac{Eh}{1-\nu^2}\left[\frac{1}{R}\frac{\partial v}{\partial \theta} + \frac{w}{R} + \nu\frac{\partial u}{\partial x}\right] \tag{10.15}$$

$$N_{x\theta} = N_{\theta x} = \int_{-h/2}^{h/2} \sigma_{x\theta} \, dz = \frac{Eh}{2(1+\nu)}\left[\frac{1}{R}\frac{\partial u}{\partial \theta} + \frac{\partial v}{\partial x}\right] \tag{10.16}$$

In general, $N_{x\theta} \neq N_{\theta x}$ even though $\sigma_{x\theta} = \sigma_{\theta x}$. The moment equations are given by

$$M_x = \int_{-h/2}^{h/2} z\sigma_x \, dz = -\frac{Eh^3}{12(1-\nu^2)}\left[\frac{\partial^2 w}{\partial x^2} + \frac{\nu}{R^2}\frac{\partial^2 w}{\partial \theta^2}\right] \tag{10.17}$$

$$M_\theta = \int_{-h/2}^{h/2} z\sigma_\theta \, dz = -\frac{Eh^3}{12(1-\nu^2)}\left[\frac{1}{R^2}\frac{\partial^2 w}{\partial \theta^2} + \nu\frac{\partial^2 w}{\partial x^2}\right] \tag{10.18}$$

$$M_{x\theta} = M_{\theta x} = \int_{-h/2}^{h/2} z\sigma_{x\theta} \, dz = -\frac{Eh^3}{12(1+\nu)}\frac{1}{R}\frac{\partial^2 w}{\partial x \partial \theta} \tag{10.19}$$

Next, we need the shear terms $Q_x$, $Q_\theta$, and $Q_{x\theta}$ in terms of the displacements. Using Equations 10.4 and 10.5, we find

$$Q_x = \frac{\partial M_x}{\partial x} + \frac{1}{R}\frac{\partial M_{\theta x}}{\partial \theta} = -\frac{Eh^3}{12(1-\nu^2)}\left[\frac{\partial^3 w}{\partial x^3} + \frac{1}{R^2}\frac{\partial^3 w}{\partial x \partial \theta^2}\right] \tag{10.20}$$

$$Q_\theta = \frac{1}{R}\frac{\partial M_\theta}{\partial \theta} + \frac{\partial M_{x\theta}}{\partial x} = -\frac{Eh^3}{12(1-\nu^2)}\left[\frac{1}{R^3}\frac{\partial^3 w}{\partial \theta^3} + \frac{1}{R}\frac{\partial^3 w}{\partial x^2 \partial \theta}\right] \tag{10.21}$$

Substituting the expressions for $N_x, \ldots, N_\theta$ in terms of the displacements $u$, $v$, and $w$ into the first three equations of motion (Equations 10.1 through 10.3) results in

$$\frac{\partial^2 u}{\partial x^2} + \left(\frac{1-\nu}{2}\right)\frac{1}{R^2}\frac{\partial^2 u}{\partial \theta^2} + \left(\frac{1+\nu}{2}\right)\frac{1}{R}\frac{\partial^2 v}{\partial x \partial \theta} + \frac{\nu}{R}\frac{\partial w}{\partial x} + \frac{p_x(1-\nu^2)}{Eh} = \frac{\rho(1-\nu^2)}{E}\frac{\partial^2 u}{\partial t^2} \tag{10.22}$$

$$\left(\frac{1+\nu}{2}\right)\frac{1}{R}\frac{\partial^2 u}{\partial x \partial \theta} + \left(\frac{1-\nu}{2}\right)\frac{\partial^2 v}{\partial x^2} + \frac{1}{R^2}\frac{\partial^2 v}{\partial \theta^2} + \frac{1}{R}\frac{\partial w}{\partial \theta} + \frac{p_\theta(1-\nu^2)}{Eh} = \frac{\rho(1-\nu^2)}{E}\frac{\partial^2 v}{\partial t^2} \tag{10.23}$$

$$\nu\frac{\partial u}{\partial x} + \frac{1}{R}\frac{\partial v}{\partial \theta} + \frac{w}{R} + \frac{h^2}{12}R\left(\frac{\partial^4 w}{\partial x^4} + \frac{2}{R^2}\frac{\partial^4 w}{\partial x^2 \partial \theta^2} + \frac{1}{R^4}\frac{\partial^4 w}{\partial \theta^4}\right) - \frac{p_z R(1-\nu^2)}{Eh} = \frac{-\rho R(1-\nu^2)}{E}\frac{\partial^2 w}{\partial t^2} \tag{10.24}$$

In order to determine the displacements $u$, $v$, and $w$, we need to establish a system of boundary conditions. The boundary conditions are generally specified on edges $x$ = constant and/or $\theta$ = constant. The typical support conditions are hinged or simply supported, clamped, or built-in and free. A physical example of a simply supported plate or shell is shown in Figure 10.6a, with clamped or built-in and free-end conditions, for $x$ = constant, are shown in Figure 10.6b and c, respectively.

For a shell with an edge hinged or simply supported, where $x$ = constant, the boundary conditions are

$$w = 0, M_x = 0, v = 0, \text{and } N_x = 0 \tag{10.25}$$

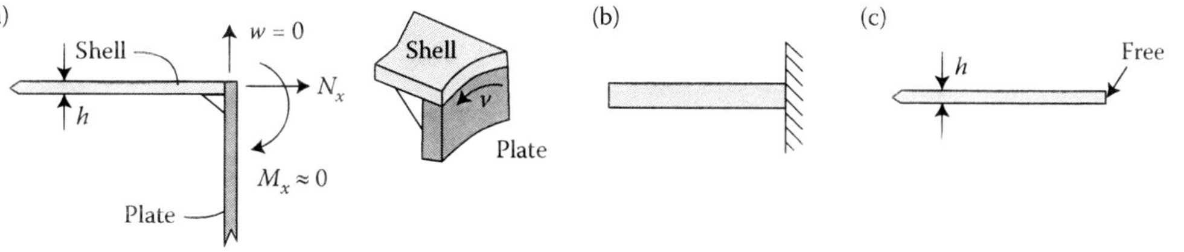


**FIGURE 10.6**
Physical example of (a) simply supported, (b) built-in, (c) free-end conditions for $x$ = constant.

For a hinged or simply supported edge, where $\theta$ = constant, the boundary conditions are

$$w = M_\theta = u = N_\theta = 0 \tag{10.26}$$

If the edge, $x$ = constant, is built-in, then the boundary conditions are

$$w = u = v = \frac{\partial w}{\partial x} = 0 \tag{10.27}$$

If the edge, $x$ = constant, is free, then the boundary conditions are

$$M_x = M_{x\theta} = N_x = N_{x\theta} = Q_x = 0 \tag{10.28}$$

We can only define four boundary conditions, so we must combine $Q_x$ and $M_{x\theta}$. Therefore,

$$Q_x + \frac{1}{R}\frac{\partial M_{x\theta}}{\partial \theta} = 0 \tag{10.29}$$

### 10.2.2 Simplified System of Equations

Assuming no loading in the $x$ and $y$ directions results in $p_x = p_y = 0$. In addition, we assume that the inertial forces in the $x$ and $y$ directions are small and can be neglected. Therefore, we have that $\rho h(\partial^2 u/\partial t^2) \approx \rho h(\partial^2 v/\partial t^2) \approx 0$. With these assumptions, we can introduce a stress function $F$ (analogous to the Airy stress function in elasticity) defined by

$$N_x = \frac{1}{R^2}\frac{\partial^2 F}{\partial \theta^2}, \quad N_\theta = \frac{\partial^2 F}{\partial x^2}, \quad N_{x\theta} = N_{\theta x} = -\frac{1}{R}\frac{\partial^2 F}{\partial x \partial \theta} \tag{10.30}$$

The equations of motion, Equations 10.1 and 10.2, involve $N_x$, $N_y$, and $N_{x\theta} = N_{\theta x}$, which are always satisfied. The third equation of motion, Equation 10.3, involves $Q_x$ and $Q_\theta$, which can be written as

$$\frac{\partial Q_x}{\partial x} + \frac{1}{R}\frac{\partial Q_\theta}{\partial \theta} - \frac{1}{R}\frac{\partial^2 F}{\partial x^2} + p_z = \rho h \frac{\partial^2 w}{\partial t^2} \tag{10.31}$$

If $Q_x$ and $Q_\theta$ are expressed in terms of $w$ as given by Equations 10.20 and 10.21, Equation 10.31 becomes

$$D\nabla^2\nabla^2 w + R\frac{\partial^2 F}{\partial \xi^2} + \rho h R^4 \frac{\partial^2 w}{\partial t^2} = R^4 p_z \tag{10.32}$$

where

$$D = \frac{Eh^3}{12(1-\nu^2)}, \quad \xi = \frac{x}{R}$$

and

$$\nabla^2 = \frac{\partial^2}{\partial \xi^2} + \frac{\partial^2}{\partial \theta^2}$$

Equation 10.32 is still in terms of two unknowns, the displacement $w$ and the stress function $F$. In order to derive the second equation simply in terms of the displacement $w$, we begin with the strain–displacement relations

$$\varepsilon_x = \frac{\partial u}{\partial x} \tag{10.33}$$

$$\varepsilon_\theta = \frac{1}{R}\frac{\partial u}{\partial \theta} + \frac{w}{R} \tag{10.34}$$

$$2\varepsilon_{x\theta} = \frac{\partial v}{\partial x} + \frac{1}{R}\frac{\partial u}{\partial \theta} \tag{10.35}$$

Next we differentiate Equation 10.33 by $(1/R^2)(\partial^2/\partial\theta^2)$, Equation 10.34 by $(\partial^2/\partial x^2)$, and Equation 10.35 by $-(1/R)$ $(\partial^2/\partial x\partial\theta)$. Then on adding the results, we obtain

$$\frac{1}{R^2}\frac{\partial^2 \varepsilon_x}{\partial \theta^2} + \frac{\partial^2 \varepsilon_\theta}{\partial x^2} - \frac{2}{R}\frac{\partial^2 \varepsilon_{x\theta}}{\partial x \partial \theta} - \frac{1}{R}\frac{\partial^2 w}{\partial x^2} = 0 \tag{10.36}$$

Then, we use Hooke's law in the form

$$\varepsilon_x = \frac{1}{Eh}(N_x - \nu N_\theta) = \frac{1}{Eh}\left(\frac{1}{R^2}\frac{\partial^2 F}{\partial \theta^2} - \nu\frac{\partial^2 F}{\partial x^2}\right) \tag{10.37}$$

$$\varepsilon_\theta = \frac{1}{Eh}(N_\theta - \nu N_x) = \frac{1}{Eh}\left(\frac{\partial^2 F}{\partial x^2} - \frac{\nu}{R^2}\frac{\partial^2 F}{\partial \theta^2}\right) \tag{10.38}$$

$$2\varepsilon_{x\theta} = \frac{2(1+\nu)}{Eh} N_{x\theta} = \frac{2(1+\nu)}{Eh}\left(-\frac{1}{R^2}\frac{\partial^2 F}{\partial x \partial \theta}\right) \tag{10.39}$$

Substituting the strains, Equations 10.37, 10.38, 10.39 into Equation 10.36 of motion, results in

$$\frac{1}{Eh}\left(\frac{\partial^4 F}{\partial x^4} + \frac{2}{R^2}\frac{\partial^4 F}{\partial x^2 \partial \theta^2} + \frac{1}{R^4}\frac{\partial^4 F}{\partial \theta^4}\right) - \frac{1}{R}\frac{\partial^2 w}{\partial x^2} = 0 \tag{10.40}$$

or

$$\frac{1}{Eh}\nabla^2\nabla^2 F - R\frac{\partial^2 w}{\partial \xi^2} = 0 \tag{10.41}$$

In order to obtain one equation, we introduce the function $\Phi(\xi, \theta, t)$ such that

$$w = \nabla^2\nabla^2\Phi(\xi,\theta,t) \tag{10.42}$$

$$F = REh\frac{\partial^2\Phi(\xi,\theta,t)}{\partial \xi^2} \tag{10.43}$$

Substituting Equations 10.42 and 10.43 into Equation 10.40 yields

$$D\nabla^2\nabla^2\nabla^2\nabla^2\Phi(\xi,\theta,t) + R^2 Eh\frac{\partial^4\Phi(\xi,\theta,t)}{\partial \xi^4} + \rho h R^4\nabla^2\nabla^2\frac{\partial^2\Phi(\xi,\theta,t)}{\partial t^2} = R^4 p_z \tag{10.44}$$

An alternate and less common form of Equation 10.44 is

$$D\nabla^2\nabla^2\nabla^2\nabla^2 w(\xi,\theta,t) + R^2 Eh\frac{\partial^4 w(\xi,\theta,t)}{\partial \xi^4} + \rho h R^4\nabla^2\nabla^2\frac{\partial^2 w(\xi,\theta,t)}{\partial t^2} = R^4 p_z \tag{10.45}$$

### 10.2.3 Solutions for Cylindrical Shells

When solving cylindrical shell problems, we can consider either a shell panel or a closed cylindrical shell as illustrated in Figure 10.7. In illustrating the solution procedure, we assume that the edges are hinged or simply supported for each case.

#### *10.2.3.1 Free Vibrations*

For free vibrations (Weingarten 1964; Chung 1981), we assume that $p_x = p_\theta = p_z = 0$. We shall consider a simply supported cylindrical shell panel similar to that shown in Figure 10.7a, with the boundary conditions: $w = v = M_x = N_x = 0$ at $x = 0$ and $x = l$, as well as $w = v = M_\theta = N_\theta = 0$ at $\theta = 0$ and $\theta = \theta_0$. We assume solutions for $F$ and $w$, which satisfy the boundary conditions of the shell element. These assumed solutions are

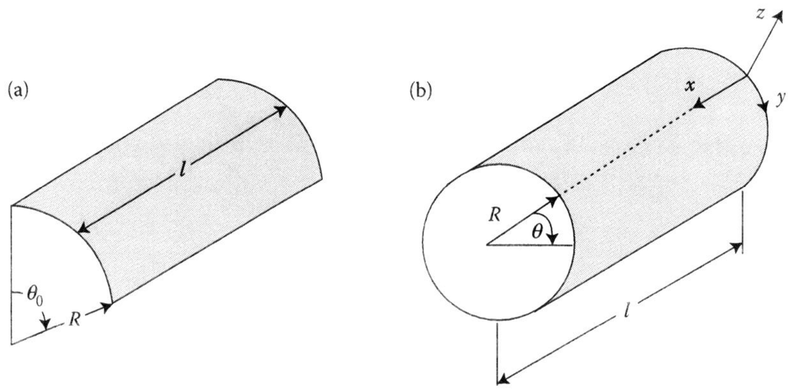


**FIGURE 10.7**
Models of simply supported (a) shell panel, (b) closed cylindrical shell.

$$F = A_{mn} \sin \alpha_m \xi \sin \beta_n \theta e^{i\omega_{mn}t}$$
$$w = B_{mn} \sin \alpha_m \xi \sin \beta_n \theta e^{i\omega_{mn}t} \tag{10.46}$$

where $\alpha_m = (m\pi R/l)$; $\beta_n = (n\pi R/l)$ with $m = 1, 2, 3, \ldots$ and $n = 1, 2, 3, \ldots$. Substituting Equation 10.46 for $F$ and $w$ into Equations 10.32 and 10.40 yields

$$B_{mn} D\left(\alpha_m^2 + \beta_n^2\right)^2 - RA_{mn}\alpha_m^2 - \rho h R^4 \omega_{mn}^2 B_{mn} = 0$$
$$A_{mn} \frac{1}{Eh}\left(\alpha_m^2 + \beta_n^2\right)^2 - B_{mn} R\alpha_m^2 = 0 \tag{10.47}$$

Solutions to Equation 10.47 for $A_{mn}$ and $B_{mn}$ exist only if the determinate of the coefficients vanishes. Hence,

$$\begin{vmatrix} -R\alpha_m^2 & D\left(\alpha_m^2 + \beta_n^2\right)^2 - \rho h R^4 \omega_{mn}^2 \\ \frac{1}{Eh}\left(\alpha_m^2 + \beta_n^2\right)^2 & R\alpha_m^2 \end{vmatrix} = 0 \tag{10.48}$$

Equation 10.48 is the frequency equation in which the only unknown is $\omega_{mn}$. Expanding the determinate gives $\omega_{mn}$ as

$$\omega_{mn}^2 = \frac{1}{h\rho R^2}\left[\frac{D}{R^2}\left(\alpha_m^2 + \beta_n^2\right)^2 + \frac{Eh\alpha_m^4}{\left(\alpha_m^2 + \beta_n^2\right)^2}\right] \tag{10.49}$$

For each $\omega_{mn}(\omega_{11}, \omega_{12}, \omega_{22}, \ldots)$, we find $A_{mn}$ and $B_{mn}$, or assume something like $B_{mn} = 1$ and solve for $A_{mn}$. This gives the mode shape corresponding to $\omega_{mn}$.

Consider now the simply supported closed cylindrical shell shown in Figure 10.7b. It is hinged or simply supported at $x = 0$, $x = l$, and we use a simplified theory.

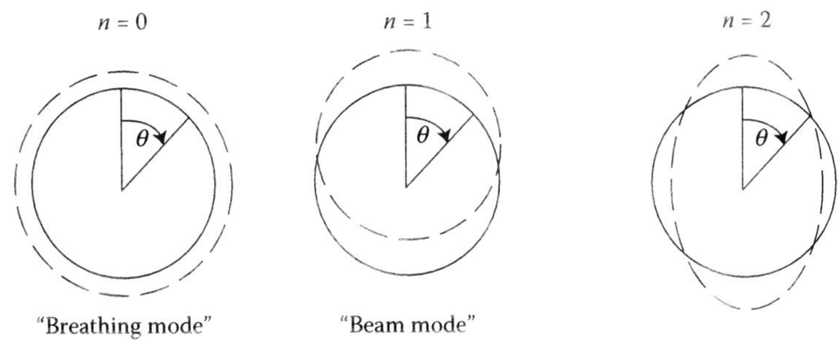


**FIGURE 10.8**
First three symmetric modes of free vibration for a closed cylindrical shell.

That is, $p_x = p_y = 0$, and $\rho h(\partial^2 u/\partial t^2) \cong 0$ and $\rho h(\partial^2 v/\partial t^2) \cong 0$. Recall that we previously defined $\Phi(\xi, \theta, t)$ such that $w(\xi, \theta, t) = \nabla^2\nabla^2\Phi(\xi, \theta, t)$ and $F = REh((\partial^2\Phi(\xi, \theta, t))/\partial\xi^2)$. Assume $\Phi(\xi,\theta,t) = C_{mn}\cos n\theta \sin \alpha_m \xi e^{i\omega_{mn}t}$, where $n = 0, 1, 2, 3, \ldots$ and $m = 1, 2, 3, \ldots$. Additionally, we note that if $n = 0$ and $m = 1, 2, 3, \ldots$, we have a condition termed axially symmetric breathing motion of the shell. Substituting into the equation for $\Phi(\xi, \theta, t)$, Equation 10.44, we obtain

$$a^2\left(\eta^2+\alpha_m^2\right)^4 C'_{mn} + \alpha_m^4 C'_{mn} - \eta^2\left(\eta^2+\alpha_m^2\right)^2 C'_{mn} = 0 \tag{10.50}$$

where $\eta^2 = (\omega^2\rho R^2/E)$ and $a^2 = (h^2/(12(1-\nu^2)R^2))$. Hence, we obtain

$$\omega_{mn}^2 = \frac{E}{\rho R^2}\frac{a^2\left(\eta^2+\alpha_m^2\right)^4 + \alpha_m^4}{\left(\eta^2+\alpha_m^2\right)^2} \tag{10.51}$$

We observe that $C'_{mn}\cos n\theta \sin\alpha_m\xi$ always produces symmetric modes with respect to $\theta = 0$, as illustrated in Figure 10.8 for $n = 0, 1, 2$. We note that $n = 0$ and 1 are referred to as the breathing and beam modes, respectively.

Another solution is obtained if we assume $\Phi(\xi,\theta,t) = C''_{mn}\sin n\theta \sin\alpha_m\xi e^{i\omega_{mn}t}$, where $n = 1, 2, 3, \ldots$ and $m = 1, 2, 3, \ldots$, the frequency $\omega_{mn}$ is the same defined by Equation 10.51. We note that $C''_{mn}$ is arbitrary. We also see that $C''_{mn}\sin n\theta \sin\alpha_m\xi$ represents antisymmetric modes with respect to $\theta = 0$, as illustrated in Figure 10.9.

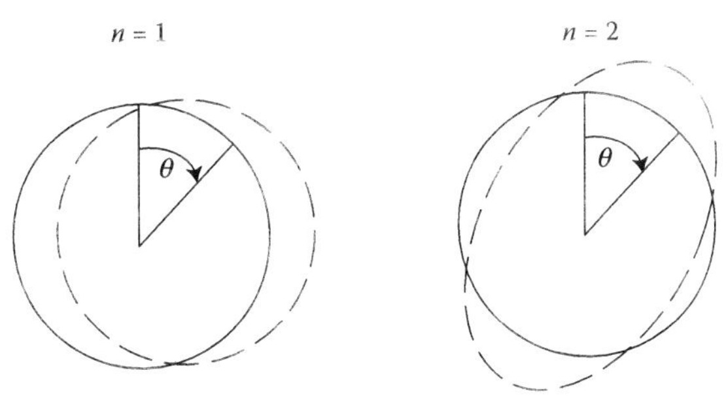


**FIGURE 10.9**
First two antisymmetric modes of free vibration for a closed cylindrical shell.

Thus, the general solution for the free vibration of a circular shell is given as

$$\Phi(\xi,\theta,t)=\sum_{m=0}^{\infty}\sum_{n=0}^{\infty}\{C'_{mn}\cos n\theta\sin\alpha_m\xi+C''_{mn}\sin n\theta\sin\alpha_m\xi\}e^{i\omega_{mn}t} \tag{10.52}$$

where $C'_{mn}$ and $C''_{mn}$ are determined from the initial conditions, which are given functions expressed as $w(\xi,\theta,0)=w_0(\xi,\theta)$ and $\dot{w}(\xi,\theta,0)=\dot{w}_0(\xi,\theta)$. If $w_o(\xi,\theta)$ and $\dot{w}_0(\xi,\theta)$ are symmetric, then $C''_{mn}=0$ and if $w_0(\xi,\theta)$ and $\dot{w}_0(\xi,\theta)$ are antisymmetric, then $C'_{mn}=0$.

#### 10.2.3.2 Forced Vibrations

For forced vibrations of a cylindrical panel as shown in Figure 10.7a, we assume $p_x=p_\theta=0$ and $p_z=p_z(x,\theta,t)$. Using modal expansion, we assume

$$p_z=\sum_m\sum_n P_{mn}(t)\sin\alpha_m\xi\sin\beta_n\theta \tag{10.53}$$

where

$$P_{mn}(t)=\frac{\int_0^{l/R}\int_0^{\theta_0}p(\xi,\theta,t)\sin\alpha_m\xi\sin\beta_n\theta\,d\xi\,d\theta}{\int_0^{l/R}\int_0^{\theta_0}(\sin\alpha_m\xi\sin\beta_n\theta)^2\,d\xi\,d\theta} \tag{10.54}$$

or, in $x$–$y$ coordinates

$$P_{mn}(t)=\frac{R\int_0^{l}\int_0^{R\theta}p(x,y,t)\sin(m\pi x/l)\sin(n\pi y/R\theta_o)dx\,dy}{\int_0^{l}\int_0^{\theta_0}(\sin(m\pi x/l)\sin(n\pi y/R\theta_o))^2\,dx\,dy} \tag{10.55}$$

The solutions for $F$ and $w$ are assumed in the form

$$\begin{aligned}F&=\sum_m\sum_n A_{mn}(t)\sin\alpha_m\xi\sin\beta_n\theta\\ w&=\sum_m\sum_n B_{mn}(t)\sin\alpha_m\xi\sin\beta_n\theta\end{aligned} \tag{10.56}$$

where $A_{mn}(t)$ and $B_{mn}(t)$ are unknown functions of time. Substituting Equation 10.56 into Equation 10.41 yields

$$A_{mn}(t)=B_{mn}(t)\frac{EhR\alpha_m^2}{\left(\alpha_m^2+\beta_n^2\right)^2} \tag{10.57}$$

Next we substitute $p_x$, $F$, and $w$ into Equation 10.32 giving

$$B_{mn}D\left(\alpha_m^2+\beta_n^2\right)^2-R\alpha_m^2A_{mn}+\rho hR^4\ddot{B}_{mn}=R^4P_{mn} \tag{10.58}$$

Using the previously established relation for $A_{mn}$, substituting Equation 10.57 in Equation 10.58 gives

$$\rho hR^4\ddot{B}_{mn}+B_{mn}D\left(\alpha_m^2+\beta_n^2\right)^2+\frac{EhR^2\alpha_m^4}{\left(\alpha_m^2+\beta_n^2\right)^2}B_{mn}=R^4P_{mn} \tag{10.59}$$

or

$$\ddot{B}_{mn}+\omega_{mn}^2B_{mn}=\frac{P_{mn}}{\rho h} \tag{10.60}$$

for each $B_{mn}(t)$.

For forced vibrations of a simply supported closed circular shell, as shown in Figure 10.7b, we assume that $p_z=p_z(x,y,t)=p_z(\xi,\theta,t)$. In addition, we assume that

$$p_z=\sum_{m=0}^{\infty}\sum_{n=0}^{\infty}\left\{P_{mn}^{(c)}(t)\cos n\theta\sin\alpha_m\xi+P_{mn}^{(s)}(t)\sin n\theta\sin\alpha_m\xi\right\}e^{i\omega_{mn}t} \tag{10.61}$$

Using the same procedures that were used for beams and plates, we find that

$$\begin{aligned}P_{mn}^{(c)}(t)&=\frac{\int_0^{2\pi}\int_0^{l/R}p_z(\xi,\theta,t)\sin\alpha_m\xi\cos n\theta\,d\xi\,d\theta}{\int_0^{2\pi}\int_0^{l/R}(\sin\alpha_m\xi\cos n\theta)^2\,d\xi\,d\theta}\\&=\frac{2R}{\pi l}\int_0^{2\pi}\int_0^{l/R}p_z\sin\alpha_m\xi\cos n\theta\,d\xi\,d\theta\end{aligned} \tag{10.62}$$

and

$$\begin{aligned}P_{mn}^{(s)}(t)&=\frac{\int_0^{2\pi}\int_0^{l/R}p_z(\xi,\theta,t)\sin\alpha_m\xi\sin n\theta d\xi d\theta}{\int_0^{2\pi}\int_0^{l/R}(\sin\alpha_m\xi\sin n\theta)^2d\xi d\theta}\\&=\frac{2R}{\pi l}\int_0^{2\pi}\int_0^{l/R}p_z\sin\alpha_m\xi\sin n\theta d\xi d\theta\end{aligned} \tag{10.63}$$

If $p_z$ is symmetric in $\theta$, then $P_{mn}^{(s)}(t)=0$, while if we have $p_z$ antisymmetric in $\theta$, then $P_{mn}^{(c)}(t)=0$. We assume a solution in the form

$$\Phi(\xi,\theta,t)=\sum_{m=0}^{\infty}\sum_{n=0}^{\infty}\left\{C_{mn}^{(c)}(t)\cos n\theta\sin\alpha_m\xi+C_{mn}^{(s)}(t)\sin n\theta\sin\alpha_m\xi\right\}e^{i\omega_{mn}t} \tag{10.64}$$

In this expression, $C_{mn}^{(c)}(t)$ and $C_{mn}^{(s)}(t)$ are unknown functions. Thus, we substitute $p_z$ and $\Phi(\xi,\theta,t)$ into Equation 10.44 and obtain two equations

$$D\left(n^2+\alpha_m^2\right)^4 C_{mn}^{(c)}+R^2Eh\alpha_m^4C_{mn}^{(c)}+\rho hR^4\left(n^2+\alpha_m^2\right)^2\ddot{C}_{mn}^{(c)}=R^4P_{mn}^{(c)} \tag{10.65}$$

for $m = 1, 2, 3, \ldots$ and $n = 0, 1, 2, \ldots$ and

$$D\left(n^2+\alpha_m^2\right)^4 C_{mn}^{(s)}+R^2Eh\alpha_m^4C_{mn}^{(s)}+\rho hR^4\left(n^2+\alpha_m^2\right)^2\ddot{C}_{mn}^{(s)}=R^4P_{mn}^{(s)} \tag{10.66}$$

for $m = 1, 2, 3, \ldots$ and $n = 0, 1, 2, 3, \ldots$. Alternately, these two equations can be expressed as

$$\ddot{C}_{mn}^{(c)}+\omega_{mn}^2C_{mn}^{(c)}=\frac{1}{\rho h\left(n^2+\alpha_m^2\right)^2}P_{mn}^{(c)} \tag{10.67}$$

and

$$\ddot{C}_{mn}^{(s)}+\omega_{mn}^2C_{mn}^{(s)}=\frac{1}{\rho h\left(n^2+\alpha_m^2\right)^2}P_{mn}^{(s)} \tag{10.68}$$

#### *10.2.3.3 Other Boundary Conditions*

Boundary conditions other than those presented in Section 10.2.1 can also be specified. The discussion of these was postponed until after solution procedures were discussed for reasons that should become clear in the following presentation. Three cases are considered:

*Case I:* This case is associated with the cylindrical panel shown in Figure 10.7a. Assume a simply supported shell along $x = 0$, $x = l$, and an arbitrary boundary condition on $\theta = 0$, $\theta = \theta_0$. For the stress function $\Phi(x, \theta, t)$, we assume

$$\Phi(x,\theta,t)=(\Psi(\theta)\sin\alpha_m\xi)e^{i\omega t} \tag{10.69}$$

Substituting Equation 10.69 into Equation 10.44 for $\Phi(x, \theta, t)$ yields an ordinary differential equation in terms of $\Psi(\theta)$. The procedure is similar to that of a plate with two edges simply supported.

*Case II:* This case is also associated with the cylindrical panel shown in Figure 10.7a. Assume a simply supported shell along $\theta = 0$, $\theta = \theta_0$ and an arbitrary boundary condition on $x = 0$, $x = l$. For the stress function $\Phi(x, \theta, t)$, we assume

$$\Phi(x,\theta,t) = \left( X_n(\xi)\sin\frac{n\pi\theta}{\theta_0} \right) e^{i\omega t} \tag{10.70}$$

Substituting into Equation 10.44 for $\Phi$, we will obtain an ordinary differential equation for $X_n(\xi)$. Proceed as before to obtain the solution.

*Case III:* This case is associated with the closed cylindrical panel shown in Figure 10.7b. Assume arbitrary boundary conditions. For the stress function $\Phi$ $(x, \theta, t)$, we assume

$$\Phi(x,\theta,t) = (C'_{mn} X_n(\xi)\cos n\theta) e^{i\omega t} \tag{10.71}$$

Substituting into Equation 10.44 for $\Phi$, we will obtain an ordinary differential equation for $X_n(\xi)$. Proceed as before. Another solution with the same $X_n(\xi)$, $w$ is given by

$$\Phi(x,\theta,t) = \left(C''_{mn} X_n(\xi)\sin n\theta\right) e^{i\omega t} \tag{10.72}$$

### 10.2.4 Membrane Theory of Cylindrical Shells

If a shell is very thin, then the rigidity appears to vanish and the shell becomes a membrane incapable of supporting bending moments and out of plane shear forces (Ventsel and Krauthammer 2001; Blaauwendraad and Hoefakker 2014). The exact equations of the bending theory (in terms of $u$, $v$, $w$) lead to complicated computations. For membrane theory to apply, we assume that only in-plane loads $N_x$, $N_\theta$ and $N_{x\theta} = N_{\theta x}$ are applied and that $M_x = M_\theta = M_{x\theta} = Q_x = Q_\theta = 0$. The equations for membrane theory are the same as the bending theory Equations 10.1 through 10.3 with $Q_x = Q_\theta = 0$, thus we have

$$\frac{\partial N_x}{\partial x} + \frac{1}{R}\frac{\partial N_{x\theta}}{\partial \theta} + p_x = \rho h \frac{\partial^2 u}{\partial t^2} \tag{10.1}$$

$$\frac{1}{R}\frac{\partial N_\theta}{\partial \theta} + \frac{\partial N_{x\theta}}{\partial x} + p_\theta = \rho h \frac{\partial^2 v}{\partial t^2} \tag{10.2}$$

and Equation 10.3 becomes

$$-\frac{1}{R} N_\theta + p_z = \rho h \frac{\partial^2 w}{\partial t^2} \tag{10.73}$$

The force–displacement relations $N_x$, $N_\theta$ and $N_{x\theta} = N_{\theta x}$ given by Equations 10.14 through 10.16 remain applicable. We then have six unknown functions ($N_x$, $N_\theta$, $N_{x\theta}$, $u$, $v$, and $w$) and six equations.

If we consider *axially symmetric deformation* in which $N_x$, $N_\theta$, $N_{x\theta}$, $u$, $v$, and $w$ do not depend on $\theta$, the equations of motion reduce to

$$
\begin{aligned}
\frac{\partial N_x}{\partial x} + p_x &= \rho h \frac{\partial^2 u}{\partial t^2} \\
\frac{\partial N_{x\theta}}{\partial x} + p_\theta &= \rho h \frac{\partial^2 v}{\partial t^2} \\
-\frac{1}{R} N_\theta + p_z &= \rho h \frac{\partial^2 w}{\partial t^2}
\end{aligned}
\tag{10.74}
$$

The force–displacement equations become

$$
\begin{aligned}
N_x &= \frac{Eh}{1-\nu^2}\left[\frac{\partial u}{\partial x} + \nu \frac{w}{R}\right] \\
N_y &= \frac{Eh}{1-\nu^2}\left[\frac{w}{R} + \nu \frac{\partial u}{\partial x}\right] \\
N_{x\theta} = N_{\theta x} &= \frac{Eh}{2(1-\nu)} \frac{\partial v}{\partial x}
\end{aligned}
\tag{10.75}
$$

Substituting the force–displacement equation (10.75) into the equations of motion (Equation 10.74) results in

$$
\begin{aligned}
\frac{\partial^2 u}{\partial x^2} + \frac{\nu}{R}\frac{\partial w}{\partial x} - C_1^{-2}\frac{\partial^2 u}{\partial t^2} &= \frac{1-\nu^2}{Eh} p_x \\
\frac{w}{R} + \frac{\nu}{R}\frac{\partial u}{\partial x} - C_1^{-2}\frac{\partial^2 w}{\partial t^2} &= \frac{1-\nu^2}{Eh} p_z
\end{aligned}
\tag{10.76}
$$

and

$$
\frac{\partial^2 v}{\partial x^2} - C_2^{-2}\frac{\partial^2 v}{\partial t^2} = -\frac{2(1+\nu)}{Eh} p_\theta \tag{10.77}
$$

where

$$
C_1^2 = \frac{E}{\rho(1-\nu^2)} \quad \text{and} \quad C_2^2 = \frac{E}{2\rho(1+\nu)} \tag{10.78}
$$

Equation 10.76 describe the axial and radial motions ($u$, $w$) and are coupled (they cannot exist independently), while Equation 10.77 describes the torsional motion of the shell ($v$—displacement) and is uncoupled from $u$ and $w$ as illustrated in Figure 10.10a and b, respectively. The solution of Equation 10.77 is the same as the axial vibration of a bar.

Considering Equation 10.76, we may only define the boundary conditions for $u$ as we do not have enough constants to define both $u$ and $w$, and it would be contradictory to our theory ($Q_x = Q_\theta = 0$). For example, suppose $w = 0$ at $x = 0$. There would have to be a shear force $Q$ in order to make $w = 0$. These situations are illustrated in Figure 10.11.

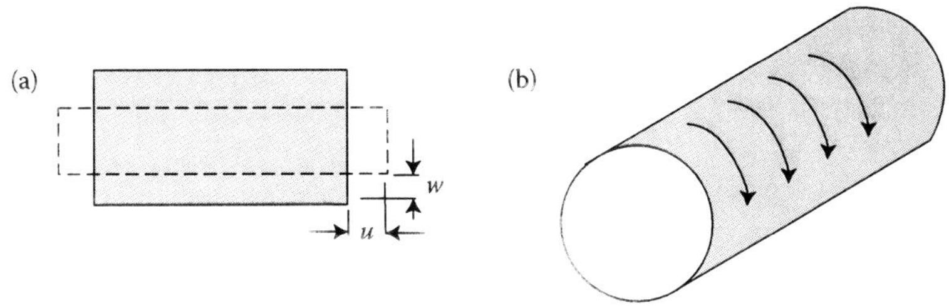


**FIGURE 10.10**
Motion described by (a) Equation 10.76, axial and radial, (b) Equation 10.77, torsional.

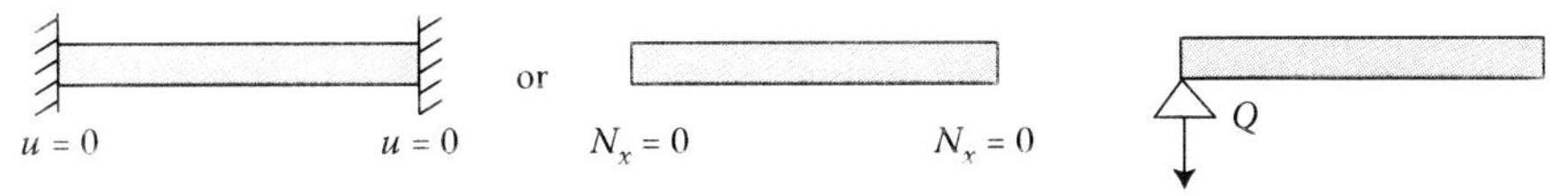


**FIGURE 10.11**
Boundary conditions for first two membranes, Equation 10.76, of motion.

Note that for loading $p_z$, we use bending theory with $f(u, w)$ and for loading $p_x$, we use membrane theory with $f(u, w)$ and then superimpose the two solutions.

#### *10.2.4.1 Free Vibrations*

There are two cases to consider when solving the first two equations of motion for free vibrations.

*Case I:* For this case both ends are free and we assume the following:

$$u = A_m \cos\frac{m\pi x}{l} e^{i\omega t} \tag{10.79}$$

and

$$w = C_m \sin\frac{m\pi x}{l} e^{i\omega t} \tag{10.80}$$

As there is no axial load applied, we know the force $N_x$ at both $x = 0$ and $l$ is

$$N_x = \frac{Eh}{1-\nu^2}\left[\frac{\partial u}{\partial x} + \nu\frac{w}{R}\right] = 0 \tag{10.81}$$

Therefore, the boundary conditions are satisfied at $x = 0$ and $x = l$. Substituting Equations 10.79 and 10.80 into the motion equation (10.76) yields

$$\begin{aligned}&-A_m\left(\frac{m\pi}{l}\right)^2 + \nu\frac{C_m}{R}\frac{m\pi}{l} + C_1^{-2}\omega^2 A_m = 0\\ &\frac{C_m}{R^2} - \nu\frac{A_m}{R}\frac{m\pi}{l} + C_1^{-2}\omega^2 C_m = 0\end{aligned} \tag{10.82}$$

where $m = 0, 1, 2, \ldots$. Equation 10.82 has two linear algebraic expressions for $A_m$ and $C_m$, which are homogeneous. For a nontrivial solution, the determinate of the coefficients $A_m$ and $C_m$ must vanish, thus

$$\begin{vmatrix} \left(\frac{m\pi}{l}\right)^2 - C_1^{-2}\omega^2 & -\nu\frac{m\pi}{Rl} \\ -\nu\frac{m\pi}{Rl} & \frac{1}{R^2} - C_1^{-2}\omega^2 \end{vmatrix} = 0 \tag{10.83}$$

This is the frequency equation for membrane theory. It is a fourth-order equation in $\omega$. For each $m$, we find two values of $\omega^2$ $(\omega_{m1}^2, \omega_{m2}^2, \ldots)$. For each $\omega_{m1}^2$, we can determine $A_{m1}$, $C_{m1}$, etc. or $A_{m1}/C_{m1}$. Similarly, for each $\omega_{m2}^2$ we can determine $A_{m2}$, $C_{m2}$, etc. or $A_{m2}/C_{m2}$. Thus, for each $m$, there are two mode shapes.

*Case II:* For this case, the ends are supported and $u = 0$ at both ends so we assume

$$u = A_m \sin\frac{m\pi x}{l} e^{i\omega t} \tag{10.84}$$

and

$$w = C_m \cos\frac{m\pi x}{l} e^{i\omega t} \tag{10.85}$$

Proceed as before to obtain similar solutions.

#### 10.2.4.2 Forced Vibrations

For forced vibrations, we assume loading conditions similar to those shown in Figure 10.12, where one end is free and the other has an applied load.

Assume an applied surface load at one end defined by $p_x = p(t)\delta(x - l)$ so that we have homogeneous boundary conditions and $p_z = 0$. We let

$$p_x(x,t) = \sum_{m=0}^{\infty} P_m(t) \cos\frac{m\pi x}{l} \tag{10.86}$$

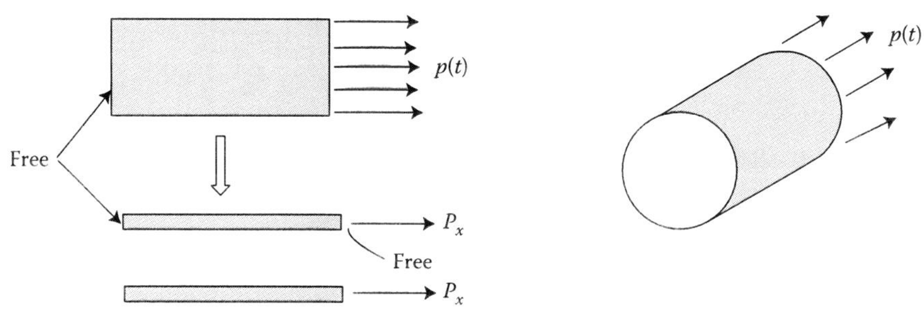


**FIGURE 10.12**
Possible loading conditions for forced vibrations of a membrane.

where

$$P_m(t) = \frac{\int_0^l p_x(x,t)\cos(m\pi x/l)dx}{\int_0^l (\cos(m\pi x/l))^2 dx} = \frac{\int_0^l p(t)\delta(x-l)\cos(m\pi x/l)dx}{\int_0^l (\cos(m\pi x/l))^2 dx} \tag{10.87}$$

$$\begin{aligned} P_m(t) &= p(t)\frac{2}{l}\cos m\pi \quad &\text{for } m = 1,2,3,\ldots \\ P_0(t) &= p(t)\frac{1}{l} \quad &\text{for } m = 0 \end{aligned} \tag{10.88}$$

Assume solutions for $u$ and $w$ of the form

$$\begin{aligned} u &= \sum_m A_m(t)\cos\frac{m\pi x}{l} \\ w &= \sum_m C_m(t)\sin\frac{m\pi x}{l} \end{aligned} \tag{10.89}$$

where $A_m(t)$ and $C_m(t)$ are unknown functions of time. Substituting Equations 10.86 and 10.89 into Equation 10.76 results in

$$\begin{aligned} -\left(\frac{m\pi}{l}\right)^2 A_m(t) + \frac{\nu}{R}\frac{m\pi}{l}C_m(t) - C_1^{-2}\ddot{A}_m(t) &= -\frac{(1-\nu^2)}{Eh}P_m(t) \\ \frac{C_m(t)}{R^2} - \frac{\nu}{R}\frac{m\pi}{l}A_m(t) + C_1^{-2}\ddot{C}_m(t) &= 0 \end{aligned} \tag{10.90}$$

Using Laplace transformations, we can solve Equation 10.90 for any $P_m(t)$. For example, if $p(t) = e^{i\omega t}$, we would assume

$$\begin{aligned} A_m(t) &= A_m^* e^{i\omega t} \\ C_m(t) &= C_m^* e^{i\omega t} \end{aligned} \tag{10.91}$$

This would lead to

$$\begin{aligned} \left[\left(\frac{m\pi}{l}\right)^2 - C_1^{-2}\omega^2\right]A_m^* - \frac{\nu}{R}\left(\frac{m\pi}{l}\right)C_m^* &= \frac{(1-\nu^2)}{Eh}\frac{l}{2}\cos m\pi \\ -\frac{\nu}{R}\left(\frac{m\pi}{l}\right)A_m^* + \left[\frac{1}{R^2} - C_1^{-2}\omega^2\right]C_m^* &= 0 \end{aligned} \tag{10.92}$$

These equations can then be solved for $A_m^*$ and $C_m^*$. Finally, the solutions frequency response functions for $u$ and $w$ are

$$\begin{aligned} u(x,t) &= \sum_{m=0}^{\infty} A_m^* \cos\frac{m\pi x}{l} e^{i\omega t} = U^*(x,i\omega)e^{i\omega t} \\ w(x,t) &= \sum_{m=1}^{\infty} C_m^* \sin\frac{m\pi x}{l} e^{i\omega t} = W^*(x,i\omega)e^{i\omega t} \end{aligned} \tag{10.93}$$

which are the frequency response functions of $u$ and $w$.

## 10.3 Shells of Revolution

When considering shells of revolution (Reissner and Wan 1969; Zingoni 1997), we categorize them as either shallow or deep shells. A shallow shells is characterized as one for which the square of the curvature is small enough that it can be treated as a linear problem. When considering deep shells, we are typically concerned with spherical shells for which membrane theory is applicable and the deformation is axisymmetric.

### 10.3.1 Spherical Shell

When developing the equations of motion for a spherical shell, we use the model shown in Figure 10.13. We assume there are no external loads so that the load vector $\mathbf{p} = 0$. In addition, we note that as there is axisymmetric deformation, $N_\theta$ is constant on the representative volume shown. The forces existing on an element ($N_\phi$ and $N_\theta$) are as shown in Figure 10.13. The area of the representative volume element in this figure is $A = Rd\phi(R \sin \phi d\theta)$ and the inertia forces in the $v$ and $w$ directions are defined as $A\rho h(\partial^2 v/\partial t^2)$ and $A\rho h(\partial^2 w/\partial t^2)$. A graphic of the geometric relations used is summing forces to establish the equations

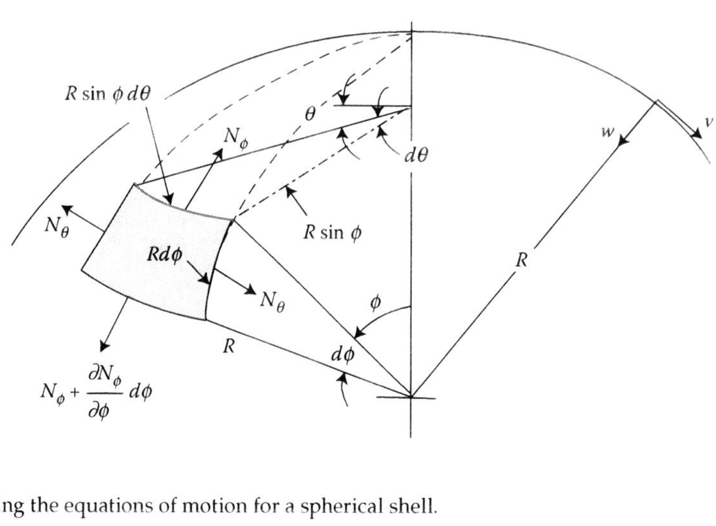


**FIGURE 10.13**
Model for defining the equations of motion for a spherical shell.

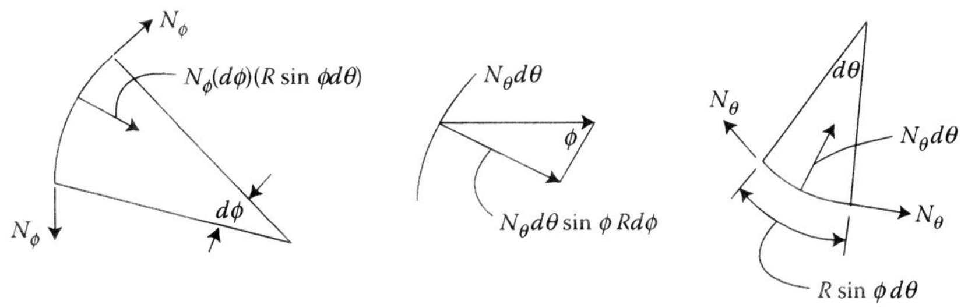


**FIGURE 10.14**
Geometric relations used to establish equations of equilibrium.

of equilibrium are shown in Figure 10.14. Summing forces in the two relevant directions (meridional and normal) results in the two equations of equilibrium shown.

$$\frac{\partial N_\phi}{\partial \phi} + N_\phi \cot\phi - N_\theta \cot\varphi = \rho h R \frac{\partial^2 v}{\partial t^2} \tag{10.94}$$

$$N_\phi + N_\theta = \rho h R \frac{\partial^2 w}{\partial t^2} \tag{10.95}$$

The strain–displacement relations are given by

$$\begin{aligned} \varepsilon_\phi &= \frac{1}{R}\frac{\partial v}{\partial \phi} - \frac{w}{R} \\ \varepsilon_\theta &= \frac{v \cot\phi}{R} - \frac{w}{R} \end{aligned} \tag{10.96}$$

Using Hooke's law, the components of stress are

$$\sigma_\phi = \frac{E}{1-\nu^2}[\varepsilon_\phi + \nu\varepsilon_\theta] = \frac{E}{R(1-\nu^2)}\left[\frac{\partial v}{\partial \phi} - w + \nu(v\cot\phi - w)\right] \tag{10.97}$$

$$\sigma_\theta = \frac{E}{1-\nu^2}[\varepsilon_\theta + \nu\varepsilon_\phi] = \frac{E}{R(1-\nu^2)}\left[v\cot\phi - w + \nu\left(\frac{\partial v}{\partial \phi} - w\right)\right] \tag{10.98}$$

The internal forces in terms of $u$ and $w$ are

$$N_\phi = \int_{-h/2}^{h/2} \sigma_\phi dz = \frac{Eh}{R(1-\nu^2)}\left[\frac{\partial v}{\partial \phi} - w + \nu(v\cot\phi - w)\right] \tag{10.99}$$

$$N_\theta = \int_{-h/2}^{h/2} \sigma_\theta dz = \frac{Eh}{R(1-\nu^2)}\left[v\cot\phi - w + \nu\left(\frac{\partial v}{\partial \phi} - w\right)\right] \tag{10.100}$$

Defining $T = ((1-\nu^2)\rho R^2)/E$ and substituting Equations 10.99 and 10.100 into Equations 10.94 and 10.95 results in

$$\frac{\partial^2 v}{\partial \phi^2} + \cot\phi \frac{\partial v}{\partial \phi} - (\nu + \cot^2\phi)v - (1+\nu)\frac{\partial w}{\partial \phi} = T\frac{\partial^2 v}{\partial t^2} \tag{10.101}$$

$$\frac{\partial v}{\partial \phi} + \cot\phi v - 2w = \frac{T}{1+\nu}\frac{\partial^2 w}{\partial t^2} \tag{10.102}$$

Consider the free vibration of a full sphere. We can assume that $v = V(\phi)e^{i\omega t}$ and $w = W(\phi)e^{i\omega t}$. Substituting into Equations 10.101 and 10.102 yields

$$\frac{\partial^2 V}{\partial \phi^2} + \cot\phi \frac{\partial V}{\partial \phi} - (\nu + \cot^2\phi)V - (1+\nu)\frac{\partial W}{\partial \phi} = -\omega^2 TV \tag{10.103}$$

$$\frac{\partial V}{\partial \phi} + \cot\phi V - 2W = -\frac{T\omega^2}{1+\nu}W \tag{10.104}$$

The solution to these equations is accomplished using the Legendre polynomials of the first kind and an associated Legendre polynomial. Therefore,

$$V_n(x) = A_n'(x) = A_n P_n'(x) = A_n \frac{(1-x^2)^{1/2}}{2^n n!}\frac{d^{n+1}(x^2-1)^n}{dx^{n+1}} \tag{10.105}$$

$$W_n(x) = C_n P_n(x) = C_n \frac{1}{2^n n!}\frac{d^n(x^2-1)^n}{dx^n} \tag{10.106}$$

where $x = \cos\phi$ and

$$P_n(x) = \text{Legendre's polynomial of the first kind}$$

$$P_n'(x) = \text{Associated Legendre's polynomial}$$

$$P_n^m(x) = \sin^m\phi \frac{d^m P_n}{dx^m}$$

Substituting $V_n(x)$ and $W_n(x)$ into the equations of motion results in two algebraic equations for $A_n$ and $C_n$, which are homogeneous. The frequency equation (second order for $\omega^2$) giving nontrivial solutions for $A_n$ and $C_n$ is obtained if the determinate of the coefficients is set to zero. For each $n$, there are two roots: $\omega_{na}^2$ and $\omega_{nb}^2$. In the case of a membrane, it is possible to obtain closed-form solutions. These are given by

$$2T\omega_{na}^2 = [n(n+1)+1+3\nu] + \{[n(n+1)+1+3\nu]^2 - 4(1-\nu^2)[n(n+1)] - 2\} \tag{10.107}$$

$$2T\omega_{nb}^2 = [n(n+1)+1+3\nu] + \{[n(n+1)+1+3\nu]^2 - 4(1-\nu^2)[n(n+1)] - 2\}^{1/2} \tag{10.108}$$

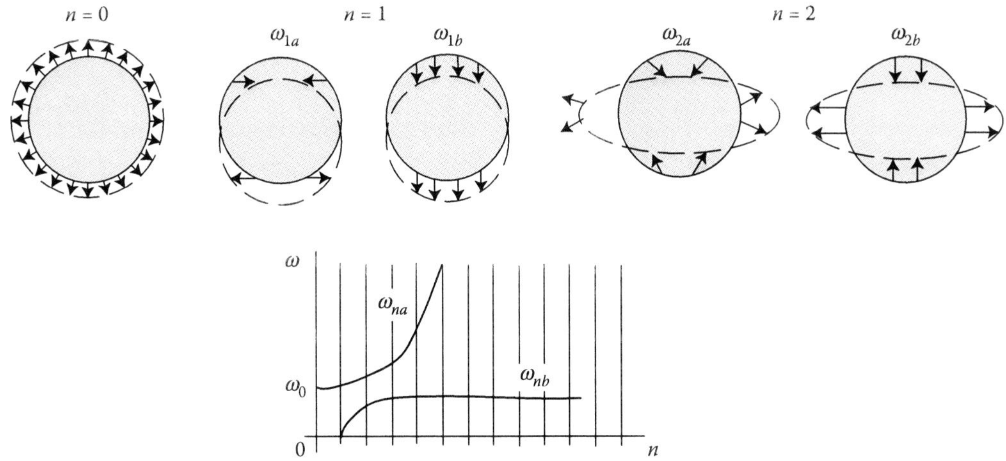


**FIGURE 10.15**
Selected mode shapes for $\omega_{na}$ and $\omega_{nb}$.

For each root of $\omega_{na}$ and $\omega_{nb}$, the mode shapes corresponding to each frequency are determined from $A_n/C_n$. The mode shape for several cases is illustrated in Figure 10.15 along with the trend for $\omega_{na}$ and $\omega_{nb}$ as $n$ increases. We note that for increasing $n$, the frequency $\omega_{nb}$ asymptotically approaches a limiting value, while $\omega_{na}$ continues to increase. The case of $n = 0$ is treated separately because there is always $V = 0$, $W = \text{constant} = C_0$ and $\omega_0^2 = 2(1+\nu)/T$.

### 10.3.2 Shallow Spherical Shells

For shallow spherical shells, we apply bending theory and note that the following assumptions on geometry are applicable: $a/R \ll 1$, $b/R \ll 1$, $h/R \ll 1$, $h/a \ll 1$, $h/b \ll 1$ and the only applied load is $p_z (p_x = p_y = 0)$. In addition to these assumptions, we assume that the difference between the curved shell and the $x$–$y$ axis is negligible. The shell geometry and representative volume element used to develop the equations of motion are shown in Figure 10.16. For the moments, see plate theory.

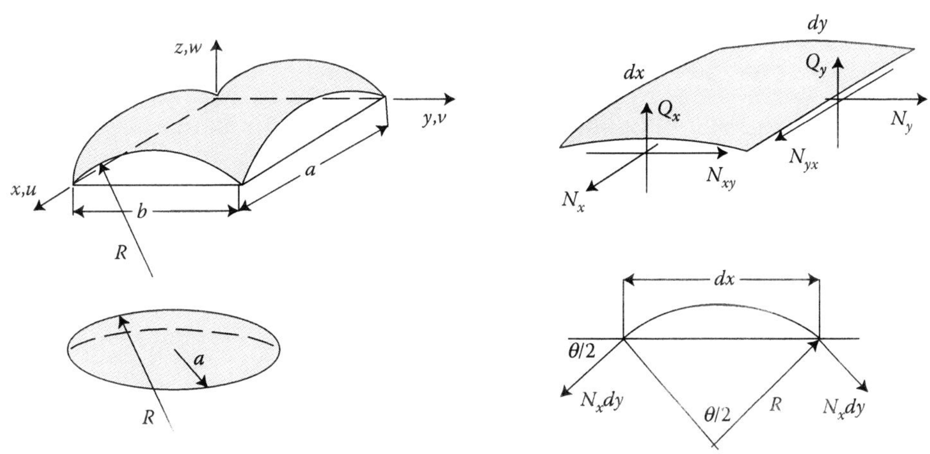


**FIGURE 10.16**
Geometry and representative volume element for a shallow shell.

The equations of motion (noting that $\rho h\partial^2 u/\partial t^2$ and $\rho h\partial^2 v/\partial t^2$ are neglected) are

$$\frac{\partial N_x}{\partial x}+\frac{\partial N_{xy}}{\partial y}=0 \tag{10.109}$$

$$\frac{\partial N_y}{\partial y}+\frac{\partial N_{xy}}{\partial x}=0 \tag{10.110}$$

$$\frac{\partial Q_x}{\partial x}+\frac{\partial Q_y}{\partial y}-\frac{N_x}{R}-\frac{N_y}{R}+p_z=\rho h\frac{\partial^2 w}{\partial t^2} \tag{10.111}$$

We note that for the shallow shell, $\tan\phi/2\approx\sin\phi/2\approx\phi/2=(x/N_x dy)\Rightarrow x=(\phi/2)N_x dy=\phi N_x dy$. In addition, we have $dx=R\phi\Rightarrow\phi=(dx/R)$. As a result, the total vertical force is $F=(N_x dydx/R)$. For the moments, we use Equations 9.10 and 9.11 and note that $M_{xy}$ is related to $N_{xy}$ and $N_{xy}$ resulting in

$$\frac{\partial M_x}{\partial x}+\frac{\partial M_{xy}}{\partial y}-Q_x=0 \tag{10.112}$$

$$\frac{\partial M_{xy}}{\partial x}+\frac{\partial M_y}{\partial y}-Q_y=0 \tag{10.113}$$

$$N_{xy}-N_{yx}-\frac{M_{xy}}{R}=0:\frac{M_{xy}}{R}\approx 0\therefore N_{xy}=N_{yx} \tag{10.114}$$

The force–displacement relations are

$$M_x=-D\left(\frac{\partial^2 w}{\partial x^2}+\nu\frac{\partial^2 w}{\partial y^2}\right) \tag{10.115}$$

$$M_y=-D\left(\frac{\partial^2 w}{\partial y^2}+\nu\frac{\partial^2 w}{\partial x^2}\right) \tag{10.116}$$

$$M_{xy}=-D(1-\nu)\frac{\partial^2 w}{\partial x\partial y} \tag{10.117}$$

$$Q_x=-D\left(\frac{\partial^3 w}{\partial x^3}+\nu\frac{\partial^3 w}{\partial x\partial y^2}\right) \tag{10.118}$$

$$Q_y=-D\left(\frac{\partial^3 w}{\partial y^3}+\nu\frac{\partial^3 w}{\partial y\partial x^2}\right) \tag{10.119}$$

The forces in the shell are defined in terms of the displacements as

$$N_x = \frac{Eh}{1-\nu^2}[\varepsilon_x + \nu\varepsilon_y] = \frac{Eh}{1-\nu^2}\left[\frac{\partial u}{\partial x} + \frac{w}{R} + \nu\left(\frac{\partial v}{\partial y} + \frac{w}{R}\right)\right] \tag{10.120}$$

$$N_y = \frac{Eh}{1-\nu^2}[\varepsilon_y + \nu\varepsilon_x] = \frac{Eh}{1-\nu^2}\left[\frac{\partial v}{\partial y} + \frac{w}{R} + \nu\left(\frac{\partial u}{\partial x} + \frac{w}{R}\right)\right] \tag{10.121}$$

$$N_{xy} = N_{yx} = \frac{Eh}{2(1-\nu)}(2\varepsilon_{xy}) = \frac{Eh}{2(1-\nu)}\left(\frac{\partial u}{\partial y} + \frac{\partial v}{\partial x}\right) \tag{10.122}$$

In order to reduce the equations of motion, we introduce a stress function, $F(x, y, t)$, such that

$$N_x = \frac{\partial^2 F}{\partial y^2}, \quad N_y = \frac{\partial^2 F}{\partial x^2}, \quad N_{xy} = N_{yx} = -\frac{\partial^2 F}{\partial x \partial y} \tag{10.123}$$

Substituting $N_x$ and $N_y$ in terms of $F$ from Equation 10.123 and $Q_x$ and $Q_y$ from Equation 10.118 and 10.119 in terms of $w$ into Equation 10.111 yields

$$D\nabla^2\nabla^2 w + \frac{1}{R}\nabla^2 F + \rho h \frac{\partial^2 w}{\partial t^2} = p_z \tag{10.124}$$

where

$$\nabla^2 = \frac{\partial^2}{\partial x^2} + \frac{\partial^2}{\partial y^2}$$

In order to derive the second equation of motion, as we have two unknowns and one equation, we need to consider the relations

$$\varepsilon_x = \frac{\partial u}{\partial x} + \frac{w}{R} \tag{10.125}$$

$$\varepsilon_y = \frac{\partial v}{\partial y} + \frac{w}{R} \tag{10.126}$$

$$2\varepsilon_{xy} = \frac{\partial u}{\partial y} + \frac{\partial v}{\partial x} \tag{10.127}$$

Next we differentiate Equation 10.125 by $(\partial^2/\partial y^2)$, Equation 10.126 by $(\partial^2/\partial x^2)$, and Equation 10.127 by $-(\partial^2/\partial x\partial y)$. Then adding the results, we obtain

$$\frac{\partial^2 \varepsilon_x}{\partial y^2} + \frac{\partial^2 \varepsilon_y}{\partial x^2} - 2\frac{\partial^2 \varepsilon_{xy}}{\partial x \partial y} - \frac{1}{R}\frac{\partial^2 w}{\partial x^2} - \frac{1}{R}\frac{\partial^2 w}{\partial y^2} = 0 \tag{10.128}$$

Expressing the strains ($\varepsilon_x$, $\varepsilon_y$, and $\varepsilon_{xy}$) in terms of the loads ($N_x$, $N_y$, and $N_{xy}$) from Equation 10.123 yields

$$\begin{aligned} \varepsilon_x &= \frac{1}{Eh}(N_x - \nu N_y) = \frac{1}{Eh}\left(\frac{\partial^2 F}{\partial y^2} - \nu\frac{\partial^2 F}{\partial x^2}\right) \\ \varepsilon_y &= \frac{1}{Eh}(N_y - \nu N_x) = \frac{1}{Eh}\left(\frac{\partial^2 F}{\partial x^2} - \nu\frac{\partial^2 F}{\partial y^2}\right) \\ 2\varepsilon_{xy} &= \frac{2(1+\nu)}{Eh}N_{xy} = \frac{2(1+\nu)}{Eh}\left(-\frac{\partial^2 F}{\partial x \partial y}\right) \end{aligned} \tag{10.129}$$

It should be noted that the normal strains come from $\varepsilon_x = (1/E)\,(\sigma_x - \nu\sigma_y)$ and $\varepsilon_y = (1/E)\,(\sigma_y - \nu\sigma_x)$, where $\sigma_x = (N_x/h)$ and $\sigma_y = (N_y/h)$, while the shear strain comes from $2\varepsilon_{xy} = (\tau_{xy}/G)$, where $\tau_{xy} = (N_{xy}/h)$. Substituting into Equation 10.128 yields

$$\frac{1}{Eh}\nabla^2\nabla^2 F - \frac{1}{R}\nabla^2 w = 0 \tag{10.130}$$

For the polar cap, we use polar coordinates for which the Del operator becomes

$$\nabla^2 = \frac{\partial^2}{\partial r^2} + \frac{1}{r}\frac{\partial}{\partial r} + \frac{\partial^2}{r^2 \partial \phi^2} \tag{10.131}$$

We introduce a single function $\Phi(r, \phi, t)$ such that $w(r, \phi, t) = \nabla^2\nabla^2\Phi(r, \phi, t)$ and $F = (Eh/R)\nabla^2\Phi(r, \phi, t)$. From Equation 10.111, we obtain

$$D\nabla^2\nabla^2\nabla^2\nabla^2\Phi(r,\phi,t) + \frac{Eh}{R}\nabla^2\nabla^2\Phi(r,\phi,t) + \rho h\nabla^2\nabla^2\frac{\partial^2\Phi(r,\phi,t)}{\partial t^2} = p_z \tag{10.132}$$

Expressed in terms of $w$, this is

$$D\nabla^2\nabla^2 w + \frac{Eh}{R}w + \rho h\frac{\partial^2 w}{\partial t^2} = p_z \tag{10.133}$$

For a plate on an elastic foundation, we have

$$D\nabla^2\nabla^2 w + Kw + \rho h\frac{\partial^2 w}{\partial t^2} = p_z \tag{10.134}$$

where $K$ is a foundation modulus. These two equations are actually the same.

In solving shallow shell problems, we note that in polar coordinates, $w$ and $F$ are found in terms of Bessel functions, while in Cartesian coordinates, $w$ and $F$ are found in terms of trigonometric and hyperbolic functions. For free vibrations of a simply supported shallow shell, the solutions are given by

$$F = \sum_m \sum_n A_{mn} \sin \alpha_m x \sin \beta_n y \, e^{i\omega_{mn}t}$$
$$w = \sum_m \sum_n B_{mn} \sin \alpha_m x \sin \beta_n y \, e^{i\omega_{mn}t} \tag{10.135}$$

where

$$\alpha_m = \frac{m\pi}{a}, \quad \beta_m = \frac{n\pi}{b}$$

and

$$\omega_{mn}^2 = \frac{1}{\rho h}\left[D\left(\alpha_m^2 + \beta_n^2\right)^2 + \frac{Eh}{R}\right]$$

## 10.4 Composite Shells

In order to provide a bridge from this text to other references pertaining to composite shells, the governing equations are presented using CLT nomenclature as outlined in Appendix A. Composite shells (Owen and Figueiras 1983; Chao and Reddy 1984; Carrera 2002; Reddy 2004) consists of multiple lamina with possibly different ply orientations and fiber/matrix combinations. The [*A*], [*B*], and [*D*] matrices, as defined by Equation A.13, are required for a complete definition of the equations of motion. In addition, using CLT, the assumed displacement fields defined by Equation A.11 are altered with $\Phi(\Phi = -\partial W_0/\partial x)$ being replaced by $\beta_x$ and $\Psi(\Psi = -\partial W_0/\partial y)$ being replaced by $\beta_\theta$. As a result, the displacement field used for composite shells is

$$U = U_0(x,y) + z\beta_x(x,y), \quad V = V_0(x,y) + z\beta_\theta(x,y), \quad W = W_0(x,y) \tag{10.136}$$

The definitions of strain remain the same as for elastic materials. The sequence of development of the equations of motion is unchanged. The assumed displacement field gives rise to strains, which in turn are related to stresses through a constitutive relationship. The stresses are related to forces and moments, which are then related to mass acceleration through Newton's laws. It is the constitutive relationship between stress and strain for a composite that makes the CLT shell equations of motion appear more complex than for a traditional material. In the equations of motion developed above, a $\rho h$ term appears on the right-hand side of the equation. To account for the possibility of different materials with different density in each lamina comprising the laminate, we define

$$(\bar{\rho}_1, \bar{\rho}_2, \bar{\rho}_3) = \sum_{k=1}^{N} \int_{h_{k-1}}^{h_k} \rho_k(1, z, z^2)\,dz \tag{10.137}$$

Note that the total thickness of the laminated shell is accounted for by the summation of the integration of the densities through each ply.

### 10.4.1 Equations of Motion

In CLT, the possibility for different surface loads on the top (surface 1) and bottom (surface 2) of the laminate is accounted for by defining $q_x = \tau_{1x} - \tau_{2x}$ and $q_\theta = \tau_{1\theta} - \tau_{2\theta}$. These two terms replace the $p_x$ and $p_\theta$ terms in the equations of motion previously derived. In addition, the surface loads in the $x$ and $\theta$ directions on the top and bottom surfaces of the laminate are separated by the laminate thickness, $h$. As a result of this, two moments are identified using CLT that are not present in the equations of motion previously defined. These two moments are $m_x = h(\tau_{1x} + \tau_{2x})/2$ and $m_\theta = h(\tau_{1\theta} + \tau_{2\theta})/2$. The other somewhat significant difference is that the applied force in the $z$ direction ($p$) above is replaced by $p_z$ in CLT. The equations of motion for composite shells using traditional CLT nomenclature are therefore,

$$
\begin{aligned}
&\frac{\partial N_x}{\partial x} + \frac{1}{R}\frac{\partial N_{x\theta}}{\partial \theta} + q_x = \bar{\rho}_1 \frac{\partial^2 U_0}{\partial t^2} + \bar{\rho}_2 \frac{\partial^2 \beta_x}{\partial t^2} \\
&\frac{1}{R}\frac{\partial N_\theta}{\partial \theta} + \frac{\partial N_{x\theta}}{\partial x} + \frac{Q_\theta}{R} + q_\theta = \bar{\rho}_1 V_{0,tt} + \bar{\rho}_2 \beta_{\theta,tt} \\
&\frac{\partial Q_x}{\partial x} + \frac{1}{R}\frac{\partial Q_\theta}{\partial \theta} - \frac{N_\theta}{R} + p_z = \bar{\rho}_1 W_{0,tt} \\
&\frac{\partial M_{x\theta}}{\partial x} + \frac{1}{R}\frac{\partial M_\theta}{\partial \theta} - (Q_\theta - m_\theta) = \bar{\rho}_2 U_{0,tt} + \bar{\rho}_3 \beta_{x,tt} \\
&\frac{1}{R}\frac{\partial M_{x\theta}}{\partial \theta} + \frac{\partial M_x}{\partial x} - (Q_x - m_x) = \bar{\rho}_2 V_{0,tt} + \bar{\rho}_3 \beta_{\theta,tt}
\end{aligned}
\tag{10.138}
$$

Using the notation $U_{0,x} = \partial U_0/\partial x$, etc., the load–displacement relations for a laminated composite shell equivalent to those given by Equations A.12 and A.14 are

$$
\begin{Bmatrix} N_x \\ N_\theta \\ N_{x\theta} \\ -- \\ M_x \\ M_\theta \\ M_{x\theta} \end{Bmatrix}
=
\begin{bmatrix}
A_{11} & A_{12} & A_{16} & | & B_{11} & B_{12} & B_{16} \\
A_{12} & A_{22} & A_{26} & | & B_{12} & B_{22} & B_{26} \\
A_{16} & A_{26} & A_{66} & | & B_{16} & B_{26} & B_{66} \\
-- & -- & -- & | & -- & -- & -- \\
B_{11} & B_{12} & B_{16} & | & D_{11} & D_{12} & D_{16} \\
B_{12} & B_{22} & B_{26} & | & D_{12} & D_{22} & D_{26} \\
B_{16} & B_{26} & B_{66} & | & D_{16} & D_{26} & D_{66}
\end{bmatrix}
\begin{Bmatrix} U_{0,x} \\ V_{0,\theta} \\ V_{0,x} + U_{0,\theta} \\ -- \\ \beta_{x,x} \\ \beta_{\theta,\theta} \\ \beta_{x,\theta}/R + \beta_{\theta,x} \end{Bmatrix}
\tag{10.139}
$$

$$
\begin{Bmatrix} Q_x \\ Q_\theta \end{Bmatrix} = \begin{bmatrix} A_{55} & A_{45} \\ A_{45} & A_{44} \end{bmatrix} \begin{Bmatrix} \beta_x + W_{0,x} \\ \beta_\theta (W_{0,\theta} - V_0)/R \end{Bmatrix}
\tag{10.140}
$$

Defining each load and moment in Equations 10.70 and 10.71 in terms of its $A_{ij}$, $B_{ij}$, $D_{ij}$, and the partial derivatives of displacement, we substitute these into Equation 10.69 to obtain the governing equations of motion for a laminated composite shell with one radius of curvature. The governing equations for shells with double radii of curvature (curvatures in the $x$ as well as the $\theta$ direction as presented herein) can also be developed. Such developments can be found in Reddy (2004). The equations of motion in terms of displacements are

$$A_{11}U_{0,xx}+2\frac{A_{16}}{R}U_{0,\theta x}+\frac{A_{66}}{R^2}U_{0,\theta\theta}+A_{16}V_{0,xx}+\frac{(A_{12}+A_{66})}{R}V_{0,\theta x}+\frac{A_{26}}{R^2}V_{0,\theta\theta}+B_{11}\beta_{x,xx}+2\frac{B_{16}}{R}\beta_{x,\theta x}$$
$$+\frac{B_{66}}{R^2}\beta_{x,\theta\theta}+B_{16}\beta_{\theta,xx}+\frac{(B_{12}+B_{66})}{R}\beta_{\theta,\theta x}+\frac{B_{26}}{R^2}\beta_{\theta,\theta\theta}+\frac{A_{12}}{R}W_{0,x}+\frac{A_{26}}{R^2}W_{0,\theta}+q_x=\bar{\rho}_1\ddot{U}_0+\bar{\rho}_2\ddot{\beta}_x$$

$$A_{16}U_{0,xx}+\frac{(A_{12}+A_{66})}{R}U_{0,\theta x}+\frac{A_{26}}{R^2}U_{0,\theta\theta}+A_{66}V_{0,xx}+2\frac{A_{26}}{R}V_{0,\theta x}+\frac{A_{22}}{R^2}V_{0,\theta\theta}+B_{16}\beta_{x,xx}$$
$$+\frac{(B_{12}+B_{66})}{R}\beta_{x,\theta x}+\frac{B_{26}}{R^2}\beta_{x,\theta\theta}+\frac{A_{45}}{R}\beta_x+B_{66}\beta_{\theta,xx}+2\frac{B_{26}}{R}\beta_{\theta,\theta x}+\frac{B_{22}}{R^2}\beta_{\theta,\theta\theta}+\frac{A_{44}}{R}\beta_\theta$$
$$+\frac{(A_{12}+A_{55})}{R}W_{0,x}+\frac{(A_{26}+A_{45})}{R^2}W_{0,\theta}+q_\theta=\bar{\rho}_1\ddot{V}_0+\bar{\rho}_2\ddot{\beta}_\theta$$

$$-\frac{A_{12}}{R}U_{0,x}-\frac{A_{26}}{R^2}U_{0,\theta}-\left(\frac{A_{26}+A_{45}}{R}\right)V_{0,x}+\left(\frac{A_{22}+A_{44}}{R^2}\right)V_{0,\theta}+\left(A_{55}-\frac{B_{12}}{R}\right)\beta_{x,x}+\left(A_{45}-\frac{B_{26}}{R}\right)\left(\frac{1}{R}\right)\beta_{x,\theta}$$
$$+\left(A_{45}-\frac{B_{26}}{R}\right)\beta_{\theta,x}+\left(A_{44}-\frac{B_{22}}{R}\right)\left(\frac{1}{R}\right)\beta_{\theta,\theta}+A_{55}W_{0,xx}+2\frac{A_{45}}{R}W_{0,\theta x}-\frac{A_{22}}{R^2}W_{0,\theta\theta}+p_z=\bar{\rho}_1\ddot{W}_0$$

(10.141)

$$B_{11}U_{0,xx}+2\frac{B_{16}}{R}U_{0,\theta x}+\frac{B_{66}}{R^2}U_{0,\theta\theta}+B_{16}V_{0,xx}+\frac{(B_{12}+B_{66})}{R}V_{0,\theta x}+\frac{B_{26}}{R^2}V_{0,\theta\theta}+D_{11}\beta_{x,xx}+2\frac{D_{16}}{R}\beta_{x,\theta x}$$
$$+\frac{D_{66}}{R^2}\beta_{x,\theta\theta}-A_{55}\beta_x+D_{16}\beta_{\theta,xx}+\frac{(D_{12}+D_{66})}{R}\beta_{\theta,\theta x}+\frac{D_{26}}{R^2}\beta_{\theta,\theta\theta}-A_{45}\beta_\theta+\left(\frac{B_{12}}{R}-A_{55}\right)W_{0,x}$$
$$+\left(\frac{B_{26}}{R}-A_{45}\right)W_{0,\theta}+m_x=\bar{\rho}_2\ddot{U}_0+\bar{\rho}_3\ddot{\beta}_x$$

$$B_{16}U_{0,xx}+\frac{(B_{12}+B_{66})}{R}U_{0,\theta x}+\frac{B_{26}}{R^2}U_{0,\theta\theta}+B_{66}V_{0,xx}+2\frac{B_{26}}{R}V_{0,\theta x}+\frac{B_{22}}{R^2}V_{0,\theta\theta}+\frac{A_{44}}{R}V_0+D_{16}\beta_{x,xx}$$
$$+\frac{(D_{12}+D_{66})}{R}\beta_{x,\theta x}+\frac{D_{26}}{R^2}\beta_{x,\theta\theta}-A_{45}\beta_x+D_{66}\beta_{\theta,xx}+2\frac{D_{26}}{R}\beta_{\theta,\theta x}+\frac{D_{22}}{R^2}\beta_{\theta,\theta\theta}-A_{44}\beta_\theta$$
$$+\left(\frac{B_{22}}{R}-A_{45}\right)W_{0,x}+\left(\frac{B_{22}}{R}-A_{44}\right)W_{0,\theta}+m_\theta=\bar{\rho}_2\ddot{V}_0+\bar{\rho}_3\ddot{\beta}_\theta$$

As the individual components of the [$A$], [$B$], and [$D$] matrices depend on material and stacking arrangement (ply orientations), these equation can reduce in complexity, depending on which terms vanish based on ply stacking arrangement. The complexity of these equations typically requires a numerical technique to obtain reliable solutions. Owen and Figueiras (1983), Chao and Reddy (1984), Carrera (2002), and Reddy (2004) can be used to identify the appropriate numerical techniques.

### 10.4.2 Free Vibrations

For free vibrations, $q_x = q_\theta = m_x = m_\theta = p_z = 0$. As was previously done, we assume solutions for the displacements $U$, $V$, and $W$ as given by Equation 10.67. These displacements include assumptions for $U_0$, $V_0$, $W_0$ as well as $\beta_x$ and $\beta_\theta$, and must satisfy the boundary conditions.

For the special case of a specially orthotropic symmetric laminated shell with $[B] = \bar{\rho}_2 = 0$ and the only nonzero terms of $[A]$ and $[D]$ are $A_{11}$, $A_{12}$, $A_{22}$, $A_{66}$, $D_{11}$, $D_{12}$, $D_{22}$, and $D_{66}$, the equations of motion become

$$\begin{aligned}
&A_{11}U_{0,xx} + \frac{A_{66}}{R^2}U_{0,\theta\theta} + \frac{(A_{12}+A_{66})}{R}V_{0,\theta x} + \frac{A_{12}}{R}W_{0,x} = \bar{\rho}_1\ddot{U}_0 \\
&\frac{(A_{12}+A_{66})}{R}U_{0,\theta x} + A_{66}V_{0,xx} + \frac{A_{22}}{R^2}V_{0,\theta\theta} + \frac{A_{12}}{R}W_{0,x} = \bar{\rho}_1\ddot{V}_0 \\
&-\frac{A_{12}}{R}U_{0,x} + \frac{A_{22}}{R^2}V_{0,\theta} - \frac{A_{22}}{R^2}W_{0,\theta\theta} = \bar{\rho}_1\ddot{W}_0 \\
&D_{11}\beta_{x,xx} + \frac{D_{66}}{R^2}\beta_{x,\theta\theta} + \frac{(D_{12}+D_{66})}{R}\beta_{\theta,\theta x} = \bar{\rho}_3\ddot{\beta}_x \\
&\frac{(D_{12}+D_{66})}{R}\beta_{x,\theta x} + D_{66}\beta_{\theta,xx} + \frac{D_{22}}{R^2}\beta_{\theta,\theta\theta} = \bar{\rho}_3\ddot{\beta}_\theta
\end{aligned} \tag{10.142}$$

For a simply supported cylindrical shell panel similar to that shown in Figure 10.7a, with boundary conditions: at $x = 0$, $l$: $W = V = M_x = N_x = 0$ and at $\theta = 0$, $\theta_0$: $W = V = M_\theta = N_\theta = 0$. We can assume displacements of the form

$$\begin{aligned}
U_0 &= U_{mn}\cos\alpha_m x \sin\beta_n\theta\, e^{i\omega_{mn}t} \\
V_0 &= V_{mn}\sin\alpha_m x \cos\beta_n\theta\, e^{i\omega_{mn}t} \\
W_0 &= W_{mn}\cos\alpha_m x \sin\beta_n\theta\, e^{i\omega_{mn}t} \\
\beta_x &= X_{mn}\cos\alpha_m x \sin\beta_n\theta\, e^{i\omega_{mn}t} \\
\beta_\theta &= Y_{mn}\sin\alpha_m x \cos\beta_n\theta\, e^{i\omega_{mn}t}
\end{aligned} \tag{10.143}$$

where

$$\alpha_m = \frac{m\pi R}{l}; \quad \beta_n = \frac{n\pi R}{l} \quad \text{with} \quad m = 1,2,3,\ldots \quad \text{and} \quad n = 1,2,3,\ldots$$

Taking the appropriate derivatives, we can express the equations of motion which become

$$\begin{aligned}
&\left(\omega_{mn}^2\bar{\rho}_1 - \alpha_m^2 A_{11} - \beta_n^2\frac{A_{66}}{R^2}\right)U_{mn} - \alpha_m\beta_n\frac{(A_{12}+A_{66})}{R}V_{mn} + \alpha_m\frac{A_{12}}{R}W_{mn} = 0 \\
&-\alpha_m\beta_n\frac{(A_{12}+A_{66})}{R}U_{mn} + \left(\omega^2\bar{\rho}_1 - \alpha_m^2 A_{66} - \beta_n^2\frac{A_{22}}{R^2}\right)V_{mn} + \alpha_m\frac{A_{12}}{R}W_{mn} = 0 \\
&\alpha_m\frac{A_{12}}{R}U_{mn} - \beta_n\frac{A_{22}}{R^2}V_{mn} + \left(\omega^2\bar{\rho}_1 + \beta_n^2\frac{A_{22}}{R^2}\right)W_{mn} = 0 \\
&\left(\omega^2\bar{\rho}_3 - \alpha_m^2 D_{11} - \beta_n^2\frac{D_{66}}{R^2}\right)X_{mn} - \alpha_m\beta_n\frac{(D_{12}+D_{66})}{R}Y_{mn} = 0 \\
&-\alpha_m\beta_n\frac{(D_{12}+D_{66})}{R}X_{mn} + \left(\omega^2\bar{\rho}_3 - \alpha_m^2 D_{66} - \beta_n^2\frac{D_{22}}{R^2}\right)Y_{mn} = 0
\end{aligned} \tag{10.144}$$

These can be put in the matrix form and as in previous cases a nontrivial solution for the fundamental frequency can be obtained provided

$$\begin{vmatrix} \omega_{mn}^2\bar{\rho}_1 - \alpha_m^2 A_{11} - \beta_n^2 \dfrac{A_{66}}{R^2} & -\alpha_m\beta_n \dfrac{(A_{12}+A_{66})}{R} & \alpha_m \dfrac{A_{12}}{R} & 0 & 0 \\ -\alpha_m\beta_n \dfrac{(A_{12}+A_{66})}{R} & \omega^2\bar{\rho}_1 - \alpha_m^2 A_{66} - \beta_n^2 \dfrac{A_{22}}{R^2} & \alpha_m \dfrac{A_{12}}{R} & 0 & 0 \\ \alpha_m \dfrac{A_{12}}{R} & -\beta_n \dfrac{A_{22}}{R^2} & \omega^2\bar{\rho}_1 + \beta_n^2 \dfrac{A_{22}}{R^2} & 0 & 0 \\ 0 & 0 & 0 & \omega^2\bar{\rho}_3 - \alpha_m^2 D_{11} - \beta_n^2 \dfrac{D_{66}}{R^2} & -\alpha_m\beta_n \dfrac{(D_{12}+D_{66})}{R} \\ 0 & 0 & 0 & -\alpha_m\beta_n \dfrac{(D_{12}+D_{66})}{R} & \omega^2\bar{\rho}_3 - \alpha_m^2 D_{66} - \beta_n^2 \dfrac{D_{22}}{R^2} \end{vmatrix} = 0 \tag{10.145}$$

### 10.4.3 Forced Vibrations

For forced vibrations, we no longer require $q_x = q_\theta = m_x = m_\theta = p_z = 0$, and depending on the loading conditions and ply orientations, we may be required to solve Equation 10.72 in its entirety. Recall that $q_x = \tau_{1x} - \tau_{2x}$, $q_\theta = \tau_{1\theta} - \tau_{2\theta}$, $m_x = h(\tau_{1x} + \tau_{2x})/2$ and $m_\theta = h(\tau_{1\theta} + \tau_{2\theta})/2$, where the subscript 1 refers to the top surface of the laminate and subscript 2 to the bottom surface. In the most common situations, the surface shear stresses do not exist, which in turn means $m_x$ and $m_\theta$ vanish. In addition, for thin laminates (small $h$), we can often treat $m_x$ and $m_\theta$ as negligible if the shear stresses are consider small. As with all forced vibration problems, the solution will contain both a homogeneous and a particular solution. In many practical applications, the only applied load will be a normal load $p_z = p_z(x, \theta, t)$ for which we can assume a modal expansion $p_z = \Sigma_m\Sigma_n P_{mn}(t) \sin \alpha_m\zeta \sin \beta_n\theta$, where, as previously illustrated,

$$P_{mn}(t) = \frac{\displaystyle\int_0^{l/R}\int_0^{\theta_0} p(\zeta,\theta,t)\sin\alpha_m\zeta \sin\beta_n\theta d\zeta d\theta}{\displaystyle\int_0^{l/R}\int_0^{\theta_0} (\sin\alpha_m\zeta \sin\beta_n\theta)^2 d\zeta d\theta}$$

The solution of such problems is best handled with a numerical technique.

## PROBLEMS

10.1 A rectangular cylindrical panel is loaded as indicated in Figure 10.17. Use a double sine series to determine the Fourier expansion coefficients for a load, which is defined as

a. A uniform pressure $P = p_z = p(t)$
b. A concentrated load $P = p(t)$ at $x = x_p$ and $\theta = \theta_p$

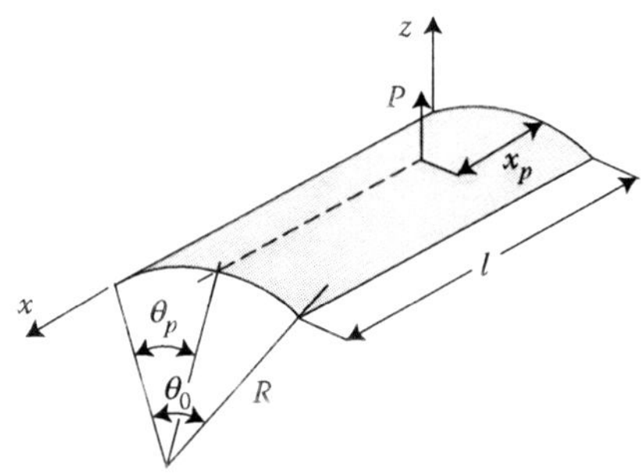


**FIGURE 10.17**
Cylindrical panel loaded with a concentrated force.

10.2 Determine the frequency response function of the deflection $w(x, \theta, t)$ for the panel in Problem 10.1 if the loads are

a. A uniform pressure $P = p_z = p(t)$
b. A concentrated load $P = p(t)$ at $x = x_p$ and $\theta = \theta_p$

10.3 Assuming $B_{mn}(0) = \dot{B}_{mn}(0) = 0$, find the response $W(x, \theta, t)$ to unit impulses for the panel in Problem 10.1 if

a. A uniform pressure $P = p_z = \delta(t)$
b. A concentrated load $P = \delta(t)$ at $x = x_p$ and $\theta = \theta_p$

10.4 Assume a closed shell with $\theta_0 = 2\pi$ and solve Problem 10.1 for

a. A uniform pressure $P = p_z = p(t)$
b. A concentrated load $P = p(t)$ at $x = x_p$ and $\theta = \theta_p$

10.5 Assume a closed shell with $\theta_0 = 2\pi$ and solve Problem 10.2 for

a. A uniform pressure $P = p_z = p(t)$
b. A concentrated load $P = p(t)$ at $x = x_p$ and $\theta = \theta_p$

10.6 Assume a closed shell with $\theta_0 = 2\pi$ and solve Problem 10.3 for

a. A uniform impulse pressure $P = p_z = \delta(t)$
b. A concentrated impulse load $P = \delta(t)$ at $x = x_p$ and $\theta = \theta_p$

## References

Blaauwendraad, J. and Hoefakker, J. H. 2014. *Structural Shell Analysis*. Dordrecht: Springer Science + Business Media.

Calladine, C. R. 1983. *Theory of Shell Structures*. Cambridge, UK: Cambridge University Press.

Carrera, E. 2002. Theories and finite elements for multilayered, anisotropic, composite plates and shells. *Arch. Comput. Meth. Eng.*, 9: 87–140.

Chao, W. C. and Reddy, J. N. 1984. Analysis of laminated composite shells using a degenerated 3-D element. *Int. J. Numer. Meth. Eng.*, 20: 1991–2007.

Chung, H. 1981. Free vibration analysis of circular cylindrical shells. *J. Sound Vib.*, 74: 331–350.

Gol'Denveizer, A. L. 1961. *Theory of Elastic Thin Shells*. Headington Hill Hall, Oxford: Pergamon Press.

Huang, H.-C. 1989. *Static and Dynamic Analysis of Plates and Shells*. Berlin, Heidelberg: Springer-Verlag.

Kil'chevskiy, N. A. 1965. *Fundamentals of the Analytical Mechanics of Shells*. Washington, DC: NASA TT F-92.

Owen, D. R. J. and Figueiras, J. A. 1983. Anisotropic elasto-plastic finite element analysis of thick and thin plates and shells. *Int. J. Numer. Meth. Eng.*, 19: 541–566.

Reddy, J. N. 2004. *Mechanics of Laminated Composite Plates and Shells*, 2nd ed. New York: CRC Press.

Reissner, E. and Wan, F. Y. M. 1969. Rotationally symmetric stress and strain in shells of revolution. *Studies Appl. Math.*, 48: 1–17.

Soedel, W. 2005. *Vibrations of Shells and Plates*, 3rd ed. Revised and Expanded. New York: Marcel Dekker, Inc.

Timoshenko, S. and Woinowsky-Krieger, S. 1959. *Theory of Plates and Shells*, 2nd ed. New York: McGraw-Hill Book Company, Inc.

Ugural, A. C. 2010. *Stresses in Beams, Plates, and Shells*, 3rd ed. Boca Raton, FL: CRC Press.

Ventsel, E. and Krauthammer, T. 2001. *Thin Plates and Shells: Theory: Analysis and Applications*. New York: Marcel Dekker, Inc.

Weingarten, V. I. 1964. Free vibration of thin cylindrical shells. *AIAA J.*, 2: 717–722.

Wempner, G. and Talaslidis, D. 2003. *Mechanics of Solids and Shells*. Boca Raton, FL: CRC Press.

Zingoni, A. 1997. *Shell Structures in Civil and Mechanical Engineering: Theory and Closed-Form Analytical Solutions*. Heron Quay, London: Thomas Telford Publishing.

# 11

# *Finite Elements and Time Integration Numerical Techniques*

## 11.1 Introduction

Direct time integration methods for single-degree-of-freedom (SDOF) systems were presented in Chapter 3. The techniques presented there are also applicable to multi-degree-of-freedom (MDOF) systems. Therefore, the response of simple systems can be obtained by solving the equations of motion together with the appropriate boundary conditions. However, in many practical situations, structures are complex and consist of an assemblage of components, such as beams, plates, shells, etc., which cannot be solved using previous established techniques. Hence, one of the fundamental problems of MDOF system is the formulation of the mass, damping, and stiffness matrices. One method that can be used to formulate the equations of motion is the *finite element method* (FEM). The method was originally developed for the static stress analysis of complex structural systems. Now it has been applied in many disciplines such as fluid mechanics, heat transfer, electromagnetics, and vibrations. The FEM is a numerical technique that can be used for an approximate solution to many vibration problems with complicated geometries, loadings, and material properties where analytical solutions cannot be obtained. The basic idea of the FEM is to determine a solution of a complex problem by replacing it with a simpler one. It is achieved by dividing the problem domain into several small subregions or elements interconnected at a finite number of *nodes*. Each element is an essentially simple unit, the behavior of which can be readily analyzed. Interpolating functions are developed in terms of unknown values of the dependent variables at the nodes. The response of a finite element is represented by a finite number of degrees of freedom at the nodes. For structural analysis, the technique is to express the relations between displacements and internal forces at the node points in terms of a set of algebraic equations. The response of the complete complex system is obtained by assembling the elements and connecting the nodes.

Several methods can be used to transform the physical formulation of the complex structural problem to a finite element discrete system. If the formulation of the problem is given as a partial differential equation, then a *Galerkin method* is the most popular method used for its finite element formulation. If the formulation of the problem can be formulated as a minimization of a functional, then a *variational* formulation of the finite element equations is most often used.

## 11.2 Basic Finite Element Approach

The FEM is a numerical method seeking an approximate solution of the distribution of field variables in the problem domain that is difficult to obtain analytically. Its main advantage is dealing with boundary conditions, which are defined over complex geometries. The application of the FEM to dynamics problems involves the following steps:

1. In the FEM, a deformable body occupying a volume $V$ is divided into a finite number of subregions $V_e$ or for a two-dimensional body occupying an area $A$ into subregions $A_e$ which are called *finite elements.* The boundary between two elements is called the *interelement boundary* which can be a plane or curved surface for a three-dimensional body, a straight, or curve line for two-dimensional body or a point for a one-dimensional body. Adjacent elements are connected at a number of discrete points, called nodes, on their common boundaries. A finite element mesh can be formed using different types of elements as long as they are compatible with connected elements as illustrated in Figure 11.1.
2. Assume a displacement field for the element. The displacement is composed of interpolation or shape functions $\mathbf{N}(x, y, z)$ and nodal displacements $\mathbf{u}^e(t)$, which are assumed functions of time. The displacement within the finite element can be interpolated as

$$\{\tilde{\mathbf{u}}(x,y,z,t)\} = \begin{Bmatrix} u(x,y,z,t) \\ v(x,y,z,t) \\ w(x,y,z,t) \end{Bmatrix} = [\mathbf{N}(x,y,z)]\{\mathbf{u}(t)\}^{(e)} \tag{11.1}$$

where $\tilde{\mathbf{u}}(x,y,z)$ is the approximation of $\mathbf{u}$. The nodal values are parameters, called *degrees of freedom* or *generalized coordinates.* The behavior of the element is characterized by these parameters. To create the interpolation or shape function for each displacement component is to approximate the value of a function between known values using an equation different from the required function itself. The known

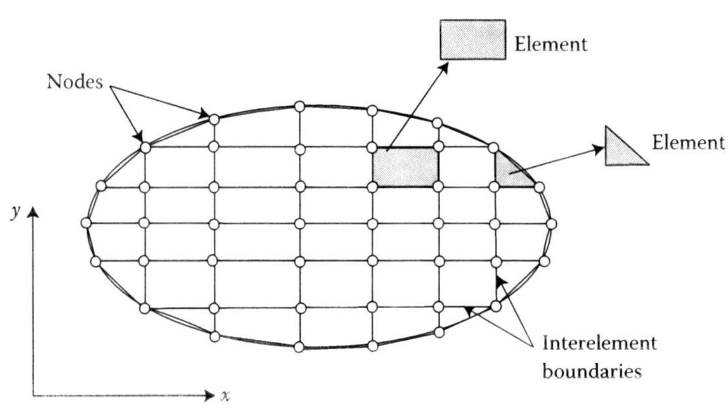


**FIGURE 11.1**
Two-dimensional system with two-dimensional elements.

values are the degrees of freedom, which are determined by solving algebraic equations which are usually approximate rather than exact. The number of shape functions is at least equal to the number of nodes. There exist a few exceptions to the last statement.

3. Develop the element characteristics. This includes the mass, structural damping, and stiffness matrices along with the load vector. This can be accomplished by using Hamilton's principle or Lagrange's equation, which were developed in Chapters 4 and 5. Equations were also developed in Chapter 5 for stresses, strains, and the stress–strain relations. The strain displacement and stress strain relations were given as

$$\varepsilon_{ij} = \frac{1}{2}(u_{i,j} + u_{j,i})$$

and

$$\sigma_{ij} = C_{ijkl}\varepsilon_{kl}$$

Using Equation 11.1, the strain displacement equation can be written as

$$\{\varepsilon\} = [\mathbf{B}]\{\mathbf{u}\}^{(e)} \tag{11.2}$$

and the stresses as

$$\{\sigma\} = [\mathbf{D}]\{\varepsilon\} = [\mathbf{D}][\mathbf{B}]\{\mathbf{u}\}^{(e)} \tag{11.3}$$

We shall use Lagrange's equation, from Section 4.3, which is given in the form

$$\frac{d}{dt}\left\{\frac{\partial L}{\partial \dot{q}}\right\} - \left\{\frac{\partial L}{\partial q}\right\} + \left\{\frac{\partial R}{\partial \dot{q}}\right\} = \{0\} \tag{11.4}$$

where $L = T - \Pi$ and is called the Lagrangian function and $R$ is the *Rayleigh dissipation function*. The kinetic energy of an element of mass $dm = \rho dV$ is

$$dT^{(e)} = \frac{1}{2}dm\left\{\dot{\mathbf{u}}\right\}^T\left\{\dot{\mathbf{u}}\right\} = \frac{1}{2}\rho\left\{\dot{\mathbf{u}}\right\}^T\left\{\dot{\mathbf{u}}\right\}dV$$

or

$$T^{(e)} = \frac{1}{2}\iiint\limits_{V^{(e)}} \rho\left\{\dot{\mathbf{u}}\right\}^T\left\{\dot{\mathbf{u}}\right\}dV \tag{11.5}$$

where $\{\dot{\mathbf{u}}\}$ is the velocity. Substituting Equation 11.1 into Equation 11.5 gives

$$T^{(e)} = \frac{1}{2}\iiint_{V^{(e)}} \{\dot{\mathbf{u}}\}^{(e)T}[\mathbf{N}]^T[\mathbf{N}]\{\dot{\mathbf{u}}\}^{(e)}\rho\, dV \tag{11.6}$$

It is noted that the nodal point velocities $\{\dot{\mathbf{u}}\}^{(e)}$ given in Equation 11.6 do not vary from point to point within the element so that Equation 11.6 can be rewritten as

$$\begin{aligned} T^{(e)} &= \frac{1}{2}\{\dot{\mathbf{u}}\}^{(e)T}\left[\iiint_{V^{(e)}} [\mathbf{N}]^T[\mathbf{N}]\rho\, dV\right]\{\dot{\mathbf{u}}\}^{(e)} \\ &= \frac{1}{2}\{\dot{\mathbf{u}}\}^{(e)T}[\mathbf{m}]^{(e)}\{\dot{\mathbf{u}}\}^{(e)} \end{aligned} \tag{11.7}$$

where

$$[\mathbf{m}]^{(e)} = \iiint_{V^{(e)}} [\mathbf{N}]^T[\mathbf{N}]\rho\, dV \tag{11.8}$$

and $[\mathbf{m}]^{(e)}$ is the element *consistent mass matrix*. The total potential energy of an element consists of the strain energy contained in elastic deformation along with the work done on the element by external forces. Thus, we have

$$\Pi^{(e)} = \frac{1}{2}\iiint_{V^{(e)}} \{\varepsilon\}^T\{\sigma\}dV - \iiint_{V^{(e)}} \{\tilde{\mathbf{u}}\}^T\{\mathbf{b}\}dV - \iint_{S_t^{(e)}} \{\tilde{\mathbf{u}}\}^T\{\mathbf{t}\}dS - \sum \{\tilde{\mathbf{u}}_i\}^T\{\mathbf{P}_i\} \tag{11.9}$$

where $\{\tilde{\mathbf{u}}\}$ is the displacement vector, $\{\mathbf{b}\}$ the vector of body forces, and $\{\mathbf{t}\}$ the vector of applied surface tractions. Substituting Equation 11.1 into Equation 11.9 gives

$$\begin{aligned} \Pi^{(e)} &= \frac{1}{2}\iiint_{V^{(e)}} \{\mathbf{u}\}^{(e)T}[\mathbf{B}]^T[\mathbf{D}][\mathbf{B}]\{u\}^{(e)}dV - \iiint_{V^{(e)}} \{\mathbf{u}\}^{(e)T}[\mathbf{N}]^T\{\mathbf{b}\}dV - \iint_{S_t^{(e)}} \{\mathbf{u}\}^{(e)T}[\mathbf{N}]^T\{\mathbf{t}\}dS \\ &\quad - \sum \{\mathbf{u}\}^{(e)T}[\mathbf{N}]_i^T\{\mathbf{P}\}_i \\ &= \frac{1}{2}\{\mathbf{u}\}^{(e)T}\left[\iiint_{V^{(e)}} [\mathbf{B}]^T[\mathbf{D}][\mathbf{B}]dV\right]\{\mathbf{u}\}^{(e)} \\ &\quad - \{\mathbf{u}\}^{(e)T}\left[\iiint_{V^{(e)}} [\mathbf{N}]^T\{\mathbf{b}\}dV + \iint_{S_t^{(e)}} [\mathbf{N}]^T\{\mathbf{t}\}dS + \sum [\mathbf{N}]_i^T\{\mathbf{P}\}_i\right] \\ &= \frac{1}{2}\{\mathbf{u}\}^{(e)T}[\mathbf{k}]^{(e)}\{\mathbf{u}\}^{(e)} - \{\mathbf{u}\}^{(e)T}\{\mathbf{f}\}^{(e)} \end{aligned} \tag{11.10}$$

where

$$[\mathbf{k}]^{(e)} = \iiint_{V^{(e)}} [\mathbf{B}]^T [\mathbf{D}][\mathbf{B}] dV \tag{11.11}$$

and

$$\begin{aligned}\{\mathbf{f}\}^e &= \iiint_{V^{(e)}} [\mathbf{N}]^T \{\mathbf{b}\} dV + \iint_{S_t} [\mathbf{N}]^T \{\mathbf{t}\} dS + \sum_i [N]_i^T \{\mathbf{P}\}_i \\ &= \{\mathbf{p}\}_b^{(e)} + \{\mathbf{p}\}_{S_t}^{(e)} + \{\mathbf{P}\}_c\end{aligned} \tag{11.12}$$

where $[\mathbf{k}]^{(e)}$ is the element stiffness matrix and $\{\mathbf{f}\}^{(e)}$ is the element load vector which is made up of

$$\{\mathbf{p}\}_b^{(e)} = \iiint_{V^{(e)}} [\mathbf{N}]^T \{\mathbf{b}\} dV \tag{11.13}$$

$$\{\mathbf{p}\}_{S_t}^{(e)} = \iint_{S_t} [\mathbf{N}]^T \{\mathbf{t}\} dS \tag{11.14}$$

$$\{\mathbf{P}\}_c = \sum_i [\mathbf{N}]_i^T \{\mathbf{P}\}_i \tag{11.15}$$

These are the vector element of nodal forces produced by the body forces $\{\mathbf{p}\}_b^{(e)}$, the vector of element nodal forces produced by the surface tractions $\{\mathbf{p}\}_{S_t}^{(e)}$, and the vector of concentrated nodal forces $\{\mathbf{P}\}_c$.

Viscous damping forces can be accounted for by assuming that the dissipative forces are proportional to the relative velocities. Thus, the dissipation function $R$ of the element can be assumed as

$$R^{(e)} = \frac{1}{2} \iiint_{V^{(e)}} \eta \left\{\dot{\mathbf{u}}\right\}^T \left\{\dot{\mathbf{u}}\right\} dV \tag{11.16}$$

where $\eta$ is known as the damping coefficient. Substituting Equation 11.1 into 11.16 yields

$$\begin{aligned}R^{(e)} &= \frac{1}{2} \iiint_{V^{(e)}} \eta \{\mathbf{u}\}^{(e)T} [\mathbf{N}]^T [\mathbf{N}] \{\mathbf{u}\}^{(e)} dV \\ &= \frac{1}{2} \{\mathbf{u}\}^{(e)T} \left[ \iiint_{V^{(e)}} \eta [\mathbf{N}]^T [\mathbf{N}] dV \right] \{\mathbf{u}\}^e \\ &= \frac{1}{2} \{\mathbf{u}\}^{(e)T} [\mathbf{c}]^{(e)} \{\mathbf{u}\}^e\end{aligned} \tag{11.17}$$

where

$$[\mathbf{c}] = \iiint\limits_{V^{(e)}} \mu [\mathbf{N}]^T [\mathbf{N}] dV \tag{11.18}$$

For the entire body (mesh of finite elements), we have the following:

$$T = \sum_{e=1}^{N_e} T^{(e)} = \frac{1}{2}\{\dot{\mathbf{u}}\}^T \left[\sum_{e=1}^{N_e} [\mathbf{m}]^{(e)}\right] \{\dot{\mathbf{u}}\} \tag{11.19}$$

$$\Pi = \sum_{e=1}^{N_e} \Pi^{(e)} = \{\mathbf{u}\}^T \left[\frac{1}{2}\sum_{e=1}^{N_e} [\mathbf{k}]^{(e)} - \{\mathbf{f}\}^{(e)}\right] \{\mathbf{u}\} \tag{11.20}$$

$$R = \sum_{e=1}^{N_e} R^{(e)} = \{\mathbf{u}\}^T \left[\frac{1}{2}\sum_{e=1}^{N_e} [\mathbf{c}]^{(e)}\right] \{\mathbf{u}\} \tag{11.21}$$

where $\{\mathbf{u}\}$ is the global nodal displacement vector and $\{\dot{\mathbf{u}}\}$ is the global nodal velocity vector.

4. Assembly of the element equations which includes the mass, structural damping and stiffness matrices, and the corresponding load vector. From theses element matrices, we develop the overall or global finite element structural system. The assembly consists of adding up or combining the contributions of all elements connected at a node. Interpreting the summation as the standard finite element assembly, we can write

$$T = \frac{1}{2}\{\dot{\mathbf{u}}\}^T [\mathbf{M}]\{\dot{\mathbf{u}}\} \tag{11.22}$$

$$\Pi = \frac{1}{2}\{\mathbf{u}\}^T [\mathbf{K}]\{\mathbf{u}\} - \{\mathbf{u}\}^T \{\mathbf{F}\} \tag{11.23}$$

$$R = \frac{1}{2}\{\dot{\mathbf{u}}\}^T [\mathbf{C}]\{\dot{\mathbf{u}}\} \tag{11.24}$$

where

$$[\mathbf{M}] = \sum_{e=1}^{N_e} [\mathbf{m}]^{(e)} \tag{11.25}$$

$$[\mathbf{K}] = \sum_{e=1}^{N_e} [\mathbf{k}]^{(e)} \tag{11.26}$$

$$[\mathbf{C}] = \sum_{e=1}^{N_e} [\mathbf{c}]^{(e)} \tag{11.27}$$

$$\{\mathbf{F}\} = \sum_{e=1}^{N_e} [\mathbf{f}]^{(e)} \tag{11.28}$$

Recall from above that $[\mathbf{m}]^{(e)}$ was called the *consistent mass* matrix. This is because the same displacement model, which is used for deriving the stiffness matrix, would be used in deriving the mass matrix. One can use simpler forms of the mass matrix. The simplest being placing concentrated masses at the node points in the assumed direction of displacement degrees of freedom. This technique is called the *lumped mass* approach.

Substituting Equations 11.22 through 11.24 into Equation 11.4 yields the equations of motion for the structure as

$$[\mathbf{M}]\{\ddot{\mathbf{u}}\} + [\mathbf{C}]\{\dot{\mathbf{u}}\} + [\mathbf{K}]\{\mathbf{u}\} = \{\mathbf{F}\} \tag{11.29}$$

For structural system with negligible damping, Equation 11.29 becomes

$$[\mathbf{M}]\{\ddot{\mathbf{u}}\} + [\mathbf{K}]\{\mathbf{u}\} = \{\mathbf{F}\} \tag{11.30}$$

5. Introduce the boundary and initial conditions. The solution of the global system of equations is only possible for structural systems that have rigid body motion removed.
6. Solve the global system of equations by consisting direct numerical integration of the equations of motion. The methods used for the direct numerical integration of SDOF systems are applicable to MDOF systems as developed using the FEM.

## 11.3 Interpolation or Shape Functions

One must select the element or elements that are most suitable for the body to be analyzed. Some typical finite elements are shown in Figure 11.2.

For an element, the generalized displacement (translation or rotation), at any point in terms of its nodal ones, was given, Equation 11.1, as

$$\{\tilde{\mathbf{u}}\} = [\mathbf{N}]\{\mathbf{u}(t)\}^{(e)}$$

where $[\mathbf{N}]$ is the interpolation or shape function. The interpolation function for a one-dimensional problem would be $[\mathbf{N}(x)]$, for two dimensions $[\mathbf{N}(x,y)]$, and for three dimensions $[\mathbf{N}(x,y,z)]$. Interpolation or shape functions can be linear, quadratic, cubic, quartic, and quantic depending on the accuracy desired or the variation needed in the field variable. Elements that have interelement displacement continuity said to have $C^0$ continuity or in some cases $C^1$

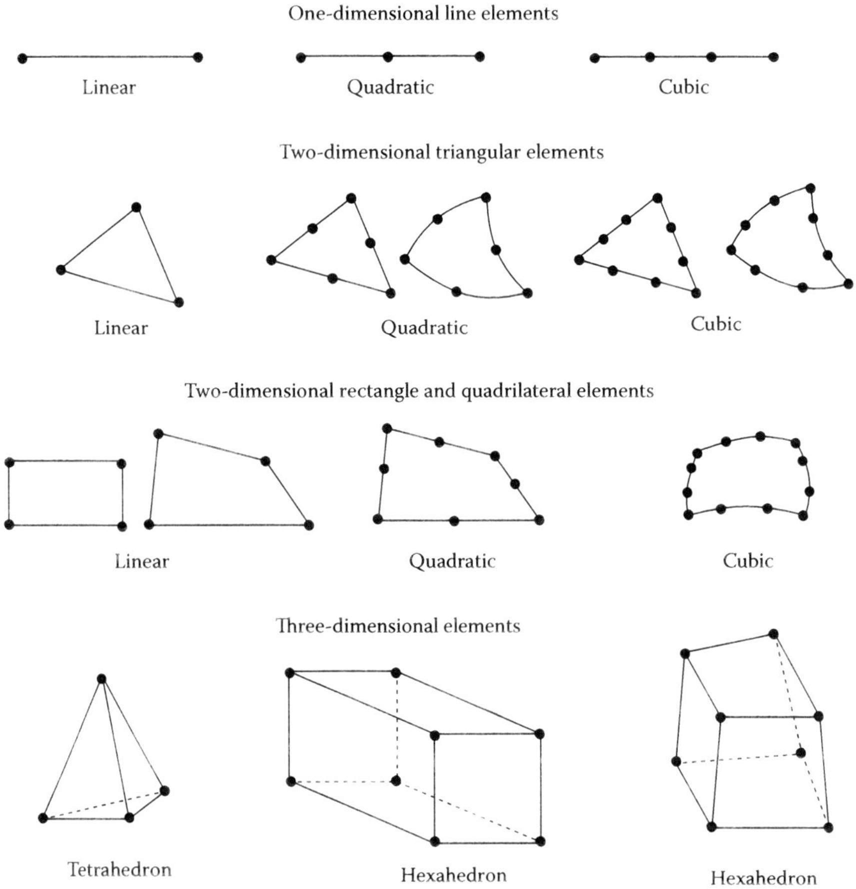


**FIGURE 11.2**
Typical one-, two-, and three-dimensional finite elements.

continuity. An element has $C^0$ continuity if its displacement field and none of its derivatives are continuous with adjacent elements and has continuous first-order derivatives within its boundary. Additionally, if the field variable and its first derivative are continuous across the element interface, and have continuous second-order derivatives within their boundary, one has $C^1$ continuity. In addition, one can have $C^2$ continuity when the field variable, its first and second derivatives maintain continuity at the interelement interface. Also $C^n$ continuity is possible in elements if the $n$th derivatives are continuous across the element interfaces.

Different types of functions that are commonly used for interpolation include polynomials, piecewise polynomials, trigonometric functions, exponential functions, and rational functions. The simplest and most common type of interpolation uses polynomials.

### 11.3.1 One-Dimensional Interpolation Formula

The polynomial expansion for a variable $\tilde{u}$ to be approximated in a one-dimensional line element can be written as

$$\tilde{u} = a_1 + a_2x + a_3x^2 + a_4x^3 + \cdots \tag{11.31}$$

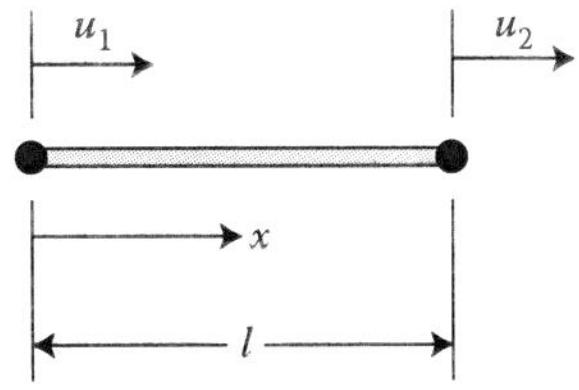


**FIGURE 11.3**
One-dimensional line element.

Since polynomials are used, they must be complete polynomials of at least degree $n$. For a linear variation of $\tilde{u}$, only two nodes are needed, one at each end of the element as shown in Figure 11.3.

Hence

$$\tilde{u} = a_1 + a_2 x \tag{11.32}$$

At $x = 0$, $\tilde{u} = u_1$ and at $x = l$, $\tilde{u} = u_2$, thus we have

$$u_1 = a_1$$
$$u_2 = a_1 + a_2 l$$

or

$$\tilde{u} = \left(1 - \frac{x}{l}\right)u_1 + \frac{x}{l}u_2 = N_1 u_1 + N_2 u_2 \tag{11.33}$$

where

$$N_1 = 1 - \frac{x}{l}$$
$$N_2 = \frac{x}{l} \tag{11.34}$$

are the interpolation or shape functions for a one-dimensional element. However, a more general approach, which is used by most finite element authors, is to use the natural coordinates, $\xi$, where the origin can be set as shown in Figure 11.4.

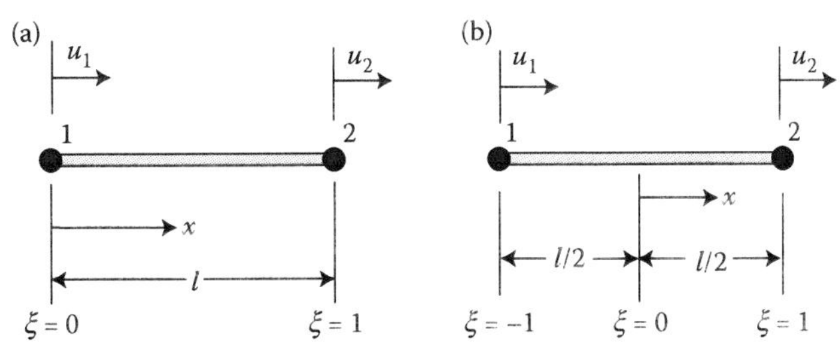


**FIGURE 11.4**
One-dimensional line element with (a) origin at end, (b) origin at center.

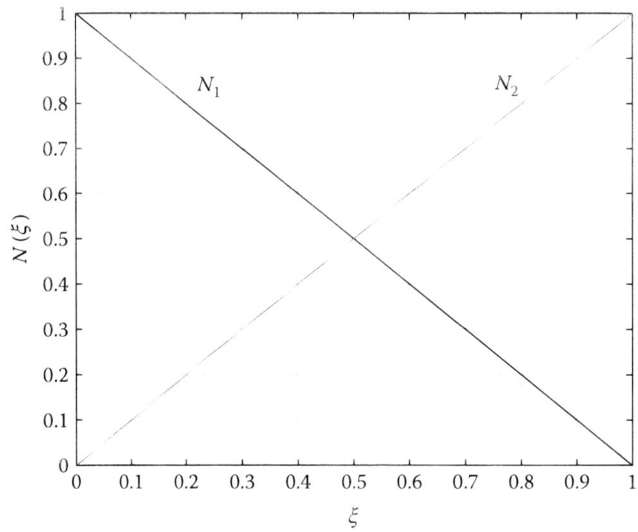


**FIGURE 11.5**
Polynomial shape functions of one-dimensional line element.

We see from Figure 11.4a that $\xi = x/l$ and from 11.4b that $\xi = x/(l/2)$. Equation 11.31 can be written as

$$\tilde{u} = a_1 + a_2\xi + a_3\xi^2 + a_4\xi^3 + \cdots \tag{11.35}$$

and for a linear two-node element we again have

$$\tilde{u} = a_1 + a_2\xi \tag{11.36}$$

As was done above, solving Equation 11.36 at each node we have for Figure 11.4a and b yields the following interpolation functions based on Figure 11.4a and b:

$$N_1 = 1 - \xi, \quad N_2 = \xi \tag{11.37}$$

$$N_1 = \frac{1}{2}(1-\xi), \quad N_2 = \frac{1}{2}(1+\xi) \tag{11.38}$$

The shape functions are shown in Figure 11.5.

#### 11.3.1.1 Lagrange's Interpolation Formula

In a similar manner, higher-order approximations can be obtained by adding the required additional nodes to the interior of the element. For example, a quadratic approximation requires an additional node which is usually chosen at the midpoint of the element. In most cases, the additional nodes are equally spaced along the element but can be placed in arbitrary locations. It is possible to find a polynomial curve fit passing through prescribed function values at specified nodes by using Lagrange's interpolation formula.

Given a set of data points $x_1, x_2, \ldots, x_n$ and the corresponding function values at the data points $u_1, u_2, \ldots, u_n$, we can write

$$\tilde{u}(x) = L_1u_1 + L_2u_2 + \cdots + L_nu_n \tag{11.39}$$

where $L_i$ are the Lagrange polynomials given by

$$L_i(x) = \prod_{m=1,m\neq i}^{n} \frac{x - x_m}{x_i - x_m} = \frac{(x - x_1)(x - x_2)\cdots(x - x_{i-1})(x - x_{i+1})\cdots(x - x_n)}{(x_i - x_1)(x_i - x_2)\cdots(x_i - x_{i-1})(x_i - x_{i+1})\cdots(x_i - x_n)} \tag{11.40}$$

and the functions $L_i(x)$ satisfy the property

$$L_i(x_j) = \begin{cases} 1 & \text{if } i = j \\ 0 & \text{if } i \neq j \end{cases} \tag{11.41}$$

Here, $x_j$ denotes the $x$ coordinate of the $j$th node in the element. Note that $L_i(x)$ given above by Equation 11.40 is just the shape functions $N_i$, therefore $N_i(x) = L_i(x)$. Using the nondimensional form $\xi = x/l$, Equation 11.40 can be written as

$$N_i(\xi) = L_i(\xi) = \prod_{m=1,m\neq i}^{n} \frac{\xi - \xi_m}{\xi_i - \xi_m} = \frac{(\xi - \xi_1)(\xi - \xi_2)\cdots(\xi - \xi_{i-1})(\xi - \xi_{i+1})\cdots(\xi - \xi_n)}{(\xi_i - \xi_1)(\xi_i - \xi_2)\cdots(\xi_i - \xi_{i-1})(\xi_i - \xi_{i+1})\cdots(\xi_i - \xi_n)} \tag{11.42}$$

Hence, we see that for Figure 11.4a and b that

$$N_1 = L_1 = \frac{\xi - \xi_2}{\xi_1 - \xi_2} = \frac{\xi - 1}{0 - 1} = 1 - \xi, \qquad N_2 = L_2 = \frac{\xi - \xi_1}{\xi_2 - \xi_1} = \frac{\xi - 0}{1 - 0} = \xi \tag{a}$$

$$N_1 = L_1 = \frac{\xi - \xi_2}{\xi_1 - \xi_2} = \frac{\xi - 1}{-1 - 1} = \frac{1}{2}(1 - \xi), \quad N_2 = L_2 = \frac{\xi - \xi_1}{\xi_2 - \xi_1} = \frac{\xi + 1}{1 + 1} = \frac{1}{2}(1 + \xi) \tag{b}$$

As noted above, the line element segments were divided into equal length segments using the expression $n = k + 1$ nodes, where $k$ is the order of approximation and $n$ is the number of nodes. Thus, for our linear variation, we have $k = 1$ and $n = 2$. For a quadratic approximation, we would have $k = 2$ and the number of nodes $n$ needed would be 3.

#### *11.3.1.2 Hermitian Interpolation Function*

As noted above, in using Lagrange's interpolation, fitting the data points meant requiring the interpolating polynomial to be equal to the functional values at the data points. On the other hand, Hermitian interpolation in addition to taking on the same value as the given function also takes the slope of the given function at specified data points. Hence, a function $u(\xi)$ is approximated by a polynomial $\tilde{u}(\xi)$ in terms of the values of $u(\xi)$ and all the first derivatives of $u(\xi)$ at $k$ discrete points as

$$\tilde{u}(\xi) = \sum_{i=1}^{k} (H_i(\xi)u(\xi_i) + u'(\xi_i)G_i(\xi)) \tag{11.43}$$

where

$$H_i(\xi) = [1 - 2(\xi - \xi_i)L_i'(\xi_i)]L_i^2(\xi)$$
$$G_i(\xi) = (\xi - \xi_i)L_i^2(\xi) \tag{11.44}$$

and where the Hermite polynomials have the properties (Scheid 1989)

$$\begin{aligned}
&H_i(\xi_j) = \delta_{ij} \\
&G_i(\xi_j) = 0 \\
&H_i'(\xi_j) = [1 - 2(\xi - \xi_i)L_i'(\xi_i)]\,2L_i(\xi)L_i'(\xi)\Big|_{\xi=\xi_j} = 0 \\
&G_i'(\xi_j) = 2(\xi - \xi_i)L_i(\xi)L_i'(\xi) + L_i^2(\xi)\Big|_{\xi=\xi_j} = \delta_{ij}
\end{aligned} \tag{11.45}$$

The values for $L_i(\xi)$ and $L_i'(\xi)$, the Lagrange polynomial and its derivative, are determined from Equation 11.40. Again the simplest and the most common Hermitian interpolation is between two data or node points, where both $u$ and $du/dx$ are taken as nodal degrees of freedom. Since we have two nodes and two degrees of freedom per node, the Hermitian interpolation function has four constants which lead to having a cubic polynomial. Thus, for a one-dimensional element with two end nodes, Equation 11.43 takes the form

$$\tilde{u}(\xi) = H_1(\xi)u_1(\xi_1) + G_1(\xi)u'(\xi_1) + H_2(\xi)u_2(\xi_2) + G_2(\xi)u'(\xi_2) \tag{11.46}$$

Equation 11.46 can also be written as

$$\tilde{u} = N_1u_1 + N_2u_1' + N_3u_2 + N_4u_2' \tag{11.47}$$

where $u' = d/d\zeta$ and the shape functions $N_1$, $N_2$, $N_3$, and $N_4$ are given as $N_1 = H_1(\xi)$, $N_2 = G_1(\xi)$, $N_3 = H_2(\xi)$, and $N_4 = G_2(\xi)$. For example, since $k = 2$, we have for $\xi_1 = 0$, $\xi_2 = 1$

$$\begin{aligned}
&L_1(\xi) = \frac{\xi - \xi_2}{\xi_1 - \xi_2} = \frac{\xi - 1}{-1} = 1 - \xi \quad && L_2(\xi) = \frac{\xi - \xi_1}{\xi_2 - \xi_1} = \frac{\xi}{1} = \xi \\
&L_1'(\xi) = -1 && L_2'(\xi) = 1
\end{aligned}$$

Therefore, we find for $H_1(\xi)$ and $G_1(\xi)$ using Equation 11.44

$$\begin{aligned}
&H_1(\xi) = [1 - 2(\xi - \xi_1)L_1'(\xi_1)]L_1^2(\xi) = [1 - 2\xi(-1)](1 - \xi)^2 = 1 - 3\xi^2 + 2\xi^3 \\
&G_1(\xi) = (\xi - \xi_1)L_1^2(\xi) = \xi(1 - \xi)^2 = \xi - 2\xi^2 + \xi^3
\end{aligned}$$

In the same manner, we determine for $H_2(\xi)$ and $G_2(\xi)$

$$\begin{aligned}
&H_2(\xi) = 3\xi^2 - 2\xi^3 \\
&G_2(\xi) = \xi^3 - \xi^2
\end{aligned}$$

The shape functions are a cubic curve fitted to the ordinance and slopes at $\xi_1 = 0$ and $\xi_2 = 1$ and are given as

$$\begin{aligned} N_1 &= H_1 = 1 - 3\xi^2 + 2\xi^3 \\ N_2 &= G_1 = \xi - 2\xi^2 + \xi^3 \\ N_3 &= H_2 = 3\xi^2 - 2\xi^3 \\ N_4 &= G_2 = -\xi^2 + \xi^3 \end{aligned} \tag{11.48}$$

If the origin is taken at the center of the beam element, we find as was done above that

$$H_1(\xi) = [1 - 2(\xi - \xi_1)L_1'(\xi_1)]L_1^2(\xi) = [1 - 2\xi(-1/2)][1/2(1-\xi)]^2 = \frac{1}{4}(2 - 3\xi + \xi^3)$$

$$G_1(\xi) = (\xi - \xi_1)L_1^2(\xi) = \frac{1}{4}(1 - \xi - \xi^2 + \xi^3)$$

and that the shape functions for $\xi_1 = -1$ and $\xi_2 = 1$ are then given as

$$\begin{aligned} N_1 &= H_1 = \frac{1}{4}(2 - 3\xi + \xi^3) \\ N_2 &= G_1 = \frac{1}{4}(1 - \xi - \xi^2 + \xi^3) \\ N_3 &= H_2 = \frac{1}{4}(2 + 3\xi - \xi^3) \\ N_4 &= G_2 = \frac{1}{4}(-1 - \xi + \xi^2 + \xi^3) \end{aligned} \tag{11.49}$$

where $[N(\zeta)] = [N_1(\zeta) \quad aN_2(\zeta) \quad N_3(\zeta) \quad aN_4(\zeta)]$. Plots of the two points first-order Hermite interpolation functions, Equations 11.48, are shown in Figure 11.6.

### 11.3.2 Two-Dimensional Interpolation Formula

For structures that vibrate in their plane, two-dimensional interpolation formula are needed. The most common element shapes used are triangles, rectangular, and quadrilaterals. These two-dimensional elements are sometimes referred to as membrane elements. For one-dimensional element, to satisfy the convergence criteria, the element displacement functions were derived from complete polynomials. The polynomial terms for one dimension are $1, x, x^2, x^3, \ldots, x^n$. One can form complete polynomial in two variables $x$ and $y$ using Pascal's triangle for the various terms of the polynomial (Table 11.1).

#### *11.3.2.1 Triangular Elements*

The simplest two-dimensional element is the triangle which can be used to idealize a flat plate of irregular shape. Figure 11.7 depicts a triangular element having three nodal points 1, 2, and 3 numbered counterclockwise from an arbitrarily selected corner. Each node has

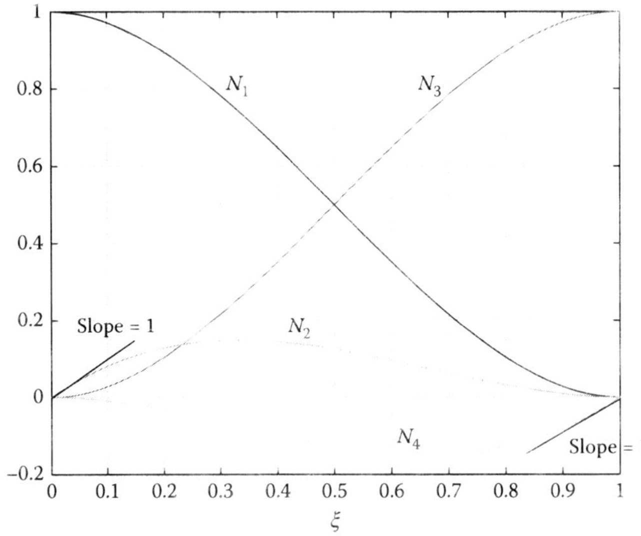


**FIGURE 11.6**
Two-point first-order Hermite interpolation function.

**TABLE 11.1**
Pascal's Triangle

| Pascal's Triangle | Degree of Complete Polynomial | Number of Terms in the Polynomial |
|---|---|---|
| $1$ | 0 | 1 |
| $x \quad y$ | 1 | 3 |
| $x^2 \quad xy \quad y^2$ | 2 | 6 |
| $x^3 \quad x^2y \quad xy^2 \quad y^3$ | 3 | 10 |
| $x^4 \quad x^3y \quad x^2y^2 \quad xy^3 \quad y^4$ | 4 | 15 |
| $x^5 \quad x^4y \quad x^3y^2 \quad x^2y^3 \quad xy^4 \quad y^5$ | 5 | 21 |

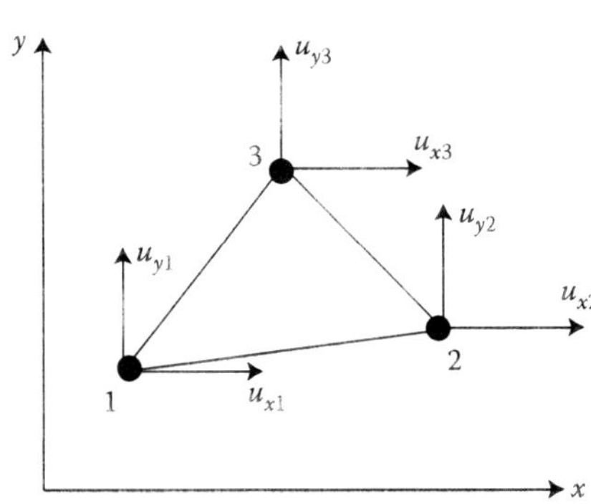


**FIGURE 11.7**
Geometry of a triangular element.

two degrees of freedom, the displacements $u_x$ and $u_y$. In the simplest form, each displacement can be represented by a linear approximation as a complete polynomial of degree 1, represented as

$$\begin{aligned} u_x(x,y) &= a_1 + a_2 x + a_3 y \\ u_y(x,y) &= a_4 + a_5 x + a_6 y \end{aligned} \tag{11.50}$$

where the linear shape functions $N_i(x,y)$ can be determined from

$$u_x(x_i, y_i) = u_{xi}, \quad i = 1,2,3 \tag{11.51}$$

Since the functions for $u_x$ and $u_y$ are of the same form, only one needs to be considered. Using Equation 11.51, we have

$$\begin{Bmatrix} u_{x1} \\ u_{x2} \\ u_{x3} \end{Bmatrix} = \begin{bmatrix} 1 & x_1 & y_1 \\ 1 & x_2 & y_2 \\ 1 & x_3 & y_3 \end{bmatrix} \begin{Bmatrix} a_1 \\ a_2 \\ a_3 \end{Bmatrix} = [\mathbf{A_e}] \begin{Bmatrix} a_1 \\ a_2 \\ a_3 \end{Bmatrix} \tag{11.52}$$

Solving Equation 11.52 gives

$$\begin{Bmatrix} a_1 \\ a_2 \\ a_3 \end{Bmatrix} = [\mathbf{A_e}]^{-1} \begin{Bmatrix} u_{x1} \\ u_{x2} \\ u_{x3} \end{Bmatrix}$$

Substituting this into Equation 11.50 yields

$$u_x(x,y) = N_1 u_{x1} + N_2 u_{x2} + N_3 u_{x3} \tag{11.53}$$

and in the same manner we find

$$u_y(x,y) = N_1 u_{y1} + N_2 u_{y2} + N_3 u_{y3} \tag{11.54}$$

where

$$N_i = \frac{1}{2A}(\alpha_i + \beta_i x + \gamma_i y), \quad i = 1,2,3 \tag{11.55}$$

and

$$\begin{array}{lll} \alpha_1 = x_2 y_3 - x_3 y_2 & \alpha_2 = x_3 y_1 - x_1 y_3 & \alpha_3 = x_1 y_2 - x_2 y_1 \\ \beta_1 = y_2 - y_3 & \beta_2 = y_3 - y_1 & \beta_3 = y_1 - y_2 \\ \gamma_1 = x_3 - x_2 & \gamma_2 = x_1 - x_3 & \gamma_3 = x_2 - x_3 \end{array} \tag{11.56}$$

The area $A$ is given by

$$2A = |\mathbf{A}_e| = \begin{vmatrix} 1 & x_1 & y_1 \\ 1 & x_2 & y_2 \\ 1 & x_3 & y_3 \end{vmatrix} = (x_2y_3 + x_3y_1 + x_1y_2 - x_2y_1 - x_3y_2 - x_1y_3) \qquad (11.57)$$

Note that the node numbers 1, 2, 3 were assigned in a counterclockwise direction. If they were assigned in a clockwise direction, then the determinant of $\mathbf{A}_e$ would be equal to $-2A$. The displacement field at an interior point of the triangular element in terms of the nodal points is given as

$$\{\mathbf{u}\} = \begin{Bmatrix} u_x(x,y) \\ u_y(x,y) \end{Bmatrix} = \begin{bmatrix} N_1 & 0 & N_2 & 0 & N_3 & 0 \\ 0 & N_1 & 0 & N_2 & 0 & N_3 \end{bmatrix} \begin{Bmatrix} u_{x1} \\ u_{y1} \\ u_{x2} \\ u_{y2} \\ u_{x3} \\ u_{y3} \end{Bmatrix} \qquad (11.58)$$

Recall that in addition to looking at the line element using Cartesian coordinates, natural coordinates were also used. This can also be applied to the triangular element by using area coordinates for the triangle as shown in Figure 11.8.

The area coordinates or natural coordinates ($L_1$, $L_2$, $L_3$) of point $P$ are defined by the following equations:

$$L_1 = \frac{A_1}{A}, \quad L_2 = \frac{A_2}{A}, \quad L_3 = \frac{A_3}{A} \qquad (11.59)$$

where $A$ is the area of the entire triangle and $A_1$, $A_2$, and $A_3$ are the areas of the subtriangles $\Delta P23$, $\Delta P13$, and $\Delta P12$. Each of the natural coordinates varies from 0 to 1. Note that since

$$A_1 + A_2 + A_3 = A \qquad (11.60)$$

then

$$L_1 + L_2 + L_3 = 1 \qquad (11.61)$$

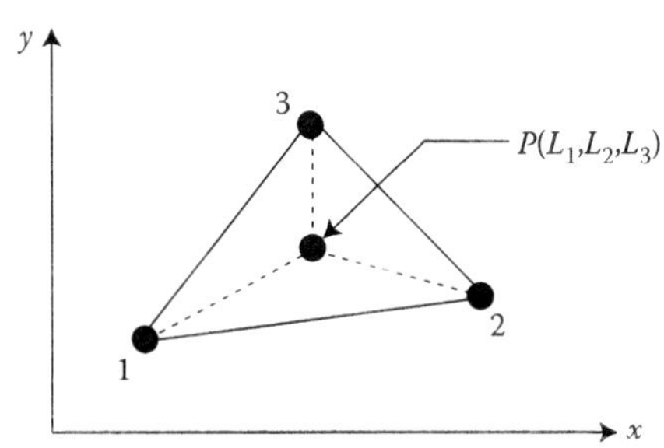


**FIGURE 11.8**
Natural coordinates for triangular element.

Consider the first coordinate given in Equation 11.59, multiply the numerator and denominator by 2 gives

$$L_1 = \frac{2A_1}{2A} \tag{11.62}$$

The area $A_1$ can be expanded in a determinant similar to Equation 11.57. Thus, we find

$$2A_1 = \begin{vmatrix} 1 & x & y \\ 1 & x_2 & y_2 \\ 1 & x_3 & y_3 \end{vmatrix} = (x_2y_3 - x_3y_2) + x(y_2 - y_3) + y(x_3 - x_2) \tag{11.63}$$

where $x$ and $y$ are the coordinates of $P$ as shown in Figure 11.8. Substituting Equation 11.63 into 11.62 yields

$$L_1 = \frac{1}{2A}[x_2y_3 - x_3y_2 + x(y_2 - y_3) + y(x_3 - x_2)] \tag{11.64}$$

It is noted that Equation 11.64 is identical to Equation 11.55 with $i = 1$, therefore, we have that

$$L_1 = N_1 \tag{11.65}$$

Similar calculations for $L_2$ and $L_3$ give

$$L_2 = N_2, \quad L_3 = N_3 \tag{11.66}$$

Thus, we have

$$\{\mathbf{u}\} = [\mathbf{N}]\{\mathbf{d}\} \tag{11.67}$$

where

$$[\mathbf{N}] = \begin{bmatrix} L_1 & L_2 & L_3 & 0 & 0 & 0 \\ 0 & 0 & 0 & L_1 & L_2 & L_3 \end{bmatrix} \tag{11.68}$$

and

$$\{\mathbf{d}\} = \{u_{x1} \quad u_{x2} \quad u_{x3} \quad u_{y1} \quad u_{y2} \quad u_{y3}\}^T \tag{11.69}$$

or

$$[\mathbf{N}] = \begin{bmatrix} L_1 & 0 & L_2 & 0 & L_3 & 0 \\ 0 & L_1 & 0 & L_2 & 0 & L_3 \end{bmatrix} \tag{11.70}$$

and

$$\{\mathbf{d}\} = \{u_{x1} \quad u_{y1} \quad u_{x2} \quad u_{y2} \quad u_{x3} \quad u_{y3}\}^T \tag{11.71}$$

A typical higher-order element is shown in Figure 11.2. For example, a quadratic element has six nodes, three at the vertices and three on the edges. The edge nodes are usually located at the midpoint between the vertices, but do not necessarily need to be. Assuming that each node has two degrees of freedom, each displacement can be represented by a quadratic approximation as a complete polynomial of degree 2, which is given as

$$\begin{aligned} u_x(x,y) &= a_1 + a_2x + a_3y + a_4xy + a_5x^2 + a_6y^2 \\ u_y(x,y) &= a_7 + a_8x + a_9y + a_{10}xy + a_{11}x^2 + a_{12}y^2 \end{aligned} \tag{11.72}$$

where again the shape functions can be determined as was done above. However, with higher-order elements it is easier to work with area or natural coordinates. Thus, we have for the six-node triangular element, where the triangular coordinates are shown in Figure 11.9, the resulting quadratic interpolation equation is given by

$$\{\mathbf{u}\} = \begin{Bmatrix} u_x \\ u_y \end{Bmatrix} = \begin{bmatrix} \{\bar{N}\}^T & \{0\}^T \\ \{0\}^T & \{\bar{N}\}^T \end{bmatrix} \begin{Bmatrix} \{d_x\} \\ \{d_y\} \end{Bmatrix} = [\mathbf{N}]\{\mathbf{d}\} \tag{11.73}$$

where

$$\{\bar{N}\}^T = [L_1(2L_1-1) \quad L_2(2L_2-1) \quad L_3(2L_3-1) \quad 4L_1L_2 \quad 4L_2L_3 \quad 4L_3L_1] \tag{11.74}$$

$$\{d_x\}^T = [u_{x1} \quad u_{x2} \quad u_{x3} \quad u_{x4} \quad u_{x5} \quad u_{x6}] \tag{11.75}$$

$$\{d_y\}^T = [u_{y1} \quad u_{y2} \quad u_{y3} \quad u_{y4} \quad u_{y5} \quad u_{y6}] \tag{11.76}$$

#### 11.3.2.2 Rectangular Elements

In some engineering structures, it is more efficient to discretize the structure such that the elements have four sides, that is, either rectangular or quadrilateral. Thus, one would use rectangular or quadrilateral elements where possible for the structure and then fill in with triangular elements when needed, usually around the boundary. Consider the

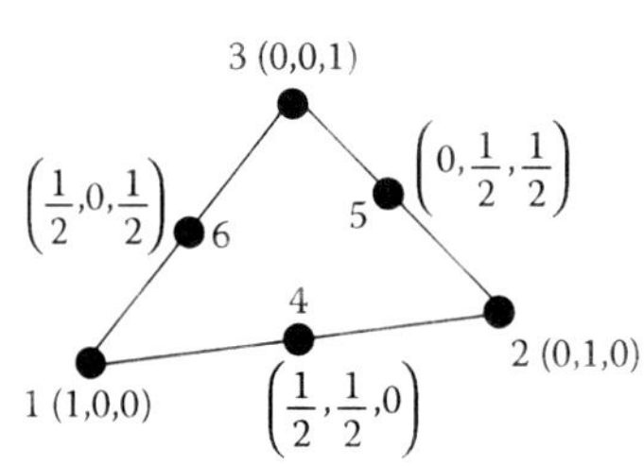


**FIGURE 11.9**
Triangular coordinates of six-node 12-dof linear strain element.

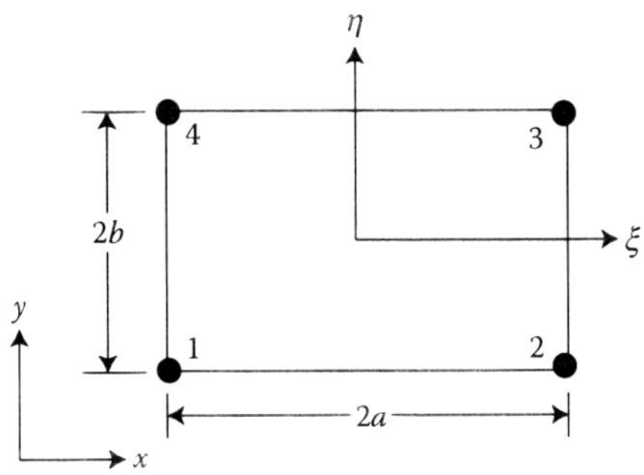


**FIGURE 11.10**
Four-node rectangular element with 8-dof.

four-node rectangular element shown in Figure 11.10 that has sides parallel to the $x$ and $y$ axes. There are two degrees of freedom at each node, the components of displacement $u_x$ and $u_y$ in the directions of the $x$ and $y$ axes.

The displacements can be represented by

$$\begin{aligned} u_x(x,y) &= a_1 + a_2x + a_3y + a_4xy \\ u_y(x,y) &= a_5 + a_6x + a_7y + a_8xy \end{aligned} \tag{11.77}$$

The eight constants $a$ can be determined, as was done for the triangular element, by using the eight nodal displacement conditions

$$\begin{aligned} u_x(x_i,y_i) &= a_1 + a_2x_i + a_3y_i + a_4x_iy_i = u_{ix} \quad i = 1,2,3,4 \\ u_y(x_i,y_i) &= a_5 + a_6x_i + a_7y_i + a_8x_iy_i = u_{iy} \quad i = 1,2,3,4 \end{aligned} \tag{11.78}$$

The eight constants are determined by substituting the eight nodal point conditions, Equation 11.78 into Equation 11.77 and then substituting the results back into Equation 11.77. This gives

$$\{\mathbf{u}\} = \begin{Bmatrix} u_x \\ u_y \end{Bmatrix} = \begin{bmatrix} N_1 & N_2 & N_3 & N_4 & 0 & 0 & 0 & 0 \\ 0 & 0 & 0 & 0 & N_1 & N_2 & N_3 & N_4 \end{bmatrix} \begin{Bmatrix} \{d_x\} \\ \{d_y\} \end{Bmatrix} = [\mathbf{N}]\{\mathbf{d}\} \tag{11.79}$$

where

$$\begin{aligned} N_1(x,y) &= \frac{(x_2-x)(y_3-y)}{(x_2-x_1)(y_3-y_1)} \\ N_2(x,y) &= \frac{(x_1-x)(y_3-y)}{(x_1-x_2)(y_3-y_2)} \\ N_3(x,y) &= \frac{(x_1-x)(y_1-y)}{(x_1-x_3)(y_1-y_3)} \\ N_4(x,y) &= \frac{(x-x_2)(y_1-y)}{(x_4-x_2)(y_1-y_4)} \end{aligned} \tag{11.80}$$

or

$$
\begin{aligned}
N_1(x,y) &= \frac{1}{4ab}(a-x)(b-y) \\
N_2(x,y) &= \frac{1}{4ab}(a+x)(b-y) \\
N_3(x,y) &= \frac{1}{4ab}(a+x)(b+y) \\
N_4(x,y) &= \frac{1}{4ab}(a-x)(b+y)
\end{aligned}
\tag{11.81}
$$

and

$$
\begin{aligned}
\{\mathbf{d_x}\}^T &= [u_{x1} \quad u_{x2} \quad u_{x3} \quad u_{x4}] \\
\{\mathbf{d_y}\}^T &= [u_{y1} \quad u_{y2} \quad u_{y3} \quad u_{y4}]
\end{aligned}
\tag{11.82}
$$

We easily see that at each node $i$ the corresponding $N_i$ have the value one, and all other values are equal to zero. Also, we have that at any point $N_1 + N_2 + N_3 + N_4 = 1$. Equation 11.79 can also be written in the form

$$
\{\mathbf{u}\} = \begin{Bmatrix} \mathbf{u_x} \\ \mathbf{u_y} \end{Bmatrix} = \begin{bmatrix} N_1 & 0 & N_2 & 0 & N_3 & 0 & N_4 & 0 \\ 0 & N_1 & 0 & N_2 & 0 & N_3 & 0 & N_4 \end{bmatrix} \begin{Bmatrix} u_{1x} \\ u_{1y} \\ u_{2x} \\ u_{2y} \\ u_{3x} \\ u_{3y} \\ u_{4x} \\ u_{4y} \end{Bmatrix}
\tag{11.83}
$$

If we use the local coordinate system $(\xi,\eta)$, the interpolation or shape functions are given as

$$
\begin{aligned}
N_1(\xi,\eta) &= \frac{1}{4}(1-\xi)(1-\eta) \\
N_2(\xi,\eta) &= \frac{1}{4}(1+\xi)(1-\eta) \\
N_3(\xi,\eta) &= \frac{1}{4}(1+\xi)(1+\eta) \\
N_4(\xi,\eta) &= \frac{1}{4}(1-\xi)(1+\eta)
\end{aligned}
\tag{11.84}
$$

One could again make use of Lagrange interpolation for quadrilateral elements as follows. The technique is depicted in Figure 11.11.

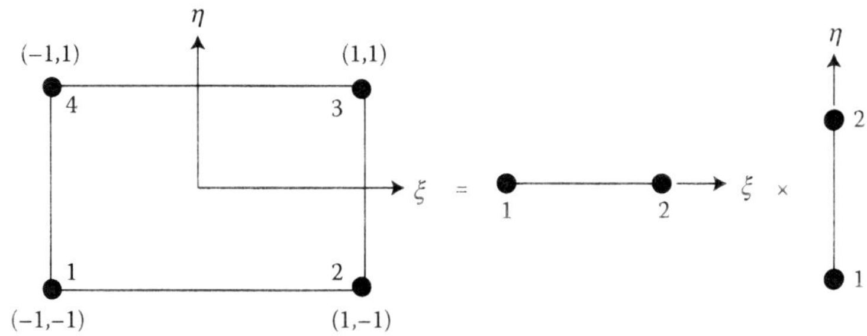


**FIGURE 11.11**
Relationship between the linear, one-dimensional shape function, and the bilinear two-dimensional shape functions.

Consider

$$L_i^{n-1}(\xi) = \prod_{m=1, m\neq i}^{n} \frac{\xi - \xi_m}{\xi_i - \xi_m} = \frac{(\xi-\xi_1)(\xi-\xi_2)\cdots(\xi-\xi_{i-1})(\xi-\xi_{i+1})\cdots(\xi-\xi_n)}{(\xi_i-\xi_1)(\xi_i-\xi_2)\cdots(\xi_i-\xi_{i-1})(\xi_i-\xi_{i+1})\cdots(\xi_i-\xi_n)} \tag{11.85}$$

then the shape function for a bilinear quadrilateral is

$$N_k(\xi,\eta) = L_k^1(\xi)L_k^1(\eta) \tag{11.86}$$

Using Equation 11.86, we find that

$$\begin{aligned}
N_1(\xi,\eta) &= L_1^1(\xi)L_1^1(\eta) = \frac{1}{4}(1-\xi)(1-\eta) \\
N_2(\xi,\eta) &= L_2^1(\xi)L_1^1(\eta) = \frac{1}{4}(1+\xi)(1-\eta) \\
N_3(\xi,\eta) &= L_2^1(\xi)L_2^1(\eta) = \frac{1}{4}(1+\xi)(1+\eta) \\
N_4(\xi,\eta) &= L_1^1(\xi)L_2^1(\eta) = \frac{1}{4}(1-\xi)(1+\eta)
\end{aligned} \tag{11.87}$$

If we introduce new variables

$$\xi_0 = \xi\xi_i \quad \eta_0 = \eta\eta_i \tag{11.88}$$

where $\xi_i$, $\eta_i$ are the normalized coordinates at node *i*. Thus, Equations 11.87 can be written as one equation

$$N_i = \frac{1}{4}(1+\xi_0)(1+\eta_0) \tag{11.89}$$

Higher-order, two-, and three-dimensional elements can be derived by taking products of the Lagrange polynomials. Consider, for example, the quadratic quadrilateral element

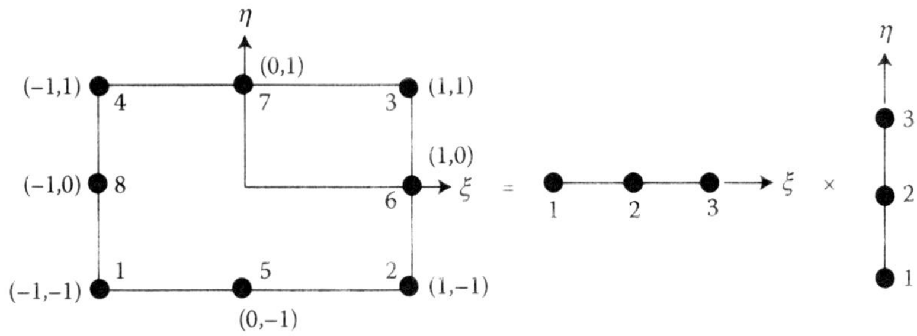


**FIGURE 11.12**
Eight-node, quadratic element with four-corner nodes and four midside nodes.

which consists of eight nodes: four corner nodes and four midside nodes (see Figure 11.12). The shape functions for a quadratic quadrilateral can be taken as

$$N_k(\xi,\eta) = L_k^2(\xi)L_k^2(\eta) \tag{11.90}$$

In the natural coordinate system ($\xi$, $\eta$), the eight shape functions are

$$\begin{aligned}
N_1 &= \frac{1}{4}(1-\xi)(\eta-1)(\xi+\eta+1) & N_5 &= \frac{1}{2}(1-\eta)(1-\xi^2)\\
N_2 &= \frac{1}{4}(1+\xi)(\eta-1)(\eta-\xi+1) & N_6 &= \frac{1}{2}(1+\xi)(1-\eta^2)\\
N_3 &= \frac{1}{4}(1+\xi)(1+\eta)(\xi+\eta-1) & N_7 &= \frac{1}{2}(1+\eta)(1-\xi^2)\\
N_4 &= \frac{1}{4}(\xi-1)(\eta+1)(\xi-\eta+1) & N_8 &= \frac{1}{2}(1-\xi)(1-\eta^2)
\end{aligned} \tag{11.91}$$

We can rewrite Equation 11.91 as

*Corner nodes:*

$$N_i = \frac{1}{4}(1+\xi_0)(1+\eta_0)(\xi_0+\eta_0-1) \tag{11.92}$$

*Midside nodes:*

$$\begin{aligned}
\xi_i = 0 \quad & N_i = \frac{1}{2}(1-\xi^2)(1+\eta_0)\\
\eta_i = 0 \quad & N_i = \frac{1}{2}(1+\xi_0)(1-\eta^2)
\end{aligned} \tag{11.93}$$

We have that $\Sigma_{i=1}^{8} N_i = 1$ at any point inside the element and the displacement field is given by

$$\{\mathbf{u}\} = \begin{Bmatrix} \mathbf{u_x} \\ \mathbf{u_y} \end{Bmatrix} = \begin{bmatrix} \{\bar{N}\}^T & \{0\}^T \\ \{0\}^T & \{\bar{N}\}^T \end{bmatrix} \begin{Bmatrix} \{d_x\} \\ \{d_y\} \end{Bmatrix} = [\mathbf{N}]\{\mathbf{d}\} \tag{11.94}$$

where

$$\{\bar{N}\}^T = [N_1 \quad N_2 \quad N_3 \quad N_4 \quad N_5 \quad N_6 \quad N_7 \quad N_8] \tag{11.95}$$

$$\{d_x\}^T = [u_{x1} \quad u_{x2} \quad u_{x3} \quad u_{x4} \quad u_{x5} \quad u_{x6} \quad u_{x7} \quad u_{x8}] \tag{11.96}$$

$$\{d_y\}^T = [u_{y1} \quad u_{y2} \quad u_{y3} \quad u_{y4} \quad u_{y5} \quad u_{y6} \quad u_{y7} \quad u_{y8}] \tag{11.97}$$

#### *11.3.2.3 Tetrahedral Elements*

For three-dimensional solids, the elements can be made up of tetrahedron or hexahedron shapes with flat or curved surfaces. Each node of the element will have three translational degrees of freedom. The element can thus deform in all three directions in space. Consider an arbitrary tetrahedron element as shown in Figure 11.13.

Natural coordinates may be defined by using any point $P$ within the element analogous to those of a triangular element. This will divide the tetrahedron into four subtetrahedra $P123$, $P234$, $P341$, and $P412$. Thus, we have for the volume coordinates

$$L_i = \frac{V_i}{V} \tag{11.98}$$

where $V_i$ is the volume of the subtetrahedra which is bound by the point $P$ and face $i$ and $V$ is the total volume. Hence, for example, $L_i$ may be defined as the ratio of the volume of the subtetrahedra $P234$ to the total volume 1234. Note that

$$V_1 + V_2 + V_3 + V_4 = V \tag{11.99}$$

therefore, we have

$$\frac{V_1}{V} + \frac{V_2}{V} + \frac{V_3}{V} + \frac{V_4}{V} = L_1 + L_2 + L_3 + L_4 = 1 \tag{11.100}$$

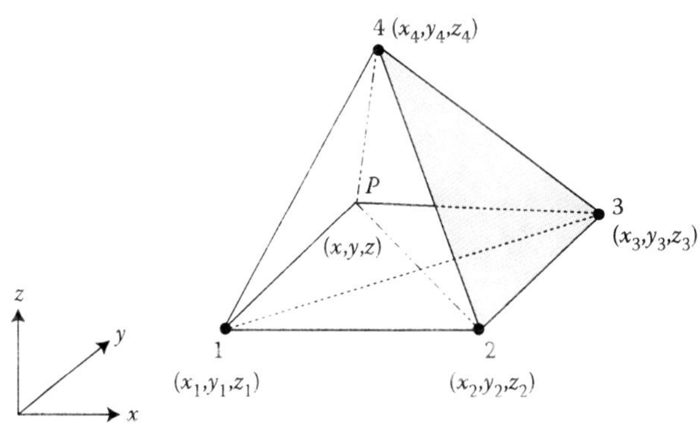


**FIGURE 11.13**
Tetrahedral element.

The total volume $V$ of the tetrahedral element is given by the determinant of the nodal coordinates as

$$V = \frac{1}{6}\begin{vmatrix} 1 & 1 & 1 & 1 \\ x_1 & x_2 & x_3 & x_4 \\ y_1 & y_2 & y_3 & y_4 \\ z_1 & z_2 & z_3 & z_4 \end{vmatrix} \tag{11.101}$$

The volume coordinates $L_1$, $L_2$, $L_3$, and $L_4$ are also the shape functions for the simple tetrahedral element. The coordinates of the tetrahedral are defined by (see Figure 11.13)

$$\begin{aligned} 1 &= L_1 + L_2 + L_3 + L_4 \\ x &= L_1x_1 + L_2x_2 + L_3x_3 + L_4x_4 \\ y &= L_1y_1 + L_2y_2 + L_3y_3 + L_4y_4 \\ z &= L_1z_1 + L_2z_2 + L_3z_3 + L_4z_4 \end{aligned} \tag{11.102}$$

or in the matrix form as

$$\begin{Bmatrix} 1 \\ x \\ y \\ z \end{Bmatrix} = \begin{bmatrix} 1 & 1 & 1 & 1 \\ x_1 & x_2 & x_3 & x_4 \\ y_1 & y_2 & y_3 & y_4 \\ z_1 & z_2 & z_3 & z_4 \end{bmatrix} \begin{Bmatrix} L_1 \\ L_2 \\ L_3 \\ L_4 \end{Bmatrix} \tag{11.103}$$

Equation 11.103 can be solved to give

$$\begin{Bmatrix} L_1 \\ L_2 \\ L_3 \\ L_4 \end{Bmatrix} = \begin{bmatrix} 1 & 1 & 1 & 1 \\ x_1 & x_2 & x_3 & x_4 \\ y_1 & y_2 & y_3 & y_4 \\ z_1 & z_2 & z_3 & z_4 \end{bmatrix}^{-1} \begin{Bmatrix} 1 \\ x \\ y \\ z \end{Bmatrix} = \frac{1}{6V}\begin{bmatrix} a_1 & b_1 & c_1 & d_1 \\ a_2 & b_2 & c_2 & d_2 \\ a_3 & b_3 & c_3 & d_3 \\ a_4 & b_4 & c_4 & d_4 \end{bmatrix} \begin{Bmatrix} 1 \\ x \\ y \\ z \end{Bmatrix} \tag{11.104}$$

where the cofactors are given as

$$\begin{aligned} a_i &= \begin{vmatrix} x_j & y_j & z_j \\ x_k & y_k & z_k \\ x_l & y_l & z_l \end{vmatrix}, \quad b_i = -\begin{vmatrix} 1 & y_j & z_j \\ 1 & y_k & z_k \\ 1 & y_l & z_l \end{vmatrix} \\ c_i &= -\begin{vmatrix} y_j & 1 & z_j \\ y_k & 1 & z_k \\ y_l & 1 & z_l \end{vmatrix}, \quad d_i = -\begin{vmatrix} y_j & z_j & 1 \\ y_k & z_k & 1 \\ y_l & z_l & 1 \end{vmatrix} \end{aligned} \tag{11.105}$$

where we have used the rule that if any two rows of a determinant are interchanged, then the value of the determinant is changed. The other constants in Equation 11.104 are obtained through cyclic permutations of subscripts *i*, *j*, *k*, and *l*. Thus, the shape functions are given as

$$N_k = L_k = \frac{1}{6V}(a_k + b_k x + c_k y + d_k z), \quad k = 1,2,3,4 \tag{11.106}$$

The displacements are given as

$$\begin{Bmatrix} u \\ v \\ w \end{Bmatrix} = [\mathbf{N}]\{\mathbf{d}\} \tag{11.107}$$

where

$$[\mathbf{N}] = \begin{bmatrix} N_1 & 0 & 0 & N_2 & 0 & 0 & N_3 & 0 & 0 & N_4 & 0 & 0 \\ 0 & N_1 & 0 & 0 & N_2 & 0 & 0 & N_3 & 0 & 0 & N_4 & 0 \\ 0 & 0 & N_1 & 0 & 0 & N_2 & 0 & 0 & N_3 & 0 & 0 & N_4 \end{bmatrix} \tag{11.108}$$

$$\{\mathbf{d}\}^T = (u_1 \quad v_1 \quad w_1 \quad u_2 \quad v_2 \quad w_2 \quad u_3 \quad v_3 \quad w_3 \quad u_4 \quad v_4 \quad w_4)$$

For a quadratic tetrahedron, we have 10 nodes that are required to define the element. This can be accomplished using the linear tetrahedron element by adding additional nodes at the edges and faces of the element. A 10-node tetrahedron element is shown in Figure 11.14.

The shape functions for the 10-noded quadratic tetrahedron element can be defined in volume coordinates as

*Corner nodes*:

$$N_i = (2L_i - 1)L_i \quad i = 1,2,3,4 \tag{11.109}$$

*Midside nodes*:

$$\begin{aligned} N_5 &= 4L_1L_2, & N_6 &= 4L_2L_3 \\ N_7 &= 4L_3L_1, & N_8 &= 4L_1L_4 \\ N_9 &= 4L_2L_4, & N_{10} &= 4L_3L_4 \end{aligned} \tag{11.110}$$

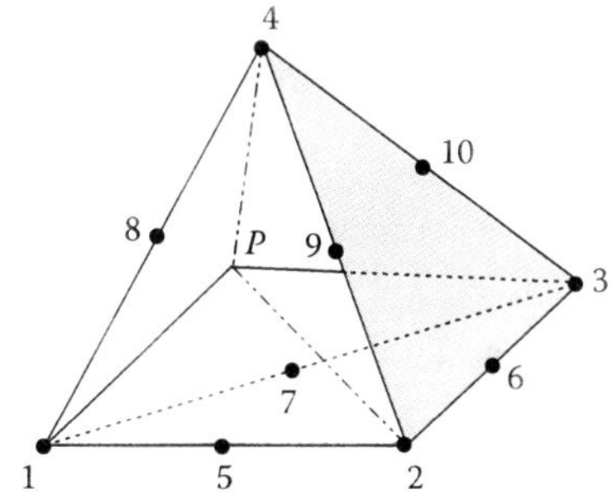


**FIGURE 11.14**
Higher-order 10-node tetrahedron element.

#### 11.3.2.4 Solid Rectangular Hexahedron Elements

In a three-dimensional domain, the system can be divided into a number of hexahedron elements with 8 nodes and 6 surfaces. The hexahedron or brick element with 8 nodes is the simplest of the three-dimensional elements. Each hexahedron element nodes are numbered 1, 2, 3, 4 and 5, 6, 7, 8 in a counterclockwise direction as shown in Figure 11.15 with dimensions $2a$, $2b$, and $2c$. There are three degrees of freedom $(u,v,w)$ at each node. For higher-order elements, the number of nodes can be increased from 8 to 20, 32, and 64.

As with the two-dimensional quadrilateral element, a natural coordinate system $(\xi,\eta,\zeta)$ can also be used for the hexahedron element as shown in Figure 11.15. Each of the displacement components can be represented by polynomials having eight terms each. To ensure geometric invariance, we need to include three quadratic terms $\xi\eta$, $\eta\zeta$, $\zeta\xi$ and a cubic term $\xi\eta\zeta$, thus we have

$$u(\xi,\eta,\zeta) = a_1 + a_2\xi + a_3\eta + a_4\zeta + a_5\xi\mu + a_6\eta\zeta + a_7\zeta\xi + a_8\xi\eta\zeta \tag{11.111}$$

The coefficients $a_1$ to $a_8$ are obtained by evaluating the $u$ value at the 8 nodes and solving the resulting equations. However, the displacement functions can be written down as done for the rectangular element above. The displacement functions are given in the form

$$\begin{Bmatrix} u \\ v \\ w \end{Bmatrix} = \sum_{i=1}^{8} N_i(\xi,\eta,\zeta) \begin{Bmatrix} u_i \\ v_i \\ w_i \end{Bmatrix} \tag{11.112}$$

The functions $N_i$ are required to have a unit value at node $i$ and zero values at all other remaining nodes. The shape functions can be expressed as a product of three linear functions based on Equation 11.111, thus, we have for the three-dimensional version of Equation 11.89

$$N_i = \frac{1}{8}(1+\xi_0)(1+\eta_0)(1+\zeta_0) \tag{11.113}$$

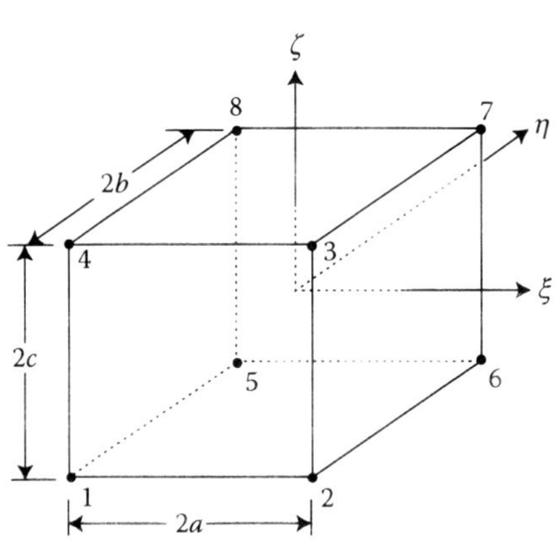


**FIGURE 11.15**
Rectangular hexahedron element with 8 nodes.

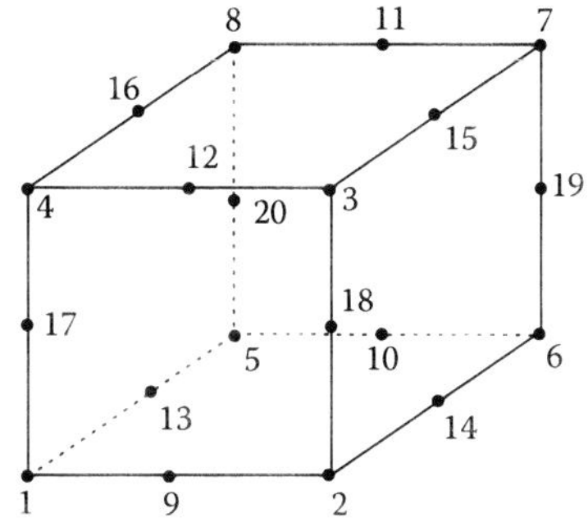


**FIGURE 11.16**
Rectangular hexahedron element with 20 nodes.

where $\xi_0 = \xi_i\xi$, $\eta_0 = \eta_i\eta$, and $\zeta_0 = \zeta_i\zeta$ and $(\xi_i, \eta_i, \zeta_i)$ are the coordinates of node $i$. One can extend this to the quadratic hexahedron element that has 20 nodes as shown in Figure 11.16.

The displacement function is given as

$$\begin{Bmatrix} u \\ v \\ w \end{Bmatrix} = \sum_{i=1}^{20} N_i(\xi, \eta, \zeta) \begin{Bmatrix} u_i \\ v_i \\ w_i \end{Bmatrix} \tag{11.114}$$

where

$$\begin{aligned} N_i(\xi,\eta,\zeta) &= \frac{1}{8}(1+\xi_0)(1+\eta_0)(1+\zeta_0)\times(\xi_0+\eta_0+\zeta_0-2) && i=1,2,\ldots,8 \\ &= \frac{1}{4}(1-\xi^2)(1+\eta_0)(1+\zeta_0) && i=9,11,17,19 \\ &= \frac{1}{4}(1-\eta^2)(1+\zeta_0)(1+\xi_0) && i=10,12,18,20 \\ &= \frac{1}{4}(1-\zeta^2)(1+\xi_0)(1+\eta_0) && i=13,14,15,16 \end{aligned} \tag{11.115}$$

#### *11.3.2.5 Isoparametric Elements*

An isoparametric element is defined as one whose displacement functions and geometry can both be described by the same matrix of shape functions. For example, for the two-dimensional element, we have for the displacement functions

$$\{\mathbf{u}\} = \begin{Bmatrix} u_x \\ u_y \end{Bmatrix} = [\mathbf{N}] \begin{Bmatrix} \mathbf{u}_{xi} \\ \mathbf{u}_{yi} \end{Bmatrix} \tag{11.116}$$

and the geometric functions are

$$\begin{Bmatrix} x \\ y \end{Bmatrix} = [\mathbf{N}] \begin{Bmatrix} \mathbf{x}_i \\ \mathbf{y}_i \end{Bmatrix} \tag{11.117}$$

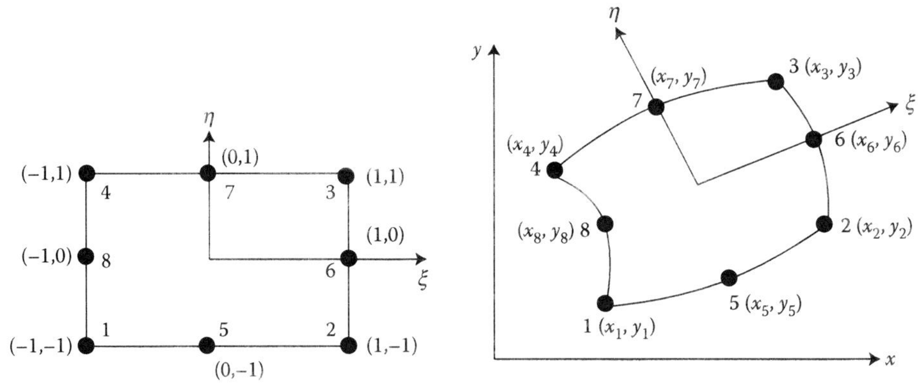


**FIGURE 11.17**
Isoparametric eight-noded quadrilateral element coordinate transformation.

Isoparametric elements are usually described by mapping nondimensional elements of regular shapes, such as triangles, rectangles, brick, etc., into actual elements of distorted shapes. The geometric functions and displacement functions are of equal orders for isoparametric elements (Zienkiewicz et al. 2005). As an example, consider an eight-noded quadrilateral element using Equations 11.116 and 11.117. The results are depicted in Figure 11.17.

### *11.3.2.6 Plate Elements*

A large number of finite elements have been proposed for plates (Hrabok and Hrudey 1984; Zienkiewicz et al. 2005). A difficulty with plate elements results because the plates are not only joined at end nodal point but also along interelement boundaries making the satisfaction of compatibility much more complex since plates carry transverse loads that lead to bending deformations. It is recalled that in one-dimensional problems the neighboring elements were joined only at the end nodal points, whereby compatibility was easily obtained. Plate elements can be triangular, rectangular, or quadrilateral in shape.

1. *Nonconforming Rectangular Element.* For convenience, we will formulate an element of rectangular shape. Consider the rectangular plate element with 12 degrees of freedom as shown in Figure 11.18.

   This element is one of the oldest and best-known elements for the analysis of plates. There are three degrees of freedom at each node. These are the displacement $w$ and the two rotations $\theta_x = \partial w/\partial y$ and $\theta_y = -\partial w/\partial x$. The displacement function is given by the polynomial having 12 terms (Melosh 1963) as

$$
\begin{aligned}
w(\xi,\eta) &= a_1 + a_2\xi + a_3\eta + a_4\xi^2 + a_5\xi\eta + a_6\eta^2 + a_7\xi^3 + a_8\xi^2\eta + a_9\xi\eta^2 + a_{10}\eta^3 + a_{11}\xi^3\eta + a_{12}\xi\eta^3 \\
&= [\mathbf{g}]\{\mathbf{a}\}
\end{aligned}
\tag{11.118}
$$

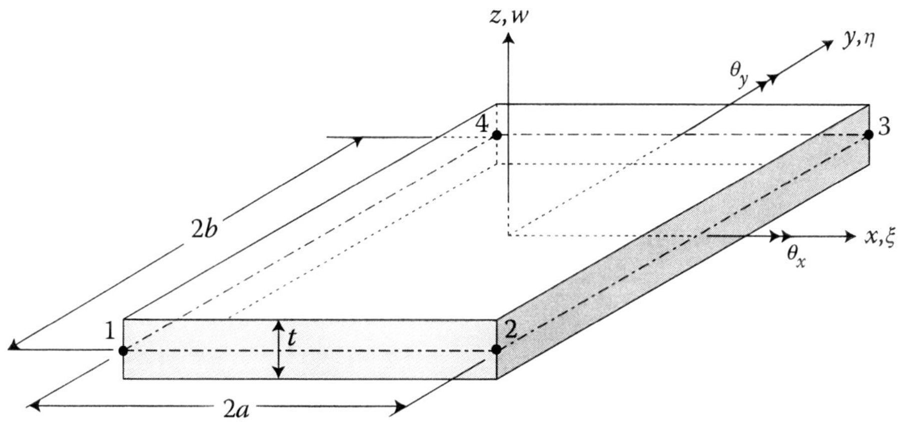


**FIGURE 11.18**
Rectangular plate element where $\xi = x/a$ and $\eta = y/b$.

where

$$[\mathbf{g}] = [1 \quad \xi \quad \eta \quad \xi^2 \quad \xi\eta \quad \eta^2 \quad \xi^3 \quad \xi^2\eta \quad \xi\eta^2 \quad \eta^3 \quad \xi^3\eta \quad \xi\eta^3] \tag{11.119}$$

$$\{\mathbf{a}\}^T = [a_1 \quad a_2 \quad a_3 \quad \ldots \quad a_{12}] \tag{11.120}$$

For the rotational terms to be in terms of $(\xi,\eta)$, we require that

$$\theta_x = \frac{1}{b}\frac{\partial w}{\partial \eta} \quad \text{and} \quad \theta_y = -\frac{1}{a}\frac{\partial w}{\partial \xi} \tag{11.121}$$

Using Equations 11.118 and 11.121, we can form the expression

$$\begin{Bmatrix} w \\ b\theta_x \\ -a\theta_y \end{Bmatrix} = \begin{bmatrix} 1 & \xi & \eta & \xi^2 & \xi\eta & \eta^2 & \xi^3 & \xi^2\eta & \xi\eta^2 & \eta^3 & \xi^3\eta & \xi\eta^3 \\ 0 & 0 & 1 & 0 & \xi & 2\eta & 0 & \xi^2 & 2\xi\eta & 3\eta^2 & \xi^3 & 3\xi\eta^2 \\ 0 & 1 & 0 & 2\xi & \eta & 0 & 3\xi^2 & 2\xi\eta & \eta^2 & 0 & 3\xi^2\eta & \eta^3 \end{bmatrix} \begin{Bmatrix} a_1 \\ a_2 \\ a_3 \\ \vdots \\ a_{12} \end{Bmatrix} \tag{11.122}$$

Evaluating Equation 11.122 at each node point $(\xi_i,\eta_i)$, where $i = 1, 2, 3, 4$, yields

$$\{\mathbf{d}\} = [\mathbf{G}]\{\mathbf{a}\} \tag{11.123}$$

where

$$\{\mathbf{d}\}^T = [w_1 \quad b\theta_{x1} \quad a\theta_{y1} \quad w_2 \quad b\theta_{x2} \quad a\theta_{y1} \quad \cdots \quad w_4 \quad b\theta_{x4} \quad a\theta_{y4}] \tag{11.124}$$

and

$$
[\mathbf{G}] = \begin{bmatrix}
1 & -1 & -1 & 1 & 1 & 1 & -1 & -1 & -1 & -1 & 1 & 1 \\
0 & 0 & 1 & 0 & -1 & -2 & 0 & 1 & 2 & 3 & -1 & -3 \\
0 & -1 & 0 & 2 & 1 & 0 & -3 & -2 & -1 & 0 & 3 & 1 \\
1 & 1 & -1 & 1 & -1 & 1 & 1 & -1 & 1 & -1 & -1 & -1 \\
0 & 0 & 1 & 0 & 1 & -2 & 0 & 1 & -2 & 3 & 1 & 3 \\
0 & -1 & 0 & -2 & 1 & 0 & -3 & 2 & -1 & 0 & 3 & 1 \\
1 & 1 & 1 & 1 & 1 & 1 & 1 & 1 & 1 & 1 & 1 & 1 \\
0 & 0 & 1 & 0 & 1 & 2 & 0 & 1 & 2 & 3 & 1 & 3 \\
0 & -1 & 0 & -2 & -1 & 0 & -3 & -2 & -1 & 0 & -3 & -1 \\
1 & -1 & 1 & 1 & -1 & 1 & -1 & 1 & -1 & 1 & -1 & -1 \\
0 & 0 & 1 & 0 & -1 & 2 & 0 & 1 & -2 & 3 & -1 & -3 \\
0 & -1 & 0 & 2 & -1 & 0 & -3 & 2 & -1 & 0 & -3 & -1
\end{bmatrix} \tag{11.125}
$$

Equation 11.123 can be solved for {**a**}, which yields

$$
\{\mathbf{a}\} = [\mathbf{G}]^{-1}\{\mathbf{d}\} \tag{11.126}
$$

where

$$
[\mathbf{G}]^{-1} = \frac{1}{8}\begin{bmatrix}
2 & 1 & -1 & 2 & 1 & 1 & 2 & -1 & 1 & 2 & -1 & -1 \\
-3 & -1 & 1 & 3 & 1 & 1 & 3 & -1 & 1 & -3 & 1 & 1 \\
-3 & -1 & 1 & -3 & -1 & -1 & 3 & -1 & 1 & 3 & -1 & -1 \\
0 & 0 & 1 & 0 & 0 & -1 & 0 & 0 & -1 & 0 & 0 & 1 \\
4 & 1 & -1 & -4 & -1 & -1 & 4 & -1 & 1 & -4 & 1 & 1 \\
0 & -1 & 0 & 0 & -1 & 0 & 0 & 1 & 0 & 0 & 1 & 0 \\
1 & 0 & -1 & -1 & 0 & -1 & -1 & 0 & -1 & 1 & 0 & -1 \\
0 & 0 & -1 & 0 & 0 & 1 & 0 & 0 & -1 & 0 & 0 & 1 \\
0 & 1 & 0 & 0 & -1 & 0 & 0 & 1 & 0 & 0 & -1 & 0 \\
1 & 1 & 0 & 1 & 1 & 0 & -1 & 1 & 0 & -1 & 1 & 0 \\
-1 & 0 & 1 & 1 & 0 & 1 & -1 & 0 & -1 & 1 & 0 & -1 \\
-1 & -1 & 0 & 1 & 1 & 0 & -1 & 1 & 0 & 1 & -1 & 0
\end{bmatrix} \tag{11.127}
$$

Substituting Equation 11.126 into 11.118 yields

$$
\begin{aligned}
w(\xi,\eta) &= [\mathbf{g}]\{\mathbf{a}\} \\
&= [\mathbf{g}][\mathbf{G}]^{-1}\{\mathbf{d}\} \\
&= [\mathbf{N}(\xi,\eta)]\{\mathbf{d}\}
\end{aligned} \tag{11.128}
$$

where

$$[\mathbf{N}(\xi,\eta)] = [[\mathbf{N}_1(\xi,\eta)] \quad [\mathbf{N}_2(\xi,\eta)] \quad [\mathbf{N}_3(\xi,\eta)] \quad [\mathbf{N}_4(\xi,\eta)]] \tag{11.129}$$

and

$$[\mathbf{N}_i(\xi,\eta)]^T = \begin{Bmatrix} (1/8)(1+\xi_0)(1+\eta_0)\left(2+\xi_0+\eta_0-\xi^2-\eta^2\right) \\ (b/8)(1+\xi_0)(\eta_i+\eta)(\eta^2-1) \\ -(a/8)(\xi_i+\xi)(\xi^2-1)(1+\eta_0) \end{Bmatrix} \tag{11.130}$$

where $(\xi_i,\eta_i)$ are the coordinates of node $i$ and $\xi_0 = \xi_i\xi, \eta_0 = \eta_i\eta$.

Note that in using the polynomial expression given in Equation 11.118, two neighboring elements will have the same values for the nodal coordinates, indicating that the continuity of the displacement $w$ will be assured along the interface of the elements. However, Equation 11.118 also shows that the derivative of $w$ normal to an interface, which varies as a cubic, only has two values of normal derivatives defined at the ends of an interelement boundary. Hence, the cubic is not uniquely defined and a discontinuity in the normal derivative can occur along the interface. Therefore, the polynomial expression given in Equation 11.118 is not a compatible one.

2. *Nonconforming Triangular Element.* We shall consider a three-node triangular element that has three degrees of freedom $w$, $\theta_x$, and $\theta_y$ at each node as shown in Figure 11.19.

   The nodal displacement vector for the nine degrees of freedom triangular element is given as

$$\{\mathbf{q}\} = \begin{Bmatrix} w_1 & \dfrac{\partial w_1}{\partial x} & \dfrac{\partial w_1}{\partial y} & w_2 & \dfrac{\partial w_2}{\partial x} & \dfrac{\partial w_2}{\partial y} & w_3 & \dfrac{\partial w_3}{\partial x} & \dfrac{\partial w_3}{\partial y} \end{Bmatrix}^T \tag{11.131}$$

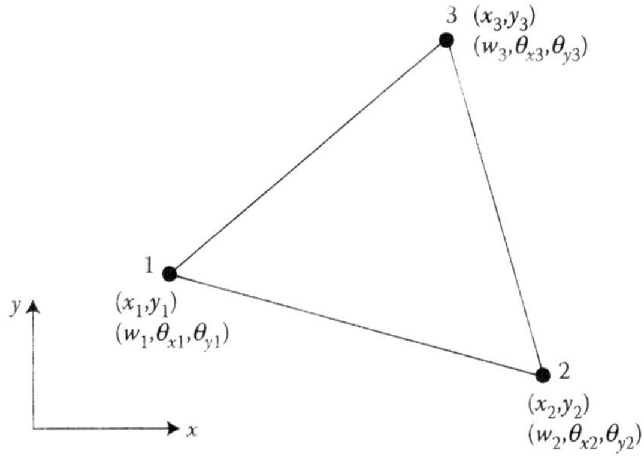


**FIGURE 11.19**
Three-node triangular element.

We shall assume that the interpolation function can be approximated by (Zienkiewicz et al. 2005)

$$w = \beta_1 L_1 + \beta_2 L_2 + \beta_3 L_3 + \beta_4 L_1^2 L_2 + \beta_5 L_1 L_2^2 + \beta_6 L_2^2 L_3 + \beta_7 L_2 L_3^2 + \beta_8 L_1 L_3^2 + \beta_9 L_1^2 L_3 + \beta_{10} L_1 L_2 L_3 \tag{11.132}$$

where $L_1$, $L_2$, and $L_3$ are the area coordinates that were used for the three-noded triangular element presented in Section 11.3.2.1 and $\beta_{10}$ corresponds to an internal degree of freedom and is approximated as

$$\beta_{10} = \frac{1}{2}(\beta_4 + \beta_5 + \beta_6 + \beta_7 + \beta_8 + \beta_9) \tag{11.133}$$

Substituting Equation 11.133 into 11.132, the interpolation function takes the form

$$w = \zeta_1 L_1 + \zeta_2 L_2 + \zeta_3 L_3 + \zeta_4\left(L_1^2 L_2 + \frac{1}{2}L_1 L_2 L_3\right) + \zeta_5\left(L_1 L_2^2 + \frac{1}{2}L_1 L_2 L_3\right) + \zeta_6\left(L_2^2 L_3 + \frac{1}{2}L_1 L_2 L_3\right) + \zeta_7\left(L_2 L_3^2 + \frac{1}{2}L_1 L_2 L_3\right) + \zeta_8\left(L_1 L_3^2 + \frac{1}{2}L_1 L_2 L_3\right) + \zeta_9\left(L_1^2 L_3 + \frac{1}{2}L_1 L_2 L_3\right) \tag{11.134}$$

Note that

$$\frac{\partial L_i}{\partial x} = \beta_i \quad \frac{\partial L_i}{\partial y} = \gamma_i \tag{11.135}$$

where

$$\left.\begin{aligned} \alpha_1 &= x_2 y_3 - x_3 y_2 \\ \beta_1 &= y_2 - y_3 \\ \gamma_1 &= x_3 - x_2 \end{aligned}\right\} \begin{array}{c} \text{other values obtained} \\ \text{by cyclic permutation} \\ \text{of suffixes 1,2,3} \end{array}$$
$$2A = (x_2 y_3 + x_3 y_1 + x_1 y_2 - x_2 y_1 - x_3 y_2 - x_1 y_3)$$
$$L_i = \frac{1}{2A}(\alpha_i + \beta_i x + \gamma_i y) \quad i = 1,2,3 \tag{11.136}$$

Substituting the nodal coordinates $(x_i, y_i)$, $i = 1,2,3$ of the three nodes of

$$w_i, \theta_{xi} = -\left(\frac{\partial w}{\partial y}\right)_i, \quad \text{and} \quad \theta_{yi} = \left(\frac{\partial w}{\partial x}\right)_i$$

in Equation 11.134, the constants and shape functions which satisfy

$$\{\mathbf{N}_1\}^T = \begin{Bmatrix} L_1 + L_1L_2(L_1 - L_2) + L_1L_3(L_1 - L_3) \\ b_3\left(L_1^2L_2 + \dfrac{1}{2}L_1L_2L_3\right) - b_2\left(L_1^2L_3 + \dfrac{1}{2}L_1L_2L_3\right) \\ b_3\left(L_1^2L_2 + \dfrac{1}{2}L_1L_2L_3\right) - b_2\left(L_1^2L_3 + \dfrac{1}{2}L_1L_2L_3\right) \end{Bmatrix} \tag{11.137}$$

The other two shape functions can be obtained by a cyclic permutation of 1, 2, and 3. We can write Equation 11.132 as

$$w = [\mathbf{N}]\{\mathbf{d}\} \tag{11.138}$$

where

$$[\mathbf{N}] = \begin{bmatrix} \{\mathbf{N}_1\} & 0 & 0 \\ 0 & \{\mathbf{N}_2\} & 0 \\ 0 & 0 & \{\mathbf{N}_3\} \end{bmatrix}$$
$$\{\mathbf{d}\} = \{w_1 \quad \theta_{1x} \quad \theta_{1y} \quad w_2 \quad \theta_{2x} \quad \theta_{2y} \quad w_3 \quad \theta_{3x} \quad \theta_{3y}\}^T \tag{11.139}$$

This element is nonconforming since it cannot develop continuous displacement $w$ and normal rotation $\partial w/\partial n$.

3. *Conforming Rectangular Element*. The difficulty encountered with the rectangular plate element presented above (*nonconforming*) can be overcome by adding additional degrees of freedom at each node. It was noted that if the twist ($\partial^2 w/\partial x\partial y$) is added, in addition to the three degrees of freedom $w$, $\partial w/\partial x$, and $\partial w/\partial y$, as a fourth degree of freedom, compatibility can be established. This implies uniqueness of the normal slope along the interelement boundaries. This implies that we have a rectangular plate element (Figure 11.20) with 16 degrees of freedom, which was proposed by Bogner et al. (1966).

   The displacement vector is written as

$$\{\mathbf{d}\}^T = \{w_1 \quad \theta_{x1} \quad \theta_{y1} \quad w_{xy1} \quad w_2 \quad \theta_{x2} \quad \theta_{y2} \quad w_{xy2} \quad \cdots \quad w_4 \quad \theta_{x4} \quad \theta_{y4} \quad w_{xy4}\}^T \tag{11.140}$$

and the interpolation function as

$$\begin{aligned} w(x,y) = {} & a_1 + a_2x + a_3y + a_4x^2 + a_5xy + a_6y^2 + a_7x^3 + a_8x^2y + a_9xy^2 + a_{10}y^3 \\ & + a_{11}x^3y + a_{12}x^2y^2 + a_{13}xy^3 + a_{14}x^3y^2 + a_{15}x^2y^3 + a_{16}x^3y^3 \end{aligned} \tag{11.141}$$

One could proceed as was done for the 12 degrees of freedom plate and solve for the coefficients and proceed to determine the shape functions. Bogner et al. determined that the shape functions can be obtained by using simple products of one-dimensional Hermite polynomials. The deflection field given by Equation

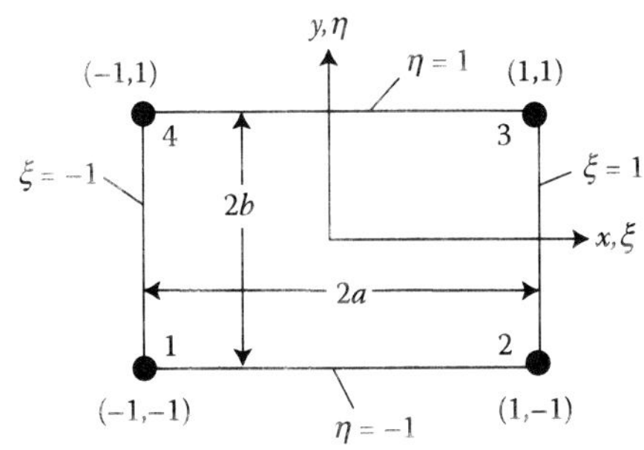


**FIGURE 11.20**
Rectangular plate element.

11.141 may be expressed in the natural coordinate system in terms of the natural coordinates $\xi$ and $\eta$ in the form

$$w(\xi,\eta) = [\mathbf{N}(\xi,\eta)]\{\tilde{\mathbf{d}}\} \tag{11.142}$$

where

$$[\mathbf{N}(\xi,\eta)] = [[\mathbf{N}_1(\xi,\eta)] \quad [\mathbf{N}_2(\xi,\eta)] \quad [\mathbf{N}_3(\xi,\eta)] \quad [\mathbf{N}_4(\xi,\eta)]] \tag{11.143}$$

and

$$\{\tilde{\mathbf{d}}\}^T = [\{\mathbf{w}\} \quad \{\theta_x\} \quad \{\theta_y\} \quad \{\mathbf{w}_{xy}\}] \tag{11.144}$$

The shape functions are given as

$$[\mathbf{N}_i(\xi,\eta)]^T = \begin{Bmatrix} g_i(\xi)h_i(\eta) \\ bg_i(\xi)h_i(\eta) \\ -ag_i(\xi)h_i(\eta) \\ abg_i(\xi)g_i(\eta) \end{Bmatrix} \tag{11.145}$$

where the Hermite functions are obtained from Equation 11.49 and given as

$$\begin{aligned} g_i(\xi) &= \frac{1}{4}\left(2+3\xi_i\xi-\xi_i\xi^3\right) \\ h_i(\xi) &= \frac{1}{4}\left(-\xi_i-\xi+\xi_i\xi^2+\xi^3\right) \end{aligned} \tag{11.146}$$

The functions of $\eta$ are obtained from Equation 11.146 by replacing $\xi$, $\xi_i$ by $\eta,\eta_i$, and $(\xi_i,\eta_i)$ are the coordinates of node $i$.

Higher-order elements can be obtained such that the generation of slope and deflection compatible elements presents less difficulty (Zienkiewicz and Taylor 2000).

4. *Conforming Triangular Element.* Compatible higher-order triangular elements have been developed which enforce continuity between finite elements. Early developments include triangular plate elements which allow only a linear variation of slope normal to an edge (Bazeley et al. 1966; Clough and Tocher 1966). These developments gave rather poor results. However, one triangular element which has 21 degrees of freedom (Argyris et al. 1968) includes all second derivatives with complete fifth-order polynomial interpolation functions along with another element (Bell 1969) which has 18 degrees of freedom but neglects the midside normal derivative nodes and gives very good accuracy.

A higher-order 3-node triangular plate element was revisited (Dasgupta and Sengupta 1990). The degrees of freedom vector at each node $i$ is

$$\{d_i\} = \left\{ w_i \quad \left(\frac{\partial w}{\partial x}\right)_i \quad \left(\frac{\partial w}{\partial y}\right)_i \quad \left(\frac{\partial^2 w}{\partial x^2}\right)_i \quad \left(\frac{\partial^2 w}{\partial x \partial y}\right)_i \quad \left(\frac{\partial^2 w}{\partial y^2}\right)_i \right\}^T \tag{11.147}$$

and the element degrees of freedom vector with nodes 1, 2, and 3 is

$$\{\mathbf{d}\} = \begin{Bmatrix} d_1 \\ d_2 \\ d_3 \end{Bmatrix} \tag{11.148}$$

Using area coordinates (see Equation 11.136), the plate deflection at any point within the element given by the polynomial is

$$\begin{aligned} w = {} & \zeta_1 L_1^5 + \zeta_2 L_2^5 + \zeta_3 L_3^5 + \zeta_4 L_1^4 L_2 + \zeta_5 L_1^4 L_3 + \zeta_6 L_2^4 L_1 + \zeta_7 L_2^4 L_3 + \zeta_8 L_3^4 L_1 + \zeta_9 L_3^4 L_2 \\ & + \zeta_{10} L_1^3 L_2^2 + \zeta_{11} L_1^3 L_3^2 + \zeta_{12} L_2^3 L_1^2 + \zeta_{13} L_2^3 L_3^2 + \zeta_{14} L_3^3 L_1^2 + \zeta_{15} L_3^3 L_2^2 + \zeta_{16} L_1^3 L_2 L_3 + \zeta_{17} L_2^3 L_1 L_3 \\ & + \zeta_{18} L_3^3 L_1 L_2 + \zeta_{19} L_1 L_2^2 L_3^2 + \zeta_{20} L_2 L_1^2 L_3^2 + \zeta_{21} L_3 L_1^2 L_2^2 \end{aligned} \tag{11.149}$$

Equation 11.149 can be solved to find the coefficients $\alpha_1$ to $\alpha_{18}$ in terms of the nodal degrees of freedom. The remaining coefficients $\zeta_{19}$, $\zeta_{20}$, and $\zeta_{21}$ are determined from the conditions imposed on the polynomial such that normal slope is constrained to have a cubic variation along the element edges. The shape functions were given as

$$w = [\mathbf{N}]\{\mathbf{d}\} \tag{11.150}$$

where

$$\begin{aligned} N_1 = {} & L_1^5 + 5L_1^4 L_2 + 5L_1^4 L_3 + 10L_1^3 L_2^2 + 10L_1^3 L_3^2 + 20L_1^3 L_3 + 30r_{21} L_1^2 L_2 L_3^2 + 30r_{31} L_1^2 L_3 L_2^2 \\ N_2 = {} & \gamma_3 L_1^4 L_2 - \gamma_2 L_1^4 L_3 + 4\gamma_3 L_1^3 L_2^2 - 4\gamma_2 L_1^3 L_3^2 + 4(\gamma_3 - \gamma_2) L_1^3 L_2 L_3 - (3\gamma_1 + 15r_{21}\gamma_2) L_1^2 L_2 L_3^2 \\ & + (3\gamma_1 + 15r_{31}\gamma_3) L_1^2 L_3 L_2^2 \\ N_3 = {} & -b_3 L_1^4 L_2 + b_2 L_1^4 L_3 - 4b_3 L_1^3 L_2^2 + 4b_2 L_1^3 L_3^2 + 4(b_2 - b_3) L_1^3 L_2 L_3 + (3b_1 + 15r_{21} b_2) L_1^2 L_2 L_3^2 \\ & - (3b_1 + 15r_{31} b_3) L_1^2 L_3 L_2^2 \end{aligned} \tag{11.151}$$

$$N_4 = \frac{\gamma_3^2}{2} L_1^3 L_2^2 + \frac{\gamma_2^2}{2} L_1^3 L_3^2 - \gamma_2\gamma_3 L_1^3 L_2 L_3 + \left(\gamma_1\gamma_2 + \frac{5}{2} r_{21}\gamma_2^2\right) L_2 L_3^2 L_1^2 + \left(\gamma_1\gamma_3 + \frac{5}{2} r_{31}\gamma_3^2\right) L_3 L_2^2 L_1^2$$

$$N_5 = -\beta_3\gamma_3 L_1^3 L_2^2 - \beta_2\gamma_2 L_1^3 L_3^2 + (\beta_2\gamma_3 + \beta_3\gamma_2) L_1^3 L_2 L_3 - (\beta_1\gamma_2 + \beta_2\gamma_1 + 5r_{21}\beta_2\gamma_2) L_2 L_3^2 L_1^2$$
$$- (\beta_1\gamma_3 + \beta_3\gamma_1 + 5r_{31}\beta_3\gamma_2) L_3 L_2^2 L_1^2$$

$$N_6 = \frac{\beta_3^2}{2} L_1^3 L_2^2 + \frac{\beta_2^2}{2} L_1^3 L_3^2 - \beta_2\beta_3 L_1^3 L_2 L_3 + \left(\beta_1\beta_2 + \frac{5}{2} r_{21}\beta_2^2\right) L_2 L_3^2 L_1^2 + \left(\beta_1\beta_3 + \frac{5}{2} r_{31}\beta_3^2\right) L_3 L_2^2 L_1^2$$

and

$$r_{ij} = -\frac{b_i b_j + c_i c_j}{b_i^2 + c_i^2}$$

The remaining 12 shape function components, $N_7$ to $N_{18}$, which correspond to the degrees of freedom at nodes 2 and 3 are obtained from Equation 11.151 by cyclically permutating the three node numbers.

### 11.3.3 Element Properties

Here we will outline the formulation of selected element stiffness and mass matrices, based on the shape functions developed above. In a static problem, inertia and damping forces are nonexistent and the displacement and force vector are independent of time. Thus, the system of equations, Equation 11.9, reduces to

$$[\mathbf{K}]\{\mathbf{u}\} = \{\mathbf{F}\} \tag{11.152}$$

1. *One-Dimensional Axial Element.* Figure 11.3 shows a one-dimensional line element, where the shape functions were given in Equation 11.34. Hence, we recall that

$$\tilde{u} = [\mathbf{N}]\begin{Bmatrix} u_1 \\ u_2 \end{Bmatrix}, \quad \text{where} \quad [\mathbf{N}] = \begin{bmatrix} 1-\dfrac{x}{l} & \dfrac{x}{l} \end{bmatrix} \tag{11.153}$$

Using Equation 11.8 with

$$\iiint_V dV = A\int_0^l dx$$

we obtain

$$[\mathbf{m}] = \rho A \int_0^l [\mathbf{N}]^T [\mathbf{N}] dx = \rho A \int_0^l \begin{bmatrix} 1-\dfrac{x}{l} \\ \dfrac{x}{l} \end{bmatrix}^T \begin{bmatrix} 1-\dfrac{x}{l} & \dfrac{x}{l} \end{bmatrix} dx$$
$$= \frac{\rho A l}{6}\begin{bmatrix} 2 & 1 \\ 1 & 2 \end{bmatrix} \tag{11.154}$$

which is called the consistent mass matrix since the shape functions used for the stiffness matrix are used to construct the mass matrix. On the other hand, one can calculate the total mass of the element as $m = \rho Al$, this is then divided into two equal parts and assigned to each node creating the so-called lumped mass matrix as

$$[\mathbf{m}] = \frac{1}{2}\rho Al \begin{bmatrix} 1 & 0 \\ 0 & 1 \end{bmatrix} \tag{11.155}$$

For the stiffness matrix, we use Equation 11.11 with $[\mathbf{B}] = d[\mathbf{N}]/dx = [-1 \;\; 1]/l$, which gives

$$[\mathbf{k}] = \int_0^l [\mathbf{B}]^T AE[\mathbf{B}]dx = \frac{AE}{l}\begin{bmatrix} 1 & -1 \\ -1 & 1 \end{bmatrix} \tag{11.156}$$

The element structural matrices $[\mathbf{k}]$ and $[\mathbf{m}]$ are evaluated in a local coordinate system $xyz$ aligned with a global coordinate system. The coordinate transformation matrices, from global to local axes, of a two-node element can be determined using Figure 11.21.

From Figure 11.21, we find

$$\begin{Bmatrix} \tilde{u}_1 \\ \tilde{u}_2 \end{Bmatrix} = \begin{bmatrix} \cos\theta & \sin\theta & 0 & 0 \\ 0 & 0 & \cos\theta & \sin\theta \end{bmatrix} \begin{Bmatrix} u_1 \\ u_2 \\ u_3 \\ u_4 \end{Bmatrix} = [\boldsymbol{\Phi}] \begin{Bmatrix} u_1 \\ u_2 \\ u_3 \\ u_4 \end{Bmatrix} \tag{11.157}$$

where $[\boldsymbol{\Phi}]$ is the transformation or rotation matrix and

$$\begin{Bmatrix} f_1 \\ f_2 \\ f_3 \\ f_4 \end{Bmatrix} = \begin{bmatrix} \cos\theta & 0 \\ \sin\theta & 0 \\ 0 & \cos\theta \\ 0 & \sin\theta \end{bmatrix} \begin{Bmatrix} \tilde{f}_1 \\ \tilde{f}_2 \end{Bmatrix} = [\boldsymbol{\Phi}]^{-1} \begin{Bmatrix} \tilde{f}_1 \\ \tilde{f}_2 \end{Bmatrix} \tag{11.158}$$

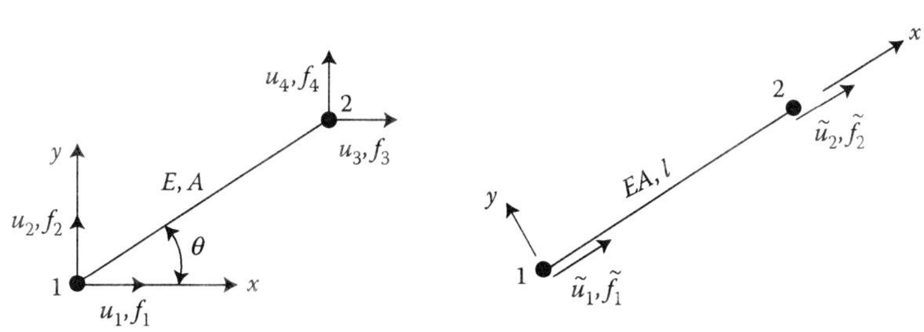


**FIGURE 11.21**
Degrees of freedom of truss element in global and local axes.

Note that $[\mathbf{\Phi}]^{-1} = [\mathbf{\Phi}]^T$. Thus, we have from Equation 11.152 for a truss element

$$[\tilde{\mathbf{k}}]\{\tilde{\mathbf{u}}\} = \{\tilde{\mathbf{f}}\}$$
$$[\tilde{\mathbf{k}}][\mathbf{\Phi}]\{\mathbf{u}\} = [\mathbf{\Phi}]\{\mathbf{f}\}$$

Premultiplying the above equation by $[\mathbf{\Phi}]^{-1}$ yields

$$\begin{aligned}[\mathbf{\Phi}]^{-1}[\tilde{\mathbf{k}}][\mathbf{\Phi}]\{\mathbf{u}\} &= [\mathbf{\Phi}]^{-1}[\mathbf{\Phi}]\{\mathbf{f}\}\\ [\mathbf{k}]\{\mathbf{u}\} &= \{\mathbf{f}\}\end{aligned} \tag{11.159}$$

where in global coordinates

$$[\mathbf{k}] = [\mathbf{\Phi}]^T[\tilde{\mathbf{k}}][\mathbf{\Phi}] = \frac{EA}{l}\begin{bmatrix} \cos^2\theta & \sin\theta\cos\theta & -\cos^2\theta & -\sin\theta\cos\theta \\ \sin\theta\cos\theta & \sin^2\theta & -\sin\theta\cos\theta & -\sin^2\theta \\ -\cos^2\theta & -\sin\theta\cos\theta & \cos^2\theta & \sin\theta\cos\theta \\ -\sin\theta\cos\theta & -\sin^2\theta & \sin\theta\cos\theta & \sin^2\theta \end{bmatrix} \tag{11.160}$$

The mass matrix in global coordinates can be found in the same manner as

$$[\mathbf{m}] = [\mathbf{\Phi}]^T[\tilde{\mathbf{m}}][\mathbf{\Phi}] = \frac{\rho A l}{6}\begin{bmatrix} 2\cos^2\theta & 2\cos\theta\sin\theta & \cos^2\theta & \cos\theta\sin\theta \\ 2\cos\theta\sin\theta & 2\sin^2\theta & \cos\theta\sin\theta & \sin^2\theta \\ \cos^2\theta & \cos\theta\sin\theta & 2\cos^2\theta & 2\cos\theta\sin\theta \\ \cos\theta\sin\theta & \sin^2\theta & 2\cos\theta\sin\theta & 2\sin^2\theta \end{bmatrix} \tag{11.161}$$

2. *Beam Finite Element*. For the beam element (see Figure 11.22), we recall that

$$\tilde{u} = N_1u_1 + aN_2u_1' + N_3u_2 + aN_4u_2' \tag{11.162}$$

where

$$\begin{aligned} N_1 &= \frac{1}{4}(2 - 3\xi + \xi^3)\\ N_2 &= \frac{1}{4}(1 - \xi - \xi^2 + \xi^3)\\ N_3 &= \frac{1}{4}(2 + 3\xi - \xi^3)\\ N_4 &= \frac{1}{4}(-1 - \xi + \xi^2 + \xi^3)\end{aligned} \tag{11.163}$$

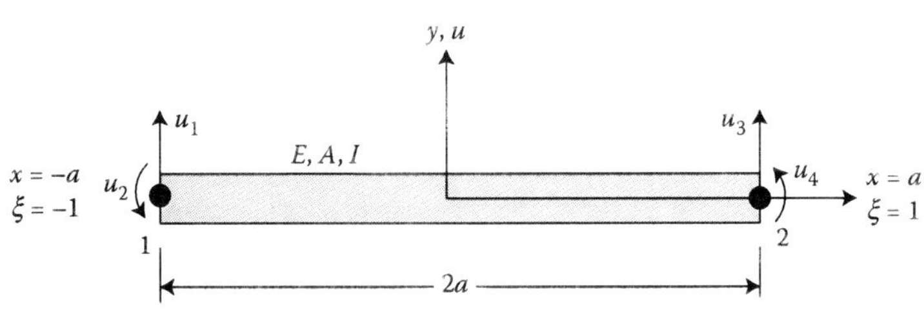


**FIGURE 11.22**
Two-node beam element with two DOF per node.

and $\zeta = x/l$. We have that $\varepsilon_x = -y(\partial^2 u/\partial x^2) = -(y/a^2)(\partial^2 u/\partial \zeta^2)$, $\sigma_x = E\varepsilon_x$, therefore $[\mathbf{B}] = -\dfrac{y}{a^2}\dfrac{\partial^2}{\partial \zeta^2}[\mathbf{N}]$ and $[D] = E$, hence

$$[\mathbf{B}] = -\frac{y}{a^2}[\mathbf{N}''] = -\frac{y}{a^2}\begin{bmatrix} N_1'' & N_2'' & N_3'' & N_4'' \end{bmatrix} \tag{11.164}$$

where

$$\begin{aligned} N_1'' &= \frac{3\zeta}{2}, \qquad N_2'' = \frac{a(-1+3\zeta)}{2} \\ N_3'' &= -\frac{3\zeta}{2}, \qquad N_4'' = \frac{a(1+3\zeta)}{2} \end{aligned} \tag{11.165}$$

The element inertia matrix is given by

$$[\mathbf{m}] = \iiint\limits_V \rho[\mathbf{N}]^T[\mathbf{N}]dV = \rho \iint\limits_A dA \int_{-a}^{a} [\mathbf{N}]^T[\mathbf{N}]dx = \rho A \int_{-1}^{1} [\mathbf{N}]^T[\mathbf{N}]a\, d\zeta$$

or

$$[\mathbf{m}] = \rho a A \int_{-1}^{1} \begin{bmatrix} N_1N_1 & N_1N_2 & N_1N_3 & N_1N_4 \\ N_2N_1 & N_2N_2 & N_2N_3 & N_2N_4 \\ N_3N_1 & N_3N_2 & N_3N_3 & N_3N_4 \\ N_4N_1 & N_4N_2 & N_4N_3 & N_4N_4 \end{bmatrix} d\zeta$$

where $A$ is the cross-sectional area of the beam. Evaluating the integrals in the above equation yields

$$[\mathbf{m}] = \frac{\rho a A}{105}\begin{bmatrix} 78 & 22a & 27 & -13a \\ 22a & 8a^2 & 13a & -6a^2 \\ 27 & 13a & 78 & -22a \\ -13a & -6a^2 & -22a & 8a^2 \end{bmatrix} \tag{11.166}$$

or using $a = l/2$ as

$$[\mathbf{m}] = \frac{\rho l A}{420}\begin{bmatrix} 156 & 22l & 54 & -13l \\ 22l & 4l^2 & 13l & -3l^2 \\ 54 & 13l & 156 & -22l \\ -13l & -3l^2 & -22l & 4l^2 \end{bmatrix} \tag{11.167}$$

The beam stiffness matrix is expressed as

$$[\mathbf{k}] = \iiint_V \int_0^l [\mathbf{B}]^T[\mathbf{D}][\mathbf{B}]dV = E\iint_A y^2 dA \int_{-a}^{a} \left[\frac{\partial^2}{\partial x^2}\mathbf{N}\right]^T \left[\frac{\partial^2}{\partial x^2}\mathbf{N}\right] dx$$

$$= EI\int_{-1}^{1} \frac{1}{a^4}\left[\frac{\partial^2}{\partial \zeta^2}(\mathbf{N})\right]^T \left[\frac{\partial^2}{\partial \zeta^2}(\mathbf{N})\right] a\,d\zeta = \frac{EI}{a^3}\int_{-1}^{1} [\mathbf{N}'']^T[\mathbf{N}'']d\zeta$$

or

$$[\mathbf{k}] = \frac{EI}{a^3}\int_{-1}^{1}\begin{bmatrix} N_1''N_1'' & N_1''N_2'' & N_1''N_3'' & N_1''N_4'' \\ N_2''N_1'' & N_2''N_2'' & N_2''N_3'' & N_2''N_4'' \\ N_3''N_1'' & N_3''N_2'' & N_3''N_3'' & N_3''N_4'' \\ N_4''N_1'' & N_4''N_2'' & N_4''N_3'' & N_4''N_4'' \end{bmatrix} d\zeta \tag{11.168}$$

where $I = \int_A y^2 dA$ is the moment of inertia of the beam cross section. Evaluating the integral values in Equation 11.168 yields

$$[\mathbf{k}] = \frac{EI}{2a^3}\begin{bmatrix} 3 & 3a & -3 & 3a \\ 3a & 4a^2 & -3a & 2a^2 \\ -3 & -3a & 3 & -3a \\ 3a & 2a^2 & -3a & 4a^2 \end{bmatrix} \tag{11.169}$$

or using $a = l/2$ as

$$[\mathbf{k}] = \frac{EI}{l^3}\begin{bmatrix} 12 & 6l & -12 & 6l \\ 6l & 4l^2 & -6l & 2l^2 \\ -12 & -6l & 12 & -6l \\ 6l & 2l^2 & -6l & 4l^2 \end{bmatrix} \tag{11.170}$$

The transformation for the beam element is similar to that obtained for the truss element. Consider the beam element in the global and local axis as shown in Figure 11.23.

The mass and stiffness matrices for the beam element are $4 \times 4$, where the transformation or rotation matrix is $6 \times 6$ and given as

$$[\mathbf{\Phi}] = \begin{bmatrix} \cos\theta & \sin\theta & 0 & 0 & 0 & 0 \\ -\sin\theta & \cos\theta & 0 & 0 & 0 & 0 \\ 0 & 0 & 1 & 0 & 0 & 0 \\ 0 & 0 & 0 & \cos\theta & \sin\theta & 0 \\ 0 & 0 & 0 & -\sin\theta & \cos\theta & 0 \\ 0 & 0 & 0 & 0 & 0 & 1 \end{bmatrix}$$

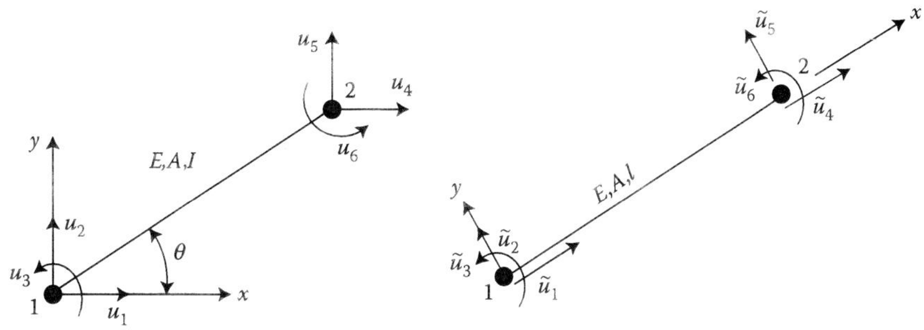


**FIGURE 11.23**
Degrees of freedom of beam element in global and local axes.

Therefore, to transform the mass and stiffness matrices, they need to be modified by adding the axial component as follows:

$$[\mathbf{k}] = \frac{EA}{l}\begin{bmatrix} 1 & 0 & 0 & -1 & 0 & 0 \\ 0 & 0 & 0 & 0 & 0 & 0 \\ 0 & 0 & 0 & 0 & 0 & 0 \\ -1 & 0 & 0 & 1 & 0 & 0 \\ 0 & 0 & 0 & 0 & 0 & 0 \\ 0 & 0 & 0 & 0 & 0 & 0 \end{bmatrix} \quad [\mathbf{m}] = \frac{\rho A l}{6}\begin{bmatrix} 2 & 0 & 0 & 1 & 0 & 0 \\ 0 & 0 & 0 & 0 & 0 & 0 \\ 0 & 0 & 0 & 0 & 0 & 0 \\ 1 & 0 & 0 & 2 & 0 & 0 \\ 0 & 0 & 0 & 0 & 0 & 0 \\ 0 & 0 & 0 & 0 & 0 & 0 \end{bmatrix}$$

Thus, the element matrices that need to be used for the transformation are

$$[\tilde{\mathbf{k}}] = \frac{EI}{l^3}\begin{bmatrix} \frac{Al^2}{I} & 0 & 0 & -\frac{Al^2}{I} & 0 & 0 \\ 0 & 12 & 6l & 0 & -12 & 6l \\ 0 & 6l & 4l^2 & 0 & -6l & 2l^2 \\ -\frac{Al^2}{I} & 0 & 0 & \frac{Al^2}{I} & 0 & 0 \\ 0 & -12 & -6l & 0 & 12 & -6l \\ 0 & 6l & 2l^2 & 0 & -6l & 4l^2 \end{bmatrix}$$

$$[\tilde{\mathbf{m}}] = \frac{\rho A l}{420}\begin{bmatrix} 140 & 0 & 0 & 70 & 0 & 0 \\ 0 & 156 & 22l & 0 & 54 & -13l \\ 0 & 22l & 4l^2 & 0 & 13l & -3l^2 \\ 70 & 0 & 0 & 140 & 0 & 0 \\ 0 & 54 & 13l & 0 & 156 & -22l \\ 0 & -13l & -3l^2 & 0 & -22l & 4l^2 \end{bmatrix}$$

The transformation or rotation matrix for the two-node beam mass and stiffness element can be obtained as

$$[\mathbf{m}] = \frac{\rho A l}{420}\begin{bmatrix} 140c^2+156s^2 & -16cs & -22ls & 70c^2+54s^2 & 16cs & 13ls \\ -16cs & 156c^2+140s^2 & 22lc & 16cs & 54c^2+70s^2 & -13lc \\ -22ls & 22lc & 4l^2 & -13ls & 13lc & -3l^2 \\ 70c^2+54s^2 & 16cs & -13ls & 140c^2+156s^2 & -16cs & 22ls \\ 16cs & 54c^2+70s^2 & 13lc & -16cs & 156c^2+140s^2 & -22lc \\ 13ls & -13lc & -3l^2 & 22ls & -22lc & 4l^2 \end{bmatrix} \tag{11.171}$$

and

$$[\mathbf{k}] = \frac{EI}{l^3}\begin{bmatrix} (Rc^2+12s^2) & (R-12)cs & -6ls & -(12s^2+Rc^2) & (12-R)cs & -6ls \\ (R-12)cs & (Rs^2+12c^2) & 6lc & (12-R)cs & -(12c^2+Rs^2) & 6lc \\ -6ls & 6lc & 4l^2 & 6ls & -6lc & 2l^2 \\ -(12s^2+Rc^2) & (12-R)cs & 6ls & (Rc^2+12s^2) & (R-12)cs & 6ls \\ (12-R)cs & -(12c^2+Rs^2) & -6lc & (R-12)cs & (Rs^2+12c^2) & -6lc \\ -6ls & 6lc & 2l^2 & 6ls & -6lc & 4l^2 \end{bmatrix} \tag{11.172}$$

where $R = Al^2/I$, $c = \cos\theta$, and $s = \sin\theta$.

3. *Constant Strain Triangle Element*. The displacements for the triangular element were given in Equation 11.67 as

$$\{\mathbf{u}\} = \begin{Bmatrix} u_x \\ u_y \end{Bmatrix} = [\mathbf{N}]\{\mathbf{d}\}$$

where [N] are the shape functions

$$[\mathbf{N}] = \begin{bmatrix} L_1 & 0 & L_2 & 0 & L_3 & 0 \\ 0 & L_1 & 0 & L_2 & 0 & L_3 \end{bmatrix}$$

and {d}, the element displacements

$$\{\mathbf{d}\}^T = \{u_{x1} \quad u_{y1} \quad u_{x2} \quad u_{y2} \quad u_{x3} \quad u_{y3}\}$$

The strain components for a two-dimensional elasticity problem having direct strains $\varepsilon_{xx}$, $\varepsilon_{yy}$ and shear strain $\gamma_{xy}$ can be found in Chapter 5 and can be given in the matrix form as

$$\{\varepsilon\} = \begin{Bmatrix} \varepsilon_{xx} \\ \varepsilon_{yy} \\ \gamma_{xy} \end{Bmatrix} = \begin{bmatrix} \dfrac{\partial}{\partial x} & 0 \\ 0 & \dfrac{\partial}{\partial y} \\ \dfrac{\partial}{\partial y} & \dfrac{\partial}{\partial x} \end{bmatrix} \begin{Bmatrix} u_x \\ u_y \end{Bmatrix} = \begin{bmatrix} \dfrac{\partial}{\partial x} & 0 \\ 0 & \dfrac{\partial}{\partial y} \\ \dfrac{\partial}{\partial y} & \dfrac{\partial}{\partial x} \end{bmatrix} [\mathbf{N}]\{\mathbf{d}\} = [\mathbf{B}]\{\mathbf{d}\} \tag{11.173}$$

Using the expression for the shape function matrix [**N**], given above, along with Equations 11.55 and 11.56 and noting that $L_i = N_i$, $i = 1,2,3$, the strain matrix [**B**] is determined as

$$[\mathbf{B}] = \frac{1}{2A} \begin{bmatrix} \beta_1 & 0 & \beta_2 & 0 & \beta_3 & 0 \\ 0 & \gamma_1 & 0 & \gamma_2 & 0 & \gamma_3 \\ \gamma_1 & \beta_1 & \gamma_2 & \beta_2 & \gamma_3 & \beta_3 \end{bmatrix} \tag{11.174}$$

Substituting Equation 11.174 into Equation 11.11 yields the stiffness matrix for the constant triangular element as

$$[\mathbf{k}] = \iint_A [\mathbf{B}]^T [\mathbf{D}][\mathbf{B}] t\, dxdy \tag{11.175}$$

where $A$ is the area of the triangular element and $t$ the thickness. For a homogeneous material, the strain matrix [**B**] is constant within the element and Equation 11.175 becomes

$$[\mathbf{k}] = tA[\mathbf{B}]^T [\mathbf{D}][\mathbf{B}] \tag{11.176}$$

Considering isotropic conditions we can consider two cases of stress: plane stress and plane strain. Each case will have a different material matrix [**D**] since the stress–strain equations are different. The plane stress equations are given in Equation 5.58 and in the matrix form the material matrix is written as

$$[\mathbf{D}] = \frac{E}{1-\nu^2} \begin{bmatrix} 1 & \nu & 0 \\ \nu & 1 & 0 \\ 0 & 0 & \dfrac{(1-\nu)}{2} \end{bmatrix} \tag{11.177}$$

For plane strain, we have from Equation 5.60

$$[\mathbf{D}] = \frac{E}{(1+\nu)(1-2\nu)} \begin{bmatrix} (1-\nu) & \nu & 0 \\ \nu & (1-\nu) & 0 \\ 0 & 0 & \left(\dfrac{1}{2}-\nu\right) \end{bmatrix} \tag{11.178}$$

Equations 11.177 and 11.178 can be put into the common matrix form so that both plane stress and strain can be analyzed. Therefore, rewrite the material matrix in the form

$$[\mathbf{D}] = \begin{bmatrix} D_1 & D_2 & 0 \\ D_2 & D_1 & 0 \\ 0 & 0 & G \end{bmatrix} \tag{11.179}$$

where

$$D_1 = \frac{E}{(1-\nu^2)} \text{ and } D_2 = \nu D_1 \text{ for the case of plane stress}$$

$$D_1 = \frac{E(1-\nu)}{(1+\nu)(1-2\nu)} \text{ and } D_2 = \frac{\nu D_1}{(1-\nu)} \quad \text{for the case of plane strain}$$

and

$$G = \frac{E}{2(1+\nu)}$$

Substituting Equation 11.178 into 11.176 yields

$$[\mathbf{k}] = \frac{t}{4A}\begin{bmatrix} \begin{pmatrix} D_1\beta_1^2 \\ +G\gamma_1^2 \end{pmatrix} & \begin{pmatrix} G\beta_1\gamma_1 \\ +D_2\beta_1\gamma_1 \end{pmatrix} & \begin{pmatrix} D_1\beta_2\beta_2 \\ +G\gamma_1\gamma_2 \end{pmatrix} & \begin{pmatrix} G\beta_2\gamma_1 \\ +D_2\gamma_1\gamma_2 \end{pmatrix} & \begin{pmatrix} D_1\beta_1\beta_3 \\ +G\gamma_1\gamma_3 \end{pmatrix} & \begin{pmatrix} G\beta_3\gamma_1 \\ +D_2\beta_1\gamma_3 \end{pmatrix} \\ & \begin{pmatrix} G\beta_1^2 \\ +D_1\gamma_1^2 \end{pmatrix} & \begin{pmatrix} G\beta_1\gamma_2 \\ +D_2\beta_1\gamma_1 \end{pmatrix} & \begin{pmatrix} G\beta_1\beta_2 \\ +D_1\gamma_1\gamma_2 \end{pmatrix} & \begin{pmatrix} G\beta_1\gamma_3 \\ +D_2\beta_3\gamma_1 \end{pmatrix} & \begin{pmatrix} G\beta_1\beta_3 \\ +D_1\gamma_1\gamma_3 \end{pmatrix} \\ & & \begin{pmatrix} D_1\beta_2^2 \\ +G\gamma_2^2 \end{pmatrix} & \begin{pmatrix} G\beta_2\gamma_2 \\ +D_2\beta_2\gamma_2 \end{pmatrix} & \begin{pmatrix} D_1\beta_2\beta_3 \\ +G\gamma_2\gamma_3 \end{pmatrix} & \begin{pmatrix} G\beta_3\gamma_2 \\ +D_2\beta_2\gamma_3 \end{pmatrix} \\ & & & \begin{pmatrix} G\beta_2^2 \\ +D_1\gamma_2^2 \end{pmatrix} & \begin{pmatrix} G\beta_2\gamma_3 \\ +D_2\beta_3\gamma_2 \end{pmatrix} & \begin{pmatrix} G\beta_2\beta_3 \\ +D_1\gamma_2\gamma_3 \end{pmatrix} \\ & \text{sym} & & & \begin{pmatrix} D_1\beta_3^2 \\ +G\gamma_3^2 \end{pmatrix} & \begin{pmatrix} G\beta_3\gamma_3 \\ +D_2\beta_3\gamma_3 \end{pmatrix} \\ & & & & & \begin{pmatrix} Gb_3^2 \\ +D_1\gamma_3^2 \end{pmatrix} \end{bmatrix} \tag{11.180}$$

The element inertia matrix is given by

$$[\mathbf{m}] = \iiint_V \rho[\mathbf{N}]^T[\mathbf{N}]dV = \iint_A \int_0^t \rho[\mathbf{N}]^T[\mathbf{N}]\,dxdA = \iint_A \rho t[\mathbf{N}]^T[\mathbf{N}]\,dA \tag{11.181}$$

For an element which has constant thickness and density, Equation 11.181 can be written as

$$[\mathbf{m}] = \rho t \iint_A \begin{bmatrix} N_1N_1 & 0 & N_1N_2 & 0 & N_1N_3 & 0 \\ 0 & N_1N_1 & 0 & N_1N_2 & 0 & N_1N_3 \\ N_2N_1 & 0 & N_2N_2 & 0 & N_2N_3 & 0 \\ 0 & N_2N_1 & 0 & N_2N_2 & 0 & N_2N_3 \\ N_3N_1 & 0 & N_3N_2 & 0 & N_3N_3 & 0 \\ 0 & N_3N_1 & 0 & N_3N_2 & 0 & N_3N_3 \end{bmatrix} dA \tag{11.182}$$

All mass terms in Equation 11.182 can be integrated in a straightforward manner analytically using the formula developed by Eisenberg and Malvern (1973), which for triangular elements is given as

$$\iint_A L_1^m L_2^n L_3^p \, dA = 2A \frac{m!n!p!}{(m+n+p+2)!} \tag{11.183}$$

whereas stated above $L_i = N_i$, $i = 1,2,3$ are the area coordinates for the triangular element. The consistent mass matrix is given as

$$[\mathbf{m}] = \frac{\rho t A}{12} \begin{bmatrix} 2 & 0 & 1 & 0 & 1 & 0 \\ 0 & 2 & 0 & 1 & 0 & 1 \\ 1 & 0 & 2 & 0 & 1 & 0 \\ 0 & 1 & 0 & 2 & 0 & 1 \\ 1 & 0 & 1 & 0 & 2 & 0 \\ 0 & 1 & 0 & 1 & 0 & 2 \end{bmatrix} \tag{11.184}$$

4. *Four-Noded Rectangular Element.* The displacement function for the four-noded, eight degrees of freedom rectangular element was given in Equation 11.83 as

$$\{\mathbf{u}\} = \begin{Bmatrix} \mathbf{u_x} \\ \mathbf{u_y} \end{Bmatrix} = \begin{bmatrix} N_1 & 0 & N_2 & 0 & N_3 & 0 & N_4 & 0 \\ 0 & N_1 & 0 & N_2 & 0 & N_3 & 0 & N_4 \end{bmatrix} \begin{Bmatrix} u_{1x} \\ u_{1y} \\ u_{2x} \\ u_{2y} \\ u_{3x} \\ u_{3y} \\ u_{4x} \\ u_{4y} \end{Bmatrix}$$

where the shape functions, Equation 11.89, in terms of nodal coordinates are expressed as

$$N_i = \frac{1}{4}(1+\xi\xi_i)(1+\eta\eta_i), \quad i = 1,2,3,4$$

The strain displacement relations in Equation 11.173 for the rectangular element can be expressed as

$$\{\varepsilon\} = \begin{Bmatrix} \varepsilon_{xx} \\ \varepsilon_{yy} \\ \gamma_{xy} \end{Bmatrix} = \begin{bmatrix} \frac{\partial}{\partial x} & 0 \\ 0 & \frac{\partial}{\partial y} \\ \frac{\partial}{\partial y} & \frac{\partial}{\partial x} \end{bmatrix} \begin{Bmatrix} u_x \\ u_y \end{Bmatrix} = \begin{bmatrix} \frac{\partial N_1}{\partial x} & 0 & \frac{\partial N_2}{\partial x} & 0 & \frac{\partial N_3}{\partial x} & 0 & \frac{\partial N_4}{\partial x} & 0 \\ 0 & \frac{\partial N_1}{\partial y} & 0 & \frac{\partial N_2}{\partial y} & 0 & \frac{\partial N_3}{\partial y} & 0 & \frac{\partial N_4}{\partial y} \\ \frac{\partial N_1}{\partial y} & \frac{\partial N_1}{\partial x} & \frac{\partial N_2}{\partial y} & \frac{\partial N_2}{\partial x} & \frac{\partial N_3}{\partial y} & \frac{\partial N_3}{\partial x} & \frac{\partial N_4}{\partial y} & \frac{\partial N_4}{\partial x} \end{bmatrix} \begin{Bmatrix} u_{1x} \\ u_{1y} \\ u_{2x} \\ u_{2y} \\ u_{3x} \\ u_{3y} \\ u_{4x} \\ u_{4y} \end{Bmatrix} \tag{11.185}$$

Differentiating the above shape function equations gives

$$\begin{aligned} \frac{\partial N_i}{\partial \xi} &= \frac{\xi_i}{4a}(1+\eta\eta_i) \\ \frac{\partial N_i}{\partial \eta} &= \frac{\eta_i}{4b}(1+\xi\xi_i) \end{aligned} \tag{11.186}$$

Substituting Equation 11.186 into 11.185 yields

$$\begin{aligned} \{\varepsilon\} &= \frac{1}{4} \begin{bmatrix} -\frac{(1-\eta)}{a} & 0 & \frac{(1-\eta)}{a} & 0 & \frac{(1+\eta)}{a} & 0 & -\frac{(1+\eta)}{a} & 0 \\ 0 & -\frac{(1-\xi)}{b} & 0 & -\frac{(1+\xi)}{b} & 0 & \frac{(1+\xi)}{b} & 0 & \frac{(1-\xi)}{b} \\ -\frac{(1-\xi)}{b} & -\frac{(1-\eta)}{a} & -\frac{(1+\xi)}{b} & \frac{(1-\eta)}{a} & \frac{(1+\xi)}{b} & \frac{(1+\eta)}{a} & \frac{(1-\xi)}{b} & -\frac{(1+\eta)}{a} \end{bmatrix} \begin{Bmatrix} u_{1x} \\ u_{1y} \\ u_{2x} \\ u_{2y} \\ u_{3x} \\ u_{3y} \\ u_{4x} \\ u_{4y} \end{Bmatrix} \\ &= [\mathbf{B}]\{\mathbf{d}\} \end{aligned} \tag{11.187}$$

where

$$[\mathbf{B}] = \frac{1}{4}\begin{bmatrix} -\frac{(1-\eta)}{a} & 0 & \frac{(1-\eta)}{a} & 0 & \frac{(1+\eta)}{a} & 0 & -\frac{(1+\eta)}{a} & 0 \\ 0 & -\frac{(1-\xi)}{b} & 0 & -\frac{(1+\xi)}{b} & 0 & \frac{(1+\xi)}{b} & 0 & \frac{(1-\xi)}{b} \\ -\frac{(1-\xi)}{b} & -\frac{(1-\eta)}{a} & -\frac{(1+\xi)}{b} & \frac{(1-\eta)}{a} & \frac{(1+\xi)}{b} & \frac{(1+\eta)}{a} & \frac{(1-\xi)}{b} & -\frac{(1+\eta)}{a} \end{bmatrix} \tag{11.188}$$

The element stiffness matrix for plane stress or strain is

$$[\mathbf{k}] = \int_{-1}^{1}\int_{-1}^{1} [\mathbf{B}]^T[\mathbf{D}][\mathbf{B}]\, d\xi d\eta \tag{11.189}$$

where $[\mathbf{D}]$ is the elasticity matrix given by Equation 11.179. Substituting Equations 11.179 and 11.188 into 11.189 yields

$$[\mathbf{k}] = [\mathbf{k}_1] + [\mathbf{k}_2] \tag{11.190}$$

where

$$[\mathbf{k}_1] = \frac{Et}{16ab(1-\nu^2)}\begin{bmatrix} 2\bar{b} & \bar{\nu} & -2\bar{b} & \bar{\nu} & -\bar{b} & -\bar{\nu} & \bar{b} & -\bar{\nu} \\ & 2\bar{a} & -\bar{\nu} & \bar{a} & -\bar{\nu} & -\bar{a} & \bar{\nu} & -2\bar{a} \\ & & 2\bar{b} & -\bar{\nu} & \bar{b} & \bar{\nu} & -\bar{b} & \bar{\nu} \\ & & & 2\bar{a} & -\bar{\nu} & -2\bar{a} & \bar{\nu} & -\bar{a} \\ & & & & 2\bar{b} & \bar{\nu} & -2\bar{b} & \bar{\nu} \\ & & & & & 2\bar{a} & -\bar{\nu} & \bar{a} \\ & \text{sym} & & & & & 2\bar{b} & -\bar{\nu} \\ & & & & & & & 2\bar{a} \end{bmatrix} \tag{11.191}$$

$$\bar{a} = \frac{8a^2}{2}, \quad \bar{b} = \frac{8b^2}{3}, \quad \bar{\nu} = 4\nu ab \tag{11.192}$$

and

$$[\mathbf{k}_2] = \frac{Et}{16ab(1+\nu)}\begin{bmatrix} 2\bar{\bar{a}} & 2ab & \bar{\bar{a}} & -2ab & -\bar{\bar{a}} & -2ab & -2\bar{\bar{a}} & 2ab \\ & 2\bar{\bar{b}} & 2ab & -2\bar{\bar{b}} & -2ab & -\bar{\bar{b}} & -2ab & \bar{\bar{b}} \\ & & 2\bar{\bar{a}} & -2ab & -2\bar{\bar{a}} & -2ab & -\bar{\bar{a}} & 2ab \\ & & & 2\bar{\bar{b}} & 2ab & \bar{\bar{b}} & 2ab & -\bar{\bar{b}} \\ & & & & 2\bar{\bar{a}} & 2ab & \bar{\bar{a}} & -2ab \\ & & & & & 2\bar{\bar{b}} & 2ab & -2\bar{\bar{b}} \\ & \text{sym} & & & & & 2\bar{\bar{a}} & -2ab \\ & & & & & & & 2\bar{\bar{b}} \end{bmatrix} \tag{11.193}$$

$$\bar{\bar{a}} = \frac{4a^2}{3}, \quad \bar{\bar{b}} = \frac{4b^2}{3} \tag{11.194}$$

The consistent mass matrix is given as

$$[\mathbf{m}] = \iint_A \rho t[\mathbf{N}]^T[\mathbf{N}]dA = \rho tab \int_{-1}^{1}\int_{-1}^{1} [\mathbf{N}]^T[\mathbf{N}]d\xi d\eta \tag{11.195}$$

where

$$[\mathbf{N}] = \begin{bmatrix} N_1 & 0 & N_2 & 0 & N_3 & 0 & N_4 & 0 \\ 0 & N_1 & 0 & N_2 & 0 & N_3 & 0 & N_4 \end{bmatrix} \tag{11.196}$$

and from Equation 11.89

$$N_i = \frac{1}{4}(1+\xi_0)(1+\eta_0)$$

where $\xi_0 = \xi\xi_i$ and $\eta_0 = \eta\eta_i$. Equation 11.195 can be written as

$$[\mathbf{m}] = \rho tab \int_{-1}^{1}\int_{-1}^{1} N_i N_j \, d\xi d\eta = \frac{\rho tab}{16}\int_{-1}^{1}(1+\xi\xi_i)(1+\xi\xi_j)d\xi \int_{-1}^{1}(1+\eta\eta_i)(1+\eta\eta_j)d\eta \tag{11.197}$$

Evaluating the integrals yields

$$[\mathbf{m}] = \frac{\rho tab}{4}\left(1+\frac{1}{3}\xi_i\xi_j\right)\left(1+\frac{1}{3}\eta_i\eta_j\right) \tag{11.198}$$

which gives

$$[\mathbf{m}] = \frac{\rho tab}{9}\begin{bmatrix} 4 & 0 & 2 & 0 & 1 & 0 & 2 & 0 \\ & 4 & 0 & 2 & 0 & 1 & 0 & 2 \\ & & 4 & 0 & 2 & 0 & 1 & 0 \\ & & & 4 & 0 & 2 & 0 & 1 \\ & & & & 4 & 0 & 2 & 0 \\ & & & & & 4 & 0 & 2 \\ & \text{sym} & & & & & 4 & 0 \\ & & & & & & & 4 \end{bmatrix} \tag{11.199}$$

5. *Tetrahedron Element*. The displacement functions for the tetrahedron element were given in Section 11.3.2.3 as

$$\{\mathbf{u}\} = \begin{Bmatrix} u_x \\ u_y \\ u_z \end{Bmatrix} = \begin{Bmatrix} u \\ v \\ w \end{Bmatrix} = [\mathbf{N}]\{\mathbf{d}\}$$

where

$$[\mathbf{N}] = \begin{bmatrix} N_1 & 0 & 0 & N_2 & 0 & 0 & N_3 & 0 & 0 & N_4 & 0 & 0 \\ 0 & N_1 & 0 & 0 & N_2 & 0 & 0 & N_3 & 0 & 0 & N_4 & 0 \\ 0 & 0 & N_1 & 0 & 0 & N_2 & 0 & 0 & N_3 & 0 & 0 & N_4 \end{bmatrix}$$

$$\{\mathbf{d}\}^T = (u_1 \quad v_1 \quad w_1 \quad u_2 \quad v_2 \quad w_2 \quad u_3 \quad v_3 \quad w_3 \quad u_4 \quad v_4 \quad w_4)$$

and $N_i = L_i$, $i = 1,2,3,4$.

For a three-dimensional element, there are six stresses $\{\sigma\} = \{\sigma_{xx} \quad \sigma_{yy} \quad \sigma_{zz} \quad \sigma_{xy} \quad \sigma_{yz} \quad \sigma_{zy}\}^T$ and six strains $\varepsilon_{ij} = \frac{1}{2}(u_{i,j} + u_{j,i})$, where in the vector form $\{\varepsilon\} = \{\varepsilon_{11} \quad \varepsilon_{22} \quad \varepsilon_{33} \quad 2\varepsilon_{12} \quad 2\varepsilon_{23} \quad 2\varepsilon_{31}\}^T = \{\varepsilon_{xx} \quad \varepsilon_y \quad \varepsilon_{zz} \quad \gamma_{xy} \quad \gamma_{yz} \quad \gamma_{zx}\}^T$. Note that the second expressions for the strains are in engineering notation. The strain–displacement relations can be written as

$$\{\varepsilon\} = \begin{Bmatrix} \varepsilon_{xx} \\ \varepsilon_{yy} \\ \varepsilon_{zz} \\ 2\varepsilon_{xy} \\ 2\varepsilon_{yz} \\ 2\varepsilon_{zx} \end{Bmatrix} = [\mathbf{L}]\{\mathbf{u}\} = [\mathbf{L}][\mathbf{N}]\{\mathbf{d}\} = [\mathbf{B}]\{\mathbf{d}\} \tag{11.200}$$

where the strain matrix $[\mathbf{B}]$ is given by

$$[\mathbf{B}] = [\mathbf{L}][\mathbf{N}] = \begin{bmatrix} \partial/\partial x & 0 & 0 \\ 0 & \partial/\partial y & 0 \\ 0 & 0 & \partial/\partial z \\ \partial/\partial y & \partial/\partial x & 0 \\ 0 & \partial/\partial z & \partial/\partial y \\ \partial/\partial z & 0 & \partial/\partial x \end{bmatrix} [\mathbf{N}] \tag{11.201}$$

Using the equation for $[\mathbf{N}]$ given above and the fact that $N_i = L_i, i = 1,2,3,4$, the strain matrix $[\mathbf{B}]$ is obtained as

$$[\mathbf{B}] = \frac{1}{6V}\begin{bmatrix} b_1 & 0 & 0 & b_2 & 0 & 0 & b_3 & 0 & 0 & b_4 & 0 & 0 \\ 0 & c_1 & 0 & 0 & c_2 & 0 & 0 & c_3 & 0 & 0 & c_4 & 0 \\ 0 & 0 & d_1 & 0 & 0 & d_2 & 0 & 0 & d_3 & 0 & 0 & d_4 \\ c_1 & b_1 & 0 & c_2 & b_2 & 0 & c_3 & b_3 & 0 & c_4 & b_4 & 0 \\ 0 & d_1 & c_1 & 0 & d_2 & c_2 & 0 & d_3 & c_3 & 0 & d_4 & c_4 \\ d_1 & 0 & b_1 & d_2 & 0 & b_2 & d_3 & 0 & b_3 & d_4 & 0 & b_4 \end{bmatrix} \tag{11.202}$$

where we note that **[B]** is constant over the element. Recall that the stress field is related to the strain field, Equation 5.40, as

$$\sigma_{ij} = C_{ijkl}\varepsilon_{kl}$$

and for a homogeneous linear isotropic material

$$\begin{Bmatrix} \sigma_{xx} \\ \sigma_{yy} \\ \sigma_{zz} \\ \sigma_{xy} \\ \sigma_{yz} \\ \sigma_{zx} \end{Bmatrix} = \frac{E}{(1+\nu)(1-2\nu)}\begin{bmatrix} 1-\nu & \nu & \nu & 0 & 0 & 0 \\ \nu & 1-\nu & \nu & 0 & 0 & 0 \\ 0 & 0 & 1-\nu & 0 & 0 & 0 \\ 0 & 0 & 0 & \frac{1}{2}-\nu & 0 & 0 \\ 0 & 0 & 0 & 0 & \frac{1}{2}-\nu & 0 \\ 0 & 0 & 0 & 0 & 0 & \frac{1}{2}-\nu \end{bmatrix}\begin{Bmatrix} \varepsilon_{xx} \\ \varepsilon_{yy} \\ \varepsilon_{zz} \\ \gamma_{xy} \\ \gamma_{yz} \\ \gamma_{zx} \end{Bmatrix} = [\mathbf{D}]\{\varepsilon\} \tag{11.203}$$

The stiffness matrix is given as

$$[\mathbf{k}] = \iiint_V [\mathbf{B}]^T[\mathbf{D}][\mathbf{B}]dV \tag{11.204}$$

Assuming that the elastic moduli do not vary over the tetrahedron element, the integrand becomes constant since matrix **[B]** is constant, and the integration is just the volume of the element, we have for the stiffness matrix

$$[\mathbf{k}] = V[\mathbf{B}]^T[\mathbf{D}][\mathbf{B}] \tag{11.205}$$

We see that the stiffness matrix is (12 × 12) and can be evaluated directly in the closed form from Equation 11.205.

The element mass matrix **[m]** for the tetrahedron can be obtained using Equation 11.8 as

$$[\mathbf{m}] = \iiint_V \rho[\mathbf{N}]^T[\mathbf{N}]dV = \iiint_V \rho\begin{bmatrix} \mathbf{N}_{11} & \mathbf{N}_{12} & \mathbf{N}_{13} & \mathbf{N}_{14} \\ \mathbf{N}_{21} & \mathbf{N}_{22} & \mathbf{N}_{23} & \mathbf{N}_{24} \\ \mathbf{N}_{31} & \mathbf{N}_{32} & \mathbf{N}_{33} & \mathbf{N}_{34} \\ \mathbf{N}_{41} & \mathbf{N}_{42} & \mathbf{N}_{43} & \mathbf{N}_{44} \end{bmatrix}dV \tag{11.206}$$

where

$$[\mathbf{N}] = \begin{bmatrix} \mathbf{N_iN_j} & 0 & 0 \\ 0 & \mathbf{N_iN_j} & 0 \\ 0 & 0 & \mathbf{N_iN_j} \end{bmatrix} \tag{11.207}$$

Using the formula developed by Eisenberg and Malvern (1973), the following, for tetrahedral elements, is given as

$$\iiint_V L_1^m L_2^n L_3^p L_4^q \, dV = \frac{m!n!p!q!}{(m+n+p+q+3)!} 6V \tag{11.208}$$

where, as above, $N_i = L_i$ are the volume coordinates for the tetrahedral element. The consistent mass matrix is obtained by evaluating the integral in Equation 11.206, which gives

$$[\mathbf{m}] = \frac{\rho V}{20} \begin{bmatrix} 2 & 0 & 0 & 1 & 0 & 0 & 1 & 0 & 0 & 1 & 0 & 0 \\ & 2 & 0 & 0 & 1 & 0 & 0 & 1 & 0 & 0 & 1 & 0 \\ & & 2 & 0 & 0 & 1 & 0 & 0 & 1 & 0 & 0 & 1 \\ & & & 2 & 0 & 0 & 1 & 0 & 0 & 1 & 0 & 0 \\ & & & & 2 & 0 & 0 & 1 & 0 & 0 & 1 & 0 \\ & & & & & 2 & 0 & 0 & 1 & 0 & 0 & 1 \\ & & & & & & 2 & 0 & 0 & 1 & 0 & 0 \\ & & & & & & & 2 & 0 & 0 & 1 & 0 \\ & & \text{sym} & & & & & & 2 & 0 & 0 & 1 \\ & & & & & & & & & 2 & 0 & 0 \\ & & & & & & & & & & 2 & 0 \\ & & & & & & & & & & & 2 \end{bmatrix} \tag{11.209}$$

6. *Hexahedron Element*. These kinds of elements are widely used in finite element simulations of three-dimensional solids. The shape functions using natural coordinates for an eight-noded hexahedron element are from Equation 11.113

$$N_i = \frac{1}{8}(1+\xi_0)(1+\eta_0)(1+\zeta_0)$$

where $\xi_0 = \xi\xi_i$, $\eta_0 = \eta\eta_i$, and $\zeta_0 = \zeta\zeta_i$. We shall assume an isoparametric element with the transforming relations from the Cartesian coordinates to the natural one $(\xi, \eta, \zeta)$ given as

$$\begin{aligned} x(\xi,\eta,\zeta) &= \sum_{i=1}^{8} N_i x_i \\ y(\xi,\eta,\zeta) &= \sum_{i=1}^{8} N_i y_i \\ z(\xi,\eta,\zeta) &= \sum_{i=1}^{8} N_i z_i \end{aligned} \tag{11.210}$$

where the shape functions $N_i$ are as given above. The properties of the shape functions are

$$\sum_{i=1}^{8} N_i(\xi,\eta,\zeta) = 1 \qquad -1 \le \begin{Bmatrix} \xi \\ \eta \\ \zeta \end{Bmatrix} \le 1$$

$$N_i(\xi_j,\eta_j,\zeta_j) = \begin{cases} 1 & \text{if} \quad i = j \\ 0 & \text{if} \quad i \ne j \end{cases}$$

The approximation of the displacement field is given as in Equation 11.112 as

$$\begin{Bmatrix} u \\ v \\ w \end{Bmatrix} = \sum_{i=1}^{8} N_i(\xi,\eta,\zeta) \begin{Bmatrix} u_i \\ v_i \\ w_i \end{Bmatrix}$$

where $u_i$, $v_i$, and $w_i$ are the displacement coordinates of the $i$th node of the hexahedron element. In the matrix form

$$\{\mathbf{u}\} = \begin{Bmatrix} u \\ v \\ w \end{Bmatrix} = \begin{bmatrix} N_1 & 0 & 0 & N_2 & 0 & 0 & \cdots & N_8 & 0 & 0 \\ 0 & N_1 & 0 & 0 & N_2 & 0 & \cdots & 0 & N_8 & 0 \\ 0 & 0 & N_1 & 0 & 0 & N_2 & \cdots & 0 & 0 & N_8 \end{bmatrix} \begin{Bmatrix} u_1 \\ v_1 \\ w_1 \\ \vdots \\ u_8 \\ v_8 \\ w_8 \end{Bmatrix} = [\mathbf{N}]\{\mathbf{d}\} \tag{11.211}$$

Using Equation 11.201, the strain matrix for the hexahedron is given as

$$[\mathbf{B}] = [\mathbf{L}][\mathbf{N}] = \begin{bmatrix} \partial/\partial x & 0 & 0 \\ 0 & \partial/\partial y & 0 \\ 0 & 0 & \partial/\partial z \\ \partial/\partial y & \partial/\partial x & 0 \\ 0 & \partial/\partial z & \partial/\partial y \\ \partial/\partial z & 0 & \partial/\partial x \end{bmatrix} \begin{bmatrix} N_1 & 0 & 0 & N_2 & 0 & 0 & \cdots & N_8 & 0 & 0 \\ 0 & N_1 & 0 & 0 & N_2 & 0 & \cdots & 0 & N_8 & 0 \\ 0 & 0 & N_1 & 0 & 0 & N_2 & \cdots & 0 & 0 & N_8 \end{bmatrix} \tag{11.212}$$

$$= \begin{bmatrix} \partial N_1/\partial x & 0 & 0 & \cdots & \partial N_8/\partial x & 0 & 0 \\ 0 & \partial N_1/\partial y & 0 & \cdots & 0 & \partial N_8/\partial y & 0 \\ 0 & 0 & \partial N_1/\partial z & \cdots & 0 & 0 & \partial N_8/\partial z \\ \partial N_1/\partial y & \partial N_1/\partial x & 0 & \cdots & \partial N_8/\partial y & \partial N_8/\partial x & 0 \\ 0 & \partial N_1/\partial z & \partial N_1/\partial y & \cdots & 0 & \partial N_8/\partial z & \partial N_8/\partial y \\ \partial N_1/\partial z & 0 & \partial N_1/\partial x & \cdots & \partial N_8/\partial z & 0 & \partial N_8/\partial x \end{bmatrix}$$

The shape functions are defined in terms of the natural coordinates $(\xi,\eta,\zeta)$, and to obtain the derivatives with respect to $(x,y,z)$ in the strain matrix, the chain rule for partial differentiation is used. Therefore, we have

$$\begin{aligned} \frac{\partial N_i}{\partial \xi} &= \frac{\partial N_i}{\partial x}\frac{\partial x}{\partial \xi} + \frac{\partial N_i}{\partial y}\frac{\partial y}{\partial \xi} + \frac{\partial N_i}{\partial z}\frac{\partial z}{\partial \xi} \\ \frac{\partial N_i}{\partial \eta} &= \frac{\partial N_i}{\partial x}\frac{\partial x}{\partial \eta} + \frac{\partial N_i}{\partial y}\frac{\partial y}{\partial \eta} + \frac{\partial N_i}{\partial z}\frac{\partial z}{\partial \eta} \\ \frac{\partial N_i}{\partial \zeta} &= \frac{\partial N_i}{\partial x}\frac{\partial x}{\partial \zeta} + \frac{\partial N_i}{\partial y}\frac{\partial y}{\partial \zeta} + \frac{\partial N_i}{\partial z}\frac{\partial z}{\partial \zeta} \end{aligned} \tag{11.213}$$

In the matrix form, Equation 11.213 becomes

$$\begin{Bmatrix} \partial N_i/\partial \xi \\ \partial N_i/\partial \eta \\ \partial N_i/\partial \zeta \end{Bmatrix} = [\mathbf{J}] \begin{Bmatrix} \partial N_i/\partial x \\ \partial N_i/\partial y \\ \partial N_i/\partial z \end{Bmatrix} \tag{11.214}$$

where [J] is the *Jacobian matrix* given as

$$[\mathbf{J}] = \begin{bmatrix} \partial x/\partial \xi & \partial y/\partial \xi & \partial z/\partial \xi \\ \partial x/\partial \eta & \partial y/\partial \eta & \partial z/\partial \eta \\ \partial x/\partial \zeta & \partial y/\partial \zeta & \partial z/\partial \eta \end{bmatrix} \tag{11.215}$$

Equation 11.214 can be written as

$$\begin{Bmatrix} \partial N_i/\partial x \\ \partial N_i/\partial y \\ \partial N_i/\partial z \end{Bmatrix} = [\mathbf{J}]^{-1} \begin{Bmatrix} \partial N_i/\partial \xi \\ \partial N_i/\partial \eta \\ \partial N_i/\partial \zeta \end{Bmatrix} \tag{11.216}$$

Equation 11.214 is used to compute the strain matrix **[B]** in Equation 11.212 by replacing the derivatives with respect to $(x,y,z)$, to those with respect to $(\xi,\eta,\zeta)$. Having the **[B]** matrix, the stiffness matrix for three-dimensional solid elements can be determined by substituting Equation 11.216 into Equation 11.11 as

$$[\mathbf{k}] = \iiint_V [\mathbf{B}]^T[\mathbf{D}][\mathbf{B}]dV = \int_{-1}^{1}\int_{-1}^{1}\int_{-1}^{1} [\mathbf{B}]^T[\mathbf{D}][\mathbf{B}]\det[\mathbf{J}]d\xi\, d\eta\, d\zeta \tag{11.217}$$

where the material matrix [**D**] is given from Equation 11.203 for a linear homogeneous isotropic material. As noted in Equation 11.217, the strain matrix [**B**] is a function of $(\xi,\eta,\zeta)$ so that evaluating the integrals can be difficult. Therefore, in most cases the integration is performed using a numerical integration technique. Using isoparametric elements, one can use numerical integration for one-, two- or three-dimensional integrations. There are many different techniques to perform numerical integration but the *Gauss quadrature* is the preferred method in finite element analysis. In finite element analysis, we encounter integrations of the following types:

$$\int f(\xi)d\xi, \quad \iint f(\xi,\eta)d\xi d\eta, \quad \iiint f(\xi,\eta,\zeta)d\xi d\eta d\zeta$$

which occur in one-, two- and three-dimensional cases. For illustration, consider the one-dimensional case where the integral can be approximated as

$$I = \int_{-1}^{1} f(\xi)d\xi = \sum_{i=1}^{n} W_i f(\xi_i) \tag{11.218}$$

where the sampling points are the specified points $\xi_i$ and $W_i$ the corresponding weights. It is noted that if a polynomial expression is assumed then for $n$ sampling points we have $2n$ unknowns, being $f_i$ and $\xi_i$, and therefore a polynomial of degree $2n$−1 could be constructed and integrated exactly (Conte and de Boor 1980). This forms the basis of Gauss quadrature. Table 11.2 shows the position and weighting factors for the first five orders.

**TABLE 11.2**

Sampling Points and Weights of the Gaussian Quadrature Equation $\int_{-1}^{1}\int f(x)dx = \sum_{i=1}^{n} W_i f(r_i)$

| $\pm r$ | | W |
|---|---|---|
| | $n = 2$ | |
| 0.57735027 | | 1.0000000 |
| | $n = 3$ | |
| 0.774596669 | | 0.5555556 |
| 0.000000000 | | 0.8888889 |
| | $n = 4$ | |
| 0.861136312 | | 0.3478548 |
| 0.339981044 | | 0.6521452 |
| | $n = 5$ | |
| 0.906179846 | | 0.2369269 |
| 0.538469310 | | 0.4786287 |
| 0.000000000 | | 0.5688889 |

The above procedure can be extended to evaluate two- and three-dimensional integrals. For these cases, the integrals are first evaluated with respect to one coordinate direction and then to the second direction and then to the third direction. For two dimensions, we have

$$\begin{aligned} I &= \int_{-1}^{1}\int_{-1}^{1} f(\xi,\eta)d\xi d\eta = \int_{-1}^{1}\sum_{i=1}^{n} W_i f(\xi_i,\eta)d\eta = \sum_{j=1}^{n} W_j\left[\sum_{i=1}^{n} W_i f(\xi_i,\eta_j)\right] \\ &= \sum_{i=1}^{n}\sum_{j=1}^{n} W_i W_j f(\xi_i,\eta_j) \end{aligned} \tag{11.219}$$

where $W_i$ and $W_j$ are the weight functions for the one-dimensional integration. The three-dimensional integral can be evaluated in the same manner, giving

$$I = \int_{-1}^{1}\int_{-1}^{1}\int_{-1}^{1} f(\xi,\eta,\zeta)d\xi d\eta d\zeta = \sum_{i=1}^{n}\sum_{j=1}^{n}\sum_{k=1}^{n} W_i W_j W_k f(\xi_i,\eta_j,\zeta_k) \tag{11.220}$$

To obtain the stiffness matrix in Equation 11.217, a $2 \times 2 \times 2$ integration scheme is usually used for linear elements and $3 \times 3 \times 3$ for quadratic elements.

The mass matrix for the hexahedron element, the shape function matrix $[\mathbf{N}]$ is substituted into Equation 11.8. This gives

$$[\mathbf{m}] = \iiint_V \rho[\mathbf{N}]^T[\mathbf{N}]dV = \int_{-1}^{1}\int_{-1}^{1}\int_{-1}^{1} \rho[\mathbf{N}]^T[\mathbf{N}]\det[\mathbf{J}]d\xi d\eta d\zeta \tag{11.221}$$

where, again, the integral is carried out using Gauss quadrature. If the hexahedron element is rectangular with dimensions $2a \times 2b \times 2c$, the determinant of the Jacobian matrix is given as

$$\det[\mathbf{J}] = 8abc = V \tag{11.222}$$

The mass or inertia matrix can be written as

$$[\mathbf{m}] = 8\rho abc\int_{-1}^{1}\int_{-1}^{1}\int_{-1}^{1} [\mathbf{N}]^T[\mathbf{N}]d\xi\, d\eta\, d\zeta \tag{11.223}$$

or expanded as

$$[\mathbf{m}] = \begin{bmatrix} \mathbf{m}_{11} & \mathbf{m}_{12} & \mathbf{m}_{13} & \mathbf{m}_{14} & \mathbf{m}_{15} & \mathbf{m}_{16} & \mathbf{m}_{17} & \mathbf{m}_{18} \\ & \mathbf{m}_{22} & \mathbf{m}_{23} & \mathbf{m}_{24} & \mathbf{m}_{25} & \mathbf{m}_{26} & \mathbf{m}_{27} & \mathbf{m}_{28} \\ & & \mathbf{m}_{33} & \mathbf{m}_{34} & \mathbf{m}_{35} & \mathbf{m}_{36} & \mathbf{m}_{37} & \mathbf{m}_{38} \\ & & & \mathbf{m}_{44} & \mathbf{m}_{45} & \mathbf{m}_{46} & \mathbf{m}_{47} & \mathbf{m}_{48} \\ & & & & \mathbf{m}_{55} & \mathbf{m}_{56} & \mathbf{m}_{57} & \mathbf{m}_{58} \\ & & & & & \mathbf{m}_{66} & \mathbf{m}_{67} & \mathbf{m}_{68} \\ & \text{sym} & & & & & \mathbf{m}_{77} & \mathbf{m}_{78} \\ & & & & & & & \mathbf{m}_{88} \end{bmatrix} \tag{11.224}$$

where

$$
\begin{aligned}
[\mathbf{m}_{\mathrm{ij}}] &= 8\rho abc \int_{-1}^{1}\int_{-1}^{1}\int_{-1}^{1} \mathbf{N}_i \mathbf{N}_j \, d\xi \, d\eta \, d\zeta \\
&= 8\rho abc \int_{-1}^{1}\int_{-1}^{1}\int_{-1}^{1} \begin{bmatrix} N_i & 0 & 0 \\ 0 & N_i & 0 \\ 0 & 0 & N_i \end{bmatrix} \begin{bmatrix} N_j & 0 & 0 \\ 0 & N_j & 0 \\ 0 & 0 & N_j \end{bmatrix} d\xi \, d\eta \, d\zeta \\
&= 8\rho abc \int_{-1}^{1}\int_{-1}^{1}\int_{-1}^{1} \begin{bmatrix} N_iN_j & 0 & 0 \\ 0 & N_iN_j & 0 \\ 0 & 0 & N_iN_j \end{bmatrix} d\xi \, d\eta \, d\zeta \\
&= \begin{bmatrix} m_{ij} & 0 & 0 \\ 0 & m_{ij} & 0 \\ 0 & 0 & m_{ij} \end{bmatrix}
\end{aligned}
\tag{11.225}
$$

The calculation for $m_{11}$ would be

$$
\begin{aligned}
m_{11} &= \frac{\rho abc}{64} \int_{-1}^{1} (1+\xi_i\xi)(1+\xi_j\xi) d\xi \int_{-1}^{1} (1+\eta_i\eta)(1+\eta_j\eta) d\eta \int_{-1}^{1} (1+\zeta_i\zeta)(1+\zeta_j\zeta) \, d\zeta \\
&= \frac{\rho abc}{8}\left(1+\frac{1}{3}\xi_i\xi_j\right)\left(1+\frac{1}{3}\eta_i\eta_j\right)\left(1+\frac{1}{3}\zeta_i\zeta_j\right) \\
&= \frac{8\rho abc}{27}
\end{aligned}
\tag{11.226}
$$

Hence, in the same manner, we find

$$
\begin{aligned}
&[\mathbf{m}_{11}] = [\mathbf{m}_{22}] = [\mathbf{m}_{33}] = [\mathbf{m}_{44}] = [\mathbf{m}_{55}] = [\mathbf{m}_{66}] = [\mathbf{m}_{77}] = [\mathbf{m}_{88}] = \frac{\rho abc}{27}\begin{bmatrix} 8 & 0 & 0 \\ 0 & 8 & 0 \\ 0 & 0 & 8 \end{bmatrix} \\
&[\mathbf{m}_{12}] = [\mathbf{m}_{14}] = [\mathbf{m}_{15}] = [\mathbf{m}_{23}] = [\mathbf{m}_{26}] = [\mathbf{m}_{34}] = [\mathbf{m}_{37}] = \\
&\qquad [\mathbf{m}_{48}] = [\mathbf{m}_{56}] = [\mathbf{m}_{58}] = [\mathbf{m}_{67}] = [\mathbf{m}_{78}] = \frac{\rho abc}{27}\begin{bmatrix} 4 & 0 & 0 \\ 0 & 4 & 0 \\ 0 & 0 & 4 \end{bmatrix} \\
&[\mathbf{m}_{13}] = [\mathbf{m}_{16}] = [\mathbf{m}_{18}] = [\mathbf{m}_{24}] = [\mathbf{m}_{25}] = [\mathbf{m}_{27}] = [\mathbf{m}_{36}] = \\
&\qquad [\mathbf{m}_{38}] = [\mathbf{m}_{45}] = [\mathbf{m}_{47}] = [\mathbf{m}_{57}] = [\mathbf{m}_{68}] = \frac{\rho abc}{27}\begin{bmatrix} 2 & 0 & 0 \\ 0 & 2 & 0 \\ 0 & 0 & 2 \end{bmatrix} \\
&\qquad [\mathbf{m}_{17}] = [\mathbf{m}_{28}] = [\mathbf{m}_{35}] = [\mathbf{m}_{46}] = \frac{\rho abc}{27}\begin{bmatrix} 1 & 0 & 0 \\ 0 & 1 & 0 \\ 0 & 0 & 1 \end{bmatrix}
\end{aligned}
\tag{11.227}
$$

7. *Rectangular Plate Element.* The shape functions for the nonconforming plate element were given in Equation 11.130 as

$$[\mathbf{N}_i(\xi,\eta)]^T = \begin{Bmatrix} (1/8)(1+\xi_0)(1+\eta_0)\left(2+\xi_0+\eta_0-\xi^2-\eta^2\right) \\ (b/8)(1+\xi_0)(\eta_i+\eta)(\eta^2-1) \\ -(a/8)(\xi_i+\xi)(\xi^2-1)(1+\eta_0) \end{Bmatrix}$$

where $(\xi_i,\eta_i)$ are the coordinates of node $i$ and $\xi_0=\xi_i\xi,\eta_0=\eta_i\eta$ as depicted in Figure 11.20 and the displacements were given as

$$\{\mathbf{d}\}^T = [w_1 \quad \theta_{x1} \quad \theta_{y1} \quad w_2 \quad \theta_{x2} \quad \theta_{y2} \quad \cdots \quad w_4 \quad \theta_{x4} \quad \theta_{y4}]$$

The linear elastic stress–strain relations in bending for a homogeneous isotropic material are defined as

$$\{\sigma\} = [\mathbf{D}]\{\varepsilon\} \tag{11.228}$$

where [**D**] is defined as

$$[\mathbf{D}] = \frac{E}{1-\nu^2}\begin{bmatrix} 1 & \nu & 0 \\ \nu & 1 & 0 \\ 0 & 0 & \dfrac{1-\nu}{2} \end{bmatrix} \tag{11.229}$$

The strain energy as given in Equation 11.9 or 11.10 is

$$\begin{aligned} \Pi &= \frac{1}{2}\iiint_V [\varepsilon]^T[\sigma]\,dV \\ &= \frac{1}{2}\iiint_V [\varepsilon]^T[\mathbf{D}][\varepsilon]\,dV \end{aligned} \tag{11.230}$$

The strains displacement relations for thin plate theory are given as

$$\{\varepsilon\} = -z\begin{Bmatrix} \dfrac{\partial^2}{\partial x^2} \\ \dfrac{\partial^2}{\partial y^2} \\ 2\dfrac{\partial^2}{\partial x\partial y} \end{Bmatrix}[\mathbf{N}]\{\mathbf{d}\} = -z\begin{Bmatrix} \dfrac{1}{a^2}\dfrac{\partial^2}{\partial \xi^2} \\ \dfrac{1}{b^2}\dfrac{\partial^2}{\partial \eta^2} \\ \dfrac{2}{ab}\dfrac{\partial^2}{\partial \xi\partial \eta} \end{Bmatrix}[\mathbf{N}(\xi,\eta)]\{\mathbf{d}\} \tag{11.231}$$

Hence

$$[\mathbf{B}] = \begin{Bmatrix} \dfrac{1}{a^2}\dfrac{\partial^2}{\partial \xi^2} \\ \dfrac{1}{b^2}\dfrac{\partial^2}{\partial \eta^2} \\ \dfrac{2}{ab}\dfrac{\partial^2}{\partial \xi \partial \eta} \end{Bmatrix} [\mathbf{N}(\xi,\eta)] \tag{11.232}$$

Substituting Equation 11.231 into the strain energy expression yields

$$\begin{aligned} \Pi &= \frac{1}{2}\iiint_V [\varepsilon]^T[\mathbf{D}][\varepsilon]\,dV \\ &= \frac{1}{2}\{\mathbf{d}\}^T \iint_A \int_{-(t/2)}^{t/2} [\mathbf{B}]^T[\mathbf{D}][\mathbf{B}]\, z^2\, dz\, dA\{\mathbf{d}\} \end{aligned} \tag{11.233}$$

The stiffness of the plate element is then obtained as

$$[\mathbf{k}] = \frac{t^3}{12}\iint_A [\mathbf{B}]^T[\mathbf{D}][\mathbf{B}]\,dA \tag{11.234}$$

Substituting the shape functions into Equation 11.232 and then into Equation 11.234, and integrating over the area yields

$$[\mathbf{k}] = \frac{Et^3}{48(1-\nu^2)ab}\begin{bmatrix} [\mathbf{k}_{11}] & [\mathbf{k}_{12}] & [\mathbf{k}_{13}] & [\mathbf{k}_{14}] \\ & [\mathbf{k}_{22}] & [\mathbf{k}_{23}] & [\mathbf{k}_{24}] \\ & & [\mathbf{k}_{33}] & [\mathbf{k}_{34}] \\ & \text{sym} & & [\mathbf{k}_{44}] \end{bmatrix} \tag{11.235}$$

where we have for the submatrices

$$[\mathbf{k}_{11}] = \begin{bmatrix} 4(\beta^2+\alpha^2)+\dfrac{2}{5}(7-2\nu) & \left\{2\alpha^2+\dfrac{1}{5}(1+4\nu)\right\}b & -\left\{2\beta^2+\dfrac{1}{5}(1+4\nu)\right\}a \\ \left\{2\alpha^2+\dfrac{1}{5}(1+4\nu)\right\}b & 4\left\{\dfrac{1}{3}\alpha^2+\dfrac{1}{15}(1-\nu)\right\}b^2 & -\nu ab \\ -\left\{2\beta^2+\dfrac{1}{5}(1+4\nu)\right\}a & -\nu ab & 4\left\{\dfrac{1}{3}\beta^2+\dfrac{1}{15}(1-\nu)\right\}a^2 \end{bmatrix} \tag{11.236}$$

$$[\mathbf{k}_{12}] = \begin{bmatrix} -\left\{(2\beta^2-\alpha^2)+\frac{2}{5}(7-2\nu)\right\} & \left\{\alpha^2-\frac{1}{5}(1+4\nu)\right\}b & -\left\{2\beta^2+\frac{1}{5}(1-\nu)\right\}a \\ \left\{\alpha^2-\frac{1}{5}(1+4\nu)\right\}b & \left\{\frac{2}{3}\alpha^2-\frac{4}{15}(1-\nu)\right\}b^2 & 0 \\ \left\{2\beta^2+\frac{1}{5}(1-\nu)\right\}a & 0 & \left\{\frac{2}{3}\beta^2-\frac{1}{15}(1-\nu)\right\}a^2 \end{bmatrix} \tag{11.237}$$

$$[\mathbf{k}_{13}] = \begin{bmatrix} -\left\{2(2\beta^2-\alpha^2)+\frac{2}{5}(7-2\nu)\right\} & \left\{\alpha^2-\frac{1}{5}(1-\nu)\right\}b & \left\{-\beta^2+\frac{1}{5}(1-\nu)\right\}a \\ \left\{-\alpha^2+\frac{1}{5}(1-\nu)\right\}b & \left\{\frac{1}{3}\alpha^2+\frac{1}{15}(1-\nu)\right\}b^2 & 0 \\ \left\{\beta^2-\frac{1}{5}(1-\nu)\right\}a & 0 & \left\{\frac{1}{3}\beta^2+\frac{1}{15}(1-\nu)\right\}a^2 \end{bmatrix} \tag{11.238}$$

$$[\mathbf{k}_{14}] = \begin{bmatrix} 2(\beta^2-2\alpha^2)-\frac{2}{5}(7-2\nu) & \left\{2\alpha^2+\frac{1}{5}(1-\nu)\right\}b & \left\{-\beta^2+\frac{1}{5}(1+4\nu)\right\}a \\ \left\{-2\alpha^2-\frac{1}{5}(1-\nu)\right\}b & \left\{\frac{2}{3}\alpha^2-\frac{1}{15}(1-\nu)\right\}b^2 & 0 \\ \left\{-\beta^2+\frac{1}{5}(1+4\nu)\right\}a & 0 & \left\{\frac{2}{3}\beta^2-\frac{4}{15}(1-\nu)\right\}a^2 \end{bmatrix} \tag{11.239}$$

where $\alpha$ and $\beta$ are the aspect ratios given as

$$\alpha = \frac{a}{b}, \quad \beta = \frac{b}{a} \tag{11.240}$$

The remaining submatrices can be found using

$$\begin{aligned} &[\mathbf{k}_{22}] = [\mathbf{I}_3]^T[\mathbf{k}_{11}][\mathbf{I}_3], \quad [\mathbf{k}_{23}] = [\mathbf{I}_3]^T[\mathbf{k}_{14}][\mathbf{I}_3], \quad [\mathbf{k}_{24}] = [\mathbf{I}_3]^T[\mathbf{k}_{23}][\mathbf{I}_3] \\ &[\mathbf{k}_{33}] = [\mathbf{I}_1]^T[\mathbf{k}_{11}][\mathbf{I}_1], \quad [\mathbf{k}_{34}] = [\mathbf{I}_1]^T[\mathbf{k}_{12}][\mathbf{I}_1] \\ &[\mathbf{k}_{44}] = [\mathbf{I}_2]^T[\mathbf{k}_{11}][\mathbf{I}_2] \end{aligned} \tag{11.241}$$

where

$$[\mathbf{I}_1] = \begin{bmatrix} -1 & 0 & 0 \\ 0 & 1 & 0 \\ 0 & 0 & 1 \end{bmatrix}, \quad [\mathbf{I}_2] = \begin{bmatrix} 1 & 0 & 0 \\ 0 & -1 & 0 \\ 0 & 0 & 1 \end{bmatrix}, \quad [\mathbf{I}_3] = \begin{bmatrix} 1 & 0 & 0 \\ 0 & 1 & 0 \\ 0 & 0 & -1 \end{bmatrix} \tag{11.242}$$

The above relationships are presented in Smith (1970).

8. *Triangular Plate Element*. The shape functions for a 9-DOF triangular element are given by Equation 11.137 as

$$w = [[\mathbf{N}_1] \quad [\mathbf{N}_2] \quad [\mathbf{N}_3]]\{\mathbf{d}\}$$

where

$$\{\mathbf{N}_1\}^T = \begin{Bmatrix} L_1 + L_1L_2(L_1 - L_2) + L_1L_3(L_1 - L_3) \\ c_3\left(L_1^2L_2 + \frac{1}{2}L_1L_2L_3\right) - c_2\left(L_1^2L_3 + \frac{1}{2}L_1L_2L_3\right) \\ b_3\left(L_1^2L_2 + \frac{1}{2}L_1L_2L_3\right) - b_2\left(L_1^2L_3 + \frac{1}{2}L_1L_2L_3\right) \end{Bmatrix}$$

$[\mathbf{N}_2]$ and $[\mathbf{N}_3]$ are obtained by cyclic permutations. The mass or inertia matrix for the triangular plate element is given by

$$[\mathbf{m}] = \iiint_{V^{(e)}} [\mathbf{N}]^T[\mathbf{N}]\rho dV = \iint_A \rho t[\mathbf{N}]^T[\mathbf{N}]dA$$
$$= \frac{\rho t A}{15{,}120}\begin{bmatrix} [\mathbf{m}_{11}] & [\mathbf{m}_{12}] & [\mathbf{m}_{13}] \\ & [\mathbf{m}_{22}] & [\mathbf{m}_{23}] \\ \text{sym.} & & [\mathbf{m}_{33}] \end{bmatrix} \quad (11.243)$$

where $A$ is the area of the triangle, $t$ the plate thickness and

$$[\mathbf{m}_{11}] = \begin{bmatrix} 2922 & 847(b_3 - b_2) & 847(c_3 - c_2) \\ & \left(225\left(b_2^2 + b_3^2\right) - 168b_2b_3\right) & (225(b_3c_3 + b_2c_2) - 84(b_2c_3 + b_3c_2)) \\ & \text{sym.} & \left(225\left(c_3^2 + c_2^2\right) - 168c_2c_3\right) \end{bmatrix} \quad (11.244)$$

$$[\mathbf{m}_{12}] = \begin{bmatrix} 1068 & (397b_1 - 643b_3) & (397c_1 - 643c_3) \\ (643b_3 - 397b_2) & \left(43b_1b_3 - 75b_1b_2 - 57b_3^2 + 43b_2b_3\right) & (43b_3c_1 - 75b_2c_1 - 57b_3c_3 + 43b_2c_3) \\ (643c_3 - 397c_2) & (43b_1c_3 - 57b_3c_3 - 75b_1c_2 + 43b_3c_2) & \left(43c_1c_3 - 75c_1c_2 - 57c_3^2 + 43c_2c_3\right) \end{bmatrix} \quad (11.245)$$

$$[\mathbf{m}_{13}] = \begin{bmatrix} 1068 & (643b_2 - 397b_1) & (643c_2 - 397c_1) \\ (397b_3 - 643b_2) & \left(43b_2b_3 - 75b_1b_3 - 57b_2^2 + 43b_1b_2\right) & (43b_3c_2 - 57b_2c_2 - 75b_3c_1 + 43b_2c_1) \\ (397c_3 - 643c_2) & (43b_2b_3 - 75b_1b_3 - 57b_2c_2 + 43b_1c_2) & \left(43c_2c_3 - 75c_1c_3 - 57c_2^2 + 43c_1c_2\right) \end{bmatrix} \quad (11.246)$$

$$[\mathbf{m}_{23}] = \begin{bmatrix} 1068 & (397b_2 - 643b_1) & (397c_2 - 643c_1) \\ (643b_1 - 397b_3) & \left(43b_1b_2 - 75b_2b_3 - 57b_1^2 + 43b_1b_3\right) & (43b_1c_2 - 75b_3c_2 - 57b_1c_1 + 43b_3c_1) \\ (643c_1 - 397c_3) & (43b_2c_1 + 57b_1c_1 - 75b_2c_3 + 43b_1c_3) & \left(43c_1c_2 - 75c_2c_3 - 57c_1^2 + 43c_1c_3\right) \end{bmatrix} \tag{11.247}$$

$[\mathbf{m}_{22}]$ and $[\mathbf{m}_{33}]$ can be obtained from $[\mathbf{m}_{11}]$ by cyclic permutation of the other subscripts in the order 1, 2, 3.

The stiffness matrix for this element is given by

$$[\mathbf{k}] = \frac{t^3}{12}\iint\limits_A [\mathbf{B}]^T[\mathbf{D}][\mathbf{B}]\,dA \tag{11.248}$$

and can be found in Cheung et al. (1968).

## 11.4 Element Assembly

Individual mass and stiffness matrices for different element types were developed in Section 11.3. However, when modeling a structure, many elements may be required and these elements need to be assembled into the overall global matrices. Recall that the global matrices were given as

$$[\mathbf{M}] = \sum_{e=1}^{N_e}[\mathbf{m}]^{(e)},\quad [\mathbf{K}] = \sum_{e=1}^{N_e}[\mathbf{k}]^{(e)},\quad [\mathbf{C}] = \sum_{e=1}^{N_e}[\mathbf{c}]^{(e)} \quad \text{and} \quad \{\mathbf{F}\} = \sum_{e=1}^{N_e}[\mathbf{f}]^{(e)}$$

When adding the element matrices, the elements need to be placed in the appropriate locations in the global matrices based on element connectivity. If elements are overlapping, the elements are simply added. In general, when using finite elements, there are two sets of node numbers. These are the local and global nodal values. In analyzing an engineering structure, the finite element mesh is generated and a global numbering system is attached to the model. In actual calculations, the global matrices can be generated by identifying the locations of the element matrices in the global matrices and adding them to existing values as the number of elements changes from 1 to $N_e$. Consider the assemblage of three one-dimensional elements as shown in Figure 11.24.

The connectivity for the above arrangement is given in Table 11.3.

The element stiffness matrices are given as

$$[\mathbf{k}]^{(1)} = \frac{A_1E_1}{L_1}\begin{matrix} 1 & 2 & \\ \left[\begin{matrix}1\\-1\end{matrix}\right. & \left.\begin{matrix}-1\\1\end{matrix}\right] & \begin{matrix}1\\2\end{matrix}\end{matrix},\quad [\mathbf{k}]^{(2)} = \frac{A_2E_2}{L_2}\begin{matrix} 2 & 3 & \\ \left[\begin{matrix}1\\-1\end{matrix}\right. & \left.\begin{matrix}-1\\1\end{matrix}\right] & \begin{matrix}2\\3\end{matrix}\end{matrix},\quad [\mathbf{k}]^{(3)} = \frac{A_3E_3}{L_3}\begin{matrix} 3 & 4 & \\ \left[\begin{matrix}1\\-1\end{matrix}\right. & \left.\begin{matrix}-1\\1\end{matrix}\right] & \begin{matrix}3\\4\end{matrix}\end{matrix} \tag{11.249}$$

Note that, for each of the elements, the position corresponding to the rows and columns of the global matrix is indicated. The global matrix is $4 \times 4$ and to be assembled from the

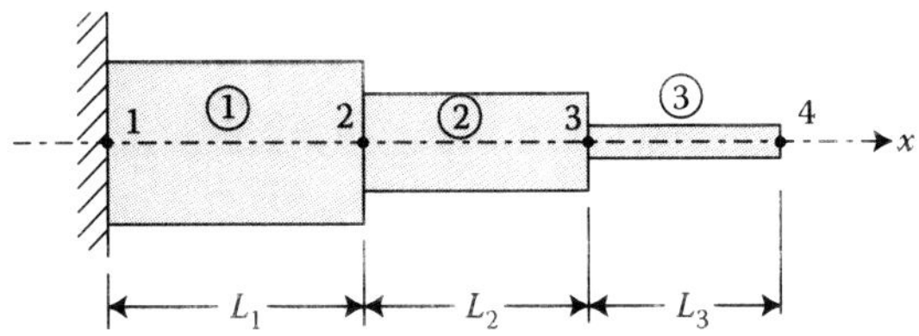


**FIGURE 11.24**
One-dimensional elements.

**TABLE 11.3**
Connectivity Matrix for One-Dimensional Axial Elements

| Element | Node 1 | Node 2 | |
|---|---|---|---|
| 1 | 1 | 2 | Global Nodes |
| 2 | 2 | 3 | |
| 3 | 3 | 4 | |

element matrices given above. After substituting, the first element matrix into the overall global matrix gives

$$[\mathbf{K}]=\begin{bmatrix} \dfrac{A_1E_1}{L_1} & -\dfrac{A_1E_1}{L_1} & 0 & 0 \\ -\dfrac{A_1E_1}{L_1} & \dfrac{A_1E_1}{L_1} & 0 & 0 \\ 0 & 0 & 0 & 0 \\ 0 & 0 & 0 & 0 \end{bmatrix}$$

Adding the second element matrix yields

$$[\mathbf{K}]=\begin{bmatrix} \dfrac{A_1E_1}{L_1} & -\dfrac{A_1E_1}{L_1} & 0 & 0 \\ -\dfrac{A_1E_1}{L_1} & \dfrac{A_1E_1}{L_1}+\dfrac{A_2E_2}{L_2} & -\dfrac{A_2E_2}{L_2} & 0 \\ 0 & -\dfrac{A_2E_2}{L_2} & \dfrac{A_2E_2}{L_2} & 0 \\ 0 & 0 & 0 & 0 \end{bmatrix}$$

Adding the third element matrix yields the final stiffness matrix in the global system as

$$[\mathbf{K}]=\begin{bmatrix} \dfrac{A_1E_1}{L_1} & -\dfrac{A_1E_1}{L_1} & 0 & 0 \\ -\dfrac{A_1E_1}{L_1} & \dfrac{A_1E_1}{L_1}+\dfrac{A_2E_2}{L_2} & -\dfrac{A_2E_2}{L_2} & 0 \\ 0 & -\dfrac{A_2E_2}{L_2} & \dfrac{A_2E_2}{L_2}+\dfrac{A_3E_3}{L_3} & -\dfrac{A_3E_3}{L_3} \\ 0 & 0 & -\dfrac{A_3E_3}{L_3} & \dfrac{A_3E_3}{L_3} \end{bmatrix} \quad (11.250)$$

The above could be written as $[\mathbf{K}]=\Sigma_{e=1}^{3}[\mathbf{K}^{(e)}]=[\mathbf{K}^{(1)}]+[\mathbf{K}^{(2)}]+[\mathbf{K}^{(3)}]$. In a similar manner, the global mass matrix can be obtained as $[\mathbf{M}]=\Sigma_{e=1}^{3}[\mathbf{M}^{(e)}]=[\mathbf{M}^{(1)}]+[\mathbf{M}^{(2)}]+[\mathbf{M}^{(3)}]$, where $[\mathbf{M}^{(1)}]$, $[\mathbf{M}^{(2)}]$, and $[\mathbf{M}^{(3)}]$ are formed using the element mass matrices $[\mathbf{m}^{(1)}]$, $[\mathbf{m}^{(2)}]$, and $[\mathbf{m}^{(3)}]$ expressed in the global coordinate system. This gives

$$[\mathbf{M}]=\frac{\rho AL}{6}\begin{bmatrix}2&1&0&0\\1&4&1&0\\0&1&4&1\\0&0&1&2\end{bmatrix} \tag{11.251}$$

where $\rho=\rho_1=\rho_2=\rho_3$, $A=A_1=A_2=A_3$, $L=L_1=L_2=L_3$.

Consider the two-element plane truss shown in Figure 11.25. For this structure, let us find the global stiffness matrix $[\mathbf{K}]$ and the consistent global mass matrix $[\mathbf{M}]$.

Information on the truss is summarized in Table 11.4

The element stiffness matrices for elements 1 and 2 can be written using Equation 11.161 and expanding to global coordinates as

$$[\mathbf{K}^{(1)}]=\frac{(A_1E_1/L_1)}{25}\begin{array}{c}\begin{matrix}1&2&3&4&5&6\end{matrix}\\\left[\begin{matrix}9&12&-9&-12&0&0\\12&16&-12&-16&0&0\\-9&-12&9&12&0&0\\-12&-16&12&16&0&0\\0&0&0&0&0&0\\0&0&0&0&0&0\end{matrix}\right]\end{array}\begin{matrix}1\\2\\3\\4\\5\\6\end{matrix}$$

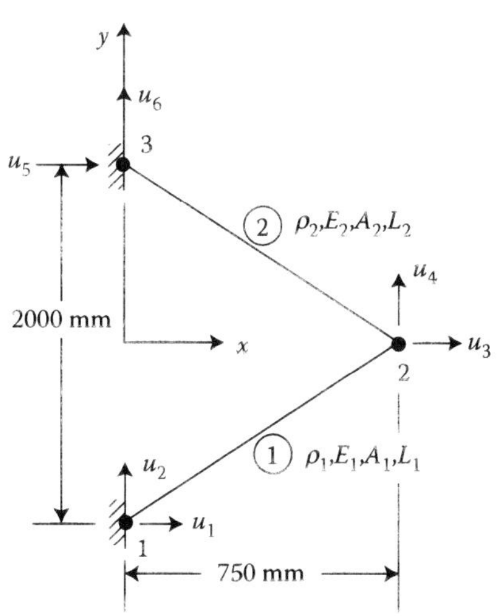


**FIGURE 11.25**
Two-element truss.

**TABLE 11.4**

Information for Truss of Figure 11.25

| Element | Node 1 | Node 2 | Area (mm²) | Length (mm) | $s = \sin\theta$ | $c = \cos\theta$ |
|---|---|---|---|---|---|---|
| 1 | 1 | 2 | 645 | 1250 | 4/5 | 3/5 |
| 2 | 2 | 3 | 645 | 1250 | 4/5 | −3/5 |

$$[\mathbf{K}^{(2)}] = \frac{(A_2E_2/L_2)}{25}\begin{array}{c}\begin{array}{cccccc}1 & 2 & 3 & 4 & 5 & 6\end{array}\\ \begin{bmatrix}0 & 0 & 0 & 0 & 0 & 0\\ 0 & 0 & 0 & 0 & 0 & 0\\ 0 & 0 & 9 & -12 & -9 & 12\\ 0 & 0 & -12 & 16 & 12 & -16\\ 0 & 0 & -9 & 12 & 9 & -12\\ 0 & 0 & 12 & -16 & -12 & 16\end{bmatrix}\end{array}\begin{array}{c}\\ 1\\ 2\\ 3\\ 4\\ 5\\ 6\end{array}$$

The global matrix then becomes

$$[\mathbf{K}] = \sum_{e=1}^{2}[\mathbf{K}^{(e)}] = [\mathbf{K}^{(1)}] + [\mathbf{K}^{(2)}]$$

$$= \frac{(AE/L)}{25}\begin{bmatrix}9 & 12 & -9 & -12 & 0 & 0\\ 12 & 16 & -12 & -16 & 0 & 0\\ -9 & -12 & 18 & 0 & -9 & 12\\ -12 & -16 & 0 & 32 & 12 & -16\\ 0 & 0 & -9 & 12 & 9 & -12\\ 0 & 0 & 12 & -16 & -12 & 16\end{bmatrix} \tag{11.252}$$

where we have set $A = A_1 = A_2$, $E = E_1 = E_2$, and $L = L_1 = L_2$. The typical consistent mass matrix for local axes of a typical member is

$$[\tilde{\mathbf{m}}] = \frac{\rho AL}{6}\begin{bmatrix}2 & 0 & 1 & 0\\ 0 & 2 & 0 & 1\\ 1 & 0 & 2 & 0\\ 0 & 1 & 0 & 2\end{bmatrix} \tag{11.253}$$

The terms in Equation 11.154 for the axial element are repeated for the $\tilde{y}$ direction since accelerations in that direction also give rise to inertia actions. For an element that is 4 × 4, the transformation or rotation matrix can be written as

$$[\mathbf{\Phi}] = \begin{bmatrix}\cos\theta & \sin\theta & 0 & 0\\ -\sin\theta & \cos\theta & 0 & 0\\ 0 & 0 & \cos\theta & \sin\theta\\ 0 & 0 & -\sin\theta & \cos\theta\end{bmatrix} \tag{11.254}$$

The transformation matrix (Equation 11.254) can be applied to the element mass matrix, Equation 11.253, as

$$[\mathbf{m}] = [\mathbf{\Phi}]^T[\tilde{\mathbf{m}}][\mathbf{\Phi}]$$

It is noted, however, that when the above equation is applied, the resulting matrix $[\mathbf{m}]$ is the same as the local mass matrix $[\tilde{\mathbf{m}}]$. Therefore, for a plane truss element, the consistent mass matrix is invariant due to a rotation of axes. Therefore, we have

$$[\mathbf{M}^{(1)}] = \frac{\rho_1 A_1 l_1}{6}\begin{array}{c} \begin{array}{cccccc} 1 & 2 & 3 & 4 & 5 & 6 \end{array} \\ \begin{bmatrix} 2 & 0 & 1 & 0 & 0 & 0 \\ 0 & 2 & 0 & 1 & 0 & 0 \\ 1 & 0 & 2 & 0 & 0 & 0 \\ 0 & 1 & 0 & 2 & 0 & 0 \\ 0 & 0 & 0 & 0 & 0 & 0 \\ 0 & 0 & 0 & 0 & 0 & 0 \end{bmatrix} \end{array} \begin{array}{c} \\ 1 \\ 2 \\ 3 \\ 4 \\ 5 \\ 6 \end{array}$$

$$[\mathbf{M}^{(2)}] = \frac{\rho_2 A_2 l_2}{6}\begin{array}{c} \begin{array}{cccccc} 1 & 2 & 3 & 4 & 5 & 6 \end{array} \\ \begin{bmatrix} 0 & 0 & 0 & 0 & 0 & 0 \\ 0 & 0 & 0 & 0 & 0 & 0 \\ 0 & 0 & 2 & 0 & 1 & 0 \\ 0 & 0 & 0 & 2 & 0 & 1 \\ 0 & 0 & 1 & 0 & 2 & 0 \\ 0 & 0 & 0 & 1 & 0 & 2 \end{bmatrix} \end{array} \begin{array}{c} \\ 1 \\ 2 \\ 3 \\ 4 \\ 5 \\ 6 \end{array}$$

The global mass matrix is given as

$$[\mathbf{M}] = \sum_{e=1}^{2}[\mathbf{M}^{(e)}] = [\mathbf{M}^{(1)}] + [\mathbf{M}^{(2)}]$$
$$= \frac{\rho AL}{6}\begin{bmatrix} 2 & 0 & 1 & 0 & 0 & 0 \\ 0 & 2 & 0 & 1 & 0 & 0 \\ 1 & 0 & 4 & 0 & 1 & 0 \\ 0 & 1 & 0 & 4 & 0 & 1 \\ 0 & 0 & 1 & 0 & 2 & 0 \\ 0 & 0 & 0 & 1 & 0 & 2 \end{bmatrix} \tag{11.255}$$

where $\rho = \rho_1 = \rho_2$, $A = A_1 = A_2$, and $L = L_1 = L_2$.

Consider now a mesh of three elements as shown in Figure 11.26. The element node numbers along with the global node numbers are shown in the figure.

For convenience, consider one degree of freedom at each node. The connectivity for the above arrangement is given in Table 11.5.

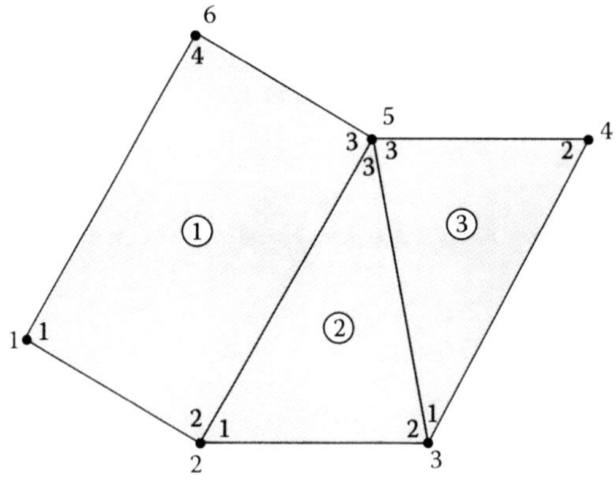


**FIGURE 11.26**
A mesh of three elements.

**TABLE 11.5**
Connectivity Matrix for Three Element Mesh

| Element | Nodes | | | |
|---|---|---|---|---|
| 1 | 1 | 2 | 5 | 6 |
| 2 | 2 | 3 | 5 | |
| 3 | 3 | 4 | 5 | |

We see that element 1 matrix is $4 \times 4$ and both elements 2 and 3 are $3 \times 3$ whereas the global matrix is $6 \times 6$. The element stiffness matrices are given as

$$[\mathbf{k}]^{(1)} = \begin{bmatrix} k_{11}^{(1)} & k_{12}^{(1)} & k_{13}^{(1)} & k_{14}^{(1)} \\ k_{21}^{(1)} & k_{22}^{(1)} & k_{23}^{(1)} & k_{24}^{(1)} \\ k_{31}^{(1)} & k_{32}^{(1)} & k_{33}^{(1)} & k_{34}^{(1)} \\ k_{41}^{(1)} & k_{42}^{(1)} & k_{43}^{(1)} & k_{44}^{(1)} \end{bmatrix} \tag{11.256}$$

$$[\mathbf{k}]^{(2)} = \begin{bmatrix} k_{11}^{(2)} & k_{12}^{(2)} & k_{13}^{(2)} \\ k_{21}^{(2)} & k_{22}^{(2)} & k_{23}^{(2)} \\ k_{31}^{(2)} & k_{32}^{(2)} & k_{33}^{(2)} \end{bmatrix} \tag{11.257}$$

$$[\mathbf{k}]^{(3)} = \begin{bmatrix} k_{11}^{(3)} & k_{12}^{(3)} & k_{13}^{(3)} \\ k_{21}^{(3)} & k_{22}^{(3)} & k_{23}^{(3)} \\ k_{31}^{(3)} & k_{32}^{(3)} & k_{33}^{(3)} \end{bmatrix} \tag{11.258}$$

Equations 11.256–11.258 are first expanded into $6 \times 6$ matrices, then summed $[\mathbf{K}] = \Sigma_{e=1}^{3}[\mathbf{K}^{(e)}] = [\mathbf{K}^{(1)}] + [\mathbf{K}^{(2)}] + [\mathbf{K}^{(3)}]$ giving the global stiffness matrix as

$$
[\mathbf{K}] = \begin{bmatrix}
k_{11}^{(1)} & k_{12}^{(1)} & 0 & 0 & k_{13}^{(1)} & k_{14}^{(1)} \\
k_{21}^{(1)} & k_{22}^{(1)} + k_{11}^{(2)} & k_{12}^{(2)} & 0 & k_{23}^{(1)} + k_{13}^{(2)} & k_{24}^{(1)} \\
0 & k_{21}^{(2)} & k_{22}^{(2)} + k_{11}^{(3)} & k_{12}^{(3)} & k_{23}^{(2)} + k_{13}^{(3)} & 0 \\
0 & 0 & k_{21}^{(3)} & k_{22}^{(3)} & k_{23}^{(3)} & 0 \\
k_{31}^{(1)} & k_{32}^{(1)} + k_{31}^{(2)} & k_{32}^{(2)} + k_{31}^{(3)} & k_{32}^{(3)} & k_{33}^{(1)} + k_{33}^{(2)} + k_{33}^{(3)} & k_{34}^{(1)} \\
k_{41}^{(1)} & k_{42}^{(1)} & 0 & 0 & k_{43}^{(1)} & k_{44}^{(1)}
\end{bmatrix} \tag{11.259}
$$

The other matrices such as the mass matrix and load vector can be developed in the same manner. It is noted that the stiffness matrix is symmetric and that the sum of the elements of each column and row is zero, which does not happen with stiffness matrices that involve rotations.

### 11.4.1 Boundary Conditions

So far, we have not considered boundary conditions. It is noted that the complete structure is capable of undergoing rigid-body motions and hence, the stiffness matrix will become singular and therefore not invertible. Therefore, it is necessary to incorporate boundary conditions to remove stiffness matrix singularity. The introduction of boundary conditions prevents rigid-body motion rendering the stiffness matrix nonsingular. There are two types of boundary conditions: displacement boundary conditions which are termed *essential* and force boundary conditions which are termed *natural*. Displacement boundary conditions can be written as

$$u_i = u_i^* \quad i = 1,2,3$$

where the star indicates prescribed values for the displacement components. Usually, the structure is supported so that the displacements are zero at a number of joints or nodes. A simple method of incorporating zero displacement boundary conditions is to eliminate the corresponding rows and columns in the mass **[M]** and stiffness **[K]** matrices and the load vector **{F}**.

Consider that in the two-element truss given in Figure 11.25 the external loads applied at node two were given as $F_{2x} = 2000\ N$ and $F_{2y} = -4500\ N$. The equations of motion of the two-element structure are then given as

$$[\mathbf{M}]\{\ddot{\mathbf{u}}\} + [\mathbf{K}]\{\mathbf{u}\} = \{\mathbf{F}\}$$

or

$$
\frac{\rho AL}{6}\begin{bmatrix}
2 & 0 & 1 & 0 & 0 & 0 \\
0 & 2 & 0 & 1 & 0 & 0 \\
1 & 0 & 4 & 0 & 1 & 0 \\
0 & 1 & 0 & 4 & 0 & 1 \\
0 & 0 & 1 & 0 & 2 & 0 \\
0 & 0 & 0 & 1 & 0 & 2
\end{bmatrix}
\begin{Bmatrix} \ddot{u}_1 \\ \ddot{v}_1 \\ \ddot{u}_2 \\ \ddot{v}_2 \\ \ddot{u}_3 \\ \ddot{v}_3 \end{Bmatrix}
+ \frac{(AE/L)}{25}\begin{bmatrix}
9 & 12 & -9 & -12 & 0 & 0 \\
12 & 16 & -12 & -16 & 0 & 0 \\
-9 & -12 & 18 & 0 & -9 & 12 \\
-12 & -16 & 0 & 32 & 12 & -16 \\
0 & 0 & -9 & 12 & 9 & -12 \\
0 & 0 & 12 & -16 & -12 & 16
\end{bmatrix}
\begin{Bmatrix} u_1 \\ v_1 \\ u_2 \\ v_2 \\ u_3 \\ v_3 \end{Bmatrix}
= \begin{Bmatrix} 0 \\ 0 \\ 2000 \\ -4500 \\ 0 \\ 0 \end{Bmatrix}
$$

Taking into account that the two-element structure is fixed at nodes 1 and 3 and incorporating the corresponding boundary conditions into the above equation yields

$$\frac{\rho AL}{6}\begin{bmatrix} 4 & 0 \\ 0 & 4 \end{bmatrix}\begin{Bmatrix} \ddot{u}_2 \\ \ddot{v}_2 \end{Bmatrix} + \frac{(AE/L)}{25}\begin{bmatrix} 18 & 0 \\ 0 & 32 \end{bmatrix}\begin{Bmatrix} u_2 \\ v_2 \end{Bmatrix} = \begin{Bmatrix} 2000 \\ -4500 \end{Bmatrix} \tag{11.260}$$

## 11.5 Free Response of Finite Element Systems (Eigenvalue Analysis)

The response of an undamped (or conservative) system with arbitrary initial conditions can be described by the global finite element system equation

$$[\mathbf{M}]\{\ddot{\mathbf{u}}\} + [\mathbf{K}]\{\mathbf{u}\} = \{\mathbf{F}\} \tag{11.261}$$

where $[\mathbf{M}]$ and $[\mathbf{K}]$ are real symmetric $n \times n$ mass (or inertia) and stiffness matrix, respectively, $\{\mathbf{u}\}$ is a vector of displacements at all the nodes, and $\{\mathbf{F}\}$ is a vector of all the nodal forces. The mass matrix $[\mathbf{M}]$ is positive definite. The usual problem is to determine the unknown response $\{\mathbf{u}(t)\}$ one method of solution of Equation 11.261 is to use the so-called direct integration method which will be discussed in Section 11.6. However, a particular solution of Equation 11.261 is obtained by setting $\{\mathbf{F}\} = \{\mathbf{0}\}$ such that the system is free of external forces and becomes

$$[\mathbf{M}]\{\ddot{\mathbf{u}}\} + [\mathbf{K}]\{\mathbf{u}\} = \{\mathbf{0}\} \tag{11.262}$$

Since the motion of a discrete system can be harmonic under certain conditions, the displacement vector $\{\mathbf{u}\}$ can be expressed as

$$\{\mathbf{u}\} = \{\boldsymbol{\Phi}\}\cos(\omega t - \varphi) \tag{11.263}$$

where $\{\boldsymbol{\Phi}\}$ is a vector containing the amplitude of nodal displacements. $\omega$ is the vibration frequency and $\varphi$ is a phase angle. Differentiating Equation 11.263 twice with respect to time and substituting into Equation 11.262 yields

$$-\omega^2[\mathbf{M}]\{\boldsymbol{\Phi}\}\cos(\omega t - \varphi) + [\mathbf{K}]\{\boldsymbol{\Phi}\}\cos(\omega t - \varphi) = \{\mathbf{0}\} \tag{11.264}$$

which must be valid for all values of $t$ and can be written as

$$([\mathbf{K}] - \omega^2[\mathbf{M}])\{\boldsymbol{\Phi}\} = \{\mathbf{0}\} \tag{11.265}$$

or

$$([\mathbf{K}] - \lambda[\mathbf{M}])\{\boldsymbol{\Phi}\} = \{\mathbf{0}\} \tag{11.266}$$

where we set $\omega^2 = \lambda$. Equations 11.265 and 11.266 represents a set of $n$ linear homogeneous algebraic equations, where $\lambda$ is a parameter of the system. The problem in finding values of $\lambda$ in which nontrivial solutions of $\{\mathbf{\Phi}\}$ exist is known as the *algebraic eigenvalue problem* or as the *characteristic value problem*. A nonzero solution for $\{\mathbf{\Phi}\}$ only exists if the determinate of the dynamic matrix vanishes, that is

$$\det([\mathbf{K}] - \omega^2[\mathbf{M}]) = \det([\mathbf{K}] - \lambda[\mathbf{M}]) = \left|[\mathbf{K}] - \lambda[\mathbf{M}]\right| = 0 \tag{11.267}$$

Equation 11.267 can be expanded leading to a polynomial of degree $n$ in $\lambda_i = \omega_i^2$ for a system of $n$ DOF. The polynomial will have at most $n$ roots, $\lambda_1, \lambda_2, \ldots, \lambda_n$, which are called eigenvalues and related to the natural frequencies of the system by $\lambda = \omega^2$. It can be shown that all $n$ roots are real and either positive or zero when the mass and stiffness matrices are symmetric and positive definite. Zero eigenvalues are possible if rigid-body movements are allowed. If the eigenvalues are not distinct, then the eigenvalue problem is said to have multiple eigenvalues. If there are $m$ coincident eigenvalues, the eigenvalue is said to be of multiplicity $m$. If all the eigenvalues are distinct, then substituting an eigenvalue $\lambda_i$ into Equation 11.266 leads to a set of nontrivial algebraic equations given as

$$([\mathbf{K}] - \lambda_i[\mathbf{M}])\{\mathbf{\Phi}_i\} = \{\mathbf{0}\} \tag{11.268}$$

where the solution $\{\mathbf{\Phi}_i\}$ is called the $i$th eigenvector or modal vector, which is associated with the $i$th vibration frequency. The eigenvector $\{\mathbf{\Phi}_i\}$ gives the shape of the structure when vibrating at frequency $\lambda_i$. To solve Equation 11.268, a supplementary equation is needed to define the vector uniquely since an eigenvector is arbitrary to the extent that a scalar multiplier is also a solution. That is, if an eigenpair $(\lambda_i, \mathbf{\Phi}_i)$ satisfies

$$[\mathbf{K}]\{\mathbf{\Phi}_i\} = \lambda_i[\mathbf{M}]\{\mathbf{\Phi}_i\} \tag{11.269}$$

then the relation

$$[\mathbf{K}]\{\alpha\mathbf{\Phi}_i\} = \lambda_i[\mathbf{M}]\{\alpha\mathbf{\Phi}_i\} \tag{11.270}$$

is also satisfied where $\alpha$ is a nonzero constant. The supplementary equation used to solve Equation 11.268 is a convenient normalization condition, where the largest element in each mode is assigned a value of unity and the remaining elements are adjusted accordingly in the normalization. Thus, we have

$$\{\mathbf{\Phi}_i\}^T[\mathbf{M}]\{\mathbf{\Phi}_i\} = 1 \tag{11.271}$$

which fixes the lengths of the eigenvectors. Premultiplying Equation 11.269 by $\{\mathbf{\Phi}_j\}^T$ and using Equation 11.271, the eigenvectors satisfy the following orthogonality conditions:

$$\{\mathbf{\Phi}_i\}^T[\mathbf{K}]\{\mathbf{\Phi}_j\} = \lambda_i\delta_{ij} \tag{11.272}$$

$$\{\mathbf{\Phi}_i\}^T[\mathbf{M}]\{\mathbf{\Phi}_j\} = \delta_{ij} \tag{11.273}$$

We say that the eigenvectors are $[\mathbf{K}]$-orthogonal and $[\mathbf{M}]$-orthogonal, respectively.

**EXAMPLE 11.1**

Determine the eigenvalues and eigenvectors for the system shown in Figure 11.27 with $k_1 = k_4 = 2$, $k_2 = k_3 = 1$, and $m_1 = m_2 = m_3 = 1$. The mass and stiffness matrices are

$$[\mathbf{M}] = \begin{bmatrix} m_1 & 0 & 0 \\ 0 & m_2 & 0 \\ 0 & 0 & m_3 \end{bmatrix}, \quad [\mathbf{K}] = \begin{bmatrix} (k_1 + k_2) & -k_2 & 0 \\ -k_2 & (k_2 + k_3) & -k_3 \\ 0 & -k_3 & (k_3 + k_4) \end{bmatrix}$$

Equation 11.268 becomes using the above values for $m$ and $k$

$$\begin{bmatrix} (3-\lambda) & -1 & 0 \\ -1 & (2-\lambda) & -1 \\ 0 & -1 & (3-\lambda) \end{bmatrix} \begin{Bmatrix} \Phi_1 \\ \Phi_2 \\ \Phi_3 \end{Bmatrix} = \begin{Bmatrix} 0 \\ 0 \\ 0 \end{Bmatrix} \tag{11.274}$$

Setting the determinate of Equation 11.274 to zero gives

$$(3-\lambda)[(2-\lambda)(3-\lambda)] - 2(3-\lambda) = 0$$

or

$$(\lambda - 1)(\lambda - 3)(\lambda - 4) = 0$$

yielding the eigenvalues as

$$\lambda = 1, 3, \text{ and } 4$$

Substituting the eigenvalues back into Equation 11.274 yields the eigenvectors. Thus, for $\lambda = 1$, we find from the first and third equations that

$$2\Phi_1 - \Phi_2 = 0$$
$$-\Phi_2 + 2\Phi_3 = 0$$

which yields

$$\Phi_2 = 2\Phi_3 \quad \text{and} \quad \Phi_1 = \Phi_3$$

Therefore, for $\lambda = 1$, the eigenvector can be written as

$$\{\mathbf{\Phi}_1\} = \begin{Bmatrix} 1 \\ 2 \\ 1 \end{Bmatrix} \Phi_3$$

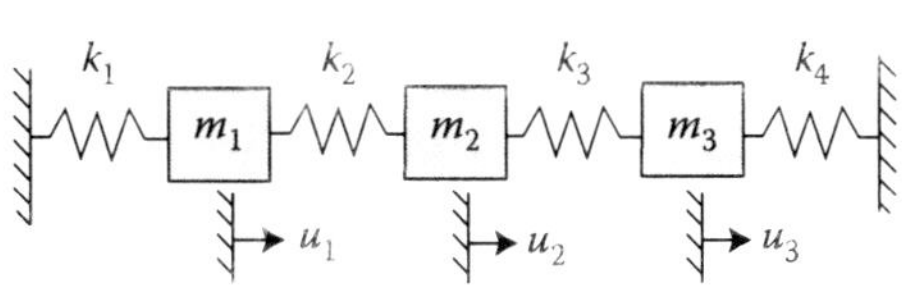


**FIGURE 11.27**
Three-degree-of-freedom spring–mass system.

In many cases, the modal vector is normalized such that Equation 11.273 is satisfied. Thus, we have

$$6\Phi_3^2 = 1$$

so that $\Phi_3 = 1/\sqrt{6}$ and the eigenvector becomes

$$\{\Phi_1\} = \begin{Bmatrix} 1/\sqrt{6} \\ 2/\sqrt{6} \\ 1/\sqrt{6} \end{Bmatrix}$$

In the same manner, the other two eigenvectors corresponding to the other two eigenvalues are

$$\lambda = 3, \quad \{\Phi_2\} = \begin{Bmatrix} -1 \\ 0 \\ 1 \end{Bmatrix} \quad \text{or} \quad \begin{Bmatrix} -1/\sqrt{2} \\ 0 \\ 1/\sqrt{2} \end{Bmatrix}$$

$$\lambda = 4, \quad \{\Phi_3\} = \begin{Bmatrix} -1 \\ 1 \\ -1 \end{Bmatrix} \quad \text{or} \quad \begin{Bmatrix} -1/\sqrt{3} \\ 1/\sqrt{3} \\ -1/\sqrt{3} \end{Bmatrix}$$

**EXAMPLE 11.2**

Consider the stepped axial bar shown in Figure 11.28 with the following values: $A_1 = 2\ \text{in}^2$, $A_2 = 1\ \text{in}^2$, $A_3 = 0.5\ \text{in}^2$, $E_i = 30\times10^6\ \text{psi}$, $\rho_i = 0.283\ \text{lb/in}^3$, $i = 1,2,3$, $L_1 = 15\ \text{in}$, $L_2 = 10\ \text{in}$, $L_3 = 5\ \text{in}$. Determine the natural frequencies and mode shapes of the axial bar.

**Solution**

Based on the degrees of freedom $u_2$, $u_3$, and $u_4$, the eigenvalue problem becomes

$$E\begin{bmatrix} \dfrac{A_1}{L_1} + \dfrac{A_2}{L_2} & -\dfrac{A_2}{L_2} & 0 \\ -\dfrac{A_2}{L_2} & \dfrac{A_2}{L_2} + \dfrac{A_3}{L_3} & -\dfrac{A_3}{L_3} \\ 0 & -\dfrac{A_3}{L_3} & \dfrac{A_3}{L_3} \end{bmatrix}\begin{Bmatrix} \Phi_1 \\ \Phi_2 \\ \Phi_3 \end{Bmatrix} = \lambda\frac{\rho}{6}\begin{bmatrix} 2A_1L_1 + 2A_2L_2 & A_2L_2 & 0 \\ A_2L_2 & 2A_2L_2 + 2A_3L_3 & A_3L_3 \\ 0 & A_3L_3 & 2A_3L_3 \end{bmatrix}\begin{Bmatrix} \Phi_1 \\ \Phi_2 \\ \Phi_3 \end{Bmatrix}$$

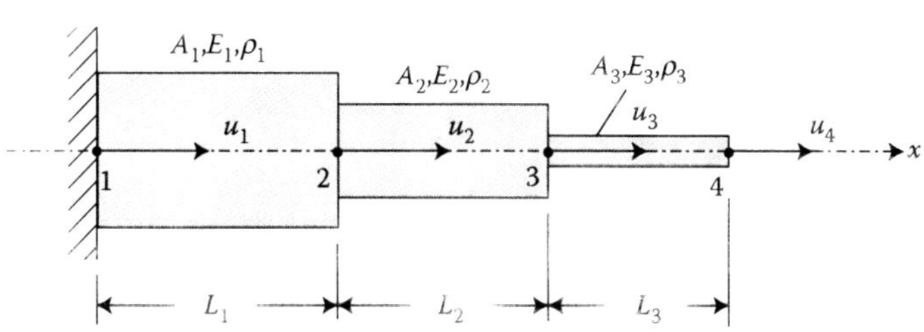


**FIGURE 11.28**
Stepped axial bar.

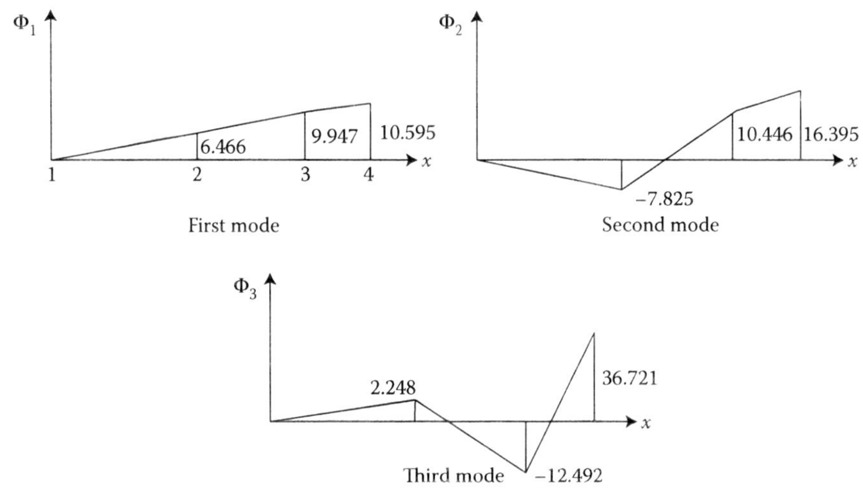


**FIGURE 11.29**
Mode shapes of stepped shaft.

Substituting the given values gives

$$30\times10^{6}\begin{bmatrix}0.2333 & -0.1 & 0\\ -0.1 & 0.2 & -0.1\\ 0 & -0.1 & 0.1\end{bmatrix}\begin{Bmatrix}\Phi_1\\ \Phi_2\\ \Phi_3\end{Bmatrix}=1.2207\times10^{-4}\lambda\begin{bmatrix}80 & 10 & 0\\ 10 & 25 & 2.5\\ 0 & 2.5 & 5.0\end{bmatrix}\begin{Bmatrix}\Phi_1\\ \Phi_2\\ \Phi_3\end{Bmatrix}$$

The eigenvalues can be determined using MATLAB. The eigenvalues are

$$\lambda_1=2.0480\times10^{8}$$
$$\lambda_2=1.3526\times10^{9}$$
$$\lambda_3=7.9376\times10^{9}$$

Since $\lambda=\omega^2$ and $\omega=2\pi f$, the frequencies in hertz are given as

$$f_1=2278\ \text{Hz}$$
$$f_2=5853\ \text{Hz}$$
$$f_3=14{,}180\ \text{Hz}$$

The corresponding eigenvectors are illustrated in Figure 11.29 and are expressed as

$$\{\boldsymbol{\Phi}_1\}^T=\{6.4657\quad 9.9466\quad 10.5954\}$$
$$\{\boldsymbol{\Phi}_2\}^T=\{-7.8252\quad 10.4461\quad 16.3952\}$$
$$\{\boldsymbol{\Phi}_3\}^T=\{2.2479\quad -12.4917\quad 36.7206\}$$

**EXAMPLE 11.3**

Consider the plane pin-jointed truss shown in Figure 11.30, which has only three members. Determine the natural frequencies and mode shapes for the plane pin-jointed truss. Take $A_1=A_2=A_3=6.45\ \text{mm}^2$, $E=207$ GPa, and $\rho=7850\ \text{kf/m}^3$.

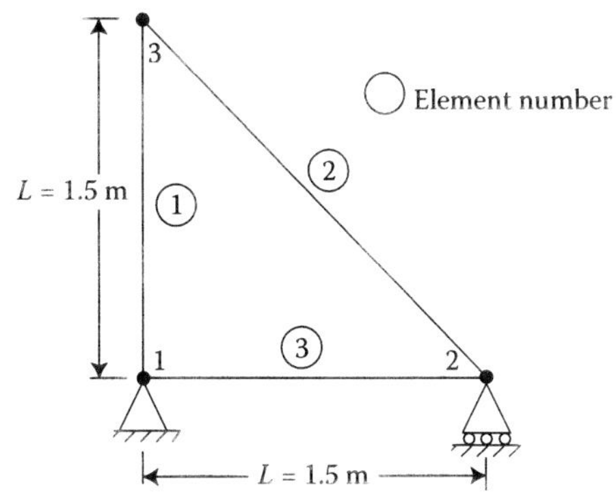


**FIGURE 11.30**
Three member pin-jointed truss.

**Solution**

The element mass and stiffness matrices are given by Equations 11.160 and 11.161. Substituting the proper given numerical values and following the rules of the direct method of assembling the system stiffness and mass matrix, we obtain

$$[\mathbf{K}] = 10^9 \begin{bmatrix} 1.2052 & -0.3148 & 0.3148 \\ -0.3148 & 0.3148 & -0.3148 \\ 0.3148 & -0.3148 & 1.2052 \end{bmatrix}$$

$$[\mathbf{M}] = \begin{bmatrix} 61.1378 & 17.9068 & 0 \\ 17.9068 & 61.1378 & 0 \\ 0 & 0 & 61.1378 \end{bmatrix}$$

Substituting the above mass and stiffness matrices into Equation 11.266 yields for the eigenvalues

$$\lambda_1 = 2.4271 \times 10^6$$
$$\lambda_2 = 1.5028 \times 10^7$$
$$\lambda_3 = 3.2751 \times 10^7$$

and the natural frequencies as

$$f_1 = 1558\,\text{Hz}$$
$$f_2 = 3877\,\text{Hz}$$
$$f_3 = 5723\,\text{Hz}$$

The corresponding eigenvectors are

$$\{\Phi_1\}^T = \{-0.03101 \quad -0.11291 \quad -0.02439\}$$
$$\{\Phi_2\}^T = \{-0.07636 \quad 0.01912 \quad 0.10495\}$$
$$\{\Phi_3\}^T = \{0.10535 \quad -0.06911 \quad 0.06889\}$$

### 11.5.1 Orthogonality for Eigenvectors of Symmetric Matrices

Consider $\{\boldsymbol{\Phi}_i\}$ and $\{\boldsymbol{\Phi}_j\}$ as two eigenvectors of symmetric matrices $[\mathbf{K}]$ and $[\mathbf{M}]$ corresponding to the eigenvalues $\lambda_i$ and $\lambda_j$ which were obtained from the expressions

$$[\mathbf{K}]\{\boldsymbol{\Phi}_i\} = \lambda_i[\mathbf{M}]\{\boldsymbol{\Phi}_i\} \tag{11.275}$$

$$[\mathbf{K}]\{\Phi_j\} = \lambda_j[\mathbf{M}]\{\Phi_j\} \tag{11.276}$$

Premultiplication of Equation 11.275 by $\{\boldsymbol{\Phi}_j\}^T$ and postmultiplication of the transpose of Equation 11.276 by $\{\boldsymbol{\Phi}_i\}$ gives

$$\{\boldsymbol{\Phi}_j\}^T[\mathbf{K}]\{\boldsymbol{\Phi}_i\} = \lambda_i\{\boldsymbol{\Phi}_j\}^T[\mathbf{M}]\{\boldsymbol{\Phi}_i\} \tag{11.277}$$

$$\{\boldsymbol{\Phi}_j\}^T[\mathbf{K}]\{\boldsymbol{\Phi}_i\} = \lambda_j\{\boldsymbol{\Phi}_j\}^T[\mathbf{M}]\{\boldsymbol{\Phi}_i\} \tag{11.278}$$

The left-hand sides of Equations 11.277 and 11.278 are equal, thus, subtracting the second equation from the first yields

$$(\lambda_i - \lambda_j)\{\boldsymbol{\Phi}_i\}^T[\mathbf{M}]\{\boldsymbol{\Phi}_j\} = 0 \tag{11.279}$$

Since $\lambda_i \neq \lambda_j$, we have that

$$\{\boldsymbol{\Phi}_j\}^T[\mathbf{M}]\{\boldsymbol{\Phi}_j\} = 0 \tag{11.280}$$

Substituting Equation 11.280 into Equations 11.277 and 11.278 gives

$$\{\boldsymbol{\Phi}_j\}^T[\mathbf{K}]\{\boldsymbol{\Phi}_j\} = 0 \tag{11.281}$$

Equations 11.280 and 11.281 represent orthogonality relations for the eigenvectors $\{\boldsymbol{\Phi}_i\}$ and $\{\boldsymbol{\Phi}_j\}$. Note that the eigenvectors are orthogonal with respect to the mass matrix $[\mathbf{M}]$ and are also orthogonal with respect to the stiffness matrix $[\mathbf{K}]$.

### 11.5.2 Rayleigh Quotient

This method has been already discussed in an earlier chapter for single-degree-of-freedom systems. Rayleigh's quotient is an important quantity in the theory and solution of eigenvalue problems. If our eigenvalue problem is expressed as

$$[\mathbf{K}]\{\boldsymbol{\Phi}\} = \omega^2[\mathbf{M}]\{\boldsymbol{\Phi}\}$$

and we premultiply by $\{\boldsymbol{\Phi}\}^T$ we obtain

$$\{\boldsymbol{\Phi}\}^T[\mathbf{K}]\{\boldsymbol{\Phi}\} = \omega^2\{\boldsymbol{\Phi}\}^T[\mathbf{M}]\{\boldsymbol{\Phi}\}$$

The mass matrix $[\mathbf{M}]$ is positive definite, thus ensuring that $\{\boldsymbol{\Phi}\}^T[\mathbf{M}]\{\boldsymbol{\Phi}\}$ is nonzero so that we can solve for $\omega^2$ giving

$$\omega^2 = \frac{\{\mathbf{\Phi}\}^T[\mathbf{K}]\{\mathbf{\Phi}\}}{\{\mathbf{\Phi}\}^T[\mathbf{M}]\{\mathbf{\Phi}\}} \tag{11.282}$$

It is noted that as $\{\mathbf{\Phi}_i\}$ becomes an increasingly good eigenvector estimate, then $\omega_i^2$ becomes an increasingly good eigenvalue estimate. The right-hand side of Equation 11.282 is known as *Rayleigh's quotient*. Let

$$\{\mathbf{\Phi}\} = A_1\{\mathbf{\Phi}_1\} + A_2\{\mathbf{\Phi}_2\} + \cdots$$

represent the arbitrary vector in terms of the normal modes of the system $\{\mathbf{\Phi}_1\}$, $\{\mathbf{\Phi}_2\}$, … , then we have

$$\{\mathbf{\Phi}\}^T[\mathbf{K}]\{\mathbf{\Phi}\} = A_1^2\{\mathbf{\Phi}_1\}^T[\mathbf{K}]\{\mathbf{\Phi}_1\} + A_2^2\{\mathbf{\Phi}_2\}^T[\mathbf{K}]\{\mathbf{\Phi}_2\} + \cdots$$

and

$$\{\mathbf{\Phi}\}^T[\mathbf{M}]\{\mathbf{\Phi}\} = A_1^2\{\mathbf{\Phi}_1\}^T[\mathbf{M}]\{\mathbf{\Phi}_1\} + A_2^2\{\mathbf{\Phi}_2\}^T[\mathbf{M}]\{\mathbf{\Phi}_2\} + \cdots$$

since $\{\mathbf{\Phi}_i\}^T[\mathbf{K}]\{\mathbf{\Phi}_j\} = 0$ and $\{\mathbf{\Phi}_i\}^T[\mathbf{M}]\{\mathbf{\Phi}_j\} = 0$. Thus, we have

$$\omega^2 = \left[\frac{A_1^2\{\mathbf{\Phi}_1\}^T[\mathbf{K}]\{\mathbf{\Phi}_1\} + A_2^2\{\mathbf{\Phi}_2\}^T[\mathbf{K}]\{\mathbf{\Phi}_2\} + \cdots}{A_1^2\{\mathbf{\Phi}_1\}^T[\mathbf{M}]\{\mathbf{\Phi}_1\} + A_2^2\{\mathbf{\Phi}_2\}^T[\mathbf{M}]\{\mathbf{\Phi}_2\} + \cdots}\right] \tag{11.283}$$

or if the modes are normalized as $\{\mathbf{\Phi}_i\}^T[\mathbf{K}]\{\mathbf{\Phi}_j\} = \omega_i^2\delta_{ij}$ and $\{\mathbf{\Phi}_i\}^T[\mathrm{M}]\{\mathbf{\Phi}_j\} = \delta_{ij}$ then Equation 11.283 becomes

$$\omega^2 = \left[\frac{A_1^2\omega_1^2 + A_2^2\omega_2^2 + \cdots}{A_1^2 + A_2^2 + \cdots}\right] \tag{11.284}$$

### 11.5.3 Reduction to Standard Form

Many textbooks such as those written by Wilkinson (1965), Bathe and Wilson (1976), and Jennings (1977) and many efficient computer subroutine programs consider the eigenproblem in the form

$$([\mathbf{A}] - \lambda[\mathbf{I}])\{\mathbf{\Phi}\} = \{\mathbf{0}\} \tag{11.285}$$

where $[\mathbf{A}]$ is a symmetric matrix and $[\mathbf{I}]$ is a unit diagonal matrix. Equation 11.285 is referred to as the *standard, symmetric form*. A simple way of reducing Equation 11.266 to the standard form is to first express the mass matrix as

$$[\mathbf{M}] = [\mathbf{L}][\mathbf{L}]^T \tag{11.286}$$

where **[L]** is a unique lower triangular matrix with positive diagonal elements. Equation 11.286 is known as the Cholesky decomposition of **[M]** (Merirovitch 1980) and the general formula factorization is given as

$$L_{jj} = \sqrt{M_{jj} - \sum_{k=1}^{j-1} L_{jk}^2}$$
$$L_{ij} = \frac{1}{L_{jj}}\left(M_{ij} - \sum_{k=1}^{j-1} L_{ik}L_{jk}\right), \quad \text{for} \quad i > j \tag{11.287}$$

Substituting Equation 11.286 into 11.266 gives

$$([\mathbf{K}] - \lambda[\mathbf{L}][\mathbf{L}]^T)\{\mathbf{\Phi}\} = \{\mathbf{0}\}$$

Premultiplying this expression by $[\mathbf{L}]^{-1}$ and using the transformation $\{\mathbf{\Phi}\} = [\mathbf{L}]^{-T}\{\mathbf{\Psi}\}$ $\{\mathbf{\Phi}\} = [\mathbf{L}]^{-T}\{\mathbf{\Psi}\}$ yields

$$([\mathbf{L}]^{-1}[\mathbf{K}][\mathbf{L}]^{-T} - \lambda[\mathbf{I}])\{\mathbf{\Psi}\} = \{\mathbf{0}\} \tag{11.288}$$

This is the form of Equation 11.282 with

$$[\mathbf{A}] = [\mathbf{L}]^{-1}[\mathbf{K}][\mathbf{L}]^{-T} \tag{11.289}$$

**[A]** is a symmetric matrix, and the eigenvalues obtained in this way are identical to those of the original system.

## 11.6 Solution Methods for Calculating Eigenvalues and Eigenvectors

There are many engineering problems that lead to the standard $[\mathbf{A}]\{\mathbf{\Phi}\} = \lambda[\mathbf{I}]\{\mathbf{\Phi}\}$ or to the generalized $([\mathbf{K}] - \lambda_i[\mathbf{M}])\ \{\mathbf{\Phi}_i\} = 0$ eigenvalue problem. Typical examples include elastic stability of a structure and structural vibrations. Here, for structural vibrations **[K]** and **[M]** denote the mass and stiffness matrices that are usually symmetric and real and at least one of the two matrices is positive definite

In many practical eigenvalue problems, the order of the mass and stiffness matrices are so large that it is impractical or too computationally expensive to obtain all the eigenvalues. However, in most cases it is accurate to consider only a partial eigensolution. A partial solution may be to determine a few of the lower eigenvalues and eigenvectors or eigenvalues and eigenvectors within a prescribed interval by solving the generalized eigenvalue equation. Finding eigenvalues is equivalent to finding the explicit evaluation of the coefficients $a_0, a_1, \ldots, a_n$ in the characteristic polynomial

$$p(\lambda) = a_n\lambda^n + a_{n-1}\lambda^{n-1} + \cdots + a_1\lambda + a_0 \tag{11.290}$$

as was employed in Chapter 4. A numerical procedure will need to be used to obtain the roots of Equation 11.290 since there is no closed form solution for $n > 4$. The number of eigenvalues required varies depending on the application and type of analysis that is needed for the particular model employed. For most structural dynamics problems, the lowest frequencies are the most desirable. The selection for a numerical procedure depends on a number of items, such as size of matrices involved, whether the matrices are symmetric or positive definite, banded or sparse. In addition, consideration is also placed on the type of solution required, such as all the frequencies are only a few needed, since in some cases not all the eigenvalues and eigenvectors are needed. Basic numerical methods for the symmetric eigenvalue computations are the vector iteration or power method, the matrix transformation method, and the polynomial root-finding method. Within each of the above three categories, a number of solution algorithms have been developed. For example, in the vector or power iteration method, one has direct iteration and inverse iteration methods. They are simple to incorporate and can be adjusted to large eigenvalue problems. It is noted that the subspace iteration method, which allows us to calculate effectively the first $m$ eigenvalues and eigenvectors of a very large generalized eigenvalue problem, is derived from the inverse iteration method. In between these iterative methods are the matrix transformation methods which include Jacobi's method, Givens method, Householder's method, and the QR method. This section is to make one aware of the different numerical solutions of the eigenproblem. More advanced information can be found in the effective computation of the eigenvalues and eigenvectors of Wilkinson (1965) and Bathe and Wilson (1972, 1973, 1976).

### 11.6.1 Vector Iteration Methods

All solution methods for eigenvalue problems are iterative in nature. As stated above, no closed form solution is available to finding the roots of the polynomial expression, Equation 11.287 for $n > 4$, thus requiring a numerical iterative solution. The equation to be solved is given in the form

$$[\mathbf{K}]\{\mathbf{\Phi}\} = \lambda[\mathbf{M}]\{\mathbf{\Phi}\}$$

where we assume that we have a dominant eigenvalue with a corresponding dominant eigenvector. Recall that for any solution $\{\mathbf{\Phi}\}$ found that any scalar multiplication will also be a solution and that the normalized conditions

$$\{\mathbf{\Phi}_i\}^T[\mathbf{K}]\{\mathbf{\Phi}_j\} = \lambda_i\delta_{ij}, \quad \{\mathbf{\Phi}_i\}^T[\mathbf{M}]\{\mathbf{\Phi}_j\} = \delta_{ij}$$

were used to circumvent this problem. Let us first factor the stiffness matrix $[\mathbf{K}]$ by using the modified Cholesky method (Strang 1988) as follows:

$$[\mathbf{K}] = [\mathbf{U}]^T[\mathbf{D}][\mathbf{U}] \tag{11.291}$$

Here $[\mathbf{D}]$ is a diagonal matrix and $[\mathbf{U}]$ is an upper triangular matrix with unit values on the main diagonal positions. Thus, we have

$$[\mathbf{K}]\{\mathbf{\Phi}\} = [\mathbf{U}]^T[\mathbf{D}][\mathbf{U}]\{\mathbf{\Phi}\} = \lambda[\mathbf{M}]\{\mathbf{\Phi}\} \tag{11.292}$$

Using an assumed initial trial vector for $\{\mathbf{\Phi}\}$, say $\{\mathbf{\Phi}_1\} = \{\mathbf{V}_1\}$, and calculate $\{\mathbf{R}_1\} = [\mathbf{M}]\{\mathbf{V}_1\}$. Therefore, for iterations $i = 1, 2, \ldots$, we have

$$[\mathbf{K}]\{\bar{\mathbf{V}}_{i+1}\} = [\mathbf{U}]^T[\mathbf{D}][\mathbf{U}]\{\bar{\mathbf{V}}_{i+1}\} = \lambda\{\mathbf{R}_i\} \tag{11.293}$$

or

$$[\mathbf{K}]\{\mathbf{V}_{i+1}\} = [\mathbf{U}]^T[\mathbf{D}][\mathbf{U}]\{\mathbf{V}_{i+1}\} = \{\mathbf{R}_i\} \tag{11.294}$$

where $\{\bar{\mathbf{V}}_{i+1}\} = \lambda\{\mathbf{V}_{i+1}\}$ is the scaled displacement solution, and we see that $\lambda$ appears as a scaling factor. Equation 11.293 or 11.294 are essentially an equilibrium equations, where $\{\mathbf{V}_{i+1}\}$ is the displacement vector and $\{\mathbf{R}_i\}$ the corresponding force vector. Iterating on the above equation, $\{\mathbf{V}_{i+1}\}$ should yield a better approximation to the eigenvector than the last calculated value $\{\mathbf{V}_i\}$. This method is termed the inverse iteration method and enables us to calculate the smallest eigenvalue $\lambda_1$ of the system.

In the above, we have stated that the iteration technique will converge leading to the smallest eigenvalue. A brief outline of convergence is given here. For a more detailed outline of convergence, see Bathe and Wilson (1976) and Bathe (1996). Since we have distinct eigenvalues $\lambda_1 < \lambda_2 < \lambda_3 < \cdots$, we also have a full set of eigenvectors $\{\mathbf{\Phi}_i\}$, $i = 1,2,3, \ldots ,n$. Thus, any arbitrary vector can be written as a linear combination of the eigenvectors as

$$\{\mathbf{V}\} = A_1\{\mathbf{\Phi}_1\} + A_2\{\mathbf{\Phi}_2\} + \cdots + A_n\{\mathbf{\Phi}_n\} = \sum_{k=1}^{n} A_k\{\mathbf{\Phi}_k\}$$

The first iteration with $i = 1$ yields

$$[\mathbf{K}]\{\bar{\mathbf{V}}_2\} = [\mathbf{M}]\{\mathbf{V}_1\}$$

and using the above expression becomes

$$\{\bar{\mathbf{V}}_2\} = \sum_{k=1}^{n} A_k[\mathbf{K}]^{-1}[\mathbf{M}]\{\mathbf{\Phi}_k\} = \sum_{k=1}^{n} \frac{1}{\lambda_k} A_k\{\mathbf{\Phi}_k\} = \frac{1}{\lambda_1}\sum_{k=1}^{n}\left(\frac{\lambda_1}{\lambda_k}\right) A_k\{\mathbf{\Phi}_k\}$$

After $i$ iterations, we have

$$\{\bar{\mathbf{V}}_{i+1}\} = \frac{1}{\lambda_1^i}\sum_{k=1}^{n}\left(\frac{\lambda_1}{\lambda_k}\right)^i A_k\{\mathbf{\Phi}_k\} \tag{11.295}$$

Based on Equation 11.295, since $\lambda_1$ is the smallest eigenvalue, then for $k > 1$ we have that $\lambda_1 < \lambda_k$ since $\lambda_1 < \lambda_2 < \cdots < \lambda_n$ and as $i \to \infty$, $(\lambda_1/\lambda_k)^i \to 0$ since the ratios $\lambda_1/\lambda_k$ are less than or equal to unity. Therefore, the $i$th iteration of vector $\{\bar{\mathbf{V}}_{i+1}\}$ converges to a vector proportional to $\{\mathbf{\Phi}_1\}$ belonging to $\lambda_1$. We see that the rate of convergence depends on the ratio $\lambda_1/\lambda_2$, the second term in the summation of Equation 11.295.

#### *11.6.1.1 Inverse Iteration*

The inverse iteration technique is used to effectively calculate the eigenvector corresponding to the lowest eigenvalue. This is accomplished by guessing a solution and repeating the iteration in the equations for the eigensystem. The procedure for the above method starts with an initial approximation iteration vector $\{\mathbf{V}_1\}$, followed by the listed steps for $i = 1,2,3,\ldots$ until convergence is established:

**Step 1:** Choose a starting vector $\{\mathbf{V}_1\}$ and determine $\{\mathbf{V}_{i+1}\}$ by solving the algebraic expression.

$$[\mathbf{K}]\{\bar{\mathbf{V}}_{i+1}\} = [\mathbf{M}]\{\mathbf{V}_i\}$$

**Step 2:** A better estimate of the eigenvalue can be obtained by using Rayleigh's quotient.

$$\lambda_{(i+1)} = \frac{\{\bar{\mathbf{V}}_{i+1}\}^T[\mathbf{K}]\{\bar{\mathbf{V}}_{i+1}\}}{\{\bar{\mathbf{V}}_{i+1}\}^T[\mathbf{M}]\{\bar{\mathbf{V}}_{i+1}\}} = \frac{\{\bar{\mathbf{V}}_{i+1}\}^T[\mathbf{M}]\{\mathbf{V}_i\}}{\{\bar{\mathbf{V}}_{i+1}\}^T[\mathbf{M}]\{\bar{\mathbf{V}}_{i+1}\}}$$

where the estimates (not the exact values) of the eigenvalues are enclosed in parentheses.

**Step 3:** Check for convergence. For a desired eigenfrequency with a precision of $2s$ digits, we must require that

$$\frac{\left|\lambda_{(i+1)} - \lambda_{(i)}\right|}{\lambda_{(i+1)}} \leq 10^{-2s}$$

where $2s$ is usually taken as 10.

**Step 4:** If convergence is not obtained, normalize $\{\bar{\mathbf{V}}_{i+1}\}$:

$$\{\mathbf{V}_{i+1}\} = \frac{\{\bar{\mathbf{V}}_{i+1}\}}{\left(\{\bar{\mathbf{V}}_{i+1}\}^T[\mathbf{M}]\{\bar{\mathbf{V}}_{i+1}\}\right)^{1/2}}$$

and return to Step 1 and continue with the iteration using the next $i$.

**Step 5:** After convergence has been achieved, the eigenvalue and eigenvector are given by

$$\lambda_1 \doteq \lambda_{(n+1)} \quad \{\Phi_1\} \doteq \frac{\{\bar{\mathbf{V}}_{n+1}\}}{\left(\{\bar{\mathbf{V}}_{n+1}\}^T[\mathbf{M}]\{\bar{\mathbf{V}}_{n+1}\}\right)^{1/2}}$$

where $n$ is the last iteration. The basic step in the iteration is the solution of the equation in Step 1.

**EXAMPLE 11. 4**

Determine the lowest eigenvalue and the corresponding eigenvector for the following system:

$$\left[k\begin{bmatrix}2 & -2 & 0\\ -2 & 4 & -2\\ 0 & -2 & 5\end{bmatrix} - \lambda m\begin{bmatrix}1 & 0 & 0\\ 0 & 1 & 0\\ 0 & 0 & 1\end{bmatrix}\right]\begin{Bmatrix}u_1\\ u_2\\ u_3\end{Bmatrix} = 0$$

where $m = 0.259$ kip s²/in and $k = 168$ kips/in.

**Solution**

Starting with an initial vector $\{\mathbf{V}_1\}^T = \{1 \quad 1 \quad 1\}$, we find the values listed in Table 11.6.

The final result is given as $\omega_1 \doteq \sqrt{312.119} = 17.667$ and $\{\mathbf{\Phi}_1\}^T = \{1.5116 \quad 1.1480 \quad 0.5081\}$.

After iterating and determining the first eigenvalue or mode shape and first eigenvalue, choosing another trial vector to determine the second mode shape will not work since a trial vector will generally contain, in some manner, the first mode shape as can be seen in the expression $\{\mathbf{V}\} = A_1\{\mathbf{\Phi}_1\} + A_2\{\mathbf{\Phi}_2\} + A_3\{\mathbf{\Phi}_3\} + \cdots + A_n\{\mathbf{\Phi}_n\}$ since $A_1$ will in general not be zero, hence . Such a trial vector will generally contain, to some extent, the first mode shape as set forth in the above equation. For the calculation of the second mode shape and second eigenvalue, a series of trial vectors whose expansions in terms of all mode shapes is such that the weighting coefficient for the first mode shape, $A_1$, should be zero, only in this way will the coefficient of the second mode dominate after a number of iterations, thus allowing convergence to the second mode. In reality, it is not possible to start with an arbitrary vector $\{\mathbf{V}\}$ and make it orthogonal to $\{\mathbf{\Phi}_1\}$. However, if the first mode has already been determined, the assumed arbitrary trial vector can be made orthogonal to $\{\mathbf{\Phi}_1\}$ by using

**TABLE 11.6**

Inverse Iteration for Lowest Eigenvalue

| Iteration | $\{\mathbf{V}_i\}$ | $\{\bar{\mathbf{V}}_{i+1}\}$ | $\lambda_{(i+1)}$ | $\{\mathbf{V}_{i+1}\}$ |
|---|---|---|---|---|
| 1 | $\begin{Bmatrix}1\\ 1\\ 1\end{Bmatrix}$ | $\begin{Bmatrix}0.0039\\ 0.0031\\ 0.0015\end{Bmatrix}$ | 317.117 | $\begin{Bmatrix}1.4646\\ 1.1717\\ 0.5858\end{Bmatrix}$ |
| 2 | $\begin{Bmatrix}1.4646\\ 1.1717\\ 0.5858\end{Bmatrix}$ | $\begin{Bmatrix}0.0048\\ 0.0037\\ 0.0017\end{Bmatrix}$ | 312.199 | $\begin{Bmatrix}1.5051\\ 1.1524\\ 0.5174\end{Bmatrix}$ |
| 3 | $\begin{Bmatrix}1.5051\\ 1.1524\\ 0.5174\end{Bmatrix}$ | $\begin{Bmatrix}0.0048\\ 0.0037\\ 0.0016\end{Bmatrix}$ | 312.120 | $\begin{Bmatrix}1.5107\\ 1.1486\\ 0.5092\end{Bmatrix}$ |
| 4 | $\begin{Bmatrix}1.5107\\ 1.1486\\ 0.5092\end{Bmatrix}$ | $\begin{Bmatrix}0.0048\\ 0.0037\\ 0.0016\end{Bmatrix}$ | 312.119 | $\begin{Bmatrix}1.5115\\ 1.1480\\ 0.5082\end{Bmatrix}$ |
| 5 | $\begin{Bmatrix}1.5115\\ 1.1480\\ 0.5082\end{Bmatrix}$ | $\begin{Bmatrix}0.0048\\ 0.0037\\ 0.0016\end{Bmatrix}$ | 312.119 | $\begin{Bmatrix}1.5116\\ 1.1480\\ 0.5081\end{Bmatrix}$ |

the *Gram–Schmidt orthogonalization* procedure (Strang 1988). The process of obtaining the third and higher modes follows the same idea as above. However, the convergence for the iteration process becomes slower when calculating the higher modes.

#### *11.6.1.2 Forward Iteration*

The forward iteration method is similar to the inverse method, but here, the method yields the eigenvector corresponding to the largest eigenvalue. In the inverse iteration method, it was assumed that **[K]** be positive definite, whereas, in the forward method, we assume that **[M]** is positive definite. The procedure for the above method starts with an initial approximation iteration vector $\{\mathbf{V}_1\}$, and we calculate for $i = 1,2,3,\ldots [\mathbf{M}]\{\bar{\mathbf{V}}_{i+1}\} = [\mathbf{K}]\{\mathbf{V}_i\}$.

**Step 1:** Choose a starting vector $\{\mathbf{Y}_1\} = [\mathbf{K}]\{\mathbf{V}_1\}$, and evaluate the algebraic expression.

$$[\mathbf{M}]\{\bar{\mathbf{V}}_{i+1}\} = \{\mathbf{Y}_i\}$$

**Step 2:** Calculate a new candidate vector.

$$\{\bar{\mathbf{Y}}_{i+1}\} = [\mathbf{K}]\{\bar{\mathbf{V}}_{i+1}\}$$

**Step 3:** Calculate the corresponding eigenvalue.

$$\lambda(\bar{\mathbf{V}}_{i+1}) = \frac{\{\bar{\mathbf{V}}_{i+1}\}^T\{\bar{\mathbf{Y}}_{i+1}\}}{\{\bar{\mathbf{V}}_{i+1}\}^T\{\mathbf{Y}_i\}}$$

**Step 4:** Check for convergence.

$$\frac{\left|\lambda_{(i+1)} - \lambda_{(i)}\right|}{\lambda_{(i+1)}} \leq 10^{-2s}$$

where $2s$ is usually taken as 10.

**Step 5:** If convergence is not obtained, normalize $\{\bar{\mathbf{Y}}_{i+1}\}$:

$$\{\mathbf{Y}_{i+1}\} = \frac{\{\bar{\mathbf{Y}}_{i+1}\}}{\left(\{\bar{\mathbf{V}}_{i+1}\}^T\{\mathbf{Y}_i\}\right)^{1/2}}$$

and return to Step 1 and continue with the iteration using the next $i$.

**Step 5:** After convergence has been achieved, the eigenvalue and eigenvector are given by

$$\lambda_n \doteq \lambda(\bar{\mathbf{V}}_{i+1}) \quad \{\mathbf{\Phi}_n\} \doteq \frac{\{\bar{\mathbf{V}}_{n+1}\}}{\left(\{\bar{\mathbf{V}}_{n+1}\}^T\{\mathbf{Y}_n\}\right)^{1/2}}$$

where $n$ is the last iteration.

**EXAMPLE 11. 5**

Determine the largest eigenvalue and the corresponding eigenvector for Example 11.4 using forward iteration.

**Solution**

Starting with an initial vector $\{\mathbf{V}_1\}^T = \{1 \quad 1 \quad 1\}$, we find the values listed in Table 11.7. The final result is given as $\omega_1 \doteq \sqrt{4481.8} = 66.946$ and $\{\Phi_1\}^T = \{0.5321 \quad -1.3062 \quad 1.3681\}$.

**TABLE 11.7**

Inverse Iteration for Largest Eigenvalue

| Iteration | $\{\bar{\mathbf{V}}_{i+1}\}$ | $\{\bar{\mathbf{Y}}_{i+1}\}$ | $\lambda(\bar{\mathbf{V}}_{i+1})$ | $\{\mathbf{Y}_{i+1}\}$ |
|---|---|---|---|---|
| 1 | $\begin{Bmatrix} 0 \\ 0 \\ 1946 \end{Bmatrix}$ | $\begin{Bmatrix} 0.0 \\ -0.6538 \\ 1.6346 \end{Bmatrix} \times 10^6$ | 3243.2 | $\begin{Bmatrix} 0 \\ -660 \\ 1651 \end{Bmatrix}$ |
| 2 | $\begin{Bmatrix} 0 \\ -2549 \\ 6373 \end{Bmatrix}$ | $\begin{Bmatrix} 0.8565 \\ -3.8543 \\ 6.2096 \end{Bmatrix} \times 10^6$ | 4048.5 | $\begin{Bmatrix} 245.2 \\ -1103.4 \\ 1777.7 \end{Bmatrix}$ |
| 3 | $\begin{Bmatrix} 946.7 \\ -4260.2 \\ 6863.7 \end{Bmatrix}$ | $\begin{Bmatrix} 1.7495 \\ -5.4872 \\ 7.1970 \end{Bmatrix} \times 10^6$ | 4343.9 | $\begin{Bmatrix} 422.7 \\ -1325.6 \\ 1738.7 \end{Bmatrix}$ |
| 4 | $\begin{Bmatrix} 1631.9 \\ -5118.2 \\ 6712.9 \end{Bmatrix}$ | $\begin{Bmatrix} 2.2680 \\ -6.2433 \\ 7.3586 \end{Bmatrix} \times 10^6$ | 4442.4 | $\begin{Bmatrix} 518.3 \\ -1426.8 \\ 1681.7 \end{Bmatrix}$ |
| 5 | $\begin{Bmatrix} 2001.3 \\ -5509.0 \\ 6493.2 \end{Bmatrix}$ | $\begin{Bmatrix} 2.5235 \\ -6.5562 \\ 7.3053 \end{Bmatrix} \times 10^6$ | 4470.9 | $\begin{Bmatrix} 566.9 \\ -1472.7 \\ 1641.0 \end{Bmatrix}$ |
| 6 | $\begin{Bmatrix} 2188.6 \\ -5686.3 \\ 6336.0 \end{Bmatrix}$ | $\begin{Bmatrix} 2.6460 \\ -6.6854 \\ 7.2328 \end{Bmatrix} \times 10^6$ | 4478.9 | $\begin{Bmatrix} 591.5 \\ -1494.4 \\ 1616.8 \end{Bmatrix}$ |
| 7 | $\begin{Bmatrix} 2283.7 \\ -5770.1 \\ 6242.5 \end{Bmatrix}$ | $\begin{Bmatrix} 2.7061 \\ -6.7423 \\ 7.1824 \end{Bmatrix} \times 10^6$ | 4481.0 | $\begin{Bmatrix} 604.1 \\ -1505.1 \\ 1603.4 \end{Bmatrix}$ |
| 8 | $\begin{Bmatrix} 2332.4 \\ -5811.2 \\ 6190.6 \end{Bmatrix}$ | $\begin{Bmatrix} 2.7363 \\ -6.7689 \\ 7.1527 \end{Bmatrix}$ | 4481.6 | $\begin{Bmatrix} 610.6 \\ -1510.5 \\ 1596.1 \end{Bmatrix}$ |
| 9 | $\begin{Bmatrix} 2357.5 \\ -5832.0 \\ 6162.7 \end{Bmatrix}$ | $\begin{Bmatrix} 2.7517 \\ -6.7891 \\ 7.1363 \end{Bmatrix} \times 10^6$ | 4481.8 | $\begin{Bmatrix} 614.0 \\ -1513.3 \\ 1592.3 \end{Bmatrix}$ |
| 10 | $\begin{Bmatrix} 2370.6 \\ -5842.7 \\ 6147.9 \end{Bmatrix}$ | $\begin{Bmatrix} 2.7597 \\ -6.7885 \\ 7.1274 \end{Bmatrix} \times 10^6$ | 4481.8 | $\begin{Bmatrix} 615.8 \\ -1514.7 \\ 1590.3 \end{Bmatrix}$ |

### 11.6.2 Vector Iteration with Shifts

The inverse iteration method can be used along with shifting of the eigenvalue scale to improve the convergence rate of the iteration process and provide a technique for determining a sequence of eigenvalues. That is, the shifting concept allows the computation of any eigenpair ($\lambda_i$, $\mathbf{\Phi}_i$) since previous calculated values can be used to estimate the required shift. In the solution of $[\mathbf{K}]\{\mathbf{\Phi}\} = \lambda[\mathbf{M}]\{\mathbf{\Phi}\}$, a shift $\mu$ is performed on the stiffness matrix [**K**].

Thus, we calculate

$$[[\mathbf{K}] - \mu[\mathbf{M}]]\{\mathbf{\Phi}\} = (\lambda - \mu)[\mathbf{M}]\{\mathbf{\Phi}\} \tag{11.296}$$

or

$$[\hat{\mathbf{K}}]\{\mathbf{\Phi}\} = \rho[\mathbf{M}]\{\mathbf{\Phi}\} \tag{11.297}$$

where

$$[\hat{\mathbf{K}}] = [\mathbf{K}] - \mu[\mathbf{M}] \quad \rho = \lambda - \mu \tag{11.298}$$

Figure 11.31 indicates the shift $\mu$ in the origin and $\rho$ as the shifted eigenvalue measured from the shifted origin.

It is noted that the eigenvalues of the original problem and shifted problem are the same. However, the eigenvalues of the shifted problem are different by the shift $\rho$ from the original problem. Inverse iteration on this system of equations will converge to the lowest eigenvalue of $\rho = \lambda - \mu$. As mentioned above, the rate of convergence depends on the ratio of the eigenvalue being sought and the next larger eigenvalue. The smaller this ratio, the faster the convergence, thus, it is desirable to choose a shift point close to the desired eigenvalue. If the shift point is located between eigenvalues $\lambda_i$ and $\lambda_{i+1}$, and $\mu$ is closer to $\lambda_i$ than $\lambda_{i+1}$, the iteration will converge to $\lambda_i$ and the rate of convergence will depend on the ratio $(\mu - \lambda_i)/(\lambda_{i+1} - \mu)$. On the other hand, it will converge to $\lambda_{i+1}$ if it is closer to $\lambda_{i+1}$ and the rate of convergence will depend on the ratio $(\lambda_{i+1} - \mu)/(\mu - \lambda_i)$. Note from the figure that if $\lambda_i < \mu$, $\rho_i$ will be negative.

**EXAMPLE 11.6**

Determine the natural frequencies and modes of vibration of Example 11.4 using inverse iteration with shifting.

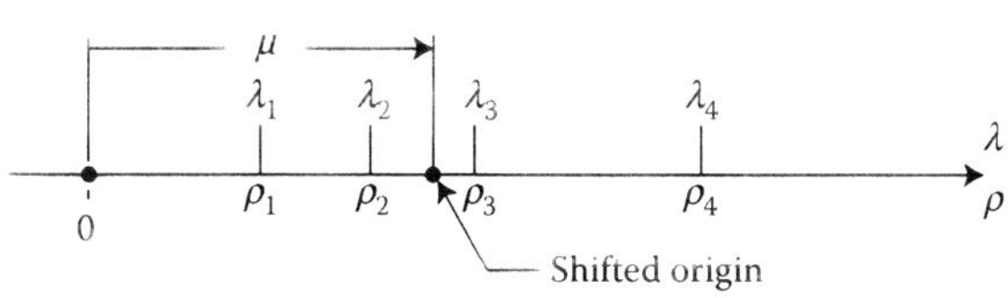


**FIGURE 11.31**
Eigenvalue spectrum with a shifted origin.

**TABLE 11.8**

Inverse Iteration for First Eigenvalue Pair Using Shifting: $\mu = 300$

| Iteration | $\{\mathbf{V}_i\}$ | $\{\bar{\mathbf{V}}_{i+1}\}$ | $\lambda_{(i+1)}$ | $\{\mathbf{V}_{i+1}\}$ |
|---|---|---|---|---|
| 1 | $\begin{Bmatrix}1\\1\\1\end{Bmatrix}$ | $\begin{Bmatrix}0.1022\\0.0778\\0.0346\end{Bmatrix}$ | 312.129 | $\begin{Bmatrix}1.5096\\1.1491\\0.5115\end{Bmatrix}$ |
| 2 | $\begin{Bmatrix}1.5096\\1.1491\\0.5115\end{Bmatrix}$ | $\begin{Bmatrix}0.1247\\0.0947\\0.0419\end{Bmatrix}$ | 312.119 | $\begin{Bmatrix}1.5116\\1.1480\\0.5081\end{Bmatrix}$ |
| 3 | $\begin{Bmatrix}1.5116\\1.1480\\0.5081\end{Bmatrix}$ | $\begin{Bmatrix}0.1247\\0.0947\\0.0419\end{Bmatrix}$ | 312.119 | $\begin{Bmatrix}1.5116\\1.1480\\0.5081\end{Bmatrix}$ |

**Solution**

Selecting a shift value of $\mu = 300$ for the first eigenvalue $[\hat{\mathbf{K}}]$ is calculated from Equation 11.298 and the inverse iteration that is outlined in Section 11.6.1.1 is applied using a starting vector $\{\mathbf{V}_1\}^T = \{1 \quad 1 \quad 1\}$. The application leads to the results listed in Table 11.8.

The final result is given as $\omega_1 \doteq \sqrt{312.119} = 17.667$ and $\{\mathbf{\Phi}_1\}^T = \{1.5116 \quad 1.1480 \quad 0.5081\}$. The results were obtained in three iterations, two iteration cycles less than Example 11.4.

For the second eigenpair, start with a shift of $\mu = 2000$ and the same starting vector $\{\mathbf{V}_1\}$. This leads to the results listed in Table 11.9.

The final result is $\omega_2 = \sqrt{2341.2} = 48.386$ and $\{\mathbf{\Phi}_2\}^T = \{-1.1369 \quad 0.9150 \quad 1.3158\}$. We see that for the value we picked for $\mu$ convergence was obtained in four iterations.

For the third eigenpair, start with a shift of $\mu = 4300$ and again use the same starting vector $\{\mathbf{V}_1\}$. This leads to the results listed in Table 11.10.

The final result is $\omega_2 = \sqrt{4481.8} = 66.946$ and $\{\mathbf{\Phi}_2\}^T = \{0.5320 \quad -1.3061 \quad 1.3683\}$. We see that for the value we picked for $\mu$ convergence was obtained in four iterations.

**TABLE 11.9**

Inverse Iteration for the Second Eigenvalue Pair Using Shifting: $\mu = 2200$

| Iteration | $\{\mathbf{V}_i\}$ | $\{\bar{\mathbf{V}}_{i+1}\}$ | $\lambda_{(i+1)}$ | $\{\mathbf{V}_{i+1}\}$ |
|---|---|---|---|---|
| 1 | $\begin{Bmatrix}1\\1\\1\end{Bmatrix}$ | $\begin{Bmatrix}-0.0029\\0.0012\\0.0025\end{Bmatrix}$ | 2252.7 | $\begin{Bmatrix}-1.4130\\0.6079\\1.2226\end{Bmatrix}$ |
| 2 | $\begin{Bmatrix}-1.4130\\0.6079\\1.2226\end{Bmatrix}$ | $\begin{Bmatrix}-0.0077\\0.0064\\0.0092\end{Bmatrix}$ | 2340.6 | $\begin{Bmatrix}-1.1113\\0.9307\\1.3266\end{Bmatrix}$ |
| 3 | $\begin{Bmatrix}-1.1113\\0.9307\\1.3266\end{Bmatrix}$ | $\begin{Bmatrix}-0.0081\\0.0065\\0.0093\end{Bmatrix}$ | 2341.2 | $\begin{Bmatrix}-1.1388\\0.9133\\1.3153\end{Bmatrix}$ |
| 4 | $\begin{Bmatrix}-1.1388\\0.9133\\1.3153\end{Bmatrix}$ | $\begin{Bmatrix}-0.0081\\0.0065\\0.0093\end{Bmatrix}$ | 2341.2 | $\begin{Bmatrix}-1.1369\\0.9150\\1.3158\end{Bmatrix}$ |

**TABLE 11.10**

Inverse Iteration for the Third Eigenvalue Pair Using Shifting: $\mu = 4300$

| Iteration | $\{\mathbf{V}_i\}$ | $\{\bar{\mathbf{V}}_{i+1}\}$ | $\lambda_{(i+1)}$ | $\{\mathbf{V}_{i+1}\}$ |
|---|---|---|---|---|
| 1 | $\begin{Bmatrix} 1 \\ 1 \\ 1 \end{Bmatrix}$ | $\begin{Bmatrix} 0.0003 \\ -0.0015 \\ 0.0009 \end{Bmatrix}$ | 4197.8 | $\begin{Bmatrix} 0.3440 \\ -1.6695 \\ 0.9774 \end{Bmatrix}$ |
| 2 | $\begin{Bmatrix} 0.3440 \\ -1.6695 \\ 0.9774 \end{Bmatrix}$ | $\begin{Bmatrix} 0.0028 \\ -0.0064 \\ 0.0074 \end{Bmatrix}$ | 4480.8 | $\begin{Bmatrix} 0.5307 \\ -1.2787 \\ 1.3943 \end{Bmatrix}$ |
| 3 | $\begin{Bmatrix} 0.5307 \\ -1.2787 \\ 1.3943 \end{Bmatrix}$ | $\begin{Bmatrix} 0.0029 \\ -0.0072 \\ 0.0075 \end{Bmatrix}$ | 4481.8 | $\begin{Bmatrix} 0.5330 \\ -1.3081 \\ 1.3659 \end{Bmatrix}$ |
| 4 | $\begin{Bmatrix} 0.5330 \\ -1.3081 \\ 1.3659 \end{Bmatrix}$ | $\begin{Bmatrix} 0.0029 \\ -0.0072 \\ 0.0075 \end{Bmatrix}$ | 4481.8 | $\begin{Bmatrix} 0.5320 \\ -1.3061 \\ 1.3683 \end{Bmatrix}$ |

### 11.6.3 Subspace Iteration

Subspace iteration (or *simultaneous iteration*) is an efficient method for approximating eigenvalues and eigenvectors based on the idea of carrying out the inverse iteration method on several trial vectors at the same time. Instead of solving the full $n \times n$ generalized eigenvalue problem, the subspace iteration method aims at finding the lowest $p$ eigenvalues and eigenvectors of a smaller or reduced problem. The iteration is performed simultaneously on a number of trial vectors $m$, where $m$ is the smaller of $2p$ and $p+8$, where $p$ is the number of modes to be determined and is usually less than $n$, the number of DOF. The general eigenvalue problem is given as

$$[\mathbf{K}]\{\mathbf{\Phi}\} = \lambda[\mathbf{M}]\{\mathbf{\Phi}\}$$

where $(\lambda_i, \{\mathbf{\Phi}_i\})$ is the $i$th eigenpair. If the order of the mass and stiffness matrices is $n$, then we have $n$ eigenpairs which we can order as

$$0 < \lambda_1 \leq \lambda_2 \leq \lambda_3 \leq \cdots \leq \lambda_n$$
$$\{\mathbf{\Phi}_1\}, \{\mathbf{\Phi}_2\}, \{\mathbf{\Phi}_3\}, \ldots, \{\mathbf{\Phi}_n\}$$

We seek a solution for the smallest $p$ eigenvalues $\lambda_i$, i = 1,2, … ,$p$ and the corresponding eigenvectors $\{\mathbf{\Phi}_i\}$, $i$ = 1,2, … ,$p$ with the ordering

$$0 < \lambda_1 \leq \lambda_2 \leq \cdots \leq \lambda_p$$

which satisfy

$$[\mathbf{K}]\{\mathbf{\Phi}_i\} = \lambda_i[\mathbf{M}]\{\mathbf{\Phi}_i\}, \quad i = 1, 2, \ldots, p$$

Along with the Kronecker delta relationships

$$\{\mathbf{\Phi}_i\}^T[\mathbf{M}]\{\mathbf{\Phi}_j\} = \delta_{ij}$$
$$\{\mathbf{\Phi}_i\}^T[\mathbf{K}]\{\mathbf{\Phi}_j\} = \lambda_i\delta_{ij}$$

Given the approximation $[\mathbf{V}_i]$ of size $n \times m$, find the next iteration vector as follows:
**Step 1:** Solve the algebraic expression.

$$[\mathbf{K}][\bar{\mathbf{V}}_{i+1}] = [\mathbf{M}][\mathbf{V}_i]$$

**Step 2:** Determine the projections of matrices **[K]** and **[M]** in an $m \times m$ space.

$$[\mathbf{K}_{i+1}] = [\bar{\mathbf{V}}_{i+1}]^T[\mathbf{K}][\bar{\mathbf{V}}_{i+1}]$$
$$[\mathbf{M}_{i+1}] = [\bar{\mathbf{V}}_{i+1}]^T[\mathbf{M}][\bar{\mathbf{V}}_{i+1}]$$

**Step 3:** Solve the reduced eigenvalue problem.

$$[\mathbf{K}_{i+1}][\mathbf{Q}_{i+1}] = [\mathbf{M}_{i+1}][\mathbf{Q}_{i+1}][\mathbf{\Lambda}_{i+1}]$$

**Step 4:** Normalize the vectors in $[Q_{i+1}]$ so that

$$[\mathbf{Q}_{i+1}]^T[\mathbf{M}][\mathbf{Q}_{i+1}] = [\mathbf{I}]$$

**Step 5:** Determine an improved estimate to the eigenvectors.

$$[\mathbf{V}_{i+1}] = [\bar{\mathbf{V}}_{i+1}][\mathbf{Q}_{i+1}]$$

Repeat Step 1.

$$[\mathbf{K}][\mathbf{Q}] = [\mathbf{M}][\mathbf{Q}][\mathbf{\Lambda}] \tag{11.299}$$

where $[\mathbf{Q}] = [\{\mathbf{\Phi}_1\} \quad \{\mathbf{\Phi}_2\} \quad \cdots \quad \{\mathbf{\Phi}_p\}]$ and $[\mathbf{\Lambda}] = \text{diag}[\lambda_p]$.

An important step in the subspace iteration method is the establishment of effective starting iteration vectors (Bathe 1996). In addition, it is usually effective to order the iteration vectors $[\mathbf{V}_i]$, from the lowest to the highest in increasing value.

**EXAMPLE 11.7**

Use the subspace iteration method to solve for the mass and stiffness **[K]** and **[M]**, which are given as

$$[\mathbf{M}] = \begin{bmatrix} 2 & 0 & 0 \\ 0 & 4 & 0 \\ 0 & 0 & 2 \end{bmatrix} \quad \text{and} \quad [\mathbf{K}] = \begin{bmatrix} 4 & -2 & 0 \\ -2 & 8 & -2 \\ 0 & -2 & 4 \end{bmatrix}$$

Assume the initial guess for the eigenvectors is given by

$$[\mathbf{V}_1] = \begin{bmatrix} 0 & 1 \\ 0.5 & 0.5 \\ 1 & 0 \end{bmatrix}$$

We find $[\bar{\mathbf{V}}_2]$ as

$$[\bar{\mathbf{V}}_2] = \begin{bmatrix} 0.2500 & 0.7500 \\ 0.5000 & 0.5000 \\ 0.7500 & 0.2500 \end{bmatrix}$$

The reduced mass and stiffness matrices become

$$\left[\mathbf{K}_2^r\right] = [\bar{\mathbf{V}}_2]^T[\mathbf{K}][\bar{\mathbf{V}}_2] = \begin{bmatrix} 2.5000 & 1.5000 \\ 1.5000 & 2.5000 \end{bmatrix}$$

$$\left[\mathbf{M}_2^r\right] = [\bar{\mathbf{V}}_2]^T[\mathbf{M}][\bar{\mathbf{V}}_2] = \begin{bmatrix} 2.2500 & 1.7500 \\ 1.7500 & 2.2500 \end{bmatrix}$$

Solution of the reduced problem

$$\left[\mathbf{K}_2^r\right][\mathbf{Q}_2] = \left[\mathbf{M}_2^r\right][\mathbf{Q}_2][\mathbf{\Lambda}_2]$$

yields

$$[\mathbf{\Lambda}_2] = \begin{bmatrix} 1 & 0 \\ 0 & 2 \end{bmatrix} \quad [\mathbf{Q}_2] = \begin{bmatrix} -0.3536 & -1.0000 \\ -0.3536 & 1.0000 \end{bmatrix}$$

The new approved eigenvectors can be obtained as

$$[\mathbf{V}_2] = [\bar{\mathbf{V}}_2][\mathbf{Q}_2] = \begin{bmatrix} -0.3536 & 0.5000 \\ -0.3536 & 0 \\ -0.3536 & -0.5000 \end{bmatrix}$$

In this example, we obtain the exact eigenvalues and eigenvectors in just one iteration. This is due to the fact that the starting vectors $[\mathbf{V}_1]$ span the subspace defined as the exact eigenvectors.

## 11.7 Transformation Methods

In solving the generalized eigenvalue problem, eigenvectors $\{\mathbf{\Phi}_k\}$ were determined that gave the following properties:

$$[\mathbf{\Phi}]^T[\mathbf{M}][\mathbf{\Phi}] = [\mathbf{I}]$$

$$[\mathbf{\Phi}]^T[\mathbf{K}][\mathbf{\Phi}] = [\mathbf{\lambda}]$$

where [I] and [$\boldsymbol{\lambda}$] are diagonal matrices and the eigenmatrix [$\boldsymbol{\Phi}$] is defined as $[\boldsymbol{\Phi}] = [\boldsymbol{\Phi}_1 \quad \boldsymbol{\Phi}_2 \quad \cdots \quad \boldsymbol{\Phi}_n]$. The idea of transformation methods is to transform the mass and stiffness matrices [M] and [K] so that the matrices tend to a diagonal forms. That is,

$$\begin{aligned}
&[\mathbf{K}_1] = [\mathbf{K}] && [\mathbf{M}_1] = [\mathbf{M}] \\
&[\mathbf{K}_2] = [\mathbf{T}_1]^T[\mathbf{K}_1][\mathbf{T}_1] && [\mathbf{M}_2] = [\mathbf{T}_1]^T[\mathbf{M}_1][\mathbf{T}_1] \\
&\qquad\vdots && \qquad\vdots \\
&[\mathbf{K}_{k+1}] = [\mathbf{T}_k]^T[\mathbf{K}_k][\mathbf{T}_k] && [\mathbf{M}_{k+1}] = [\mathbf{T}_k]^T[\mathbf{M}_k][\mathbf{T}_k]
\end{aligned} \tag{11.300}$$

with

$$[\mathbf{K}_1] = [\mathbf{K}] \quad \text{and} \quad [\mathbf{M}_1] = [\mathbf{M}]$$

It is noted that if the matrices $[\mathbf{T}_k]$ are properly selected we have

$$[\mathbf{K}_{k+1}] \to [\boldsymbol{\lambda}] \quad \text{and} \quad [\mathbf{M}_{k+1}] \to [\mathbf{I}] \quad \text{as} \quad k \to \infty$$

Assume that $p$ is the last iteration; we can represent the eigenmatrix as

$$[\boldsymbol{\Phi}] = [\{\mathbf{T}_1\} \quad \{\mathbf{T}_2\} \quad \ldots \quad \{\mathbf{T}_p\}]$$

Since we are performing numerical calculations, we will not be able to achieve exact diagonalization but will have a so-called effective diagonalization such that

$$\lambda_i = \frac{K_{ii}^d}{M_{ii}^d} \tag{11.301}$$

There are a number of different methods that use the ideas described above such as Householder's, LR, and QR transformation methods. Here, only the Jacobi method will be discussed. The general Jacobi method yields simultaneously all the eigenvalues of a symmetric matrix. Jacobi's method is characterized by its simplicity and exceptional stability.

### 11.7.1 Generalized Jacobi Method

As noted above, reducing the initial symmetric matrix to the diagonal form requires a sequence of orthogonal transformations. Thus, each matrix $[\mathbf{T}_k]$ is chosen such that nondiagonal nonzero terms are zero after transformations. For matrix $[\mathbf{T}_k]$, the following form is used:

$$[\mathbf{T}_k] = \begin{bmatrix} 1 & & & & & & 0 \\ & \ddots & & & & & \\ & & 1 & & \alpha & & \\ & & & \ddots & & & \\ & & \beta & & 1 & & \\ & & & & & \ddots & \\ 0 & & & & & & 1 \end{bmatrix} \begin{matrix} \\ \\ -i\text{th row} \\ \\ -j\text{th row} \\ \\ \\ \end{matrix} \tag{11.302}$$

$$\qquad\qquad\quad |\qquad\quad\; |$$
$$\qquad\qquad i\text{th column} \quad j\text{th column}$$

where the constants $\alpha$ and $\beta$ are selected such that the off-diagonal elements $(i,j)$ of $[\mathbf{K}_k] = [\mathbf{M}_k] = 0$. Therefore, we have that the values of $\alpha$ and $\beta$ are functions of $k_{ij}^{(k)}, k_{ii}^{(k)}, k_{jj}^{(k)}, m_{ij}^{(k)}, m_{ii}^{(k)}$, and $m_{jj}^{(k)}$, where the superscript $(k)$ denotes the $k$th iteration. Performing the multiplications

$$[\mathbf{K}_{k+1}] = [\mathbf{T}_k]^T[\mathbf{K}_k][\mathbf{T}_k] \quad [\mathbf{M}_{k+1}] = [\mathbf{T}_k]^T[\mathbf{M}_k][\mathbf{T}_k]$$

And applying the conditions that $k_{ij}^{(k+1)}$ and $m_{ij}^{(k+1)}$ shall become zero leads to

$$(1+\alpha\beta)k_{ij}^{(k)} + \alpha k_{ii}^{(k)} + \beta k_{jj}^{(k)} = 0 \tag{11.303}$$

$$(1+\alpha\beta)m_{ij}^{(k)} + \alpha m_{ii}^{(k)} + \beta m_{jj}^{(k)} = 0 \tag{11.304}$$

Equations 11.303 and 11.304 are of the same form and can be solved for the constants $\alpha$ and $\beta$. From Equation 11.304, we have

$$\alpha = -\left(\frac{m_{ij}^{(k)} + \beta m_{jj}^{(k)}}{m_{ii}^{(k)} + \beta m_{ij}^{(k)}}\right)$$

Substituting this expression into Equation 11.303 yields

$$A\beta^2 - B\beta - C = 0 \tag{11.305}$$

where

$$A = k_{jj}^{(k)}m_{ij}^{(k)} - k_{ij}^{(k)}m_{jj}^{(k)}, \quad B = k_{ii}^{(k)}m_{jj}^{(k)} - k_{jj}^{(k)}m_{ii}^{(k)}, \quad C = k_{ii}^{(k)}m_{ij}^{(k)} - m_{ii}^{(k)}k_{ij}^{(k)}$$

We have

$$D = \frac{B}{2} + \operatorname{sign}(B)\sqrt{\left(\frac{B}{2}\right)^2 + AC}$$

Therefore,

$$\alpha = \frac{A}{D}, \quad \beta = -\frac{C}{D} \tag{11.306}$$

When $[\mathbf{M}]$ is positive definite, full or banded, the coefficient $(B/2)^2 + AC$ is always positive. When $D = 0$, $\alpha$ and $\beta$ are given by

$$\alpha = 0, \quad \beta = \frac{-k_{ij}^{(k)}}{k_{jj}^{(k)}} \tag{11.307}$$

The generalized Jacobi procedure can also be adopted for the case when the mass matrix [**M**] is a diagonal matrix, with or without zero diagonal elements. Convergence is obtained by comparing successive eigenvalue approximations. Assuming that $l$ is the last iteration, then convergence is achieved if

$$\frac{\left|\lambda_m^{(l+1)} - \lambda_m^{(l)}\right|}{\lambda_m^{(l+1)}} \leq 10^{-2s}, \quad \text{for } m = 1, 2, \ldots, n \tag{11.308}$$

where

$$\lambda_m^{(l)} = \frac{k_{mm}^{(l)}}{m_{mm}^{(l)}}, \quad \lambda_m^{(l+1)} = \frac{k_{mm}^{(l+1)}}{m_{mm}^{(l+1)}} \tag{11.309}$$

and

$$\left[\frac{\left(k_{ij}^{(l+1)}\right)^2}{k_{ii}^{(l+1)} k_{jj}^{(l+1)}}\right]^{1/2} \leq \varepsilon, \quad \left[\frac{\left(m_{ij}^{(l+1)}\right)^2}{m_{ii}^{(l+1)} m_{jj}^{(l+1)}}\right]^{1/2} \leq \varepsilon, \quad \text{for all } i, j \text{ with } i < j,\ j = 1, 2, \ldots, n \tag{11.310}$$

where $\varepsilon$ is the convergence tolerance usually taken as $10^{-2m}$. The steps in using the generalized Jacobi method are given as follows:

**Step 1:** Initialize variables

a. Initialize sweep counter $m$ and rotation counter $k$: $m = 0, k = 1$

b. Initialize eigenvalues and eigenvectors

$$\lambda_i^{(1)} = \frac{k_{ii}^{(1)}}{m_{ii}^{(1)}}, \quad [\mathbf{\Phi}] = [\mathbf{I}]$$

c. Select convergence tolerance $\varepsilon_{tol}$

**Step 2:** Increment sweep counter: $m = m+1$

a. Define dynamic tolerance $\varepsilon_m$:$\varepsilon_m = 10^{-2m}$

**Step 3:** Loop over $k = 1, n(n-1)/2$, all $(i,j)$ off-diagonal elements, $i = 1, 2, \ldots, n-1$, $j = i+1, i+2, \ldots, n$ and evaluate

$$\left(\frac{\left(k_{ij}^{(k)}\right)}{k_{ii}^{(k)} k_{jj}^{(k)}}\right)^{1/2} \leq \varepsilon_m \quad \text{and} \quad \left(\frac{\left(m_{ij}^{(k)}\right)}{m_{ii}^{(k)} m_{jj}^{(k)}}\right)^{1/2} \leq \varepsilon_m$$

If both thresholds are satisfied, do not do a rotation and look to the next element. If the thresholds are not satisfied, proceed with transformation.

**Step 4:** Determine $\alpha$ and $\beta$ using

$$A = k_{jj}^{(k)} m_{ij}^{(k)} - k_{ij}^{(k)} m_{jj}^{(k)}, \quad B = k_{ii}^{(k)} m_{jj}^{(k)} - k_{jj}^{(k)} m_{ii}^{(k)}, \quad C = k_{ii}^{(k)} m_{ij}^{(k)} - m_{ii}^{(k)} k_{ij}^{(k)}$$

$$D = \frac{B}{2} + \text{sign}(B)\sqrt{\left(\frac{B}{2}\right)^2 + AC}, \quad \alpha = \frac{A}{D}, \quad \beta = -\frac{C}{D}$$

and construct transformation matrix, Equation 11.275.

**Step 5:** Determine transformed matrices (only the rows and columns $(i,j)$ are affected)

$$[\mathbf{K}_{k+1}] = [\mathbf{T}_k]^T [\mathbf{K}_k][\mathbf{T}_k] \quad [\mathbf{M}_{k+1}] = [\mathbf{T}_k]^T [\mathbf{M}_k][\mathbf{T}_k]$$

**Step 6:** Update eigenvectors

$$[\boldsymbol{\Phi}_{k+1}] = [\boldsymbol{\Phi}_k][\mathbf{T}_k]$$

**Step 7:** End loop over $k$

**Step 8:** Update the eigenvalues

$$\lambda_i^{(m+1)} = \frac{k_{ii}^{(k)}}{m_{ii}^{(k)}}, \quad i = 1, 2, \ldots, n$$

**Step 9:** Check convergence on eigenvalues $\lambda_i^{(m+1)}$

$$\frac{\left|\lambda_i^{(m+1)} - \lambda_i^{(m)}\right|}{\left|\lambda_i^{(m+1)}\right|} \leq \varepsilon_{tol}, \quad i = 1, 2, \ldots, n$$

**Step 10:** Computation of coupling factors

$$\left(\frac{\left(k_{ij}^{(k)}\right)^2}{k_{ii}^{(k)} k_{jj}^{(k)}}\right)^{1/2} \leq \varepsilon_{tol}, \quad \left(\frac{\left(m_{ij}^{(k)}\right)^2}{m_{ii}^{(k)} m_{jj}^{(k)}}\right)^{1/2} \leq \varepsilon_{tol}, \quad i, j = 1, 2, \ldots, n, i < j$$

If neither of these is satisfied for any off-diagonal term, go to Step 2 to begin another complete sweep. Otherwise convergence has been obtained.

**Step 11:** End sweeps

**Step 12:** The approximate eigenvalue and eigenvectors

$$\lambda_i \approx \lambda_i^{(m+1)}, \quad i = 1, 2, \ldots, n \quad \text{(where } m \text{ is the number of sweeps)}$$

$$[\Phi] \approx [\Phi_k]\left(\frac{1}{\sqrt{m_{ii}^k}}\right) \quad \text{(where } k-1 \text{ is the number of rotations)}$$

**EXAMPLE 11.8**

Use the generalized Jacobi method to calculate the eigensystem of the problem $[\mathbf{K}]\{\boldsymbol{\Phi}\} = \lambda[\mathbf{M}]\{\boldsymbol{\Phi}\}$ where

$$[\mathbf{K}] = \begin{bmatrix} 4 & 3.5 \\ 3.5 & 4 \end{bmatrix}, \quad [M] = \begin{bmatrix} 1 & 0 \\ 0 & 2 \end{bmatrix}$$

Determine $\alpha$ and $\beta$ using

$$A = k_{22}^{(1)} m_{12}^{(1)} - k_{12}^{(1)} m_{22}^{(1)} = 4 * (0) - 3.5 * 2 = -7.0$$
$$B = k_{11}^{(1)} m_{22}^{(1)} - k_{22}^{(1)} m_{11}^{(1)} = 4 * 2 - 4 * 1 = 4.0$$
$$C = k_{11}^{(1)} m_{12}^{(1)} - k_{12}^{(1)} m_{11}^{(1)} = 4 * (0) - 3.5 * 1 = -3.5$$
$$D = \frac{B}{2} + \text{sign}(B)\sqrt{\left(\frac{B}{2}\right)^2 + AC} = 2 + \sqrt{4 + (7) * (3.5)} = 7.3385$$

which yields

$$\alpha = \frac{A}{D} = \frac{-7}{7.3385} = -0.9539, \quad \beta = -\frac{C}{D} = -\frac{-3.5}{7.3385} = 0.4769$$

The rotation matrix $[\mathbf{T}_k]$ becomes and

$$[\mathbf{T}_1]^T [\mathbf{K}] [\mathbf{T}_1] = \begin{bmatrix} 1 & 0.4769 \\ -0.9539 & 1 \end{bmatrix} \begin{bmatrix} 4 & 3.5 \\ 3.5 & 4 \end{bmatrix} \begin{bmatrix} 1 & -0.9539 \\ 0.4769 & 1 \end{bmatrix} = \begin{bmatrix} 8.2484 & 0 \\ 0 & 0.9624 \end{bmatrix}$$
$$[\mathbf{T}_1]^T [\mathbf{M}] [\mathbf{T}_1] = \begin{bmatrix} 1 & 0.4769 \\ -0.9539 & 1 \end{bmatrix} \begin{bmatrix} 1 & 0 \\ 0 & 2 \end{bmatrix} \begin{bmatrix} 1 & -0.9539 \\ 0.4769 & 1 \end{bmatrix} = \begin{bmatrix} 1.4549 & 0 \\ 0 & 2.9099 \end{bmatrix}$$

The eigenvalues and eigenvectors are obtained from

$$\lambda_i = \frac{k_{ii}}{m_{ii}}, \quad [\boldsymbol{\Phi}] = [\mathbf{T}_1]\text{diag}\left(\frac{1}{\sqrt{m_{ii}}}\right)$$

Thus, we have

$$\lambda_1 = \frac{k_{11}}{m_{11}} = \frac{8.2484}{1.4549} = 5.6693, \quad \lambda_2 = \frac{k_{22}}{m_{22}} = \frac{0.9624}{2.9099} = 0.3307$$

and

$$[\boldsymbol{\Phi}] \approx \begin{bmatrix} 1 & -0.9539 \\ 0.4769 & 1 \end{bmatrix} \begin{bmatrix} 1/\sqrt{1.4549} & 0 \\ 0 & 1/\sqrt{2.9099} \end{bmatrix} = \begin{bmatrix} 0.8293 & -0.5589 \\ 0.3955 & 0.5862 \end{bmatrix}$$

Sorting the eigenvalues in ascending order yields

$$[\boldsymbol{\Lambda}] = \begin{bmatrix} 0.3307 & \\ & 5.6693 \end{bmatrix}; \quad [\boldsymbol{\Phi}] = \begin{bmatrix} -0.5589 & 0.8293 \\ 0.5862 & 0.3955 \end{bmatrix}$$

## 11.8 Multiple-Degrees-of-Freedom Numerical Techniques

The numerical integration of SDOF systems was presented in Chapter 3. The techniques and comments made are applicable to MDOF systems. The equation for MDOF systems is given as

$$[\mathbf{M}]\{\ddot{\mathbf{u}}\} + [\mathbf{C}]\{\dot{\mathbf{u}}\} + [\mathbf{K}]\{\mathbf{u}\} = \{\mathbf{f}\} \tag{11.311}$$

with initial conditions

$$\{\mathbf{u}\}_0 = \{\mathbf{u}(0)\} \quad \text{and} \quad \{\dot{\mathbf{u}}\}_0 = \{\dot{\mathbf{u}}(0)\} \quad \text{at } t = 0 \tag{11.312}$$

where

$[\mathbf{M}]$ = inertia matrix
$[\mathbf{K}]$ = stiffness matrix
$[\mathbf{C}]$ = damping matrix
$\{\mathbf{u}\}$ = column matrix of nodal displacements
$\{\mathbf{f}\}$ = column matrix of equivalent nodal forces

As in SDOF systems, different methods of solution are used to solve the above equation depending on whether the applied forces are harmonic, periodic, transient, or random in nature.

The mass matrix, as mentioned before, can be considered consistent or lumped. Consistent mass matrices are nondiagonal and tend to yield more accurate frequencies for flexural structural elements, whereas lumped mass matrices are diagonal. Often a diagonal mass matrix gives eigenvalues which converge from below as you refine the mesh, while consistent mass matrices converge from above.

In the following section, a brief presentation of several approaches used for the numerical integration of forced MDOF vibration problems will be discussed. Direct integration of the equation of motion provides the response of the MDOF system at discrete interval of time. Response of the system requires the computation of the displacement, velocity, and acceleration. The algorithms described in this section can be applied to many practical problems in structural dynamics.

### 11.8.1 Central Difference Method

In Chapter 3, the central difference method was developed for an SDOF system. This method can be extended to MDOF systems where the velocity and acceleration equations can be expressed in vector notation as

$$\{\dot{\mathbf{u}}\}_i = \frac{1}{2h}[\{\mathbf{u}\}_{i+1} - \{\mathbf{u}\}_{i-1}] \tag{11.313}$$

$$\{\ddot{\mathbf{u}}\}_i = \frac{1}{h^2}[\{\mathbf{u}\}_{i+1} - 2\{\mathbf{u}\}_i + \{\mathbf{u}\}_{i-1}] \tag{11.314}$$

where $h = \Delta t = t_{i+1} - t_i$ and $t_i = ih$. At time $t_i$, Equation 11.311 becomes

$$[\mathbf{M}]\{\ddot{\mathbf{u}}\}_i + [\mathbf{C}]\{\dot{\mathbf{u}}\}_i + [\mathbf{K}]\{\mathbf{u}\}_i = \{\mathbf{f}\}_i$$

Substituting Equations 11.313 and 11.314 and rearranging terms yields

$$\left(\frac{1}{h^2}[\mathbf{M}] + \frac{1}{2h}[\mathbf{C}]\right)\{\mathbf{u}\}_{i+1} + \left(\frac{1}{h^2}[\mathbf{M}] - \frac{1}{2h}[\mathbf{C}]\right)\{\mathbf{u}\}_{i-1} + \left([\mathbf{K}] - \frac{2}{h^2}[\mathbf{M}]\right)\{\mathbf{u}\}_i = \{\mathbf{f}\}_i$$

or

$$\left(\frac{1}{h^2}[\mathbf{M}] + \frac{1}{2h}[\mathbf{C}]\right)\{\mathbf{u}\}_{i+1} = \{\mathbf{f}\}_i - \left([\mathbf{K}] - \frac{2}{h^2}[\mathbf{M}]\right)\{\mathbf{u}\}_i - \left(\frac{1}{h^2}[\mathbf{M}] - \frac{1}{2h}[\mathbf{C}]\right)\{\mathbf{u}\}_{i-1} \quad (11.315)$$

Equation 11.315 can be written as

$$[\hat{\mathbf{K}}]\{\mathbf{u}\}_{i+1} = \{\hat{\mathbf{f}}\}_i \quad (11.316)$$

where

$$[\hat{\mathbf{K}}] = \frac{1}{h^2}[\mathbf{M}] + \frac{1}{2h}[\mathbf{C}] \quad (11.317)$$

and

$$\{\hat{\mathbf{f}}\}_i = \{\mathbf{f}\}_i - \left(\frac{1}{h^2}[\mathbf{M}] - \frac{1}{2h}[\mathbf{C}]\right)\{\mathbf{u}\}_{i-1} - \left([\mathbf{K}] - \frac{2}{h^2}[\mathbf{M}]\right)\{\mathbf{u}\}_i \quad (11.318)$$

are the effective stiffness matrix and effective load vector, respectively.

The solution vector $\{\mathbf{u}\}_{i+1}$ can be found using Equation 11.315 provided we know the previous displacements $\{\mathbf{u}\}_i$ and $\{\mathbf{u}\}_{i+1}$ along with the current applied external force $\{\mathbf{f}\}_i$. When $i = 0$, $\{\mathbf{u}\}_0$ and $\{\mathbf{u}\}_{-1}$ are needed to compute $\{\mathbf{u}\}_1$. Using the initial conditions only gives the values $\{\mathbf{u}\}_0$ and $\{\dot{\mathbf{u}}\}_0$; hence, a special starting technique is needed to determine $\{\mathbf{u}\}_0$ and $\{\mathbf{u}\}_{-1}$. Using Equations 11.311 through 11.314, evaluated at $i = 0$, gives

$$[\mathbf{M}]\{\ddot{\mathbf{u}}\}_0 + [\mathbf{C}]\{\dot{\mathbf{u}}\}_0 + [\mathbf{K}]\{\mathbf{u}\}_0 = \{\mathbf{f}\}_0 \quad (11.319)$$

$$\{\dot{\mathbf{u}}\}_0 = \frac{1}{2h}[\{\mathbf{u}\}_1 - \{\mathbf{u}\}_{-1}] \quad (11.320)$$

$$\{\ddot{\mathbf{u}}\}_0 = \frac{1}{h^2}[\{\mathbf{u}\}_1 - 2\{\mathbf{u}\}_0 + \{\mathbf{u}\}_{-1}] \quad (11.321)$$

Equation 11.319 gives the initial acceleration vector as

$$\{\ddot{\mathbf{u}}\}_0 = [\mathbf{M}]^{-1}[\{\mathbf{f}\}_0 - [\mathbf{C}]\{\dot{\mathbf{u}}\}_0 - [\mathbf{K}]\{\mathbf{u}\}_0] \quad (11.322)$$

**TABLE 11.11**

Algorithm for Central Difference Method for MDOF System

*Initial Computations:*

1. Set the initial conditions at $t = 0$: $\{\mathbf{u}\}_0 = \{\mathbf{u}(0)\}$ and $\{\dot{\mathbf{u}}\}_0 = \{\dot{\mathbf{u}}(0)\}$ and calculate

$$\{\ddot{\mathbf{u}}\}_0 = [\mathbf{M}]^{-1}[\{\mathbf{f}\}_0 - [\mathbf{C}]\{\dot{\mathbf{u}}\}_0 - [\mathbf{K}]\{\mathbf{u}\}_0]$$

2. Select a time step $h$, $h \le h_{cr} = T_i/\pi$
3. Calculate

$$a_0 = \frac{1}{h^2};\quad a_1 = \frac{1}{2h};\quad a_2 = 2a_0;\quad a_3 = \frac{1}{a_2}$$

4. Calculate

$$\{\mathbf{u}\}_{-1} = \{\mathbf{u}\}_0 - h\{\dot{\mathbf{u}}\}_0 + a_3\{\ddot{\mathbf{u}}\}_0$$

5. Calculate effective stiffness matrix $[\hat{\mathbf{K}}]$: $[\hat{\mathbf{K}}] = a_0[\mathbf{M}] + a_1[\mathbf{C}]$
6. Triangularize $[\hat{\mathbf{K}}]$:$[\hat{\mathbf{K}}] = [\mathbf{L}][\mathbf{D}][\mathbf{L}]^T$

*Calculate for each time step i: $i = 0, 1, 2, \ldots, t_f$*

7. Calculate effective load vector $\{\hat{\mathbf{f}}\}_i$ at time $t_i$

$$\{\hat{\mathbf{f}}\}_i = \{\mathbf{f}\}_i - (a_0[\mathbf{M}] - a_1[\mathbf{C}])\{\mathbf{u}\}_{i-1} - ([\mathbf{K}] - a_2[\mathbf{M}])\{\mathbf{u}\}_i$$

8. At time $t_{i+1}$, calculate the displacement vector $\{\mathbf{u}\}_{i+1}$

$$[\mathbf{L}][\mathbf{D}][\mathbf{L}]^T\{\mathbf{u}\}_{i+1} = \{\hat{\mathbf{f}}\}_i$$

9. Calculate the velocity and acceleration vectors

$$\{\dot{\mathbf{u}}_i\}_i = a_1[\{\mathbf{u}_{i+1}\}_{i+1} - \{\mathbf{u}_{i-1}\}_{i-1}]$$
$$\{\ddot{\mathbf{u}}\}_i = a_0[\{\mathbf{u}\}_{i+1} - 2\{\mathbf{u}\}_i + \{\mathbf{u}\}_{i-1}]$$

and Equations 11.320 and 11.321 give after eliminating $\{\mathbf{u}\}_1$

$$\{\mathbf{u}\}_{-1} = \{\mathbf{u}\}_0 - h\{\dot{\mathbf{u}}\}_0 + \frac{h^2}{2}\{\ddot{\mathbf{u}}\}_0 \tag{11.323}$$

where $\{\ddot{\mathbf{u}}\}_0$ is given by Equation 11.292. The numerical stability for the central difference technique depends upon the time step chosen. The technique will become *unstable* if the time step is not chosen small enough. The method requires that the time step $h$ be smaller than a critical time step $h_{cr}$ (Hughes and Belytschko 1983) given by

$$h \le h_{cr} = \frac{T_i}{\pi} = 0.3183T_i \tag{11.324}$$

where $T_i$ is the smallest fundamental period of vibration of the physical system under study. The algorithm used in the central difference method is given in Table 11.11.

**EXAMPLE 11.9**

Using the central difference method, calculate the displacement response of the two degree, second-order system

$$[\mathbf{M}]\{\ddot{\mathbf{u}}\} + [\mathbf{C}]\{\dot{\mathbf{u}}\} + [\mathbf{K}]\{\mathbf{u}\} = \{\mathbf{f}\}$$

with initial conditions

$$\{\mathbf{u}\}_0 = \{\mathbf{u}\{0\}\} \quad \text{and} \quad \{\dot{\mathbf{u}}\}_0 = \{\dot{\mathbf{u}}(0)\} \quad \text{at } t = 0$$

where

$$[\mathbf{M}] = \begin{bmatrix} 1 & 0 \\ 0 & 2 \end{bmatrix}; \quad [\mathbf{C}] = \begin{bmatrix} 0 & 0 \\ 0 & 0 \end{bmatrix}; \quad [\mathbf{K}] = \begin{bmatrix} 6 & -2 \\ -2 & 8 \end{bmatrix}; \quad \{\mathbf{f}\} = \begin{Bmatrix} 0 \\ 10 \end{Bmatrix}$$

**Solution**

The integration constants are

$$a_0 = \frac{1}{h^2} = 100, \quad a_1 = \frac{1}{2h} = 5, \quad a_2 = 2a_0 = 200, \quad a_3 = \frac{1}{a_2} = 0.005$$

Substituting $\{\mathbf{u}\}_0$ and $\{\dot{\mathbf{u}}\}_0$ into the system gives

$$[\mathbf{M}]\{\ddot{\mathbf{u}}\}_0 = \begin{Bmatrix} 0 \\ 10 \end{Bmatrix} \quad \{\ddot{\mathbf{u}}\}_0 = \begin{Bmatrix} 0 \\ 5 \end{Bmatrix}$$

The central difference algorithm, with $h = \Delta t = 0.1$

$$[\hat{\mathbf{K}}]\{\mathbf{u}\}_{i+1} = \{\hat{\mathbf{f}}\}_i$$

$$\{\hat{\mathbf{f}}\}_i = \{\mathbf{f}\}_i - (a_0[\mathbf{M}] - a_1[\mathbf{C}])\{\mathbf{u}\}_{i-1} - ([\mathbf{K}] - a_2[\mathbf{M}])\{\mathbf{u}\}_i$$

$$\{\hat{\mathbf{f}}\}_i = \{\mathbf{f}\}_i + [\mathbf{M}](200\{\mathbf{u}\}_i - 100\{\mathbf{u}\}_{i-1}) + 5[\mathbf{C}]\{\mathbf{u}\}_{i-1} - [\mathbf{K}]\{\mathbf{u}\}_i$$

$$[\hat{\mathbf{K}}] = 100[\mathbf{M}] + 5[\mathbf{C}] = \begin{bmatrix} 100 & 0 \\ 0 & 200 \end{bmatrix}$$

With $t = 0$, we have

$$\{\mathbf{u}\}_{-1} = \{\mathbf{u}\}_0 - h\{\dot{\mathbf{u}}\}_0 + a_3\{\ddot{\mathbf{u}}\}_0 = \begin{Bmatrix} 0 \\ 0.025 \end{Bmatrix}$$

The results are given in Table 11.12.

### 11.8.2 The Houbolt Method

The Houbolt method has been applied successfully to a variety of problems in structural dynamics (Houbolt 1950; Johnson and Greif 1966). The method is similar to the central

**TABLE 11.12**

Displacement Results for Example 11.9

| Time $t$ | $u_1$ | $u_2$ |
|---|---|---|
| 0 | 0.000000 | 0.000000 |
| 0.1 | 0.000000 | 0.025000 |
| 0.2 | 0.000500 | 0.099000 |
| 0.3 | 0.002950 | 0.219045 |
| 0.4 | 0.009604 | 0.380358 |
| 0.5 | 0.023289 | 0.576552 |
| 0.6 | 0.047107 | 0.799917 |
| 0.7 | 0.084098 | 1.041757 |
| 0.8 | 0.136877 | 1.292767 |
| 0.9 | 0.207300 | 1.543436 |
| 1.0 | 0.296153 | 1.784440 |
| 1.5 | 0.948432 | 2.575961 |
| 2.0 | 1.440928 | 2.360525 |
| 2.5 | 1.036949 | 1.434676 |
| 3.0 | −0.106143 | 0.561995 |
| 3.5 | −0.834902 | 0.307710 |
| 4.0 | −0.291522 | 0.756325 |

difference method. The equations are determined by fitting a third-order polynomial through four consecutive displacement time points and differentiating the expression twice to arrive at an approximation for the velocity and acceleration. The approximations are (Houbolt 1950)

$$\{\dot{\mathbf{u}}\}_{i+1} = \frac{1}{6h}[11\{\mathbf{u}\}_{i+1} - 18\{\mathbf{u}\}_i + 9\{\mathbf{u}\}_{i-1} - 2\{\mathbf{u}\}_{i-2}] \tag{11.325}$$

$$\{\ddot{\mathbf{u}}\}_{i+1} = \frac{1}{h^2}[2\{\mathbf{u}\}_{i+1} - 5\{\mathbf{u}\}_i + 4\{\mathbf{u}\}_{i-1} - \{\mathbf{u}\}_{i-2}] \tag{11.326}$$

The equation approximations are based on two backward difference equations with errors of $h^2$. Consider Equation 11.311 written at time $(t + \Delta t)$, thus we have

$$[\mathbf{M}]\{\ddot{\mathbf{u}}\}_{i+1} + [\mathbf{C}]\{\dot{\mathbf{u}}\}_{i+1} + [\mathbf{K}]\{\mathbf{u}\}_{i+1} = \{\mathbf{f}\}_{i+1} \tag{11.327}$$

Substituting Equations 11.325 and 11.326 into Equation 11.327 and rearranging terms yields

$$\begin{aligned}\left[[\mathbf{K}] + \frac{2}{h^2}[\mathbf{M}] + \frac{11}{6h}[\mathbf{C}]\right]\{\mathbf{u}\}_{i+1} = \{\mathbf{f}\}_{i+1} &+ [\mathbf{M}]\left[\frac{5}{h^2}\{\mathbf{u}\}_i - \frac{4}{h^2}\{\mathbf{u}\}_{i-1} + \frac{1}{h^2}\{\mathbf{u}\}_{i-2}\right] \\ &+ [\mathbf{C}]\left[\frac{3}{h}\{\mathbf{u}\}_i - \frac{3}{2h}\{\mathbf{u}\}_{i-1} + \frac{1}{3h}\{\mathbf{u}\}_{i-2}\right]\end{aligned} \tag{11.328}$$

Equation 11.328 can be written as

$$[\hat{\mathbf{K}}]\{\mathbf{u}\}_{i+1} = \{\hat{\mathbf{f}}\}_{i+1} \tag{11.329}$$

where

$$[\hat{\mathbf{K}}] = [\mathbf{K}] + \frac{2}{h^2}[\mathbf{M}] + \frac{11}{6h}[\mathbf{C}] \tag{11.330}$$

and

$$\begin{aligned}\{\hat{\mathbf{f}}\}_{i+1} = \{\mathbf{f}\}_{i+1} + [\mathbf{M}]\left[\frac{5}{h^2}\{\mathbf{u}\}_i - \frac{4}{h^2}\{\mathbf{u}\}_{i-1} + \frac{1}{h^2}\{\mathbf{u}\}_{i-2}\right] \\ + [\mathbf{C}]\left[\frac{3}{h}\{\mathbf{u}\}_i - \frac{3}{2h}\{\mathbf{u}\}_{i-1} + \frac{1}{3h}\{\mathbf{u}\}_{i-2}\right]\end{aligned} \tag{11.331}$$

It is noted that one needs to determine $\{\mathbf{u}\}_i$, $\{\mathbf{u}\}_{i-1}$, and $\{\mathbf{u}\}_{i-2}$ to obtain the solution $\{\mathbf{u}\}_{i+1}$. Hence, we see that, like the central difference method, a special starting procedure is required. Several methods have been used for the starting procedure. Houbolt initially used, at time equal to zero, the values $\{\mathbf{u}\}_{-1} = \{\mathbf{u}\}_{-2} = 0$. After calculating $\{\mathbf{u}\}_1$, the following values were used:

$$\begin{aligned}\{\ddot{\mathbf{u}}\}_0 &= [\mathbf{M}]^{-1}[\{\mathbf{f}(t_0)\} - [\mathbf{C}]\{\dot{\mathbf{u}}\}_0 - [\mathbf{K}]\{\mathbf{u}\}_0] \\ \{\mathbf{u}\}_{-1} &= h^2\{\ddot{\mathbf{u}}\}_0 - \{\mathbf{u}\}_1 + 2\{\mathbf{u}\}_0 \\ \{\mathbf{u}\}_{-2} &= 6h\{\dot{\mathbf{u}}\}_0 + 6h^2\{\ddot{\mathbf{u}}\}_0 - 8\{\mathbf{u}\}_1 + 9\{\mathbf{u}\}_0\end{aligned} \tag{11.332}$$

One could also employ the central difference method to start the method. Starting with the initial conditions $\{\mathbf{u}\}_0 = \{\mathbf{u}(t=0)\}$ and $\{\dot{\mathbf{u}}\}_0 = \{\dot{\mathbf{u}}(t=0)\}$, calculate $\{\ddot{\mathbf{u}}\}_0$ using Equation 11.322. Then, compute $\{\mathbf{u}\}_{-1}$ using Equation 11.323. Starting with $i = 0$, compute $\{\mathbf{u}\}_1$ and $\{\mathbf{u}\}_2$ using Equation 11.316. We can then proceed solving Houbolt's Equation 11.329 starting at $i = 2$. The algorithm for using the Houbolt method is given in Table 11.13.

**EXAMPLE 11.10**

Determine the response of the two degrees of freedom, second-order system considered in Example 11.9 using the Houbolt method and using the central difference method to start the method.

**Solution**

Use $h = \Delta t = 0.1$. The value of $\{\ddot{\mathbf{u}}\}_0$ is determined from Equation 11.322

$$[\mathbf{M}]\{\ddot{\mathbf{u}}\}_0 = \begin{Bmatrix}0\\5\end{Bmatrix} \quad \{\ddot{\mathbf{u}}\}_0 = \begin{Bmatrix}0\\5\end{Bmatrix}$$

**TABLE 11.13**

Algorithm for Houbolt Method for MDOF System

*Initial Computations:*

1. Set the initial conditions at $t = 0$: $\{\mathbf{u}\}_0 = \{\mathbf{u}(0)\}$ and $\{\dot{\mathbf{u}}\}_0 = \{\dot{\mathbf{u}}(0)\}$ and calculate

$$\{\ddot{\mathbf{u}}\}_0 = [\mathbf{M}]^{-1}[\{\mathbf{f}\}_0 - [\mathbf{C}]\{\dot{\mathbf{u}}\}_0 - [\mathbf{K}]\{\mathbf{u}\}_0]$$

2. Select a time step $h$
3. Calculate

$$a_0 = \frac{2}{h^2};\quad a_1 = \frac{11}{6h};\quad a_2 = \frac{5}{h^2};\quad a_3 = \frac{3}{h}$$
$$a_4 = \frac{4}{h^2};\quad a_5 = \frac{3}{2h};\quad a_6 = \frac{1}{h^2};\quad a_7 = \frac{1}{3h}$$

4. Using some method, calculate $\{\mathbf{u}\}_1$ and $\{\mathbf{u}\}_2$
5. Calculate effective stiffness matrix $[\hat{\mathbf{K}}]$: $[\hat{\mathbf{K}}] = [\mathbf{K}] + a_0[\mathbf{M}] + a_1[\mathbf{C}]$
6. Triangularize $[\hat{\mathbf{K}}]$: $[\hat{\mathbf{K}}] = [\mathbf{L}][\mathbf{D}][\mathbf{L}]^T$

*Calculate for each time step i: $i = 2, 3, \ldots, t_f$*

7. Calculate effective load vector $\{\hat{\mathbf{f}}\}_i$ at time $t_i$

$$\{\hat{\mathbf{f}}\}_{i+1} = \{\mathbf{f}\}_{i+1} + [\mathbf{M}][a_2\{\mathbf{u}\}_i - a_4\{\mathbf{u}\}_{i-1} + a_6\{\mathbf{u}\}_{i-2}] + [\mathbf{C}][a_3\{\mathbf{u}\}_i - a_5\{\mathbf{u}\}_{i-1} + a_7\{\mathbf{u}\}_{i-2}]$$

8. At time $t_{i+1}$, calculate the displacement vector $\{\mathbf{u}\}_{i+1}$

$$[\mathbf{L}][\mathbf{D}][\mathbf{L}]^T\{\mathbf{u}\}_{i+1} = \{\hat{\mathbf{f}}\}_i$$

9. Calculate the velocity and acceleration vectors

$$\{\dot{\mathbf{u}}\}_{i+1} = a_1\{\mathbf{u}\}_{i+1} - a_3\{\mathbf{u}\}_i + a_5\{\mathbf{u}\}_{i-1} - a_7\{\mathbf{u}\}_{i-2}$$

$$\{\ddot{\mathbf{u}}\}_{i+1} = a_0\{\mathbf{u}\}_{i+1} - a_2\{\mathbf{u}\}_i + a_4\{\mathbf{u}\}_{i-1} - a_6\{\mathbf{u}\}_{i-2}$$

and $\{\mathbf{u}\}_{-1}$ from Equation 11.323

$$\{\mathbf{u}\}_{-1} = \{\mathbf{u}\}_0 - h\{\dot{\mathbf{u}}\}_0 + a_3\{\ddot{\mathbf{u}}\}_0 = \begin{Bmatrix} 0 \\ 0.025 \end{Bmatrix}$$

Determine $\{\mathbf{u}\}_1$ and $\{\mathbf{u}\}_2$ using Equation 11.311 or 11.312

$$\{\mathbf{u}\}_1 = \begin{Bmatrix} 0.00000 \\ 0.02500 \end{Bmatrix},\quad \{\mathbf{u}\}_2 = \begin{Bmatrix} 0.000500 \\ 0.099000 \end{Bmatrix}$$

The results are given in Table 11.14.

### 11.8.3 Newmark Method

Newmark's method (Newmark 1959) for time integration was presented in Chapter 3 for SDOF systems. The method can easily be extended to matrix format for MDOF systems. The velocity and displacement at times $t_i$ and $t_{i+1}$ become, where $t_{i+1} = t_i + h$,

$$\{\dot{\mathbf{u}}\}_{i+1} = \{\dot{\mathbf{u}}\}_i + (1-\gamma)h\{\ddot{\mathbf{u}}\}_i + \gamma h\{\ddot{\mathbf{u}}\}_{i+1} \tag{11.333}$$

**TABLE 11.14**

Displacement Results for Example 11.10

| Time $t$ | $u_1$ | $u_2$ |
|---|---|---|
| 0 | 0.000000 | 0.000000 |
| 0.1 | 0.000000 | 0.025000 |
| 0.2 | 0.000500 | 0.099000 |
| 0.3 | 0.003332 | 0.218154 |
| 0.4 | 0.010779 | 0.377390 |
| 0.5 | 0.025475 | 0.570388 |
| 0.6 | 0.050186 | 0.789722 |
| 0.7 | 0.087550 | 1.027121 |
| 0.8 | 0.139784 | 1.273774 |
| 0.9 | 0.208408 | 1.520684 |
| 1.0 | 0.293997 | 1.759012 |
| 1.5 | 0.913187 | 2.563815 |
| 2.0 | 1.400766 | 2.397693 |
| 2.5 | 1.082904 | 1.502937 |
| 3.0 | 0.031940 | 0.607498 |
| 3.5 | −0.762674 | 0.299526 |
| 4.0 | −0.424410 | 0.698924 |

$$\{\mathbf{u}\}_{i+1} = \{\mathbf{u}\}_i + h\{\dot{\mathbf{u}}\}_i + \left[(1-2\beta)\{\ddot{\mathbf{u}}\}_i + 2\beta\{\ddot{\mathbf{u}}\}_{i+1}\right]\frac{h^2}{2} \tag{11.334}$$

where $0 < \gamma < 1$ and $0 < \beta < 1/2$ are parameters that can be determined to obtain accuracy and stability. These parameters establish how much acceleration is allowed into the velocity and displacement equations at the end of each time interval $h$. Typically, $\gamma$ is set to 1/2, which corresponds to zero artificial damping and $\beta$ is set to a value between 1/6 and 1/4. For average acceleration, the parameters are $\gamma = 1/2$ and $\beta = 1/4$, where for linear acceleration the parameters are $\gamma = 1/2$ and $\beta = 1/6$. Since the right-hand side of Equations 11.333 and 11.334 depend on the acceleration at time $t_{i+1}$, the method is implicit and stable provided that $\gamma \geq 1/2$ and $\beta \geq (\gamma+1/2)^2/4$ (Hilber et al. 1977). The equation of motion at time point $t_{i+1}$ is given as

$$[\mathbf{M}]\{\ddot{\mathbf{u}}\}_{i+1} + [\mathbf{C}]\{\dot{\mathbf{u}}\}_{i+1} + [\mathbf{K}]\{\mathbf{u}\}_{i+1} = \{\mathbf{f}\}_{i+1} \tag{11.335}$$

The finite difference expressions for the Newmark beta method are obtained using Equations 11.333 and 11.334, thus we have

$$\{\ddot{\mathbf{u}}\}_{i+1} = \frac{1}{\beta h^2}(\{\mathbf{u}\}_{i+1} - \{\mathbf{u}\}_i) - \frac{1}{\beta h}\{\dot{\mathbf{u}}\}_i - \left(\frac{1}{2\beta} - 1\right)\{\ddot{\mathbf{u}}\}_i \tag{11.336}$$

and

$$\{\dot{\mathbf{u}}\}_{i+1} = \frac{\gamma}{\beta h}(\{\mathbf{u}\}_{i+1} - \{\mathbf{u}\}_i) + \left(1 - \frac{\gamma}{\beta}\right)\{\dot{\mathbf{u}}\}_i + h\left(1 - \frac{\gamma}{2\beta}\right)\{\ddot{\mathbf{u}}\}_i \tag{11.337}$$

Substituting Equations 11.336 and 11.337 into Equation 11.335 yields

$$[\hat{\mathbf{K}}]\{\mathbf{u}\}_{i+1} = \{\hat{\mathbf{f}}\}_{i+1} \tag{11.338}$$

where

$$[\hat{\mathbf{K}}] = [\mathbf{K}] + \frac{1}{\beta h^2}[\mathbf{M}] + \frac{\gamma}{\beta h}[\mathbf{C}] \tag{11.339}$$

and

$$\begin{aligned}\{\hat{\mathbf{f}}\}_{i+1} = \{\mathbf{f}\}_i + [\mathbf{M}]\left[\frac{1}{\beta h^2}\{\mathbf{u}\}_i + \frac{1}{\beta h}\{\dot{\mathbf{u}}\}_i + \left(\frac{1}{2\beta} - 1\right)\{\ddot{\mathbf{u}}\}_i\right] \\ + [\mathbf{C}]\left[\frac{\gamma}{\beta h}\{\mathbf{u}\}_i + \left(\frac{\gamma}{\beta} - 1\right)\{\dot{\mathbf{u}}\}_i + h\left(\frac{\gamma}{2\beta} - 1\right)\{\ddot{\mathbf{u}}\}_i\right]\end{aligned} \tag{11.340}$$

Solution of Equation 11.338 gives $\{\mathbf{u}\}_{i+1}$, which, when substituted into Equations 11.336 and 11.337, gives the velocity and acceleration $\{\dot{\mathbf{u}}\}_{i+1}$ and $\{\ddot{\mathbf{u}}\}_{i+1}$.

The Newmark numerical method is unconditional stable (Hughes and Belytschko 1983) when

$$h \le h_{cr} = \frac{T}{2\pi\sqrt{\gamma/2 - \beta}} \tag{11.341}$$

where $T$ is the period of vibration. The algorithm for using the Newmark beta method is given in Table 11.15.

**EXAMPLE 11.11**

Determine the response of the two degrees of freedom, second-order system considered in Example 11.9 using the Newmark beta method.

**Solution**

The initial conditions at $t = 0$ were determined in Example 11.9:

$$\{\mathbf{u}\}_0 = \begin{Bmatrix}0\\0\end{Bmatrix}, \quad \{\dot{\mathbf{u}}\}_0 = \begin{Bmatrix}0\\0\end{Bmatrix}, \quad \{\ddot{\mathbf{u}}\}_0 = \begin{Bmatrix}0\\5\end{Bmatrix}$$

Select the integration method: linear acceleration method with $\beta = 1/2$ and $\gamma = 1/6$. Determine the integration constants

$$a_0 = \frac{1}{\beta h^2} = 600 \quad a_1 = \frac{\gamma}{\beta h} = 30 \quad a_2 = \frac{1}{\beta h} = 60 \quad a_3 = \frac{1}{2\beta} - 1 = 2$$

$$a_4 = \frac{\gamma}{\beta} - 1 = 2 \quad a_5 = h\left(\frac{\gamma}{2\beta} - 1\right) = 0.05 \quad a_6 = (1-\gamma)h = 0.05 \quad a_7 = \gamma h = 0.05$$

**TABLE 11.15**

Algorithm for Newmark Beta Method for MDOF Systems

*Initial Computations:*

1. Set the initial conditions $\{\mathbf{u}\}_0 = \{\mathbf{u}(0)\}$ and $\{\dot{\mathbf{u}}\}_0 = \{\dot{\mathbf{u}}(0)\}$ and calculate

$$[\mathbf{M}]\{\ddot{\mathbf{u}}\}_0 = [\{\mathbf{f}\}_0 - [\mathbf{C}]\{\dot{\mathbf{u}}\}_0 - \mathbf{K}\{\mathbf{u}\}_0]$$

2. Select $\gamma$ and $\beta$

$$\text{Average acceleration method } \left(\gamma = \frac{1}{2}, \beta = \frac{1}{4}\right)$$

$$\text{Linear acceleration method } \left(\gamma = \frac{1}{2}, \beta = \frac{1}{6}\right)$$

3. Select a time step $h$: $h \le h_{cr} = \dfrac{T}{2\pi\sqrt{\gamma/2-\beta}}$, based on $\gamma$ and $\beta$

4. Calculate constants

$$a_0 = \frac{1}{\beta h^2}; \quad a_1 = \frac{\gamma}{\beta h}; \quad a_2 = \frac{1}{\beta h}; \quad a_3 = \frac{1}{2\beta} - 1;$$

$$a_4 = \frac{\gamma}{\beta} - 1; \quad a_5 = h\left(\frac{\gamma}{2\beta} - 1\right); \quad a_6 = (1-\gamma)h; \quad a_7 = \gamma h$$

5. Calculate the effective stiffness matrix $[\hat{\mathbf{K}}]$: $[\hat{\mathbf{K}}] = [\mathbf{K}] + a_0[\mathbf{M}] + a_1[\mathbf{C}]$

6. Triangularize $[\hat{\mathbf{K}}]$: $[\hat{\mathbf{K}}] = [\mathbf{L}][\mathbf{D}][\mathbf{L}]^T$

*Calculate for each time step: $i = 0, 1, 2, \ldots, t_f$*

7. Increment time: $t_{i+1} = t_i + h$

8. Calculate effective forces $\{\hat{\mathbf{f}}\}_{i+1}$

$$\{\hat{\mathbf{f}}\}_{i+1} = \{\mathbf{f}\}_{i+1} + [\mathbf{M}][a_0\{\mathbf{u}\}_i + a_2\{\dot{\mathbf{u}}\}_i + a_3\{\ddot{\mathbf{u}}\}_i] + [\mathbf{C}][a_1\{\mathbf{u}\}_i + a_4\{\dot{\mathbf{u}}\}_i + a_5\{\ddot{\mathbf{u}}\}_i]$$

9. At time $t_{i+1}$, calculate the displacement vector $\{\mathbf{u}\}_{i+1}$

$$[\mathbf{L}][\mathbf{D}][\mathbf{L}]^T\{\mathbf{u}\}_{i+1} = \{\hat{\mathbf{f}}\}_{i+1}$$

10. Determine the velocity and acceleration vectors at time $t_{i+1}$

$$\{\ddot{\mathbf{u}}\}_{i+1} = a_0(\{\mathbf{u}\}_{i+1} - \{\mathbf{u}\}_i) - a_2\{\dot{\mathbf{u}}\}_i - a_3\{\ddot{\mathbf{u}}\}_i$$

$$\{\dot{\mathbf{u}}\}_{i+1} = \{\dot{\mathbf{u}}\}_i + a_6\{\ddot{\mathbf{u}}\}_i + a_7\{\ddot{\mathbf{u}}\}_{i+1}$$

Calculate the effective stiffness matrix

$$[\hat{\mathbf{K}}] = [\mathbf{K}] + a_0[\mathbf{M}] + a_1[\mathbf{C}]$$

$$= \begin{bmatrix} 6 & -2 \\ -2 & 8 \end{bmatrix} + 600\begin{bmatrix} 1 & 0 \\ 0 & 2 \end{bmatrix} = \begin{bmatrix} 606 & -2 \\ -2 & 1208 \end{bmatrix}$$

Factorization of $[\hat{\mathbf{K}}]$ and then perform the step-by-step integration $n = 0, 1, 2, \ldots, t_f$. Determine the effective load vector

$$\{\hat{\mathbf{f}}\}_{i+1} = \{\mathbf{f}\}_{i+1} + [\mathbf{M}]\left[a_0\{\mathbf{u}\}_i + a_2\{\dot{\mathbf{u}}\}_i + a_3\{\ddot{\mathbf{u}}\}_i\right] + [\mathbf{C}]\left[a_1\{\mathbf{u}\}_i + a_4\{\dot{\mathbf{u}}\}_i + a_5\{\ddot{\mathbf{u}}\}_i\right]$$

$$= \begin{Bmatrix} 0 \\ 10 \end{Bmatrix} + \begin{bmatrix} 1 & 0 \\ 0 & 2 \end{bmatrix}\left[600\{\mathbf{u}\}_i + 60\{\dot{\mathbf{u}}\}_i + 2\{\ddot{\mathbf{u}}\}_i\right]$$

**TABLE 11.16**
Displacement Results for Example 11.11

| Time $t$ | $u_1$ | $u_2$ |
|---|---|---|
| 0 | 0.000000 | 0.000000 |
| 0.1 | 0.000082 | 0.024835 |
| 0.2 | 0.000812 | 0.098353 |
| 0.3 | 0.003591 | 0.217644 |
| 0.4 | 0.010603 | 0.377998 |
| 0.5 | 0.024585 | 0.573118 |
| 0.6 | 0.048545 | 0.795395 |
| 0.7 | 0.085433 | 1.036240 |
| 0.8 | 0.137796 | 1.286452 |
| 0.9 | 0.207447 | 1.536613 |
| 1.0 | 0.295172 | 1.777477 |
| 1.5 | 0.939328 | 2.574584 |
| 2.0 | 1.433807 | 2.370006 |
| 2.5 | 1.050773 | 1.448098 |
| 3.0 | −0.077577 | 0.568663 |
| 3.5 | −0.828216 | 0.304104 |
| 4.0 | −0.326013 | 0.744480 |

For each time step, calculate the displacement vector

$$[\mathbf{L}][\mathbf{D}][\mathbf{L}]^T\{\mathbf{u}\}_{i+1} = \{\hat{\mathbf{f}}\}_{i+1}$$

The results are given in Table 11.16.

### 11.8.4 Wilson-Theta Method

The Wilson-theta method is a modification of the Newmark linear acceleration methods where the acceleration is assumed to vary in a linear manner in the extended time interval from time $t_i$ to time $t_{i+\theta} = t_i + \theta h$, where $\theta \geq 1.0$ (see Figure 11.32). The parameter $\theta$ is selected to yield the desired characteristics of stability and accuracy. This is an implicit scheme and does not require any special starting procedures.

Using dynamic equilibrium equations at time $t_{i+\theta} = t_i + \theta h$, our equation of motion gives

$$[\mathbf{M}]\{\ddot{\mathbf{u}}\}_{i+\theta} + [\mathbf{C}]\{\dot{\mathbf{u}}\}_{i+\theta} + [\mathbf{K}]\{\mathbf{u}\}_{i+\theta} = \{\mathbf{f}\}_{i+\theta} \tag{11.342}$$

where the force vector is given by

$$\{\mathbf{f}\}_{i+\theta} = (1-\theta)\{\mathbf{f}\}_i + \theta\{\mathbf{f}\}_{i+1} \tag{11.343}$$

Since $\{\ddot{\mathbf{u}}(t)\}$ varies linearly between $t_i$ and $t_{i+\theta}$, as shown in Figure 11.32, the linear change in acceleration $\{\ddot{\mathbf{u}}(t)\}$, at any time $t_i + \tau$ where $0 \leq \tau \leq \theta h$, and $\tau$ the integration variable, can be expressed as

$$\{\ddot{\mathbf{u}}(t_i + \tau)\} = \{\ddot{\mathbf{u}}\}_i + \frac{\tau}{\theta h}(\{\ddot{\mathbf{u}}\}_{i+\theta} - \{\ddot{\mathbf{u}}\}_i) \tag{11.344}$$

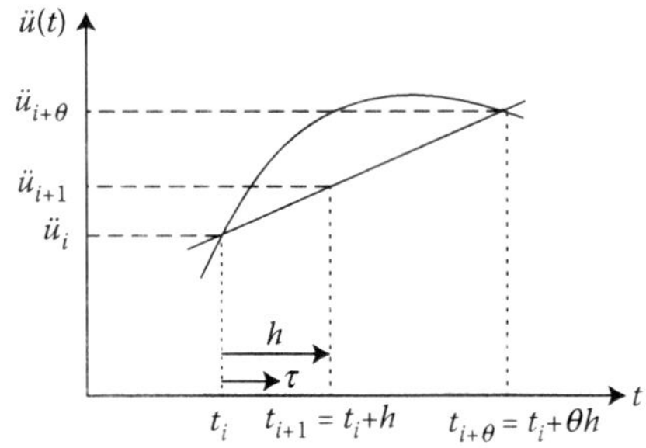


**FIGURE 11.32**
Wilson-theta method.

Integrating over time variable $\tau$ yields the velocity

$$\{\dot{\mathbf{u}}(t_i+\tau)\} = \{\dot{\mathbf{u}}\}_i + \{\ddot{\mathbf{u}}\}_i\tau + \frac{\tau^2}{2\theta h}(\{\ddot{\mathbf{u}}\}_{i+\theta} - \{\ddot{\mathbf{u}}\}_i) \tag{11.345}$$

Integrating again gives the displacement as

$$\{\mathbf{u}(t_i+\tau)\} = \{\mathbf{u}\}_i + \{\dot{\mathbf{u}}\}_i\tau + \frac{\tau^2}{2}\{\ddot{\mathbf{u}}\}_i + \frac{\tau^3}{6\theta h}(\{\ddot{\mathbf{u}}\}_{i+\theta} - \{\ddot{\mathbf{u}}\}_i) \tag{11.346}$$

At time $t_{i+\theta} = t_i + \theta$, $\tau = \theta h$, thus, Equations 11.345 and 11.346 become

$$\{\dot{\mathbf{u}}\}_{i+\theta} = \{\dot{\mathbf{u}}\}_i + \frac{\theta h}{2}(\{\ddot{\mathbf{u}}\}_{i+\theta} + \{\ddot{\mathbf{u}}\}_i) \tag{11.347}$$

$$\{\mathbf{u}\}_{i+\theta} = \{\mathbf{u}\}_i + \theta h\{\dot{\mathbf{u}}\}_i + \frac{(\theta h)^2}{6}(\{\ddot{\mathbf{u}}\}_{i+\theta} + 2\{\ddot{\mathbf{u}}\}_i) \tag{11.348}$$

Equations 11.347 and 11.348 can be solved for $\{\ddot{\mathbf{u}}\}_{i+\theta}$ and $\{\dot{\mathbf{u}}\}_{i+\theta}$ in terms of $\{\mathbf{u}\}_{i+\theta}$. Thus, from Equation 11.348, we have

$$\{\ddot{\mathbf{u}}\}_{i+\theta} = \frac{6}{(\theta h)^2}(\{\mathbf{u}\}_{i+\theta} - \{\mathbf{u}\}_i) - \frac{6}{(\theta h)}\{\dot{\mathbf{u}}\}_i - 2\{\ddot{\mathbf{u}}\}_i \tag{11.349}$$

and substituting Equation 11.349 into Equation 11.347 we have for the velocity at $t_{i+\theta}$

$$\{\dot{\mathbf{u}}\}_{i+\theta} = \frac{3}{\theta h}(\{\mathbf{u}\}_{i+\theta} - \{\mathbf{u}\}_i) - 2\{\dot{\mathbf{u}}\}_i - \frac{\theta h}{2}\{\ddot{\mathbf{u}}\}_i \tag{11.350}$$

An expression in terms of $\{\mathbf{u}\}_{i+\theta}$ can be obtained by substituting Equations 11.349 and 11.350 into Equation 11.342. Hence, substituting and collecting terms yields

$$[\hat{\mathbf{K}}]\{\mathbf{u}\}_{i+\theta} = \{\hat{\mathbf{f}}\}_{i+\theta} \tag{11.351}$$

$$[\hat{\mathbf{K}}] = [\mathbf{K}] + \frac{6}{(\theta h)^2}[\mathbf{M}] + \frac{3}{\theta h}[\mathbf{C}] \tag{11.352}$$

and

$$\begin{aligned}\{\hat{\mathbf{f}}\}_{i+\theta} &= \{\mathbf{f}\}_i + \theta(\{\mathbf{f}\}_{i+1} - \{\mathbf{f}\}_i) + [\mathbf{M}]\left(\frac{6}{(\theta h)^2}\{\mathbf{u}\}_i + \frac{6}{\theta h}\{\dot{\mathbf{u}}\}_i + 2\{\ddot{\mathbf{u}}\}_i\right) \\ &+ [\mathbf{C}]\left(\frac{3}{\theta h}\{\mathbf{u}\}_i + 2\{\dot{\mathbf{u}}\}_i + \frac{\theta h}{2}\{\ddot{\mathbf{u}}\}_i\right)\end{aligned} \tag{11.353}$$

After determining $\{\mathbf{u}\}_{i+\theta}$ from Equation 11.351, it can be substituted into Equation 11.349 yielding $\{\ddot{\mathbf{u}}\}_{i+\theta}$, this is then in turn used in Equations 11.346, evaluated at $\tau = h$ to give the acceleration at time $t_{i+1}$.

$$\{\ddot{\mathbf{u}}\}_{i+1} = \frac{6}{\theta^3 h^2}(\{\mathbf{u}\}_{i+\theta} - \{\mathbf{u}\}_i) - \frac{6}{\theta^2 h}\{\dot{\mathbf{u}}\}_i + \left(1 - \frac{3}{\theta}\right)\{\ddot{\mathbf{u}}\}_i \tag{11.354}$$

The velocity can be calculated at time $t_{i+1}$ knowing that the acceleration is linear in time or using Equation 11.347 with $\theta = 1$ leads to

$$\{\dot{\mathbf{u}}\}_{i+1} = \{\dot{\mathbf{u}}\}_i + \frac{h}{2}(\{\ddot{\mathbf{u}}\}_{i+1} + \{\ddot{\mathbf{u}}\}_i) \tag{11.355}$$

Substituting Equation 11.355 into Equation 11.354 yields

$$\{\dot{\mathbf{u}}\}_{i+1} = \left(1 - \frac{3}{\theta^2}\right)\{\dot{\mathbf{u}}\}_i + \left(1 - \frac{3}{2\theta}\right)h\{\ddot{\mathbf{u}}\}_i + \frac{3}{\theta^3 h}(\{\mathbf{u}\}_{i+\theta} - \{\mathbf{u}\}_i) \tag{11.356}$$

For the displacement, we have using Equation 11.348 with $\theta = 1$

$$\{\mathbf{u}\}_{i+1} = \{\mathbf{u}\}_i + h\{\dot{\mathbf{u}}\}_i + \frac{h^2}{6}(\{\ddot{\mathbf{u}}\}_{i+1} + 2\{\ddot{\mathbf{u}}\}_i) \tag{11.357}$$

or again using Equation 11.354

$$\{\mathbf{u}\}_{i+1} = \{\mathbf{u}\}_i + \left(1 - \frac{1}{\theta^2}\right)h\{\dot{\mathbf{u}}\}_i + \left(1 - \frac{1}{\theta}\right)\frac{h^2}{2}\{\ddot{\mathbf{u}}\}_i + \frac{1}{\theta^3}(\{\mathbf{u}\}_{i+\theta} - \{\mathbf{u}\}_i) \tag{11.358}$$

For $\theta = 1$, the Wilson-theta method reduces to the linear acceleration method. In linear problems, the method is unconditionally stable for $\theta \geq 1.37$ (Bathe and Wilson 1976). The value of $\theta = 1.4$ is usually used. Using $\theta = 1$, $\gamma = 1/2$, and $\beta = 1/4$ leads to the Newmark average acceleration method. For $\theta = 1$, $\gamma = 1/2$, and $\beta = 1/6$, we obtain the Newmark linear acceleration method. The algorithm for the Wilson-theta method is given in Table 11.17.

**TABLE 11.17**

Algorithm for Wilson's Theta Method

| |
|---|
| *Initial Computations:* |
| 1. Set the initial conditions $\{\mathbf{u}\}_0 = \{\mathbf{u}(0)\}$ and $\{\dot{\mathbf{u}}\}_0 = \{\dot{\mathbf{u}}(0)\}$ and calculate $$\{\ddot{\mathbf{u}}\}_0 = [\mathbf{M}]^{-1}[\{\mathbf{f}\}_0 - [\mathbf{C}]\{\dot{\mathbf{u}}\}_0 - [\mathbf{K}]\{\mathbf{u}\}_0]$$ |
| 2. Selection of the integration method<br>Newmark average acceleration method: $\theta = 1$, $\gamma = 1/2$, and $\beta = 1/4$<br>Newmark linear acceleration method: $\theta = 1$, $\gamma = 1/2$, and $\beta = 1/6$<br>Wilson-theta method: $\theta > 1$, $\gamma = 1/2$, and $\beta = 1/6$ (unconditionally stable for $\theta \geq 1.37$) |
| 3. Select time step $h \leq h_{cr} = \dfrac{T}{2\pi\sqrt{\gamma/2-\beta}}$, for $\gamma = 1/2$ and $\beta = 1/6$ |
| 4. Calculate the integration constants $$a_0 = \frac{1}{\beta(\theta h)^2}, \quad a_1 = \frac{\gamma}{\beta\theta h}, \quad a_2 = \frac{1}{\beta\theta h}, \quad a_3 = \frac{1}{2\beta} - 1; \quad a_4 = \frac{\gamma}{\beta} - 1$$ $$a_5 = \left(\frac{\gamma}{\beta} - 2\right)\frac{\theta h}{2}, \quad a_6 = (1-\gamma)h, \quad a_7 = \lambda h, \quad a_8 = \left(\frac{1}{2} - \beta\right)h^2, \quad a_9 = \beta h^2$$ |
| 5. Determine the effective stiffness matrix $[\hat{\mathbf{K}}]$:$[\hat{\mathbf{K}}] = [\mathbf{K}] + a_0[\mathbf{M}] + a_1[\mathbf{C}]$ |
| 6. Triangularize $[\hat{\mathbf{K}}]$: $[\hat{\mathbf{K}}] = [\mathbf{L}][\mathbf{D}][\mathbf{L}]^T$ |
| *Calculate for each time step: for i = 1, 2, ..., $t_f$* |
| 7. Determine the effective load vector $$\{\hat{\mathbf{f}}\}_{i+\theta} = \{\mathbf{f}\}_i + \theta(\{\mathbf{f}\}_{i+1} - \{\mathbf{f}\}_i) + [\mathbf{M}](a_0\{\mathbf{u}\}_i + a_2\{\dot{\mathbf{u}}\}_i + a_3\{\ddot{\mathbf{u}}\}_i) + [\mathbf{C}](a_1\{\mathbf{u}\}_i + a_4\{\dot{\mathbf{u}}\}_i + a_5\{\ddot{\mathbf{u}}\}_i)$$ |
| 8. Solve for the displacements at time $t_{n+\theta}$ $$[\mathbf{L}][\mathbf{D}][\mathbf{L}]^T\{\mathbf{u}\}_{i+\theta} = \{\hat{\mathbf{f}}\}_{i+\theta}$$ |
| 9. Calculate the displacements, velocities, and acceleration at time $t_{i+1}$ $$\{\ddot{\mathbf{u}}\}_{i+\theta} = a_0(\{\mathbf{u}\}_{i+\theta} - \{\mathbf{u}\}_i) - a_2\{\dot{\mathbf{u}}\}_i - a_3\{\ddot{\mathbf{u}}\}_i$$ $$\{\ddot{\mathbf{u}}\}_{i+1} = \{\ddot{\mathbf{u}}\}_i + \frac{1}{\theta}(\{\ddot{\mathbf{u}}\}_{i+\theta} - \{\ddot{\mathbf{u}}\}_i)$$ $$\{\dot{\mathbf{u}}\}_{i+1} = \{\dot{\mathbf{u}}\}_i + a_6\{\ddot{\mathbf{u}}\}_i + a_7\{\ddot{\mathbf{u}}\}_{i+1}$$ $$\{\mathbf{u}\}_{i+1} = \{\mathbf{u}\}_i + h\{\dot{\mathbf{u}}\}_i + a_8\{\ddot{\mathbf{u}}\}_i + a_9\{\ddot{\mathbf{u}}\}_{i+1}$$ |

**EXAMPLE 11.12**

With the Wilson-theta method, calculate the displacement response of Example 11.9.

**Solution**

The initial conditions at $t = 0$ that were determined in Example 11.9:

$$\{\mathbf{u}\}_0 = \begin{Bmatrix} 0 \\ 0 \end{Bmatrix}, \quad \{\dot{\mathbf{u}}\}_0 = \begin{Bmatrix} 0 \\ 0 \end{Bmatrix}, \quad \{\ddot{\mathbf{u}}\}_0 = \begin{Bmatrix} 0 \\ 5 \end{Bmatrix}$$

Select the integration method for the Wilson-theta method: $\gamma = 1/2$, $\beta = 1/6$, $\theta = 1.41$.

Select the time step: $h = 0.1$ s.
Determine the integration constants:

$$a_0 = \frac{1}{\beta(\theta h)^2} = 301.78 \quad a_1 = \frac{\gamma}{\beta\theta h} = 21.28 \quad a_2 = \frac{1}{\beta\theta h} = 42.55$$

$$a_3 = \frac{1}{2\beta} - 1 = 2 \qquad a_4 = \frac{\gamma}{\beta} - 1 = 2 \qquad a_5 = \left(\frac{\gamma}{\beta} - 2\right)\frac{\theta h}{2} = 0.071$$

$$a_6 = (1-\gamma)h = 0.05 \qquad a_7 = \lambda h = 0.05 \qquad a_8 = \left(\frac{1}{2} - \beta\right)h^2 = 0.0033$$

$$a_9 = \beta h^2 = 0.0017$$

Determine the effective stiffness matrix $[\hat{\mathbf{K}}]$:

$$\begin{aligned}[\hat{\mathbf{K}}] &= [\mathbf{K}] + a_0[\mathbf{M}] + a_1[\mathbf{C}]\\ &= \begin{bmatrix} 6 & -2 \\ -2 & 8 \end{bmatrix} + 301.78\begin{bmatrix} 1 & 0 \\ 0 & 2 \end{bmatrix}\\ &= \begin{bmatrix} 307.78 & -2 \\ -2 & 611.59 \end{bmatrix}\end{aligned}$$

Triangularize the effective stiffness matrix $[\hat{\mathbf{K}}]$ and then perform the step-by-step integration $n = 0, 1, 2, \ldots, t_f$ incrementing the time: $t_{i+1} = t_i + h, t_0 = 0$. Determine the effective load vector

$$\begin{aligned}\{\hat{\mathbf{f}}\}_{i+\theta} &= \{\mathbf{f}\}_i + \theta(\{\mathbf{f}\}_{i+1} - \{\mathbf{f}\}_i) + [\mathbf{M}](a_0\{\mathbf{u}\}_i + a_2\{\dot{\mathbf{u}}\}_i + a_3\{\ddot{\mathbf{u}}\}_i)\\ &\quad + [\mathbf{C}](a_1\{\mathbf{u}\}_i + a_4\{\dot{\mathbf{u}}\}_i + a_5\{\ddot{\mathbf{u}}\}_i)\\ &= \begin{Bmatrix} 0 \\ 10 \end{Bmatrix} + 1.41\left(\begin{Bmatrix} 0 \\ 10 \end{Bmatrix} - \begin{Bmatrix} 0 \\ 10 \end{Bmatrix}\right) + \begin{bmatrix} 1 & 0 \\ 0 & 2 \end{bmatrix}(301.78\{\mathbf{u}\}_i + 42.55\{\dot{\mathbf{u}}\}_i + 2\{\ddot{\mathbf{u}}\}_i)\end{aligned}$$

Determine the displacement $\{\mathbf{u}\}_{i+\theta}$ at time $t_{i+\theta}$

$$[\hat{\mathbf{K}}]\{\mathbf{u}\}_{i+\theta} = \{\hat{\mathbf{f}}\}_{i+\theta}$$

$$\begin{bmatrix} 307.78 & -2 \\ -2 & 611.59 \end{bmatrix}\{\mathbf{u}\}_{i+\theta} = \{\hat{\mathbf{f}}\}_{i+\theta}$$

Determine the acceleration, velocity, and displacement vectors at time $t_{i+1}$

$$\{\ddot{\mathbf{u}}\}_{i+\theta} = 301.78(\{\mathbf{u}\}_{i+\theta} - \{\mathbf{u}\}_i) - 42.55\{\dot{\mathbf{u}}\}_i - 2\{\ddot{\mathbf{u}}\}_i$$

$$\{\ddot{\mathbf{u}}\}_{i+1} = \{\ddot{\mathbf{u}}\}_i + 0.7092(\{\ddot{\mathbf{u}}\}_{i+\theta} - \{\ddot{\mathbf{u}}\}_i)$$

$$\{\dot{\mathbf{u}}\}_{i+1} = \{\dot{\mathbf{u}}\}_i + 0.05\{\ddot{\mathbf{u}}\}_i + 0.05\{\ddot{\mathbf{u}}\}_{i+1}$$

$$\{\mathbf{u}\}_{i+1} = \{\mathbf{u}\}_i + 0.1\{\dot{\mathbf{u}}\}_i + 0.0033\{\ddot{\mathbf{u}}\}_i + 0.0017\{\ddot{\mathbf{u}}\}_{i+1}$$

The results are given in Table 11.18.

**TABLE 11.18**

Displacement Results for Example 11.12

| Time $t$ | $u_1$ | $u_2$ |
|---|---|---|
| 0 | 0.000000 | 0.000000 |
| 0.1 | 0.000114 | 0.024768 |
| 0.2 | 0.001035 | 0.097878 |
| 0.3 | 0.004194 | 0.216311 |
| 0.4 | 0.011702 | 0.375398 |
| 0.5 | 0.026183 | 0.568953 |
| 0.6 | 0.050500 | 0.789512 |
| 0.7 | 0.087459 | 1.028654 |
| 0.8 | 0.139475 | 1.277353 |
| 0.9 | 0.208267 | 1.526357 |
| 1.0 | 0.294574 | 1.766563 |
| 1.5 | 0.926239 | 2.570361 |
| 2.0 | 1.420164 | 2.383476 |
| 2.5 | 1.067816 | 1.470773 |
| 3.0 | −0.031366 | 0.582297 |
| 3.5 | −0.807890 | 0.300193 |
| 4.0 | −0.373527 | 0.725767 |

### 11.8.5 HHT-Alpha Method

A precursor of the HHT method is the Newmark method (Newmark 1959), in which a family of integration formulas that depend on two parameters $\beta$ and $\gamma$ was defined. The HHT method (also called the $\alpha$ method) was introduced by Hilber et al. (1977) as a generalization of the Newmark beta method to control dissipation of the higher frequency modes. It is an implicit technique. The method uses the Newmark difference (Equations 11.333 and 11.334) but uses a modified time-discrete equation of motion in the following form:

$$[\mathbf{M}]\{\ddot{\mathbf{u}}\}_{i+1}+(1+\alpha)[\mathbf{C}]\{\dot{\mathbf{u}}\}_{i+1}-\alpha[\mathbf{C}]\{\dot{\mathbf{u}}\}_i+(1+\alpha)[\mathbf{K}]\{\mathbf{u}\}_{i+1}-\alpha[\mathbf{K}]\{\mathbf{u}\}_i=(1+\alpha)\{\mathbf{f}\}_{i+1}-\alpha\{\mathbf{f}\}_i \tag{11.359}$$

where $\alpha$ is the parameter that controls numerical damping. If $\alpha = 0$, Equation 11.359 would reduce to the Newmark beta method. Selecting $-1/3 \le \alpha \le 0$, $\gamma = 1/2 - \alpha$, and $\beta = (1-\alpha)^2/4$ results in an unconditionally stable second-order accuracy.

Substituting Equations 11.333 and 11.334 into Equation 11.359 gives after collecting terms

$$\begin{aligned}&\left[(1+\alpha)[\mathbf{K}]+(1+\alpha)\frac{\gamma}{\beta h}[\mathbf{C}]+\frac{1}{\beta h^2}[\mathbf{M}]\right]\{\mathbf{u}\}_{i+1}=(1+\alpha)\{\mathbf{f}\}_{i+1}-\alpha\{\mathbf{f}\}_i\\&+[\mathbf{M}]\left[\frac{1}{\beta h^2}\{\mathbf{u}\}_i+\frac{1}{\beta h}\{\dot{\mathbf{u}}\}_i+\left(\frac{1}{2\beta}-1\right)\{\ddot{\mathbf{u}}\}_i\right]\\&+[\mathbf{C}]\left[(1+\alpha)\frac{\gamma}{\beta h}\{\mathbf{u}\}_i+(1+\alpha)\left(\frac{\gamma}{\beta}-1\right)\{\dot{\mathbf{u}}\}_i+(1+\alpha)\left(\frac{\gamma}{2\beta}-1\right)h\{\ddot{\mathbf{u}}\}_i\right]+\alpha[\mathbf{C}]\{\dot{\mathbf{u}}\}_i+\alpha[K]\{\mathbf{u}\}_i\end{aligned} \tag{11.360}$$

or

$$[\hat{\mathbf{K}}]\{\mathbf{u}\}_{i+1} = \{\hat{\mathbf{f}}\}_{i+1} \tag{11.361}$$

where

$$[\hat{\mathbf{K}}] = \left[(1+\alpha)[\mathbf{K}] + (1+\alpha)\frac{\gamma}{\beta h}[\mathbf{C}] + \frac{1}{\beta h^2}[\mathbf{M}]\right] \tag{11.362}$$

and

$$\{\hat{\mathbf{f}}\}_{i+1} = (1+\alpha)\{\mathbf{f}\}_{i+1} - \alpha\{\mathbf{f}\}_i + [\mathbf{M}]\left[\frac{1}{\beta h^2}\{\mathbf{u}\}_i + \frac{1}{\beta h}\{\dot{\mathbf{u}}\}_i + \left(\frac{1}{2\beta} - 1\right)\{\ddot{\mathbf{u}}\}_i\right]$$
$$+ [\mathbf{C}]\left[(1+\alpha)\frac{\gamma}{\beta h}\{\mathbf{u}\}_i + (1+\alpha)\left(\frac{\gamma}{\beta} - 1\right)\{\dot{\mathbf{u}}\}_i + (1+\alpha)\left(\frac{\gamma}{2\beta} - 1\right)h\{\ddot{\mathbf{u}}\}_i\right] + \alpha[\mathbf{C}]\{\dot{\mathbf{u}}\}_i + \alpha[\mathbf{K}]\{\mathbf{u}\}_i \tag{11.363}$$

Equation 11.361 is used to calculate the displacement $\{\mathbf{u}\}_{i+1}$ and Equations 11.336 and 11.333 are used to calculate $\{\ddot{\mathbf{u}}\}_{i+1}$ and $\{\dot{\mathbf{u}}\}_{i+1}$. The algorithm for using the HHT-$\alpha$ method is given in Table 11.19.

**EXAMPLE 11.13**

Using the HHT-$\alpha$ method, calculate the displacement response of the system of Example 11.9.

**Solution**

The initial conditions at $t = 0$ that were determined in Example 11.9:

$$\{\mathbf{u}\}_0 = \begin{Bmatrix} 0 \\ 0 \end{Bmatrix}, \quad \{\dot{\mathbf{u}}\}_0 = \begin{Bmatrix} 0 \\ 0 \end{Bmatrix}, \quad \{\ddot{\mathbf{u}}\}_0 = \begin{Bmatrix} 0 \\ 5 \end{Bmatrix}$$

Select the integration method for the HHT-$\alpha$ method:

$$\alpha = -0.3, \quad \gamma = \left(\frac{1}{2} - \alpha\right) = 0.8, \quad \frac{\beta = (1-\alpha)^2}{4} = 0.4225$$

Select the time step: $h = 0.1$ s:
Determine the integration constants:

$$a_0 = \frac{1}{\beta h^2} = 236.69 \qquad a_1 = \frac{\gamma(1+\alpha)}{\beta h} = 13.25 \qquad a_2 = \frac{1}{\beta h} = 23.67$$

$$a_3 = \left(\frac{1}{2\beta} - 1\right) = 0.1834 \quad a_4 = (1+\alpha)\left(\frac{\gamma}{\beta} - 1\right) = 0.6254 \quad a_5 = (1+\alpha)\left(\frac{\gamma}{2\beta} - 1\right)h = -0.0037$$

$$a_6 = (1-\gamma)h = 0.02 \qquad a_7 = \gamma h = 0.08$$

**TABLE 11.19**

Algorithm for HHT-$\alpha$ method

*Initial Computations:*

1. Initial conditions $\{\mathbf{u}\}_0 = \{\mathbf{u}(0)\}$ and $\{\dot{\mathbf{u}}\}_0 = \{\dot{\mathbf{u}}(0)\}$ and determine $\{\ddot{\mathbf{u}}\}_0$

$$\{\ddot{\mathbf{u}}\}_0 = [\mathbf{M}]^{-1}[\{\mathbf{f}\}_0 - [\mathbf{C}]\{\dot{\mathbf{u}}\}_0 - [\mathbf{K}]\{\mathbf{u}\}_0]$$

2. Specify the integration method
   Newmark average acceleration method $\gamma = 1/2,\quad \beta = 1/4\quad \alpha = 0$
   Newmark linear acceleration method $\gamma = 1/2,\quad \beta = 1/6\quad \alpha = 0$
   HHT-$\alpha$ method $\gamma = 1/2 - \alpha\quad \beta = (1-\alpha)^2/4\quad -1/3 \le \alpha \le 0$
3. Select time step $h$: $h \le h_{cr} = 0.5513\,T$, for $\gamma = 1/2$ and $\beta = 1/6$ ($T$ is the smallest period of the system)
4. Determine the integration constants

$$a_0 = \frac{1}{\beta h^2},\qquad a_1 = \frac{\gamma(1+\alpha)}{\beta h},\qquad a_2 = \frac{1}{\beta h},\qquad a_3 = \left(\frac{1}{2\beta} - 1\right)$$

$$a_4 = (1+\alpha)\left(\frac{\gamma}{\beta} - 1\right),\quad a_5 = (1+\alpha)\left(\frac{\gamma}{2\beta} - 1\right)h,\quad a_6 = (1-\gamma)h,\quad a_7 = \gamma h$$

5. Determine the effective stiffness matrix $[\hat{\mathbf{K}}]$

$$[\hat{\mathbf{K}}] = (1+\alpha)[\mathbf{K}] + a_1[\mathbf{C}] + a_0[\mathbf{M}]$$

6. Triangularize $[\hat{\mathbf{K}}]$: $[\hat{\mathbf{K}}] = [\mathbf{L}][\mathbf{D}][\mathbf{L}]^T$

*Calculate for each time step: for $n = 0, 1, 2, \ldots, t_f$*

7. Increment the time: $t_{i+1} = t_i + h, t_0 = 0$.
8. Determine the effective load vector at time $t_{i+1}$

$$\{\hat{\mathbf{f}}\}_{i+1} = (1+\alpha)\{\mathbf{f}\}_{i+1} - \alpha\{\mathbf{f}\}_i + [\mathbf{M}][a_0\{\mathbf{u}\}_i + a_2\{\dot{\mathbf{u}}\}_i + a_3\{\ddot{\mathbf{u}}\}_i] \\ + [\mathbf{C}][a_1\{\mathbf{u}\}_i + a_4\{\dot{\mathbf{u}}\}_i + a_5\{\ddot{\mathbf{u}}\}_i] + \alpha[\mathbf{C}]\{\dot{\mathbf{u}}\}_i + \alpha[\mathbf{K}]\{\mathbf{u}\}_i$$

9. Determine the displacement vector $\{\mathbf{u}\}_{i+1}$ at time $t_{i+1}$

$$[\mathbf{L}][\mathbf{D}][\mathbf{L}]^T\{\mathbf{u}\}_{i+1} = \{\hat{\mathbf{f}}\}_{i+1}$$

10. Determine the acceleration and velocity vectors at time $t_{i+1}$

$$\{\ddot{\mathbf{u}}\}_{i+1} = a_0(\{\mathbf{u}\}_{i+1} - \{\mathbf{u}\}_i) - a_2\{\dot{\mathbf{u}}\}_i - a_3\{\ddot{\mathbf{u}}\}_i$$

$$\{\dot{\mathbf{u}}\}_{i+1} = \{\dot{\mathbf{u}}\}_i + a_6\{\ddot{\mathbf{u}}\}_i + a_7\{\ddot{\mathbf{u}}\}_{i+1}$$

Determine the effective stiffness matrix $[\hat{\mathbf{K}}]$:

$$[\hat{\mathbf{K}}] = (1+\alpha)[\mathbf{K}] + a_1[\mathbf{C}] + a_0[\mathbf{M}]$$

$$= (1-0.3)\begin{bmatrix} 6 & -2 \\ -2 & 8 \end{bmatrix} + 236.69\begin{bmatrix} 1 & 0 \\ 0 & 2 \end{bmatrix}$$

$$= \begin{bmatrix} 240.89 & -1.4 \\ -1.4 & 478.97 \end{bmatrix}$$

Triangularize the effective stiffness matrix $[\hat{\mathbf{K}}]$ and then perform the step-by-step integration $n = 0, 1, 2, \ldots, t_f$ incrementing the time: $t_{i+1} = t_i + h$, $t_0 = 0$. Determine the effective load vector.

**TABLE 11.20**

Displacement Results for Example 11.13

| Time $t$ | $u_1$ | $u_2$ |
|---|---|---|
| 0 | 0.000000 | 0.000000 |
| 0.1 | 0.000144 | 0.024708 |
| 0.2 | 0.001063 | 0.097823 |
| 0.3 | 0.004140 | 0.216432 |
| 0.4 | 0.011492 | 0.375889 |
| 0.5 | 0.025774 | 0.569978 |
| 0.6 | 0.049905 | 0.791186 |
| 0.7 | 0.086749 | 1.031025 |
| 0.8 | 0.138780 | 1.280401 |
| 0.9 | 0.207767 | 1.529988 |
| 1.0 | 0.294483 | 1.770617 |
| 1.5 | 0.930824 | 2.572688 |
| 2.0 | 1.426221 | 2.378843 |
| 2.5 | 1.062764 | 1.461534 |
| 3.0 | −0.049678 | 0.575919 |
| 3.5 | −0.818892 | 0.300996 |
| 4.0 | −0.357221 | 0.733004 |

$$\{\hat{\mathbf{f}}\}_{i+1} = (1+\alpha)\{\mathbf{f}\}_{i+1} - \alpha\{\mathbf{f}\}_i + [\mathbf{M}]\left[a_0\{\mathbf{u}\}_i + a_2\{\dot{\mathbf{u}}\}_i + a_3\{\ddot{\mathbf{u}}\}_i\right]$$
$$+ [\mathbf{C}][a_1\{\mathbf{u}\}_i + a_4\{\dot{\mathbf{u}}\}_i + a_5\{\ddot{\mathbf{u}}\}_i] + \alpha[\mathbf{C}]\{\dot{\mathbf{u}}\}_i + \alpha[\mathbf{K}]\{\mathbf{u}\}_i$$
$$= 0.7\begin{Bmatrix} 0 \\ 10 \end{Bmatrix} + 0.3\begin{Bmatrix} 0 \\ 10 \end{Bmatrix} + \begin{bmatrix} 1 & 0 \\ 0 & 2 \end{bmatrix}(236.69\{\mathbf{u}\}_i + 23.67\{\dot{\mathbf{u}}\}_i + 0.1834\{\ddot{\mathbf{u}}\}_i) - 0.3\begin{bmatrix} 6 & -2 \\ -2 & 8 \end{bmatrix}\{\mathbf{u}\}_i$$

Determine the displacement $\{\mathbf{u}\}_{i+1}$ at time $t_{i+1}$:

$$[\hat{\mathbf{K}}]\{\mathbf{u}\}_{i+1} = \{\hat{\mathbf{f}}\}_{i+1}$$

$$\begin{bmatrix} 240.89 & -1.4 \\ -1.4 & 478.97 \end{bmatrix}\{\mathbf{u}\}_{i+1} = \{\hat{\mathbf{f}}\}_{i+1}$$

Determine the acceleration, velocity, and displacement vectors at time $t_{i+1}$:

$$\{\ddot{\mathbf{u}}\}_{i+1} = 236.69(\{\mathbf{u}\}_{i+1} - \{\mathbf{u}\}_i) - 23.67\{\dot{\mathbf{u}}\}_i - 0.1834\{\ddot{\mathbf{u}}\}_i$$

$$\{\dot{\mathbf{u}}\}_{i+1} = \{\dot{\mathbf{u}}\}_i + 0.02\{\ddot{\mathbf{u}}\}_i + 0.08\{\ddot{\mathbf{u}}\}_{i+1}$$

The results are given in Table 11.20.

## PROBLEMS

11.1 Using Lagrange's interpolation formula, determine the quadratic shape function for a one-dimensional linear element.

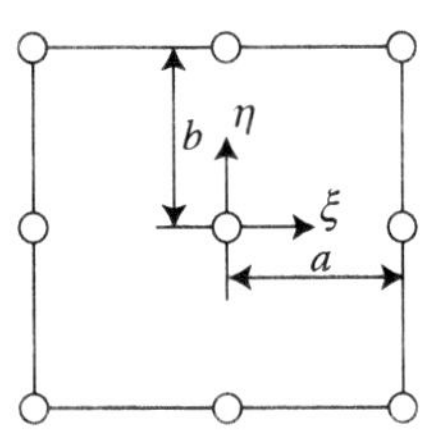


**FIGURE 11.33**
Nine-noded element.

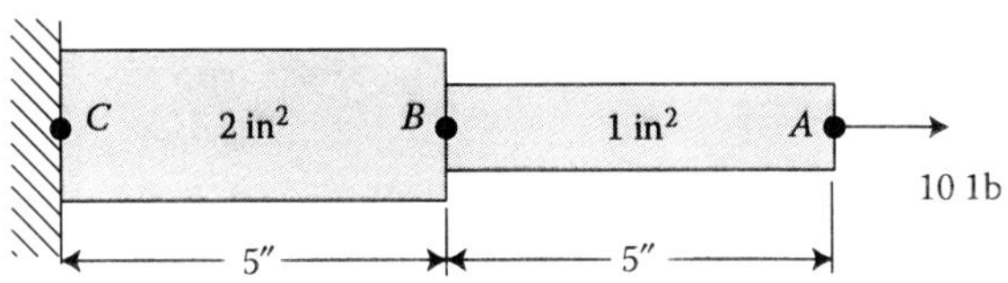


**FIGURE 11.34**
Stepped steel shaft.

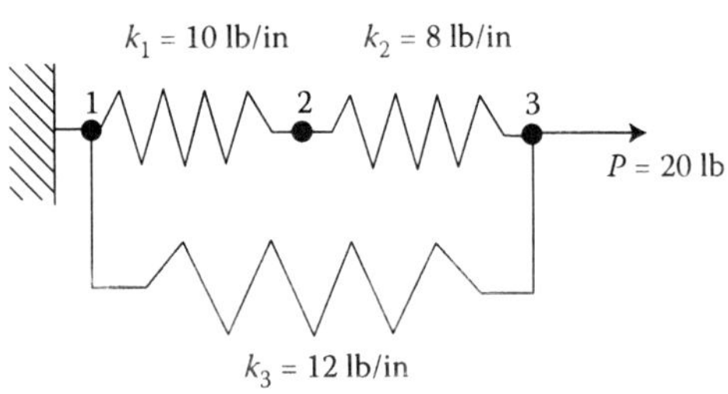


**FIGURE 11.35**
Three-spring system.

11.2 Apply the Lagrange's interpolation quadratic formula, used in Problem 11.1, in the $\xi$ and $\eta$ directions to determine the shape function for the nine-noded element as shown in Figure 11.33.

11.3 The stepped steel ($E = 30 \times 10^6$ psi) shaft in Figure 11.34 is subjected to a 10 lb axial force as shown. Assume each 5″ segment of the shaft can be treated as an element and determine (using hand calculations) the displacement at points $A$ and $B$.

11.4 The system of springs shown in Figure 11.35 can be modeled as three elements. The element stiffness matrix for each spring element is

$$[K]^e = \begin{bmatrix} k_i & -k_i \\ -k_i & k_i \end{bmatrix}$$

Using hand calculations, determine the force in each spring.

11.5 For the plane truss shown in Figure 11.36, each element has a cross-sectional area of $A = 5$ in² and an elastic modulus of $E = 1 \times 10^6$ psi. Using hand calculations, determine the displacement of node 2 knowing that the element stiffness matrix for a truss element is given by

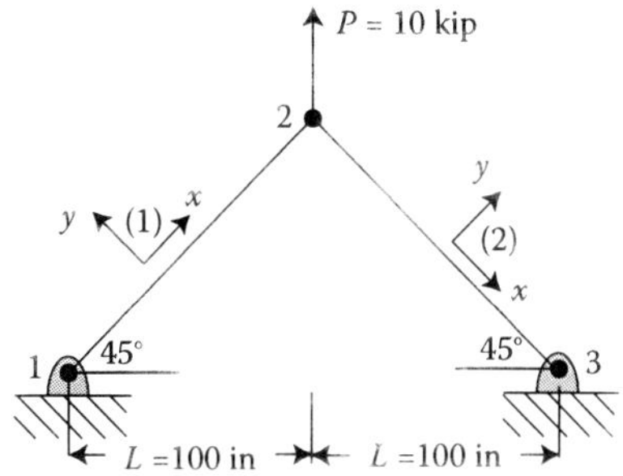


**FIGURE 11.36**
Plane truss.

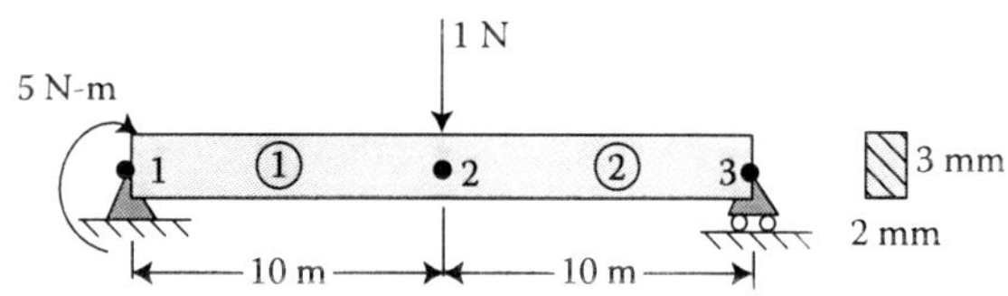


**FIGURE 11.37**
Simply supported beam.

$$[K]^e = \frac{AE}{L}\begin{bmatrix} \cos^2\theta & \sin\theta\cos\theta & -\cos^2\theta & -\sin\theta\cos\theta \\ \sin\theta\cos\theta & \sin^2\theta & -\sin\theta\cos\theta & -\sin^2\theta \\ -\cos^2\theta & -\sin\theta\cos\theta & \cos^2\theta & \sin\theta\cos\theta \\ -\sin\theta\cos\theta & -\sin^2\theta & \sin\theta\cos\theta & \sin^2\theta \end{bmatrix}$$

11.6 A 20 m long simply supported beam with an elastic modulus of $E = 1$ GPa can be modeled by two elements as illustrated in Figure 11.37. A 1 N downward force and a 5 N m moment are applied as indicated. . Using hand calculations, determine the displacement and slope at all three nodes knowing the element stiffness matrix is given by

$$[K]^e = \frac{EI}{L^3}\begin{bmatrix} 12 & 6L & -12 & 6L \\ 6L & 4L^2 & -6L & 2L^2 \\ -12 & -6L & 12 & -6L \\ 6L & 2L^2 & -6L & 4L^2 \end{bmatrix}$$

11.7 The 6 m long fixed-fixed steel ($E = 210$ GPa, $I = 4 \times 10^{-4}$ m$^4$) in Figure 11.38 can be modeled using two 3 m long elements. Using hand calculations, determine the displacement and slope at node 2 knowing the element stiffness matrix is given by

$$[K]^e = \frac{EI}{L^3}\begin{bmatrix} 12 & 6L & -12 & 6L \\ 6L & 4L^2 & -6L & 2L^2 \\ -12 & -6L & 12 & -6L \\ 6L & 2L^2 & -6L & 4L^2 \end{bmatrix}$$

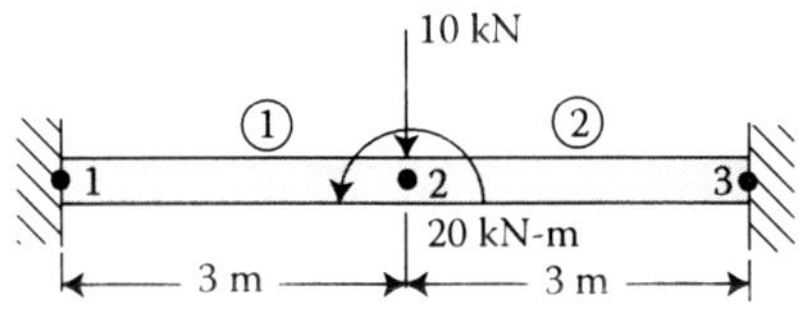


**FIGURE 11.38**
Fixed-fixed steel beam.

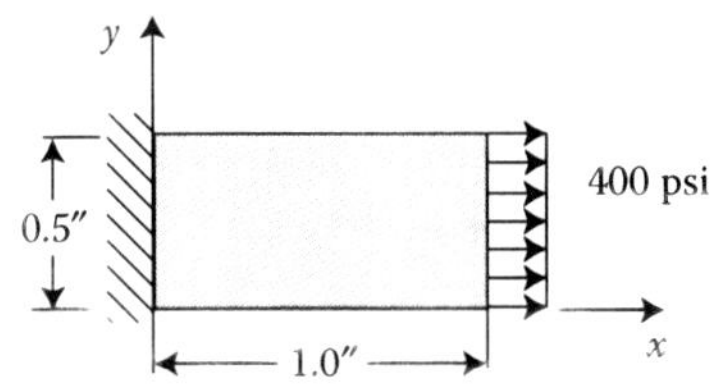


**FIGURE 11.39**
Rectangular plate with axial load.

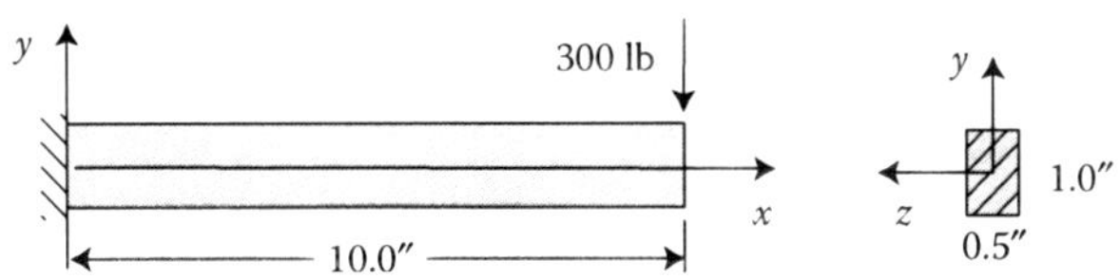


**FIGURE 11.40**
Cantilevered steel beam.

11.8 The rectangular steel ($E = 30 \times 10^6$ psi, $\nu = 0.30$) plate in Figure 11.39 is 0.25 in thick and is loaded with an axial stress of 400 psi. Use two triangular plane stress elements to approximate the deflections at the end of the plate subjected to the applied load. Assume one node at the fixed end is on rollers. Hand calculations are to be used.

11.9 The cantilevered steel ($E = 30 \times 10^6$ psi, $\nu = 0.30$) beam in Figure 11.40 has a cross section as indicated and is subjected to a 300 lb vertical lend load. The theoretical vertical deflection at the free end of the beam is approximately 0.08 in downward ($\nu = -0.08''$). Using an appropriate finite element and a finite element code of your choice, construct several meshes with an increasing number of elements in each mesh to determine the vertical deflection at the free end. Use at least three meshes and not the change in accuracy as the number of elements increases.

11.10 A plate of width $w$ has a hole of diameter $d$ in the center. A force $F$ is applied as shown in Figure 11.41. The hole creates a stress concentration factor ($K$), which is a function of geometry ($d/w$) and loading condition. A quarter model of the plate (as shown) can be used to approximate the stress concentration factor. Find a table of stress concentration factors and select a ratio of $d/w$ to conduct a finite element analysis. Define appropriate boundary conditions for the model, select an appropriate finite element (and finite element code), define several (at least three) mesh refinements, and observe the difference in stress concentration approximations based on the meshes selected.

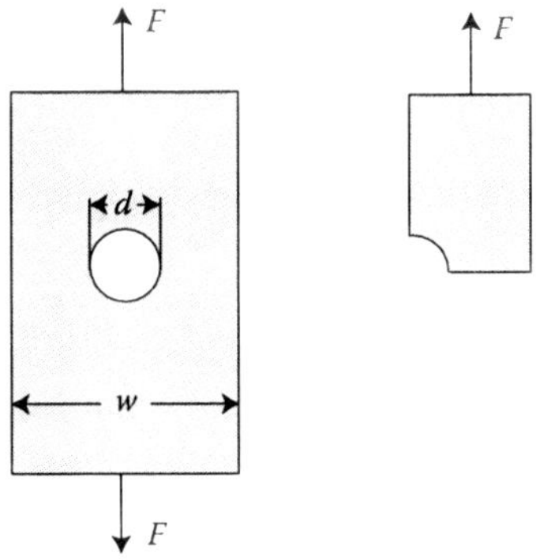


**FIGURE 11.41**
Plate with a hole.

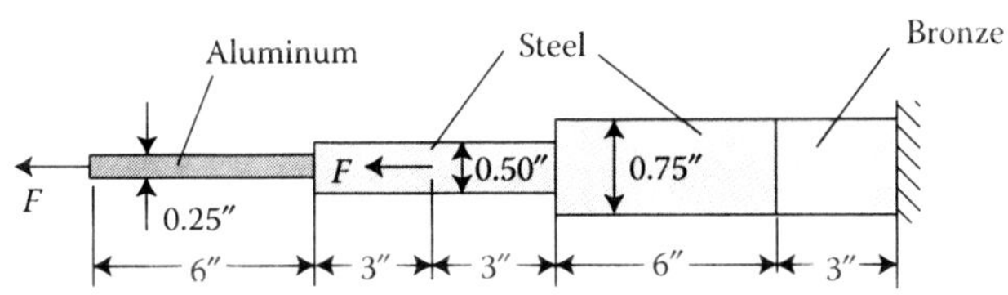


**FIGURE 11.42**
Stepped shaft.

11.11 Using the plate of Figure 11.39, select an appropriate element (and finite element code) and construct a mesh at least eight elements in order to determine the horizontal deflection of the free end. The result obtained in Problem 11.8 for the free end was $1.333 \times 10^{-5}$ in. Your results may vary from this.

11.12 A stepped shaft is made from three materials and has three different diameters as illustrated in Figure 11.42. A force of $F = 3000$ lb is applied at two different location as shown. Select appropriate elements and an appropriate finite element code to determine the displacements at all interfaces between materials and diameters, the reaction force at the wall, the force, strain, and stress in each element.

11.13 The cross section of a structure with an applied pressure of 600 kN is shown in Figure 11.43. The structure is made from a material with $E = 40$ GPa, $G = 17$ GPa, and $\nu = 0.15$. Select an appropriate finite element code and perform a finite element analysis to obtain a plot of the deformed shape, including displacements and stresses. A mesh using 8-node quadrilateral elements is suggested but not necessary.

11.14 The statically indeterminate beam shown in Figure 11.44 has a uniform wall thickness of 0.0625 in. Using appropriate elements and an appropriate finite element code, determine the reactions at the supports, the deflection and rotation at the force application positions, and the stress and strain at several key locations along the beam. You are free to select any material you wish, as long as beam failure does not occur.

11.15 Write a MATLAB program to determine all eigenpairs of a generalized eigenvalue problem using the generalized Jacobi method.

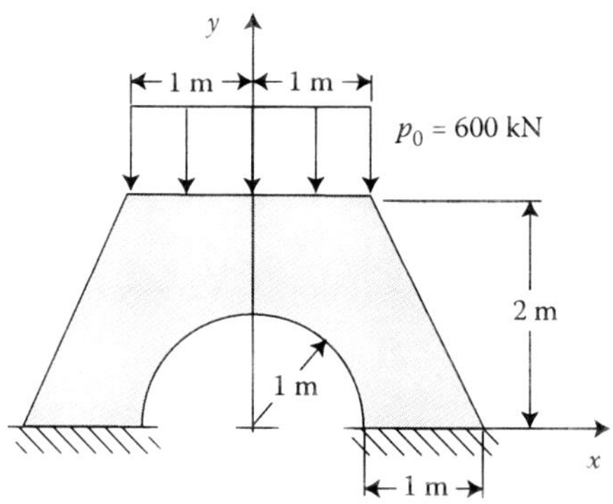


**FIGURE 11.43**
Structure with applied pressure.

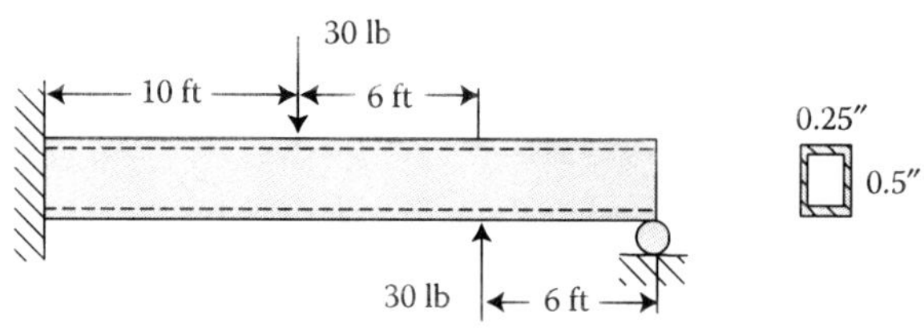


**FIGURE 11.44**
Statically indeterminate beam.

11.16 Write a MATLAB program using the subspace iteration method and use it to solve for the eigenvalues and eigenvectors of the following system:

$$[M]=\begin{bmatrix}259 & 0 & 0\\ 0 & 259 & 0\\ 0 & 0 & 259\end{bmatrix}\frac{\text{lb s}^2}{\text{in}},\quad [K]=1000\begin{bmatrix}336 & -186 & 0\\ -186 & 336 & -186\\ 0 & -186 & 186\end{bmatrix}\frac{\text{lb}}{\text{in}}$$

11.17 Solve Problem 11.16 using the forward iteration method.

11.18 Solve Problem 11.16 using the inverse iteration method.

11.19 Determine the response of a three-degree-of-freedom system when $f_1(t) = 0$, $f_2(t) = 0$, and $f_3(t) = 150\ N$ using the central difference method. The mass, stiffness, and damping matrices are given as

$$[M]=30\begin{bmatrix}1 & 0 & 0\\ 0 & 2 & 0\\ 0 & 0 & 1\end{bmatrix}\text{kg},\quad [C]=1\times e^3\begin{bmatrix}3 & -1 & 0\\ -1 & 4 & -3\\ 0 & -3 & 3\end{bmatrix}\frac{\text{N-s}}{\text{m}},\quad [K]=1\times e^5\begin{bmatrix}3 & -1 & 0\\ -1 & 4 & -3\\ 0 & -3 & 3\end{bmatrix}\frac{\text{N}}{\text{m}}$$

11.20 Solve Problem 11.19 using the Houbolt method.

11.21 Solve Problem 11.19 using the Newmark beta method using $\gamma = 1/2$, $\beta = 1/6$.

11.22 Solve Problem 11.19 using the Wilson-theta method with $\theta = 1.41$ and $\gamma = 1/2$, $\beta = 1/6$.

11.23 Solve Problem 11.19 using the HHT-alpha method using $\alpha = -0.3$, $\gamma = ((1/2) - \alpha)$, $\beta = (1 - \alpha)^2/4$ and select a time step to ensure stability.

## References

Argyris, J. H., Fried, I., and Scharpf, D. W. 1968. The TUBA family of plate elements for the matrix displacement method. *J. Royal Aeronaut. Soc.*, 72: 701–709.

Bathe, K.-J. 1996. *Finite Element Procedures*. Englewood Cliffs, NJ: Prentice-Hall.

Bathe, K. J. and Wilson, E. L. 1972. Large eigenvalue problem in dynamic analysis. *ASCE, J. Eng. Mech. Div.*, EM6: 1471–1485.

Bathe, K.-J. and Wilson, E. L. 1973. Solution methods for eigenvalue problems in structural mechanics. *Int. J. Num. Meth. Eng.*, 6: 213–226.

Bathe, K. J. and Wilson, E. L. 1976. *Numerical Methods in Finite Element Analysis*. Englewood Cliffs, NJ: Prentice-Hall.

Bazeley, G. P., Cheung, Y. K., Irons, B. M., and Zienkiewicz, O. C. 1966. Triangular elements in plate bending-conforming and non-conforming solutions. *Matrix Meth. Struct. Mech.*, AFFDL-TR-66-80: 547–576.

Bell, K. 1969. A refined triangular plate bending finite element. *Int. J. Numer. Methods Eng.*, 1: 101–122.

Bogner, F. K., Fox, R. L., and Schmit, L. A. 1966. The generation of inter-element compatible stiffness and mass matrices by the use of interpolation formulas. *Matrix Meth. Struct. Mech.*, AFFDL-TR-66-80: 397–443.

Cheung, Y. K., King, I. P., and Zienkiewicz, O. C. 1968. Slab bridges with arbitrary shape and support conditions: A general method of analysis based on finite elements. *Proc. Inst. Civil Eng.*, 40: 9–36.

Clough, R. W. and Tocher, J. L. 1966. Finite element stiffness matrices for analysis of plate bending. *Matrix Meth. Struct. Mech.*, AFFDL-TR-66-80: 515–545.

Conte, S. D. and de Boor, C. 1980. *Elementary Numerical Analysis*, 3rd ed. New York: McGraw-Hill Book Company.

Dasgupta, S. and Sengupta, D. 1990. A higher-order triangular plate bending element revisted. *Int. J. Numer. Methods Eng.*, 30: 419–430.

Eisenberg, M. A. and Malvern, L. E. 1973. On finite element integration in natural coordinates. *Int. J. Num. Meth. Eng.*, 7: 574–575.

Hilber, H. M., Hughes, T. J. R., and Taylor, R. L. 1977. Improved numerical dissipation for time integration algorithms in structural dynamics. *Earthquake Eng. Struc. Dyn.*, 5: 282–292.

Houbolt, J. C. 1950. A recurrence matrix solution for the dynamic response of elastic aircraft. *J. Aeronautical Sci.*, 17: 540–550.

Hrabok, M. M. and Hrudey, T. M. 1984. A review and catalogue of plate bending elements. *Comput. Struct.*, 19: 479–495.

Hughes, T. J. R. and Belytschko, T. 1983. A precis of developments in computational methods for transient analysis. *J. Appl. Mech.*, 50: 1033–1041.

Jennings, A. 1977. *Matrix Computation for Engineers and Scientists*. Chichester: John Wiley & Sons.

Johnson, D. E. and Greif, R. 1966. Dynamic response of a cylindrical shell: Two numerical methods. *AIAA J.*, 4: 486–494.

Melosh, R. H. 1963. Basis for derivation of matrices by the direct stiffness method. *AIAA J.*, 1: 1631–1637.

Merirovitch, L. 1980. *Computational Methods in Structural Dynamics*. Rijn, The Netherlands: Sijthoff & Noordhoff.

Newmark, N. M. 1959. A method of computation for structural dynamics. *J. Eng. Mech. Div., ASCE,* 85: 67–94.

Scheid, F. 1989. *Numerical Analysis*, 2nd ed. Schaum's Outline Series. New York: McGraw-Hill.

Smith, C. V. 1970. *Finite Element Model, with Applications to Buildings and K-33 and K-31*. Union Carbide Corporation Report No. CTC-29.

Strang, G. 1988. *Linear Algebra and Its Application*, 3rd ed. Pacific Grove, CA: Brooks-Cole Publishing.

Wilkinson, J. H. 1965. *The Algebraic Eigenvalue Problem*. Oxford: Clarendon Press.

Zienkiewicz, O. C. and Taylor, R. L. 2000. *The Finite Element Method for Solid and Structural Mechanics*. Vol. 2. 5th ed. Jordan Hill, Oxford: Butterworth-Heinemann.

Zienkiewicz, O. C., Taylor, R. L., and Zhu, Z. 2005. *The Finite Element Method for Solid and Structural Mechanics*, 6th ed. Jordan Hill, Oxford: Elsevier Butterworth-Heinemann.

# 12

# Shock Spectra

## 12.1 Introduction

Shock waves are propagating disturbances that result from explosions, earthquakes, powerful electric discharges, etc. Mechanical shock pulses are often analyzed in terms of shock response spectra. The shock response spectrum assumes that the shock pulse is applied as a common base input to an array of independent single-degree-of-freedom systems. The shock response spectrum gives the peak response of each system with respect to the natural frequency. The shock response spectrum is particularly suited for analyzing pyrotechnic shock. It may also be used for evaluating classical pulses, such as a half-sine pulse. A majority of shock-related topics are associated with random vibrations, which are adequately described in various texts including Yang (1986), (Gupta 1990), Lutes and Sarkani (2004), and Bendat and Piersol (2010). Additionally, the overall subject of shock and its subsequent analysis is extensively addressed in handbooks (Harris and Piersol 2010). The intent of this chapter is to introduce selected topics relevant to shock spectra.

## 12.2 One-Degree-of-Freedom System

Consider a simple one-degree-of-freedom system without damping, whose equation of motion is $\ddot{y} + \omega^2 y = f(t)$. In this equation the forcing function $f(t)$ is arbitrary, complicated, and nonperiodic in time. In some cases numerical integration is required to evaluate $f(t)$. Once the forcing function is known, we determine the shock spectrum for the maximum displacement $y_{max}$. This is sequentially illustrated in Figure 12.1.

The maximum $y$ is typically determined by:

1. Numerical integration. Using numerical integration, one assumes different values of $\omega$ and integrates to obtain $y(t)$.
2. Computer simulation. Using commercially available programs such as MATLAB, one can develop codes to obtain $y(t)$.
3. Experimental technique where one could set up an experiment similar to that illustrated in Figure 12.2. Applying a known $f(t)$ to a base with a known mass attached, the maximum displacement for a known frequency ($\omega$) can be measured.

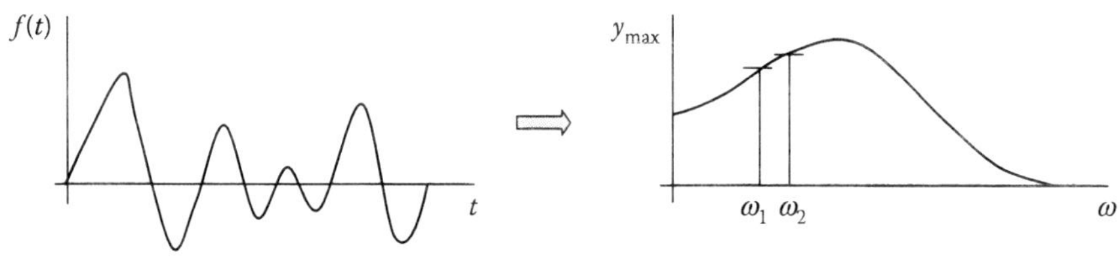


**FIGURE 12.1**
Sequence for forcing function $f(t)$ yielding shock spectrum $y_{max}(\omega)$.

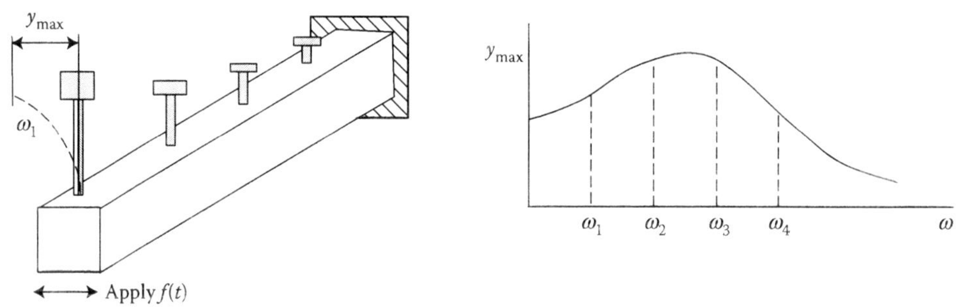


**FIGURE 12.2**
Illustration of an experiment to determine a shock spectrum.

## 12.3 Several Degrees of Freedom Systems

Multiple-degree-of-freedom systems pose a slightly more complex approach. Suppose we have a structure such as that shown in Figure 12.3. This structure could simulate a three-story building with rigid floors and three degrees of freedom. Associated with each floor is a unique displacement $u_i$. Assume the input $f(t)$ to this structure is applied at the base. We want to determine the natural frequencies $\omega_1$, $\omega_2$, $\omega_3$ or $f_1$, $f_2$, $f_3$ (in cycles per second) for all three floors.

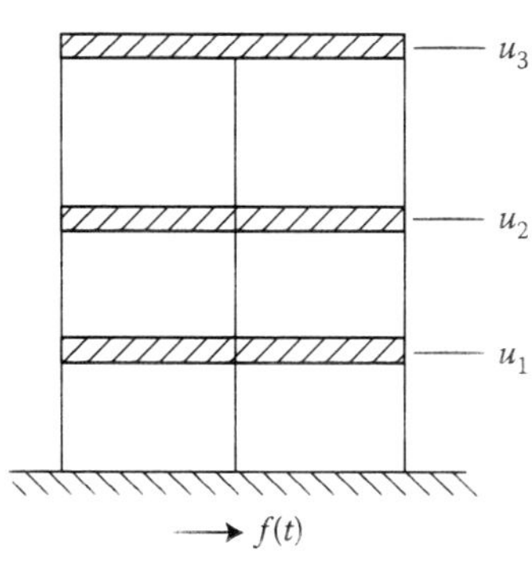


**FIGURE 12.3**
Model of a three-degree-of-freedom structure.

We determine the modal matrix for all three frequencies. For $\omega_1$, we have $u_1 = \phi_1^{(1)}$, $u_2 = \phi_2^{(1)}$, $u_3 = \phi_3^{(1)}$. Similarly, for $\omega_2$ we have $u_1 = \phi_1^{(2)}$, $u_2 = \phi_2^{(2)}$, $u_3 = \phi_3^{(2)}$, and for $\omega_3$ we have $u_1 = \phi_1^{(3)}$, $u_2 = \phi_2^{(3)}$, $u_3 = \phi_3^{(3)}$. Assume that for the input $f(t)$, we have

$$
\begin{aligned}
u_1 &= g_1\phi_1^{(1)} + g_2\phi_1^{(2)} + g_3\phi_1^{(3)} \\
u_2 &= g_1\phi_2^{(1)} + g_2\phi_2^{(2)} + g_3\phi_2^{(3)} \\
u_3 &= g_1\phi_3^{(1)} + g_2\phi_3^{(2)} + g_3\phi_3^{(3)}
\end{aligned}
\tag{12.1}
$$

The $g_r$ terms come from the equation $\ddot{g}_r + \omega_r^2 g_r = C_r f(t)$, where $r = 1, 2, 3$. From the shock spectrum for $f_1, f_2$, and $f_3$, we read $y_1$, $y_2$, and $y_3$ which gives

$$
g_{1-\max} = C_1 y_1, \quad g_{2-\max} = C_2 y_2, \quad g_{3-\max} = C_3 y_3
$$

Then we get the conservative estimates

$$
\begin{aligned}
u_{1-\max} &= g_{1-\max}\phi_1^{(1)} + g_{2-\max}\phi_1^{(2)} + g_{3-\max}\phi_1^{(3)} \\
u_{2-\max} &= g_{1-\max}\phi_2^{(1)} + g_{2-\max}\phi_2^{(2)} + g_{3-\max}\phi_2^{(3)} \\
u_{3-\max} &= g_{1-\max}\phi_3^{(1)} + g_{2-\max}\phi_3^{(2)} + g_{3-\max}\phi_3^{(3)}
\end{aligned}
\tag{12.2}
$$

This method can be modified slightly. The modification is not considered to be conservative and is given by

$$
\begin{array}{ccccccc}
(u_{1-\max})^2 & = & \left(g_{1-\max}\phi_1^{(1)}\right)^2 & + & \left(g_{2-\max}\phi_1^{(2)}\right)^2 & + & \left(g_{3-\max}\phi_1^{(3)}\right)^2 \\
\vdots & & \vdots & & \vdots & & \vdots \\
\vdots & & \vdots & & \vdots & & \vdots
\end{array}
\tag{12.3}
$$

Velocity is $v = \dot{y}$, and its equation is $\ddot{v} + \omega v = \dot{f}(t)$. Find $v_{\max}$ corresponding to $\dot{f}(t)$ or find $v_{\max} = \dot{y}_{\max}$, determine $y$ and $\dot{y}$.

## 12.4 Random Loading

Random loadings are typically random in both time and space. Recall that random loading in time only is as depicted in Figure 12.4.

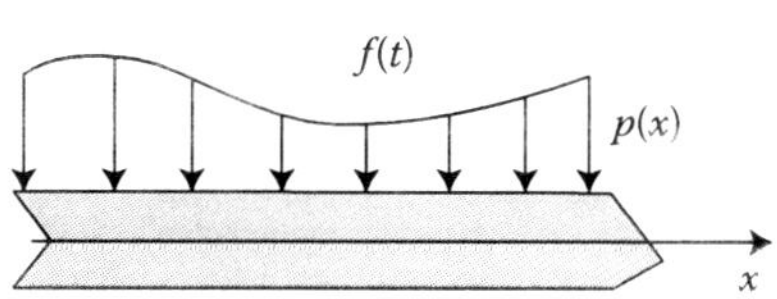


**FIGURE 12.4**
Random loading in time only.

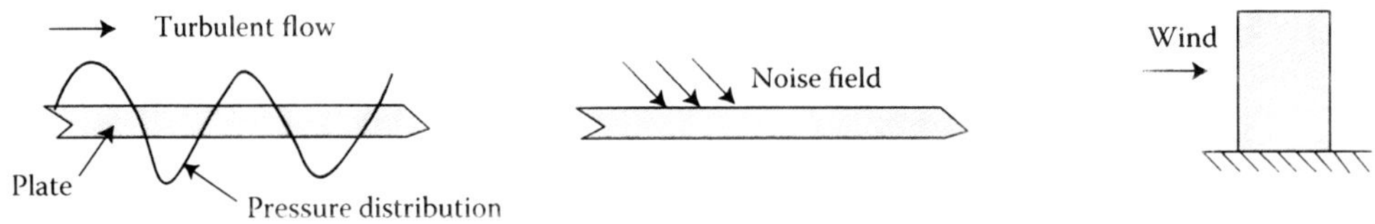


**FIGURE 12.5**
Examples of random loading in time and space.

The distribution in space is given by $p(x)$ = deterministic function. The time dependence is described by $f(t)$ = random function. Then,

$$S_W(\omega) = |\Phi^*(i\omega)|^2 \, S_f(\omega) \tag{12.4}$$

where

$S_W(\omega)$ = spectral density of the output

$\Phi^*(x, i\omega)$ = frequency response function = response to the loading, $p(x)e^{i\omega t}$

$S_f(\omega)$ = spectral density of the function $f(t)$

Examples of random loading in space and time are illustrated in Figure 12.5.

In order to describe a loading that is random in time and space, we need a number of observations. We would require observations of the load in time (say for $t = t_1$ and $t_2$) and in space (say for $x = x_1$ and $x_2$) as illustrated in Figure 12.6.

The correlation function of $p$ (or the cross-correlation function of $p$) is

$$R_p(x_1,t_1;x_2,t_2) = E[p(x_1,t_1)\cdot p(x_2,t_2)] \cong \langle p(x_1,t_1)p(x_2,t_2)\rangle \tag{12.5}$$

where $\langle p(x_1, t_1) \cdot p(x_2, t_2)\rangle$ = ensemble average.

Assume you have $n$ observations for positions and times 1 and 2. Constructing a table similar to that shown in Table 12.1 provides the necessary information to evaluate the spectrum.

Then we identify $\langle p(x_1, t_1) \cdot p(x_2, t_2)\rangle$ = (summ. of terms)/$n$. Depending on the type of loading, we can define a correlation function in several ways. For a stationary random $p$ (pressure), we define

$$R_p(x_1,t_1;x_2,t_2) = R_p(x_1,x_2,t_2-t_1) = R_p(x_1,x_2,\tau) \tag{12.6}$$

For a homogenous random function, we have

$$R_p(x_1,t_1;x_2,t_2) = R_p(x_2-x_1,t_1,t_2) = R_p(\varsigma,t_1,t_2) \tag{12.7}$$

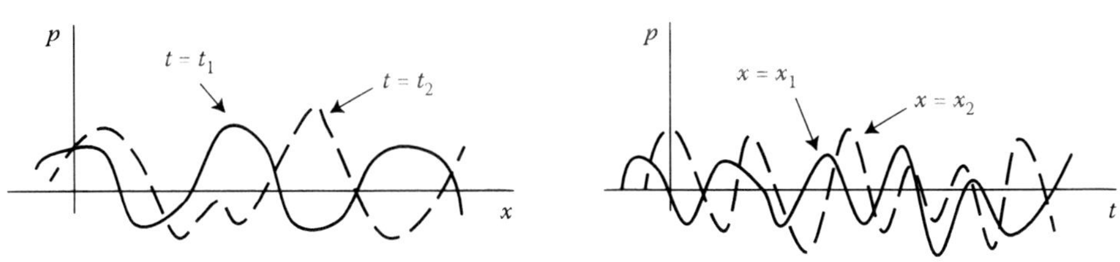


**FIGURE 12.6**
Observations of random loading in time and space.

**TABLE 12.1**

Observed Events for Random Time and Space Loading

| Observation | $p(x_1, t_1) \cdot p(x_2, t_2)$ |
|---|---|
| 1 | $p^{(1)}(x_1, t_1) \cdot p^{(1)}(x_2, t_2)$ |
| 2 | $p^{(2)}(x_1, t_1) \cdot p^{(2)}(x_2, t_2)$ |
| ⋮ | ⋮ |
| ⋮ | ⋮ |
| $n$ | $p^{(n)}(x_1, t_1) \cdot p^{(n)}(x_2, t_2)$ |
| | Summation of terms |

It does not matter where you are. All you need is the difference $x_2 - x_1$.

For a stationary and homogeneous, we have

$$R_p(x_1,t_1;x_2,t_2) = R_p(x_2 - x_1,t_2,-t_1) = R_p(\varsigma,\tau) \tag{12.8}$$

Assume we have a correlation function $R_p = Ae^{-|x_2-x_1|/l_c}e^{-|t_2-t_1|/t_c}$, where $A$, $l_c$, and $t_c$ are constants. We can extract certain information from $R_p$:

a. With $x_1 = x_2 = x$ and $t_1 = t_2 = t$ we find
   $R_p(x_1, t_1; x_2, t_2) = E[p^2(x,t)]$ = mean square value of $x$, $t$.
b. With $x_1 = x_2 = x$
   $R_p(x_1, t_1; x_2, t_2)$ = correlation function of $p$ at some $x$.
   This describes the time behavior of $p$ at a fixed point.
c. With $t_1 = t_2 = t$
   $R_p(x_1, t_1; x_2, t_2)$ = describes the spatial correlation of $p$.

Assume $p$ is stationary and its mean value is zero, that is, $E[p] = 0$, as illustrated in Figure 12.7.

The cross-spectral density of $p$ is

$$S_p(x_1,x_2,\omega) = \int_{-\infty}^{\infty} R_p(x_1,x_2,\tau)e^{-i\omega\tau}\, d\tau$$

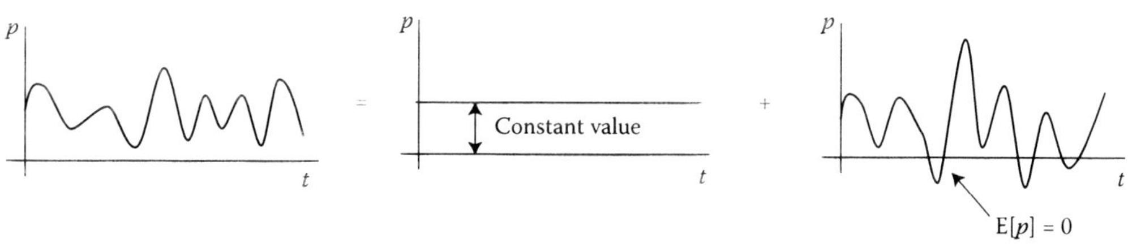


**FIGURE 12.7**
Stationary wave with mean value $E[p] = 0$.

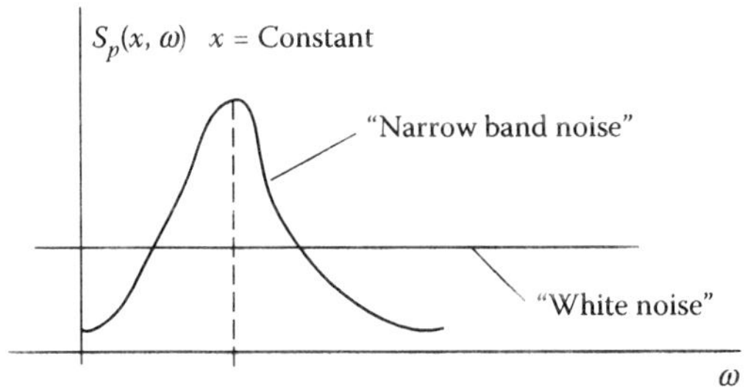


**FIGURE 12.8**
Narrow band and white noise.

Thus

$$R_p(x_1, x_2, \tau) = \frac{1}{2\pi} \int_{-\infty}^{\infty} S_p(x_1, x_2, \omega) e^{i\omega\tau} \, d\omega$$

If $x_1 = x_2 = x$, then $S_p(x, \omega)$ = spectral density of $p$ at some point $x$. For $x$ = constant, we can generate "white noise" or "narrow band noise" as illustrated in Figure 12.8.

Note that $(1/2\pi)\int_{-\infty}^{\infty} S_p(x, \omega)\, d\omega = E[p^2(x)]$ = mean square value of $p$ at $x$.

## 12.5 Input–Output Relations

If we know the correlation function or the spectral density for an event (pressure, or stress, or deflection, etc.), we can predict the number of times the event will occur in a given time interval. For example, assume we have a plate with given pressure in terms of either $R_p$ or $S_p$. We wish to determine the output $q$ (deflection, a component of stress, etc.) in terms of $R_p$ or $S_q$. If we let $q$ be a component of stress, then knowing $R_q(x, \tau)$ or $S_p(x, \omega)$ we can predict how many times in a specified time interval a given stress level is reached.

Suppose we have a beam with a loading $p = \delta(t)$ at some arbitrary location $\zeta$ as shown in Figure 12.9. We wish to define an output $q = q(x, t)$. The initial step is to determine the response of $p = \delta(t)$ at $x = \zeta$.

The response is $Q = (x, \zeta, t)$, which has a Laplace transformation of $\bar{Q}(x, \varsigma, s)$, where $s$ is the transformation parameter. Its Fourier transformation is $Q^*(x, \zeta, i\omega)$. If the loading at $\zeta$ is $e^{i\omega t}$, then the response is $Q^*(x, \zeta, i\omega)e^{i\omega t}$.

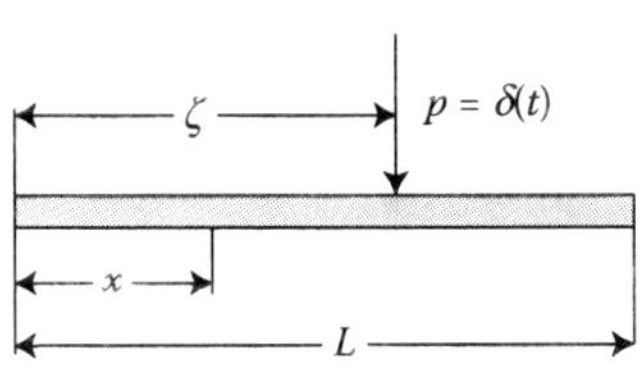


**FIGURE 12.9**
Beam with load $p = \delta(t)$ at $x$.

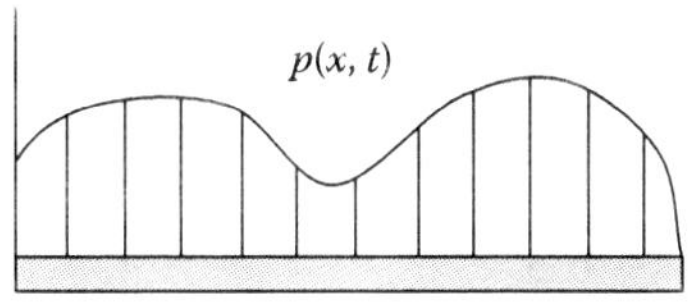


**FIGURE 12.10**
Beam with an arbitrary load distribution.

For an arbitrary loading $p(x,t)$ shown in Figure 12.10, the output is given as

$$q(x,t) = \int_0^L \int_{-\infty}^t Q(x,\varsigma,t-\theta)p(\varsigma,\theta)\,d\theta\,d\varsigma \tag{12.9}$$

The proof of Equation 12.9 used the model shown in Figure 12.11, which shows a concentrated force $p(\zeta, t)$ acting at $\zeta$, but arbitrary in time.

The response to this input is

$$\sum_{\varsigma}\left(\int_{-\infty}^{t} Q(x,\varsigma,t-\theta)p(\varsigma,\theta)d\theta\right)d\varsigma$$

We note that this is similar to a spring with a forcing function applied. In this model $u(t) = \int_{-\infty}^{t} U(t-\theta)p(\theta)d\theta$, where $U(t)$ is the response to $p(t) = \delta(x)$. To take into account the $p$ distributed force, we use

$$\int_0^L\left(\int_{-\infty}^{t} Q(x,\varsigma,t-\theta)p(\varsigma,\theta)d\theta\right)d\varsigma$$

We have $R_q(x_1,t_1;x_2,t_2) = E[q(x_1,t_1)\, q(x_2,t_2)]$, where

$$q(x_1,t_1) = \int_0^L \int_{-\infty}^{t_1} Q(x_1,t_1-\theta_1)p(\varsigma_1,\theta_1)d\theta_1\,d\varsigma_1$$

$$q(x_2,t_2) = \int_0^L \int_{-\infty}^{t_2} Q(x_2,t_2-\theta_2)p(\varsigma_2,\theta_2)d\theta_2\,d\varsigma_2$$

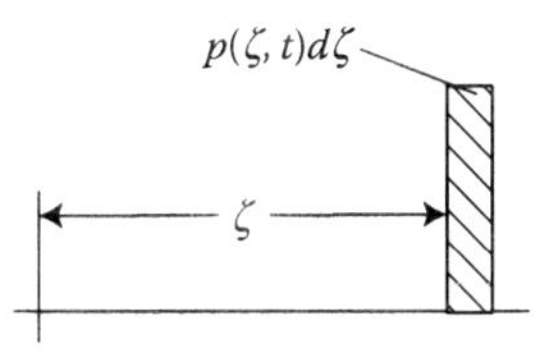


**FIGURE 12.11**
Model for the proof of Equation 12.9.

Multiplying the two and taking the mean value

$$E[q(x_1,t_1)q(x_2,t_2)] = \int_0^L \int_0^L \int_{-\infty}^{t_1} \int_{-\infty}^{t_2} Q(x_1,\varsigma_1,t_1-\theta_1)Q(x_2,\varsigma_2,t_2-\theta_2)E[p(\varsigma_1,\theta_1)p(\varsigma_2,\theta_2)]d\theta_1\, d\theta_2\, d\varsigma_1\, d\varsigma_2$$

or

$$R_q(x_1,x_2,t_1,t_2) = \int_0^L \int_0^L \int_{-\infty}^{t_1} \int_{-\infty}^{t_2} Q(x_1,\varsigma_1,t_1-\theta_1)Q(x_2,\varsigma_2,t_2-\theta_2)R_p(\varsigma_1,\varsigma_2,\theta_1,\theta_2)d\theta_1\, d\theta_2\, d\varsigma_1\, d\varsigma_2 \tag{12.10}$$

We are interested in $R_q(x_1,t_1,t_2)$; $x_1 = x_2 = x$

$$E[q(x,t)] = R_q(x,x;t,t)$$

This can be simplified for the stationary process to

$$R_q(x,t_1,t_2) = R_q(x,t_2-t_1) = R_q(x,\tau) \quad \text{for} \quad x_1 = x_2 = x \tag{12.11}$$

and

$$R_p(\varsigma_1,\varsigma_2,\theta_1,\theta_2) = R_p(\varsigma_1,\varsigma_2,\theta_2-\theta_1) = R_p(\varsigma_1,\varsigma_2,u) \tag{12.12}$$

Use Equation 12.12 in Equation 12.10 and multiply both sides by $e^{i\omega\tau}/2\pi$. Then, integrate from $-\infty$ to $+\infty$ in $\tau$. We find

$$S_q(x,\omega) = \int_0^L \int_0^L Q^*(x,\varsigma_1,i\omega)Q^*(x,\varsigma_2,-i\omega)S_p(\varsigma_1,\varsigma_2,\omega)d\varsigma_1\, d\varsigma_2 \tag{12.13}$$

**EXAMPLE 12.1**

Assume a load on a beam as shown in Figure 12.10, where $p(x)$ is stationary and homogeneous. In addition, assume $R_p(x_1,x_2,t_1,t_2) = R_p(x_2-x_1,t_2-t_1)$ and the distributions for $R_p$ shown in Figure 12.12 are applicable. Determine an expression for $R_q(x,t)$.

**Solution**

For this type of $R_p$, for the purposes of integration, we can assume

$$R_p = D\delta(x_2-x_1)\delta(t_2-t_1) \quad \text{or} \quad R_p = D\delta(\varsigma_2-\varsigma_1)\delta(\theta_2-\theta_1)$$

Let the output $q(x,t)$ be the deflection, $w$. Since deflection is a required output, we need to find $Q$ using a beam model as shown in Figure 12.13.

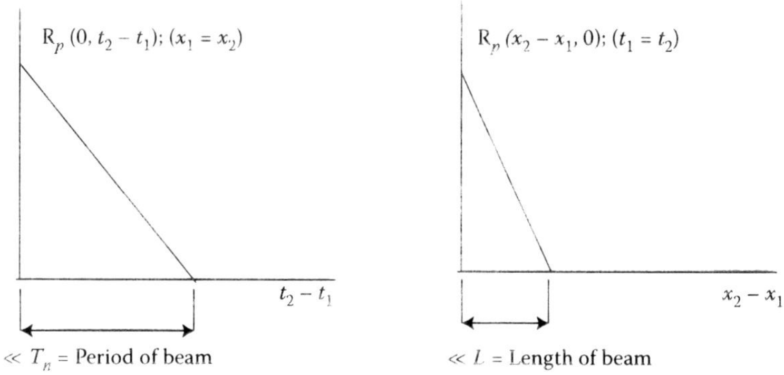


**FIGURE 12.12**
Distributions of $R_i$.

Since we are dealing with a beam deflection, we need to use the beam equation of motion, which is

$$EI\frac{\partial^4 q}{\partial x^4} + \beta\frac{\partial q}{\partial t} + A\rho\frac{\partial^2 q}{\partial t^2} = p(x,t) = \delta(x-\varsigma)\delta(t)$$

$$p(x,t) = \sum P_m(t)X_m(\varsigma)$$

where

$$P_m(t) = \frac{\int_0^L p(x,t)X_m\,dx}{\int_0^L X_m^2\,dx} = \frac{X_m(\varsigma)\delta(t)}{K_m} \quad \text{and} \quad K_m = \int_0^L X_m^2\,dx$$

In addition, $q = Q(x,\zeta,t) = \sum C_m(t)X_m(x)$. Substituting into the beam equation of motion, we can write

$$EIX_m^{IV} = \omega_m^2 A\rho X_m$$

From this, we obtain

$$\omega_m^2 C_m + \frac{\beta}{A\rho}\dot{C}_m + \ddot{C}_m = \frac{X_m(\varsigma)}{A\rho K_m}\delta(t)$$

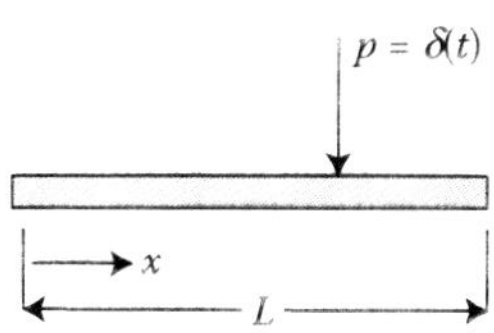


**FIGURE 12.13**
Beam model for determining deflection.

The solution to this is

$$C_m(t) = \frac{X_m(\varsigma)}{K_m A\rho\omega'_m} e^{-\gamma t} \sin\omega'_m t$$

where $\gamma = (\beta/2A\rho)$ and $\omega'_m = \sqrt{\omega_m^2 - \gamma^2}$ with the natural frequency of the beam (without damping) being $\omega_m$

$$Q(x,\varsigma,t) = \sum_m \frac{X_m(\varsigma)}{K_m A\rho\omega'_m} e^{-\gamma t} \sin\omega'_m t X_m(x)$$

For $x_1 = x_2 = x$, $t_1 = t_2 = t$

$$R_q(x,t) = \int_0^L \int_0^L \int_{-\infty}^{t_1} \int_{-\infty}^{t_2} Q(x_1,\varsigma_1,t_1-\theta_1)Q(x_2,\varsigma_2,t_2-\theta_2)D\delta(\varsigma_2-\varsigma_1)\delta(\theta_2-\theta_1)d\theta_1\,d\theta_2\,d\varsigma_1\,d\varsigma_2$$

$$\int_0^L \int_0^L X_m(\varsigma_1)X_m(\varsigma_2)\delta(\varsigma_2-\varsigma_1)d\varsigma_1\,d\varsigma_2 = \int_0^L X_m(\varsigma_1)X_m(\varsigma_2)d\varsigma_1 = K_m$$

$$\int_{-\infty}^{t} \int_{-\infty}^{t} e^{-\gamma(t-\theta_1)} \sin\omega'_m(t-\theta_1)e^{-\gamma(t-\theta_2)} \sin\omega'_m(t-\theta_2)\delta(\theta_2-\theta_1)d\theta_1\,d\theta_2$$

$$= \int_{-\infty}^{t} e^{-\gamma(t-\theta_1)} \sin\omega'_m(t-\theta_1)e^{-\gamma(t-\theta_1)} \sin\omega'_m(t-\theta_1)d\theta_1$$

$$= \int_{-\infty}^{t} e^{-2\gamma(t-\theta_1)} \sin^2\omega'_m(t-\theta_1)d\theta_1 = \frac{\omega'^2_m}{4\gamma\omega_m^2}$$

Thus,

$$R_q(x,t) = E[q^2(x,t)] = \frac{D}{4A^2\rho^2\gamma} \sum_m \frac{1}{K_m\omega_m^2} X'_m(x)$$

## References


Bendat, J. S. and Piersol, A. G. 2010. *Random Data Analysis and Measurement Procedures.* New York: John Wiley and Sons.

Gupta, A. K. 1990. *Response Spectrum Method in Seismic Analysis and Design of Structures.* Brookline Village, MA: Blackwell Scientific Publications, Inc.

Harris, C. M. and Piersol, A. G. 2010. *Harris' Shock and Vibration Handbook,* 6th ed. New York: McGraw-Hill.

Lutes, L. D. and Sarkani, S. 2004. *Random Vibrations Analysis of Structural and Mechanical Systems,* Burlington, MA: Elsevier Butterworth-Heinemann.

Yang, Y. C. 1986. *Random Vibration of Structures.* New York: John Wiley and Sons.

# Appendix A: Introduction to Composite Materials

## A.1 Introduction

As a background to the analysis of composite materials, one needs to be familiar with orthotropic material behavior. A complete presentation of the topics related to composite materials is beyond the scope of this text. Numerous references are readily available regarding the general topic of laminated composite materials, including Ashton, Halpin and Petit (1969), Jones (1975), Agarwal and Broutman (1980), Vinson and Sierakowski (1986), and Staab (2015). The fundamental building block for a traditional continuous fiber composite plate is the lamina. Once a lamina's response to loads can be assessed, one can consider a laminate, which consists of individual lamina arranged with their fiber oriented at selected angles with respect to a reference axis. The intent of this appendix is to provide the reader with sufficient information to make them familiar enough with composite material terminology so that they can more fully appreciate sections of the text pertaining to laminated composite materials.

## A.2 Orthotropic Composite Lamina

A lamina is typically an arrangement of unidirectional (or woven) fibers suspended in a matrix material. It is generally assumed to be orthotropic, and its thickness depends on the material from which it is made. The matrix is the binder material which supports, separates, and protects the fibers. It provides a path by which load is both transferred to the fibers and redistributed among the fibers in the event of fiber breakage. The matrix typically has a lower density, stiffness, and strength than the fibers. Matrices can be brittle, ductile, elastic, or plastic. They can have either linear or nonlinear stress–strain behavior. The behavior of a lamina to applied loads is characterized by defining directions relative to the fiber orientation. These directions as illustrated in Figure A.1 are: 1—principal fiber direction; 2— in-plane direction perpendicular to fibers; and 3—out-of-plane direction perpendicular to fibers.

In order to evaluate the response of a lamina to applied loads, each component of a stiffness matrix $[C]$ must be determined. The stress–strain relationships needed to define $[C]$ are obtained by experimental procedures. The stiffness matrix is established by first developing the compliance matrix $[S]$ and inverting it to obtain $[C]$. The lamina is orthotropic so extension and shear are uncoupled in the principal material directions. Figure A.2 shows the directions of normal load application required to establish the normal stress components of $[S]$ in each direction.

There is no shear–extension coupling, so the relationship between each normal strain and the applied stress in the 1-direction is

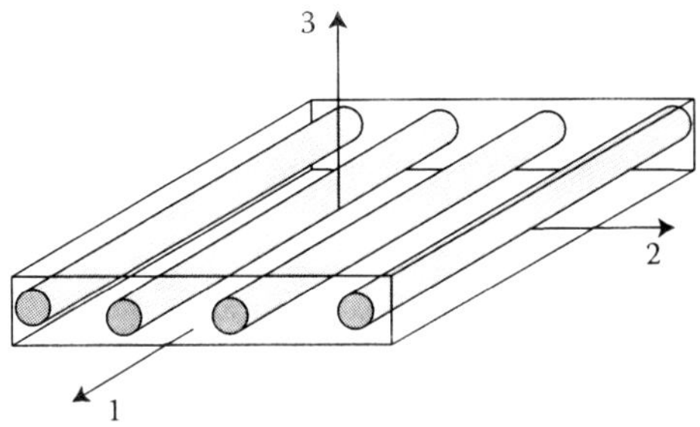


**FIGURE A.1**
Schematic of actual and modeled lamina.

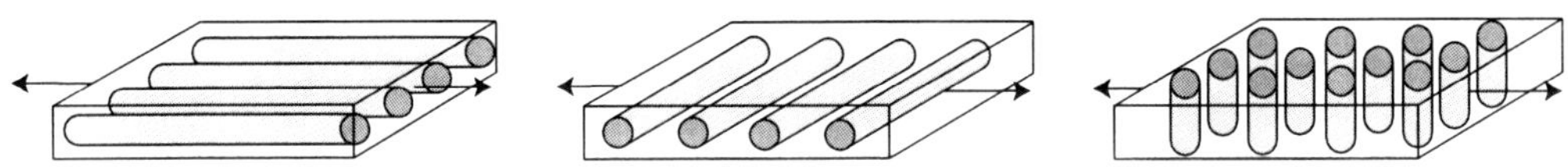


**FIGURE A.2**
Schematic of applied stress for determining components of an orthotropic compliance matrix.

$$\varepsilon_1 = \frac{\sigma_1}{E_1} \quad \varepsilon_2 = -\frac{\nu_{12}\sigma_1}{E_1}$$

where
$E_1$ = elastic modulus in the 1-direction (parallel to the fibers).
$v_{12}$ = Poisson's ratio in the 2-direction when the lamina is loaded in the 1-direction.
$v_{13}$ = Poisson's ratio in the 3-direction when the lamina is loaded in the 1-direction.

Similarly, by application of a normal stresses $\sigma_2$ and $\sigma_3$ with all other stresses zero results in

$$\varepsilon_1 = -\frac{\nu_{21}\sigma_2}{E_2} \quad \varepsilon_2 = \frac{\sigma_2}{E_2} \quad \varepsilon_3 = -\frac{\nu_{23}\sigma_2}{E_2}$$

$$\varepsilon_1 = -\frac{\nu_{31}\sigma_3}{E_3} \quad \varepsilon_2 = -\frac{\nu_{32}\sigma_3}{E_3}$$

Combining these results, and noting Hook's law applies with $\{\varepsilon_i\} = [S_{ij}]\{\sigma_j\}$, the extension terms are

$$\begin{aligned} S_{11} &= \frac{1}{E_1} \quad S_{12} = -\frac{\nu_{21}}{E_2} \quad S_{13} = -\frac{\nu_{31}}{E_3} \\ S_{21} &= -\frac{\nu_{12}}{E_1} \quad S_{22} = \frac{1}{E_2} \quad S_{23} = -\frac{\nu_{32}}{E_3} \\ S_{31} &= -\frac{\nu_{13}}{E_1} \quad S_{32} = -\frac{\nu_{23}}{E_2} \quad S_{33} = \frac{1}{E_3} \end{aligned} \tag{A.1}$$

The elastic modulus and Poisson's ratio can be expressed as

$E_i$ = elastic modulus in the $i$-direction, with a normal stress applied in the $i$-direction.

$v_{ij}$ = Poisson's ratio for transverse strain in the $j$-direction, with a normal stress applied in the $i$-direction.

Since the compliance matrix is symmetric, a simplifying relationship exists between Poisson's ratio and the elastic moduli.

$$\frac{\nu_{ij}}{E_i} = \frac{\nu_{ji}}{E_j} \quad (i, j = 1, 2, 3) \tag{A.2}$$

The relationship defined by Equation A.2 can be used to express (A.1) as

$$\begin{matrix} S_{11} = \dfrac{1}{E_1} & S_{12} = -\dfrac{\nu_{12}}{E_1} & S_{13} = -\dfrac{\nu_{13}}{E_1} \\ S_{21} = -\dfrac{\nu_{12}}{E_1} & S_{22} = \dfrac{1}{E_2} & S_{23} = -\dfrac{\nu_{23}}{E_2} \\ S_{31} = -\dfrac{\nu_{13}}{E_1} & S_{32} = -\dfrac{\nu_{23}}{E_2} & S_{33} = \dfrac{1}{E_3} \end{matrix} \tag{A.3}$$

In addition to the normal components of the compliance matrix, the shear terms $S_{44}$, $S_{55}$, and $S_{66}$ must be determined. In principle this is a simple matter, since there in no shear–extension coupling. By application of a pure shear on the 2–3, 1–3, and 1–2 planes, the relationships between shear stress and strain are

$$S_{44} = \frac{1}{G_{23}}, \quad S_{55} = \frac{1}{G_{13}}, \quad S_{66} = \frac{1}{G_{12}} \tag{A.4}$$

where $G_{ij}$ is the shear modulus corresponding to a shear stress applied to the $ij$-plane.

The stiffness matrix is obtained by inverting the compliance matrix. The stiffness matrix is, by convention, expressed as [Q] instead of [C]. The lamina stress–strain relationship (commonly referred to as a *specially orthotropic* lamina) is

$$\begin{Bmatrix} \sigma_1 \\ \sigma_2 \\ \sigma_3 \\ \sigma_4 \\ \sigma_5 \\ \sigma_6 \end{Bmatrix} = \begin{bmatrix} Q_{11} & Q_{12} & Q_{13} & 0 & 0 & 0 \\ Q_{12} & Q_{22} & Q_{23} & 0 & 0 & 0 \\ Q_{13} & Q_{23} & Q_{33} & 0 & 0 & 0 \\ 0 & 0 & 0 & Q_{44} & 0 & 0 \\ 0 & 0 & 0 & 0 & Q_{55} & 0 \\ 0 & 0 & 0 & 0 & 0 & Q_{66} \end{bmatrix} \begin{Bmatrix} \varepsilon_1 \\ \varepsilon_2 \\ \varepsilon_3 \\ \varepsilon_4 \\ \varepsilon_5 \\ \varepsilon_6 \end{Bmatrix} \tag{A.5}$$

The individual components of the stiffness matrix [Q] are expressed in terms of the elastic constants as

$$
\begin{aligned}
Q_{33} &= \frac{E_3(1-\nu_{12}\nu_{21})}{\Delta} \\
Q_{12} &= \frac{E_1(\nu_{21}+\nu_{31}\nu_{23})}{\Delta} = \frac{E_2(\nu_{12}+\nu_{32}\nu_{13})}{\Delta} \\
Q_{13} &= \frac{E_1(\nu_{31}+\nu_{21}\nu_{32})}{\Delta} = \frac{E_3(\nu_{13}+\nu_{12}\nu_{23})}{\Delta} \\
Q_{23} &= \frac{E_2(\nu_{32}+\nu_{12}\nu_{31})}{\Delta} = \frac{E_3(\nu_{23}+\nu_{21}\nu_{13})}{\Delta} \\
Q_{66} &= G_{12}
\end{aligned}
\tag{A.6}
$$

where $\Delta = 1 - \nu_{12}\nu_{21} - \nu_{23}\nu_{32} - \nu_{31}\nu_{13} - 2\nu_{13}\nu_{21}\nu_{32}$.

Under appropriate conditions, these expressions can be simplified. For example, the elastic moduli in the 2- and 3-directions are generally assumed to be the same ($E_2 = E_3$), which implies $\nu_{23} = \nu_{32}$ and $\nu_{21}\nu_{13} = \nu_{12}\nu_{31}$. Simplifications to these equations can be made by assumptions such as plane stress.

Equation A.5 is based on elastic constants for the special case of the $x$-axis coinciding with the 1-axis of the material (an *on-axis* configuration). In this arrangement, the $x$(1)-direction is associated with the maximum lamina stiffness, while the $y$(2)-direction corresponds to the direction of minimum lamina stiffness. It is not always practical for the $x$-axis of a lamina to correspond to the 1-axis of the material. An orientation in which $x$–$y$ and 1–2 do not coincide is an *off-axis* configuration. Both configurations are illustrated in Figure A.3.

The on-axis stress–strain relationship of Equation A.5 is not adequate for the analysis of an off-axis configuration. Relating stresses and strains in the $x$–$y$ system to the constitutive relations developed for the 1–2 system requires the use of stress and strain transformation matrices as described in undergraduate strength of materials courses. Applying these transformations results in

$$
\begin{Bmatrix} \sigma_x \\ \sigma_y \\ \sigma_z \\ \tau_{yz} \\ \tau_{xz} \\ \tau_{xy} \end{Bmatrix}
=
\begin{bmatrix}
\bar{Q}_{11} & \bar{Q}_{12} & \bar{Q}_{13} & 0 & 0 & \bar{Q}_{16} \\
\bar{Q}_{12} & \bar{Q}_{22} & \bar{Q}_{23} & 0 & 0 & \bar{Q}_{26} \\
\bar{Q}_{13} & \bar{Q}_{23} & \bar{Q}_{33} & 0 & 0 & \bar{Q}_{36} \\
0 & 0 & 0 & \bar{Q}_{44} & \bar{Q}_{45} & 0 \\
0 & 0 & 0 & \bar{Q}_{45} & \bar{Q}_{55} & 0 \\
\bar{Q}_{16} & \bar{Q}_{26} & \bar{Q}_{36} & 0 & 0 & \bar{Q}_{66}
\end{bmatrix}
\begin{Bmatrix} \varepsilon_x \\ \varepsilon_y \\ \varepsilon_z \\ \gamma_{yz} \\ \gamma_{xz} \\ \gamma_{xy} \end{Bmatrix}
\tag{A.7}
$$

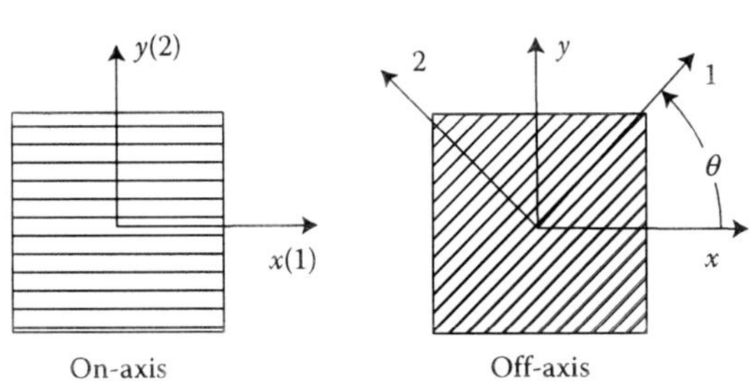


**FIGURE A.3**
On- and off-axis configurations.

Each element of $[\bar{Q}]$ is defined as

$$\bar{Q}_{11} = Q_{11}m^4 + 2(Q_{12} + 2Q_{66})m^2n^2 + Q_{22}n^4$$

$$\bar{Q}_{12} = (Q_{11} + Q_{22} - 4Q_{66})m^2n^2 + Q_{12}(m^4 + n^4)$$

$$\bar{Q}_{13} = Q_{13}m^2 + Q_{23}n^2$$

$$\bar{Q}_{16} = -Q_{22}mn^3 + Q_{11}m^3n - (Q_{12} + 2Q_{66})mn(m^2 - n^2)$$

$$\bar{Q}_{22} = Q_{11}n^4 + 2(Q_{12} + 2Q_{66})m^2n^2 + Q_{22}m^4$$

$$\bar{Q}_{23} = Q_{13}n^2 + Q_{23}m^2$$

$$\bar{Q}_{26} = -Q_{22}m^3n + Q_{11}mn^3 - (Q_{12} + 2Q_{66})mn(m^2 - n^2) \quad \text{(A.8)}$$

$$Q_{33} = Q_{33}$$

$$\bar{Q}_{36} = (Q_{13} - Q_{23})mn$$

$$\bar{Q}_{44} = Q_{44}m^2 + Q_{55}n^2$$

$$\bar{Q}_{45} = (Q_{55} - Q_{44})mn$$

$$\bar{Q}_{55} = Q_{55}m^2 + Q_{44}n^2$$

$$\bar{Q}_{66} = (Q_{11} + Q_{22} - 2Q_{12})m^2n^2 + Q_{66}(m^2 - n^2)^2$$

where $m = \cos\theta$ and $n = \sin\theta$. Equations A.7 and A.8 allow for the analysis of an off-axis lamina provided an on-axis constitutive relationship exists.

For a condition of plane stress, Equations A.5 and A.7 reduce since one normal and two shear components of stress are zero. As with the case of an isotropic material, the elimination of stress components does not imply that strain components become zero. In the case of plane stress, we assume that in the material coordinate system $\sigma_3 = \sigma_4 = \sigma_5 = 0$. The stiffness matrix for plane stress is termed the *reduced stiffness matrix*. The on-axis form of the reduced stiffness matrix is similar to the $[Q]$ of Equation A.5 and is

$$[Q] = \begin{bmatrix} Q_{11} & Q_{12} & 0 \\ Q_{12} & Q_{22} & 0 \\ 0 & 0 & Q_{66} \end{bmatrix} \quad \text{(A.9)}$$

where the individual terms are

$$Q_{11} = \frac{E_1}{1-\nu_{12}\nu_{21}}, \quad Q_{22} = \frac{E_2}{1-\nu_{12}\nu_{21}}, \quad Q_{12} = \frac{\nu_{12}E_2}{1-\nu_{12}\nu_{21}} = \frac{\nu_{21}E_1}{1-\nu_{12}\nu_{21}}, \quad Q_{66} = G_{12}$$

Although both out-of-plane shear strains are zero for this case, the normal strain ($\varepsilon_3$) exists, and is easily derived from Equation A.5.

The off-axis form of the reduced stiffness is formulated using the stress and strain transformations of Chapter 2 along with Equation A.9. Following the same procedures as before results in

$$\begin{Bmatrix} \sigma_x \\ \sigma_y \\ \tau_{xy} \end{Bmatrix} = \begin{bmatrix} \bar{Q}_{11} & \bar{Q}_{12} & \bar{Q}_{16} \\ \bar{Q}_{12} & \bar{Q}_{22} & \bar{Q}_{26} \\ \bar{Q}_{16} & \bar{Q}_{26} & \bar{Q}_{66} \end{bmatrix} \begin{Bmatrix} \varepsilon_x \\ \varepsilon_y \\ \gamma_{xy} \end{Bmatrix} \tag{A.10}$$

where the corresponding terms from Equation A.8 remain applicable for the plane stress case given by Equation A.10.

## A.3 Orthotropic Composite Laminates

A *laminate* is a collection of lamina arranged in a specified manner. Within a specific laminate, there are $N$ lamina. Adjacent lamina may be of the same or different materials and their fiber orientations with respect to a reference axis (by convention, the $x$-axis). The $[\bar{Q}]$ for each lamina is denoted by $[\bar{Q}]_k$, indicating that it is associated with the $k$th lamina. The stacking arrangement is defined with respect to the midplane of the laminate, with lamina number 1 being on the bottom and lamina $n$ being at the top, as illustrated in Figure A.4.

The conventional analysis of a laminates is based on classical lamination theory (CLT). The approach used in formulating CLT is similar to that used in developing load–stress relationships in elementary strength of materials courses. The governing load–displacement relationships are developed with respect to the midsurface strains $\{\varepsilon^0\}$ and curvatures $\{k\}$, which in turn are based on midplane displacements and curvatures, which are based on the assumed displacement fields.

$$U = U_0(x,y) + z\Phi(x,y), \quad V = V_0(x,y) + z\Psi(x,y), \quad W = W_0(x,y) \tag{A.11}$$

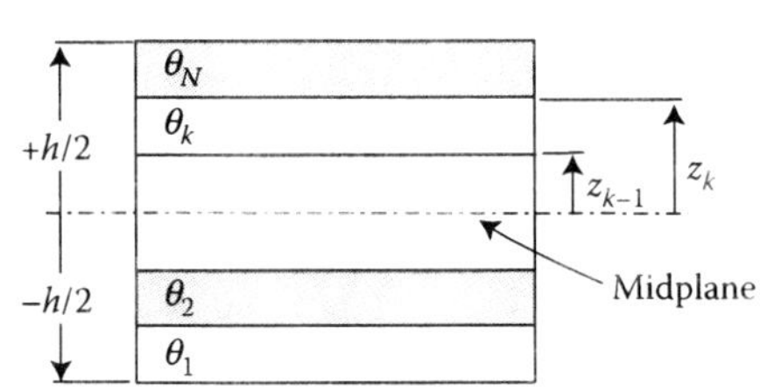


**FIGURE A.4**
Laminate stacking arrangement.

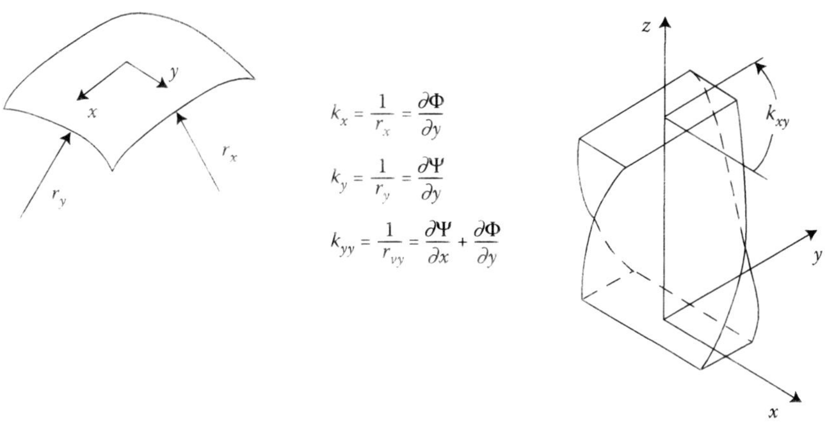


**FIGURE A.5**
Plate curvatures for classical lamination theory.

The resulting midsurface strains and curvatures (illustrated in Figure A.5) are defined as

$$\{\varepsilon^0\} = \begin{Bmatrix} \varepsilon_x^0 \\ \varepsilon_y^0 \\ \gamma_{xy}^0 \end{Bmatrix} = \begin{Bmatrix} \partial U_0/\partial x \\ \partial V_0/\partial y \\ \partial V_0/\partial x + \partial U_0/\partial y \end{Bmatrix} \quad \{\kappa\} = \begin{Bmatrix} \kappa_x \\ \kappa_y \\ \kappa_{xy} \end{Bmatrix} = \begin{Bmatrix} \partial \Phi/\partial x \\ \partial \Psi/\partial y \\ \partial \Psi/\partial x + \partial \Phi/\partial y \end{Bmatrix}$$

The stress in the *k*th lamina each is related to the strains and curvatures by

$$\begin{Bmatrix} \sigma_x \\ \sigma_y \\ \tau_{xy} \end{Bmatrix}_k = [\bar{Q}]_k \left( \begin{Bmatrix} \varepsilon_x^0 \\ \varepsilon_y^0 \\ \gamma_{xy}^0 \end{Bmatrix} + z \begin{Bmatrix} \kappa_x \\ \kappa_y \\ \kappa_{xy} \end{Bmatrix} \right)$$

The applied in-plane loads, moments, and shear forces on the laminate are related to the stresses in each lamina by summing the contribution of each lamina. These are expressed as

$$\begin{Bmatrix} N_x \\ N_y \\ N_{xy} \end{Bmatrix} = \sum_{k=1}^{N} \int_{z_{k-1}}^{z_k} \begin{Bmatrix} \sigma_x \\ \sigma_y \\ \tau_{xy} \end{Bmatrix}_k dz \quad \begin{Bmatrix} M_x \\ M_y \\ M_{xy} \end{Bmatrix} = \sum_{k=1}^{N} \int_{z_{k-1}}^{z_k} z \begin{Bmatrix} \sigma_x \\ \sigma_y \\ \tau_{xy} \end{Bmatrix}_k dz \quad \begin{Bmatrix} Q_x \\ Q_y \end{Bmatrix} = \sum_{k=1}^{N} \int_{z_{k-1}}^{z_k} \begin{Bmatrix} \tau_{xz} \\ \tau_{yz} \end{Bmatrix}_k dz$$

Substituting the equations for the stresses in *k*th lamina and neglecting the $Q_x$ and $Q_y$ terms results in

$$\begin{Bmatrix} M_x \\ M_y \\ M_{xy} \end{Bmatrix} = \sum_{k=1}^{N} [\bar{Q}]_k \left( \int_{z_{k-1}}^{z_k} z \begin{Bmatrix} \varepsilon_x^0 \\ \varepsilon_y^0 \\ \gamma_{xy}^0 \end{Bmatrix} dz + \int_{z_{k-1}}^{z_k} z^2 \begin{Bmatrix} \kappa_x \\ \kappa_y \\ \kappa_{xy} \end{Bmatrix} dz \right)$$

The end result is that the loads and moments are related to the strains and curvatures (which are related to displacements) through the matrix expression

$$\begin{Bmatrix} N_x \\ N_y \\ N_{xy} \\ -- \\ M_x \\ M_y \\ M_{xy} \end{Bmatrix} = \begin{bmatrix} A_{11} & A_{12} & A_{16} & | & B_{11} & B_{12} & B_{16} \\ A_{12} & A_{22} & A_{26} & | & B_{12} & B_{22} & B_{26} \\ A_{16} & A_{26} & A_{36} & | & B_{16} & B_{26} & B_{66} \\ -- & -- & -- & | & -- & -- & -- \\ B_{11} & B_{12} & B_{16} & | & D_{11} & D_{12} & D_{16} \\ B_{12} & B_{22} & B_{26} & | & D_{12} & D_{22} & D_{26} \\ B_{16} & B_{26} & B_{66} & | & D_{16} & D_{26} & D_{66} \end{bmatrix} \begin{Bmatrix} \varepsilon_x^0 \\ \varepsilon_y^0 \\ \gamma_{xy}^0 \\ -- \\ \kappa_x \\ \kappa_y \\ \kappa_{xy} \end{Bmatrix} \quad \text{(A.12)}$$

Each term of the $[A]$, $[B]$, and $[D]$ is defined by

$$\begin{aligned} [A_{ij}] &= \sum_{k=1}^{N} [\bar{Q}_{ij}]_k (z_k - z_{k-1}) \\ [B_{ij}] &= \frac{1}{2} \sum_{k=1}^{N} [\bar{Q}_{ij}]_k \left(z_k^2 - z_{k-1}^2\right) \\ [D_{ij}] &= \frac{1}{3} \sum_{k=1}^{N} [\bar{Q}_{ij}]_k \left(z_k^3 - z_{k-1}^3\right) \end{aligned} \quad \text{(A.13)}$$

These matrices are termed

$[A_{ij}] =$ extensional stiffness matrix
$[B_{ij}] =$ extension-bending coupling matrix
$[D_{ij}] =$ bending stiffness matrix

When determining these matrices, it is useful to note that the region between $z_{k-1}$ and $z_k$ is the thickness of the $k$th lamina, $t_k = z_k - z_{k-1}$. In addition the centroid of each lamina, $\bar{z}_k$, can be defined with respect to the midsurface of the laminate (noting that below the midplane it is negative and it is positive above the midplane). With these definitions in place: $(z_k - z_{k-1}) = t_k$, $(1/2)\left(z_k^2 - z_{k-1}^2\right) = t_k \bar{z}_k$, $(1/3)\left(z_k^3 - z_{k-1}^3\right) = t_k \bar{z}_k^2 + \left(z_k^3/12\right)$. This simplifies the computation of the $[A]$, $[B]$, and $[D]$ matrices.

The shear forces $Q_x$ and $Q_y$ are treated differently since they are not expressed in terms of midsurface strains and curvatures. The stress–strain relationship for shear in the $k$th layer is expressed as

$$\begin{Bmatrix} \tau_{xz} \\ \tau_{yz} \end{Bmatrix}_k = \begin{bmatrix} \bar{Q}_{55} & \bar{Q}_{45} \\ \bar{Q}_{45} & \bar{Q}_{44} \end{bmatrix}_k \begin{Bmatrix} \gamma_{xz} \\ \gamma_{yz} \end{Bmatrix}_k$$

Subsequently the forces $Q_x$ and $Q_y$ are

$$\begin{Bmatrix} Q_x \\ Q_y \end{Bmatrix}_k = \sum_{k=1}^{N} \int_{z_{k-1}}^{z_k} \begin{bmatrix} \bar{Q}_{55} & \bar{Q}_{45} \\ \bar{Q}_{45} & \bar{Q}_{44} \end{bmatrix}_k \begin{Bmatrix} \gamma_{xz} \\ \gamma_{yz} \end{Bmatrix}_k dz$$

It is generally assumed that the transverse shear stresses have a parabolic distributed over the laminate thickness. This can be represented by a weighting function $f(z) = c[1 - (2z/h)^2]$. The coefficient $c$ is commonly termed the shear correction factor. The numerical value of $c$ depends upon the cross-sectional shape of the laminate. For a rectangular section, generally of interest in laminate analysis, $c = 6/5$ (1.20). The derivation of this can be found in many strength of materials texts. The expression for $Q_x$ and $Q_y$ can be written as

$$\begin{Bmatrix} Q_x \\ Q_y \end{Bmatrix} = \begin{bmatrix} A_{55} & A_{45} \\ A_{45} & A_{44} \end{bmatrix}_k \begin{Bmatrix} \gamma_{xz} \\ \gamma_{yz} \end{Bmatrix}_k \tag{A.14}$$

where

$$A_{ij} = c\sum_{k=1}^{N}[\bar{Q}_{ij}]\left\{(z_k - z_{k-1}) - \frac{4}{3h^2}\left(z_k^3 - z_{k-1}^3\right)\right\}$$

where $i, j = 4, 5$ and $h$ is the total laminate thickness.

## References

Agarwal, B. D. and Broutman, L. J. 1980. *Analysis and Performance of Fiber Composites*. New York: John Wiley and Sons.

Ashton, J. E., Halpin, J. C., and Petit, P. H. 1969. *Primer on Composite Materials*. Westport, CT: Technomic.

Jones, R. M. 1975. *Mechanics of Composite Materials*. New York: Hemisphere Publishing.

Staab, G. H. 2015. *Laminar Composites*, 2nd ed. Oxford: Elsevier.

Vinson, J. R. and Sierakowski, R. L. 1986. *The Behavior of Structures Composed of Composite Materials*. Dordrecht, The Netherlands: Martinus Nijhoff.

# Appendix B: Additional References

Appendix B contains general references that are applicable to the subject matter of this book, but were not cited in any of the chapters.


Anderson, J. C. and Naeim, F. 2012. *Basic Structural Dynamics*. New York: John Wiley & Sons, Inc.

Banakh, L. Y. and Kempner, M. L. 2010. *Vibrations of Mechanical Systems with Regular Structure*. New York: Springer-Verlag.

Beards, C. F. 1995. *Engineering Vibration Analysis with Application to Control Systems*. London: Edward Arnold.

Beards, C. F. 1996. *Structural Vibration: Analysis and Damping*. New York: Halstead Press.

Benaroya, H. and Nagurka, M. L. 2010. *Mechanical Vibration—Analysis, Uncertainties, and Control*, 3rd ed. Boca Raton: CRC Press/Taylor & Francis Group.

Biggs, J. M. 1964. *Introduction to Structural Dynamics*. New York: McGraw-Hill.

Bottega, W. J. 2006. *Engineering Vibrations*. Boca Raton: CRC Press/Taylor & Francis Group.

Buchholdt, H. A. and Nejad, S. E. M. 2012. *Structural Dynamics for Engineers*, 2nd ed. London: ICE Publishing.

Cheng, F. Y. 2000. *Matrix Analysis of Structural Dynamics—Applications and Earthquake Engineering*. New York: Marcel Dekker, Inc.

Cheng, F. Y. 2001. *Matrix Analysis of Structural Dynamics*. New York: Marcel Dekker, Inc.

Cho, W. S. T. 2014. *Stochastic Structural Dynamics Application of Finite Element Methods*. New York: John Wiley & Sons, Inc.

Chopra, A. K. 2001. *Dynamics of Structures Theory and Applications to Earthquake Engineering*. Upper Saddle River, NJ: Prentice-Hall.

Chopra, A. K. 2012. *Dynamics of Structures Theory and Application to Earthquake Engineering*, 4th ed. Upper Saddle River, NJ: Prentice-Hall.

Chowdhury, I. and Dasgupta, S. 2009. *Dynamics of Structure and Foundation—A Unified Approach 1. Fundamentals*. Boca Raton: CRC Press/Taylor & Francis Group.

Church, A. 1948. *Elementary Mechanical Vibrations*. New York: Pitman Publishing Corporation.

Clough, R. W. and Penzein, J. 1975. *Dynamics of Structures*. New York: McGraw-Hill.

Clough, R. W. and Penzien, J. 2003. *Dynamics of Structures*, 3rd ed. Berkeley, CA: Computer & Structures, Inc.

Craig, R. R. Jr. and Kurdila, A. J. 2006. *Fundamentals of Structural Dynamics*, 2nd ed. Hoboken: John Wiley & Sons, Inc.

De Silva, C. W. 1999. *Vibration: Fundamentals and Practice*. Boca Raton: CRC Press.

De Silva, C. W. 2005. *Computer Techniques in Vibration*. Boca Raton: CRC Press/Taylor & Francis Group.

Den Hartog, J. P. 1985. *Mechanical Vibrations*, 4th ed. New York: Dover Publications, Inc.

Donaldson, B. H. 2006. *Introduction to Structural Dynamics*. Cambridge: Cambridge University Press.

Donea, J. ed. 1980. *Advanced Structural Dynamics*. Essex: Applied Science Publishers Ltd.

Drozdov, A. D. 1998. *Viscoelastic Structures*. San Diego: Academic Press.

Dukkipati, R. V. 2007. *Solving Vibration Analysis Problems Using MATLAB*. New Delhi: New Age International Limited.

Freberg, C. R. and Kemler, E. N. 1949. *Elements of Mechanical Vibration*, 2nd ed. New York: John Wiley & Sons, Inc.

Gatti, P. L. and Ferrari, V. 2003. *Applied Structural and Mechanical Vibrations—Theory, Methods and Measuring Instrumentation*. New York: Routledge.

Gawronski, W. K. 2004. *Advanced Structural Dynamics and Active Control of Structures*. New York: Springer-Verlag.

Geradin, M. and Rixen, D. J. 2015. *Mechanical Vibrations Theory and Application to Structural Dynamics*, 3rd ed. New York: John Wiley & Sons, Inc.

Girard, A. and Roy, N. 2008. *Structural Dynamics in Industry*. Hoboken: John Wiley & Sons, Inc.

Hagedron, P. and DasGupta, A. 2007. *Vibrations and Waves in Continuous Mechanical Systems*. West Sussex: John Wiley & Sons Ltd.

Hansen, H. M. and Chenea, P. F. 1952. *Mechanics of Vibration*. New York: John Wiley & Sons, Inc.

Hatch, M. R. 2001. *Vibration Simulation Using MATLAB and ANSYS*. Boca Raton: Chapman & Hall/CRC Press.

Hodges, D. H. and Pierce, G. A. 2002. *Introduction to Structural Dynamics and Aeroelasticity*. Cambridge: Cambridge University Press.

Hopkins, B. 1910. *Vibrations of Systems Having One Degree of Freedom*. Cambridge: Cambridge University Press.

Humar, J. L. 2001. *Dynamics of Structures*, 2nd ed. Tokyo: A. A. Balkema Publishers.

Inman, D. 2014. *Engineering Vibration*, 4th ed. Upper Saddle River: Pearson Education, Inc.

Jacobsen, L. S. and Ayre, R. S. 1958. *Engineering Vibrations with Applications to Structures and Machinery*. New York: McGraw-Hill Company.

Jazar, R. N. 2013. *Advanced Vibrations—A Modern Approach*. New York: Springer Science + Business.

Kachapi, S. H. H. and Ganji, D. D. 2014. *Dynamics and Vibrations—Progress in Nonlinear Analysis*. New York: Springer Science + Business.

Kappos, A. J. 2002. *Dynamic Loading and Design of Structures*. London: Spon Press/Taylor & Francis Group.

Karnovsky, I. A. and Lebed, O. I. 2001. *Formulas for Structural Dynamics: Tables, Graphs and Solutions*. New York: McGraw-Hill.

Kelly, G. S. 2000. *Fundamentals of Mechanical Vibrations*, 2nd ed. Boston: McGraw-Hill.

Kelly, G. S. 2012. *Mechanical Vibrations Theory and Applications*. Stamford: Cengage Learning.

Krysinski, T. and Malburet, F. 2007. *Mechanical Vibrations Active and Passive Control*. Chippenham: Anthony Rowe Ltd.

Lee, K. C. 2011. *Energy Methods in Dynamics*. Heidelberg: Springer-Verlag.

Leissa, A. W. and Quatu, M. S. 2011. *Vibration of Continuous Systems*. New York: McGraw-Hill Companies, Inc.

Lewandowski, R., Bartkowiak, A., and Maciejewski, H. 2012. Dynamic analysis of frames with viscoelastic dampers: A comparison of Samper models. *Struct. Eng. Mech.*, 41:113–137.

Li, J. and Chen, J. 2009. *Stochastic Dynamics of Structures*. Singapore: John Wiley & Sons (Asia) Pte Ltd.

Lutes, L. D. and Sarkani, S. 2004. *Random Vibrations Analysis of Structural and Mechanical Systems*. Oxford: Elsevier Butterworth-Heinemann.

Lynch, S. 2014. *Dynamical Systems with Applications Using MATLAB*, 2nd ed. New York: Springer.

McLachlan, N. W. 1951. *Theory of Vibrations*. New York: Dover Publications.

Meirovitch, L. 1975. *Elements of Vibration Analysis*, 2nd ed. New York: McGraw-Hill.

Meirovitch, L. 1980. *Computational Methods in Structural Dynamics*. Rijn: Sijhoff & Noordgoff.

Meirovitch, L. 1986. *Elements of Vibration Analysis*, 2nd ed. New York: McGraw-Hill.

Meirovitch, L. 2001. *Fundamentals of Vibrations*. Boston: McGraw-Hill Company.

Mobley, R. K. 1999. *Vibration Fundamentals*. Boston: Newnes.

Morrill, B. 1957. *Mechanical Vibrations*. New York: Roland Press Company.

Newland, D. E. 1993. *An Introduction to Random Vibrations, Spectral & Wavelet Analysis*, 3rd ed. New York: Prentice Hall.

Papadrakakis, M., Stefanou, G., and Papadopoulos, V. ed. 2011. *Computational Methods in Stochastic Dynamics*. New York: Springer Science + Business.

Paultre, P. 2010. *Dynamics of Structures*. Hoboken, NJ: John Wiley & Sons, Inc.

Paultre, P. 2011. *Dynamics of Structures*. Hoboken, NJ: John Wiley & Sons, Inc.

Paz, M. and Leigh, W. 2004. *Structural Dynamics Theory and Computation*, 5th ed. Boston: Kluwer Academic Publishers.

Rajasekaran, S. 2009. *Structural Dynamics of Earthquake Engineering Theory and Application Using MATHEMATICA and MATLAB*. Boca Raton: CRC Press/Woodhead Publishing.

Rao, S. S. 2007. *Vibration of Continuous Systems*. Hoboken: John Wiley & Sons, Inc.

Rao, S. S. 2011. *Mechanical Vibrations*, 5th ed. Upper Saddle River, NJ: Prentice-Hall.

Rogers, G. L. 1959. *An Introduction to the Dynamics of Framed Structures*. New York: John Wiley & Sons, Inc.

Roy, D. and Rao, G. V. 2012. *Elements of Structural Dynamics A New Perspective*. New York: John Wiley & Sons, Inc.

Schmitz, T. L. and Smith, K. S. 2012. *Mechanical Vibrations Modeling and Measurement*. New York: Springer-Verlag.

Sinha, A. 2010. *Vibration of Mechanical Systems*. Cambridge: Cambridge University Press.

Stojanovic, V. and Kozic, P. 2015. *Vibrations and Stability of Complex Beam Systems*. New York: Springer International Publishing Switzerland.

Strommen, E. N. 2014. *Structural Dynamics*. New York: Springer-Verlag.

Svetlitsky, V. A. 2005. *Dynamics of Rods*. Berlin: Springer-Verlag.

Thompson, W. T. and Dahleh, M. D. 2005. *Theory of Vibration with Applications*, 5th ed. Singapore: Pearson Education Asia Limited and Tsinghua University Press.

Thomson, W. T. 1993. *Theory of Vibration with Applications*, 4th ed. Englewood Cliffs, NJ: Prentice-Hall.

Thorby, D. 2008. *Structural Dynamics and Vibration in Practice an Engineering Handbook*. Oxford: Elsevier Butterworth-Heinemann.

Thornton, D. L. 1940. *Mechanics Applied to Vibrations and Balancing*. New York: John Wiley and Sons, Inc.

Timoshenko, S. 1937. *Vibration Problems in Engineering*, 2nd ed. New York: D. Van Nostrand Company.

Tong, K. N. 1960. *Theory of Mechanical Vibration*. Hoboken: John Wiley & Sons, Inc.

Tongue, B. H. 2002. *Principles of Vibration*, 2nd ed. New York: Oxford University Press.

Tse, F. S., Morse, I. E., and Hinkle, R. T. 1963. *Mechanical Vibrations*. Boston: Allyn and Bacon, Inc.

Tse, F. S., Morse, I. E., and Hinkle, R. T. 1978. *Mechanical Vibrations, Theory and Applications*, 2nd ed. Englewood Cliffs, NJ: Prentice Hall.

Veselic, K. 2011. *Damped Oscillations of Linear Systems—A Mathematical Introduction*. Berlin: Springer-Verlag.

Vierck, R. K. 1979. *Vibration Analysis*, 2nd ed. New York: Harper & Row Publishers.

Virgin, L. N. 2007. *Vibration of Axially Loaded Structures*. Cambridge: Cambridge University Press.

Warburton, G. B. 1976. *The Dynamical Behavior of Structures*, 2nd ed. Oxford: Pergamon Press.

Weaver, W. Jr., Timoshenko, S. P., and Young, D. H. 1990. *Vibration Problems in Engineering*, 5th ed. New York: Wiley-Interscience.

Xie, W.-C. 2006. *Dynamic Stability of Structures*. New York: Cambridge University Press.

Yang, C. Y. 1986. *Random Vibration of Structures*. New York: John Wiley & Sons, Inc.

Yuen, K.-V. 2010. *Bayesian Methods for Structural Dynamics and Civil Engineering*. Singapore: John Wiley & Sons (Asia).

# Index


A

Algebraic Eigenvalue problem, 497
α method, *see* HHT-method
Alternate forms of Fourier series, 42
Amplification factor, 23, 28, 32, 34
Amplitude of vibration, 8
Amplitude ratio, 23, 28, 32
Angular frequency, 10
Anisotropic material, 229, 233
Arbitrary forcing function, 186, 210–211
  asymptotic step forcing function, 66–67, 68
  Duhamel integral, 59
  exponentially decaying function, 65–66
  half-cycle sine impulse, 74–77
  integral transformations, 77–90
  rectangular impulse, 70–73
  response to, 58
  response to external force, 63–65
  response to ramp-step function, 67–70
  step response of undamped system, 60–63, 64
  triangular impulse, 73–74, 75
Arbitrary viscous damping, 213–214
Asymptotic step forcing function, 66–67, 68
Autocorrelation function, 103, 124
Average acceleration method, 148–149
Average value, *see* Mean value
Axial force
  effect, 332–334
  vibrations of frames with axial forces, 345–348
Axially symmetric deformation, 409–410

B

Bars, 265
  with concentrated end load and associated expressions, 294–296
  forced vibrations, 287–289
  free vibrations, 280–287
  longitudinal vibrations, 279–280
Beam(s), 299, 339
  finite element, 466–470
  modes, 405
Beam vibrations; *see also* Random vibrations
  Euler–Bernoulli theory, 301–325
  shear beam, 299–301
  Timoshenko beam theory, 326–334
Beating phenomenon, 25, 26
Bending theory, 409, 411, 417
Bernoulli beam theory, *see* Euler–Bernoulli beam theory
Body forces, 217, 236, 433
Boundary conditions, 4, 234–235, 266–267
  of cylindrical shells, 408–409
  element assembly, 495–496
Breathing modes, 405
Bridges, 1
Buildings, 1
Bulk modulus, 231

C

Cartesian coordinates, 168, 373
Central difference method, 142–145, 521
  algorithm for, 523
  displacement results, 525
  example, 523–524
  solution vector, 522
Central Limit Theorem, 97–98
Characteristic equation, 176, 199
Characteristic values, 176, 199
  problem, 497
Characteristic vector, 199, 200
Cholesky decomposition, 504
Circular plates, 373; *see also* Sandwich plates
  Cartesian coordinates, 373–374
  equations of motion, 374–378
  forced vibrations, 378
Classical beam theory, *see* Euler–Bernoulli beam theory
Classical fourth-order Runge–Kutta method, 141
Classical lamination theory (CLT), 369, 421, 422, 562, 563
Classical plate theory (CPT), 351
Closed-form solutions, 127, 391–392, 416–417
CLT, *see* Classical lamination theory
Complementary function, 22
Complementary solution, 27, 277
Compliance matrix, 163, 229, 559
Composite materials, 557
  orthotropic composite lamina, 557–562
  orthotropic composite laminates, 562–565

Composite plates, 369; *see also* Sandwich plates
forced vibrations, 373
free vibrations of simply supported plate, 370–373
relationship between loads and displacements, 369–370
Composite shells, 421
equations of motion, 422–423
forced vibrations, 425
free vibrations, 423–425
Computer simulation, 547
Concentrated force, 293–294, 314
Configuration space, 213
Conservative system, 175
Consistent mass matrix, 432, 435, 465, 521
Constant acceleration method, 147–148
Constant strain triangle element, 470–473
Continuous beams, 339
slope–deflection method, 339–344
vibrations of frames with axial forces, 345–348
Convergence, 507, 518
Correlation function, 103, 123, 552
ensemble average. *vs.* time average, 106
even function, 109
Fourier transformation, 107
multiple events, 103
practical white noise, 111
spectral density and correlation function reflect apparent periodicity, 108
stationary process correlation function depends, 104
stationary random function, 110
weak and strong correlation functions, 105
CPT, *see* Classical plate theory
Creep, 52
Cubic dilatation, 231
Curvilinear coordinates, 395
Cylindrical shells, 395
coordinates and applied loads, 396
equations of motion, 396–401
forced vibrations, 406–408
free vibrations, 403–406
membrane theory of, 409–414
other boundary conditions, 408–409
simplified system of equations, 401–403
solutions for, 403

D

d'Alembert's principle, 240
d'Alembert's solution, *see* Traveling wave solution
Damped forced vibration, 191–194, 195
Damped free vibration(s), 13, 186–191
actual damping in vibrating system, 14
damped pendulum, 21
displacement of critically and overdamped system, 16, 17
hyperbolic sine and cosine, 15
period and frequency of underdamped case, 19
plane model, 20
underdamped case with initial displacement, 18
Damping
coefficient, 433
effects, 51
forces, 211
material with, 289
Degrees of freedom (DOF), 430–431, 446
damped forced vibration, 191–194, 195
damped free vibrations, 186–191
equations of motion, 159–166
forced damped MDOF systems, 211–214
forced undamped vibrations, 208–211
free undamped MDOF systems, 198–208
free vibrations, 175–181
frequency response function, 181–186
general theory of MDOF, 194–198
Lagrange's equations, 166–171
MDOF, 159
potential energy, 171–175
systems, 548–549
two-node beam element with two DOF per node, 466
Degrees of freedom systems, 548–549
Del operator (∇), 356, 374, 420
Deterministic loading, 90
Diagonalization, 516
Dirac delta filtering property, 56
Dirac delta function, 56–57, 115, 293
response to, 57–58
Dirac functions, 81–82, 83
Direct integration method, 496
Direct time integration methods, 429
Discontinuous strings, 274–277
Discrete-time equation of motion, 155
Discrete systems, 1
Displacement and Strain Relations, 224
Displacement equations for elastic bodies, 233–234
Displacement transmissibility, 35, 36, 37
Dissipated energy, 48, 49
DOF, *see* Degrees of freedom
Donnell's type theory, 395

Duhamel integral, 59, 128–129, 296
Dummy index, 221
Dynamic analysis, 1
Dynamic characteristics of linear systems, 115–117
Dynamic load, 1
  factor, 60
Dynamic response of SDOF systems using numerical methods
  finite differences, 133–145
  HHT-Alpha method, 153, 155–157
  interpolating excitation function, 128–133, 134
  Newmark method, 145–150
  time-stepping method, 128
  Wilson-Theta method, 150–153, 154

E

Earthquake, 1, 93, 547
Eigenmatrix, 516
Eigenvalue(s), 199
  analysis, 496–504
  spectrum with shifted origin, 511
  inverse iteration for first and second eigenvalue pair using shifting, 512
  inverse iteration for third eigenvalue pair using shifting, 513
  solution methods for calculating, 504
  subspace iteration, 513–515
  vector iteration methods, 505–510
  vector iteration with shifts, 511–513
Eigenvector(s), 199, 504–513
Einstein's convention, 221
Elastic beam, 243, 258
Elastic bodies, displacement equations for, 233–234
Elastic modulus, 51, 559
Elastic–plastic materials, 50–51
Elastic potential energy, 172–173
Elastic system, 4
Element matrix, 490–491
Energy, 235
  kinetic energy, 238
  strain energy, 236–237
Energy dissipation, work performing by, 47–55
  elastic–plastic materials, 50–51
  material damping, 49–50
  viscoelastic materials, 51–53
  viscoelastic relations for harmonic functions, 53–55
Ensemble averaging, 105, 106
Equations of motion, 2, 159–166, 191, 326–329, 351
  boundary conditions, 357–358
  circular plates, 374–378
  composite plates, 369–373
  composite shells, 422–423
  cylindrical shells, 396–401
  differential equation, 355–356
  forced vibrations, 365–369
  mass of element, 354–355
  plane stress, 353–354
  for plates vibrations, 351
  rectangular plate, 351–353
  sandwich plates, 383–388
  solutions for other boundary conditions, 361–365
  solution to plate equations, 358–361
Equations of motion of continuous systems
  boundary conditions, 234–235
  displacement and strain relations, 224–228
  displacement equations for elastic bodies, 233–234
  equations of equilibrium, 219–221
  forces and stresses, 217–219
  Galerkin's method, 258–261
  general energy theorem, 247–248
  Hamilton's principle, 241–247
  plane stress, 222–224
  principle of virtual work, 238–241
  Rayleigh's method, 248–253
  Ritz method, 253–257
  stress–strain relations, 228–233
  work and energy, 235–238
Equilibrium equations, 219–221
Equivalent spring constant, 4–7
Ergodic property, 104
Euler beam theory, *see* Euler–Bernoulli beam theory
Euler–Bernoulli beam theory, 299, 301–302, 326
Euler–Bernoulli theory, 301; *see also* Timoshenko beam theory
  application of Laplace transformations, 318–319
  axial force effect, 324–325
  forced vibrations, 312–318
  free vibrations, 303–312
  frequency response function, 319–324
  out of plane displacements, 302
Euler method, 135–138
Event, 93
Expected value, *see* Mean value
Experimental technique, 547
Explicit method, 143
Exponentially decaying function, 65–66
Extensional strain, 227

External force(s), 208–209, 217
elastic–plastic materials, 50–51
material damping, 49–50
response to, 63–65
three-dimensional body acted by external forces, 218
viscoelastic materials, 51–53
viscoelastic relations for harmonic functions, 53–55
work performing by, 47
External harmonic force, 22

**F**

FEM, *see* Finite element method
Finite differences, 133
central difference method, 142–145
computational grid for finite difference, 134
Euler method, 135–138
forward and backward difference equations, 135
modified Euler method, 138–139
Runge–Kutta method, 139–142, 143
Finite element method (FEM), 1, 429, 430
Finite elements, 429, 430
assembly of element equations, 434–435
boundary conditions, 495–496
convenience, 493–494
degrees of freedom, 430–431
developing element characteristics, 431–434
direct time integration methods, 429
element assembly, 489
element matrix, 490–491
free response of finite element systems, 496–504
global matrices, 489–490
interpolation or shape functions, 435–489
multiple-degrees-of-freedom numerical techniques, 521–539
solution methods for calculating eigenvalues and eigenvectors, 504–515
stiffness matrix, 494–495
transformation matrix, 492–493
transformation methods, 515–520
two-element truss, 491
Finite length strings
forced vibrations of, 277–279
free vibrations of, 271–277
First mode of vibration, *see* Fundamental mode
Fixed-fixed ends, 283
Fixed-free ends, 282
Flexibility coefficient, 178–179
Flexural rigidity, 354
Force(s), 217
components of stress tensor, 219
three-dimensional body acted by external forces, 218
transmitting to base, 37
Forced damped harmonic motion, 27–30
using complex format, 30–32
Forced damped MDOF systems, 211
arbitrary viscous damping, 213–214
proportional damping, 212–213
Forced harmonic motion of support, 32–36
Force–displacement equations, 410
Forced undamped harmonic motion, 22–27
Forced undamped vibrations, 208
arbitrary forcing function, 210–211
steady-state harmonic motion, 208–210
Forced vibration(s), 289–293, 312–318, 334, 406–408
of bars, 287–289
circular plates, 378
composite plate, 373
composite shells, 425
of finite length strings, 277–279
forced damped harmonic motion, 27–32
forced harmonic motion of support, 32–36
forced undamped harmonic motion, 22–27
force transmitting to base, 37
under harmonic force, 21
membrane theory of cylindrical shells, 412–414
plates vibrations, 365–369
sandwich plates, 389–391
for slope–deflection method, 344
Forcing function, 37
impulse function, 56–57
response to, 55
response to Dirac delta function, 57–58
Forcing response to periodic loading, 37
alternate forms of Fourier series, 42
complex form of Fourier series, 42–44
response to general periodic forces, 44–47
trigonometric functions and Fourier series, 38–41
Forward iteration, 509–510
Four-noded rectangular element, 473–476
Fourier coefficients, 42, 46
Fourier series, 38–41
alternate forms of, 42
complex form of, 42–44
Fourier transform(s), 78
application to ODOF systems with material damping, 82–86
Fourier transformation(s), 107

application to viscoelastic relations, 80
solution, 268–271
Frames, 339
slope–deflection method, 339–344
vibrations of frames with axial forces, 345–348
Free-edge condition, 357
Free ends, 281
Free response of finite element systems, 496–504
mass matrix, 496–497
nontrivial algebraic equations, 497
orthogonality for eigenvectors of symmetric matrices, 502
Rayleigh quotient, 502–503
reduction to standard form, 503
Free undamped MDOF systems, 198
free vibrations with initial conditions, 202–208
normalized modes, 200–202
orthogonality of natural modes, 199–200
Free vibrations, 7, 175, 303–309, 329, 403–406
amplitude of vibration, 8
bars, 280
beam deformations, 331
built in beam, 330
car, trailer system, 12
composite shells, 423–425
discontinuous strings, 274–277
of finite length strings, 271–274
fixed-fixed ends, 283
fixed-free ends, 282
flexibility coefficient, 178–179
free ends, 281
fundamental mode, 177, 180
individual contributions of term in Equation, 9
with initial conditions, 202–208, 285–287, 309–312
inverted pendulum, 11
membrane theory of cylindrical shells, 411–412
natural frequencies, 13, 176–177
node, 181
orthogonality of natural modes, 283–285
sandwich plates, 388–389
effect of shear deformation and rotary inertia, 332–334
simple harmonic motion, 10
of simply supported plate, 370–373
spring mass system, 175
Frequency, 10
characteristics, 85
determinant, 198
equation, 176, 199
ratio, 34, 36
response curves, 28
Frequency response function (F.R.F.), 32, 85, 116, 181–184, 319–324
application of, 184
arbitrary forcing function, 186
transfer function, 184–186
Function of time, 93
Fundamental mode, 177, 180

G

Galerkin's method, 258–261, 382, 429
Gaussian distribution, 97
Gauss quadrature, 482
General energy theorem, 247–248
Generalized coordinates, 167–168, 430
Generalized forces, 169–171
Generalized Jacobi method, 516
convergence, 518
example, 520
off-diagonal elements, 517
General periodic forces, response to, 44–47
Global matrices, 489–490
Gram–Schmidt orthogonalization procedure, 509
Gravitational forces, 171–172
Green's theorem, 239

H

Hamiltonian mechanics, 247
Hamilton's principle, 241–247, 431
Harmonic function, 33
viscoelastic relations for, 53–55
Harmonic motion, steady-state, 208–210
Harmonic stress and strain of unit, 55
Heaviside functions, 81–82, 83
Hermite functions, 461–462
Hermitian interpolation function, 439–441
Heun's method, 137, 138–139
Hexahedron element, 479–484
HHT-Alpha method, 153, 155–157; *see also* Newmark method; Wilson-Theta method
HHT-alpha method, 536
algorithm, 538
displacement results, 539
example, 537
modified time-discrete equation of motion, 536
HHT method, 153–154
Holonomic constraints, 167

Holonomic system, 167
Homogeneous solutions, 277
Homogenous random function, 550
Hooke's law, 228–229, 402, 415
Houbolt method, 524; *see also* Newmark method
  algorithm for, 527
  approximations, 525
  displacement results, 528
  example, 526–527
  starting procedure, 526
Hydrostatic pressure, 231

## I

Idealized linear elastic material, 50
Impulse function, 56–57
In-plane displacements, 302
Index notation, 219, 221
Indicial notation, 219, 221
Input–output relationship, 552–556
  nonstationary random processes, 121–124
  for stationary random processes, 117–120
Integral transformations, 77
  application of Fourier transformations to viscoelastic relations, 80
  application of Laplace and Fourier transforms to ODOF systems, 82–86
  application of Laplace transformations to viscoelastic relations, 80–81
  application of $U^*$, 86–87
  dirac and heaviside functions, 81–82, 83
  experimental determination of $U^*$, 87–90
  Laplace transform and Fourier transform, 78
  time derivatives of Laplace and Fourier transformations, 79
Interelement boundary, 430
Interpolating excitation function, 128–133, 134
Interpolating functions, 429, 435
  beam finite element, 466–470
  constant strain triangle element, 470–473
  element properties, 464–489
  finite elements, 435–436
  four-noded rectangular element, 473–476
  hexahedron element, 479–484
  one-dimensional axial element, 464–466
  one-dimensional interpolation formula, 436–441
  rectangular plate element, 485–487
  tetrahedron element, 477–479
  triangular plate element, 488–489
  two-dimensional interpolation formula, 441–464
Inverse iteration, 507–509
Isoparametric elements, 455–456
Isotropic material, 229–230
Iteration technique, 506

## J

Jacobian matrix, 481
Joint probability distribution function, 99

## K

Kelvin model, 53, 54, 81
Kelvin–Voigt model, 51
Kinetic energy, 238
Kronecker delta relationships, 514

## L

Lagrange's equation, 166, 195–196, 431
  application, 174–175
  generalized coordinates, 167–168
  generalized forces, 169–171
Lagrange's interpolation formula, 438–439
Lagrangian mechanics, 247
Lame's constants, 231
Lamina stress–strain relationship, 559
Laminate, 557, 562
Laplace transform, 78
  application to ODOF systems, 82–86
Laplace transformations, 413
  application, 318–319
  application to viscoelastic relations, 80–81
L'Hospital's rule, 25
Linear acceleration method, 149–150
Linear damping coefficient, 13
Linear elastic materials, 49
Linear stress–strain behavior, 557
Linear system, 10
  dynamic characteristics of, 115–117
Logarithmic decrement, 19
Longitudinal strain, 227
Longitudinal vibrations of bars, 279–280
Lumped mass
  approach, 435
  matrix, 465

## M

Magnification factor, 23, 28, 32
Mass matrix, 496, 502–503, 521
Material damping, 49–50, 289
  Laplace and Fourier transforms application to ODOF systems with, 82–86

Matrix notation for MDF systems, 196–198
Maxwell model, 51, 53, 54, 81
MDF systems, matrix notation for, 196–198
MDOF systems, *see* Multi-degree-of-freedom systems
Mean, 96–98
Membrane theory of cylindrical shells, 409
  axially symmetric deformation, 409–410
  forced vibrations, 412–414
  free vibrations, 411–412
  motion, 411
Midplane of plate, 351, 352, 383
Modal damping ratio, 213
Modal expansion method, 365
Modal matrix, 201
Modal vector, 177, 497
Mode shape, 199
Modified Cholesky method, 505
Modified Euler method, 137, 138–139
Monoclinic material, 229
Multi-degree-of-freedom systems (MDOF systems), 2, 127, 159, 429
  central difference method, 521–524
  forced damped, 211–214
  free undamped, 198–208
  general theory, 194
  HHT-alpha method, 536–539
  Houbolt method, 524–527
  Lagrange's equation, 195–196
  mass matrix, 521
  matrix notation for MDF systems, 196–198
  Newmark method, 527–531
  numerical techniques, 521
  Wilson-theta method, 531–536

N

Natural circular frequencies, 199
Natural coordinates, 451
Natural frequencies, 176–177, 199
Natural mode, 199
  expansion method, 289–293
  orthogonality of, 199–200, 283–285
Newmark beta method, 147, 148, 528
Newmark method, 145, 527; *see also* HHT-Alpha method; Houbolt method; Wilson-Theta method
  algorithm for, 530
  average acceleration method, 148–149
  constant acceleration method, 147–148
  displacement results, 531
  equation for velocity and displacement, 146
  example, 529–531
  finite difference expressions, 528
  linear acceleration method, 149–150
  Newmark beta method, 147, 148
Newtonian mechanics, 247
Newton's law, 1, 195, 219, 421
Newton's second law, 13, 302, 354
Nine-DOF triangular element Ψ9-Ψ, 488
Node, 181, 429, 430
Nonholonomic system, 167
Nonhomogeneous second-order differential equation, 22
Nonlinear stress–strain behavior, 557
Nonstationary process, 104
Nonstationary random processes, input–output relations for, 121–124
Normal distribution, 97–98
Normal force, 217
Normalization, 200
Normalized modes, 200–202
Normal modes, 199, 200
Numerical analysis, 1
Numerical integration, 128, 482, 521, 547
Numerical simulation, 127

O

ODOF system, *see* One-degree-of-freedom system
Off-axis configuration, 560
On-axis configuration, 560
One-degree-of-freedom system (ODOF system), 83, 90, 115, 547–548
  application of Laplace and Fourier transforms to, 82–86
One-dimensional
  axial element, 464–466
  Hooke's law, 327
  wave equation, 266, 280
One-dimensional interpolation formula, 436; *see also* Two-dimensional interpolation formula
  Hermitian interpolation function, 439–441
  Lagrange's interpolation formula, 438–439
  one-dimensional line element, 436–438
Orthogonality
  of natural modes, 199–200, 283–285
  for symmetric matrix eigenvectors, 502
Orthotropic composite lamina, 557
  actual and modeled lamina, 558
  applied stress for components of orthotropic compliance matrix, 558
  elastic modulus and Poisson's ratio, 559
  on-and off-axis configurations, 560
  reduced stiffness matrix, 561–562

Orthotropic composite laminates, 562–565
Orthotropic material, 229
Out of plane displacements, 302

P

Partial solution, 504
Particular solution, 22, 27, 30, 33, 277, 316, 366
Pascal's Triangle, 442
Phase angle, 29, 36
Piecewise-constant interpolation, 128–129
Piecewise-linear excitation interpolation, 128–129
Plane state of stress, 222
Plane strain, 232
Plane stress, 222–224, 232
Plate elements, 456
  conforming rectangular element, 461–462
  conforming triangular element, 463–464
  nonconforming rectangular element, 456–459
  nonconforming triangular element, 459–461
Plates vibrations, 351; *see also* Shells vibration
  approximate solutions, 379
  circular plates, 373–378
  equations for plates of variable thickness, 391–392
  equations of motion, 351–373
  Galerkin's method, 382
  Rayleigh method, 380–382
  Ritz method, 382
  sandwich plates, 382–391
  strain energy, 379–380
Plate theory, 354
Poisson's ratio, 559
Potential energy, 167, 171
  application of Lagrange's equation, 174–175
  elastic, 172–173
  gravitational forces, 171–172
Predictor-corrector technique, 138
Principal modes, 199
Principle of conservation of energy, 248
Principle of minimum potential energy, 243
Probability, 93–96
  distribution, 93–96
Probability density, 93–96
  function, 94, 95, 98
Proportional damping, 212–213

Q

Quadratic interpolation equation, 446

R

Ramp-step function, response to, 67–70
Random functions, 101
  combinations of random processes, 111–112
  correlation function and spectral densities, 103–111
  different random functions with identical distributions, 102
  ensemble of random process, 102
  level crossings, 112–115
Random loading, 549–552
Random processes, combinations of, 111–112
Random variable, 94
Random vibrations, 93; *see also* Beam vibrations
  combined probabilities, 98–101
  dynamic characteristics of linear systems, 115–117
  input–output relations for nonstationary random processes, 121–124
  input–output relations for stationary random processes, 117–120
  mean, variance, standard deviation, and distributions, 96–98
  probability, probability distribution, and probability density, 93–96
  random functions, 101–115
  typical time history for system, 94
Rayleigh damping, 212
Rayleigh dissipation function, 431
Rayleigh's method, 248–253, 380–382
Rayleigh's quotient, 249, 502–503
Rectangular elements, 446–451
Rectangular impulse, 70–73
Rectangular plate element, 485–487
Reduced stiffness matrix, 561–562
Reissner–Mindlin theory, 383
Residual error, 258
Resonance, 24, 26, 48
Response
  to arbitrary forcing function, 58–77
  to Dirac delta function, 57–58
  to general forcing function, 55
  to general periodic forces, 44–47
  to ramp-step function, 67–70
  to unit impulse, 115
Revolution, shells of, 414–421
Rheonomic constraints, 167
Ritz method, 253, 382
  approach to, 257
  displacements, 253–254
  property of, 256–257
  system of linear homogeneous equations, 255

Rotary inertia effect, 332–334
Rotational motion model, 4
Runge–Kutta method, 139–142, 143

S

Sandwich plates, 382–383; *see also* Circular plates; Composite plates
  equations of motion, 383–388
  forced vibrations, 389–391
  free vibrations, 388–389
Scleronomic constraints, 167
SDOF system, *see* Single-degree-of-freedom system
Second-order nonhomogeneous differential equation, 27
Shape functions, 435
  beam finite element, 466–470
  constant strain triangle element, 470–473
  element properties, 464–489
  finite elements, 435–436
  four-noded rectangular element, 473–476
  hexahedron element, 479–484
  one-dimensional axial element, 464–466
  one-dimensional interpolation formula, 436–441
  rectangular plate element, 485–487
  tetrahedron element, 477–479
  triangular plate element, 488–489
  two-dimensional interpolation formula, 441–464
Shear beam, 299–301
Shear correction factor, 326
Shear deformation effect, 332–334
Shearing strains, 227
Shell(s), 395
Shells vibration, 395; *see also* Plates vibrations
  composite shells, 421–425
  cylindrical shells, 395–414
  shallow spherical shells, 417–421
  shells of revolution, 414
  spherical shell, 414–417
Shell theories, 395
Shock response spectrum, 547
Shock spectra, 547
  degrees of freedom systems, 548–549
  input–output relationship, 552–556
  one-degree-of-freedom system, 547–548
  random loading, 549–552
Shock waves, 547
Simple beam theory, 301–302, 326
Simultaneous iteration, *see* Subspace iteration
Single-degree-of-freedom system (SDOF system), 2, 127, 142, 429
  damped free vibration, 13–21
  equivalent spring constant, 4–7
  forced vibration under harmonic force, 21–37
  forcing response to periodic loading, 37–47
  free vibrations, 7–13
  one degree of freedom, 2–4
  response to arbitrary forcing function, 58–77
  response to general forcing function, 55–58
  work performing by external forces and energy dissipation, 47–55
Sinusoidal function, 22
Slope–deflection method, 339
  bending moment and shear force, 341
  deflection, rotation, and loads at two joints, 340
  forced vibrations, 344
  shear-force equation, 343
  three-membered frames, 342
Solid rectangular hexahedron elements, 454–455
Specially orthotropic lamina, *see* Lamina stress–strain relationship
Specific strain energy, 237
Spectral densit(ies), 103, 552
  ensemble average *vs.* time average, 106
  even function, 109
  Fourier transformation, 107
  matrix, 111–112
  multiple events, 103
  practical white noise, 111
  spectral density and correlation function reflect apparent periodicity, 108
  stationary process correlation function depends, 104
  stationary random function, 110
  weak and strong correlation functions, 105
Spherical shell, 414–417; *see also* Cylindrical shells
Spring mass system, 4, 6
Standard deviation, 96–98
Standardization, 200
Standard solid model, 54, 81
State space, 213
Static equilibrium, 3
Static loads, 1
Stationary ergodic process, 106
Stationary functions, 104
Stationary random
  function, 110
  input–output relations for process, 117–120
  processes, 104

Steady-state
amplitudes, 194
displacement, 35
harmonic motion, 208–210
response, 24, 55
solution, 22
vibration, 22
Step load, 60
Step response of undamped system, 60–63
response to constant acceleration, 64
Stiffness matrix, 163, 229, 494–495, 557
Stochastic process, 101–102
Strain, 232
energy, 236–237
Stress(es), 217–219, 421
resultants, 222
tensor, 219
vector, 217
Stress–strain relations, 228
anisotropic materials, 233
bulk modulus, 231
isotropic material, 229–230
plane stress and strain, 232
rectangular parallelepiped, 230
uniaxial stress, 232–233
Strings, 265
forced vibrations of finite length, 277–279
free vibrations of finite length, 271–277
transverse string vibration, 265–267
Structural analysis, 429
Structural components, 395
Structural dynamics, 1
Structural elements, 351
Subspace iteration, 513–515
Subspace iteration method, 505
Surface forces, 217
Symmetric form, 503
Symmetric matrix, 503
orthogonality for symmetric matrix eigenvectors, 502
System of equations, 220, 364
inverse iteration on, 511
simplified, 401–403

T

Tangential force, 217
Taylor's theorem, 133
Ten-node tetrahedron element Ψ10Ψ, 453
Tetrahedron element, 477–479
Thick beam theory, 326
Thin plate theory, 351
Time-averaging procedure, 104
Time-dependent coordinates, 2
Time-stepping method, 128
Time average, 106
Time integration method, 145, 527–528; *see also* Finite elements
Timoshenko beam model, 326
Timoshenko beam theory, 299, 326; *see also* Euler–Bernoulli theory
equations of motion, 326–329
forced vibration, 334
free vibrations, 329–334
Timoshenko theory, 302
Total potential energy, 242, 432
Traction vector, 217
Transfer function, 184–186
Transformation, 78; *see also* Integral transformations
Transformation matrix, 492–493
Transformation methods, 515
eigenmatrix, 516
generalized Jacobi method, 516–520
Transient solution, 22
Transverse loads, 299
Transversely isotropic material, 229
Transverse shear stresses, 565
Transverse string vibration, 265
elastic string, 266
initial and boundary conditions, 266–267
Traveling wave solution, 268
Triangular impulse, 73–74, 75, 76
Triangular plate element, 488–489
Trigonometric functions, 38–41
Two-degree-of-freedom, 159, 160
Two-dimensional interpolation formula, 441; *see also* One-dimensional interpolation formula
isoparametric elements, 455–456
plate elements, 456–464
rectangular elements, 446–451
solid rectangular hexahedron elements, 454–455
tetrahedral elements, 451–453
triangular elements, 441–446
Two-dimensional plate theory, 351

U

Undamped system
response to constant acceleration, 64
step response of, 60–63
Uniaxial stress, 232–233
Unit diagonal matrix, 503

Unit impulse, 89
 function, 56
 response to, 184–186
Unit vectors, 218

V

Variance, 96–98
Vector equation, 160
Vector iteration
 forward iteration, 509–510
 inverse iteration, 507–509
 methods, 505
 with shifts, 511–513
Vibration(s), 9, 305
 approximate solutions, 379–382
 circular plates, 373–378
 equations for plates of variable thickness, 391–392
 equations of motion, 351–373
 of frames with axial forces, 345–348
 of plates, 351
 response, 211
 sandwich plates, 382–391
Vibration of shells, 395
 composite shells, 421–425
 cylindrical shells, 395–414
 shells of revolution, 414–421
Vibration of strings and bars
 bar with concentrated end load and associated expressions, 294–296
 concentrated force, 293–294
 forced vibrations and natural mode expansion method, 289–293
 forced vibrations of bars, 287–289
 forced vibrations of finite length strings, 277–279
 free vibrations of bars, 280–287
 free vibrations of finite length strings, 271–277
 general solution of wave equation, 267–271
 longitudinal vibrations of bars, 279–280
 material with damping, 289
 transverse string vibration, 265–267
Virtual displacements, 238
Virtual strains, 238
Virtual work principle, 238–241
Viscoelastic materials, 51–53
Viscoelastic relations
 Fourier transformations application to, 80
 for harmonic functions, 53–55
 Laplace transformations application to, 80–81
Viscous damping forces, 433
Volumetric strain, 231

W

Wave equation
 Fourier transformation solution, 268–271
 general solution of, 267
 traveling wave solution, 268
Wave propagation velocity, 266
"Weighted average", 259
White noise, 108
 narrow band and, 552
 practical, 111
Wide sense, 104
Wiener–Khinchine relations, 109
Wilson-theta method, 150–153, 154, 531, 532; *see also* HHT-Alpha method; Newmark method
 algorithm for, 534
 displacement results, 536
 dynamic equilibrium equations, 531
 example, 534–535
 velocity calculation, 533
Work, 235–238

Z

Zener models, 52